Microbial Polymers

Anukool Vaishnav •
Devendra Kumar Choudhary
Editors

Microbial Polymers

Applications and Ecological Perspectives

Springer

Editors
Anukool Vaishnav
Department of Biotechnology
GLA University
Mathura, India

Devendra Kumar Choudhary
Amity Institute of Microbial Technology
Amity University
Noida, India

ISBN 978-981-16-0044-9 ISBN 978-981-16-0045-6 (eBook)
https://doi.org/10.1007/978-981-16-0045-6

This Springer imprint is published by the registered company Springer Nature Singapore Pte Ltd.
The registered company address is: 152 Beach Road, #21-01/04 Gateway East, Singapore 189721, Singapore

Preface

The trend in polymer sciences during early twentieth century was using synthetic plastic, oil-derived plastics, and their products in burgeoning chemical industry. However, discarded synthetic polymers contribute to the pollution of the environment, and the bulk ends up on beaches, in the oceans, or clogging landfill sites. Recently, there has been an increased awareness of and a sense of responsibility for pollution. This has led to a determined search for biodegradable alternatives to synthetic polymers. Polymers of natural origin (biopolymers) have become increasingly important for both scientific and industrial sectors due to their renewable nature, unique properties, and wide availability. Increased bio-based polymer production and application has been positioned as one of the most promising ways to meet the sustainable development goal of replacing traditional petroleum polymers in several industrial sectors. In this context, microbe-based polymers, including exopolysaccharides, chitin, polyhydroxybutyrates (PHB), polyhydroxyalkanoates (PHAs), fungal mycelium, bacterial pigments and biofilms, have attracted a significant amount of interest recently, especially for industrial, medical, packaging, and agricultural sectors. Microorganisms efficiently convert different carbon sources into a diverse range of polymers, i.e., polysaccharides, polyesters, polyamides, and inorganic polyanhydrides with varying chemical and material properties. Most of the polymers secreted by microbes are extracellular, hence easy to extract them through fermentation. Some of the polymers from different organisms have same functions, whereas other polymers are specific for certain microbial species.

Recent discoveries in molecular tools and our understanding about microbial synthesis of polymers have gained interest in rational engineering in microbes towards production of desired polymers with specific high value application at viable economic cost. The fungal mycelium has attracted increasing interests recently as a new form of low energy bio-fabrication and waste upcycling. Fungal mycelium binds organic matter tightly in their hyphal filamentous network that can be exploited to produce packaging and composite materials from agricultural and industrial waste. In agriculture, some microbial polymers such as PHAs are used as a carrier material for controlled and targeted pesticides delivery. Cell encapsulation method is also used for microbial inoculants delivery into soil. In this method, microbial cells such as plant growth-promoting rhizobacteria (PGPR) and arbuscular mycorrhizae fungi (AMF) encapsulate in gel matrix. More than 80% of bio-encapsulation processes are carried out using various types of alginates. Alginates, linear polymers of b (1,4)-D-mannuronic acid and a(1,4)-L-guluronic acid monomers, are produced by brown algae. and their composition varies depending on the algal source. Likewise, several microbial polymers are already produced commercially including Chitosan, Curdlan, Beta-Glucan, Xanthan Gum, Agar Gum, Gallan Gum, etc. for different industrial purposes. Therefore, significant attention has been paid on the improvement of those sustainable polymers, overall their production, properties, and performance.

Hence, in this book, editors compiled researches carried out on microbe-based polymers and their application in diverse sectors including agriculture, food, and medicine.

Mathura, India — Anukool Vaishnav
Noida, India — Devendra Kumar Choudhary

Contents

Part I Diversity of Microbial Polymers

1 **The Production and Applications of Microbial-Derived Polyhydroxybutyrates** . . . 3
S. I. Magagula, M. Mohapi, J. S. Sefadi, and M. J. Mochane

2 **Fungal Exopolysaccharides: Types, Production and Application** . . . 45
Ashim Debnath, Bimal Das, Maimom Soniya Devi, and Ratul Moni Ram

3 **Isolation and Purification of Microbial Exopolysaccharides and Their Industrial Application** . . . 69
Veena S. More, Allwin Ebinesar, A. Prakruthi, P. Praveen, Aneesa Fasim, Archana Rao, Farhan Zameer, K. S. Anantharaju, and Sunil More

4 **A Review on Properties and Applications of Xanthan Gum** . . . 87
Surabhi Chaturvedi, Sanchita Kulshrestha, Khushboo Bhardwaj, and Rekha Jangir

5 **Biosynthesis and Characterization of Poly-(3)-hydroxyalkanoic Acid by *Bacillus megaterium* SF4 Using Different Carbohydrates** . . . 109
Temitope O. Fadipe, Nazia Jamil, and Adekunle K. Lawal

6 **Mushroom Mycelia-Based Material: An Environmental Friendly Alternative to Synthetic Packaging** . . . 131
Abhik Mojumdar, Himadri Tanaya Behera, and Lopamudra Ray

7 **An Overview of Microbial Derived Polyhydroxybutyrate (PHB): Production and Characterization** . . . 143
Monika Sharma and Harish Kumar Dhingra

8 **Insight of Biopolymers and Applications of Polyhydroxyalkanoates** . . . 177
Rishabh Agrahari, Gargi Sarraf, Naveen Chandra Joshi, Swati Mohapatra, and Ajit Varma

9 **Microbial Pigments and Their Application** 193
Selvaraju Vishnupriya, Sundaresan Bhavaniramya, Dharmar Baskaran, and Arulselvam Karthiayani

Part II Microbial Polymers in Agriculture

10 **Extracellular Polymeric Substances from Agriculturally Important Microorganisms** 217
Valeria Valenzuela Ruiz, Roel Alejandro Chávez Luzania, Fannie Isela Parra Cota, Gustavo Santoyo, and Sergio de los Santos Villalobos

11 **Significance of Bacterial Polyhydroxyalkanoates in Rhizosphere** . . . 235
Sundaresan Bhavaniramya, Selvaraju Vishnupriya, and Dharmar Baskaran

12 **Role of Microbial Biofilms in Agriculture: Perspectives on Plant and Soil Health** 251
Anupam Gogoi, Mandeep Poudel, Jagajjit Sahu, and Geetanjali Baruah

13 **Biological Soil Crusts to Keep Soil Alive, Rehabilitate Degraded Soil, and Develop Soil Habitats** 289
Atoosa Gholamhosseinian, Adel Sepehr, Behnam Asgari Lajayer, Nasser Delangiz, and Tess Astatkie

14 **Fungal Chitosan: The Importance and Beneficiation of this Biopolymer in Industrial and Agricultural Process** 311
Allwin Ebinesar, Veena S. More, D. L. Ramya, G. R. Amrutha, and Sunil S. More

15 **Role of Microbial Extracellular Polymeric Substances in Soil Fertility** 341
Alok Bharadwaj

16 **Microbes Derived Exopolysaccharides Play Role in Salt Stress Alleviation in Plants** 355
Purnima Singh, Vibha Pandey, and Prerana Parihar

Part III Microbial Polymers in Industrial Sectors

17 **Microbial Exopolysaccharides: Structure and Therapeutic Properties** 375
Hiran Kanti Santra and Debdulal Banerjee

18 **Microbial Biopolymers: Pharmaceutical, Medical, and Biotechnological Applications** 421
Rohit Godbole, Asha Goutam, and Aniket Mali

19 **Mycobacterium Biofilms Synthesis, Ultrastructure, and Their Perspectives in Drug Tolerance, Environment, and Medicine** 465
Kundan Kumar Chaubey, Mohd. Abdullah, Saurabh Gupta, Manthena Navabharath, and Shoor Vir Singh

20 **A Comprehensive Review on Different Microbial-Derived Pigments and Their Multipurpose Activities** 479
Archana S. Rao, Sidhartha Pratim Deka, Sunil S. More, Ajay Nair, Veena S. More, and K. S. Ananthjaraju

21 **Microbial Polysaccharides with Potential Industrial Applications: Diversity, Synthesis, and Their Applications** 521
Himadri Tanaya Behera, Abhik Mojumdar, Chiranjib Mohapatra, and Lopamudra Ray

22 **Eco-friendly Microbial Biopolymers: Recent Development, Biodegradation, and Applications** 547
Chandrabose Selvaraj and Sanjeev Kumar Singh

Part IV Advances in Microbial Polymers

23 **Microbial Biopolymers as an Alternative Construction Binder** 581
Gye-Chun Cho, Ilhan Chang, and Jooyoung Im

24 **Genetic Engineering Approaches for High-End Application of Biopolymers: Advances and Future Prospects** 619
Saurabh Gupta, Kundan Kumar Chaubey, Vishal Khandelwal, Tarubala Sharma, and Shoor Vir Singh

25 **Microbial Pigments: Secondary Metabolites with Multifaceted Roles** 631
Himadri Tanaya Behera, Abhik Mojumdar, Suchismita Nivedita, and Lopamudra Ray

26 **Bio-fermentative Production of Xanthan Gum Biopolymer and Its Application in Petroleum Sector** 655
Nanthakumar Kuppanan, Paul Jeyaseelan, Shishram Chahar, Subhasis Das, Veeranna Channashettar, and Banwari Lal

27 **A Comparative Study on Biodegradable Packaging Materials: Current Status and Future Prospects** 675
Archana S. Rao, Ajay Nair, Sunil S. More, Arpita Roy, Veena S. More, and K. S. Anantharaju

28 **Environmental Implications of Microbial Bioplastics for a Sustainable Future** 695
A. Mbotho, S. I. Magagula, K. M. Moloantoa, J. S. Sefadi, and M. J. Mochane

About the Editors

Dr. Anukool Vaishnav is working as an Assistant professor in Department of Biotechnology, GLA University, Mathura. He operated SERB-National Post-doctoral Fellowship (NPDF) project as a Principal Investigator at Department of Mycology and Plant Pathology, Institute of Agricultural Sciences, Banaras Hindu University, Varanasi, India. He has 7 years of research experience in Agriculture Microbiology. His research is mainly focuses on microbial mediated plant protection against biotic and abiotic stress, characterization of signaling molecules and secondary metabolites (soluble and VOCs) in plant holobionts. As an active researcher, he has published more than 20 research and review articles along with 13 book chapters for reputed journals and edited books. He has filed 9 Indian patents in association with his research group. He is an editorial member of Current Genomics Journal published by Bentham Science. He has been awarded with Young Scientist Award from reputed societies.

Dr. Devendra Kumar Choudhary is selected member-National Academy of Sciences, India (NASI) and shows his presence at Amity University Uttar Pradesh, Noida. Recently he was associated with PDM and worked as Professor (Ad hoc) for three months. He is active researcher and operated major projects with worth amount 10 million sponsored by DBT, DST and SERB, New Delhi, India, as principal investigator and co-investigator. He has involved and contributed in NAAS policy paper number 36 in the year 2006. He has published more than 80 research/review articles along with several book chapters for reputed journals and edited books (Springer, Elsevier and Taylor Francis/Wiley-Blackwell). He has supervised 08 research scholars for their doctoral programme as supervisor and co-supervisor. In addition, he is recipient of Indian National Science Academy (INSA) visiting and summer research fellowship-2014 and his team received DR. RS Rana Memorial best research award 2013 sponsored by Association of Microbiologists of India. His team filed 3 patents and submitted bacterial strains to public domain wherein incurred two accession numbers from MTCC, IMTECH for submitted bacterial cultures (MTCC, 12057 & 12058) along with one MCC no 2607.

Part I

Diversity of Microbial Polymers

The Production and Applications of Microbial-Derived Polyhydroxybutyrates 1

S. I. Magagula, M. Mohapi, J. S. Sefadi, and M. J. Mochane

Abstract

For the last few decades, PHBs have been used to address issues around environmental pollution. This is due to their biodegradable nature when compared with conventional plastics. Moreover, their 100% biodegradability combined with biocompatibility has secured a place for PHBs in many industrial applications such as biomedical, packaging, tissue engineering and drug delivery. However, properties such as brittleness, thermal instability and narrow processing window limit the application of PHBs in many industrial applications. Therefore, in order to overcome this setback, PHBs are usually blended with other polymers, fibres or additives to obtain PHB-based composites with optimum properties. PHBs are a group of PHAs that are synthesized and stored by many microorganisms as a source of carbon and energy. With the growing environmental concerns, there is an increasing demand for the production of PHB-based materials on a larger scale. However, the major problem facing by large-scale production of PHB right now is the high cost of production. In spite of the many attempts to reduce the cost of production, there is still a need for more ways of significantly reducing the production cost. Some of the ways are briefly mentioned in this chapter. This chapter is aimed at providing a detailed discussion on the production and applications of microbial PHBs.

S. I. Magagula · M. Mohapi · M. J. Mochane (✉)
Department of Life Sciences, Central University of Technology, Bloemfontein Campus, Bloemfontein, South Africa
e-mail: mmochane@cut.ac.za

J. S. Sefadi (✉)
School of Natural and Applied Sciences, Sol Plaatje University, Kimberly, South Africa
e-mail: jeremia.sefadi@spu.ac.za

A. Vaishnav, D. K. Choudhary (eds.), *Microbial Polymers*,
https://doi.org/10.1007/978-981-16-0045-6_1

Keywords

Polyhydroxybutyrate · Bacteria · Cyanobacteria · Biodegradable plastic

1.1 Introduction

Polyhydroxybutyrates (PHBs) are biopolymers that are synthesized and accumulated by a wide range of microorganisms as energy reserves, especially when the microbes are grown under stressful conditions (Galia 2010; Getachew and Woldesenbet 2016; Ramadas et al. 2009). PHBs are produced via bacterial fermentation processes where different strains of bacteria are used to synthesize and accumulate PHBs in the form of intracellular granules (Ramadas et al. 2009). The bacteria is usually cultured under limited conditions where one of the nutrients like phosphorus, nitrogen, oxygen or sulphur is absent and the presence of carbon is in abundance (Sudesh et al. 2000; Kessler and Witholt 2001). PHBs form a major part of a group of microbial-derived, lipid-like, water-insoluble polyester molecules called polyhydroxyalkanoates (PHAs) (Raza et al. 2019). PHAs are the major building blocks of a group of biopolymers called biodegradable plastics or biopolymers (Carpine et al. 2020). Figure 1.1 shows a classification of the different types of biodegradable plastics.

PHBs have processing properties that are similar to those of synthetic thermoplastics such as polypropylene (Raza et al. 2019; Faria and Martins-Franchetti 2010). They are also biocompatible, hydrophobic and 100% biodegradable (Bugnicourt et al. 2014; Lasprilla et al. 2012). Due to their 100% biodegradability, thermoplastic processability and biocompatibility, PHBs have been used to replace synthetic thermoplastics in several industrial applications such as in the packaging industry (Mousavioun et al. 2012), agricultural industry (Alejandra et al. 2012) as well as in the biomedical (Ke et al. 2017; Ho et al. 2014) and pharmaceutical fields (Villanova et al. 2010). Figure 1.2 shows the general structures of biodegradable and non-biodegradable plastics.

Due to their non-degradable nature, synthetic thermoplastics have caused some irreparable damages to the environment (Sharma et al. 2007). Most of the synthetic thermoplastics accumulating in the environment as plastic waste end up in our oceans (Anderson et al. 2018; Maes et al. 2017; Do Sul and Costa 2014), our freshwater (Eerkes-Medrano et al. 2015; Jambeck et al. 2015; Mani et al. 2015), our air, our soil and food (Carpine et al. 2020). This causes devastating damages to the ecosystem. For example, most of the synthetic thermoplastics that end up as micro plastics in the ocean are ingested by aquatic creatures that depend on the ocean to survive such as fish (Anderson et al. 2018; Maes et al. 2017). The micro plastics have been discovered in the digestive systems of many aquatic species (Rummel et al. 2016; Van Cauwenberghe and Janssen 2014). So far, clarity is required on how the ingested plastics affect the organisms (Mintenig et al. 2019). However, so far, it has been reported that ingested micro plastics can induce gut inflammations, but their transportation to other organs has not been reported (Bouwmeester et al. 2015).

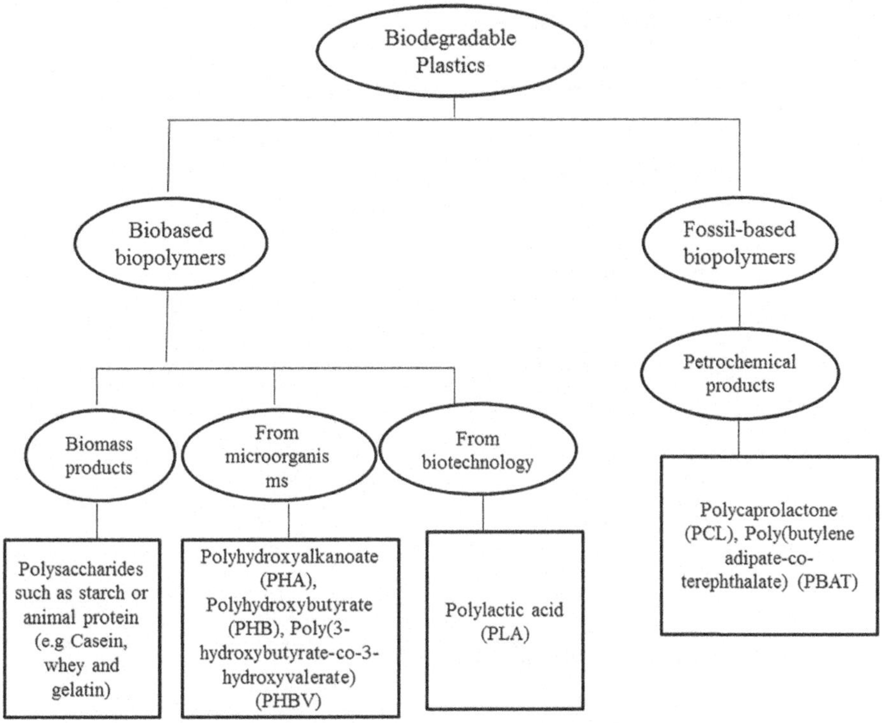

Fig. 1.1 Classification of the different types of biodegradable plastics

Micro plastics have been found in seafood (Tanaka and Takada 2016), salt (Iñiguez et al. 2017), honey (Liebezeit and Liebezeit 2015) and tap water (Kosuth et al. 2017). Furthermore, micro plastic particles have also been found in bottled drinking water, and it was suspected that the packaging might also be a contributor to the contamination (Schymanski et al. 2018). This could cause problems for the human health and ecosystem in the long run.

Currently, the global consumption of synthetic thermoplastics is approximately 140 million tons per annum (Carpine et al. 2020). This explains the rate at which plastic waste is accumulating around the world. According to EU officials, plastic waste accounts for about 70% of the trash found in the oceans (Carpine et al. 2020). This makes the plastic waste a global concern. Now, several methods of plastic waste disposal have been used over the years. However, these methods have always been faced with serious challenges. For instance, the degradability of the thermoplastics disposed in landfills has been found to be extremely slow. The incineration of the thermoplastics has been found to generate toxic by-products. For those who have tried recycling, they have found recycling to be time-consuming. In addition to that, recycling has also been found to change the properties of the original plastic material (Costa et al. 2019).

Biodegradable plastics

Poly-ε-caprolactone(PCL)

Polylactid (PLA)

Polybutylene succinate (PBS)

Poly(butylene succinate-co adipate) (PBSA)

Poly(3-hydroxybutyrateco-3-hydroxyvalerate)(PHBV)

Non-biodegradable plastic

Polyethylene (PE)

Polyethylene terephthalate (PET)

Polypropylene (PP)

Polystyrene (PS)

Polyvinyl chloride (PVC)

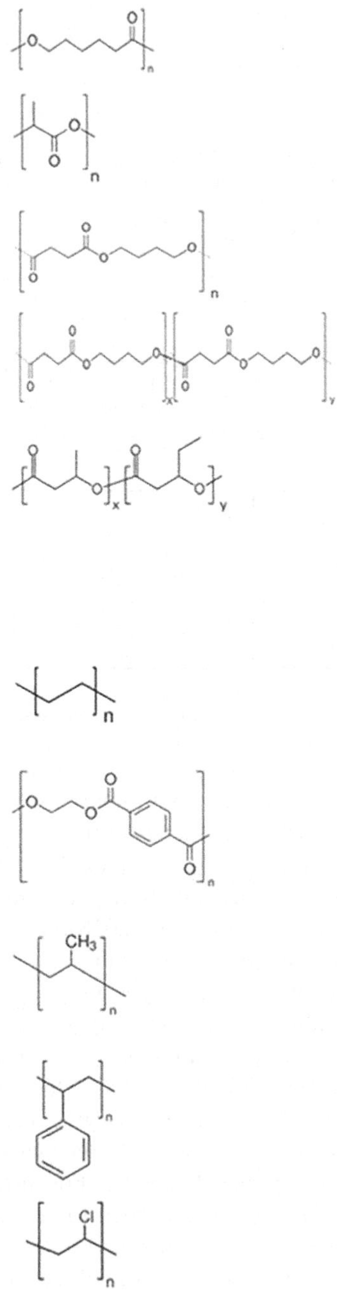

Fig. 1.2 General structures of biodegradable and non-biodegradable plastics. (Springer open access Urbanek et al. (2018))

As a result of their eco-friendliness, PHBs have become solution of choice when it comes to protecting the environment against all environmental hazards that have been caused by the synthetic thermoplastics. PHBs are also the only biodegradable plastics that are 100% biodegradable as compared to other biodegradable plastics. This qualifies them even more as ideal candidates for solving the global environmental concerns (Carpine et al. 2020).

Currently, bioplastics, with PHBs included, account for 1% of the plastics produced globally per annum. In spite of this, the bioplastic global market continues to grow at approximately 20–100% per annum. In 2019, the production capacity of bioplastics recorded by European Bioplastics and the Research Institute Nova-Institute was 2.11 million tons, and it is expected to increase to 2.43 million tons by 2024 (Carpine et al. 2020).

One of the major drawbacks in the extensive production and commercialization of bioplastics, especially PHBs is their high production cost as compared to synthetic thermoplastics (Carpine et al. 2020). As a result, most research has been geared towards reducing the cost of producing PHBs. This has been achieved by developing new strains of bacteria (Bistué Rovira et al. 2019; Peña et al. 2014a) as well as optimizing fermentation and PHB recovery processes (Peña et al. 2014a; Al-Battashi et al. 2019; Lopez-Arenas et al. 2017; Garcia-Gonzalez et al. 2015; Reddy et al. 2019; Carpine et al. 2020). The increase in the cost of production of PHBs is as a result the high carbon substrate cost as well as the high cost of the processes used to recover the PHB granules from the bacterial cell cytoplasm (Carpine et al. 2020).

The high production cost together with properties like thermal instability and poor mechanical performance limit the use of PHBs in many industrial applications (Carpine et al. 2020). Therefore, more effort has been focused on reducing the production cost and improving the mechanical properties as well as the thermal stability of the PHBs (Raza et al. 2019). In order to improve their mechanical and thermal properties for packaging, biomedical and tissue engineering applications, PHBs are incorporated with other materials or biopolymers (Raza et al. 2019). The aim of this chapter is to review the production methods used to produce microbial PHBs and their applications in the biomedical field, packaging industry and in tissue engineering.

1.2 A Brief History of Polyhydroxybutyrates (PHBs)

PHBs were first discovered, isolated and characterized in 1925 by a scientist named Maurice Lemoigne (López-Cortés et al. 2008). Due to their 100% biodegradability, biocompatibility and non-toxicity, PHBs have become some of the most important and commonly used type of PHAs (Bugnicourt et al. 2014; Lasprilla et al. 2012). Now, PHBs are aliphatic polyesters characterized by a linear polymer chain (Raza et al. 2019) (see Fig. 1.3). The linear polymer chain of PHBs is composed of 3-hydroxybutyrate monomers consisting of a short chain length of four to five carbon atoms. Each 3-hydroxybutyrate is composed of a β-hydroxybutyric acid consisting of carboxylic acid (-COOH) and alcoholic (-OH) functional groups

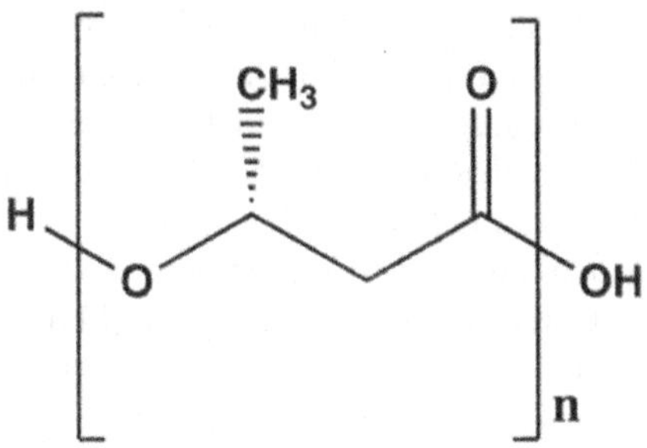

Fig. 1.3 A typical structure of microbial polyhydroxybutyrates (PHBs) (Carpine et al. 2020). (MDPI Open access)

Table 1.1 Comparison between the physical properties of polyhydroxybutyrates (PHB) and polypropylene (PP) (Verlinden et al. 2007; Hankermeyer and Tjeerdema 1999)

Physical property	PHB	PP
Crystalline melting point (°C)	177	176
Crystallinity (%)	60	50–70
Molecular weight (Daltons)	5×10^5	2×10^5
Glass transition temperature (°C)	2	−10
Density (g/cm^3)	1.25	0.905
Young's modulus (GPa)	3.5	1.7
Tensile strength (MPa)	40	38
Extension to break (%)	6	400
UV resistance	Good	Poor

(Raza et al. 2019). The PHB polymer chain structure is composed of a carbonyl group that is chromophoric, which gives the PHBs an optically active nature (Raza et al. 2018). However, even though PHBs are optically active, it is important to note that PHBs are stable towards (UV) radiation (Raza et al. 2019) (see Table 1.1). Furthermore, the polymer chain structure of PHBs is also isotactic and semi-crystalline with an R-configuration (Raza et al. 2019).

PHBs have a molecular mass that varies between 50,000 and 1 million Da. The reason for this wide molecular mass range is because the molecular mass of PHBs depends on production conditions such as the carbon source, bacteria growth conditions and the PHB recovery methods used (Dos Santos et al. 2017; Rai et al. 2011). The molecular weight, melting point, crystallinity and tensile strength of PHBs are comparable with those of polypropylene (PP) as shown by Table 1.1. Therefore, PHBs can be used in the place of polypropylene in many industrial applications (Verlinden et al. 2007). PHBs have a glass transition temperature (Tg) of about 2–5 °C (Raza et al. 2019; Carpine et al. 2020) and a melting temperature (Tm) of about 177–180 °C (Raza et al. 2019, Carpine et al. 2020), which means they are capable of overcoming the harsh processing conditions associated with injection, blowing and extrusion techniques. PHBs also have a density of about 1.26 g/cm^3 as well as a specific gravity of about 1.18 g/cm^3 (Raza et al. 2019), which makes them lightweight materials suitable for any industrial application.

PHBs have superior barrier properties against oxygen than those of polyethylene terephthalate (PETE) and polypropylene (PP) (Rajan et al. 2018). In addition to that, the water vapour barrier properties as well as the barrier properties against odour and

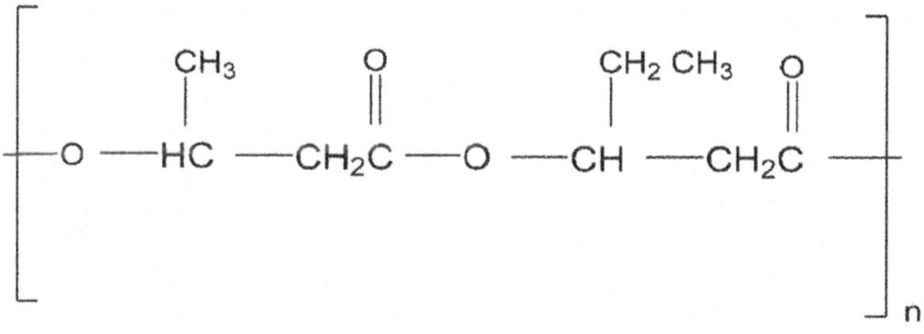

Fig. 1.4 General structural formula of polyhydroxybutyrate-co-valerate (PHBV) (Rivera-Briso and Serrano-Aroca 2018). (MDPI Open access)

fat of PHBs are also superior to those of polypropylene (Hankermeyer and Tjeerdema 1999). These excellent barrier properties make PHBs suitable for use in food packaging. Furthermore, PHB has been reported to have a mammalian lethal dose (LD_{50}) of more than 500 mg/kg. Therefore, they have been considered to be non-toxic and suitable for use in the denitrification of drinking water (Hankermeyer and Tjeerdema 1999).

Regardless of their superior properties, the applications of PHBs are limited by a few major setbacks. One is that PHBs have poor impact properties (Hankermeyer and Tjeerdema 1999). Their extension at break was about 6% in comparison to the 400% of polypropylene (see Table 1.1). Another major setback is that PHBs decompose 10 °C above their melting temperature of 177 °C, which makes their processing difficult. However, this problem is usually solved by incorporating long pendant hydroxyl acid monomer unit groups, such as hydroxyvalerate. This results in a polyhydroxybutyrate-co-valerate (PHBV) copolymer with superior strength and flexibility than PHB (Hankermeyer and Tjeerdema 1999). Figure 1.4 shows the general structural formula of PHBV.

1.3 Mechanism for the Biosynthesis of PHB in Microorganisms

PHBs have been reported to be synthesized and accumulated by over 300 microorganisms (Koller et al. 2010; Alarfaj et al. 2015). The most commonly investigated microorganisms for PHB accumulation include genera of *Alcaligenes*, *Azotobacter*, *Bacillus*, *Rhizobium*, *Rhodospirillum* and *Pseudomonas* (Chandani et al. 2014; Jiang et al. 2011; Bhagowati et al. 2015). These microorganisms synthesize and accumulate PHBs as energy and carbon sources as reserves for use when experiencing extremely harsh growth conditions. Therefore, the synthesis of PHBs is usually carried out under unbalanced cellular growth conditions. That is, in the presence of excessive amounts of a suitable carbon source, and a limited supply of essential nutrients such as nitrogen, phosphorus, sulphur or oxygen (Carpine et al. 2020). Table 1.2 shows some of the microorganisms used in the production of PHB, the used feedstock and the type of polymer recovered.

Table 1.2 Some strains of bacteria used for the synthesis and accumulation of PHB

Type of bacterial strain	Source of carbon	Type of biopolymer (s) produced	Amount of PHB produced (%)	Refs
Aeromonas hydrophila 4AK4 mutant	Oleic acid, lauric acid	MCL-PHA	64	Han et al. (2004)
Alcaligenes latus	Malt, milk waste, soy waste	PHB	70	Wong et al. (2004)
Azotobacter chroococcum H23	Wastewater obtained from olive oil mills	PHB	80	Pozo et al. (2002)
Bacillus cereus UW85	Glucose	PHB	9	Łabużek and Radecka (2001)
Bacillus megaterium ATCC 6748	Molasses, corn steep liquor	PHB	43	Chaijamrus and Udpuay (2008)
Bacillus spp. 87I	Glucose	PHB	67	Thirumala et al. (2010)
Burkholderia sacchari sp. IPT101	Glucose	PHB PHBV	68	Gent (2001)
Burkholderia cepacia IPT 048	Bagasse	PHB	62	Silva et al. (2004)
Caulobacter crescentus DSM 4727	Glucose	PHB	18	Qi and Rehm (2001)
Cupriavidus necator DSM 545	Corn syrup	PHB	30	Daneshi et al. (2010)
Cupriavidus necator DSM 545	Waste glycerol	PHB	62	Cavalheiro et al. (2009)
Enterobacter aerogenes 12Bi	Wastewater	PHB	43	Ceyhan and Ozdemir (2011)
Escherichia coli mutants	Xylose	PHB	27	Nikel et al. (2006)
Halomonas boliviensis LC1	Hydrolysate, starch, maltose, maltohexaose and maltotetraose	PHB	56	Quillaguaman et al. (2005)
Legionella pneumophila 74/81	Nutrient broth	PHB	16	James et al. (1999)
Methylocystis sp. GB 25 DSM 7674	Methane	PHB	51	Wendlandt et al. (2005)
Microlunatus phosphovorus DSM 10555	Glucose, acetate	PHB	30	Akar et al. (2006)
Mixed microbial culture	Lactate	PHB	90	By et al. (2013)
Pandoraea sp.	Crude glycerol	PHB	63	de Paula et al. (2017)

(continued)

Table 1.2 (continued)

Type of bacterial strain	Source of carbon	Type of biopolymer (s) produced	Amount of PHB produced (%)	Refs
Pseudomonas aeruginosa NCIB 40045	Agro-industrial oily wastes	MCL-PHA	66	Fernández et al. (2005)
Pseudomonas hydrogenovora DSM 1749	Dairy whey	MCL-PHA	21	Koller et al. (2008)
Pseudomonas putida CA-3	Petrochemical plastic waste	PHB	30	Goff et al. (2007)
Pseudomonas fluorescens A2a5	Sugar cane liquor	PHB	70	Jiang et al. (2008)
Rhodopseudomonas palustris SP5212	Acetate	PHB	34	Mukhopadhyay et al. (2005)
Ralstonia pickettii 61A6	Sugar cane liquor	PHB	10	Bonatto et al. (2004)

MCL-PHA medium-chain-length-polyhydroxyalkanoates, *PHB* polyhydroxybutyrate, *PHBV* poly (3-hydroxybutyrate-co-3-hydroxyvalerate)

Now, PHB-producing microorganisms synthesize PHBs via three consecutive enzymatic reactions. The first step is the condensation of two acetyl-CoAs to acetoacetyl-CoA by β-ketothiolase (PhaA). The second step involves reducing acetoacetyl-CoA to 3-hydroxybutyril-CoA by an acetoacetyl-CoA reductase (PhaB). Finally, the third step involves the polymerization of 3-hydroxybutyril-CoA monomers to PHB using PHB synthase (PhaC and PhaE) (Sagong et al. 2018) (see Fig. 1.5).

Once the PHB is polymerized, it is packed into PHB granules called carbonosomes (see Fig. 1.6). These carbonosomes primarily act as carbon and energy storage materials situated inside the cells of various bacterial strains (Nikel et al., 2006, Asada et al., 1999, Qi and Rehm, 2001, Zhang et al., 2015, Angermayr et al., 2015, Carpine et al., 2020). The carbonosomes consist of PHBs covered by granule-associated proteins (GAPs) (Jendrossek 2009; Pfeiffer and Jendrossek 2012). Now, there are four main types of GAPs reported in literature: PHB depolymerases (PhaZ), PHB synthases (PhaC), phasins (PhaP) and regulators (PhaR or PhaM) (Tirapelle et al. 2013; De Los Angeles Martínez-Martínez et al. 2019). The synthases and depolymerases are responsible for the accumulation and utilization of PHB (Juengert et al. 2018; Mezzolla et al. 2018). Phasins are responsible for coating and stabilizing the PHB in the cytoplasm (Sznajder et al. 2015). In addition to that, phasins are also responsible for regulating the number and size of carbonosomes (Pfeiffer and Jendrossek 2012; Hauf et al. 2015). Furthermore, the regulators (PhaR) are responsible for regulating the synthesis of phasins (Maehara et al. 2002; Pötter et al. 2002) and the localization of the PHB granules (PhaM) (Pfeiffer et al. 2011; Pfeiffer and Jendrossek 2014).

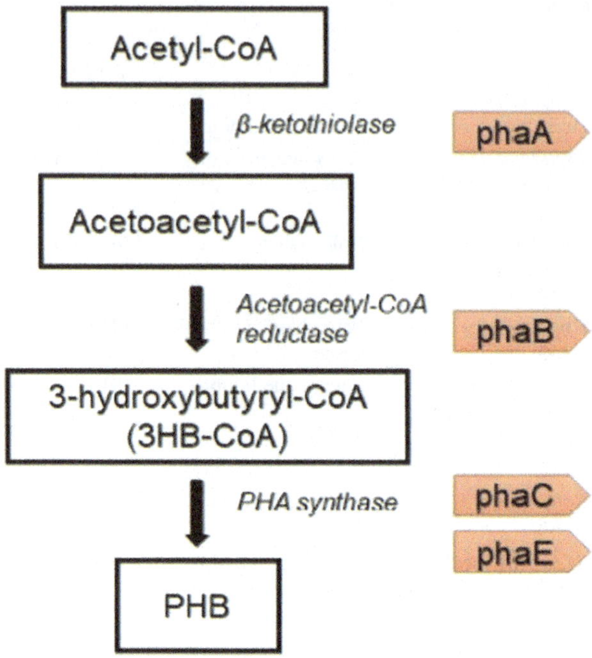

Fig. 1.5 Mechanism for the biosynthesis of PHB (Carpine et al. 2020). (MDPI Open access)

Phasins have been regarded as the most important type of GAPs (Sznajder et al. 2015; De Los Angeles Martínez-Martínez et al. 2019). This is mostly due to the fact that they are able to interact with both PHB depolymerases and PHB polymerases. This depends on the type of microorganism used (Ushimaru et al. 2014; Tian et al. 2005; Handrick et al. 2004a; Handrick et al. 2004b). For example, studies on *Aeromonas hydrophila* revealed a two-fold increase of the PHB content when only PhaP was overexpressed (Ushimaru et al. 2014). Changes in the composition of the monomers of the poly(3-hydroxybutyrate-co-3-hydrohexanoate) [P(HB-co-HHx)] copolyester were also observed (Tian et al. 2005). Furthermore, studies on *Rhodospirillum rubrum* revealed that the presence PhP also promoted the degradability of PHB (Handrick et al. 2004b; De Los Angeles Martínez-Martínez et al. 2019). These studies showed that the interaction of the phasins with PHB polymerases and PHB depolymerases has a positive effect on the synthesis and degradability of PHB (De Los Angeles Martínez-Martínez et al. 2019).

1.4 Production of PHB

With the increasing impact of plastics on the environment, there is an increasing need for the production of bioplastics on a larger scale. Currently, the production of bioplastics on a smaller scale in comparison to the production of petroleum-derived plastics (Carpine et al. 2020). From literature, the production of PHB by Brazilian,

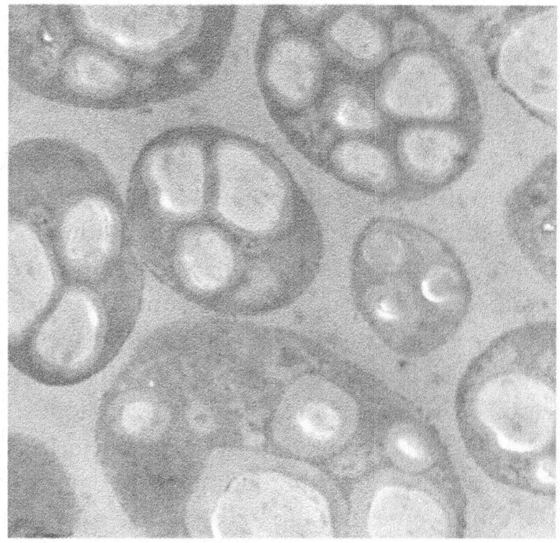

Fig. 1.6 PHB granules (white inclusions) shown on a transmission electron micrograph of a thin section belonging to *A. vinelandii* (Peña et al., 2014a). Open Access

British and Chinese bioplastic companies was reported to be 0.1, 0.3 and 10 kilotonnes per year, which is quite low when compared with petroleum-based plastics (Carpine et al. 2020). The question is why? This question will be answered in a moment. Now, due to their 100% biodegradability, eco-friendliness and biocompatibility, PHBs have gained enormous attention in research and numerous commercial applications (Ramadas et al. 2009; Raza et al. 2019; Carpine et al. 2020). The production of PHBs is normally done via the commonly known bacterial fermentation processes (Carpine et al. 2020; Ramadas et al. 2009). The fermentation process can be divided into two phases. In the first phase, all the nutrients (e.g. carbon source, NH_3, O_2) needed for the growth of the biomass are added into the fermenter in order to raise the biomass concentration to the desired level. In the second phase, the supply of nitrogen is cut off so that only the carbon source and O_2 are added into the fermenter. This restricts the biomass from growing so that the excess carbon source is redirected towards PHB production (Rathore 2014).

Over the years, the fermentation processes have used a bacterial system called the chemoheterotrophic bacterial system. The chemoheterotrophic bacterial system (as shown in Table 1.2) uses agro-based carbon source materials as feedstock (Carpine et al. 2020). Now, this brings us back to our previous question about why the production of bioplastics such as PHB is done on a small scale. The agro-based carbon source materials used as feedstock for the production of PHB by the chemoheterotrophic bacteria are expensive (Carpine et al. 2020). They occupy 50% of the total cost of production of PHB (Ramadas et al. 2009). Therefore, this limits the production of PHB on a larger scale (Carpine et al. 2020).

A lot of work, including the use of waste material as a carbon source (Al-Battashi et al. 2019; Reddy et al. 2019) and improving the fermentation strategy (Lopez-Arenas et al. 2017; Garcia-Gonzalez et al. 2015) is still being conducted for reducing the cost of production of PHBs. Unfortunately, most of the work reported so far is still based on the chemoheterotrophic bacterial system which uses an expensive feedstock. Therefore, a bacterial system called the photoautotrophic cyanobacteria system has been alternatively used for reducing the cost of production of PHBs (Carpine et al. 2020). The photoautotrophic cyanobacteria system uses CO_2 and sunlight as feedstock instead of the expensive carbon source materials. This bacterial system uses different strains of cyanobacteria for the production of PHB. Judging from the previous history of PHB production, no one ever thought that photoautotrophic bacteria like cyanobacteria could ever be used in producing PHB (Carpine et al. 2020). Table 1.3 shows a summary of some of the cyanobacteria strains that are used in producing PHB. Using cheap feedstock such as CO_2 and sunlight for the production of PHB by cyanobacteria has attracted a lot of research attention (Carpine et al. 2020). Therefore, different techniques have been developed for the improvement of PHB production by cyanobacteria. These strategies include: (1) designing cyanobacteria growth conditions; (2) genetic engineering of the cyanobacteria for the improvement of the PHB storage as shown in Table 1.4; (3) developing suitable models as tools for identifying the challenges faced by the microbial production of PHB; and (4) facilitating the operation of photobioreactors in order to get a maximum yield (Carpine et al. 2020). The synergy between these strategies is very important for the designing of an optimum PHB production process using cyanobacteria (Carpine et al. 2020). The production of PHB using cyanobacteria has a few challenges as well. The first challenge is that the recovery process for recovering the PHB homopolymer from the bacterial cells is expensive (Carpine et al. 2020). The second challenge is that the chemicals that are used during the recovery process are not environmentally friendly (Carpine et al. 2020). Therefore, a cheaper and eco-friendly recovery process needs to be investigated.

Just like their chemoheterotrophic counterparts, cyanobacteria have all the qualities of bacteria (Carpine et al. 2020). However, there is one feature that distinguishes cyanobacteria from their chemoheterotrophic counterparts. That feature is their ability to use CO_2 and sunlight energy, instead of the expensive carbon source materials, to convert them into a product of interest such as PHB (Carpine et al. 2020). The use of cyanobacteria for PHB production has two advantages. The first is that cyanobacteria do not use agro-food market products for their feedstock (Kadiyala 2014; Gupta et al. 2013; Wijffels et al. 2013) and the second is that cyanobacteria use captured CO_2, thus reducing the pressure of releasing greenhouse gases into the atmosphere (Carpine et al. 2020).

Cyanobacteria produce and accumulate PHB when they are exposed to photoautotrophic conditions under a limited supply of nutrients such as nitrate, phosphorus, calcium and magnesium, as well as macro-nutrients such as manganese, ferrous, cobalt, zinc and copper (Balaji et al. 2013; Costa et al. 2018). The accumulated PHB is usually used as carbon and energy storages by the cyanobacteria (Wu et al. 2001). Now, *Synechocystis sp.* PCC 6803 is one of the most commonly used as well as characterized cyanobacteria strain for the production of PHB. This is because this strain of the cyanobacteria grows very fast. Its doubling time has been reduced to less than 7 h. The other reason for the cyanobacteria strain's common use is that this

Table 1.3 Summary of some of the cyanobacteria strains used for the production of PHB

Type of cyanobacterial strain	Cyanobacteria growth conditions	Type of produced biopolymer	Amount of produced biopolymer (%)	Refs
Arthrospira (Spirulina) platensis	Photoautotrophic	PHB	6	Campbell et al. (1982)
Anabaena cylindrica 10C	Propionate	P(3HB-co-3HV)	2	Lama et al. (1996)
Aulosira fertilissima CCC 444	Citrate, acetate and K_2HPO_4	PHB	85	Samantaray and Mallick (2012)
Aulosira fertilissima CCC 444	Fructose and valerate	P(3HB-co-3HV)	77	Ceyhan and Ozdemir (2011)
Gloeothece sp. PCC 6909	Acetate	PHB	9	Stal et al. (1990)
Nostoc muscorum Agardh	Deficiency of nitrogen, acetate, glucose and valerate	P(3HB-co-3HV)	71	Cavalheiro et al. (2009)
Nostoc muscorum CCAP 1453/9	CO_2	PHB	22	Akar et al. (2006)
Scytonema geitleri Bharawaja	30 mM acetate	PHB	7	Singh et al. (2019)
Synechococcus sp. PCC7942	Acetate and deficiency of nitrogen	PHB	26	Nikel et al. (2006)
Synechocystis sp. PCC 6803	CO_2; concentration of nitrate is the half of the optimal concentration	PHB	8	Wendlandt et al. (2005)
Synechocystis sp. PCC 6803	Deficiency of nitrate; CO_2	PHB	4	Qi and Rehm (2001)
Synechocystis sp. PCC 6803	Deficiency of phosphate, gas exchange limitation and acetate	PHB	38	Goff et al. (2007)
Spirulina sp. LEB 18	Sodium bicarbonate, nitrogen and phosphorus deficiency	PHB	31	Quillaguaman et al. (2005)
Spirulina platensis UMACC 161	CO_2 and acetate	PHB	10	Toh et al. (2008)

strain of the cyanobacteria does not have specific nutritional requirements (Angermayr et al. 2009). It is flexible.

Table 1.4 Genetic engineering approaches used to improve PHB accumulation in cyanobacteria

Type of cyanobacteria strain	Genetic engineering approach	Source of carbon	Cyanobacteria growth conditions	Type of reactor	Amount of PHB accumulated (% dcw)	Refs
Synechocystis sp. PCC 6803 mutant strain	Insertion of the *agp* gene	0.035% CO_2, 5 mM acetate	Deprived of nitrogen	Flask	18.6	Gupta et al. (2013)
Synechococcus sp. PCC7942 mutant strain	Expression of a PHB synthesizing enzyme obtained from *A. eutrophus* in a plasmid	5% CO_2, 10 mM acetate	Deprived of nitrogen	Flask	25	Nikel et al. (2006)
Synechocystis sp. PCC 6803 mutant strain	Expression of PHB synthase from *C. necator* in a plasmid	1% CO_2, 10 mM acetate	Deprived of nitrogen	Flask	11	Gopi et al. (2014)
Synechocystis sp. PCC 6803 mutant strain	Expression of native *sigE* from the chromosome	1% CO_2	Deprived of nitrogen	Flask	1.4	Blankenship (2010)
Synechocystis sp. PCC 6803 mutant strain	Expression of PHA biosynthetic operon obtained from *M. aeruginosa* in a plasmid	2% CO_2	Deprived of nitrogen	Flask	7	Wijffels et al. (2013)
Synechocystis sp. PCC 6803 mutant strain	Native *PhaAB* from the chromosome was overexpressed	0.035% CO_2, 4 mM acetate	Deprived of nitrogen	Flask	35	Poirier et al. (1992)
Synechocystis sp. PCC 6803 mutant strain	Expression of *xfpk* from *B. breve* in a double *ach* and *pta* knock out	2% CO_2	BG11	PBR	12	Vermaas (2001)

1.5 Factors Affecting the PHB Accumulation in Bacteria

1.5.1 Effect of the Bacteria Strain

A lot of bacteria strains which have the natural capacity to produce different amounts of PHBs have been used over the years. However, most of these bacteria strains have shown limited industrial applicability during the industrial scale production of PHBs. This is because these bacteria strains produce low yields of PHB with reference to the yields required on an industrial scale. Therefore, to expand PHB production to an industrial scale, more efforts have been geared towards the genetic modification of the bacteria strains. The genetic modification of the bacteria strains was achieved in several ways which include the modification of their metabolism pathways so that they favour the synthesis of PHB, the regulation of the regulatory systems that regulate the synthesis of PHB and expression of the recombinant *phb* gene (Peña et al. 2014a).

The PHB biosynthetic pathways in bacteria contend for the same precursors with the bacteria's central metabolic pathways. These central metabolic pathways include the tricarboxylic acid (TCA) cycle, degradation of fatty acid (β-oxidation) and biosynthesis of fatty acid. These PHB biosynthetic pathways also contend with many other biosynthetic pathways with which they share common precursors (Peña et al. 2014a). Three examples of PHB synthesis by metabolic pathway modification in *A. vinelandii* have been reported (Peña et al. 2014a). The first example involved the mutation of the respiratory NDH oxidase to give a new strain (UWD). This strain was able to accumulate PHB in the exponential growth phase without depriving it of nutrients (Page and Knosp 1989). The second instance involved inactivating pyruvate carboxylate, which is the enzyme responsible for catalysing the reactions that replenish the TCA cycle (Segura and Espín 2004). This type of mutation increased the production of PHB three times more than what was produced by the natural PHB-producing strain (*A. vinelandii* UW136). This was attributed to the blocked flux of acetyl-CoA into the TCA cycle and hence increasing its availability for PHB synthesis. The third example involved the mutation that blocks the synthesis of the alginate exopolysaccharide. This mutation increased the production of PHB five times more than what is produced by the natural species (UTCC9046). This mutation did not only increase the content of PHB per biomass unit produced by the bacterium, but it also increased the bacterium growth rate. As a result, the volumetric polymer production was increased by 10-fold (Segura et al. 2003).

A few examples involving PHB synthesis by the modification of the regulatory systems that regulate the synthesis of PHB in *A. vinelandii* have also been reported. PHB synthesis in *A. vinelandii* is regulated by two systems (Peña et al. 2014a). The first system is called the nitrogen-related phosphotransferase system, in which the expression of the PHB biosynthetic operon is suppressed by the IIA^{Ntr} protein in a non-phosphorylated form (Noguez et al. 2008). The second system is called the post-transcriptional regulatory system (RsmZ/Y- A). In this PHB synthesis system, the translation of mRNAs from *phbBAC* biosynthetic operon and the *phbR* that codes for the RsmA protein's transcriptional activator is suppressed by the RsmA protein (Hernandez-Eligio et al. 2012). In order to improve PHB production, the IIA^{Ntr} and RsmA-negative regulators were first identified. *A. vinelandii* was used in the production of PHB by inactivating the gene coding in its IIA^{Ntr} (ptsN) protein. As a result of the mutation, a 77% PHB yield was achieved which was equivalent to 4.1 g/L in the genetically modified strain and 3.5 g/L in the natural strain. Therefore, this means that the obtained yield was increased by 36% when compared to the original strain depending on the substrate consumed (Peña et al. 2014b). To further increase PHB production, both the IIA^{Ntr} and RsmA-negative regulators were inactivated. This mutation together with a suitable fermentation process further increased the PHB production up to 27 g/L (García et al. 2014).

Another example involving the synthesis of PHB by the modification of the regulatory systems that regulate PHB synthesis in cyanobacterium Synechocystis sp. PCC 6803 was reported. This is also shown in Table 1.4. Here, the overexpression of the sigma factor SigE, which is responsible for the activation of many catabolic genes of sugar and increasing the acetyl-CoA levels, increased PHB production by two- or three-fold compared to the natural strain (Osanai et al. 2013).

The successful production of PHB is usually determined by the fast growth rate of the bacterium and an increased cell density consisting of an increased PHB content in the culture media. *E. coli*, a well-investigated bacterium using several state-of-the-art technologies used in genetic engineering has attracted a lot of research attention for its use in the production of PHB.*E. coli* is a non-producer of PHB. However, in 1988 Slater et al. (Slater et al. 1988) cloned for the first time the genes of the PHB producer strain, *C. necator* H16 in *E. coli*. The experiment was a success and it resulted in the production of PHB by *E. coli*. Since then, a lot of genetic modifications have been performed on *E. coli* for improving PHB accumulation in these microorganisms. Most of the strategies for genetic modification have been reported by Li et al. (2007) and Wang et al. (2013). Most studies have shown that the use of recombinant *E. coli* strains increases the cell concentration in cultures up to 200 g dry cell weight (DCW)/L, cell growth rate and PHB content up to 90% of the DCW than in natural species (Ahn et al. 2000; Reddy et al. 2003).

1.5.2 Effect of the Carbon Source Materials

The cost of producing PHB is usually influenced by several factors which include fermentation substrate costs, polymer extraction from the bacteria cell cytoplasm as well as the handling the extraction and fermentation wastes (Chen 2010). However, of all the costs involved in the production of PHBs, the carbon source materials cost is the highest (Peña et al. 2014a). The carbon source materials used in the production of PHB can be categorized into two categories; the agro-based and non-agro materials such as CO_2 and sunlight. The agro-based carbon source materials are used by most chemoheterotrophic bacterial systems such as the ones shown in Table 1.2 and the non-agro materials are used by most photoautotrophic bacterial systems such as cyanobacteria as shown by Table 1.3. The use of agro-based carbon source materials is very expensive because it competes with the agro-food market (Carpine et al. 2020). Therefore, new strategies which use waste materials (Al-Battashi et al. 2019; Reddy et al. 2019), CO_2 and sunlight (Carpine et al. 2020) as carbon source materials have been introduced as new ways of reducing the production cost of PHBs. For example, the use of several bacteria strains to convert wastewater (Reddy et al. 2019), paper waste (Al-Battashi et al. 2019), CO_2 and sunlight (Carpine et al. 2020) into PHB has been considered a cheaper PHB synthetic route. When a suitable carbon source material is selected, the standard of selection should not be based only on its market price but on its availability and global price as well (Chanprateep 2010). Table 1.2 shows a summary of some of the agro-based carbon sources used in the production of PHB. From the information presented in the table, it is clear that the agro-based carbon source material used as the fermentation substrate has a significant effect on the amount of PHB produced. For example, when corn syrup and waste glycerol were used in the production of PHB using *Cupriavidus necator* DSM 545 bacteria, 30 and 62% of PHB were accumulated. Contrastingly, Table 1.4 shows a summary of some of the non-agro carbon sources used in the production of PHB. From the table, it was also observed that the non-agro carbon source materials significantly influenced the PHB content accumulated by cyanobacteria. For example, the use of 2% CO_2 as well as 0.035% CO_2 and 4 mM acetate as carbon source materials in the presence of *Synechocystis* sp. PCC 6803 mutant strain under nitrogen-deprived conditions, gave PHB yields of 7% and 35%. This was an indication that both agro-based and non-agro carbon source materials had an influence on the amount of PHB accumulated by bacteria.

1.5.3 Effect of the Fermentation Process

Over the last few decades, very few bacteria strains including *C. necator* and recombinant *E. coli* have been employed in the industrial scale synthesis of PHB (Khanna and Srivastava 2005). Surprisingly, high PHB content accumulating bacteria strains such as *A. vinelandii* and *A. chrooccum* have not been used in the industrial scale production of PHB (Peña et al. 2014a). As a result, more bacteria strains have been added to the list of bacteria strains used in the industrial scale production of PHB. Many fermentation processes which use fed-batch, batch and continuous cultures are used in PHBs production (Peña et al. 2014a; Sirohi et al. 2020).

Fermentation processes using batch cultures for PHB production have been the most commonly used as a result of their low cost of operation and flexibility. However, their disadvantage is that batch cultures produce a low PHB yield (Peña et al. 2014a). For example, the production of PHB in batch cultures consisting of *C. necator* ATCC 17699 and acetic acid as the source of carbon was investigated by Wang and Yu (Wang and Yu 2001). In their research, only a PHB content of 50% (w/w) and productivity rate of 0.046 g/Lh was achieved with a carbon/nitrogen (C/N) ratio of 76 in their batch culture conditions.

Fermentation processes involving the use of fed-batch cultures have been used to obtain increased yields of PHB (Kulpreecha et al. 2009). During fed-batch cultivation systems, one or several nutrients are fed into the bioreactor. The obtained products and other constituents of the system are left inside the bioreactor until the fermentation process is done. This allows materials to only flow in and not out. As a result of this, the contents of the bioreactor keep changing with time (Mejía et al. 2010). In order to regulate the carbon source materials concentration during the fed-batch process, a good substrate feeding strategy is required. There are several feeding strategies used which include, pulse feeding, continuous feeding with a defined feeding rate, exponential feeding, using pH control (pH-stat mode) and dissolved oxygen (DO) concentration to control the nutrient feed (DO-stat mode) (Cruz et al. 2019). From previous studies, it has been shown that the feeding strategy used in fed-batch cultures has a significant effect on the amount of PHB accumulated as shown by Table 1.5.

Fermentation processes involving the use of a continuous culture system usually produce PHB in two or three stages (Ruan et al. 2003; Rocha et al. 2008). The first stage is for biomass accumulation and the second stage is for PHB production. The continuous culture PHB production process has attracted a lot of commercial interest. This is due to its high PHB productivity rate especially when strains such as *R. eutropha* characterized by an increased specific rate of growth are used. For example, Du et al. (2001) obtained a maximum PHB yield of 72.1% at a dilution rate of 0.075 h^{-1} when a two-stage continuous culture of *R. eutropha* was used in the presence of glucose as the carbon source. In another study, Tan et al. (2011) obtained a maximum PHB yield of 65–70% after 14 days when an open unsterile continuous culture of *Halomonas*TD01 was used in the presence of a nitrogen-deficient glucose salt medium.

1.5.4 Effect of Culture Conditions

The effect of culture conditions such as pH, medium composition and oxygen availability on the molecular weight of PHB has been reported by several authors (Peña et al. 2014a). Table 1.6 shows a summary of the effect of different culture conditions on PHB molecular weight. The PHB molecular weight is an important feature that determines its medical applications. Therefore, the molecular weight of PHB is highly considered during the commercial production of PHB (Peña et al. 2014a). Moreover, properties such as the elastic behaviour and mechanical resistance of the PHB are controlled by its molecular weight. For instance, PHB fibres with a molecular weight of about 3.0×10^2 kDa are characterized by a tensile strength of about 190 MPa as well as an elongation at break of about 5%. Contrastingly, PHB fibres with a higher molecular weight of about 5.3×10^3 can experience a seven-fold increase of their tensile strength up to 1320 MPa and an increase of their elongation at break up to 57% (Iwata 2005).

Table 1.5 Effect of different feeding strategies on the PHB production in fed-batch cultures

Bacteria strain	Carbon source	Feeding strategy	Dry cell weight (DCW) ($g\ L^{-1}$)	PHB concentration ($g\ L^{-1}$)	PHB productivity	PHB content (% wt)	Refs
B. megaterium SRKP-3	Dairy waste + rice bran + seawater	DO-stat mode	n.m.	11.32	n.m.	n.m.	Pandian et al. (2010)
E. Coli WΔcscR mutant	Sucrose	pH-stat mode	n.m.	47.7 ± 3.1	17.5 ± 0.01 $g\ L^{-1}\ h^{-1}$	45.8 ± 3.6	Arifin et al. (2011)
E. Coli W	Glucose		n.m.	36.0 ± 0.4	1.87 ± 0.05 $g\ L^{-1}\ h^{-1}$	40.1 ± 0.4	
E. Coli W	Sucrose		n.m.	31.7 ± 1.4	1.40 ± 0.04 $g\ L^{-1}\ h^{-1}$	36.2 ± 1.2	
Methylobacterium sp. ZP24	Cheese whey	Intermittant ± DO-stat	n.m.	3.54	0.09 $g\ L^{-1}\ h^{-1}$	64	Nath et al. (2008)
C. necator	Corn stover alkaline pretreatment liquor	Single pulse feeding (300 mL) with supplements	n.m.	4.02	0.09 $g\ L^{-1}\ h^{-1}$	n.m.	Li and Wilkins (2020)
		4 times pulses feeding (each 75 mL) with supplements	n.m.	4.47	0.08 $g\ L^{-1}\ h^{-1}$	n.m.	
C. necator DSM 428	Used cooking oil	Exponential feeding	21.3 ± 0.9	17.9 ± 0.9	4.5 ± 0.2 $g\ L^{-1}\ day^{-1}$	84.0 ± 4.5	Cruz et al. (2019)
		DO-stat	27.2 ± 0.5	19.8 ± 1.8	12.6 ± 0.8 $g\ L^{-1}\ day^{-1}$	77.0 ± 5.7	
B. megaterium BA-019	Sugar cane molasses	pH-stat	72.6	30.5	1.27 $g\ L^{-1}\ h^{-1}$	42	Kulpreecha et al. (2009)
		Intermittent	90.7	41.6	1.73 $g\ L^{-1}\ h^{-1}$	46	Kanjanachumpol et al. (2013)

(continued)

Table 1.5 (continued)

Bacteria strain	Carbon source	Feeding strategy	Dry cell weight (DCW) ($g\ L^{-1}$)	PHB concentration ($g\ L^{-1}$)	PHB productivity	PHB content (% wt)	Refs
C. necator	Glycerol	Pulses	75	53	$0.92\ g\ L^{-1}\ h^{-1}$	71	Tanadchangsaeng and Yu (2012))
	Soybean oil	Pulses	83	67.2	$2.5\ g\ L^{-1}\ h^{-1}$	81	da Cruz Pradella et al. (2012)
	Glycerol	Pulses	82.5	51.2	$1.52\ g\ L^{-1}\ h^{-1}$	62	Cavalheiro et al. (2009)
	Glucose	Exponential ± coupled to alkali addition monitoring ± constant with N2 limitation	164	125	$2.03\ g\ L^{-1}\ h^{-1}$	76.2	Mozumder et al. (2014)
E. coli CGSC 4401	Whey	pH-stat	119.5	96.2	$2.57\ g\ L^{-1}\ h^{-1}$	80	Ahn et al. (2000)
A. vinelandii	Sucrose and yeast extract	Exponential ± pulses	37.2	27.3	$0.5\ g\ L^{-1}\ h^{-1}$	73.3	García et al. (2014)

n.m. stands for not mentioned

1.6 Extraction of PHB

The production of PHB usually involves the fermentation, extraction and purification of PHB. Now, the commercial PHB production and industrial PHB application is limited by its expensive production cost when compared to petroleum-derived plastics. The high production cost is mostly influenced by the cost of the fermentation process and the extraction/purification technologies. In addition to the fermentation cost, the PHB recovery process also increases the overall production cost quite significantly (Aramvash et al. 2015). In order to reduce the PHB extraction costs, solvents such as sodium hydroxide and sulphuric acid have been used during the extraction process (López-Abelairas et al. 2015). Unfortunately, the use of these solvents has been considered not environmentally friendly. Therefore, an environmentally friendly, more efficient and cheaper PHB extraction method is needed.

In order to develop a suitable extraction process, more details on the formation and accumulation of PHB need to be provided. PHB is a compound belonging to a class of polyesters that are produced by several bacteria strains and then stored as a carbon and energy reserve. The PHB is accumulated in the cell cytoplasm of the microorganisms in the form of inclusion bodies or PHB granules. The PHB granules appear as sack-like structures wrapped up by layer made up of a lot of proteins (see Fig. 1.4). The PHB granules have supramolecular complexes-like structures and specific functions which make them to be more than just carbon and energy storage materials (Jendrossek and Pfeiffer 2014).

The mechanism for the formation of PHB granules in bacteria is a topic that is currently under intense discussion. In order to understand the mechanisms that lead to the formation of PHB granules, three models were developed. These models were developed from theoretical manipulations and experiments performed on cultures of *Ralstonia eutropha*. The first model is called the micelle model. This model is based on the assumption that the soluble (cytoplasmic) PHB synthase can only start producing PHB once the substrate (3-hydroxybutyryl-CoA) concentration has increased sufficiently (Gerngross et al. 1993). During the PHB production, the nascent PHB polymer chains form hydrophobic and less soluble aggregates which form micelle structures in the cell cytoplasm. According to the micelle model, PHB granules assume a random distribution within the cell (Jendrossek and Pfeiffer 2014; Stubbe et al. 2005). The second model is called the budding model. According to this model, the PHB synthase is situated in the cytoplasm membrane and the PHB granules that have been newly formed are found in the membrane bilayer (Das et al. 1997). The third model is called the scaffold model. According to this model, it is assumed that the PHB synthase of nascent PHB granules attaches itself on a scaffold molecule that is yet unknown in the cell (Jendrossek and Pfeiffer 2014). This forms a granule-scaffold complex that is attached to the bacteria nucleoid and consists of complexes that initiate PHB granule formation. According to the scaffold model, the subcellular localization of the PHB granules is determined by the localization and nature of the scaffold molecules consisting of the cells accumulating PHB (Jendrossek and Pfeiffer 2014).

Table 1.6 Effect of different culture conditions on the molecular weight and content of PHB accumulated

Bacteria strain	Carbon source	Culture conditions	Mw (kDa)	PHB content (%)	Refs
C. necator	Acetic acid	C/N ratio = 4	820	50	Wang and Yu (2001)
		C/N ratio = 72	520		
A. lata	Sucrose	C/N ratio = 20, C/P ratio = 1.9, D.O. concentration = 20%, culture residence time in stationary phase = 1 h	2576	15	Penloglou et al. (2012)
		C/N ratio = 8, C/P ratio = 1.9, D.O. concentration = 20%, culture residence time in the stationary phase = 1 h	596	35	
		C/P ratio = 8, C/N ratio = 10, D.O. concentration = 20%, culture residence time in the stationary phase = 1 h	2076	27	
		C/N ratio = 10, C/P ratio = 1.9, D.O. concentration = 20%, culture residence time in the stationary phase = 1 h	894	46	
R. eutropha	Glucose	Oxygen concentration = 21%	102.1	n.m	
		Oxygen concentration = 50%	134.4	80.8	
		Oxygen concentration = 65%	111.35	99.8	
		Oxygen concentration = 75%	125.52	51.2	
		Oxygen concentration = 100%	71.9	3.4	
A. cchroccoccum 7B	Sucrose	2% (w/v) sucrose only and sucrose supplemented with molasses	1200–1600	74–79	Myshkina et al. (2008)
	Sucrose + Molasses		590	60	
E. coli XL-1	Glucose	pH of between 6.0 and 7.0	2000–2500	32–35	Bocanegra et al. (2013)
*A. vinelandii*OPN	Sucrose	When aeration is low	2020	67	Peña et al. (2014b)
		When aeration is high	1010	62	
A. vinelandii OPNA	Sucrose	C/N ratio = 10 in a batch culture	2300	85	Castillo et al. (2017)
		C/N ratio = 18 in a batch culture	6600	85	

		C/N ratio = 10 in a fed-batch culture with a one pulse feeding strategy	1900	60	
		C/N ratio = 14 in in a fed-batch culture with a one pulse feeding strategy	3700	74	
		C/N ratio = 18 in a fed-batch culture with a one pulse feeding strategy	5300	80	
		C/N ratio = 14 in a fed-batch culture with a two pulse feeding strategy	3700	71	
		C/N ratio = 18 in a fed-batch culture with a two pulse feeding strategy	3100	77	

n.m stands for not mentioned

Regardless of how the PHB is formed or accumulated, the extraction of PHB from the cell cytoplasm involves the rupture of the bacteria cell and the subsequent elimination of the layer of protein wrapped around the PHB granules. However, other PHB extraction methods which involve selectively dissolving PHB granules using a good solvent and subsequently precipitating the dissolved PHB have been alternatively used. This method of extraction is called the solvent extraction method (Kunasundari and Sudesh 2011).

Solvent extraction is one of the most commonly used technique for extracting of PHB from cellular biomass. This is a result of the simplicity, rapidity and the ability of the technique to allow for the assessment of PHB accumulation. The solvent extraction method involves a two-step sequential procedure. The first step involves changing the permeable nature of the cell membrane to improve the release and dissolution of the PHB. In the second step, non-solvents are used to precipitate the PHB that has been dissolved in the first step (Jacquel et al. 2008). The dissolution step of the solvent extraction method uses solvents such as chlorinated hydrocarbons like chloroform and 1,2-dichloroethane (Ramsay et al. 1994) as well as cyclic carbonates like 1,2-propylene carbonate and ethylene carbonate (Carpine et al. 2020). Contrastingly, the precipitation step of the solvent extraction method uses non-solvents such as ethanol and methanol (Ramsay et al. 1994).

The solvent extraction method has a few advantages as compared to other extraction methods. One of its main advantages amongst others is its extraction efficiency. Another advantage is that the solvent extraction method is capable of eliminating bacterial endotoxin and therefore eliminates the chances of significant polymer degradation (Jacquel et al. 2008). Therefore, this method gives the possibility of obtaining PHB with high purity and significantly high molecular weights. Unfortunately, the application of the solvent extraction method on a larger scale is limited by the use of solvents which are not environmentally friendly (Carpine et al. 2020). Table 1.7 shows a summary of all the solvents that are used during the solvent extraction method. The PHB yield and purity results obtained after the use of the different solvents is also shown. Furthermore, the digestion method has been used as an alternative extraction method to the solvent extraction technique (Carpine et al. 2020). According to the digestion method, the extraction of PHB is achieved by the dissolution of the cellular material wrapped around the PHB. Table 1.8 shows a summary of the digestion methods applied during the extraction of PHB. The purity and the PHB yield are also shown in the table.

In order to prevent environmental pollution, a simple, practical, efficient, cost-effective and environmentally friendly PHB extraction method that uses non-halogenated solvents is required (Carpine et al. 2020). As a result, several PHB extraction methods have been reported in this regard. For instance, the use of ionic liquids (ILs) to recover PHB accumulated in cyanobacteria by dissolving the cyanobacteria and retaining the PHB has been reported in literature (Kobayashi et al. 2015). Now, ILs are gaining much attention for use as novel green solvents. Their melting point is below 100 °C. ILs also have interesting properties such as high thermal stability and high ionic conductivity which are not found in conventional organic solvents (Kobayashi et al. 2015). Other authors have also reported the use of

Table 1.7 Summary of the solvents used by the PHB solvent extraction method

Solvents used	Bacteria strain	Results	Refs
Chloroform	*Bacillus cereus* SPV	Purity = 92%; Yield = 31%	Valappil et al. (2007)
Chloroform	*Cupriavidus necator* DSM 545	Purity = 95%; Yield = 96%	Fiorese et al. (2009)
1,2-propylene carbonate	*Cupriavidus necator* DSM 545	Purity = 84%; Yield = 95%	Fiorese et al. (2009)
Methyl tert-butyl ether	*Pseudomonas putida* KT2440	Yield = 80–85%	Wampfler et al. (2010)
Butyl acetate	*C. necator*	Purity = 99; Yield: 96	Aramvash et al. (2015)
Non-halogenated acetone/ ethanol/propylene carbonate	*C. necator*	Purity = 93%; Yield = 92%	Fei et al. (2016)
Acetone	*P. putida* GPo1	Yield = 94%	Elbahloul and Steinbüchel (2009)

the genetic engineering approach to liberate intracellular PHB granules. Jung et al. (Jung et al. 2005) developed an *Escherichia coli* strain that was capable of producing sufficiently high yields of PHB by manipulating the inoculum size and medium composition. After a spontaneous disintegration of the bacteria cells, the PHB granules were delivered into the culture medium. The use of solvents that are not chlorinated such as anisole (Rosengart et al. 2015) or solvents that are not halogenated such as butyl acetate (Wang et al. 2015) has also been considered as environmentally friendly PHB extraction method. Furthermore, the use of a solvent-free technique such as enzyme digestion has also been considered as an environmentally friendly route for the extraction of PHB. Martino et al. (Martino et al. 2014) used a combination of the enzyme Alcalase and an aqueous medium of sodium dodecyl sulphate (SDS) and ethylenediaminetetraacetic acid (EDTA) for the extraction of PHB granules from the biomass of *Cupriavidus necator* DSM 428 where used cooking oil (UCO) was used as a source of carbon. The extracted PHB granules had a purity of >90% and no crystallization.

1.7 Applications of PHB

The most outstanding feature of PHBs is that they are 100% biodegradable. However, the biodegradability occurs when the materials are exposed to biologically active environments such as in soils, seawater or freshwater as well as in aerobic and anaerobic compost (Telles et al. 2011). Now, biologically active environments are environments where the PHB materials come into contact with degrading microorganisms (Woolnough et al. 2010). In addition, environments such as in activated sludge (Corrêa et al. 2008) and in sanitary landfills, in which the materials can be discarded without causing any harm to the environment, are also considered as biologically active (Casarin et al. 2013).

Table 1.8 A summary of the digestion methods for the extraction of PHB

Digestion extraction method	Materials used	Bacteria strain	Results	Refs
Surfactant	SDS	Recombinant *Escherichia coli*	Purity = 99%; Yield = 89%	Choi and Lee (1999)
	Palmitoyl carnitine	*C. necator*, *Alcaligenes latus*	Degree of lysis = 56–78%	Lee et al. (1993)
Surfactant-sodium hypochlorite	SDS-sodium hypochlorite	*Azotobacter chroococcum* G-3	Purity = 98%; Yield = 87%	Dong and Sun (2000)
	Triton X-100-sodium hypochlorite	*C. necator* DSM 545	Purity = 98%	Ramsay et al. (1990)
Sodium hypochlorite	Sodium hypochlorite	*C. necator* recombinant *Escherichia coli*	Purity = 86%; Purity = 93%	Hahn et al. (1995)
	Sodium hypochlorite	*Staphylococcus epidermidis*		Marjadi and Dharaiya (2014)
Surfactant chelate	Triton X-100-EDTA	*Sinorhizobium meliloti*	Purity = 68%	Lakshman and Shamala (2006)
	Betaine-EDTA disodium salt	*C. necator* DSM 545	Purity ≥ 96%; Yield = 90%	Chen et al. (2001)
Dispersion of chloroform and sodium hypochlorite	Chloroform-sodium hypochlorite	*B. cereus* SPV	Purity = 95%; Yield = 30%	Valappil et al. (2007)
	Chloroform-sodium hypochlorite	*C. necator* recombinant *Escherichia coli*	Purity ≥ 98%	Hahn et al. (1995)
Proton-selective dissolution	Sulphuric acid	*C. necator*	Purity ≥ 97%; Yield ≥ 95%	Yu and Chen (2006)
Enzymatic digestion	Combination of enzyme with SDS-EDTA	*P. putida*	Purity = 93%	Kathiraser et al. (2007)
	Alcalase mixed with EDTA and SDS	*Cupriavidus necator* DSM 428	Purity ≥ 90%; Yield ≥ 90%	Koksharova and Wolk (2002)
Mechanical disruption	Bead mill	*A. latus*		Tamer et al. (1998)
	High-pressure homogenization	*A. latus*		Tamer et al. (1998)
	SDS-high-pressure homogenization	*Methylobacterium* sp. V49	Purity = 95%; Yield = 98%	Ghatnekar et al. (2002)
Spontaneous liberation		*E. coli*	80% autolysis	Toh et al. (2008)

Furthermore, biocompatibility is another important feature of PHBs. Now, biocompatibility refers to the suitable interaction between a particular material and its biological environment. In simpler terms, the substance is considered biocompatible if it is not rejected by the body and non-toxic waste is generated during assimilation (Dos Santos et al. 2017). PHB is considered a biocompatible material for two reasons. The first reason is that, it is found in the form of low molar mass PHB in the bloodstream. The second reason is that it produces 3-hydroxybutyric acid upon degradation, which is a common metabolite found in living things (Dos Santos et al. 2017).

Due to their biodegradability and biocompatibility, PHBs find many applications in the biomedical, packaging, tissue engineering and drug delivery fields (Dos Santos et al. 2017). Some of the applications of PHBs are shown in Table 1.9. PHBs have other features that limit their applications in various fields. For instance, PHBs have a physical ageing effect which leads to their brittle behavior (Dos Santos et al. 2017; Raza et al. 2019). PHBs are also thermally unstable and are therefore susceptible to thermal degradation during processing. This can be attributed to their narrow processing window, which is characterized by the close proximity between their melting and degradation temperatures. Another major drawback is that PHBs flow during processing (Dos Santos et al. 2017, Raza et al. 2019). Therefore, several approaches aimed at improving their mechanical properties and reduction of the production cost of PHB has been reported in literature. These approaches involve the preparation of PHB blends, PHB copolymers and insertion of additives to develop novel PHB-based composites with optimum properties (Dos Santos et al. 2017, Raza et al. 2019). Some of the PHB-based composites used in PHB applications together with their preparation methods are shown in Table 1.9.

1.8 Conclusion

Polyhydroxybutyrates (PHBs) are a class of biopolymers that are synthesized and accumulated by several bacterial strains as a reserve for energy and carbon. The microorganisms produce the PHB under stressful conditions where there is an excess of carbon and an absence of an essential nutrient such as nitrogen, phosphorus, sulphur or oxygen. Furthermore, PHBs are part of a large group of polyesters called polyhydroxyalkanoates (PHAs) which have microbial-derived, lipid-like and water-insoluble properties. The world is faced with an environmental pollution problem. Petroleum-based plastics are the main contributors to this environmental pollution problem. Due to their 100% biodegradability, PHBs have been used to replace the non-degradable petroleum-based plastics. Their biodegradability combined with their biocompatibility has earned PHBs a place in many industrial applications such as in the biomedical, packaging, tissue engineering and drug delivery fields. Due their high production cost, PHBs are still produced on a small scale. The hike in the production cost is mainly due to the cost of the carbon source materials and extraction processes. The high cost of the carbon source materials is due to their competition with the agro-food market. In order to reduce the carbon source material

Table 1.9 A summary of the preparation, applications and properties of PHB-based composed

PHB-based composite	Preparation of composites	Properties of prepared composites	Applications	Refs
TCP/PHBV, HA/PHBV	Compounding, compression molding	Reduced degradation Temperature and crystallinity	Biomedical	Chen and Wang (2002)
PHB/ cellulose	Grafting by extrusion mechanism	Reduced crystallinity and brittleness	Packaging	Wei et al. (2015)
PHB/PLA/ CNF	Melt mixing, injection molding and compounding	Decreased elongation at break and increased elastic modulus	Packaging	Kiziltas et al. (2016)
PHB/CNCs	Solution casting	Increased tensile strength and Young's modulus	Packaging	Seoane et al. (2016)
PHB/ cellulose cardboard	Solvent casting and compression molding	Increased Young's modulus	Packaging	Seoane et al. (2015)
PHBV/ recycled cellulose fibres	Melt mixing	Increased storage modulus and tensile strength	Packaging	Bhardwaj et al. (2006)
PHB/HA	Compression molding	Increased crystallinity and decreased decomposition temperature	Bone tissue engineering	Shishatskaya et al. (2006)
PHB/HA	Ball milling and hot pressing	Improved bending strength of 46.6 MPa	Tissue engineering	Zhuo et al. (2010)
PHB/starch	Melt compounding	Improved thermal stability	Packing	Zhang and Thomas (2010)
PHB/maize starch	Compression molding	Decreased elongation at break and tensile strength	Packing	Thiré et al. (2006)
PHB/starch	Hot pressing and melt mixing	Increased tensile strength (5.9 ± 0.1 MPa) and tear strength (44.1 ± 0.2 kJ/m^2)	Packing	Lai et al. (2006)
PHB/chitin	Solution casting	Melting temperature of 166–170 °C	Biomedical	Reddy et al. (2015)
PHB/α-chitin/ chitosan	Solution casting	Melting temperature of 155–170 °C	Biomedical	Ikejima et al. (1999)
PHB/chitosan	Precipitation method	Tm of 173.2 °C and ΔHm of 91.5 J/g, respectively	Tissue engineering	Chen et al. (2005)

(continued)

Table 1.9 (continued)

PHB-based composite	Preparation of composites	Properties of prepared composites	Applications	Refs
PHB/chitosan	Freeze drying and emulsion blending	Elastic modulus (8.7 ± 1.6 to 4.9 ± 0.6 MPa) and swelling capability (141.1 ± 5.9 to 119.3 ± 4.3%)	Tissue engineering	Cao et al. (2005)
PHB/chitosan	Injection molding	Young's modulus as (1044–2499 MPa), tensile strength as 11 ± 0.2 to 8.5 ± 0.3 MPa and impact strength as 1595 ± 5 to 159 ± 3 kJ/m^2, respectively	Drug delivery	Rajan et al. (2012)
PHB/PCL	Solution casting	Increase in elongation at break (300%) and tensile strength (28.23 MPa)	Bones regeneration	Saeb et al. (2014)
PHB/PCL	Compression molding and compounding	Elastic modulus (1643 ± 59 MPa) and tensile strength (21.4 ± 0.3 MPa)	Bones regeneration	Garcia-Garcia et al. (2016)
PHB/silk	Electrospinning technique	Elongation at break as 17.10% and tensile strength as 3.81 ± 0.1 MPa	Tissue engineering	Karahaliloğlu (2017)
PHBV/silk	Electrospinning technique	Tensile strength as 1.31 ± 0.20 to 5.82 ± 0.50 MPa, Young's modulus as 56.5 ± 5.7 to 67.7 ± 5.2 MPa and elongation at break as 1.31 ± 0.20 to 5.82 ± 0.50 MPa	Tissue engineering	Yang et al. (2011)
PHB/PLA	Melt mixing	58% decrease in tensile strength, ~50-fold increase in elongation at break and Tg as −23 to 19 °C	Food packing	D'Amico et al. (2016)
PHB/PLA	Melt compounding	Decreased tensile strength and elongation at break	Food packing	Zhang and Thomas (2011)

(continued)

Table 1.9 (continued)

PHB-based composite	Preparation of composites	Properties of prepared composites	Applications	Refs
PHB/flax fibre	Compression molding and injection molding	Reduced tensile properties	Biocomposites development	Barkoula et al. (2010)
PHB/kenaf fibre	Compression molding	Increased elongation at break and tensile strength	Packaging	Kuciel and Liber-Kneć (2011)
PHB/PEG	Salt leaching method	Young's modulus as 35 ± 1 to 110 ± 2 kPa and bulk density as 1.08 g/cm^3	Biomedical	Castellanos et al. (2003)

CNF cellulose nanofibres, *CNC* cellulose nanocrystals, *PCL* poly(caprolactone), *HA* hydroxyapatite, *PHB* poly(hydroxybutyrate), *PEG* poly(ethylene glycol), *TCP* tricalcium phosphate, *PLA* poly (lactic acid)

cost, waste materials, CO_2 and sunlight have been used as carbon source materials. However, more work needs to be done to reduce the cost of the carbon source materials. For the extraction of PHB, the cheaper solvent extraction method is used. Unfortunately, the solvent extraction method is not environmentally friendly on a large scale. However, the digestion method is therefore used as an alternative to the solvent extraction method. Therefore, in order to reduce the cost of the extraction process, a process that is environmentally friendly, cheaper, simple, efficient and uses non-halogenated solvents needs to be investigated. Furthermore, the industrial application of PHB is limited by its brittleness and narrow processing window which is caused by the close proximity between their melting and degradation temperatures. This limitation is overcome by blending PHBs with other polymers, fibres or additives to form PHB-based composites with optimum properties.

References

Ahn WS, Park SJ, Lee SY (2000) Production of poly (3-hydroxybutyrate) by fed-batch culture of recombinant Escherichia coli with a highly concentrated whey solution. Appl Environ Microbiol 66:3624–3627

Akar A, Akkaya EU, Yesiladali SK, Çelikyilmaz G, Çokgor EU, Tamerler C, Orhon D, Çakar ZP (2006) Accumulation of polyhydroxyalkanoates by Microlunatus phosphovorus under various growth conditions. J Ind Microbiol Biotechnol 33:215–220

Alarfaj AA, Arshad M, Sholkamy EN, Munusamy MA (2015) Extraction and characterization of polyhydroxybutyrates (PHB) from Bacillus thuringiensis KSADL127 isolated from mangrove environments of Saudi Arabia. Braz Arch Biol Technol 58:781–788

Al-Battashi H, Annamalai N, Al-Kindi S, Nair AS, Al-Bahry S, Verma JP, Sivakumar N (2019) Production of bioplastic (poly-3-hydroxybutyrate) using waste paper as a feedstock: optimization of enzymatic hydrolysis and fermentation employing Burkholderia sacchari. J Clean Prod 214:236–247

Alejandra R-C, Margarita C-M, Soledad MCM (2012) Enzymatic degradation of poly (3-hydroxybutyrate) by a commercial lipase. Polymer Degrad Stability 97:2473–2476

Anderson ZT, Cundy AB, Croudace IW, Warwick PE, Celis-Hernandez O, Stead JL (2018) A rapid method for assessing the accumulation of microplastics in the sea surface microlayer (SML) of estuarine systems. Sci Rep 8:1–11

Angermayr SA, Hellingwerf KJ, Lindblad P, Teixeira De Mattos MJ (2009) Energy biotechnology with cyanobacteria. Curr Opin Biotechnol 20:257–263

Angermayr SA, Rovira AG, Hellingwerf KJ (2015) Metabolic engineering of cyanobacteria for the synthesis of commodity products. Trends Biotechnol 33:352–361

Aramvash A, Gholami-Banadkuki N, Moazzeni-Zavareh F, Hajizadeh-Turchi S (2015) An environmentally friendly and efficient method for extraction of PHB biopolymer with non-halogenated solvents. J Microbiol Biotechnol 25:1936–1943

Arifin Y, Sabri S, Sugiarto H, Krömer JO, Vickers CE, Nielsen LK (2011) Deletion of cscR in Escherichia coli W improves growth and poly-3-hydroxybutyrate (PHB) production from sucrose in fed batch culture. J Biotechnol 156:275–278

Asada Y, Miyake M, Miyake J, Kurane R, Tokiwa Y (1999) Photosynthetic accumulation of poly-(hydroxybutyrate) by cyanobacteria—the metabolism and potential for CO_2 recycling. Int J Biol Macromol 25:37–42

Balaji S, Gopi K, Muthuvelan B (2013) A review on production of poly β hydroxybutyrates from cyanobacteria for the production of bio plastics. Algal Res 2:278–285

Barkoula N, Garkhail S, Peijs T (2010) Biodegradable composites based on flax/polyhydroxybutyrate and its copolymer with hydroxyvalerate. Ind Crop Prod 31:34–42

Bhagowati P, Pradhan S, Dash HR, Das S (2015) Production, optimization and characterization of polyhydroxybutyrate, a biodegradable plastic by Bacillus spp. Biosci Biotechnol Biochem 79:1454–1463

Bhardwaj R, Mohanty AK, Drzal L, Pourboghrat F, Misra M (2006) Renewable resource-based green composites from recycled cellulose fiber and poly (3-hydroxybutyrate-co-3-hydroxyvalerate) bioplastic. Biomacromolecules 7:2044–2051

Bistué Rovira M, Martínez Martínez C, Maurino Reyes M, Pérez Rubio P (2019) Bioprocess design of a PHB production plant

Blankenship RE (2010) Early evolution of photosynthesis. Plant Physiol 154:434–438

Bocanegra JK, Da Cruz Pradella JG, Da Silva LF, Taciro MK, Gomez JGC (2013) Influence of pH on the molecular weight of poly-3-hydroxybutyric acid (P3HB) produced by recombinant Escherichia coli. Appl Biochem Biotechnol 170:1336–1347

Bonatto D, Matias F, Lisbôa MP, Bogdawa HM, Henriques JAP (2004) Production of short side chain-poly[Hydroxyalkanoate] by a newly isolated Ralstonia Pickettii strain. World J Microbiol Biotechnol 20:395–403

Bouwmeester H, Hollman PC, Peters RJ (2015) Potential health impact of environmentally released micro-and nanoplastics in the human food production chain: experiences from nanotoxicology. Environ Sci Technol 49:8932–8947

Bugnicourt E, Cinelli P, Lazzeri A, Alvarez VA (2014) Polyhydroxyalkanoate (PHA): review of synthesis, characteristics, processing and potential applications in packaging. eXPRESS Polym Lett 8(11):791–808

By PUC, Roland-Holst D, Triolo R, Heft-Neal S, Bayrami B (2013) Bioplastics in California

Campbell J, Stevens SE, Balkwill DL (1982) Accumulation of poly-beta-hydroxybutyrate in Spirulina platensis. J Bacteriol 149:361

Cao W, Wang A, Jing D, Gong Y, Zhao N, Zhang X (2005) Novel biodegradable films and scaffolds of chitosan blended with poly (3-hydroxybutyrate). J Biomater Sci Polym Ed 16:1379–1394

Carpine R, Olivieri G, Hellingwerf KJ, Pollio A, Marzocchella A (2020) Industrial production of poly-β-hydroxybutyrate from CO_2: can cyanobacteria meet this challenge? PRO 8:323

Casarin SA, Agnelli JA, Malmonge SM, Rosário F (2013) Blendas PHB/copoliésteres biodegradáveis: biodegradação em solo. Polímeros 23:115–122

Castellanos IJ, Crespo R, Griebenow K (2003) Poly (ethylene glycol) as stabilizer and emulsifying agent: a novel stabilization approach preventing aggregation and inactivation of proteins upon encapsulation in bioerodible polyester microspheres. J Control Release 88:135–145

Castillo T, Flores C, Segura D, Espín G, Sanguino J, Cabrera E, Barreto J, Díaz-Barrera A, Peña C (2017) Production of polyhydroxybutyrate (PHB) of high and ultra-high molecular weight by Azotobacter vinelandii in batch and fed-batch cultures. J Chem Technol Biotechnol 92:1809–1816

Cavalheiro JM, De Almeida MCM, Grandfils C, Da Fonseca M (2009) Poly (3-hydroxybutyrate) production by Cupriavidus necator using waste glycerol. Process Biochem 44:509–515

Ceyhan N, Ozdemir G (2011) Poly-hydroxybutyrate (PHB) production from domestic wastewater using Enterobacter aerogenes 12Bi strain. Afr J Microbiol Res 5:690–702

Chaijamrus S, Udpuay N (2008) Poly (3-Hydroxybutyrate) production with high productivity and high polymer content by a fed-batch culture of alcaligenes latus under nitrogen limitation. Agric Eng Int CIGRE J 27

Chandani N, Mazumder P, Bhattacharjee A (2014) Production of polyhydroxybutyrate (biopolymer) by Bacillus tequilensis NCS-3 isolated from municipal waste areas of Silchar, Assam. Int J Sci Res 3:198–203

Chanprateep S (2010) Current trends in biodegradable polyhydroxyalkanoates. J Biosci Bioeng 110:621–632

Chen GQ (2010) Plastics completely synthesized by bacteria: polyhydroxyalkanoates. In: Plastics from bacteria. Springer

Chen L, Wang M (2002) Production and evaluation of biodegradable composites based on PHB–PHV copolymer. Biomaterials 23:2631–2639

Chen Y, Yang H, Zhou Q, Chen J, Gu G (2001) Cleaner recovery of poly (3-hydroxybutyric acid) synthesized in Alcaligenes eutrophus. Process Biochem 36:501–506

Chen C, Zhou X, Zhuang Y, Dong L (2005) Thermal behavior and intermolecular interactions in blends of poly (3-hydroxybutyrate) and maleated poly (3-hydroxybutyrate) with chitosan. J Polym Sci B Polym Phys 43:35–47

Choi JI, Lee SY (1999) Efficient and economical recovery of poly (3-hydroxybutyrate) from recombinant Escherichia coli by simple digestion with chemicals. Biotechnol Bioeng 62:546–553

Corrêa M, Rezende M, Rosa D, Agnelli J, Nascente P (2008) Surface composition and morphology of poly (3-hydroxybutyrate) exposed to biodegradation. Polym Test 27:447–452

Costa JAV, Moreira JB, Lucas BF, Braga VDS, Cassuriaga APA, Morais MGD (2018) Recent advances and future perspectives of PHB production by cyanobacteria. Ind Biotechnol 14:249–256

Costa SS, Miranda AL, De Morais MG, Costa JAV, Druzian JI (2019) Microalgae as source of polyhydroxyalkanoates (PHAs)—a review. Int J Biol Macromol 131:536–547

Cruz MV, Gouveia AR, Dionísio M, Freitas F, Reis MA (2019) A process engineering approach to improve production of P (3HB) by Cupriavidus necator from used cooking oil. Int J Polym Sci 2019:2191650

D'Amico DA, Montes MI, Manfredi LB, Cyras VP (2016) Fully bio-based and biodegradable polylactic acid/poly(3-hydroxybutirate) blends: use of a common plasticizer as performance improvement strategy. Polym Test 49:22–28

Da Cruz Pradella JG, Ienczak JL, Delgado CR, Taciro MK (2012) Carbon source pulsed feeding to attain high yield and high productivity in poly (3-hydroxybutyrate)(PHB) production from soybean oil using Cupriavidus necator. Biotechnol Lett 34:1003–1007

Daneshi A, Younesi H, Ghasempouri SM, Sharifzadeh M (2010) Production of poly-3-hydroxybutyrate by Cupriavidus necator from corn syrup: statistical modeling and optimization of biomass yield and volumetric productivity. J Chem Technol Biotechnol 85:1528–1539

Das S, Lengweiler UD, Seebach D, Reusch RN (1997) Proof for a nonproteinaceous calcium-selective channel in Escherichia coli by total synthesis from (R)-3-hydroxybutanoic acid and inorganic polyphosphate. Proc Natl Acad Sci U S A 94:9075–9079

De Los Angeles Martínez-Martínez M, González-Pedrajo B, Dreyfus G, Soto-Urzúa L, Martínez-Morales LJ (2019) Phasin PhaP1 is involved in polyhydroxybutyrate granules morphology and in controlling early biopolymer accumulation in Azospirillum brasilense Sp7. AMB Express 9:155

De Paula FC, Kakazu S, De Paula CBC, Gomez JGC, Contiero J (2017) Polyhydroxyalkanoate production from crude glycerol by newly isolated Pandoraea sp. J King Saud Univ Sci 29:166–173

Do Sul JAI, Costa MF (2014) The present and future of microplastic pollution in the marine environment. Environ Pollut 185:352–364

Dong Z, Sun X (2000) A new method of recovering polyhydroxyalkanoate fromAzotobacter chroococcum. Chin Sci Bull 45:252–256

Dos Santos AJ, Oliveira Dalla Valentina LV, Hidalgo Schulz AA, Tomaz Duarte MA (2017) From obtaining to degradation of PHB: material properties. Part I. Ingeniería y Ciencia 13:269–298

Du G, Chen J, Yu J, Lun S (2001) Continuous production of poly-3-hydroxybutyrate by Ralstonia eutropha in a two-stage culture system. J Biotechnol 88:59–65

Eerkes-Medrano D, Thompson RC, Aldridge DC (2015) Microplastics in freshwater systems: a review of the emerging threats, identification of knowledge gaps and prioritisation of research needs. Water Res 75:63–82

Elbahloul Y, Steinbüchel A (2009) Large-scale production of poly (3-hydroxyoctanoic acid) by Pseudomonas putida GPo1 and a simplified downstream process. Appl Environ Microbiol 75:643–651

Faria AUD, Martins-Franchetti SM (2010) Biodegradação de filmes de polipropileno (PP), poli (3-hidroxibutirato)(PHB) e blenda de PP/PHB por microrganismos das águas do Rio Atibaia. Polímeros 20:141–147

Fei T, Cazeneuve S, Wen Z, Wu L, Wang T (2016) Effective recovery of poly-β-hydroxybutyrate (PHB) biopolymer from Cupriavidus necator using a novel and environmentally friendly solvent system. Biotechnol Prog 32:678–685

Fernández D, Rodríguez E, Bassas M, Viñas M, Solanas A, Llorens J, Marqués A, Manresa A (2005) Agro-industrial oily wastes as substrates for PHA production by the new strain Pseudomonas aeruginosa NCIB 40045: effect of culture conditions. Biochem Eng J 26:159–167

Fiorese ML, Freitas F, Pais J, Ramos AM, De Aragão GM, Reis MA (2009) Recovery of polyhydroxybutyrate (PHB) from Cupriavidus necator biomass by solvent extraction with 1, 2-propylene carbonate. Eng Life Sci 9:454–461

Galia MB (2010) Isolation and analysis of storage compounds. In: Handbook of hydrocarbon and lipid microbiology

García A, Segura D, Espín G, Galindo E, Castillo T, Peña C (2014) High production of poly-β-hydroxybutyrate (PHB) by an Azotobacter vinelandii mutant altered in PHB regulation using a fed-batch fermentation process. Biochem Eng J 82:117–123

Garcia-Garcia D, Ferri J, Boronat T, López-Martínez J, Balart R (2016) Processing and characterization of binary poly (hydroxybutyrate)(PHB) and poly (caprolactone)(PCL) blends with improved impact properties. Polym Bull 73:3333–3350

Garcia-Gonzalez L, Mozumder MSI, Dubreuil M, Volcke EI, De Wever H (2015) Sustainable autotrophic production of polyhydroxybutyrate (PHB) from CO_2 using a two-stage cultivation system. Catal Today 257:237–245

Gent L (2001) Burkholderia sacchari sp. nov., a polyhydroxyalkanoate-accumulating bacterium isolated from soil of a sugar-cane plantation in Brazil. Int J Syst Evol Microbiol 51:1709–1713

Gerngross T, Reilly P, Stubbe J, Sinskey A, Peoples O (1993) Immunocytochemical analysis of poly-beta-hydroxybutyrate (PHB) synthase in Alcaligenes eutrophus H16: localization of the synthase enzyme at the surface of PHB granules. J Bacteriol 175:5289–5293

Getachew A, Woldesenbet F (2016) Production of biodegradable plastic by polyhydroxybutyrate (PHB) accumulating bacteria using low cost agricultural waste material. BMC Res Notes 9:1–9

Ghatnekar MS, Pai JS, Ganesh M (2002) Production and recovery of poly-3-hydroxybutyrate from Methylobacterium sp V49. J Chem Technol Biotechnol 77:444–448

Goff M, Ward PG, O'Connor KE (2007) Improvement of the conversion of polystyrene to polyhydroxyalkanoate through the manipulation of the microbial aspect of the process: a nitrogen feeding strategy for bacterial cells in a stirred tank reactor. J Biotechnol 132:283–286

Gopi K, Balaji S, Muthuvelan B (2014) Isolation purification and screening of biodegradable polymer PHB producing cyanobacteria from marine and fresh water resources. Iran J Energy Environ 5:94–100

Gupta V, Ratha SK, Sood A, Chaudhary V, Prasanna R (2013) New insights into the biodiversity and applications of cyanobacteria (blue-green algae)—prospects and challenges. Algal Res 2:79–97

Hahn SK, Chang YK, Lee SY (1995) Recovery and characterization of poly (3-hydroxybutyric acid) synthesized in Alcaligenes eutrophus and recombinant Escherichia coli. Appl Environ Microbiol 61:34–39

Han J, Qiu Y-Z, Liu D-C, Chen G-Q (2004) Engineered Aeromonas hydrophila for enhanced production of poly (3-hydroxybutyrate-co-3-hydroxyhexanoate) with alterable monomers composition. FEMS Microbiol Lett 239:195–201

Handrick R, Reinhardt S, Schultheiss D, Reichart T, Schüler D, Jendrossek V, Jendrossek D (2004a) Unraveling the function of the Rhodospirillum rubrum activator of polyhydroxybutyrate (PHB) degradation: the activator is a PHB-granule-bound protein (phasin). J Bacteriol 186:2466–2475

Handrick R, Technow U, Reichart T, Reinhardt S, Sander T, Jendrossek D (2004b) The activator of the Rhodospirillum rubrum PHB depolymerase is a polypeptide that is extremely resistant to high temperature (121 C) and other physical or chemical stresses. FEMS Microbiol Lett 230:265–274

Hankermeyer CR, Tjeerdema RS (1999) Polyhydroxybutyrate: plastic made and degraded by microorganisms. In: Reviews of environmental contamination and toxicology. Springer

Hauf W, Watzer B, Roos N, Klotz A, Forchhammer K (2015) Photoautotrophic polyhydroxybutyrate granule formation is regulated by cyanobacterial phasin PhaP in Synechocystis sp. strain PCC 6803. Appl Environ Microbiol 81:4411–4422

Hernandez-Eligio A, Moreno S, Castellanos M, Castaneda M, Nunez C, Muriel-Millan LF, Espín G (2012) RsmA post-transcriptionally controls PhbR expression and polyhydroxybutyrate biosynthesis in Azotobacter vinelandii. Microbiology 158:1953–1963

Ho MH, Li SY, Ciou CY, Wu TM (2014) The morphology and degradation behavior of electrospun poly (3-hydroxybutyrate)/magnetite and poly (3-hydroxybutyrate-co-3-hydroxyvalerate)/magnetite composites. J Appl Polym Sci 131:41070

Ikejima T, Yagi K, Inoue Y (1999) Thermal properties and crystallization behavior of poly (3-hydroxybutyric acid) in blends with chitin and chitosan. Macromol Chem Phys 200:413–421

Iñiguez ME, Conesa JA, Fullana A (2017) Microplastics in Spanish table salt. Sci Rep 7:1–7

Iwata T (2005) Strong fibers and films of microbial polyesters. Macromol Biosci 5:689–701

Jacquel N, Lo C-W, Wei Y-H, Wu H-S, Wang SS (2008) Isolation and purification of bacterial poly (3-hydroxyalkanoates). Biochem Eng J 39:15–27

Jambeck JR, Geyer R, Wilcox C, Siegler TR, Perryman M, Andrady A, Narayan R, Law KL (2015) Plastic waste inputs from land into the ocean. Science 347:768–771

James BW, Mauchline WS, Dennis PJ, Keevil CW, Wait R (1999) Poly-3-hydroxybutyrate in Legionella pneumophila, an energy source for survival in low-nutrient environments. Appl Environ Microbiol 65:822–827

Jendrossek D (2009) Polyhydroxyalkanoate granules are complex subcellular organelles (carbonosomes). J Bacteriol 191:3195–3202

Jendrossek D, Pfeiffer D (2014) New insights in the formation of polyhydroxyalkanoate granules (carbonosomes) and novel functions of poly (3-hydroxybutyrate). Environ Microbiol 16:2357–2373

Jiang Y, Song X, Gong L, Li P, Dai C, Shao W (2008) High poly(β-hydroxybutyrate) production by Pseudomonas fluorescens A2a5 from inexpensive substrates. Enzym Microb Technol 42:167–172

Jiang Y, Marang L, Kleerebezem R, Muyzer G, Van Loosdrecht MC (2011) Polyhydroxybutyrate production from lactate using a mixed microbial culture. Biotechnol Bioeng 108:2022–2035

Juengert JR, Patterson C, Jendrossek D (2018) Poly (3-hydroxybutyrate)(PHB) polymerase PhaC1 and PHB depolymerase PhaZa1 of Ralstonia eutropha are phosphorylated in vivo. Appl Environ Microbiol 84:e00604–e00618

Jung IL, Phyo KH, Kim KC, Park HK, Kim IG (2005) Spontaneous liberation of intracellular polyhydroxybutyrate granules in Escherichia coli. Res Microbiol 156:865–873

Kadiyala G (2014) Isolation purification and screening of biodegradable polymer PHB producing cyanobacteria from marine and fresh water resources. Iran J Energy Environ 5:94–100

Kanjanachumpol P, Kulpreecha S, Tolieng V, Thongchul N (2013) Enhancing polyhydroxybutyrate production from high cell density fed-batch fermentation of Bacillus megaterium BA-019. Bioprocess Biosyst Eng 36:1463–1474

Karahaliloğlu Z (2017) Cell-compatible PHB/silk fibroin composite nanofiber mat for tissue engineering applications. Turk J Biol 41:503–513

Kathiraser Y, Aroua MK, Ramachandran KB, Tan IKP (2007) Chemical characterization of medium-chain-length polyhydroxyalkanoates (PHAs) recovered by enzymatic treatment and ultrafiltration. J Chem Technol Biotechnol 82:847–855

Ke Y, Zhang X, Ramakrishna S, He L, Wu G (2017) Reactive blends based on polyhydroxyalkanoates: preparation and biomedical application. Mater Sci Eng C 70:1107–1119

Kessler B, Witholt B (2001) Factors involved in the regulatory network of polyhydroxyalkanoate metabolism. J Biotechnol 86:97–104

Khanna S, Srivastava AK (2005) Recent advances in microbial polyhydroxyalkanoates. Process Biochem 40:607–619

Kiziltas A, Nazari B, Kiziltas EE, Gardner DJ, Han Y, Rushing TS (2016) Method to reinforce polylactic acid with cellulose nanofibers via a polyhydroxybutyrate carrier system. Carbohydr Polym 140:393–399

Kobayashi D, Fujita K, Nakamura N, Ohno H (2015) A simple recovery process for biodegradable plastics accumulated in cyanobacteria treated with ionic liquids. Appl Microbiol Biotechnol 99:1647–1653

Koksharova O, Wolk C (2002) Genetic tools for cyanobacteria. Appl Microbiol Biotechnol 58:123–137

Koller M, Bona R, Chiellini E, Fernandes EG, Horvat P, Kutschera C, Hesse P, Braunegg G (2008) Polyhydroxyalkanoate production from whey by Pseudomonas hydrogenovora. Bioresour Technol 99:4854–4863

Koller M, Salerno A, Dias M, Reiterer A, Braunegg G (2010) Modern biotechnological polymer synthesis: a review. Food Technol Biotechnol 48:255–269

Kosuth M, Wattenberg EV, Mason SA, Tyree C, Morrison D (2017) Synthetic polymer contamination in global drinking water. Orb Media

Kuciel S, Liber-Kneć A (2011) Biocomposites based on PHB filled with wood or kenaf fibers. Polimery 56:218–223

Kulpreecha S, Boonruangthavorn A, Meksiriporn B, Thongchul N (2009) Inexpensive fed-batch cultivation for high poly (3-hydroxybutyrate) production by a new isolate of Bacillus megaterium. J Biosci Bioeng 107:240–245

Kunasundari B, Sudesh K (2011) Isolation and recovery of microbial polyhydroxyalkanoates. Express Polym Lett 5:620

Łabużek S, Radecka I (2001) Biosynthesis of PHB tercopolymer by Bacillus cereus UW85. J Appl Microbiol 90:353–357

Lai SM, Don TM, Huang YC (2006) Preparation and properties of biodegradable thermoplastic starch/poly (hydroxy butyrate) blends. J Appl Polym Sci 100:2371–2379

Lakshman K, Shamala TR (2006) Extraction of polyhydroxyalkanoate from Sinorhizobium meliloti cells using Microbispora sp. culture and its enzymes. Enzym Microb Technol 39:1471–1475

Lama L, Nicolaus B, Calandrelli V, Manca MC, Romano I, Gambacorta A (1996) Effect of growth conditions on endo- and exopolymer biosynthesis in Anabaena cylindrica 10 C. Phytochemistry 42:655–659

Lasprilla AJ, Martinez GA, Lunelli BH, Jardini AL, Maciel Filho R (2012) Poly-lactic acid synthesis for application in biomedical devices—A review. Biotechnol Adv 30:321–328

Lee KM, Chang HN, Chang YK, Kim BS, Hahn SK (1993) The lysis of Gram-negative Alcaligenes eutrophus and Alcaligenes latus by palmitoyl carnitine. Biotechnol Tech 7:295–300

Li M, Wilkins M (2020) Fed-batch cultivation and adding supplements to increase yields of polyhydroxybutyrate production by Cupriavidus necator from corn stover alkaline pretreatment liquor. Bioresour Technol 299:122676

Li R, Zhang H, Qi Q (2007) The production of polyhydroxyalkanoates in recombinant Escherichia coli. Bioresour Technol 98:2313–2320

Liebezeit G, Liebezeit E (2015) Origin of synthetic particles in honeys. Polish J Food Nutr Sci 65:143–147

López-Abelairas M, García-Torreiro M, Lú-Chau T, Lema J, Steinbüchel A (2015) Comparison of several methods for the separation of poly (3-hydroxybutyrate) from Cupriavidus necator H16 cultures. Biochem Eng J 93:250–259

Lopez-Arenas T, González-Contreras M, Anaya-Reza O, Sales-Cruz M (2017) Analysis of the fermentation strategy and its impact on the economics of the production process of PHB (polyhydroxybutyrate). Comput Chem Eng 107:140–150

López-Cortés A, Lanz-Landázuri A, García-Maldonado JQ (2008) Screening and isolation of PHB-producing bacteria in a polluted marine microbial mat. Microb Ecol 56:112–120

Maehara A, Taguchi S, Nishiyama T, Yamane T, Doi Y (2002) A repressor protein, PhaR, regulates polyhydroxyalkanoate (PHA) synthesis via its direct interaction with PHA. J Bacteriol 184:3992–4002

Maes T, Jessop R, Wellner N, Haupt K, Mayes AG (2017) A rapid-screening approach to detect and quantify microplastics based on fluorescent tagging with Nile Red. Sci Rep 7:1–10

Mani T, Hauk A, Walter U, Burkhardt-Holm P (2015) Microplastics profile along the Rhine River. Sci Rep 5:1–7

Marjadi D, Dharaiya N (2014) Recovery and characterization of poly (3-Hydroxybutyric acid) synthesized in Staphylococcus epidermidis. Afr J Environ Sci Technol 8:319–329

Martino L, Cruz MV, Scoma A, Freitas F, Bertin L, Scandola M, Reis MA (2014) Recovery of amorphous polyhydroxybutyrate granules from Cupriavidus necator cells grown on used cooking oil. Int J Biol Macromol 71:117–123

Mejía M, Segura D, Espín G, Galindo E, Peña C (2010) Two-stage fermentation process for alginate production by Azotobacter vinelandii mutant altered in poly-β-hydroxybutyrate (PHB) synthesis. J Appl Microbiol 108:55–61

Mezzolla V, D'urso OF, Poltronieri P (2018) Role of PhaC type I and type II enzymes during PHA biosynthesis. Polymers 10:910

Mintenig S, Löder M, Primpke S, Gerdts G (2019) Low numbers of microplastics detected in drinking water from ground water sources. Sci Total Environ 648:631–635

Mousavioun P, George GA, Doherty WO (2012) Environmental degradation of lignin/poly (hydroxybutyrate) blends. Polym Degrad Stab 97:1114–1122

Mozumder MSI, De Wever H, Volcke EI, Garcia-Gonzalez L (2014) A robust fed-batch feeding strategy independent of the carbon source for optimal polyhydroxybutyrate production. Process Biochem 49:365–373

Mukhopadhyay M, Patra A, Paul AK (2005) Production of poly(3-hydroxybutyrate) and poly (3-hydroxybutyrate-co-3-hydroxyvalerate) by Rhodopseudomonas palustris SP5212. World J Microbiol Biotechnol 21:765–769

Myshkina V, Nikolaeva D, Makhina T, Bonartsev A, Bonartseva G (2008) Effect of growth conditions on the molecular weight of poly-3-hydroxybutyrate produced by Azotobacter chroococcum 7B. Appl Biochem Microbiol 44:482–486

Nath A, Dixit M, Bandiya A, Chavda S, Desai A (2008) Enhanced PHB production and scale up studies using cheese whey in fed batch culture of Methylobacterium sp. ZP24. Bioresour Technol 99:5749–5755

Nikel PI, Pettinari MJ, Galvagno MA, Méndez BS (2006) Poly (3-hydroxybutyrate) synthesis by recombinant Escherichia coli arcA mutants in microaerobiosis. Appl Environ Microbiol 72:2614–2620

Noguez R, Segura D, Moreno S, Hernandez A, Juarez K, Espín G (2008) Enzyme INtr, NPr and IIANtr are involved in regulation of the poly-β-hydroxybutyrate biosynthetic genes in Azotobacter vinelandii. J Mol Microbiol Biotechnol 15:244–254

Osanai T, Numata K, Oikawa A, Kuwahara A, Iijima H, Doi Y, Tanaka K, Saito K, Hirai MY (2013) Increased bioplastic production with an RNA polymerase sigma factor SigE during nitrogen starvation in Synechocystis sp. PCC 6803. DNA Res 20:525–535

Page WJ, Knosp O (1989) Hyperproduction of poly-β-hydroxybutyrate during exponential growth of Azotobacter vinelandii UWD. Appl Environ Microbiol 55:1334–1339

Pandian SR, Deepak V, Kalishwaralal K, Rameshkumar N, Jeyaraj M, Gurunathan S (2010) Optimization and fed-batch production of PHB utilizing dairy waste and sea water as nutrient sources by Bacillus megaterium SRKP-3. Bioresour Technol 101:705–711

Peña C, Castillo T, García A, Millán M, Segura D (2014a) Biotechnological strategies to improve production of microbial poly-(3-hydroxybutyrate): a review of recent research work. Microb Biotechnol 7:278–293

Peña C, López S, García A, Espín G, Romo-Uribe A, Segura D (2014b) Biosynthesis of poly-β-hydroxybutyrate (PHB) with a high molecular mass by a mutant strain of Azotobacter vinelandii (OPN). Ann Microbiol 64:39–47

Penloglou G, Kretza E, Chatzidoukas C, Parouti S, Kiparissides C (2012) On the control of molecular weight distribution of polyhydroxybutyrate in Azohydromonas lata cultures. Biochem Eng J 62:39–47

Pfeiffer D, Jendrossek D (2012) Localization of PHB granule associated proteins during PHB granule formation and identification of two new phasins, PhaP6 and PhaP7, in Ralstonia eutropha H16. J Bacteriol 194:5909

Pfeiffer D, Jendrossek D (2014) PhaM is the physiological activator of poly (3-hydroxybutyrate) (PHB) synthase (PhaC1) in Ralstonia eutropha. Appl Environ Microbiol 80:555–563

Pfeiffer D, Wahl A, Jendrossek D (2011) Identification of a multifunctional protein, PhaM, that determines number, surface to volume ratio, subcellular localization and distribution to daughter cells of poly (3-hydroxybutyrate), PHB, granules in Ralstonia eutropha H16. Mol Microbiol 82:936–951

Poirier Y, Dennis DE, Klomparens K, Somerville C (1992) Polyhydroxybutyrate, a biodegradable thermoplastic, produced in transgenic plants. Science 256:520–523

Pötter M, Madkour MH, Mayer F, Steinbüchel A (2002) Regulation of phasin expression and polyhydroxyalkanoate (PHA) granule formation in Ralstonia eutropha H16. Microbiology 148:2413–2426

Pozo C, Martınez-Toledo M, Rodelas B, Gonzalez-Lopez J (2002) Effects of culture conditions on the production of polyhydroxyalkanoates by Azotobacter chroococcum H23 in media containing a high concentration of alpechin (wastewater from olive oil mills) as primary carbon source. J Biotechnol 97:125–131

Qi Q, Rehm BH (2001) Polyhydroxybutyrate biosynthesis in Caulobacter crescentus: molecular characterization of the polyhydroxybutyrate synthase. Microbiology 147:3353–3358

Quillaguaman J, Hashim S, Bento F, Mattiasson B, Hatti-Kaul R (2005) Poly (β-hydroxybutyrate) production by a moderate halophile, Halomonas boliviensis LC1 using starch hydrolysate as substrate. J Appl Microbiol 99:151–157

Rai R, KeshaVarz T, Roether J, Boccaccini AR, Roy I (2011) Medium chain length polyhydroxyalkanoates, promising new biomedical materials for the future. Mater Sci Eng R Rep 72:29–47

Rajan R, Sreekumar P, Joseph K, Skrifvars M (2012) Thermal and mechanical properties of chitosan reinforced polyhydroxybutyrate composites. J Appl Polym Sci 124:3357–3362
Rajan K, Thomas S, Gopanna A, Chavali M (2018) Polyhydroxybutyrate (PHB): a standout biopolymer for environmental sustainability. In: Martínez LMT, Kharissova OV, Kharisov BI (eds) Handbook of ecomaterials, pp 1–23
Ramadas NV, Singh SK, Soccol CR, Pandey A (2009) Polyhydroxybutyrate production using agro-industrial residue as substrate by Bacillus sphaericus NCIM 5149. Braz Arch Biol Technol 52:17–23
Ramsay J, Berger E, Ramsay B, Chavarie C (1990) Recovery of poly-3-hydroxyalkanoic acid granules by a surfactant-hypochlorite treatment. Biotechnol Tech 4:221–226
Ramsay J, Berger E, Voyer R, Chavarie C, Ramsay B (1994) Extraction of poly-3-hydroxybutyrate using chlorinated solvents. Biotechnol Tech 8:589–594
Rathore DP (2014) Bioprospects of PHB: a review. IJETST 1(4):529–532
Raza ZA, Riaz S, Banat IM (2018) Polyhydroxyalkanoates: properties and chemical modification approaches for their functionalization. Biotechnol Prog 34:29–41
Raza ZA, Noor S, Khalil S (2019) Recent developments in the synthesis of poyyly (hydroxybutyrate) based biocomposites. Biotechnol Prog 35:e2855
Reddy C, Ghai R, Kalia VC (2003) Polyhydroxyalkanoates: an overview. Bioresour Technol 87:137–146
Reddy KR, Sato H, Takahashi I, Ozaki Y (2015) Intermolecular hydrogen bondings in the poly (3-hydroxybutyrate) and chitin blends: their effects on the crystallization behavior and crystal structure of poly (3-hydroxybutyrate). Polymer 75:141–150
Reddy MV, Mawatari Y, Onodera R, Nakamura Y, Yajima Y, Chang Y-C (2019) Bacterial conversion of waste into polyhydroxybutyrate (PHB): a new approach of bio-circular economy for treating waste and energy generation. Bioresour Technol Rep 7:100246
Rivera-Briso AL, Serrano-Aroca Á (2018) Poly (3-Hydroxybutyrate-co-3-Hydroxyvalerate): enhancement strategies for advanced applications. Polymers 10:732
Rocha RC, Da Silva LF, Taciro MK, Pradella JG (2008) Production of poly (3-hydroxybutyrate-co-3-hydroxyvalerate) P (3HB-co-3HV) with a broad range of 3HV content at high yields by Burkholderia sacchari IPT 189. World J Microbiol Biotechnol 24:427–431
Rosengart A, Cesário MT, De Almeida MCM, Raposo RS, Espert A, De Apodaca ED, Da Fonseca MMR (2015) Efficient P (3HB) extraction from Burkholderia sacchari cells using non-chlorinated solvents. Biochem Eng J 103:39–46
Ruan W, Chen J, Lun S (2003) Production of biodegradable polymer by A. eutrophus using volatile fatty acids from acidified wastewater. Process Biochem 39:295–299
Rummel CD, Löder MG, Fricke NF, Lang T, Griebeler E-M, Janke M, Gerdts G (2016) Plastic ingestion by pelagic and demersal fish from the North Sea and Baltic Sea. Mar Pollut Bull 102:134–141
Saeb A, Alshammari AS, Al-Brahim H, Al-Rubeaan KA (2014) Production of silver nanoparticles with strong and stable antimicrobial activity against highly pathogenic and multidrug resistant bacteria. Sci World J 2014:704708
Sagong H-Y, Son HF, Choi SY, Lee SY, Kim KJ (2018) Structural insights into polyhydroxyalkanoates biosynthesis. Trends Biochem Sci 43:790–805
Samantaray S, Mallick N (2012) Production and characterization of poly-β-hydroxybutyrate (PHB) polymer from Aulosira fertilissima. J Appl Phycol 24:803–814
Schymanski D, Goldbeck C, Humpf H-U, Fürst P (2018) Analysis of microplastics in water by micro-Raman spectroscopy: release of plastic particles from different packaging into mineral water. Water Res 129:154–162
Segura D, Espín G (2004) Inactivation of pycA, encoding pyruvate carboxylase activity, increases poly-β-hydroxybutyrate accumulation in Azotobacter vinelandii on solid medium. Appl Microbiol Biotechnol 65:414–418
Segura D, Guzmán J, Espín G (2003) Azotobacter vinelandii mutants that overproduce poly-β-hydroxybutyrate or alginate. Appl Microbiol Biotechnol 63:159–163

Seoane IT, Manfredi LB, Cyras VP (2015) Properties and processing relationship of polyhydroxybutyrate and cellulose biocomposites. Procedia Mater Sci 8:807–813

Seoane IT, Fortunati E, Puglia D, Cyras VP, Manfredi LB (2016) Development and characterization of bionanocomposites based on poly (3-hydroxybutyrate) and cellulose nanocrystals for packaging applications. Polym Int 65:1046–1053

Sharma L, Singh AK, Panda B, Mallick N (2007) Process optimization for poly-β-hydroxybutyrate production in a nitrogen fixing cyanobacterium, Nostoc muscorum using response surface methodology. Bioresour Technol 98:987–993

Shishatskaya E, Khlusov I, Volova T (2006) A hybrid PHB–hydroxyapatite composite for biomedical application: production, in vitro and in vivo investigation. J Biomater Sci Polym Ed 17:481–498

Silva L, Taciro M, Ramos MM, Carter J, Pradella J, Gomez J (2004) Poly-3-hydroxybutyrate (P3HB) production by bacteria from xylose, glucose and sugarcane bagasse hydrolysate. J Ind Microbiol Biotechnol 31:245–254

Singh MK, Rai PK, Rai A, Singh S, Singh JS (2019) Poly-β-hydroxybutyrate production by the cyanobacterium Scytonema geitleri Bharadwaja under varying environmental conditions. Biomol Ther 9:198

Sirohi R, Pandey JP, Gaur VK, Gnansounou E, Sindhu R (2020) Critical overview of biomass feedstocks as sustainable substrates for the production of polyhydroxybutyrate (PHB). Bioresour Technol 311:123536

Slater SC, Voige W, Dennis D (1988) Cloning and expression in Escherichia coli of the Alcaligenes eutrophus H16 poly-beta-hydroxybutyrate biosynthetic pathway. J Bacteriol 170:4431–4436

Stal LJ, Heyer H, Jacobs G (1990) Occurrence and role of poly-hydroxy-alkanoate in the cyanobacterium Oscillatoria limosa. In: Novel biodegradable microbial polymers. Springer

Stubbe J, Tian J, He A, Sinskey AJ, Lawrence AG, Liu P (2005) Nontemplate-dependent polymerization processes: polyhydroxyalkanoate synthases as a paradigm. Annu Rev Biochem 74:433–480

Sudesh K, Abe H, Doi Y (2000) Synthesis, structure and properties of polyhydroxyalkanoates: biological polyesters. Prog Polym Sci 25:1503–1555

Sznajder A, Pfeiffer D, Jendrossek D (2015) Comparative proteome analysis reveals four novel polyhydroxybutyrate (PHB) granule-associated proteins in Ralstonia eutropha H16. Appl Environ Microbiol 81:1847–1858

Tamer IM, Moo-Young M, Chisti Y (1998) Disruption of Alcaligenes latus for recovery of poly (β-hydroxybutyric acid): comparison of high-pressure homogenization, bead milling, and chemically induced lysis. Ind Eng Chem Res 37:1807–1814

Tan D, Xue Y-S, Aibaidula G, Chen G-Q (2011) Unsterile and continuous production of polyhydroxybutyrate by Halomonas TD01. Bioresour Technol 102:8130–8136

Tanadchangsaeng N, Yu J (2012) Microbial synthesis of polyhydroxybutyrate from glycerol: gluconeogenesis, molecular weight and material properties of biopolyester. Biotechnol Bioeng 109:2808–2818

Tanaka K, Takada H (2016) Microplastic fragments and microbeads in digestive tracts of planktivorous fish from urban coastal waters. Sci Rep 6:34351

Telles MR, Saran LM, Unêda-Trevisolli SH (2011) Produção, propriedades e aplicações de bioplástico obtido a partir da cana-de-açúcar. Ciência e Tecnologia 2:52–63

Thiré RM, Ribeiro TA, Andrade CT (2006) Effect of starch addition on compression-molded poly (3-hydroxybutyrate)/starch blends. J Appl Polym Sci 100:4338–4347

Thirumala M, Reddy SV, Mahmood S (2010) Production and characterization of PHB from two novel strains of Bacillus spp. isolated from soil and activated sludge. J Ind Microbiol Biotechnol 37:271–278

Tian S-J, Lai W-J, Zheng Z, Wang H-X, Chen G-Q (2005) Effect of over-expression of phasin gene from Aeromonas hydrophila on biosynthesis of copolyesters of 3-hydroxybutyrate and 3-hydroxyhexanoate. FEMS Microbiol Lett 244:19–25

Tirapelle EF, Müller-Santos M, Tadra-Sfeir MZ, Kadowaki MA, Steffens MB, Monteiro RA, Souza EM, Pedrosa FO, Chubatsu LS (2013) Identification of proteins associated with polyhydroxybutyrate granules from Herbaspirillum seropedicae SmR1-old partners, new players. PLoS One 8:e75066

Toh PS, Jau MH, Yew SP, Abed RM, Sudesh K (2008) Comparison of Polyhydroxyalkanoates biosynthesis, mobilization and the effects on cellular morphology in Spirulina platensis and Synechocystis Sp. Uniwg J Biosci 19:21–38

Urbanek AK, Rymowicz W, Mirończuk AM (2018) Degradation of plastics and plastic-degrading bacteria in cold marine habitats. Appl Microbiol Biotechnol 102:7669–7678

Ushimaru K, Motoda Y, Numata K, Tsuge T (2014) Phasin proteins activate Aeromonas caviae polyhydroxyalkanoate (PHA) synthase but not Ralstonia eutropha PHA synthase. Appl Environ Microbiol 80:2867–2873

Valappil SP, Misra SK, Boccaccini AR, Keshavarz T, Bucke C, Roy I (2007) Large-scale production and efficient recovery of PHB with desirable material properties, from the newly characterised Bacillus cereus SPV. J Biotechnol 132:251–258

Van Cauwenberghe L, Janssen CR (2014) Microplastics in bivalves cultured for human consumption. Environ Pollut 193:65–70

Verlinden RA, Hill DJ, Kenward M, Williams CD, Radecka I (2007) Bacterial synthesis of biodegradable polyhydroxyalkanoates. J Appl Microbiol 102:1437–1449

Vermaas W (2001) Encyclopedia of life sciences. In: Photosynthesis and respiration in cyanobacteria

Villanova JC, Oréfice RL, Cunha AS (2010) Aplicações farmacêuticas de polímeros. Polímeros 20:51–64

Wampfler B, Ramsauer T, Rezzonico S, Hischier R, Kohling R, Thony-Meyer L, Zinn M (2010) Isolation and purification of medium chain length poly (3-hydroxyalkanoates)(mcl-PHA) for medical applications using nonchlorinated solvents. Biomacromolecules 11:2716–2723

Wang J, Yu J (2001) Kinetic analysis on formation of poly (3-hydroxybutyrate) from acetic acid by Ralstonia eutropha under chemically defined conditions. J Ind Microbiol Biotechnol 26:121–126

Wang Q, Zhuang Q, Liang Q, Qi Q (2013) Polyhydroxyalkanoic acids from structurally-unrelated carbon sources in Escherichia coli. Appl Microbiol Biotechnol 97:3301–3307

Wang G, Gislum M, Filippov G, Montgomery S (2015) An environment-friendly and efficient method for extraction of PHB biopolymer by non-halogenated solvents. J Microbiol Biotechnol 31:785–794

Wei L, Mcdonald AG, Stark NM (2015) Grafting of bacterial polyhydroxybutyrate (PHB) onto cellulose via in situ reactive extrusion with dicumyl peroxide. Biomacromolecules 16:1040–1049

Wendlandt K-D, Geyer W, Mirschel G, Hemidi FAH (2005) Possibilities for controlling a PHB accumulation process using various analytical methods. J Biotechnol 117:119–129

Wijffels RH, Kruse O, Hellingwerf KJ (2013) Potential of industrial biotechnology with cyanobacteria and eukaryotic microalgae. Curr Opin Biotechnol 24:405–413

Wong PA, Cheung MK, Lo WH, Chua H, Yu PHF (2004) Investigation of the effects of the types of food waste utilized as carbon source on the molecular weight distributions and thermal properties of polyhydroxybutyrate produced by two strains of microorganisms. e-Polymers 4

Woolnough CA, Yee LH, Charlton T, Foster LJR (2010) Environmental degradation and biofouling of 'green' plastics including short and medium chain length polyhydroxyalkanoates. Polym Int 59:658–667

Wu GF, Wu QY, Shen ZY (2001) Accumulation of poly-β-hydroxybutyrate in cyanobacterium Synechocystis sp. PCC6803. Bioresour Technol 76:85–90

Yang HX, Sun M, Zhang Y, Zhou P (2011) Degradable PHBHHx modified by the silk fibroin for the applications of cardiovascular tissue engineering. ISRN Mater Sci 2011

Yu J, Chen LX (2006) Cost-effective recovery and purification of polyhydroxyalkanoates by selective dissolution of cell mass. Biotechnol Prog 22:547–553

Zhang M, Thomas NL (2010) Preparation and properties of polyhydroxybutyrate blended with different types of starch. J Appl Polym Sci 116:688–694

Zhang M, Thomas NL (2011) Blending polylactic acid with polyhydroxybutyrate: the effect on thermal, mechanical, and biodegradation properties. Adv Polym Technol 30:67–79

Zhang D, Dechatiwongse P, Hellgardt K (2015) Modelling light transmission, cyanobacterial growth kinetics and fluid dynamics in a laboratory scale multiphase photo-bioreactor for biological hydrogen production. Algal Res 8:99–107

Zhuo ZQ, Dong LM, Wang C, Zan QF, Tian JM (2010) Preparation and mechanical properties of HA/PHB composites. Adv Mater Res 105-106:104–107

Fungal Exopolysaccharides: Types, Production and Application

2

Ashim Debnath, Bimal Das, Maimom Soniya Devi, and Ratul Moni Ram

Abstract

Exopolysaccharides (EPSs) may be regarded as extracellular metabolites of living organisms (bacteria, fungi, algae, plants and animals, etc.) associated with adaptation, survival and functionalities. Exopolysaccharides (EPSs) produced from fungi have been documented as high worth molecules for the recent few decades. A wide array of fungal exopolysaccharides is reported to exhibit numerous applications in pharmaceuticals, industries, medicine, foods and other sectors, viz. scleroglucan, botryosphaeran, pestan, pullulan, etc. Although fungal EPSs are exceptionally relevant, to date data concerning fungal biosynthesis is scant and a broad chase for new fungal species which could create novel EPSs is as yet required. As a rule, the molecular weight disparities and sugar syntheses of fungal EPSs are needy to culture medium synthesis and distinctive states of being given during maturation. An accentuation is likewise given to drilling down various parasitic strains that can deliver effective EPSs. The variable synthetic and biochemical engineering that describes an EPS preset its organic usefulness and potential biotechnological benefits. It is agreeable to hereditary, biotechnological

A. Debnath
Department of Genetics and Plant Breeding, A. N. D. University of Agriculture and Technology, Ayodhya, Uttar Pradesh, India

B. Das
Department of Genetics and Plant Breeding, Uttar Banga Krishi Vishwavidyalaya, Dakshin Dinajpur, West Bengal, India

M. S. Devi
Department of Entomology, Rani Lakshmi Bai Central Agricultural University, Jhansi, Uttar Pradesh, India

R. M. Ram (✉)
Department of Plant Pathology, SGT University, Chandu Budhera, Gurugram, Haryana, India

A. Vaishnav, D. K. Choudhary (eds.), *Microbial Polymers*,
https://doi.org/10.1007/978-981-16-0045-6_2

and biochemical mobility for sought bioactivity or application during their production and extraction. A brief idea regarding fungal EPS is an insight into this chapter. The overall chapter describes the concept of fungal EPS, their production confirmation and wide applications.

Keywords

Fungal exopolysaccharide · Scleroglucan · Botryosphaeran

2.1 Introduction

Polysaccharides of high molecular weight have been utilized as additives for emulsifiers (a substance that stabilizes an emulsion, in particular, an additive used to stabilize processed foods), flocculants (clarifying agents removing suspended solids from liquids by inducing flocculation), coatings or films (direct coatings to food into a film used as a food wrap without changing the original ingredients), gelling agents (substances providing texture to foods through gel formation), stabilizers (additives helping in the preservation of foods structure) and thickening agents (substances which increases fluid viscosity without substantially changing its other properties) used in many food products. Polysaccharides obtained from plants cellulose or pectin and seaweeds carrageenan or alginate (Kleerebezem et al. 1999). Interest in polysaccharides from fungi has risen in recent years due to the high productivity and low production costs. Polysaccharides bioactive molecule is produced by many microorganisms such as bacteria, fungi and yeast (Wang et al. 2012). Among the different microbial polysaccharides, one of the most important types is exopolysaccharides which have many advantages over intracellular polysaccharides and its isolation process is easy in a short period from the microorganism (Mahapatra and Banerjee 2013). Exopolysaccharides (EPS) are mainly extracellular polymers of carbohydrate which are synthesis by microorganisms and utilize them in cell outside. They do not serve as reserve materials, making them different from various other intracellular polymeric inclusions. Exopolysaccharides are fitted to the cell surfaces and play a major role in protecting against phagocytosis and phage attack. They also act as barriers to diffusion and bind the metal ions (Cu, Pb, Zn, Co, etc.) which help microorganism under adverse conditions. Microbial origin EPSs, including fungal EPSs, are a justifiable alternative to marketed polysaccharides. The microbial-based polysaccharides are almost identical to those currently used gums (Sutherland 1996a, b). These polysaccharides have achieved importance from the last few decades based on their different applications in various fields.

The term 'exopolysaccharide' was first described as high molecular weight polymers of carbohydrate produced by marine bacteria (Sutherland 1982). Exopolysaccharides composed of repeated sugar moieties may be associated with other molecules like proteins, organic, inorganic compounds, lipids, DNA and metal ions. The definite actions of EPS and exact function depend on ecological niches and structural units of the concerned microorganisms (Mishra and Jha 2013). Substantial progress is being carried forward to discover and develop new microbial EPSs with novel and interesting industrial applications (Nicolaus et al. 2010). EPS are used as emulsifiers, binders, stabilizers, coagulants, gelling agents, lubricants, thickening and film-forming agents in food and pharmaceutical industries. Microbial exopolysaccharides served many functions in the microbial cells and are differentiated into three main types mainly structural, intracellular and extra or exopolysaccharides. Chitin was the first polysaccharide of carbohydrate polymer discovered in edible mushrooms and the production of extracellular polysaccharide by microorganisms first reported in 1861. However, microorganisms are known for their ability to synthesize exopolysaccharides (EPS), both homo- and heteropolymers with different structural complexities (Seviour et al. 1992; Sutherland 1994). These microbial exopolysaccharides represent an attractive alternative to synthetic polymers and represent biodegradable polymers (Dave et al. 2016) which is ideal to replace petro-based polymers (Meybod and Mohammadifar 2015; Castillo et al. 2015; Moscovici 2015; Schmid et al. 2016). Similarly, fungal exopolysaccharides far better alternative to exopolysaccharides produced by plant or macroalgae (Freitas et al. 2011; Moscovici 2015; Meybod and Mohammadifar 2015).

Several metabolites have been reported to be produced from fungi. The important among these metabolites are enzymes, exopolysaccharides, antibiotics, etc. Interest in these natural metabolites is significant because of their medical, industrial and agricultural importance. Different fungal EPS with interesting biological activities, from *Agaricus blazi*, *Ganoderma lucidum*, *Cordyceps* sp., *Grifola frondosa* and *Lentinus edodes* through submerged cultures (Yang and He 2008). A glucan was extracted from aqueous extract of fruiting bodies of *Pleurotus florida* (Rout et al. 2005). EPS of medicinal importance has been extracted from *Pleurotus* (Fr.) P. Karst and *Lentinus edodes* (Berk.), *Phellinus rimosus*, *Ganoderma lucidum*, *Pleurotus pulmonaris* and *Pleurotus florida* and are found to possess profound antioxidant and antitumor activities (Elisashvili et al. 2009; Ajith and Janardhanan 2007). Biologically active extracellular exopolysaccharides and intracellular polysaccharides in fruit bodies, cultured broth and cultured mycelium (Wasser 2010) of many fungi are useful to pharmacological industries in the aspects of the antioxidant property, hypoglycemia, anti-inflammatory, antitumour, immuno-stimulating and ROS activity (Chen et al. 2012; Hu et al. 2006a, b; Wu et al. 2010. Despite having immense potential for EPS production, only medicinal aspects of these fungi have been explored. They are capable of producing EPS of high commercial value and can be used for various industrial applications.

2.2 Sources of Fungal Exopolysaccharides

Fungal EPSs have been identified as highly useful bio-macromolecules. EPSs like scleroglucan, botryosphaeran and pullulan have many uses in food, medicine, pharmaceuticals industries, etc. The fungal EPS are extremely pertinent and identification of novel fungal species for the biosynthesis of potential EPSs is still required.

Table 2.1 Source of fungal exopolysaccharide

Sl. no.	Fungus species	Exopolysaccharide
1	Filamentous fungi, *Candida albicans, Zygosaccharomyces rouxii*	Chitin and chitosan
2	*Sclerotium rolfsii*	Scleroglucan
3	*Schizophyllum commune*	Schizophyllan
4	*Botryosphaeria rhodina*	Botryosphaeran
5	*Cryptococcus neoformans, Aspergillus fumigatus*	Galactosaminoglucan
6	*Aureobasidium pullulans, Pullularia pullulans*	Pullulan
7	*Pleurotus ostreatus*	Pleuran

Variations in molecular mass and composition of sugar in fungal EPSs depends on the composition of culture media and various physical conditions during fermentation. EPS production through submerged fermentation by different fungal strains, including *Agaricus blazi*, *Ganoderma lucidum*, *Cordyceps* sp., *Grifola frondosa* and *Lentinus edodes* had been reported, with unique biological activities. Presently, a large number of fungi including higher basidiomycetes, lower filamentous fungi and yeasts from various ecological habitats are found to produce EPSs in laboratory conditions. However, many remain unexplored or under-investigated. Some of the important fungal exopolysaccharides sources (Table 2.1) and their structure are presented in the following:

2.2.1 Chitin/Chitosan

The cellulose analogue chitin constitutes one of the most abundant glycans. It occurs in different crystalline polymorphic forms. Among them α-chitin is on the most common forms of chitin. Chitin is a long-chain polymer of a derivative of glucose, *N*-acetylglucosamine. The primary source of chitin is cell walls of fungi and exoskeletons of insects and different arthropods like lobsters, shrimps, crabs, etc. Chitins are chemically modified to be used in food processing as edible films and as thickeners and stabilizers chitin and chitosan have been used in paper strengthening (Hosokawa et al. 1990). Chitosan is fully or partially N-deacetylated derivatives of cellulose. Chitosan is made up of β (1-4)-linked 2-acetamido-2-deoxy-β-D-glucose (*N*-acetylglucosamine) (Fig. 2.1). Chitosan has received considerable attention recently (Sanford and Hutchings 1987). Chitosans are the major elements derived from arthropods shells including that of shrimps, crabs, lobsters and insects. They are also formed extracellularly by the cell walls of fungi and brown algae. Chitosan-derived biomaterials have received considerable attention as an antimicrobial, functional, renewable, non-toxic, biocompatible, bioabsorbable and biodegradable biopolymer agent (Biagini et al. 1991; Zhang et al. 2010).

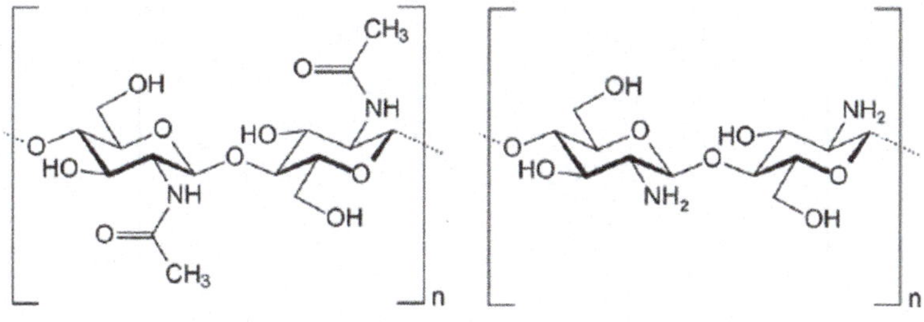

Fig. 2.1 Structure of (**a**) chitin showing *N*-acetylglucosamine linked through β-(1→4)-linkage and (**b**) chitosan comprising of β (1-4)-linked D-glucosamine as deacetylated unit and N-acetyl-D-glucosamine as acetylated unit

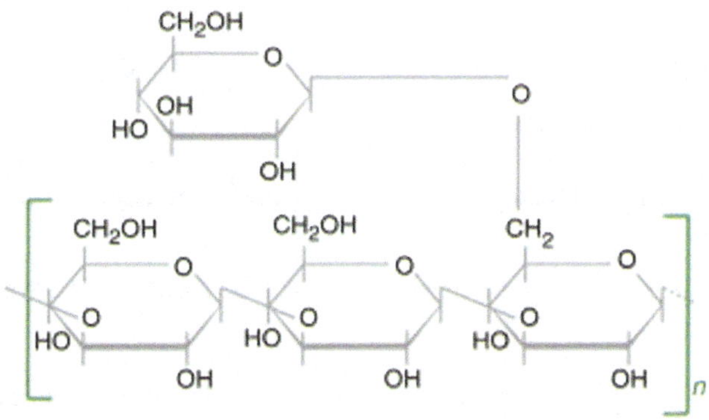

Fig. 2.2 Structure of scleroglucan made by a β-(1, 3) backbone linked with D-glucosyl groups linked by β-(1, 6) linkages

2.2.2 Scleroglucan

It is a natural polysaccharide produced by the fungus *Sclerotium rolfsii.* Structurally is a β-(1, 3) linked with D-glucosyl groups linked via β-(1,6) linkages (Fig. 2.2) and is stable at low pH and high temperature. Scleroglucan is used for the improvement of drug delivery system (Coviello et al. 2005). It is acceptable in making of sustained-release tablets and ocular formulations. It shows antiviral property against Herpes Virus and is used in formulations for agriculture. This polysaccharide is mainly used as an industrial viscosifier for the enhanced oil recovery. It has been reported for its antiviral and antitumor activity.

2.2.3 Schizophyllan

The main producer of this polysaccharide is *Schizophyllan commune.* The average molecular weight is about 500,000 Da (Itou et al. 1986). Schizophyllan is a β-1,

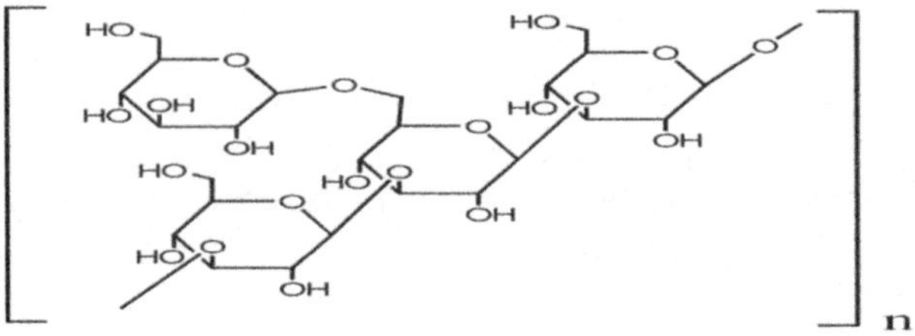

Fig. 2.3 Structure of schizophyllan showing a β-1, 3 glucan with β-1, 6 branching

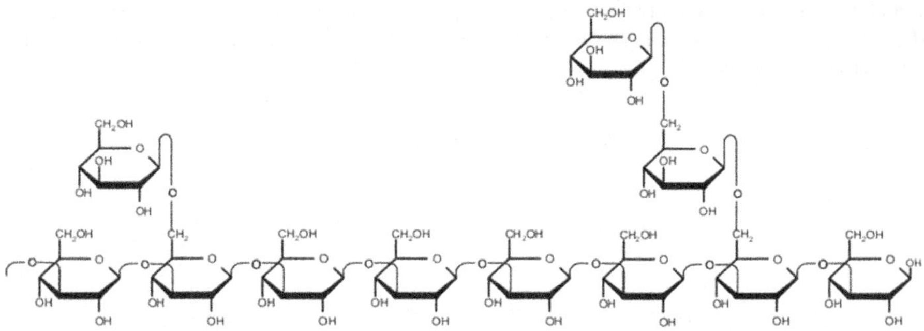

Fig. 2.4 Structure of botryosphaeran comprises of β (1, 3) backbone and β (1, 6) branched residues of glucose

3 glucan with β-1, 6 branching structure (Fig. 2.3). This polysaccharide is known for its capability for stimulating the immune system, hold metals in water, help in delivering drugs and making of nanofibers.

2.2.4 Botryosphaeran

Botryosphaeran is an exopolysaccharide which is produced by *Botryosphaeria rhodina* a ligninolytic and ascomycetous fungus that makes the culture medium-high in viscosity when grown on glucose as the main source of carbon. It is capable of forming a strong gel. It is a potent immunomodulatory and shows an antimutagenic effect. The triple-helical conformation of the polysaccharide is stabilized by hydrogen bonding between hydroxyl groups of the polysaccharide chain, which is the main reason for the stable behaviour of these macromolecules (Fig. 2.4).

2.2.5 Glucuronoxylomannan

Glucuronoxylomannan is a major capsular fungal polysaccharide produced by *Cryptococcus neoformans*. Glucuronoxylomannan is a high molecular weight polysaccharide that is shed during cryptococcosis and very important for immune

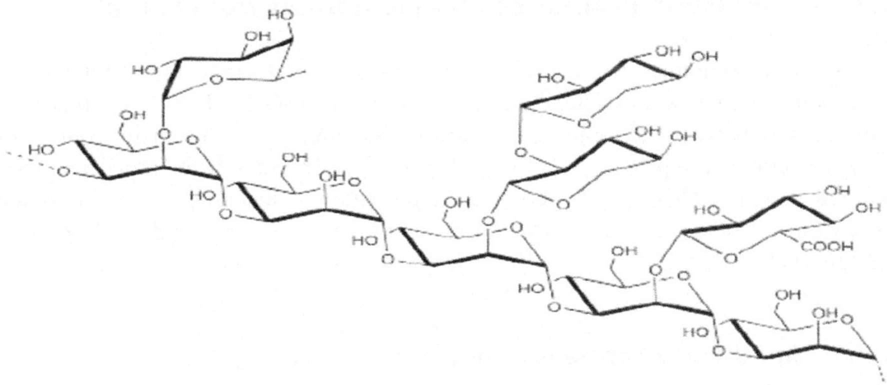

Fig. 2.5 Structure of glucuronoxylomannan consisting of two repeating polysaccharide units

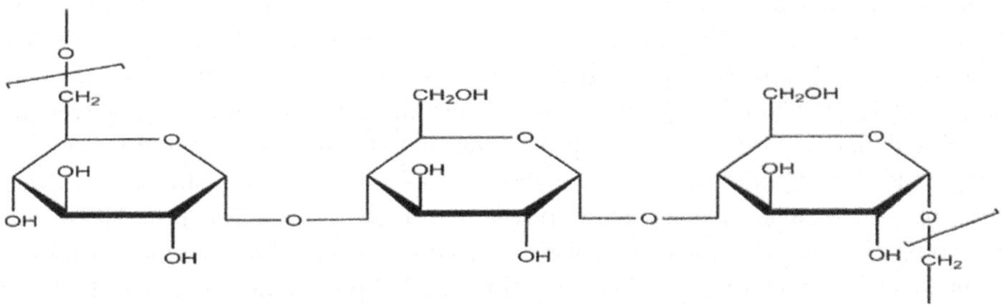

Fig. 2.6 Structure of pullulan consisting of maltotriose units with alternating α-(1, 4) and α-(1,6) linkages

response. It also inhibits the entrance of HIV in TZM-bl cell lines. The interaction of glucuronoxylomannan with monocytes or macrophages regulates the killing process, resulting in the reduction of superoxide anion (Monari et al. 2003) and decreased proinflammatory cytokines (Vecchiarelli et al. 1995); alongwith supression of IL-12 (Retini et al. 2001) (Fig. 2.5).

2.2.6 Pullulan

Pullulan is a linear water-soluble polymer isolated from *Pullularia pullulans* and *Aureobasidium pullulans*. The molecular weight of pullulan varies from 10–400 × 103 Da and highly soluble than amylose. Pullulan is a soluble homopolysaccharide comprises of maltotriose units with alternating α-(1, 4) and α-(1, 6) linkages (Fig. 2.6). Generally, pullulan is used in the agri-food industry as viscosity stabilizers, thickeners and biodegradable preparation, edible plastic materials and non-toxic. Since pullulan is only partially degraded by the human amylases, it is utilized as dietary fibre or a prebiotic substance.

2.3 Production Process of Polysaccharides from Fungi

Among all the parameters affecting polysaccharides production, the techniques or the production process has a huge impact on producing polysaccharides qualitatively and quantitatively. The outer cell surfaces of fungi are the main source of polysaccharides. For production of fungal exopolysaccharides different fermentation method are practised under controlled condition. Generally, two important processors are deployed for maximum fungal biomass production and for the quality end product.

2.3.1 Solid-State Fermentation (SSF)

This technique provides good results mostly for the biotechnological fungal products, for example, enzymes or other secondary metabolites (Saqib et al. 2010). One of the important factors for choosing this method is that they are relatively cheaper when the raw material is used as the main subtract (Castilho et al. 2000). Singhania et al. (2009) suggested that this substrate provides the best growing condition for the fungi to grow for producing the polysaccharides. However, he suggested that the limitation of this process lies with the biomass separation for most fungi. Nie et al. (2005) proposed a method for polysaccharides quantification using high-performance liquid chromatography with refractive index detection (HPLC-RI) for comparison of glycoproteins collected from tea species from various locations. Solid-state fermentation overall is the best method of producing fungal biomass as it is a cost-effective process and can yield maximum quantities of fungal biomass. Also for the production of foods that contain polysaccharides derived from fungi such as glucomannans produced from *A. subrufescens,* this procedure can be utilized to consume as whole biomass obtained through the extension of functional foods.

2.3.2 Submersed Fermentation

This method generally provides more productivity of polysaccharides than the solid-state fermentation as the procedure is subjected to those fungi which have the potentialities to produce a large volume of biomass compared to others and thereby producing a large number of polysaccharides. This method is mostly used for the pharmaceutical area as they require more precise fungal growth condition such as physical, chemical, nutritional factor, a uniform yield of biomass and quality polysaccharides yield (Fazenda et al. 2008).

Generally, two types of bioreactors are used for cultivation of fungal biomass in submersed fermentation method, viz. stirred-tank bioreactor which is applied under the aerobic condition and is mostly used in the industrial sector. The other one is airlift and bubble column bioreactor which is operated without mechanical agitators. The disadvantage that lies with the stirred-tank is the stirring requirement to reach ideal biomass of fungi sometimes causes damage to the cells. Apart from that they are more expensive and have much lesser resistant than later. The work on the airlift bioreactor is done by Camelini et al. (2013) on *A. subrufescens* and bubble column by Kawagoe et al. (2004) on the same species was 8 times lower average productivity (0.095 g/L-day) on the latter than the former which has 0.76 g/L-day subject to placement at same airflow rates.

The production of polysaccharides mostly depends on right species selection, cultivation method, extraction and purification for final successful results of interest. With the gaining of popularity in today's world, this technology will help future research with the production of quality product needed by different sectors.

2.4 Compositions of EPS Produced by Different Fungi

The fungal EPSs composition differs from pure sugars to sugars combined with a second unit. This second unit can be a protein, phosphate, sulphate or amine. In homopolysaccharides, glucose is the only monomer but in case of Ascomycota and Basidiomycota EPSs, they are mainly heteropolysaccharides where different types of sugar units were found in fungal EPSs such as glucose, galactose, mannose, fucose, xylose and rhamnose. Different fungi synthesize same monosaccharide units which are the composition of EPSs (Table 2.2). The chemical properties and structures of EPS such as the molecular weight, monosaccharide composition, linkage type and the conformation chain were evaluated with the help of different chromatography technology such as GLC, TLC, HPLC and also with various spectrum analysis like FTIR, 1D and 2D NMR spectroscopy and GLC-MS (Kozarski et al. 2012; Osinska-Jaroszuk et al. 2014; Silveira et al. 2015).

2.5 Parameters Affecting Polysaccharides Production

Polysaccharides production and usage in today's world have gained a tremendous market due to its wide range of beneficial applications. They form an important constituent in several industrial purposes in the form of feedstock, different types of chemicals material production and also energy. Apart from these various sources of polysaccharides viz., plant, animals and microorganisms, the former contributes most for increasing demand of the market. It has been estimated to produce 90% of polysaccharides only through the vegetable biomass. Among the microorganism producing polysaccharides, fungi are the dominant producers and are the most studied area for the production of polysaccharides. Polysaccharides production from fungal biomass has gained a lot of interest among the researcher as they are most predominant on the cell wall and shares equal relevant physico-chemical and structural properties that have wide range of applications. However, the production process of polysaccharides from fungal biomass varies with that of the other sources and several factors affect the production process.

Table 2.2 Compositions of some fungal exopolysaccharide

Organism	EPS composition	References
Aspergillus niger	Glucose	Sutherland (1996a, b)
Nigrospora oryzae var. *glucanicum*	Glucose	Sudhakaran and Shewale (1988)
Aspergillus alliaceus	Glucose, Galactosamine, Galactose, Acetate	Miranda and Leal (1981)
Drechslera spicifera	Glucose	Aouadi et al. (1992)
Aspergillus parasiticus	Glucose, Galactose, Acetate, Galactosamine, Phosphate	Ruperez and Lal (1981)
Aspergillus sp. Y16	Galactose, Mannose	Chen et al. (2011)
Candida boidinii	Glucose, Mannose	Petersen et al. (1990)
Cordyceps sphecocephala	Glucose, Galactose, Mannose, Protein	Oh et al. (2007)
Cryphonectria parasitica	1st type—Glucose, Galactose, Mannose; 2nd type—Glucose	Forabosco et al. (2006)
Cryptococcus laurentii AL100	Glucose, Galactose, Arabinose, Mannose, Rhamnose	Pavlova et al. (2011)
Cryptococcus neoformans	1st—Glucuronic acid, Xylose, Mannose; 2nd—Galactose, Xylose, Mannose; 3rd—Mannose, Protein	
Stemphylium sp.	Glucose, Mannose	Mahapatra and Banerjee (2012)
Antrodia camphorata	–	Shu and Lung (2004)
Elsinoe leucospila	Glucose	Sutherland (1996a, b)
Epicoccum nigrum Ehrenb. ex Schlecht	Glucose	Schmid et al. (2001)
Fusarium solani SD5	Galactose, Rhamnose	Mahapatra and Banerjee (2012)
Isaria farinosa BO5	Glucose, Mannose, Galactose, Uronic acid, Protein (4.40%)	Jiang et al. (2008)
Lachnum sp. YM261	Glucose	
Cyttaria harioti	Glucose	Sutherland (1996a, b)
Phomopsis foeniculi	1st type—Rhamnose, Mannose, Galactose; 2nd type—Mannose	Corsaro et al. (1998)
Penicillium varians	Glucose, Galactose	Jansson and Lindberg (1980)
Pestalotia sp. 815	Glucose	Misaki et al. (1984)
Phanerochaete chrysosporium	Glucose	Buchala and Leisola (1987)

(continued)

Table 2.2 (continued)

Organism	EPS composition	References
Trichosporon asahii	Mannose, Xylose, Glucuronic acid	Fonseca et al. (2009)
Pleurotus tuber-regium	Glucose, Mannose	Zhang and Cheung (2011)
Sporobolomyces sp.	Galactose	Pavlova et al. (2004)
Pleurotus sp.	Glucose, Galactose, Mannose, Protein	Gutiérrez et al. (1996)
Tremella fuciformis	Mannose, Xylose, Fucose, Protein	Cho et al. (2006)
Acremonium diospyri Crandall	Glucose	Seviour and Hensgen (1983)
Botryosphaeria rhodina RCYU30101	Glucose	Weng et al. (2011)

2.5.1 Nutrient Source

One of the major factor affecting fungal growth and development is the wide range of nutrient sources. The effect of a nutrient on mycelium can lead to overgrowth or even restrict the growth with negative results. It has been found by various researchers that there has been a direct co-relation of polysaccharides production from fungal biomass and the nutrients supplied. Among the nutrient sources studied, most focus has been given to the effect of nitrogen and carbon sources (Wu et al. 2012a, b; Mahapatra and Banerjee 2013).

2.5.1.1 Nitrogen Source

The nitrogen source in the growing media has a great influence on fungal growth, development and metabolite productions. Both the organic and inorganic source of nitrogen can be effectively used. However, organic sources have been seen to produce relatively higher polysaccharides than later. $NH_4H_2PO_4$ when used as nitrogen sources produce a much higher yield of polysaccharides than KNO_3 from *Tricholoma margolicum*. Some of the well-known inorganic sources are ammonium chloride, ammonium sulphate, potassium nitrate, sodium nitrate, diammonium oxalate monohydrate and urea (Mahapatra and Banerjee 2013). Of these ammonium salts are found to produce effective results in polysaccharide production. Few reports have suggested sodium and urea as a better source of nitrogen when *Epicoccum nigrum* and *S.rolfsii* fungi are used for polysaccharides production (Farina et al. 1998; Schmid et al. 2001). Even though a very less amount of nitrogen are required by the fungi (1–10 g/L) to produce polysaccharides, on the one hand, their unavailability, on the other hand, may result in a subsequent reduction in fungal growth and thereby less production on the polysaccharides.

2.5.1.2 Carbon Source

In contrast to nitrogen which is needed in a very minute amount, carbon is very much important in polysaccharides production from fungi as they are the important sources that have a direct effect on growth and development of high fungi. The polysaccharides production intensity is much dependent on the types of carbohydrates used, i.e. on carbon sources. Many sources of carbon have been studied by the researchers: glucose, sucrose, fructose, lactose, maltose, galactose, cellobiose, xylose, sorbitol, mannitol, xylitol, etc. Among which glucose, maltose and sucrose have been mostly studied and have given the most influential results in the production of the polysaccharide from fungal biomass. However, apart from these sources, it has been found that different fungi respond to other carbon sources for their growth and development with the production of the desired end product. Different fungi which have been studied for different carbon sources are presented in the following Table 2.3.

Besides the sources of carbon, the concentrations of carbon also play a vital role in the growth and development of fungal biomass and ultimately on polysaccharide production. On average, it has been found that 30–60 g/L carbon was best to produce polysaccharide from fungi. Few reports have also suggested that combined carbon source can enhance the polysaccharide where 2 g/L glucose combined with 30 g/L starch increased production of polysaccharide when produced from *Sorangium cellulosum*.

2.5.2 pH

The pH is one of the important factors affecting the fungal biomass growth that is required for effective polysaccharide production. Their growth varies with the change of pH, with slight variation in it can lead to overgrowth or undergrowth of fungal biomass within a given environment condition. In general, fungi mostly prefer pH ranging between 3.0 and 6.5 (Pavlova et al. 2011, Luo et al. 2009, Wu

Table 2.3 List of fungi studied for different carbon sources

Fungi	Carbon source	References
1. *G. lucidum*	Glucose	Elisashvili et al. (2009)
2. *Inonotus levis*	Glucose	
3. *Phellinus robustus*	Glucose	
4. *Cerrena maxima*	Maltose	
5. *Phellinus igniarius*	Maltose	
6. *Trametes versicolor*	Maltose	
7. *Agaricus nevei*	Mannitol	Bae et al. (2001)
8. *Paecilomyces japonica*	Mannitol	
9. *L. edodes*	Sodium gluconate	Elisashvili et al. (2009)
10. *Pleurotus* sp.	Sodium gluconate	
11. *Sclerotium rolfsii* (ATCC2011126)	Sucrose	Farina et al. (1998)

et al. 2012a, b). Besides these neutral pH loving fungi, several other fungi prefer mostly neutral or alkaline pH for polysaccharide production (Kim and Yun 2005; Pokhrel and Ohga 2007). Once the study has suggested a direct relation of polysaccharide production to initial pH where the mycelia growth at an earlier stage was seen faster, however on the later stage there was no clustering of mycelia which resulted in a decrease in mycelia biomass formation and final accumulation of polysaccharides. An example to the effect of pH on the production of polysaccharide can be seen on the studies of Shu and Lung (2004) where they found the variation in polysaccharide production with different molecular weight (Mw) by *Antrodia camphorata* with variation in pH. They found that at lower pH medium-high Mw polysaccharide was produced in low amount and at high pH, medium-low Mw polysaccharide was produced with high yield. Another study on *Mucorrouxii* by Abdel-Aziz et al. (2012) showed that the polysaccharide production gradually increased with increase in pH. A work on *A. subrufescens* showed that the yield of mycelium was maximum when placed in the pH initially at 6.06 and incubated at a temperature of 27 °C (Lin and Yang 2006). A slight change in pH with increased polysaccharide production was also reported by Fan et al. (2007) where pH 6.1 and temperature 30 °C was suitable for mycelial yield. However, some of the reports also suggested that the pH of about 5.6 and 6.9 and a temperature of about 24.6 °C and 20 °C was best for maximum polysaccharides production (Hamedi et al. 2007).

2.5.3 Temperature

Temperature is relatively an important factor required for fungal growth and development which in turn affects the polysaccharide production. Different fungi respond to varying temperature for the production of polysaccharides where maximum polysaccharide production from fungal strain lies within the range of 22 °C to 30 °C (Mahapatra and Banerjee 2013). Apart from these few fungal strain have been reported to respond well at 20 °C for polysaccharide production. Xiotong et al. (2012) found that at 25 °C *T. mongolium* responded well and yield maximum polysaccharide. However, with slight variation in above temperature, the mycelia did not respond well to grow resulting in low yield of polysaccharides production.

2.5.4 Size and Age of Fungal Inoculum

The different physiological properties of fungus also play a significant effect on the production of the polysaccharide. Among them, the size of the plant and the age of the fungus in the culture have a tremendous effect on the fungal growth and development (Glazebrook et al. 1992). In other words, if the size of the inoculum is taken smaller it may lengthen the fermentation period, which on the other larger size inoculum may overcome disadvantage because the large inoculum may consume the nutrient quickly which helps in the growth of fungal biomass and thereby producing the polysaccharides preventing the bacterial infection during the fermentation process. Xiaotong in his experiment concluded that out of different inoculum, viz., 5, 8, 10, 12 and 15% v/v, only 10–12% inoculum size was seen to produce better polysaccharides production. Age, on the other hand, have a direct effect on the fungal mycelial growth, as over-aged fungal colony may take more time to reach the log phage as compared to the fresh culture provided the same nutrient media and temperature.

2.5.5 Fungal Material Preservation

Storage of fungal material plays an important role in the production of quality polysaccharides. As the condition of the fungal material before and after the preservation directly influence the growth of the fungi. Different type of techniques has been used by various researchers previously but for fungal storage, one can store them depending on the duration of time as follows:

2.5.5.1 Short-Term Preservation

It includes maintenance and preservation of culture for a period of short time (less than 1 year). This method is simple, cost-effective and labour intensive. Besides these, it has several disadvantages of frequent checking of contamination, morphological or physiological change of fungi with time. For fungal short term storage, the use of mineral oil as well as distilled water is most popular and widely used for the preservation (Richter 2008; Croan et al. 1999).

2.5.5.2 Long-Term Preservation

This technique is mainly used to preserve the fungal culture for long period (1–3 years or even more duration). This method generally overcomes the disadvantages of short term storage. The main logic behind this technique is to arrest growth and metabolic activities for long-term preservation. For fungal long-term storage lyophilization and cryopreservation in liquid nitrogen are mostly used (Croan et al. 1999; Colauto et al. 2012a, b).

The choice for opting the method of preservation depends upon several factors. After preservation, the fungal culture must be contamination-free apart from producing a pure mycelium as preserved it must be able to grow and multiply producing successful biomass for polysaccharide production. Also, the factors such as biological activity in the end product are one of the major concern before opting for any method of preservation as some preservative chemicals or substracts may interact with the chemical structure of the polysaccharides present in the cell wall of mycelia which may alter or change its main structural chain also its lateral end chain or the arrangement of its branching degree, which may have an impact in quality polysaccharide production (Lavi et al. 2010; Mantovani et al. 2008). Few examples in case of *Agaricus subrufescens*. While preserved in cryopreserving with DMSO, it was found to respond differently, only 40% was recovered when frozen and maintained at −196 °C for 1.5 years (Colauto et al. 2012a, b). Attenuation of fungal activity is one of the important factors as some of the fungi have the capabilities to produce a large number of secondary metabolites as agaritine when their growth are arrested during preservation placed in the mycelia for growth may affect the fungal species through their mutagenic effect (Walton et al. 1997). Maheshwari and Navaraj (2008) found that the storage of most fungi can be done at a temperature ranging between 4 °C and 12 °C for 6–8 months; however, the difficulties pertaining this can be the induction of senescence which may lead to the failure of progressive potential mycelial growth, induction of the mutation, hindrance of growth which may lead to fungal death. Some researchers also suggest that the use of activated charcoal for preserving *A. subrufescens* is a quick, easy and cost-effective method as it was found that it did not affect much on morphological or genetical changes even when stored for 1 year.

2.5.6 Additives

There have been several reports that the addition of some additives resulted in maximum EPS production. Such finding has been presented with the addition of fatty acid, vitamins, vegetable oils and surfactants (Yang and He 2008; Xiao et al. 2010; Zhang and Cheung 2011; Yang et al. 2000). In addition to these use of glutamic acid, biotin, thiamine (Lee et al. 2001), soybean oil (0:1) in culture media, palmitic acid and oleic acid showed increased production of EPS when *Shiraria bambusicola* was used. Apart from these some of the surfactant which has widely been tested such as Tween 80 and Span 80 induced maximum EPS production when 3 g/L is added. However, Hsieh et al. (2008) found that the same surfactants even though induced the fungal growth but EPS production was reduced significantly in *G. frondora*. He also found that at a stationary phase when olive oil (0.5%) is added the fungus was able to produce maximum EPS. Furthermore, addition vitamins like Vit A & D in media, 0.1% (v/v) thiamin & nicotinic acid at the rate of 0.1% (v/v) helped to increase exopolysaccharides production by *Oudemansiella radicata* and *Antrodiac cinnamomea,* respectively (Lin and Chen 2007; Xiao et al. 2010).

2.6 Different Applications of Fungal Polysaccharides

Fungal EPSs being one of the prime components of microbial EPSs have gained wide significance in the course of a recent couple of decades, as few studies demonstrated their applicability in various fields. Besides this, production of EPSs is simpler and a large number of these products can be developed in a shorter period in comparison with products of plant or algae. Fungal EPSs have got few applications in the food, cosmetics and drug enterprises beside others. Fungal EPSs, like pullulan, scleroglucan and botryosphaeran, has many important applications in fields of food and pharmaceutical industries. Among the various applications, EPSs are used as biosurfactants, bioemulsifiers and bioflocculants due to their structural and functional diversity. EPS-based bioflocculants are mostly

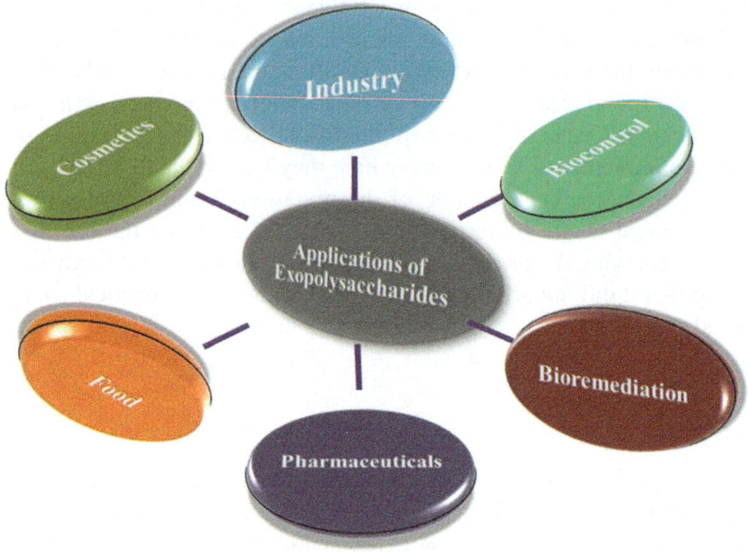

Fig. 2.7 Multiple applications of fungal EPs in different fields

composed of polysaccharides, proteins, glycoproteins, nucleic acids and lipids. The process where stable aggregates are formed by extracellular polymers released by living cells is termed as bioflocculation. Exopolysaccharides as bioflocculants have gained increased attention as they possess ionizable functional groups like amine, carboxyl, acetate, sulphate and hydroxyl groups which make these biopolymers to be effective in the removal of suspended solids, heavy metals and in reducing the turbidity of different types of industrial wastewater effluents. Their use has been well documented in textiles, detergents, raw water treatment, adhesives, petroleum refinery effluent, oil recovery, paper industry and settling of sludge in aerobic and anaerobic treatment systems.

An overview of various applications of fungal EPSs is presented in Fig. 2.7. Among the diverse fungal EPSs scleroglucan, botryosphaeran and pullulan are notable for exhibiting multiple roles in various fields. However, a few fungal EPSs have been described for understanding the extent of vivid applications of macromolecules.

Pullulan, an EPS of *A. pullulans*, can be utilized as a thickener, a viscosity stabilizer in the food business, and for designing of non-toxic, biodegradable, palatable plastic materials (Paul et al. 1986).

Fungal EPS scleroglucan is utilized mostly for the Enhanced Oil Recovery (Holzwarth 1985). Scleroglucan often is known as schizophyllan under the trade name BIOVIS with bacterial polysaccharides, xanthan gums are usually used in making of drilling liquids with low mud impurities (Hamed and Belhadri 2009). Farwick et al. (2009) reported that scleroglucan possess the characteristic feature of high-water adhering ability on epithelial cells. Other modern utilizations of this EPS were in the manufacture of glues, water hues, printing inks and formation of animal feed (Halleck 1969). It is also being applied in the manufacture of cosmetics and different skin health management items, lotion and creams (Halleck 1970, 1972).

EPS of *Byssochlamys* demonstrated the property of Kaolin flocculating conduct. Gomoiu and Catley proposed that the EPS can be utilized for deposition of filaments in downstream expulsion of paper strands from water effluent in paper enterprises and subsequently has reduced efficacy in the paper industry and rivers wherever mechanical effluents were disposed of (Catley 1971). Five yeast strain EPSs and from one yeast-like organism demonstrated drag-lessening action (Petersen et al. 1990). Exopolymer obtained from *G. lucidum* can enhance the swimming durability of mice by around 10 min and decreased the glycogen content in muscle and liver depletion by 18.5% and 67.2%, respectively (Yang et al. 2001). Fungal-glucans (both intracellular and extracellular) are compelling in advancing health, safeguard from mutations and therapy of ailments like disease, microbial contaminations, diabetes and hypercholesterolaemia (Chen and Seviour 2007; Mantovani et al. 2008). Pestan, a contagious EPS obtained from *Pestalotiopsis* sp. KCTC 8637, has been used in the treatment of wastewater as a biosorbent of zinc and lead. Every gram of pestan can absorb 60 mg zinc and 120 mg lead (Moon et al. 2006). Similarly, EPS of *Aspergillus fumigatus* likewise demonstrated sorption proficiency of two substantial metal particles, i.e. lead and copper (Yin et al. 2011).

Botryosphaeran produced from *Botryosphaeria rhodina* appeared to diminish altogether the clastogenic impact of cyclophosphamide-instigated micronucleus arrangement in polychromatic erythrocytes of bone marrow and reticulocytes infringe blood in mice (Miranda et al. 2008). It is also known to be a strong immunomodulator (Weng et al. 2011). EPSs delivered by *Akanthomyces pistillariiformis* BCC2694, *P. tenuipes* BCC2656, *Cordyceps dipterigena* BCC2073 and *Phytocordyceps* sp. BCC2744 demonstrated to be biocompatible and have the probability to act as a wound dressing material (Madla et al. 2005). EPS of *Phellinus baumii* Pilat demonstrated direct safe invigorating action on splenocyte proliferative reaction and corrosive phosphatase movement in peritoneal macrophages of mice (Luo et al. 2009). Wang et al. (2011) described the immunomodulatory exercises of EPS delivered by *C. sinensis* Cs-HK1.

Water solvent EPS of *Isaria farinosa* BO5 demonstrated antitumor and cell reinforcement movement in Kunming mice (Weng et al. 2011). EPS of *Fomes fomentarius* and *Hypsizygus marmoreus* demonstrated antitumor action inspected on human gastric disease cells SGC-7901 (Chen et al. 2008; Zhang et al. 2012). EPS of *C. sinensis* Cs-HK1 has moderate cancer prevention potential.57 EPSs of *Gomphidius rutilus*, *Aspergillus* sp. Y16 and *Ganoderma resinaceum* appeared in vitro opponent of oxidant movement (Kim et al. 2006; Chen et al. 2011; Gao et al. 2012). *Fusarium oxysporum* DzF17 has been accounted to be an EPS-producing fungus. The EPS indicated elicitor exercises on the development and diosgenin formation in cell suspension culture of *Dioscorea zingiberensis* (Li et al. 2011). *F. solani*SD5 was accounted for to create an extracellular rhamnogalactan that indicated calming and hostile to hypersensitive action in vitro. (Mahapatra and Banerjee 2012). *G. frondosa* HB0071 EPS demonstrated inhibitory impact on grid metallo proteinase-1 articulation in UV-lighted human dermal fibroblasts, and this manner may add to inhibitory activity in photograph maturing skin by diminishing framework metallo proteinase-1-related network debasement framework (Bae et al. 2005). EPS produced by *Stropharia rugosoannulata* and *P. baumii* show hypoglycemic movement in streptozotocin actuated high sugar rodents (Zhai et al. 2012). Primer assessments with EPS formed by *R. glutinins* are hostile to oxidant, antitumor and antiviral exercises (Ibrahim et al. 2012).

2.7 Conclusion

Hurtley et al. (2001) stated that the science and science of carbohydrate research resembles a 'Cinderella field', however till date, it doesn't receive much appreciation as that of genomics and proteomics. This acknowledgement is touched with authenticity, even though microorganisms have been widely employed since the 1940s for the development of assorted biomaterials. On account of EPS creation, notwithstanding, the endeavours were not acceptable to extinguish the hunger for information in this field as of not long ago. Significant increments in the examination of fungal EPS production or their physio-synthetic portrayals by specialists were seen uniquely in the course of the most recent 20 years. By the by, the discoveries are amazing from both the logical edge furthermore, pertinence. There is no uncertainty that EPS and its derivatives have promising possibilities that could be investigated for assorted additions, including public and unfamiliar trade benefits through connecting with worldwide biopolymer market. The expectation is that this pattern will proceed to increment and will enhance both logical information base and will open new avenues in different sectors for encouraging our livelihood.

References

Abdel-Aziz SM, Hamed HA, Mouafi FE, Gad AS (2012) Acidic shock induces the production of an exopolysaccharide by the fungus *Mucor rouxii*. Util Beet Molasses 5:52–61

Ajith TA, Janardhanan KK (2007) Indian medicinal mushrooms as a source of antioxidant and antitumor agents. J Clin Biochem Nutr 40(3):157–162

Aouadi S, Heyraud A, Seigle-Murandi F, Steiman R, Fournet B (1992) Structural analysis and rheological behaviour of an extracellular polysaccharide from *Drechslera spicifera*. Carbohydr Polym 17(3):177–183

Bae JT, Park JP, Song CH, Yu CB, Park MK, Yun JW (2001) Effect of carbon sources on the mycelial growth and exo-biopolymer production by submerged culture of *Paecilomyces japonica*. J Biosci Bioeng 91(5):522–524

Bae JT, Sim GS, Lee DH (2005) Production of exopolysaccharide from mycelial culture of *Grifola frondosa* and its inhibitory effect on matrix metalloproteinase-1 expression in UV-irradiated human dermal fibroblasts. FEMS Microbiol Lett 251(2):347–354

Biagini G, Bertani A, Muzarelli RAA, Damadei A, Benedetto G, Di Belligolli A, Riccoti G, Zucchini C, Rizzoli C (1991) Wound management with *N*-carboxybutyl chitosan. Biomaterials 12:281–286

Buchala AJ, Leisola M (1987) Structure of the β-glucan secreted by *Phanerochaete chrysosporium* in continuous culture. Carbohydr Res 165:146–149

Camelini CM, Gomes A, Cardozo-Sousa FTG, Simões CMO, Rossi MJ, Giachini AJ, Petrus JCC, Mendonça MM (2013) Production of polysaccharide from *Agaricus subrufescens* Peck on solid-state fermentation. Appl Microbiol Biotechnol 97(1):123–133

Castilho LR, Polato CMS, Baruque EA, Sant'Anna GLJ, Freire DMG (2000) Economic analysis of lipase production by *Penicillium restrictum* in solid-state and submerged fermentations. Biochem Eng J4(3):239–247

Castillo AN, Valdez AL, Julia IF (2015) Microbial production of scleroglucan and downstream processing. Front Microbiol 6:1106

Catley BJ (1971) Role of pH and nitrogen limitation in the elaboration of the extracellular polysaccharide pullulan by *Pullularia pullulans*. Appl Microbiol 22(4):650–654

Chen J, Seviour R (2007) Medicinal importance of fungal β-(1-3), (1-6)-glucans. Mycol Res 111:635–652

Chen W, Zhao Z, Chen SF, Li YQ (2008) Optimization for the production of exopolysaccharide from *Fomes fomentarius* in submerged culture and its antitumor effect in vitro. Bioresour Technol 99(8):3187–3194

Chen Y, Mao W, Tao H (2011) Structural characterization and antioxidant properties of an exopolysaccharide produced by the mangrove endophytic fungus *Aspergillus* sp. Y16. Bioresour Technol 102(17):8179–8184

Chen J, Mao D, Yong Y, Li J, Wei H, Lu L (2012) Hepatoprotective and hypolipidemic effects of water-soluble polysaccharidic extract of *Pleurotus eryngii*. Food Chem 130:687–694

Cho EJ, Oh JY, Chang HY, Yun JW (2006) Production of exopolysaccharides by the submerged mycelial culture of a mushroom *Tremella fuciformis*. J Biotechnol 127(1):129–140

Colauto NB, Eira AF, Linde GA (2012a) Cryopreservation of *Agaricus blazei* in liquid nitrogen using DMSO as cryoprotectant. Biosci J 28(6):1034–1037

Colauto NB, Cordeiro FA, Geromini KVN, Lima TG, Lopes AD, Nunes RAR, Roratto FB, Tanaka HS, Zaghi LL Jr, Linde GA (2012b) Viability of *Agaricus blazei* after long-term cryopreservation. Ann Microbiol 62:871–876

Corsaro MM, Castro CD, Evidente A (1998) Chemical structure of two phytotoxic exopolysaccharides produced by *Phomopsis foeniculi*. Carbohydr Res 308(3–4):349–357

Coviello T, Palleschi A, Grassi M, Matricardi P, Bocchinfuso G, Alhaique F (2005) Scleroglucan: a versatile polysaccharide for modified drug delivery. Molecules 10(1):6–33

Croan SC, Burdsall HH Jr, Rentmeester RM (1999) Preservation of tropical wood-inhabiting basidiomycetes. Mycologia 91(5):908–916

Dave SR, Vaishnav AM, Upadhyay KH, Tipre D (2016) Microbial exopolysaccharide—An inevitable product for living beings and environment. J Bacteriolmycol Open Access 2(4):00034

Elisashvili VI, Kachlishvili ET, Wasser SP (2009) Carbon and nitrogen source effects on basidiomycetes exopolysaccharide production. Appl Biochem Micro 45(5):531–535

Fan L, Soccol AT, Pandey A, Soccol CR (2007) Effect of nutritional and environmental conditions on the production of exopolysaccharide of *Agaricus brasiliensis* by submerged fermentation and its antitumor activity. LWT 40:30–35

Farina JI, Sineriz F, Molina OE, Perotti NI (1998) High scleroglucan production by *Sclerotium rolfsii*: influence of medium composition. Biotechnol Lett 20(9):825–831

Farwick M, Lersch P, Schmitz G, Müllner S, Wattenberg A (2009) "Skin-omics": use of genomics, proteomics and lipidomics to assess effects of low molecular weight scleroglucan. Cos Sci Technol E Indus, 100–5

Fazenda ML, Seviour R, Mcneil B, Harvey LM (2008) Submerged culture fermentation of "higher fungi": the macrofungi. Adv Appl Microbiol 63:33–103

Fonseca FL, Frases S, Casadevall A, Gompertz OF, Nimrichter L, Rodrigues ML (2009) Structural and functional properties of the *Trichosporon asahii* glucuronoxylomannan. Fungal Genet Biol 46(6–7):496–505

Forabosco A, Bruno G, Sparapano L, Liut G, Marino D, Delben F (2006) Pullulans production by strains of *Cryphonectria parasitica*-I. Production and characterization of the exopolysaccharides. Carbohydr Polym 63(4):535–544
Freitas F, Alves VD, Reis MAM (2011) Advances in bacterial exopolysaccharides: from production to biotechnological applications. Trends Biotechnol 29:388–398
Gao C, Wang Z, Su T, Zhang J, Yang X (2012) Optimization of exopolysaccharide production by *Gomphidius rutilus* and its antioxidant activities in vitro. Carbohydr Polym 87(3):2299–2305
Glazebrook MA, Vining LC, White RL (1992) Growth morphology of *Streptomyces akiyoshiensis* in submerged culture: influence of pH, inoculums, and nutrients. Can J Microbiol 38:98–103
Gutiérrez A, Prieto A, Martínez AT (1996) Structural characterization of extracellular polysaccharides produced by fungi from the genus *Pleurotus*. Carbohydr Res 281(1):143–154
Halleck FE (1969) Paint composition containing polysaccharides. US Patent 3447940
Halleck FE (1970) Wave set composition containing a polysaccharide. US Patent 3507290
Halleck FE (1972) Cosmetic composition employing water-soluble polysaccharide. US Patent 3659025
Hamed SB, Belhadri M (2009) Rheological properties of biopolymers drilling fluids. J Pet Sci Eng 67:84–90
Hamedi A, Vahid H, Ghanati F (2007) Optimization of the medium composition for the production of mycelia biomass and exopolysaccharide by *Agaricus blazei* Murill DPPh 131 using response-surface methodology. Biotechnology 6(4):456–464
Holzwarth G (1985) Xanthan and scleroglucan: structure and use in enhanced oil recovery. Dev Ind Microbiol 26:271–280
Hosokawa J, Nishiyama M, Yoshihara K, Kubo T (1990) Biodegradable film derived from chitosan and homogenized cellulose. Indus Eng Chem Res 29(5):800–805
Hsieh C, Wang HL, Chen CC, Hsu TH, Tseng MH (2008) Effect of plant oil and surfactant on the production of mycelial biomass and polysaccharides in submerged culture of *Grifola frondosa*. Biochem Eng J 38(2):198–205
Hu SH, Liang ZC, Chia YC, Lien JL, Chen KS, Lee MY, Wang JC (2006a) Antihyperlipidemic and antioxidant effects of extracts from *Pleurotus citrinopileatus*. J Agric Food Chem 54:2103–2110
Hu SH, Wang JC, Lien JL, Liaw ET, Lee MY (2006b) Antihyperglycemic effect of polysaccharide from fermented broth of *Pleurotus citrinopileatus*. Appl Microbiol Biotechnol 70:107–113
Hurtley S, Service R, Szuromi P (2001) Cinderella's coach is ready. Science 291(5512):2337
Ibrahim GS, Mahmoud MG, Asker MMS, Ghazy EA (2012) Production and biological evaluation of exopolysaccharide from isolated *Rhodotorula glutinins*. Aust J Basic Appl Sci 6:401–408
Itou T, Teramoto A, Matsuo T, Sug H (1986) Macromolecules 19:1234–1240
Jansson PE, Lindberg B (1980) Structural studies of varianose. Carbohydr Res 82:97–102
Jiang Y, Jiang X, Wang P, Mou H, Hu X, Liu S (2008) The antitumor and antioxidative activities of polysaccharides isolated from *Isaria farinosa* BO5. Microbiol Res 163(4):424–430
Kawagoe M, Nagaoka Y, Araki M, Yamagaki K, Naoe K, Noda H (2004) Submerged culture of *Agaricus blazei* mycelium in a bubble column fermentor. J Chem Eng Jpn 37(8):1056–1061
Kim HO, Yun JW (2005) A comparative study on the production of exopolysaccharides between two entomopathogenic fungi *Cordyceps militaris* and *Cordyceps sinensis* in submerged mycelial cultures. J Appl Microbiol 99(4):728–738
Kim HM, Kim SW, Hwang HJ (2006) Influence of agitation intensity and aeration rate on production of antioxidative exopolysaccharides from submerged mycelial culture of *Ganoderma resinaceum*. J Microbiol Biotechnol 16(8):1240–1247
Kleerebezem M, van Kranenburg R, Tuinier R, Boels IC, Zoon P, Looijesteijn E, Hugenholtz J, de Vos WM (1999) Exopolysaccharides produced by Lactococcus lactis: from genetic engineering to improved rheological properties? Anton van Leeuwenhoek 76(1–4):357–365
Kozarski M, Klaus A, Niksic M, Vrvic MM, Todorovic N, Jakovljevic D, Van Griensven LJLD (2012) Antioxidative activities and chemical characterization of polysaccharide extracts from

the widely used mushrooms *Ganoderma applanatum*, *Ganoderma lucidum*, *Lentinus edodes* and *Trametes versicolor*. J Food Compost Anal 26(1–2):144–153
Lavi I, Levinson D, Peri I, Tekoah Y, Hadar Y, Schwartz B (2010) Chemical characterization, antiproliferative and antiadhesive properties of polysaccharides extracted from *Pleurotus pulmonarius* mycelium and fruiting bodies. Appl Microbiol Biotechnol 85:1977–1990
Lee JH, Kim JH, Zhu IH (2001) Optimization of conditions for the production of pullulan and high molecular weight pullulan by *Aureobasidium pullulans*. Biotechnol Lett 23(10):817–820
Li P, Mou Y, Shan T (2011) Effects of polysaccharide elicitors from endophytic *Fusarium oxysporium* Dzf17 on growth and diosgenin production in cell suspension culture of *Dioscorea zingiberensis*. Molecules 16(11):9003–9016
Lin ES, Chen YH (2007) Factors affecting mycelial biomass and exopolysaccharide production in submerged cultivation of *Antrodia cinnamomea* using complex media. Bioresour Technol 98 (13):2511–2517
Lin JH, Yang SS (2006) Mycelium and polysaccharide production of *Agaricus blazei* Murril by submerged fermentation. J Microbiol Immunol Infect 39:98–108
Luo J, Liu J, Ke C (2009) Optimization of medium composition for the production of exopolysaccharides from *Phellinus baumii* Pilát in submerged culture and the immuno-stimulating activity of exopolysaccharides. Carbohydr Polym 78(3):409–415
Madla S, Methacanon P, Prasitsil M, Kirtikara K (2005) Characterization of biocompatible fungi-derived polymers that induce IL-8 production. Carbohydr Polym 59(3):275–280
Mahapatra S, Banerjee D (2012) Structural elucidation and bioactivity of a novel exopolysaccharide from endophytic *Fusarium solani* SD5. Carbohydr Polym 90(1):683–689
Mahapatra S, Banerjee D (2013) Fungal exopolysaccharide: production, composition and applications. Microbiol Insights 6:1–16
Maheshwari R, Navaraj A (2008) Senescence in fungi: the view from Neurospora. FEMS Microbiol Lett 280(2):135–143
Mantovani MS, Bellini MF, Angeli JPF, Oliveira RJ, Silva AF, Ribeiro LR (2008) β-Glucans in promoting health: prevention against mutation and cancer. Mutat Res 658:154–161
Meybod NM, Mohammadifar MA (2015) Microbial exopolysaccharides: a review of their function and application in food sciences. J Food Qual Hazard Control 2:112–117
Miranda BG, Leal JA (1981) Extracellular and cell wall polysaccharides of *Aspergillus alliaceus*. Trans Br Mycol Soc 76(2):249–253
Miranda CCBO, Dekker RFH, Serpeloni JM, Fonseca EAI, Cólus IMS, Barbosa AM (2008) Anticlastogenic activity exhibited by botryosphaeran, a new exopolysaccharide produced by *Botryosphaeria rhodina* MAMB-05. Int J Biol Macromol 42(2):172–177
Misaki A, Kawaguchi K, Miyaji H (1984) Structure of pestalotan, a highly branched (1-3)-β-D-glucan elaborated by *Pestalotia* sp 815 and the enhancement of its antitumour activity by polyol modification of the side chains. Carbohydr Res 129:209–227
Mishra A, Jha B (2013) Microbial exopolysacchrides. In: Rosenberg E, DeLong EF, Thompson F, Lory S, Stackebrandt E (eds) The prokaryotes: applied bacteriology and biotechnology, 4th edn. Springer, Berlin, pp 179–192
Monari C, Retini C, Casadevall A, Netski D, Bistoni F, Kozel TR, Vecchiarelli A (2003) Differences in outcome of the interaction between *Cryptococcus neoformans* glucuronoxylomannan and human monocytes and neutrophils. Eur J Immunol 33:1041–1051
Moon SH, Park CS, Kim YJ, Park YI (2006) Biosorption isotherms of Pb (II) and Zn (II) on Pestan, an extracellular polysaccharide, of *Pestalotiopsis* sp. KCTC 8637P. Process Biochem 41 (2):312–316
Moscovici M (2015) Present and future medical applications of microbial exopolysaccharides. Front Microbiol 6:1012
Nicolaus B, Kambourova M, Oner ET (2010) Exopolysaccharides from extremophiles: from fundamentals to biotechnology. Environ Technol 31:1145–1158
Nie S, Xie M, Wang Y (2005) Preparation of tea glycoprotein and its application as a calibration standard for the quantification and molecular weight determination of tea glycoprotein in

different tea samples by high-performance gel permeation chromatography. Anal Bioanal Chem 383:680–686

Oh JY, Cho EJ, Nam SH, Choi JW, Yun JW (2007) Production of polysaccharide-peptide complexes by the submerged mycelial culture of an entomopathogenic fungus *Cordyceps sphecocephala*. Process Biochem 42(3):352–362

Osinska-Jaroszuk M, Jaszek M, Mizerska-Dudka M, Blachowicz A, Rejczak T, Janusz G, Wydrych J, Polak J, Jarosz-Wilkołazka A, Kandefer-Szerszen M (2014) Exopolysaccharide from *Ganoderma applanatum* as a promising bioactive compound with cytostatic and antibacterial properties. Biomed Res Int 2014:1–10

Paul F, Morin A, Monsan P (1986) Microbial polysaccharides with actual potential industrial applications. Biotechnol Adv 4(2):245–259

Pavlova K, Koleva L, Kratchanova M, Panchev I (2004) Production and characterization of an exopolysaccharide by yeast. World J Microbiol Biotechnol 20(4):435–439

Pavlova K, Rusinova-Videva S, Kuncheva M, Kratchanova M, Gocheva M, Dimitrova S (2011) Synthesis and characterization of an exopolysaccharide by Antarctic yeast strain *Cryptococcus laurentii* AL. Appl Biochem Biotechnol 163(8):1038–1052

Petersen GR, Schubert WW, Richards GF, Nelson GA (1990) Yeasts producing exopolysaccharides with drag-reducing activity. Enzyme Microb Technol 12(4):255–259

Pokhrel CP, Ohga S (2007) Submerged culture conditions for mycelial yield and polysaccharides production by *Lyophyllum decastes*. Food Chem 105(2):641–646

Retini C, Kozel TR, Pietrella D, Monari C, Bistoni F, Vecchiarelli A (2001) Interdependency of interleukin-10 and interleukin12 in the regulation of T-cell differentiation and effector function of monocytes in response to stimulation with *Cryptococcus neoformans*. Infect Immun 69:6064–6073

Richter DL (2008) Revival of saprotrophic and mycorrhizal basidiomycete cultures after 20 years in cold storage in sterile water. Can J Microbiol 54(8):595–599

Rout D, Mondal S, Chakraborty I, Pramanik M, Islam SS (2005) Chemical analysis of a new (1--> 3)-, (1-->6)-branched glucan from an edible mushroom, *Pleurotus florida*. Carbohydr Res 340:2533–2539

Ruperez P, Lal JA (1981) Extracellular galactosaminogalactan from *Aspergillus parasiticus*. Trans Br Mycol Soc 77(3):621–625

Sanford PA, Hutchings GP (1987) Chitosan—a natural, cationic biopolymer: commercial applications. In: Yapalma M (ed) Industrial polysaccharides: genetic engineering, structure/ property relations and applications. Elsevier Science Publishers B.V., Amsterdam, pp 363–376

Saqib AAN, Hassan M, Khan NF, Baig S (2010) Thermostability of crude endoglucanase from *Aspergillus fumigatus* grown under solid-state fermentation (SSF) and submerged fermentation (SmF). Process Biochem 45(5):641–646

Schmid F, Stone BA, McDougall M (2001) Structure of epiglucan, a highly side-chain/branched (1-3;1-6)-β-glucan from the micro fungus *Epicoccum nigrum* Ehrenb. ex Schlecht. Carbohydr Res 331(2):163–171

Schmid J, Farina J, Rehm B, Sieber V (2016) Microbial exopolysaccharides: from genes to applications. Frontiers Media SA 161

Seviour RJ, Hensgen K (1983) Exocellular glucan production by *Acremonium diospyri*. FEMS Microbiol Lett 16(2–3):343–347

Seviour RJ, Stasinopoulos SJ, Auer DPF, Gibbs PA (1992) Production of pullulan and other exopolysaccharides by filamentous fungi. Crit Rev Biotechnol 12:279–298

Shu CH, Lung MY (2004) Effect of pH on the production and molecular weight distribution of exopolysaccharide by *Antrodia camphorata* in batch cultures. Process Biochem 39(8):931–937

Silveira ML, Smiderle FR, Agostini F, Pereira EM, Bonatti-Chaves M, Wisbeck E, Ruthes AC, Sassaki GL, Cipriani TR, Furlan SA, Iacomini M (2015) Exopolysaccharide produced by *Pleurotus sajor-caju*: its chemical structure and anti-inflammatory activity. Int J Biol Macromol 75:90–96

Singhania RR, Patel AK, Soccol CR, Pandey A (2009) Recent advances in solid-state fermentation. Biochem Eng J 44(1):13–18
Sudhakaran VK, Shewale JG (1988) Exopolysaccharide production by *Nigrospora oryzae* var. *glucanicum*. Enzyme Microb Technol 10(9):547–551
Sutherland IW (1982) Biosynthesis of microbial exopolysaccharides. Adv Microb Phys 23:79–150
Sutherland IW (1994) Structure-function relationships in microbial exopolysaccharides. Biotech Adv 12(2):393–448
Sutherland IW (1996a) Extracellular polysaccharides. In: Rhem HJ, Reed G (eds) Biotechnology, vol 6. Wiley-VCH, Weinheim, pp 615–657
Sutherland IW (1996b) Extracellular polysaccharides. In: Rehm HJ, Reed G (eds) Biotechnology: products of primary metabolism, vol 6, 2nd edn. Wiley-VCH Verlag GmbH, Weinheim, pp 613–657
Vecchiarelli A, Retini C, Pietrella D, Monari C, Tascini C, Beccari T, Kozel TR (1995) Downregulation by cryptococcal polysaccharide of tumour necrosis factor-alpha and interleukin1 beta secretion from human monocytes. Infect Immun 63:2919–2923
Walton K, Coombs MM, Walker R, Ioannides C (1997) Bioactivation of mushroom hydrazines to mutagenic products by mammalian and fungal enzymes. Mutat Res 381:131–139
Wang ZM, Peng X, Lee KLD, Tang JC, Cheung PCK, Wu JY (2011) Structural characterization and immunomodulatory property of an acidic polysaccharide from mycelial culture of *cordyceps sinensis* fungus Cs-HK1. Food Chem 125(2):637–643
Wang X, Tao F, Gai Z, Tang H, Xu P (2012) Genome sequence of the welan gum producing strain *Sphingomonas* sp. ATCC 31555. J Bacteriol 194:5989–5990
Wasser SP (2010) Medicinal mushroom science: history, current status, future trends, and unsolved problems. Int J Med Mushrooms 12(1):1–16
Weng BBC, Lin YC, Hu CW (2011) Toxicological and immunomodulatory assessments of botryosphaeran (β-glucan) produced by *Botryosphaeria rhodina* RCYU 30101. Food Chem Toxicol 49(4):910–916
Wu SH, Nilsson HR, Chen CT, Yu SY, Hallenberg N (2010) The white rotting genus Phanerochaete is polyphyletic and distributed throughout the phleboid clade of the Polyporales (Basidiomycota). Fungal Divers 42:107–118
Wu S, Chen J, Pan S (2012a) Optimization of fermentation conditions for the production of pullulan by a new strain of *Aureobasidium pullulans* isolated from sea mud and its characterization. Carbohydr Polym 87(2):1696–1700
Wu X, Xu R, Ren Q, Bai J, Zhao J (2012b) Factors affecting extracellular and intracellular polysaccharide production in submerged cultivation of *Tricholoma mongolicum*. Afr J Microbiol Res 6(5):909–916
Xiao JH, Xiao DM, Xiong Q, Liang ZQ, Zhong JJ (2010) Nutritional requirements for the hyperproduction of bioactive exopolysaccharides by submerged fermentation of the edible medicinal fungus *Cordyceps taii*. Biochem Eng J 49:241–249
Yang H, He G (2008) Influence of nutritional conditions on exopolysaccharide production by submerged cultivation of the medicinal fungus *Shiraia bambusicola*. World J Microbiol Biotechnol 24(12):2903–2907
Yang FC, Ke YF, Kuo SS (2000) Effect of fatty acids on the mycelial growth and polysaccharide formation by *Ganoderma lucidum* in shake flask cultures. Enzyme Microbiol Tech 27 (3–5):295–301
Yang BK, Jeong SC, Park JB (2001) Swimming endurance capacity of mice after administration of exo-polymer produced from submerged mycelia culture of *Ganoderma lucidum*. J Microbiol Biotechnol 11(5):902–905
Yin Y, Hu Y, Xiong F (2011) Sorption of Cu(II) and Cd(II) by extracellular polymeric substances (EPS) from *Aspergillus fumigatus*. Int Biodeter Biodegr 65(7):1012–1018
Zhai X, Zhao A, Geng L, Xu C (2012) Fermentation characteristics and hypoglycemic activity of an exopolysaccharide produced by submerged culture of *Stropharia rugosoannulata* 2. Ann Microbiol 4(1):754–772

Zhang BB, Cheung CK (2011) Use of stimulatory agents to enhance the production of bioactive exopolysaccharide from *Pleurotus tuber-regium* by submerged fermentation. J Agric Food Chem 59(4):1210–1216

Zhang J, Xia W, Liu P, Cheng Q, Tahirou T, Gu W (2010) Chitosan modification and pharmaceutical/biomedical applications. Mar Drugs 8:1962–1987

Zhang B, Yan P, Chen H, He J (2012) Optimization of production conditions for mushroom polysaccharides with high yield and antitumor activity. Carbohydr Polym 87(4):2569–2575

Isolation and Purification of Microbial Exopolysaccharides and Their Industrial Application

3

Veena S. More, Allwin Ebinesar, A. Prakruthi, P. Praveen, Aneesa Fasim, Archana Rao, Farhan Zameer, K. S. Anantharaju, and Sunil More

Abstract

Exopolysaccharide is long-chain high molecular weight polymeric carbohydrates composed of monosaccharide units bound together by glycosidic linkage, which are secreted extracellularly by the secreting microscopic cell or organism. These exopolysaccharides can be either homopolysaccharide or heteropolysaccharide in nature. Exopolysaccharide comprises repeated units of sugar moieties attached to the carrier lipid and can be associated with proteins, lipids, organic and inorganic compounds, and metal ions. Bacteria, archaea, yeast, filamentous fungi, and single cell of eukaryotes produce microbial exopolysaccharides. Microbial synthesis of polysaccharides is greatly influenced by environmental factors such as temperature, pH, pressure, salinity, toxicity, and radiation levels across their ecological niche. Due to the very less production time of exopolysaccharides and its simple purification process, these have found various successful applications in various industrial sectors such as pharmacology, diagnostics, nutraceuticals, functional foods, cosmetics, herbicides and insecticides, bioremediation, biotechnology, petrochemicals, dairy industry, and paint industry. Some of the most important exopolysaccharides and its industrial applications are Acetan (as preservative), Alternan (as commercial gum Arabic), Biodispersion (as remediation of oil spills), Cellulose (as non-indigestible fiber), Curdlan (as a food additive and gelling agent), Dextran (as moisturizers), Emulsan (as crude oil recoveree), Gellan (as stabilizer and microencapsulation matrix), Hyaluronan

V. S. More (✉) · A. Ebinesar · P. Praveen
Department of Biotechnology, Sapthagiri College of Engineering, H. Ghatta, Bangalore, India

A. Prakruthi · A. Fasim · A. Rao · F. Zameer · S. More
School of Basic and Applied Sciences, Dayananda Sagar University, Bangalore, India

K. S. Anantharaju
Department of Chemistry, Dayananda Sagar College of Engineering, Bangalore, India

A. Vaishnav, D. K. Choudhary (eds.), *Microbial Polymers*,
https://doi.org/10.1007/978-981-16-0045-6_3

(as for the removal of dead cells in skincare products), Kefiran (as anti-cancer agents), Levan (as viscosifier), Mutan (as adhesives), Succinoglucan (as emulsion stabilizers), Welan (as thickening agents in high-temperature industries), Xanthan (as an emulsifier and suspending agent), and Pullulan (as orally consumable films). These exopolysaccharides are produced by bacterial species such as Acetobacter, Pseudomonas, Leuconostoc, Alteromonas, Alcaligenes, Lactobacillus, Zymomonas, Xanthomonas, Aureobasidium (fungus). These compounds are purified by techniques such as membrane filtration, dialysis, precipitation, various types of column chromatography, lyophilization, distillation, and rotatory vaporization. This chapter describes different isolation and purification techniques for microbial exopolysaccharides.

Keywords

Exopolysaccharide · Bacteria · Chromatography · Purification

3.1 Introduction

Microbes being one among the most abundant live forms in this biosphere are known to produce many extracellular metabolic products, commonly known as extracellular products or exopolysaccharides. Microbes are found in all forms of habitable or non-habitable ecological conditions. Microbes consist of domains such as archaea, mycota, and fungi. Microbial exopolysaccharides are found in extracellular space, long chain, and high molecular weight polymeric carbohydrates composed of monosaccharide units bound together by glycosidic linkage. These exopolysaccharides can be found in two forms, homopolysaccharide and heteropolysaccharide. Homopolysaccharides are those that consist of same repeating units of monomeric molecules and heteropolysaccharides are those that consist of different repeating units of monomeric compounds. Microorganisms are living in very harsh environmental conditions such as arid deserts, geographical poles, volcanoes, deep trench, rich ecological niche, acidic environments, and many more commonly synthesized exopolysaccharides. These exopolysaccharides are highly influenced by the factors of nature such as temperature, pH, humidity, toxicity, pressure, and radiation levels. Exopolysaccharides consist of repeated units of sugar moiety attached to the carrier lipid and can be found associated with proteins, lipids, organic or inorganic compounds, and metal ions. Microorganisms such as bacteria, archaea, yeast, filamentous fungi, and single cell of eukaryotes produce exopolysaccharides. Exopolysaccharides are usually found attached to the cell membrane of microbes, or as free molecules in the extracellular space. Exopolysaccharides are usually synthesized by microorganisms for their protection from pathogens and virulent organisms, as removal of endotoxins for the microorganisms, as by-products of metabolism, for symbiotic relationship with the host. As these are found in outer space of microorganisms, exopolysaccharides are very easy to isolate, purify, and find varied applications in industrial sector.

In this chapter, we will discuss some of the most important and commonly used microbial exopolysaccharides such as alternan, cellulose, curdlan, dextran, gellan, hyaluronan, kefiran, levan, welan, xanthan gum, pullan, and also their isolation and purification techniques such as salting out method, ultracentrifugation method, ultrafiltration method, and various chromatography techniques such as anion exchange column chromatography, affinity, and chromatography. Industrial applications of above-mentioned various enzymes are also discussed.

3.2 Various Exopolysaccharides

3.2.1 Xanthan

Xanthan gum was first discovered in 1960s and in the late 1970s it was commercialized (Barcelos et al. 2020). Xanthan is produced by pure cultures of *Xanthomonas campestris* (Frese et al. 2014). Xanthan is a long-chain polysaccharide consisting of D-glucose, D-mannose, and D-glucuronic acid as building blocks in molecular ratio of 3:3:2 with high number of trisaccharide side chains. Average molecular weight of xanthan is about 2000 kDa (Jindal and Khattar 2018). Xanthan gum is a water-soluble fast hydrating hydrocolloid, which easily dissolves at room temperature. Stirring and continuous mixing increases the dissolution process of Xanthan gum. Generally, greater the ionic strength more is the dissolvation time. Xanthan solution can function well in the pH range of 3–10. Xanthan gum (XG) is recommended due to its high solubility in cold and hot water, stable viscosity at lower concentrations. To add on the helical conformation possessed by XG displays intense pseudoplastic performance and helps in instant recovery.

3.2.1.1 Pharmaceutical Applications

XG is currently being used in pharmaceutical preparations due to its branched polymeric structure. The cohesive and adhesive characteristics is majorly because of the special structure it possesses. Additionally, it finds its application as a thickening, stabilizing, gelling, and emulsifying agent along with being used as tablet binder. Viscoelastic properties of xanthan solutions form a weak gel, giving characteristic suspending in liquids. Xanthan solutions are more stable than other thickeners and even resistant to enzymes, such as amylases, proteases, and cellulases (Wüstenberg 2014).

A study was conducted to understand the encapsulations of β-carotene for stabilizing pickering emulsions of wheat gluten nanoparticles by using xanthan gum. It was established that the utilization of xanthan helped in preventing the chemical degradation of β-carotene during storage for 1 month. It could help in sustaining at higher pH and excessive salt concentration as well.

In another study, XG was supplied to CMXG, and further cross-linked as a hydrophilic polymer with Ca^{+2} ion (in situ) for the research of prednisolone matrix tablets. The usage of XG assisted in the reduction of immediate release of the drug in designated regions like upper gastrointestinal tract.

3.2.1.2 Personal Hygiene Products

In products like toothpaste, shampoos, and liquid soap, XG assists in boosting the flow properties by this means enabling the easy extrusion from pump dispensers and tubes. The effect of XG displaying a pseudoplastic behavior also helps in offering an increased level of smoothness and softness in creams, lotions, and gels.

Xanthan was employed in rice bran oil formulation for its utilization in sun protection agents. The concentration of XG was varied and tested to check its emulsification properties. The droplet size varied and the zeta potential values were augmented; however, the pH remained consistent throughout. It was thus concluded that XG can be used to enhance the physicochemical properties of the oil which can further be used in different fields.

3.2.1.3 Oil Industry

The stability factor of XG makes it easily utilized in petroleum industries as an additive in cleaning pipes, drilling fluids, and in enhancing the recovery of oil. Hydrolyzed polyacrylamide has been replaced with xanthan in oil recovery process as it is inadequate to reservoirs with moderate or low hardness brine solution. It cannot be frequently employed even during lower temperatures as their viscosities get highly affected by the presence of electrolytes. This is possible due to its tolerance towards high salt concentration and the capacity to maintain viscosity of fluids that are water based. XG is also known to be used in micellar-polymer flooding in the recovery of oils. The oils that are trapped in tiny gaps of small sandstone need to be pumped by using polymer-based salt solutions. XG helps in easy recovery of residual oil and helps in maintaining the efficiency of oil extraction.

3.2.1.4 Food Processing Industries

Plastic materials which are used in food packaging are being replaced by polysaccharides like xanthan due to its contribution in increasing the film forming capacity due to its stability in hot and cold water. It acts as an excellent UV light shielding component along with maintaining the moister content of the food. (Abu Elella et al. 2020).

3.2.1.5 Other Applications

Another interesting contribution of XG is in the field of catalytic application, due to its active functional groups which include carboxyl and hydroxyl moieties. The presence of these groups increases the dispersal property of catalytic nanoparticles such as Pd, Ag, and aid in the decreased size for enhancing their catalytic accomplishment.

Reports also explain XG being used in the treatment of contaminated water. XG-based hydrogels and nanocomposites are used as they exhibit high surface area (specific), gelation capabilities, and thermal stability along with great mechanical strength properties.

Thus, xanthan can be used as an alternative in the above-mentioned fields and exploratory research of its other applications is yet to be investigated.

3.2.2 Gellan

Gellan gum is exopolysaccharide produced by *Sphingomonas elodea* by fermentation (Bajaj et al. 2007; Wang et al. 2006). Gellan is linear, anionic heteropolysaccharide, straight chain consisting building blocks of D-glucose, L-rhamnose, and D-glucuronic acid in molecular ratio of 1.5:1:1. The chain consists of tetrasaccharide repeating unit in which beta (1–4) linked glucose, glucuronic acid, glucose, and rhamnose in alpha (1–3) linkage are bonded together. The native product is partially esterified. In its native or high-acyl form, two acyl substituents acetate and glycerate are present. One glycerate for one repeat and one acetate for every two repeating units are observed. Gellan is suspected to adopt two- or three-fold helical structure after heating or cooling (Ferreira et al. 2016). Gellan gum easily dissolves in hot and cold water. In high gellan gum concentration in solutions, a network-like structure is observed called demoldable gel. In lower gellan gum concentration in solutions, molecules are associated closely but system remains fluid, hence known as fluid gel (Nitta and Nishinari 2005; Coviello et al. 2007).

3.2.2.1 Pharmaceutical Applications

Gellan gum is used in biomedical and pharmaceuticals research as a protein carrier. It also finds its application in gene transfection and gene therapy. Gellan gum combined with chitosan was effectively used in the manufacture of nasal insert which aided in the enhancement of drug release and water up-taking capacity. The combinational product was a replacement to other polymers which displayed deficiencies with their efficacy. The drug release at lower pH is not possible by compounds like chitosan along, hence when polysaccharides like gellan are used along. The facilitation of drug release is improved.

Nasal cavity is handled mostly for the local treatment of nasal infections and congestion. It was studied that gellan amended the epithelial transport further successfully than the usually employed isotonic D-mannitol solution. Furthermore, the dwelling time in the nasal cavity was up to 4 h and there was no side effects noted (Osmałek et al. 2014).

3.2.2.2 Tissue Engineering

Gellan gum has been lately examined in the arena of tissue engineering, as a substantial for cartilage restoration. The ability to self-repair after a damage to tissue is very inadequate. Hence, it is important to note that the ability of cartilage to self-repair after degeneration or mechanical damage is limited. Therefore, biomaterials to repair damaged tissue is essential. Gellan-based structures are presently discovered as injectable carriers for numerous autologous cells like bone marrow cells and chondrocytes. *In vivo* results of viability and proliferation were promising when conducted (Osmałek et al. 2014).

3.2.2.3 Food Industry

A study reported that when gellan was combined with pectin (methoxy pectin) the aroma compounds present in food products, wine, and essential oils helped in retaining the aroma for an extended duration. Also the same combinational product was utilized in the preparation of dough, a nutritious acidic milk prepared from yogurt. After the addition, it was noted that the usability of the product was improved by increasing the stability (Zia et al. 2018).

3.2.3 Pullan

Pullan being an extracellular glucan by fungal strain *Aureobasidium pullulans* commonly known as black yeast through fermentation (Choi et al. 2014). Pullan is linear homopolysaccharide consisting of maltotriose as monomers. Three glucose units of maltotriose are linked through alpha (1–4) glycosidic linkage, and maltotriose units are linked through alpha (1–6) linkage. Molecular weight of pullan is between 10 and 400 kDa (Mano et al. 2007; Navard 2012). Pullan forms a stable, viscous fluid when dissolved in hot or cold water but does not form a gel (Imeson 2011). Pullan is insoluble in organic solvents except dimethyl sulfoxide and dimethylformamide. Viscosity of pullan remains not much affected in the pH range of 2–11. In addition, it remains stable in the presence of ions (Cui 2005). Pullan is an odorless and tasteless white powder. Pullan is heat resistant and it readily forms film, which is oil repellent, transparent, edible, and readily soluble in water (Shit and Shah 2014).

3.2.3.1 Food Processing Industries

Due to its non-toxic, non-immunological, and non-carcinogenic properties pullan is extensively used in food edible coatings, as foaming agents and flocculants (Zia et al. 2018). Edible coatings assist in retarding dehydration which in turn recover the quality of the food medium by retaining the volatile flavor complexes and also decrease bacterial growth. Pullan-based films can interact with the food by enriching the safety and quality by releasing essential compounds.

Pullan is used for preparation of coating along with whey protein. Chestnuts were dipped in these film-forming solutions to reduce the crop wastage because of high water wastage. SEM results were conducted to check the adherence of the coated films and the results conveyed the active properties of pullan-based coated films. Further experiments also demonstrated that moisture loss was reduced and the shelf life of the chestnuts were improved (Trinetta and Cutter 2016).

3.2.4 Dextran

Dextran is an extracellular product of *Streptococcus* or *Leuconostoc* by fermentation of sucrose at optimum pH 6.5–7 and optimum temperature of 25–30 °C. The commercial dextran is produced by *L. mesenteroides* and *L. dextranicum* (Wüstenberg 2014). Dextran is synthesized using dextransucrase in two pathways.

- By hydrolyzing the sucrose and binding the glycosyl moiety.
- Building up the dextran by the insertion mechanism.

Dextran is highly branched, neutral, high molecular weight homopolysaccharide made up of glucose monomers. Glucose monomers are linked together through alpha (1–4) linkage. Dextrans have 10–50,000 kDa molecular weight (Kapoor et al. 2013). Dextran is readily soluble in water or glycerin at room temperature. Dextran solution is highly viscous or slimy fluid. Dextran size and structure greatly depends on dextransucrase enzyme produced by the different microbial strains. This significantly justifies the difference in the structure of dextran isolated from diverse bacterial species. Dextran prevents crystallization, improves moisture retention, and increases viscosity in the solution. Dextrans are odorless, tasteless, and non-toxic, and these characteristics are favorable over other stabilizers. Humans can easily digest dextran, and hence it serves as high-energy source (Bhavani and Nisha 2010; Kothari et al. 2014).

3.2.5 Curdlan

Curdlan is a secondary extracellular polysaccharide high molecular weight polymer of glucose produced by a non-pathogenic bacterium *Agrobacterium biobar* and mutants of *Alcaligenes faecalis* (Zhang and Edgar 2014). It is a linear beta (1–3) glucan. It is partially esterified with succinic acid and appears as succinoglucan (Wüstenberg 2014). Curdlan dissolves in aqueous alkaline solution and is an insoluble exopolysaccharide in water and alcohol. Curdlan is known to form strong thermos irreversible high set elastic gel at temperatures above 54 °C. Gel strength is known to be between agar and gelatin (Saha and Bhattacharya 2010). Curdlan when heated with solvent forms an irreversible gel. Gel strength increases with an increase in temperature. Curdlan forms water-insoluble films. These gels are known to be biodegradable, impermeable to oxygen, and edible. Curdlan is an odorless white powder (Ghanbarzadeh and Almasi 2013).

3.2.5.1 Food Processing Industry

Curdlan displays thermal gelling and finds its applications in various food industries and is legally permitted to be used as a food additive in countries like the USA, Japan, Taiwan, and Korea.

Curdlan comprises dietary fibers which are highly insoluble and are thus used in the production of low-calorie foods. It is likewise used as a stabilizer, texturizer in jelly-based food and in meat products, dairy products, respectively. Potato starch and other starch are extensively used in the noodle industry due to the need of higher content of amylose. Addition of curdlan improved the viscidness and solidity along with the gradual decrease in the stickiness of the noodles (Verma et al. 2020).

3.2.5.2 Biomedical Applications

Curdlan also illustrates potential applications in the field of biomedical and pharmaceuticals. Gel encapsulation of indomethacin, prednisolone was performed using curdlan to check for better drug release. When the drug release system was compared and analyzed, curdlan-gels were found to remain intact and improved the drug release location. This also helped in avoiding drug uptake into portal vein (Zhang and Edgar 2014).

3.2.6 Levan

Levan is a fructose-based polysaccharide, which is extracellularly secreted by many microorganism species. Some of the microorganisms are *Acetobacter, Bacillus, Brenneria, Geobacillus, Halomonas, Lactobacillus, Zymomonas,* and *Saccharomyces* (Han 1990). Levan is composed of beta D-fructofuranose with beta (2–6) linkages between fructose rings. Levansucrase synthesizes levan by directly converting sucrose into a levan polymer. Levan is highly soluble in water and oils, however is insoluble in many organic solvents such as methanol, ethanol, and isopropanol. Large amount of tensile strength and cohesiveness of levan is due to its branching state of existence. Adhesiveness with molecules is achieved by the presence of hydroxyl groups. Levan has a very low viscosity and is non-toxic in nature. Levan has high value of adhesivity, strong biocompatibility, and film forming ability (Ullrich 2009).

3.2.6.1 Tissue Engineering

Reports discuss the application of levan as a bioadhesive agent in drug delivery systems. In tissue engineering, levan is successfully used as a material for scaffold preparation as a remedy for wound healing.

3.2.6.2 Biotechnological Applications

Due to its adhesive nature and water retracting properties, levan is advantageous in assisting microbes and plants to survive in regions with limited source of water. Thereby increasing the water availability in hypersaline environments as well which forms the major portion of earth ecosystem.

3.2.6.3 Other Applications

In the manufacturing of adhesive membranes, film casting a conventional technique. In the process, levan from microbial sources were supplemented with chitosan and polyethylene (PEO) as a casting solution. Levan facilitated an improvement with the biocompatibility of the films. Furthermore, mechanical and thermal characteristics were maintained by suppressing the crystallinity of PEO.

It is an ideal biopolymer for many industrial applications due to its wide range of potential characteristics. (Combie and Öner 2018).

3.2.7 Welan

Welan is a novel fermented exopolysaccharide product of *Sphingomomonas* species, a Gram-negative bacterium and by *Alcaligenes* species, a mutant species. Of the two species, *Alcaligenes* species is known to produce welan in a large amount of quantity (Kaur et al. 2014). Welan is an acidic heteropolysaccharide and has an approximate molecular weight of 1.0*10^6 g/mol (Kaur et al. 2014). The extracellular polysaccharide welan gum is a series of oligosaccharides, which consists of L-mannose, L-rhamnose, D-glucose, and D-glucuronic acid as backbone in the molar ratio of 1.0:4.5:3.1:2.3. Welan has a single side chain containing either L-rhamnose units or L-mannose units substituted on C3 of every 1,4 glucose repeating units (O'Neill et al. 1986). The presence of side chains greatly reduce the gelling behavior of the welan. Welan in aqueous solutions has a weak gel property. Welan gum polymer produces a high viscous aqueous solution exhibiting higher viscosity retention at high temperature (up to 150 °C) and wide pH range of (2–12) (Jansson et al. 1985).

3.2.7.1 Cement Industries

The utilization of welan gum in cement configurations has a diverse implication. It decreases fluid loss of the cement compositions and upsurges the suspension characteristics of cement. Also is highly operative in low concentration as well. It is widely used in oilfield processes which include wellbore cleanup, hydraulic fracturing, and drilling for viscosity enhancement. It has its application in oil well drilling as a spacer fluid.

3.2.7.2 Other Applications

Welan gum has certain qualities like binding, thickening, and also display emulsifying properties. Hence, it finds its application as an constituent in beverages like citric acid centered drinks and dairy food products like yoghurt and ice creams (Kaur et al. 2014).

Oil recovery is one such field where welan is used because of its gelling properties. The decent thermal insulating characteristics and rheology of the gel make them feasible as an insulating material for pipeline bundles so that the temperature (above 30 °C) is maintained and is above sea level.

3.2.8 Kefiran

Kefiran is a capsular polysaccharide synthesized extracellularly by slime forming rod-shaped *Lactobacillus* species. Kefiran is a fermented product obtained from kefir grains (Farnworth 2006). Kefiran is soluble in water and in aqueous solutions containing ethanol forms gels. Branched glucogalactan comprises equal amount of

D-glucose and D-galactose residues. Kefiran is resistant to hydrolysis. Kefiran has a structure of hexasaccharide or heptasaccharide repeating units which itself composed of pentasaccharide units to which one or two residues are randomly linked. Primary structure of kefiran is known to consist of galactose and glucose in the molar ratio of 1.1:0.9.

3.2.8.1 Food Industry

Kefiran has been reported to possess antibacterial, antifungal, and antitumor properties (Rodrigues et al. 2005). Kefiran has been shown to enhance the production of beta cortisol and noradrenaline in human cell lines. It also finds its application in food industry (Agoub et al. 2007). Studies have reported that kefiran combined with whey protein isolate was used to manufacture bio-nanocomposites with nanofillers which exhibited greater efficiency in acting as oxygen and moisture barriers. Hence, they can be used in food packaging industries. Recent studies also suggest that when kefiran was added to the films for food packaging, it also assisted in the acting as a UV-blocking. UV-blocking is a very vital factor in edible food packaging sector. Kefiran is considered to be a stress-reducing food supplement due to its competence in increasing the production of interferon B-cortisol and noradrenaline in human cell lines.

3.2.8.2 Medical Applications

Research was conducted to determine and demonstrate the efficiency of kefiran in treating *Candida albicans* by improving the protective outcome to dermal connective tissue and encouraging wound healing. This study was conducted in vivo to justify the outcome. In human intestinal system immune system, kefiranis found to act as a stimulant by altering the immune cells (Tan et al. 2020).

3.2.9 Hyaluronan

Hyaluronan also known as hyaluronic acid is a linear high molecular weight heteropolysaccharide consisting of alternate beta 1, 3 N-acetylglucosamine and beta 1, 4-glucuronic acid linked through beta-(1–4) and beta-(1–3) glycosidic bonds (Chen et al. 2014a, b). Hyaluronan is synthesized by the bacterial species of streptococcus such as *S.equisimilis, S.pyogenes, S.thermophilus, and S.equi* (Shiedlin et al. 2004). A recombinant strain of Bacillus subtilis is known to produce hyaluronan at laboratory scale. Hyaluronan can be approximately 20,000–25,000 disaccharide repeats in length (Chen et al. 2014a, b). Hyaluronan has a very high rate of water holding capacity, high biocompatibility, and high viscosity. Hence, it is considered as a natural lubricant. It is energetically stable, in part because of the stereochemistry of its component disaccharides. Bulky groups on each sugar molecule are in sterically favored position; likewise, the small hydrogen groups assume the less favorable axial position (Izawa et al. 2009, 2011). Due to these various properties of hyaluronan, it finds applications in various industries which include in the manufacture of ophthalmic products, treatment of ulcers and surgical procedures.

3.2.9.1 Medical Applications

Hyaluronan plays a dynamic role in the growth of cartilage and in the sinovial fluid maintenance for tendon regeneration. Hyaluronan is thus present in greater concentrations in the ECM of all adult joint tissues. Thus, HA acts in the joint as a lubricant and shock absorber and can externally supplied to improve bone health.

Study reports that hyaluronan hydrogels have accelerated the healing of wounds by providing highly hydrated non-immunogenic environment that assisted in faster healing process. Hyaluronan has been extensively used in the field of cosmetics due to its property of hydration.

Hyaluronan has also proven to be operative for intensifying the blood compatibilities of cardiovascular implants such as stents and vascular grafts. Biomaterial surfaces treated with cross-linked hyaluronan have been associated with reduced platelet adhesion and thrombus formation. HA derivatives with higher degrees of sulfation are connected with augmented aptitudes to avoid blood coagulation (Necas et al. 2008)

3.2.10 Alternan

Alternan is a microbial high molecular weight exopolysaccharide consisting of alternating sequence of alpha (1–3) and alpha (1–6) linked D-glucose units (Misaki et al. 1980). Some single extracellular enzymes of certain strains of lactic acid bacterium Leuconostoc mesenteroides produce alternan. Alternan are synthesized as extracellular slimes by lactic acid bacteria (Cote et al. 2002). Due to an unusual linkage structure, alternan possesses resistance to many hydrolytic enzymes that degrade starch, dextran, and other alpha D-glucan. Alternan possesses low viscosity than those posed by other food gums. Alternan is readily soluble in water (Leathers et al. 2003). Due to similarity of alternan in structure with dextran, alternan finds applications in the artificial sweeteners in the sugar confectionary industries.

3.2.11 Cellulose

Cellulose is an exopolysaccharide synthesized by various species of bacteria such as *Gluconacetobacter, Agrobacterium, Rhizobium, Salmonella, and Sarcina*. Among these, *Gluconacetobacter* is considered as the most cellulose producing species (Glenn et al. 2009). Cellulose is a polysaccharide composed of carbon, hydrogen, and oxygen. Cellulose is a straight chain polymer whose base units of glucose beta linkages hold together. Each polymeric chain is asymmetric containing reducing end and non-reducing end. The reducing end has carbonyl functional group whereas non-reducing end has hydroxyl functional group (de Souza Lima and Borsali 2004). Cellulose are found in different physical states such as crystalline, amorphous, sheets, and fibers. Cellulose fibrils are insoluble, inelastic, and have tensile strength comparable to that of steel. They possess high mechanical strength, water absorption capacity, temperatures, crystallinity, pure fine fiber network, and are highly stable to chemicals. Cellulose sheets are known to be resistant to higher stability of pressure (Glenn et al. 2009; Hokkanen et al. 2016). Cellulose is tasteless, odorless, hydrophobic, chiral, renewable, biodegradable, and is insoluble in water and organic solvents. Due to all these varied properties, cellulose finds applications in medicinal, food, commercial, and industrial fields.

3.2.11.1 Food Industry

Cellulose nanocrystals (CNC) are used to enhance the mechanical characteristics of polymers which include soy protein, rubber latex, and starch-based matrixes. They find their applications in food packaging industries as coating materials. They act as excellent oxygen barriers and also enhance the biodegradable properties.

In specific, nanocellulose can be used as a low calorie substitute for carbohydrate extracts as thickeners, flavor carriers, and suspension stabilizers in a wide variety of food products. It is useful for producing crushes, fillings, chips, soups, wafers, gravies, etc.

3.3 Isolation and Purification Techniques

For microbial exopolysaccharides, the various downstream processes of isolation and purification do not have a clear distinguishing difference because sometimes the isolation process is a part of purification process itself and vice versa. Hence, in this chapter we discuss some of the most important and common techniques of purification and isolation of the microbial exopolysaccharide.

3.3.1 Ultracentrifugation Method

Ultracentrifugation method is one of analytical techniques, which involves ultracentrifuge with optical monitoring systems. Various polysaccharides possess different molecular weight and different sedimentation speed in strong centrifugal force. Using this property, various polysaccharides can be purified. Ultracentrifugation is of two types, one is differential centrifugation and other is density gradient zonal centrifugation method (Shi 2016).

Different molecular weight polysaccharides are separated in batches by gradual addition of speed in the differential centrifugation method. The high molecular weight polysaccharides are separated in the low speed whereas the low molecular weight polysaccharides are separated at higher speeds.

Density gradient zonal centrifugation is used to detect the homogeneity of the polysaccharides. The rationale of this method is when polysaccharides are centrifuged in the inert gradient media and reach equilibrium. Various polysaccharides can be collected and distributed to certain specific positions within gradient and form different zones. These zones are separated and different polysaccharide fractions are obtained. Commonly used inert media are water, NaCl, CsCl solution, and many more (Shi 2016).

3.3.2 Ultrafiltration Method

Ultrafiltration works on the principle of pressure-induced separation of solutes from solvent through semi-permeable membrane. The relationship between the applied pressure on the solution to be separated and the flux through the membrane is most commonly described by Darcy equation. Suspended solids and solutes of high molecular weight are retained in retentate, while water and low molecular weight exopolysaccharides pass through the membrane in permeate or filtrate. Ultrafiltration principle is also defined as molecular sieve, which mainly works on size exclusion or particle capture mechanism in the membrane. Ultrafiltration membranes are defined by molecular weight cut-off value of the membrane used. Ultrafiltration is applied in cross flow or dead end mode (Schachman 2013).

3.3.3 Salting Out Method

Salting out is also known as salt-induced precipitation, salt fractionation, anti-solvent crystallization, precipitation crystallization, or drowning out. Salting out method is based on the electrolyte–nonelectrolyte interaction, in which the non-electrolyte could be less soluble at high concentration. This method is used for purification of protein and to prevent protein denaturation due to excessively diluted samples. Salt compounds dissociate in the aqueous solutions. When the salt concentration is increased, some of the water molecules are attracted by the salt ions, which decreases the number of water molecules available for the interaction with the charged part of exopolysaccharides.

The principle of this method is that different molecular weight polysaccharides have different solubility in the salt solutions of certain concentrations. When neutral salt is added to the polysaccharide solution and when it reaches equilibrium with the solution, the exopolysaccharide precipitates out from the solution. This precipitate can be separated by centrifugation.

3.3.4 Anion Exchange Column Chromatography

Anion exchange column chromatography is the widely used analytical technique for polysaccharide isolation and purification. It is used for bulk quantity isolation of polysaccharide. Concentration and preliminary purification of polysaccharide solution is possible in this method. The most widely used anion exchanger so far is DEAE-cellulose, DEAE-sephadex, and DEAE-sepharose. The separation mechanism of anion exchange column chromatography is ion exchange and adsorption-desorption. Hence, anion chromatography can be used in separation of neutral polysaccharide, acidic polysaccharide, and mucopolysaccharide. Acidic polysaccharide is adsorbed to the exchanger whereas neutral polysaccharides cannot be absorbed. Then the buffers, which have the same pH value but different values, can be used to elute these acidic polysaccharides, respectively. The adsorption ability generally increases with an increase in acidic groups in the polysaccharide; the adsorption ability of straight chain polysaccharide is greater than branched chain polysaccharide. The height of column bed and flow rate of the polysaccharide also greatly affects this method of isolation and purification (Sober et al. 1956).

3.3.5 Affinity Chromatography

Affinity is generally known as the binding ability between two different molecules. Two specific molecules can dissociate after they bind. Using this principle, polysaccharides can be isolated, purified by the process of binding-dissociation. The affinity column is eluted using polysaccharide solution as mobile phase. As polysaccharide solution is a mixture of multiple polysaccharides, only specific polysaccharide will bind to the affinity column and are adsorbed whereas the remaining polysaccharide will elute from the column. The ionic strength and pH value of the mobile phase can be changed to dissociate the polysaccharide fraction combined with ligand. Hence, purified polysaccharide fraction can be obtained. The advantage of the affinity chromatography is that it has high efficiency and is easy to operate. The most important aspect and advantage of affinity chromatography usage that is several hundred to several thousand fold of concentration purity can be achieved (Cuatrecasas and Anfinsen 1971).

3.4 Industrial Applications of Exopolysaccharides

Polysaccharide	Industrial applications	References
Xanthan	Emulsification and gelationation in food industry, thickener, suspension stabilizer in pharmaceuticals creams, oil recovery, mineral ore processing, paper manufacturing, agriculture, cosmetics, ice creams, puddings, cheese spreads, toothpaste, textile, and agriculture	Born et al. (2002), Morris and Hardling (2009), Saha and Bhattacharya (2010)
Gellan	Solidification agent in beverage industry, excipient in oral, ophthalmic, and nasal drug formulations, tablet dis-integration, jams, jellies, microencapsulation	Felt et al. (2002), Hagerstrom (2003), Osmałek et al. (2014), Mariod and Fadul (2013)), Saha and Bhattacharya (2010))
Pullan	Tablet granulation and coating, binder, and oxygen impermeable film forming, non-animal capsules, oral, and wound care products, food	Ferreira et al. (2016), Shit and Shah (2014)

(continued)

	additive in low calorie foods, preservative, food packing industry	
Dextran	Blood plasma extender, cholesterol-lowering agent, microcarrier in tissue and cell culture techniques, artificial sugars	Nwodo et al. (2012), Bhavani and Nisha (2010), Kothari et al. (2014)
Curdlan	Gelling agent, ice creams, yogurts, cheese, salads, preservatives	Chien and Iwata (2018)
Levan	Blood plasma extender, cholesterol-lowering agent, prebiotic, fodder for animals	Patel and Prajapati (2013), Kang et al. (2009)
Welan	Stabilizer and viscosifier in food industry, jellies, beverages, dairy products, cosmetics, capsule coating of tablets	Asubiaro and Shah (2008), Ma et al. (2012)
Kefiran	Gelatination and viscoelasticizer in dairy industry	
Hyaluronan	Moisturizers in skin care industry, synovial fluid substitute in stem cell industry, chronic difficult wound healing, eye surgery, disinfectant	Kogan et al. (2006), Necas et al. (2008)
Alternan	Blood flow improving agent, cholesterol-lowering agent	
Cellulose	Natural non-indigestible fiber, temporary artificial skin, acoustic membranes in audiovisual equipment	Fang and Catchmark (2015), Chawla et al. (2008)

3.5 Conclusion

As evident from the review of the numerous EPS produced from the different microorganisms remains in the chapter remains to be an unexplored field of study which finds it applications in industries. The eco-friendly exopolysaccharides discussed here are mainly carbohydrates and determines their functions in respective fields. The various isolation and purification techniques mentioned here can be optimized and used in large-scale production of EPS due to its extensive application purposes. The outline of the chapter mainly holds to initiate a proper characterization and documentation of EPS composition and functionality to understand its applications in distinguished fields.

References

Abu Elella MH, Goda ES, Gab-Allah MA, Hong SE, Pandit B, Lee S, Gamal H, ur Rehman A, Yoon KR, 2020 Xanthan gum-derived materials for applications in environment and eco-friendly materials: a review. J Environ Chem Eng, 104702. https://doi.org/10.1016/j.jece.2020.104702

Agoub AA, Smith AM, Giannouli P, Richardson RK, Morris ER (2007) "Melt-in-the-mouth" gels from mixtures of xanthan and konjac glucomannan under acidic conditions: A rheological and calorimetric study of the mechanism of synergistic gelation. Carbohydr Polym 69(4):713–724

Asubiaro A, Shah SN (2008) Rheological and hydraulic properties of welan gum fluids in straight and coiled tubings. J Fluids Eng 130:081506

Bajaj IB, Survase SA, Saudagar PS, Singhal RS (2007) Gellan gum: fermentative production, downstream processing and applications. Food Technol Biotechnol 45(4):341–354

Barcelos MC, Vespermann KA, Pelissari FM, Molina G (2020) Current status of biotechnological production and applications of microbial exopolysaccharides. Crit Rev Food Sci Nutr 60 (9):1475–1495

Bhavani AL, Nisha J (2010) Dextran—the polysaccharide with versatile uses. Int J Pharm Biol Sci 1 (4):569–573

Chawla P, Bajaj IB, Survase S, Singhal RS (2008) Microbial cellulose: fermentative production and application. Biotechnology 47:107–124

Chen X, Gao H, Ploehn HJ (2014a) Montmorillonite–levan nanocomposites with improved thermal and mechanical properties. Carbohydr Polym 101:565–573

Chen X, Siu KC, Cheung YC, Wu JY (2014b) Structure and properties of a (1→3)-β-d-glucan from ultrasound-degraded exopolysaccharides of a medicinal fungus. Carbohydr Polym 106:270–275

Chien C-Y, Iwata T (2018) Synthesis and characterization of regioselectively substituted curdlan hetero esters with different ester groups on primary and secondary hydroxyl groups. Carbohydr Polym 181:300–306

Choi S, Choi W, Kim S, Lee SY, Noh I, Kim CW (2014) Purification and biocompatibility of fermented hyaluronic acid for its applications to biomaterials. Biomater Res 18(1):6

Combie J, Öner ET (2018) From healing wounds to resorbable electronics, levan can fill bioadhesive roles in scores of markets. Bioinspir Biomim 14:011001. https://doi.org/10.1088/1748-3190/aaed92

Cote Gl, Alternan Vandamme EJ, De baets S, Steinbuchel A (eds) (2002) Biopolymers, Wiley-vch, weinheim, chapter 13. P. 323–350

Coviello T, Matricardi P, Marianecci C, Alhaique F (2007) Polysaccharide hydrogels for modified release formulations. J Control Release 119(1):5–24

Cuatrecasas P, Anfinsen CB (1971) Affinity chromatography. In: Methods in enzymology, vol 22. Academic, Cambridge, pp 345–378

Cui SW (ed) (2005) Food carbohydrates: chemistry, physical properties, and applications. CRC Press, Boca Raton

de Souza Lima MM, Borsali R (2004) Rodlike cellulose microcrystals: structure, properties, and applications. Macromol Rapid Commun 25(7):771–787

Fang L, Catchmark JM (2015) Characterization of cellulose and other exopolysaccharides produced from Gluconacetobacter strains. Carbohydr Polym 115:663–669

Farnworth ER (2006) Kefir–a complex probiotic. Food Sci Technol Bull 2(1):1–17

Ferreira AR, Alves VD, Coelhoso IM (2016) Polysaccharide-based membranes in food packaging applications. Membranes 6(2):22

Frese M, Schatschneider S, Voss J, Vorhölter FJ, Niehaus K (2014) Characterization of the pyrophosphate-dependent 6-phosphofructokinase from Xanthomonas campestris pv. Campestris. Arch Biochem Biophys 546:53–63

Ghanbarzadeh B, Almasi H (2013) Boidegradable polymers. In: Chamy R, Rosenkranz F (eds) Biodegradation- life of science. InTech Publications, Croatia, pp 141–186

Glenn GM, Holtman KM, Ludvik C, Chiou BS, Imam SH, Orts WJ, Wood D (2009) Green composites derived from natural fibers. In: Natural fibre reinforced polymer composites: from macro to nanoscale

Han YW (1990) Microbial levan. In: Advances in applied microbiology, vol 35. Academic, Cambridge, pp 171–194

Hokkanen S, Bhatnagar A, Sillanpää M (2016) A review on modification methods to cellulose-based adsorbents to improve adsorption capacity. Water Res 91:156–173

Imeson A (ed) (2011) Food stabilisers, thickeners and gelling agents. Wiley, Hoboken
Izawa N, Hanamizu T, Iizuka R, Sone T, Mizukoshi H, Kimura K, Chiba K (2009) Streptococcus thermophilus produces exopolysaccharides including hyaluronic acid. J Biosci Bioeng 107 (2):119–123
Izawa N, Serata M, Sone T, Omasa T, Ohtake H (2011) Hyaluronic acid production by recombinant Streptococcus thermophilus. J Biosci Bioeng 111(6):665–670
Jansson PE, Lindberg B, Widmalm G, Sandford PA (1985) Structural studies of an extracellular polysaccharide (S-130) elaborated by Alcaligenes ATCC 31555. Carbohydr Res 139:217–223
Jindal N, Khattar JS (2018) Microbial polysaccharides in food industry. In: Biopolymers for food design. Academic, Cambridge, pp 95–123
Kang, S. A., Jang, K. H., Seo, J. W., Kim, K. H., Kim, Y. H., Rairakhwada, D., ... & Rhee, S. K. (2009). Levan: applications and perspectives. In: Microbial production of biopolymers and polymer precursors, pp 145–162
Kapoor M, Khandal RK, Seshadri G, Aggarwal S, Kumar Khandal R (2013) Novel hydrocolloids: preparation and applications–a review. IJRRAS 16(3):432–482
Kaur V, Bera MB, Panesar PS, Kumar H, Kennedy JF (2014) Welan gum: microbial production, characterization, and applications. Int J Biol Macromol 65:454–461
Kogan G, Šoltés L, Stern R, Gemeiner P (2006) Hyaluronic acid: a natural biopolymer with a broad range of biomedical and industrial applications. Biotechnol Lett 29:17–25
Kothari D, Das D, Patel S, Goyal A (2014) Dextran and food application. In: Polysaccharides. Springer
Leathers TD, Nunnally MS, Ahlgren JA, Côté GL (2003) Characterization of a novel modified alternan. Carbohydr Polym 54(1):107–113
Ma L, Zhao Q, Yao C, Zhou M (2012) Impact of welan gum on tricalcium aluminate–gypsum hydration. Mater Charact 64:88–95
Mano JF, Silva GA, Azevedo HS, Malafaya PB, Sousa RA, Silva SS et al (2007) Natural origin biodegradable systems in tissue engineering and regenerative medicine: present status and some moving trends. J R Soc Interface 4(17):999–1030
Mariod AA, Fadul H (2013) Gelatin, source, extraction and industrial applications. Acta Scientiarum Polonorum Technologia Alimentaria 12(2):135–147
Misaki A, Torii M, Sawai T, Goldstein IJ (1980) Structure of the dextran of Leuconostoc mesenteroides B-1355. Carbohydr Res 84(2):273–285
Navard P (ed) (2012) The European Polysaccharide Network of Excellence (EPNOE): research initiatives and results. Springer
Necas J, Bartosikova L, Brauner P, Kolar J (2008) Hyaluronic acid (hyaluronan): a review. Veterinarni Medicina 53(8):397–411
Nitta Y, Nishinari K (2005) Gelation and gel properties of polysaccharides gellan gum and tamarind xyloglucan. J Biol Macromol 5(3):47–52
Nwodo U, Green E, Okoh A (2012) Bacterial exopolysaccharides: functionality and prospects. IJMS 13:14002–14015. https://doi.org/10.3390/ijms131114002
O'Neill MA, Selvendran RR, Morris VJ, Eagles J (1986) Structure of the extracellular polysaccharide produced by the bacterium Alcaligenes (ATCC 31555) species. Carbohydr Res 147 (2):295–313
Osmałek T, Froelich A, Tasarek S (2014) Application of gellan gum in pharmacy and medicine. Int J Pharm 466:328–340. https://doi.org/10.1016/j.ijpharm.2014.03.038
Patel A, Prajapati J (2013) Food and health applications of exopolysaccharides produced by lactic acid bacteria. Adv Dairy Res 1:1–7
Rodrigues KL, Caputo LRG, Carvalho JCT, Evangelista J, Schneedorf JM (2005) Antimicrobial and healing activity of kefir and kefiran extract. Int J Antimicrob Agents 25(5):404–408
Saha D, Bhattacharya S (2010) Hydrocolloids as thickening and gelling agents in food: a critical review. J Food Sci Technol 47(6):587–597
Schachman HK (2013) Ultracentrifugation in biochemistry. Elsevier

Shi L (2016) Bioactivities, isolation and purification methods of polysaccharides from natural products: a review. Int J Biol Macromol 92:37–48

Shiedlin A, Bigelow R, Christopher W, Arbabi S, Yang L, Maier RV et al (2004) Evaluation of hyaluronan from different sources: Streptococcus zooepidemicus, rooster comb, bovine vitreous, and human umbilical cord. Biomacromolecules 5(6):2122–2127

Shit SC, Shah PM (2014) Edible polymers: challenges and opportunities. J Polym 2014:1

Sober HA, Gutter FJ, Wyckoff MM, Peterson EA (1956) Chromatography of proteins. II. Fractionation of serum protein on anion-exchange cellulose. J Am Chem Soc 78(4):756–763

Tan K-X, Chamundeswari VN, Loo SCJ (2020) Prospects of kefiran as a food-derived biopolymer for agri-food and biomedical applications. RSC Adv 10:25339–25351. https://doi.org/10.1039/D0RA02810J

Trinetta V, Cutter CN (2016) Pullulan. In: Antimicrobial food packaging. Elsevier, pp 385–397. https://doi.org/10.1016/B978-0-12-800723-5.00030-9

Ullrich M (ed) (2009) Bacterial polysaccharides: current innovations and future trends. Horizon Scientific Press, Poole

Verma DK, Niamah AK, Patel AR, Thakur M, Singh Sandhu K, Chávez-González ML, Shah N, Noe Aguilar C (2020) Chemistry and microbial sources of curdlan with potential application and safety regulations as prebiotic in food and health. Food Res Int 133:109136. https://doi.org/10.1016/j.foodres.2020.109136

Wang A, Lewus R, Rathore AS (2006) Comparison of different options for harvest of a therapeutic protein product from high cell density yeast fermentation broth. Biotechnol Bioeng 94 (1):91–104

Wüstenberg T (2014) Cellulose and cellulose derivatives in the food industry: fundamentals and applications. Wiley, Hoboken

Zhang R, Edgar KJ (2014) Properties, chemistry, and applications of the bioactive polysaccharide curdlan. Biomacromolecules 15(4):1079–1096

Zia KM, Tabasum S, Khan MF, Akram N, Akhter N, Noreen A, Zuber M (2018) Recent trends on gellan gum blends with natural and synthetic polymers: a review. Int J Biol Macromol 109:1068–1087. https://doi.org/10.1016/j.ijbiomac.2017.11.099

A Review on Properties and Applications of Xanthan Gum

4

Surabhi Chaturvedi, Sanchita Kulshrestha, Khushboo Bhardwaj, and Rekha Jangir

Abstract

Xanthan gum is a hetero-polysaccharide formed by the strains of *Xanthomonas* spp. It is a naturally found polysaccharide molecule of high molecular weight maily formed by various fermentation processes. Its extraordinary rheological properties make it a very useful stabilizing agent for water-based systems. It has enormous applications ranging from the food industry to oil drilling. It is typically used in food industry in salad coverings, sauces, milk products, gravies, sweets, and low calorie foods in general. Xanthan gum is also used in making cleansers, varnishes, polishes, and in agricultural flowables. This chapter describes the extraction procedure of xanthan gum from microbes, factors affecting production, and application in different sectors.

Keywords

Xanthan gum · Oil drilling · Bakery · Food industry

S. Chaturvedi (✉) · S. Kulshrestha · R. Jangir
Lab No. 10, Department of Botany, University of Rajasthan, Jaipur, Rajasthan, India

K. Bhardwaj
Lab of Lifesciences, JECRC University, Jaipur, Rajasthan, India

A. Vaishnav, D. K. Choudhary (eds.), *Microbial Polymers*,
https://doi.org/10.1007/978-981-16-0045-6_4

4.1 Introduction

4.1.1 History

In 1950, it was first learnt about xanthan gum (XG) at the National Center for Agricultural Utilization of the United Sates Department of Agriculture (USDA). Its first production by industries started in 1960. After dextran, it was the second polysaccharide, which was isolated from microbes and was adapted for industrial commercialization (Palaniraj and Jayaraman 2011). Xanthan gum is a hetero-polysaccharide produced by the capsules of strains of Xanthomonas spp.which is usually associated with a loose bond with the bacterial cells which provides the best substitute of gluten. Its primary structure has repeats of pentasaccharide units containing two glucose, two mannose, and one glucuronic acid units with main chain comprising of β-d-glucose units interconnected at the first and fourth positions (Sworn 2009; Sharma et al. 2014). Galactomannan combinations made with xanthan gum are main ingredients to prepare gellies or thickeners of foods. It is also used as stabilizer in cattle-feed products, herbicides, fungicides, pesticides, fertilizers, and in toothpastes (Flores et al. 2010). It is permitted as food packing material in paper and wooden industries, mining ores, paints, and polishes. Additional major applications include oil recovery; gelatinization of xanthan in pumping operations, etc. (Kandra et al. 2018).

Industrial production of xanthan gum is performed by spp. of Xanthomonas such as X. arboricola, X. axonopodis, X. campestris, X. citri, X. fragaria, X. gummisudans, X. juglandis, X. phaseoli, X. Vasculorium, and X. campestris (Petri 2015).

Xanthomonas is a genus of Gram-negative tiny rod-shaped aerobes coming under the family *Pseudomonadaceae*. Many members of this genus are producers of xanthan gum, but most of them are phytopathogens (Vendruscolo et al. 2000). At commercial level, 70% of the substrate can be converted into the gum through aerobic fermentation at a temperature between 27 and 30 °C, by using organic acids as stimulants. (Lopes et al. 2015).

As a general practice, commercially exopolysaccharides are produced by batch fermentation and precipitation and centrifugation are the main techniques used for recovery of exopolysaccharides from the fermentation broth. Similar process is also applied in the xanthan gum production process. Studies have shown that the parameters and variables like composition of nutrients, techniques of feeding, temperature, pH, agitation, and foaming reduction can affect conversion of a raw material in xanthan (Rosalam et al. 2008).

4.1.2 Properties

Xanthan gum has the following important properties:

1. **High solubility**: It has high solubility in both cold and hot water but barely soluble in maximum organic solvents. Highest solubility of xanthan gum is in 25% phosphoric acid and then in acetic acid, sulfuric acid, nitric acid, and sodium hydroxide.
   This propertiy varies according to the end use of the product; for instance, the bacteriological loads and heavy metal content for food application have to be very low although this is not the case for applications such as those in other industrial applications.
2. **Highly viscous**: It is very viscous even at low concentrations; even at a concentration of 1% it seems almost gel, yet pours readily and easy to mix and pumping. The viscosity is determined by temperature, the biopolymer concentration, amount of salts, and pH.
3. **Emulsifying nature**: It has notably high emulsifying, stabilizing, and suspending abilities. Xanthan gum exhibits great compatibility with thickeners present in the market such as cellulose derivatives, starch, pectin, gelatin, dextrin, alginate, and carrageenan. By combining various gums with xanthan gum in altered proportions. It performs better after modification with enzymatic treatments whch results in the removal of galactose residues (Kuppuswami 2014).
4. **Low calorie:** Similarly like other gums, xanthan gum is non-digestible in human gut and thus provides lower calories and easily passes through the gastric tract.
5. **Pseudoplasticity**: In solution, it has pseudoplastic nature; that is, the viscosity inversely changes with shear rate of a xanthan solution. This property enhances sensory potentials (flavor release, mouth feel) in food products and promises a high level of mixing, pumping, and pouring properties (Sharma et al. 2014).
6. **High resistance to pH**: It has a high tolerance to deviations in the pH range of 2–12 which makes it suitable to foods. Exceptional constancy is seen at low pH over long episodes of time. This property is vital for cleaners as well as for acidic food presentations such as salad dressings and fruit systems.
7. **Thermostability**: High resistance to temperature changes, even in the presence of acids and salts. It has brilliant freeze–thaw ability, viscosity which has stability up to 100 °C, and even after heat treatment processes of sterilization, viscosity is maintained after chilling. It is evident that the pyruvate content of xanthan gum affects its thermal stability.

Structure: The primary structure of xanthan gum, shown in Fig. 4.1, has a linear (1–4)-linked-D-glucose backbone with a trisaccharide side chain on every other glucose at C-3, containing a glucuronic acid residue linked (1–4) to a terminal mannose unit and (1–2) to a second mannose that connects to the backbone (Sanderson 1982). There is a repeated disachharide making backbone with right handed helix to form fiber. This conformation gives stability to the structure as the trisaccharide side chains are lined up with the backbone strengthened by non-covalent and hydrogen bonds. In solution, these side chains cover the backbone all over; in this manner, the (1–4) bonds are protected from attacks. It is believed that this shield is key to the excellent stability of the gum under hostile situations. Xanthan gum undergoes a conformational change on heating which is related with

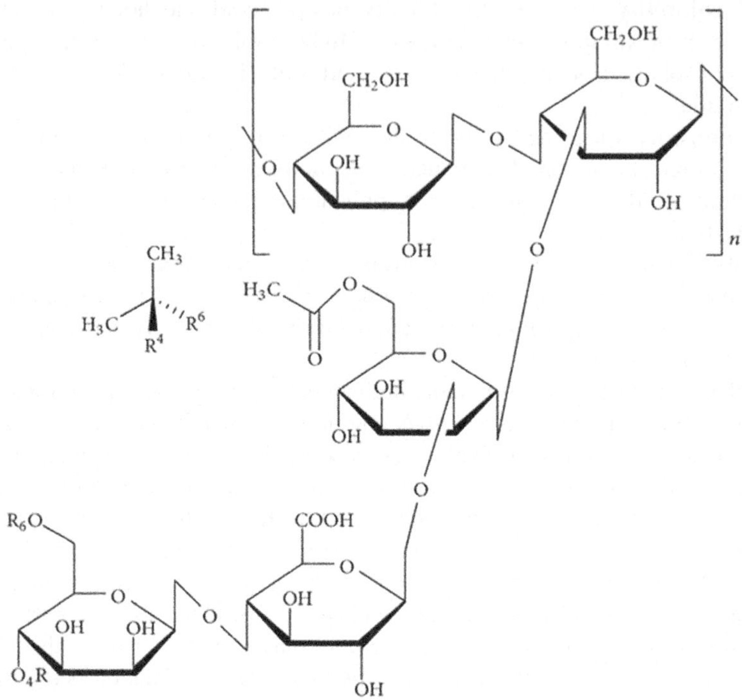

Fig. 4.1 Primary structure of xanthan gum

the conversion from an inflexible systematic state at low temperature to a more bendable and disordered state as the temperature increases (Pelletier et al. 2001).

4.2 Microbial Production of Xanthan Gum

The major operations involved in the microbial production of xanthan are as follows:

1. Organism and inoculum preparation
2. Media preparation
3. Fermentation

4.2.1 Organism and Inoculum Preparation

Industrial production of xanthan gum is performed by spp. of Xanthomonas such as X. arboricola, X. axonopodis, X. campestris, X. citri, X. fragaria, X. gummisudans, X. juglandis, X. phaseoli, X. Vasculorium, and X. campestris (Petri 2015) as shown in Fig. 4.2.

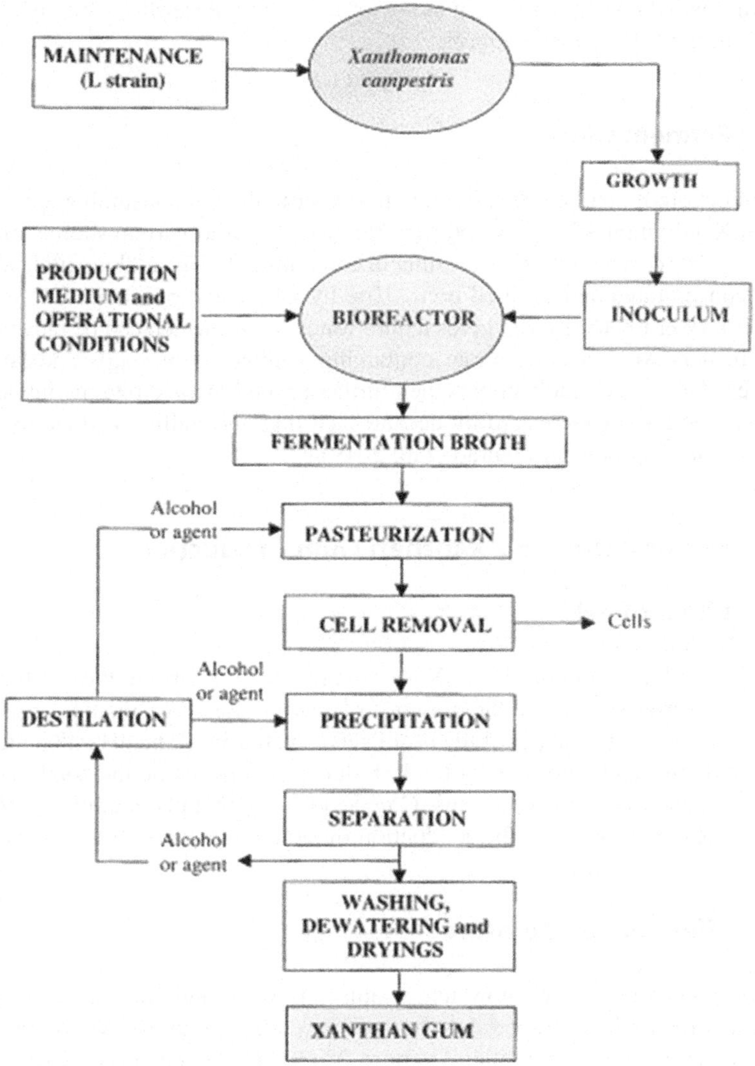

Fig. 4.2 Generalized schematic for industrial production of xanthan gum

4.2.2 Media Preparation

Mostly naturally found raw materials are utilized for industrial production of xanthan gum. Media is diluted with tap water and carbon source generally used are glucose, sucrose, or starch and sometimes acid whey, up to a concentration of 5% as greater concentrations could suppress both growth and final production. To reduce the cost of media, its components are priorly sterilized. It is also beneficial to simplify the procedure, increase flexibility as modern process and equipment are used. Suspended constituents, such as soy protein, are not sterilized simply and thus cause fouling of the heated surfaces.

4.2.3 Fermentation

The production of xanthan gum starts with the fermentation consuming any suitable strain of Xanthomonas bacteria and substrate with sugar as carbon source, nitrogen, and salts. The fermentation broth results in containing 2–3% xanthan gum, which is further concentrated and purified up to 10% by UF. Thus, generated UF permeate stream if further treated by RO gives a pure water stream which can be recycled into the fermenter. Also, a concentrate containing xanthan gum, salts, and sugars is obtaained. The concentrated gum is then further refined by precipitating in methanol and the crude gum is recovered by decanter. Drying and milling is done to make a white free-flowing powder (Murad et al. 2019).

4.3 Factors Affecting Xanthan Gum Production

4.3.1 Effect of pH

It is assumed that optimum pH for XG growth is neutral. During the production of XG, pH decreases to 5 due to the presence of acid groups. Investigations reveal that after a fixed time duration pH of the broth keeps increasing. The pH value increased from 7 to 8 after 24 h and 8–9.5 after 48 h duration depends on the combination of temperature and agitation conditions. Conclusions say that pH conditions affect the growth of Xanthomonas not the production of xanthan (Psomas et al. 2007).

4.3.2 Effect of Temperature

It has been broadly studied how temperature affects production of xanthan gum. Gumus and researchers studied that the optimum temperature for XG production is 25–34 °C while culture conditions require 28–30 °C (Gumus et al. 2010). It was studied by Cadmus and researchers if temperature is increased for culture production, pyruvate content decreases. It is clear from the studies of Shu et al. that for the production of XG medium play as main pivotal factor for optimal temperature (Shu and Yang 1990).

4.3.3 Effect of Pressure

The pseudoplastic behavior and intrinsic viscosity of XG are reduced due to microfluidization. The effect of dynamic high pressure was studied by Laneuville and coworkers on XG in which they found that gel aggregates convert into solution with high shear force if applied once. When applied again, double-helical structure of XG was degraded into single chain. If the applied pressure is removed, XG will gain its original structure (Funahashi et al. 1987).

4.3.4 Effect of Carbon Sources

Carbon is the necessary component of culture medium. Every cell needs optimum concentration of nutrients and growth factors. Variation into nutrient and substrate concentration affects the side chain structure, molecular mass, and yield and not the backbone. Glucose and sucrose are the most common carbon sources in any medium. The carbon concentration for XG production is preffered to be 2–4%; concentration more than it prevents growth (Funahashi et al. 1987). Deacylation effect on rheological properties of xanthan–guar interaction is investigated by Khouryieh et al. and revealed that deacylated xanthan–guar gum mixture behave like gel whereas it is liquid like if the mixture is native. The reason behind that is destabilization of double-helical structure increased flexibility of side chain. Moreover, the deacylated XG contains more intrinsic viscosity than native. (De Vuyst et al. 1987).

4.3.5 Effect of Polymer Concentration and Salts

The XG solutions are more viscous if the concentration of XG is higher in the solution. The viscosity is produced due to its intermolecular bonding and reticulated structure. It has come into observation that the viscosity of XG solution may be changed by salt. Addition of small amount of salt lowers the viscosity of low concentrated solution of XG. This happens due to reduced electrostatic forces which otherwise causes an increase in viscosity resulting into less availability of molecules (Smith and Page 1982). Addition of salt into solution of high XG concentration causes more interaction of XG and salt molecules which result in higher viscosity (Palaniraj and Jayaraman 2011; Garcia-Ochoa et al. 2000).

4.3.6 Effect of Viscosity on Xanthan Gum in the Presence of Galactomannan

XG and galactomannan interaction causes an increase in viscosity of XG solution and also affected majorly by temperature. The dissolving temperature of galactomannan affects the viscosity and mesh size of the gel (Garcıa-Ochoa et al. 2000). It determines the maximum size of any passing molecule which affects the diffusion. Measurement of diffusion rate helps in its utilization for estimating the release kinetics.

4.4 Applications of Xanthan Gum

Xanthan gum is reported to have enormous applications in many industries because it has some extraordinary properties such as longer stability of emulsions in acidic, alkaline, and saline mediums; thermal stability and pseudoplasticity (Fig. 4.3, Table 4.1). It has wide use in chemical industries such as xanthan and locust bean gum are used for manufacturing of deodorant gels. Due to high dispersion stability and low viscosity it is used to give a proper uniformity to the toothpaste (Palaniraj and Jayaraman 2011). By adding a minimum quantity of bivalent salts, xanthan gum can improve the color depth (K/S value) and acuity of the printed fabrics (Cao et al. 2019).

4.4.1 Pharmaceutical Applications

Xanthan gum is a stabilizer for suspensions of many unsolvable ingredients like barium sulfate (X-ray diagnoses), complexed dextromethorphan (for cough preparations) and thiabendazole. It is able to maintain homogeneity of liquids (Palaniraj and Jayaraman 2011), and it is reported to provide sustained release of

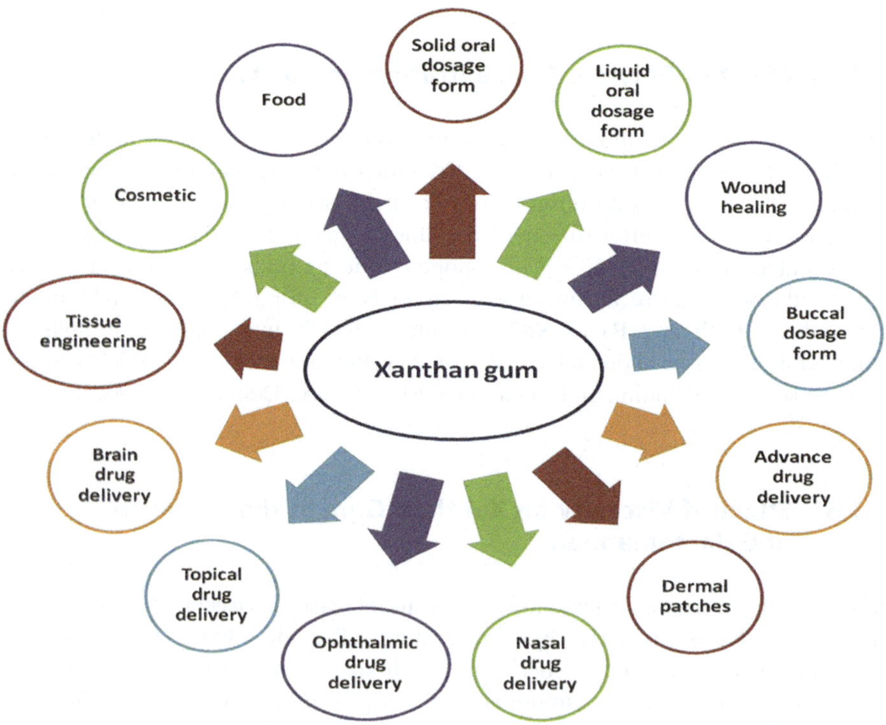

Fig. 4.3 Applications of xanthan gum in various industries

Table 4.1 Summarizing different applications of xanthan gum

Food application	Usage in %	Function	References
Salad dressing Bakery Beverage Preparation of food Soups, sausage, and gravies Dairy products Meat products	0.1–0.05 0.05–0.3 0.05–0.2 0.1–0.3 0.05–0.5 0.05–0.2 0.5–0.2	Easy pouring and good cling Improves texture Enhances taste Mixes fruit pulp Stabilizer Makes it tempting. Stabilize Stabilizer and emulsifier Holds water	Sharma et al. (2006) Rosalam and England (2006)
Personal care Toothpaste Creams and lotion Shampoo	0.7–0.1 0.2–0.5 0.2–0.5	Easy pumping and good texture Stabilizer Improves solubility	Rosalam and England (2006)
Industrial application Agricultural products Cleansers Polishes Aqueous paints Textile and flooring print Glues Paper firms Ceramic glases Drilling of oil	0.1–0.3 0.2–0.7 0.2–0.7 0.1–0.3 0.2–0.5 0.1–0.3 0.1–0.2 0.3–0.5 0.1–0.4	Proper mixing Provide stable pH range Mix abrasive components Control liquidity Improved color Control rheology and penetration Suspends particles Suspends solid Stabilizer Function recovery agent	Flickinger and Draw (1999) Katzbauer (1998)
Biomedical Tablets	0.1–0.5 0.1–3.0	Provide excellent stability and good flow Retards drug release	Bindu and Ashok (2012)

drugs in the human system (Patel and Patel 2007), for example, XG-based pills of drug aceclofenac has been made by Ramasamy and colleagues for targeted drug delivery in colon. These tablets were stable at acidic pH of stomach and were degraded and absorbed in colon only (Ramasamy et al. 2011). Scientists have also formulated gastroretentive tablets using different hydrocolloids including XG using direct techniques with controlled physicochemical properties. These preparations confirmed to have sustained release for more than 20 h and stayed resilient for long. Similarly, Butani and colleagues formulated venlafaxine multifaceted matrix tablets using XG with rate controlling agent. Hence, it has been demonstrated by many authors that bursting of the tablets can be controlled by adding XG (Butani 2013). Scientists have established an antacid and antiulcer suspension using herbal extractions of Glycyrrhiza glabra, Terminalia chebula, Terminalia bellirica, Emblica officinalis, etc. and stabilized with various quantities of XG (Roopa et al. 2015).

4.4.2 Food Industries

Xanthan gum is mainly applied as suspending and thickening agent in food industries of fruit juices and chocolate. It has already been approved by the US food and drug administration (FDA) for its use in food products. Most of the food items, nowadays need improvement of texture, look, flavor enhancement, viscidness, and liquidity. Xanthan gum is able to improve all these properties and additionally maintains the rheological performance of the foodstuffs. Also, it is useful to maintain pseudoplasticity and tempers the "gummy" taste than other gums (Palaniraj and Jayaraman 2011).

4.4.3 Dairy

When mixed with other polysaccharides such as galactomannans, carrageenan and guar gum are finer stabilizing agents for a lot of industrialized dairy foodstuffs such as milkshakes, ice milk, and ice cream. It has been used in manufacturing of frozen dairy products such as dessert puddings, acidified milk gels by mixing methyl-carboxy methyl cellulose and CMC. These dairy stuffs are popular for being more stable, gentle, fresh in taste, viscidness, and tempting flavor (Flores-Candia and Deckwer 1999; Rosalam et al. 2008). With some other gums it is crucial for firmness, texture, and flavor of creamy cheese (Frank and Somkuti 1979) and karish cheese (Ahmed et al. 2005).

4.4.4 Bakery Products

Xanthan gum has been utilized for manufacturing bakery products as it lifts up the water holding capability throughout baking process and storage. It is also being used to increase shelf life of baked foods, as a replacement of eggs in some of bakery products, improves softness, air absorption, and holding in bread blends, biscuits, and cakes. The foods supplemented with xanthan gum are known to lessen calorie and gluten content in breads and increases shelf permanency, freezing and thawing stability in creams and fruit fillings (Murad et al. 2019). Due to its capability of producing highly stable emulsions, it is functional to make oil and non-oil-based sauces and ketchups (Koocheki et al. 2009). As it has stability to withstand in acidic, alkaline, and salty mediums and also possess thermostability, it is used to make products for a long shelf life.

Besides having textural properties, the capability to maintain flavor for a longer period and defrosting constancy of xanthan gum has been exploited to derive very successful products (Bylaite et al. 2005). It is also applied in dairy goods due to stabilizing properties and protective effect against heat shock and prevent formation of icy crystals (Hemar et al. 2001).

There are reports that adding xanthan gum was good to improve the firmness of cassava flour, taste, hardness, and crumbliness of the biscuits (Lu et al. 2019).

Monthe et al. (2019) have demonstrated xanthan gum as a gluten substitute to make GF bread with a mixed flour prepared from fermented cassava, sorghum, and sweet potato and assessed their properties under varying parameters. Mohammadi et al. (2014) mixed xanthan gum and CMC to the blend of GF bread and found improved texture and taste. Kaur et al. (2015) reported that biscuits made from buckwheat flour with gums have more moisture, weight, size, and chunkiness and low breakage strength in comaprison to regular biscuits. The xanthan gum could considerably improve the look, color, flavor, and complete appropriateness of biscuits.

4.4.5 Beverages

In the processing of beverages, xanthan gum has been used as a bodying and dangling agent in drinks. It gives decent look and touch along with improvement in mouth feel to have pleasant taste, fast solubility, and compatibility with most components. In reconstituted beverage, it is used as taste enhancer which improves viscosity also (Murad et al. 2019).

4.4.6 Biomedical Application

Latest research shows that the xanthan gum is mixed with glycerides for improvement of performance of drilling fluids (Nunes et al. 2014). It is also mixed with magnetic nanoparticles and used as enhancer for the adsorption and removal of pollutant dyes from water (Mittal et al. 2014). It has appeared as an important tool in the biomedical field such as the intra-articular injection of xanthan are applied for treatment of osteoarthritis as it improves elasticity and provides lubrication between the joints (Han et al. 2012).

Its high molecular weight is useful to build up physical and chemical networks, which is further used as carriers for drugs delivery and as scaffolds in biomedical tissue engineering. Due to acidic tolerance, it is used with some more polymers and mixed as excipient in tablets or as supportive hydrogels for drug release (Petri 2015).

4.4.7 Nanoparticle

In human lung cancer treatment, xatham gum is used to synthesize gold nanoparticle and these EPSs perform good to transport doxorubicin hydrochloride. (Pooja et al. 2014). Silver nanoparticles with increased catalytic and antiseptic properties for Escherichia coli and Staphylococcus aureus have been prepared with xanthan gum. (Xu et al. 2014). With other metals EPSs have also been used to synthesize nanoparticles because xathan gum reduces and stabilizes the reaction to produce palladium nanoparticles (Pd-NPs) with better catalytic activity through out the reaction of 4-nitrophenol to 4-aminophenol by sodium borohydride (Santoshi et al. 2015. For remediation of polychlorinated biphenyl- (PCB-) from contaminated soil, xantham gum is used with surfactant BriJ35 to stabilize iron, palladium NPs. (Fan et al. 2013).

In food and non-food industries, xathan gum is commonly used for an increase in thickness and viscosity due to its softness. Xatham gum is also used as stabilizers in emulsion,suspensions, and foarms (Michel et al. 2003). It is also used with chitosan membranes in treatment of dermo-epidermal wounds (Michel et al. 2003; Bellini and Caliari 2014) and many pharmaceuticals (Gardin and Pauss 2001; Jagdale and Pawar 2014; Kuo et al. 2014; Razavi et al. 2014) and cosmetic industries (Jianlon and Yi 1999; Jamshidian et al. 2014).

4.4.8 Drug Delivery

Xanthan gum has enormous role in healthcare and drug delivery systems (Fig. 4.4, Table 4.2). It can be explained in the following points:

Liposomes: Liposomes are spherical vehicles of lipid bilayer designed artificially to deliver both hydrophobic and hydrophilic compounds as they can protect entrapped compounds from degradation and release it on the targeted sites. Lipophilic barriers around the entrapped compound help in protecting it from enzymes, free radicals, and digestive juices. Stability of liposomes can be improved

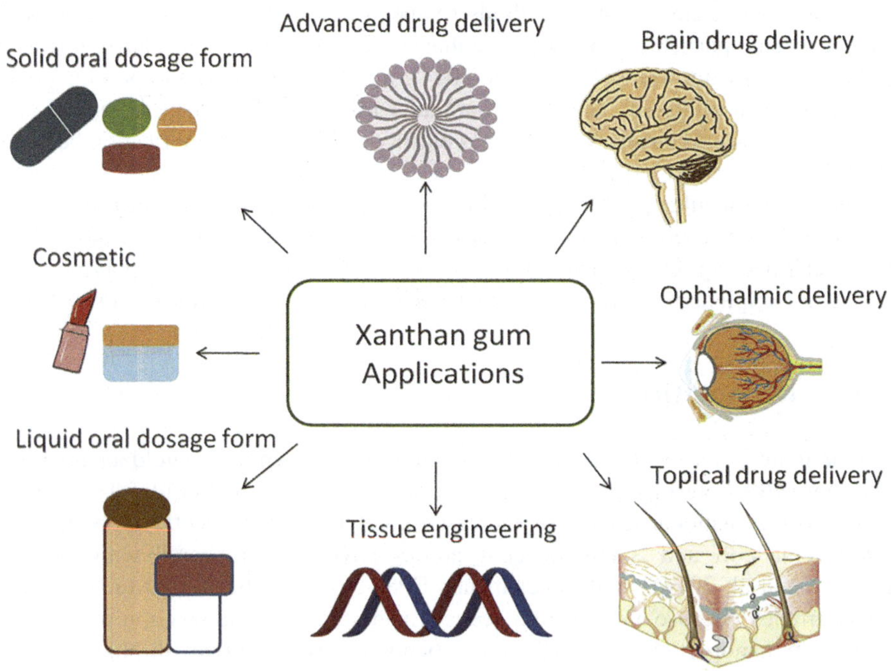

Fig. 4.4 Potential role of xanthan gum in healthcare systems

Table 4.2 Application of xanthan gum in drugs

Application	Drug	Application	References
Treatment of posttraumatic corneal abrasions	Netilmicin	An eye gel with the antibiotic netilmicin	Faraldi et al. (2012)
Buccal drug delivery	Zolmitriptan	Xanthum gum-based bilayer mucoadhesive	Shiledar et al. (2014)
Treatment of sialorrhea		Modified xanthum gum for treating sialorrhea	Laffleur and Michalek (2017)
Advanced drug delivery	Lamivudine	Drug delivery	Harika et al. (2011)
Brain drug delivery	Carbamazepine	Carbamazepine nanoemulgel for brain	Samia et al. (2012)
Wound healing	Metalloprotenase-9	Wound healing by matrix metalloprotenase-9	Reiss et al. (2010)
Dermal patches	Paracetamol	For treating dermitis	Gorle et al. (2017)
Nasal drug delivery	Loratadine	Ph-dependent in situ nasal gel of loratadine using XG as a polymer	Saudagar and Badhe (2016)
Cosmetic cream	Caffeine	Skin gel	Parente et al. (2015)

by using suitable polymers such as chitosan (Akbarzadeh et al. 2013). A positive effect of liposomal formulation was observed for pulmonary delivery by the liposome coating with polyelectrolyte complex formed by complexation of xanthan gum and chitosan (Sandolo et al. 2007; Manca et al. 2011).

Hydrogel: Hydrogels are hydrophilic polymeric network chains which help in imbibing huge amount of water just like a biological tissue. Porosity of a hydrogel determines the drug loading capacity and the succeeding drug release. Porous structure can be regulated by cross-linking agents and polymer concentrations (Bindu and Ashok 2012). Superporous hydrogels are water-loving polymers with cross-linkage networks which improves their puffiness and absorption properties. Permeation and capillary action are the main modes of quick water absorption in a macromolecular structure. Superporous hydrogels can be prepared by using xanthan gum, 2 hydroxyethylmethacrylate and acrylic acid (Santos et al. 2005; Gils et al. 2009). It is observed that xanthan gum does not form hydrogels very fast instead these are only shaped when the aqueous solutions are taken to a particular temperature and immediately cooled (Sharma et al. 2006; Iijima et al. 2007).

Matrix Systems: Matrix systems are composed of polymeric compounds which swell in the presence of water or biological fluids and can be used in controlled release of active drug ingredients. In matrix system, drug powder is evenly distributed in polymeric material matrix and compressed to form a tablet. The enlargement property of these polymers is helpful in the fragmentation of tablets and improvement of the termination rate of drug. According to Verhoeven et al.

(2006) high enlargement and erosion rate, quicker drug release, and high absorption rate are the qualities of xanthan gum-mediated drug delivery in sustained release matrices. In colon drug delivery also when xanthan gum-mixed tablets show higher drug retarding capability than other formulations (Tiwari and Kumar 2009; Jackson and Ofoefule 2011). Xanthan gum is used in gum-based sustained release tablets and provides compatibility and inertness (Shinde and Kanojiya 2014). According to Vendruscolo et al. (2005) xanthan gum has increased the ability to hold back the drug release in comparison with synthetic cellulose derivatives.

Niosomes: Niosomes are vesicular type of drug carrier transport system and have similarity with structure of liposome, but they are better in stability and are cheap in comparison to liposome. These possess both hydrophilic and hydrophobic parts so that both hydrophobic and hydrophilic drugs may be conveyed to the targeted sites. These can be used in eye drugs where their function is to carry hemoglobin and peptides (Ashutosh et al. 2012).

Niosomes prepared with xanthan gum have changed particle size and good spreadability in comparison to the formulation without xanthan gum and the niosomal preparation have pseudoplasticity. The use of xanthan gum as a gelling agent to form serratiopeptidase niosome gel is proved experimentally (Vishakha et al. 2012; Shinde and Kanojiya 2014).

Nanoparticles: Nanoparticles perform adsorbtion or encapsulation of the drug molecule for its protection from various biological degradation process, thus are very useful in drug delivery. The small size of nanoparticles [1–100 nm] makes them easily absorbable in comparison to larger molecules. Surface characters and the particle size of nanoparticles can be efficiently changed (Mohanraj and Chen 2006). Covalent, ionic, and poly bonds are the main types of linkages in the formation of polysaccharide nanoparticles (Liu et al. 2008; Vishakha et al. 2012). Gold nanoparticles synthesized using xanthan gum are nontoxic and biocompatible and also used to improve drug loading, stabilization, and increment of cytotoxicity in lung cancer cells (Wang et al. 2002; Pooja et al. 2014). According to a study, micro- and nanoscale iron particles can be stabilized by viscoelastic gel made of xanthan gum and guar gum mixtures (Williams et al. 1991; Xue and Sethi 2012).

Microspheres: Microspheres are small spherical particles (1–1000 μm) used in the regulated delivery of unstable drugs. They can be applied in gene delivery, nasal drug delivery, oral drug delivery, and gastrointestinal drug delivery (Sahil et al. 2011). A 3D interpenetrating polymer network is developed by cross-linking of xanthan gum and polyvinyl alcohol and potentially delivered drug in a much sustained release pattern (Bhattacharya et al. 2013). According to a research, hydrophilic gums like xanthan gum and locust bean gum can be useful in delaying the drug release and enhance the duration of drug release (Deshmukh et al. 2009; Xue and Sethi 2012).

4.4.9 Food Applications

XG is used as an additive in the food industry. It is biosynthetic and edible gum and at present additives are widely used for baking. Serum cholesterol is found to be lowered by the consumption of guar gum-like edible gums. In a study subjects of diabetes and without diabetes were chosen to assess the acceptability of XG (present in muffins) with a daily dose of 12 g in diet. Postload and fasting glucose level was found to be lowered after the XG consumption thus can be suggested that XG use into diet may help in the initial treatment of diabetes mellitus (Osilesi et al. 1985). In a study conducted by Preichardt et al., role of XG in maintaining the quality of gluten-free cakes was assessed for celiac patients. Three different concentrations of XG were taken for the formulations.Some authors analyzed the chemical, physical, and sensory features which revealed that XG combination improves quality of the characteristics such as enhanced and uniform features, delayed staling, and an increase in specific volume whereas both control and sample cake appeared same physically (Preichardt et al. 2011).

In a study conducted for the examination of baked food quality having four different hydrocolloids, viz. Arabic gum, guar gum, methyl 2-hydroxyl ethyl cellulose, and XG. Evaluation of certain parameters such as degree of softening, dough development time, mixing tolerance index, and water absorbing capacity. Crumb hardness, baked food quality, and sensory evaluation were assessed. XG containing dough was found to be more water absorbing due to the presence of hydroxyl group in studies of water absorption capacity. In addition, loaves which contain guar gum were softer than that which were containing XG after 72 h of storage (Hui 2005).

Another reason for the use of XG in food industry is its high viscosity even at a very low concentration which enables it to be used as a thickner. It shows pseudoplastic behavior in aqueous solution helping in pouring, filling, mixing and pumping, etc. (Geremia and Rinaudo 2005; Milas et al. 1990). Highly stable at variable pH, ionic strength, temperature and under shear when processed and packed (Hui 2005). For the prevention of water migration into pastry from the baked foods, xanthan gum is also used in bakery fillings. Recently, due to the elasticity of dough lent by XG, it is being added to gluten-free baked goods where its concentration range between approximately 0.05 and 0.7 wt % (Zhou et al. 2014).

4.4.10 Cosmetics

In cosmetic industry also XG is used as an emulsion stabilizer, thickener, and texture enhancer. Due to the ability to make gel structure into water, XG is also used as viscosity modifier such as guar gum. XG is patented as fixative for the composition of cosmetics by Cao et al. In the study, it was revealed that heat treated XG containing composition performed better than the carbomer hair gel/polyvinyl pyrrolidone in wet and dry combs and stiffness, etc. (Cao et al. 2003). An eyeshadow containing XG mixed into dyestuff and silicate aqueous medium has been patented by Collin and his team. It can be stored at 45 °C and remain homogenous for

2 months due to the presence of gelling agent in the composition. Due to fluid-like texture, spreading on the eye lids as well as removal is easy and stays long (Collin et al. 2003). Formulation of bioadhesive using caffeine and XG as model drug and secondary polymer for skin application was carried out by Parente and coworkers which was characterized by adhesion, rheological, spreading capacity, and in vitro release studies. The revelation of in vitro studies was that release time for approximately two third of drug was very less. Other physicochemical studies were helpful in selecting optimum formulation (Parente et al. 2015). In a research by Saharudin along with coworkers examined what effect XG shows on some properties of rice bran oil such as physicochemical and rheological. By the use of emulsification process, six formulations were prepared. In studies, it was revealed that droplet size and zeta potential were increased with an increase in XG concentration. Author concluded that improvement in physicochemical properties of formulation can be done by using XG (Saharudin et al. 2016). Amnuaikit and his coworkers prepared a sunscreen lotion by using XG which was found to be having high sun protection factor (SPF) as well as both UV filters (organic and inorganic). In the formulation, XG was used in aqueous phase whereas UV filters were titanium dioxide and anisotriazine. In the evaluation results of prepared creams, in vitro SPF value was found to be more when the above-mentioned combination was used (Amnuaikit and Boonme 2013). Meanwhile at present, XG is present in various cosmetic products as well as toothpaste and some dermatological products (Singhvi et al. 2019).

4.4.11 Oil Industry

Xanthan is also used in petroleum industry in extraction of oil (Katzbauer 1998) fracturing, pipeline cleaning, and work-over and completion. In thermal degradation, xanthan gum has good coordination with salt and tolerance that is why it is also used as an ingredient in drilling fluids (Khalil and Jan 2012). In micellar-polymer, xanthan gum is used as a tertiary oil recovery agent. Xantham gum is also used to fill the rock of sand in which oil is poured to make sure the proper extraction of oil. Xanthan gum is used in formulation of drilling fluid, starch modifications, and calcium products because of its high specific gravity. The drilling fluid's thickness is managed by weighting the supplements which to make sure that hydrostatic pressure is same and fluid is not lost (Xia et al. 2020).

4.5 Conclusion and Future Prospective

XG is a natural gum which has an extraordinarily miscellaneous role in different branches of science. Particularly, it is capable of undertaking a range of chemical amendments with personalized properties. Because of such qualities, it is extensively used to make adhesives, gums, coatings, emulsions, and binders as it has been approved as a nontoxic gelling and stabilizing agent with wide industrial applications. It is a nonsensitizing compound which makes it suitable for use in drugs. On the ground of such properties, xanthan is approved by the FDA for use as a food stabilizer without any precise magnitude limitations and in pharmaceuticals with certain guidelines. It is shown that XG along with herbal powders can give better uniformity to tablets. It gives wide growth to industries; besides being natural, it has tolerance to generally practiced processing protocols and offers the advantages of biodegradability and low toxicity.

References

Ahmed NH, Soda ME, Hassan AN, Frank JF (2005) Improving the textural properties of an acid-coagulated (Karish) cheese using exopolysaccharide producing cultures. LWT-Food Sci Technol 38:843–847

Akbarzadeh A, Rezaei-Sadabady R, Davaran S, Joo SW, Zarghami N, Hanifehpour Y (2013) Liposome: classification, preparation, and applications. Nanoscale Res Lett 8:1–9

Amnuaikit T, Boonme P (2013) Formulation and characterization of sunscreen creams with synergistic efficacy on SPF by combination of UV filters. J Appl Pharm Sci 3(8):1–9

Ashutosh L, Suman R, Sidhyartha S, Altaf S, Vandana YM (2012) Review article a novel drug delivery system: niosomes review. J Drug Deliv Ther 2:129–135

Bellini MZ, Caliari C (2014) Combining xanthan and chitosan membranes to multipotent mesenchymal stromal cells as bioactive dressings for dermo-epidermal wounds. J Biomater Appl 29:1155–1166. https://doi.org/10.1177/0885328214553959

Bhattacharya SS, Mazahir F, Banerjee S, Verma A, Ghosh A (2013) Preparation and in vitro evaluation of xanthan gum facilitated superabsorbent polymeric microspheres. Carbohydr Polym 98:64–72

Bindu S, Ashok V (2012) As a review on hydrogels as drug delivery in the pharmaceutical field. Int J Pharm Chem Sci 1(2):642–661

Butani SB (2013) Development and optimization of venlafaxine hydrochloride sustained release triple layer tablets adopting quality by design approach. Pharmacol Pharm 4:9–16. https://doi.org/10.4236/pp.2013.43A002

Bylaite E, Adler-Nissen J, Meyer AS (2005) Effect of xanthan on flavor release from thickened viscous food model systems. J Agric Food Chem 53:3577–3583

Cao H, Martino G, Richardson P, Inventors; National Starch, Chemical Investment Holding Corp, Assignee (2003) Use of xanthan gum as a hair fixative [United States patent application US 09/932,435]

Cao H, Li A, Yang Z, Zhu Y (2019) Application of xanthan gum as a pre-treatment and sharpness evaluation for inkjet printing on polyester. Polymers 11:1504

Collin N, Bichon Y, Pays K, Inventors; L'Oreal SA; Assignee. Eyeshadow comprising a xanthan gum and a mixed silicate; August 21, 2003 [United States patent application US 10/337,904]

De Vuyst L, Vermiere A, Van Loo J, Vandamme EJ (1987) Two-step fermentation process for improved xanthan production by Xanthomonas campestris NRRL B-1459. J Chem Technol Biotechnol 39:263–273

Deshmukh VN, Jadhav JK, Masirkar VJ, Sakarkar DM (2009) Formulation optimization and evaluation of controlled release alginate microspheres using synergy gum blends. Res J Pharm Tec 2:324–327

Fan G, Cang L, Qin W, Zhou C, Gomes HI, Zhou D (2013) Surfactants-enhanced electrokinetic transport of xanthan gum stabilized nanoPd/Fe for the remediation of PCBs contaminated soils. Sep Purif Technol 114:64–72

Faraldi F, Papa V, Santoro D, Rasá D, Mazza AL, Rabbione MM, Russo SA (2012) New eye gel containing sodium hyaluronate and xanthan gum for the management of post-traumatic corneal abrasions. Clin Ophthalmol 6:727–731

Flickinger, FC, Draw, SW, (1999) Encyclopedia of Bioprocess Technology: Fermentation. vol 5, pp 2706–2707
Flores-Candia JL, Deckwer WD (1999) Xanthan gum. In: Flickinger MC, Drew SW (eds) Fermentation, biocatalysis, and bioseparation, vol 1-5. Wiley, Hoboken, pp 2695–2711
Flores SK, Costa D, Yamashita F, Gerschenson LN, Grossmann MV (2010) Mixture design for evaluation of potassium sorbate and xanthan gum effect on properties of tapioca starch films obtained by extrusion. Mater Sci Eng 30:196–202
Frank JF, Somkuti GA (1979) General properties of beta-galactosidase of *Xanthomonas campestris*. Appl Environ Microbiol 38:554–556
Funahashi H, Yoshida T, Taguchi H (1987) Effect of glucose concentration on xanthan gum production by Xanthomonas campestris. J Ferment Technol 65:603–606
Garcia-Ochoa F, Santos VE, Casas JA, Gomez E (2000) Xanthan gum: production, recovery and properties. Biotechnol Adv 18:549–579. https://doi.org/10.1016/S0734-9750(00)00050-1
Gardin H, Pauss A (2001) κ-carrageenan/gelatin gel beads for the co-immobilization of aerobic and anaerobic microbial communities degrading 2,4,6-trichlorophenol under air-limited conditions. Appl Microbiol Biotechnol 56:517–523. https://doi.org/10.1007/s002530000581
Geremia R, Rinaudo M (2005) In: Dimitriu S (ed) Polysaccharides: structural diversity and functional versalitity. Dekker M, New York, pp 411–430
Gils PS, Ray D, Sahoo PK (2009) Characteristics of xanthan gum-based biodegradable superporous hydrogel. Int J Biol Macromol 45:364–371
Gorle A, Pawara I, Achaliya A (2017) Design development and evaluation of transdermal drug delivery system of antipyretic agent. Int J Pharm Res Health Sci 5(4):1743–1749
Gumus T, Demirci AS, Mirik M, Arici M, Aysan Y (2010) Xanthan gum production of Xanthomonas spp. isolated from different plants. Food Sci Biotechnol 19:201–206
Han GY, Wang G, Zhu X, Shao H, Liu F, Yang P, Ying Y, Wang F, Ling P (2012) Preparation of xanthan gum injection and its protective effect on articular cartilage in the development of osteoarthritis. Carbohydr Polym 87:1837–1842
Harika MJ, Kumar M, Sinhal A, Saifi A (2011) Recent development in novel drug delivery systems of herbal drugs. Int J Green Pharm 5(2):87–94
Hemar Y, Tamehana M, Munro P, Singh H (2001) Viscosity, microstructure and phase behavior of aqueous mixtures of commercial milk protein products and xanthan gum. Food Hydrocoll 15:565–574
Hui YH (ed) (2005) Handbook of food science, technology and engineering, vol 4. CRC Taylor and Francis, Boca Raton, p 3632
Iijima M, Shinozaki M, Hatakeyama T, Takahashi M, Hatakeyama H (2007) AFM studies on gelation mechanism of xanthan gum hydrogels. Carbohydr Polym 68:701–707
Jackson C, Ofoefule S (2011) Use of xanthan gum and ethylcellulose in formulation of metronidazole for colon delivery. J Chem Pharm Res 3:11–20
Jagdale SC, Pawar CR (2014) Application of design of experiment for polyox and xanthan gum coated floating pulsatile delivery of sumatriptan succinate in migraine treatment. Biomed Res Int 2014:1–11. https://doi.org/10.1155/2014/547212
Jamshidian M, Savary G, Grisel M, Picard C (2014) Stretching properties of xanthan and hydroxypropyl guar in aqueous solutions and in cosmetic emulsions. Carbohydr Polym 112:334–341. https://doi.org/10.1016/j.carbpol.2014
Jianlon W, Yi Q (1999) Microbial degradation of 4-chlorophenol by microorganisms entrapped in carrageenan-chitosan gels. Chemosphere 38:3109–3117. https://doi.org/10.1016/S0045-6535(98)00516-5
Kandra P, Challa MM, Chinthala R (2018) Biopolymers for food design: consumer-friendly natural ingredients. In: Alexandru MG, Alina MH (eds) Handbook of food bioengineering, biopolymers for food design, 3rd edn. Academic, Cambridge, pp 1–32
Katzbauer B (1998) Properties and applications of xanthan gum. Polym Degrad Stab 59 (1–3):81–84

Kaur M, Sandhu KS, Arora A, Sharma A (2015) Gluten free biscuits prepared from buckwheat flour by incorporation of various gums: physicochemical and sensory properties. LWT Food Sci Technol 62(1):628–632

Khalil M, Jan BM (2012) Herschel-Bulkley rheological parameters of a novel environmentally friendly lightweight biopolymer drilling fluid from Xanthan gum and starch. J Appl Polym Sci 124:595–606

Koocheki A, Ghandi A, Razavi S, Mortazavi SA, Vasiljevic T (2009) The rheological properties of ketchup as a function of different hydrocolloids and temperature. Int J Food Sci Technol 44:596–602

Kuo SM, Chang SJ, Wang HY, Tang SC (2014) Evaluation of the ability of xanthan gum/gellan gum/hyaluronan hydrogel membranes to prevent the adhesion of postrepaired tendons. Carbohydr Polym 114:230–237. https://doi.org/10.1016/j.carbpol.2014.07.049

Kuppuswami GM (2014) Production of xanthan gum. In: Encyclopedia of Food Microbiology, vol. 1. https://doi.org/10.1016/B978-0-12-384730-0.00110-5

Laffleur F, Michalek M (2017) Modified xanthan gum for buccal deliveryda promising approach in treating sialorrhea. Int J Biol Macromol 102:1250–1256

Liu Z, Jiao Y, Wang Y, Zhou C, Zhang Z (2008) Polysaccharides-based nanoparticles as drug delivery systems. Adv Drug Deliv Rev 60:1650–1662

Lopes BM, Lessa VL, Silva BM, La Cerda LG (2015) Xanthan gum: properties, production conditions, quality and economic perspective. J Food Nutr Res 54(3):185–194

Lu H, Guo L, Zhang L et al (2019) Study on quality characteristics of cassava flour and cassava flour short biscuits. Food Sci Nutr 8:521–533

Manca ML, Manconi M, Valenti D, Lai F, Loy G, Matricardi P (2011) Liposomes coated with chitosan—xanthan gum (chitosomes) as potential carriers for pulmonary delivery of rifampicin. J Pharm Sci 101(2):566–575

Michel G, Helbert W, Kahn R, Dideberg O (2003) The structural bases of the processive degradation of ι-carrageenan, a main cell wall polysaccharide of red algae. J Mol Biol 334:421–433. https://doi.org/10.1016/j.jmb.2003.09.056

Milas M, Rinaudo M, Knipper M, Schuppiser JL (1990) Macromolecules 23:2506

Mittal H, Parashar V, Mishra SB, Mishra AK (2014) Fe3O4 MNPs and gum xanthan based hydrogels nanocomposites for the efficient capture of malachite green from aqueous solution. Chem Eng J 255:471

Mohammadi M, Sadeghnia N, Azizi MH, Neyestani TR, Mortazavian AM (2014) Development of gluten-free flat bread using hydrocolloids: xanthan and CMC. J Ind Eng Chem 20(4):1812–1818

Mohanraj VJ, Chen Y (2006) Nanoparticles—a review. Trop J Pharm Res 5:561–573

Monthe OC, Grosmaire L, Nguimbou RM, Dahdouh L, Ricci J, Tran T, Ndjouenkeu R (2019) Rheological and textural properties of gluten-free doughs and breads based on fermented cassava, sweet potato and sorghum mixed flours. LWT Food Sci Technol 101:575–582

Murad HA, Abo-Elkhair AG, Azzaz HH (2019) Production of xanthan gum from nontraditional substrates with perspective of the unique properties and wide industrial applications. JSMC Microbiol 1:6

Nunes DG, da Silva APM, Cajaiba J, Perez-Gramatges A, Lachter ER (2014) Nascimento RSVJ. Appl Polym Sci 131:41085

Osilesi O, Trout DL, Glover EE, Harper SM, Koh ET, Behall KM, O'Dorisio TM, Tartt J (1985) Use of xanthan gum in dietary management of diabetes mellitus. Am J Clin Nutr 42(4):597–603

Palaniraj A, Jayaraman V (2011) Production, recovery and applications of xanthan gum by *Xanthomonas campestris*. J Food Eng 106:1–12

Parente ME, Ochoa Andrade A, Ares G, Russo F, Jimenez-Kairuz A (2015) Bioadhesive hydrogels for cosmetic applications. Int J Cosmet Sci 37(5):511–518

Patel VF, Patel NM (2007) Statistical evaluation of influence of xanthan gum and guar gum blends on dipyridamole release from floating matrix tablets. Drug Dev Ind Pharm 33:327–334

Pelletier E, Viebke C, Meadows J, Williams PA (2001) A rheological study of the order–disorder conformational transition of xanthan gum. Biopolymers 59:339–346

Petri DFS (2015) Xanthan gum: a versatile biopolymer for biomedical and technological applications. J Appl Polym Sci 132:1–13

Pooja D, Panyaram S, Kulhari H, Rachamalla SS, Sistla R (2014) Xanthan gum stabilized gold nanoparticles: characterization, biocompatibility, stability and cytotoxicity. Carbohydr Polym 110:1–9

Preichardt LD, Vendruscolo CT, Gularte MA, Moreira AD (2011) The role of xanthan gum in the quality of gluten free cakes: improved bakery products for coeliac patients. Int J Food Sci Technol 46(12):2591–2597

Psomas SK, Liakopoulou-Kyriakides M, Kyriakidis DA (2007) Optimization study of xanthan gum production using response surface methodology. Biochem Eng J 35:273–280. https://doi.org/10.1016/j.bej.2007.01.036

Ramasamy T, Kandhasami UDS, Ruttala H, Shanmugam S (2011) Formulation and evaluation of xanthan gum based aceclofenac tablets for colon targeted drug delivery. Braz J Pharm Sci 47:2

Razavi M, Nyamathulla S, Karimian H (2014) Hydrogel polysaccharides of tamarind and xanthan to formulate hydrodynamically balanced matrix tablets of famotidine. Molecules 19:13909–13931. https://doi.org/10.3390/molecules1909.13909

Reiss MJ, Han YP, Garcia E, Goldberg M, Yu H, Garner WL (2010) Matrix metalloproteinase-9 delays wound healing in a murine wound model. Surgery 147(2):295–302

Roopa G, Bhat RS, Dakshina MS (2015) Formulation and evaluation of an antacid and anti-ulcer suspension containing herbal drugs. Biomed Pharmacol J 3(1):01–06

Rosalam S, England R (2006) Review of xanthan gum production from unmodified starches by *Xanthomonas camprestris* sp. Enzym Microb Technol 39:197–207. https://doi.org/10.1016/j.enzmictec.2005.10.019

Rosalam S, Krishnaiah D, Bono A (2008) Cell free xanthan gum production using continuous recycled packed fibrous-bed bioreactor-membrane. Malays J Microbiol 4:1–5

Saharudin SH, Ahmad Z, Basri M (2016) Role of xanthan gum on physicochemical and rheological properties of rice bran oil emulsion. Int Food Res J 23(4):1361–1366

Sahil K, Akanksha M, Premjeet S, Bilandi A, Kapoor B (2011) Microsphere: a review. Int J Res Pharm Chem 1:1184–1198

Samia O, Hanan R, Kamal ET (2012) Carbamazepine mucoadhesive nanoemulgel (MNEG) as brain targeting delivery system via the olfactory mucosa. Drug Deliv 19(1):58–67

Sanderson GR (1982) The interactions of xantham gum in food systems. Prog Food Nutr Sci 6:77–87

Sandolo C, Coviello T, Matricardi P, Alhaique F (2007) Characterization of polysaccharide hydrogels for modified drug delivery. Eur Biophys J 36(7):693–700

Santos H, Veiga F, Pina ME, Sousa JJ (2005) Compaction, compression and drug release properties of diclofenac sodium and ibuprofen pellets comprising xanthan gum as a sustained release agent. Int J Pharm 295:15–27

Santoshi AK, Venkatesham M, Ayodhya D, Veerabhadram G (2015) Green synthesis, characterization and catalytic activity of palladium nanoparticles by xanthan gum. Appl Nanosci 5 (3):315–320

Saudagar RB, Badhe KP (2016) Development and evaluation of ph dependent in situ nasal gel of loratadine. Eur J Biomed Pharm Sci 1(3):233–238

Sharma BR, Naresh L, Dhuldhoya NC, Merchant SU, Merchant UC (2006) Xanthan gum—a boon to food industry. Food Promotion Chronicle 1(5):27–30

Sharma A, Gautam S, Wadhawan S (2014) Xanthomonas. In: Batt CA, Tortorello M (eds) Encyclopedia of food microbiology, 2nd edn. Academic, Amsterdam, pp 811–817

Shiledar RR, Tagalpallewar AA, Kokare CR (2014) Formulation and in vitro evaluation of xanthan gum-based bilayered mucoadhesive buccal patches of zolmitriptan. Carbohydr Polym 101:1234–1242

Shinde UA, Kanojiya SS (2014) Serratiopeptidase niosomal gel with potential in topical delivery. J Pharm 2014:1–9

Shu CH, Yang ST (1990) Effects of temperature on cell growth and xanthan production in batch cultures of Xanthomonas campestris. Biotechnol Bioeng 35:454–468. https://doi.org/10.1002/bit.260350503

Singhvi G, Hans N, Shiva N, Kumar Dubey S (2019) Xanthan gum in drug delivery applications. In: Natural polysaccharides in drug delivery and biomedical applications, pp 121–144. https://doi.org/10.1016/b978-0-12-817055-7.00005-4

Smith JH, Page GW (1982) Recovery of microbial polysaccharides. J Chem Technol Biotechnol 32:119–129. https://doi.org/10.1002/jctb.5030320116

Sworn G (2009) Xanthan gum. In: Phillips GO, Williams PA (eds) Handbook of hydrocolloids, 2nd edn. Woodhead Publishing, Cambridge, pp 186–203

Tiwari A, Kumar S (2009) Natural polymer in colon targeting. Int J Pharm Clin Res 1:43–46

Vendruscolo C T, Moreira AS, Souza AS, Zambiazi R, Scamparini, ARP (2000) Heteropolysaccharides produced by Xanthomonas campestris pv pruni. vol 24, pp 187–191. https://doi.org/10.1016/B978-044450178-3/50022-6

Vendruscolo C, Andreazza I, Ganter J, Ferrero C, Bresolin T (2005) Xanthan and galactomannan (from M. Scabrella) matrix tablets for oral controlled delivery of theophylline. Int J Pharm 296 (1–2):1–11

Verhoeven E, Vervaet C, Remon JP (2006) Xanthan gum to tailor drug release of sustained-release ethylcellulose mini-matrices prepared via hot-melt extrusion: in vitro and in vivo evaluation. Eur J Pharm Biopharm 63:320–330

Vishakha K, Kishor B, Sudha R (2012) Natural polymers—a comprehensive review. Int J Res Pharm Biomed Sci 3:1597–1613

Wang F, Wang YJ, Sun Z (2002) Conformational role of xanthan in its interaction with locust bean gum. J Food Sci 67(7):2609–2614

Williams P, Clegg S, Day D, Phillips G, Nishinari K (1991) Mixed gels formed with konjac mannan and xanthan gum. In: Dickinson E (ed) Food polymers, gels and colloids. Royal Society of Chemistry, Cambridge, pp 339–348

Xia S, Zhang L, Davletshin A, Li Z, You J, Tan S (2020) Application of polysaccharide biopolymer in petroleum recovery. Polymers 12:1860. https://doi.org/10.3390/polym12091860

Xu W, Jin W, Lin L et al (2014) Green synthesis of xanthan conformation-based silver nanoparticles: antibacterial and catalytic application. Carbohydr Polym 101:961–967

Xue D, Sethi R (2012) Viscoelastic gels of guar and xanthan gum mixtures provide long-term stabilization of iron micro and nanoparticles. J Nanopart Res 14(11):1–14

Zhou W, Therdthai N, Hui YH (2014) Bakery products science and technology. Wiley, West Sussex, pp 1–16

Biosynthesis and Characterization of Poly-(3)-hydroxyalkanoic Acid by *Bacillus megaterium* SF4 Using Different Carbohydrates

5

Temitope O. Fadipe, Nazia Jamil, and Adekunle K. Lawal

Abstract

The non-degradability of synthetic polymers of plastics has been associated with ecosystem imbalance. This requires the development of eco-friendly biopolymers. Poly-(3)-hydroxyalkanoic acids (PHAs) are microbially derived biopolymers produced by bacteria and have physical and chemical properties analogous to synthetic polymers thus making them candidates for bioplastic production. We share results from the use of isolate SF4 with potential for sustainable PHA production. The isolate was screened on PHA detection agar supplemented Nile red and Nile blue A and with Sudan Black B staining. It was biochemically characterized and identified by 16SrRNA sequencing. Production of PHA was done in a nitrogen-limiting medium containing 2% carbon source over 96 h. Extraction of PHA was by sodium hypochlorite/chloroform method and PHA was characterized by FT-IR and GC-MS. The PHA synthase genes, *PhaC* and *PhaR*, of the isolate were also partially amplified and sequenced. Orange and yellow fluorescence of PHA were observed for all carbon sources used. Blue-black intracellular inclusions of PHA were also observed after Sudan Black B staining. The isolate was identified as *Bacillus megaterium* SF4 sharing closest homology with *B. megaterium* NBRC15308 = ATCC 14581. Growth

T. O. Fadipe (✉)
Biotechnology Department, Federal Institute of Industrial Research, Lagos, Nigeria

Department of Microbiology and Molecular Genetics, University of the Punjab, Quaid-e-Azam Campus, Lahore, Pakistan

N. Jamil
Department of Microbiology and Molecular Genetics, University of the Punjab, Quaid-e-Azam Campus, Lahore, Pakistan

A. K. Lawal
Biotechnology Department, Federal Institute of Industrial Research, Lagos, Nigeria

A. Vaishnav, D. K. Choudhary (eds.), *Microbial Polymers*,
https://doi.org/10.1007/978-981-16-0045-6_5

curves showed highest biomass accumulation in 2% starch and PHA production of 26.53 ± 0.91%. FT-IR spectra showed peaks corresponding to the presence of P3HB and P3HB3HV. Sequences of 16SrRNA, *PhaC,* and *PhaR* have been accessioned in the NCBI-GenBank as KY855376.1, KY855378.1, and MF947449.1, respectively. The results show that *B. megaterium* SF4 has great capacity for PHA production in varied carbohydrates.

Keywords

Poly-(3)-hydroxyalkanoic acid · Biopolymers · Bioplastics · *Bacillus megaterium* SF4 · Sustainable

5.1 Introduction

5.1.1 Plastics

Plastics derived from petrochemicals have found broad applications in day-to-day human activities and in the industries because of their favourable mechanical and thermal properties (Ojumu et al. 2004). Their durability, flexibility, lightness, and resistance to degradation makes them more desirable than glass and wood (Aarthi and Ramana 2011). They are made from synthetic polymers such as polypropylene, polyvinylchloride, and polyethylene (Pathak et al. 2014). The annual production of petroleum-based plastics was estimated at over 300 million tons with over 150 million tons of synthetic plastics and plastic-derived materials utilized globally on an annual basis.

The production of synthetic plastics is expected to increase to an estimated amount of 810 million tons by 2050 (Muhammadi et al. 2015; Kumar et al. 2020). These plastics are recalcitrant and resistant to microbial degradation and are reported to accumulate in the environment at the rate of 28 million tons yearly (Muhammadi et al. 2015). Their accumulation and persistence in the environment as well as very slow degradation have been associated with serious environmental pollution of land and water bodies. It has been reported that up to ten million tons of synthetic plastics find their way into the oceans yearly causing detrimental effects to marine life forms and disruption of the oceanic environment (Kumar et al. 2020). In addition, several health problems such as cancer and nervous system-related diseases have been associated with toxic by-products of synthetic plastics (Pathak et al. 2014; Muhammadi et al. 2015).

In order to reduce these negative environmental effects and overcome these problems, synthetic polymers can be replaced with biopolymers such as polylactic acid, polysaccharides polymers, aliphatic polyesters, and poly-(3)-hydroxyalkanoic acid which have similar mechanical, thermal, and chemical properties as synthetic polymers (Muhammadi et al. 2015; Kumar et al. 2020). These biopolymers are becoming increasingly popular because they are biodegradable, environmentally friendly, and have widespread applications in the industries. In addition, the global bioplastic market is expected to reach a production volume of 2.44 million tons by 2022 (Kumar et al. 2020)

5.1.2 Poly-(3)-hydroxyalkanoic Acid: Discovery, Structure, and Classification

Poly-(3)-hydroxyalkanoic acids (PHAs) are a group of polymers synthesized as intracellular carbon and energy reserves by archaea, several bacterial species (Gram positive and negative), fungi, and plants (McCool and Cannon 2001; Kung et al. 2007; Mahmoud 2012; Muhammadi et al. 2015; Kumar et al. 2020). They can be accumulated to levels as high as 90–97% of dry cell weight in these organisms (Muhammadi et al. 2015). They are biodegradable and biocompatible and can be sustainably produced (Kung et al. 2007; Alkotaini et al. 2015). When disposed in the natural environment, they can be degraded by indigenous microbes within 4–6 weeks (Ali and Jamil 2016). The production of the first PHA, poly-(3)-hydroxybutyric acid (PHB), as intracellular inclusions in *Bacillus megaterium* was described by Maurice Lemoigne and co-workers in 1926 (Verlinden et al. 2007; Muhammadi et al. 2015; Kumar et al. 2020).

PHAs have been extensively researched and commercialized by some companies in countries such as the USA (Procter & Gamble), Germany (Biomer Inc.), China (Lianyi Biotech), and Brazil (PHB Industrial Company) (Suchada 2010) though its production cost is higher compared with other synthetic polymers. As a result, cheap carbon substrates are being exploited for PHAs production to reduce overall bioplastics production cost by up to 60% (Du et al. 2012; Naheed and Jamil 2014).

A typical PHA granule is coated on the surface with a layer of phospholipid and proteins called phasins. These proteins are the major compounds in the interface of the PHA molecule. The layer of phasin proteins stabilizes and prevents coalescence of PHA granules in the cytosol. The phasins influence the number and size of PHA granule (Verlinden et al. 2007; Kabilan et al. 2012). A PHA molecule consists of 600–35,000 R(−)-3-hydroxyalkanoic acid monomers ranging from 3 to more than 16 carbon atoms. These monomers could have extra side chains of saturated, unsaturated, straight, or branched compounds containing aromatic or aliphatic side groups (Tan et al. 2014; Muhammadi et al. 2015; Kumar et al. 2020). During the formation of the PHA polymer, the carboxyl group of one monomer forms an ester bond with the hydroxyl group of the adjourning monomer.

There are over 150 types of hydroxyalkanoic acids of PHAs are known, and this number continues to increase with various research progress. The molecular weight of PHA ranges from 2×10^5 and 3×10^6 Da which depends on the number of constituent monomeric unit, bacterial strain used, and growth conditions (Muhammadi et al. 2015). Typically. PHAs are classified based on the number of carbon atoms in the side chain length of a PHA monomer (Kabilan et al. 2012). Short chain length PHA (SCL-PHA) contains 3–5 carbon in their side chain (Kumar et al. 2020). Poly-(3)-hydroxybutyric acid (PHB) and poly-(3)-hydroxyvaleric acid (PHV) belong to this class (Verlinden et al. 2007). Medium chain length PHA (MCL-PHA) contains 6–16 carbon atoms and long chain length PHA (LCL-PHA) contains more than 16 carbon atoms) in their side chains. These classes of PHA are accumulated by a wide variety of bacteria (Kung et al. 2007). In addition, some bacteria such as Aeromonas strains are accumulate copolymers of poly-(3)-hydroxybutyric and hexanoic acid (Muhammadi et al. 2015). Thus, PHAs have found broad range application properties because of the variations in their physical, chemical, and mechanical properties (Kabilan et al. 2012).

5.1.3 Biosynthesis of Poly-(3)-hydroxyalkanoic Acid

Over 90 microbial genera including 300 bacteria species, isolated from different ecosystems, that are capable of accumulating PHAs granules intracellularly have been studied (Aneesh et al. 2017; Kumar et al. 2020). In addition, some PHA-producing bacteria are constructed as genetic recombinants (Du et al. 2012). The production of PHAs from bacterial systems is preferred to production from other living systems including fungi and plants because of their high accumulation ability (Kumar et al. 2020). Desirable qualities in any PHA-synthesizing bacteria are high growth rate, ability to use inexpensive carbon substrates, and high accumulation capacity (Naheed and Jamil 2014). Some known factors that affect PHA composition and yield are bacterial strain type, substrate (type and concentration), and growth parameters used during production. Thus to produce PHA cost-effectively, the use of an efficient bacterial strain is critical and has been at the centre of several research efforts (Valappil et al. 2007; Babruwad et al. 2015). The properties of PHAs including elasticity, biodegradability, and renewability strongly depend on the pathway of biosynthesis, monomer composition, and chemical structure (Kumar et al. 2020).

Broadly, bacteria are divided into two major groups based on culture conditions required for PHA synthesis (Muhammadi et al. 2015). The first group of bacteria requires an essential nutrient limitation such as phosphorous, nitrogen, sulphur, or magnesium in the presence of excess carbon for PHA synthesis. Examples of bacteria in this group are *Ralstonia eutropha*, *Pseudomonas oleovorans,* and *Protomonas extorquens* and members of the *Bacillus* genera. The second group of bacteria does not require a nutrient limiting condition for PHA synthesis. The PHA polymer is accumulated during their growth phase. Examples of bacteria in this category are *Alcaligenes latus*, a mutant strain of *Azotobacter vinelandii* and recombinant *Escherichia coli* (Muhammadi et al. 2015).

The major pathway of PHA biosynthesis in bacteria is catalysed by three enzymes. In the first reaction, two acetyl coenzyme A (CoA) molecules are condensed into acetoacetyl-CoA by β-ketothiolase encoded by *phaA* gene. The second reaction involves the reduction of acetoacetyl-CoA to (R)-3-hydroxybutyryl-CoA by the NADPH-dependent enzyme, acetoacetyl-CoA reductase/dehydrogenase encoded by *phaB* gene. In the last step, (R)-3-hydroxybutyryl-CoA monomers are used as precursors for the polymerization of poly-(3)-hydroxybutyric acid (PHB) by a PHA polymerase encoded by *phaC* gene in an esterification reaction. The genes involved in this major pathway are assembled into the phaCAB operon (Sheu et al. 2000; Muhammadi et al. 2015; Kumar et al. 2020). In addition, some can also synthesize MCL-PHA through de novo fatty acid biosynthesis and β-oxidation pathways. This has been reported in members of the *Pseudomonas* genera (Muhammadi et al. 2015; Kumar et al. 2020). Over 20 other PHA biosynthetic operons have been cloned and analyzed in bacteria till date (Sheu et al. 2000).

Based on the type of PHA polymerase, bacteria are grouped into four classes, namely class I made by *Ralstonia eutropha*, class II by *Pseudomonas* species, class III by *Allochromatium vinosum,* and class IV by *Bacillus* species (Rehm and Steinbüchel 1999; Ali and Jamil 2016). Class IV PHA polymerases are encoded by *phaC* and *phaR* genes which are both important for polymerizing activity in vivo and in vitro (McCool and Cannon 2001; Nayak et al. 2013).

Several bacterial species belonging to the *Bacillus* genera including strains of *Bacillus megaterium* are known to utilize varieties of carbon sources for growth and PHA production as reported by several authors. They are able to use pure carbon substrates such as glucose, lactose, fructose, sucrose, (Shamala et al. 2003; Krueger et al. 2012; Okwuobi and Ogunjobi 2013; Medjeber et al. 2015), pure glycerol, waste glycerol (from biodiesel production), castor oil, waste frying oil, whey (Cardozo et al. 2016), sugarcane molasses, corn steep liquor (Gouda et al. 2001), beet molasses (Medjeber et al. 2015), and pure and waste cassava starch (Krueger et al. 2012; Medjeber et al. 2015; Aneesh et al. 2016).

This study investigates the potential of a novel *Bacillus megaterium* strain isolated from cassava dumpsite for PHA sustainable production.

5.2 Materials and Methods

5.2.1 Detection of PHA Production by Isolate SF4

Bacterial isolate SF4, previously isolated from sugar farm site soils by Salaam et al. (2016), was used in this study. Poly-(3)-hydroxyalkanoic acid (PHA) production by isolate SF4 was detected following the viable colony staining protocol in a nitrogen-limiting and carbon-rich medium (2% of each carbon source: glucose, sugarcane molasses, glycerol, and starch). Isolate SF4 was first refreshed on nutrient agar for 24 h and pure colonies were cultured on PHA detection agar (PDA) containing potassium dihydrogen phosphate—1.33 g, ammonium sulfate—0.2 g, citric acid—0.17 g, magnesium sulfate heptahydrate—0.12 g, agar—1.5 g, 1 mL trace elements solution supplemented with 0.5 μg/mL of Nile red or Nile blue A stains in a final volume of 100 mL. The petri-plates were incubated at 37 °C for 24 h. For fluorescence detection of PHA, 24 h petri-plates were observed under an ultraviolet transilluminator and documented (Spiekermann et al. 1999; Fadipe et al. 2019a). Further screening for detection of intracellular PHA inclusions was done by Sudan Black B staining as described by Lee and Choi (1999). This was done by preparing heat-fixed smears of the isolate grown on PDA followed by staining with 0.3%

Sudan Black B solution. Microscope slides were observed for blue-black intracellular inclusions under a compound light microscope and photomicrographs were taken.

5.2.2 Characterization of Isolate SF4

Isolate SF4 was streaked on freshly prepared nutrient agar plates and incubated for 24 h after which its colonial morphology was documented. Gram staining was done using heat-fixed smear of the isolate and its cellular morphology was observed under compound light microscope. Biochemical profile of the isolate was determined using the analytical profile index kit (BioMerieux). Other tests including catalase, oxidase, and starch hydrolysis were also done. Genomic DNA was extracted from 16–18 h bacterial broth culture according to the protocol in the Exgene™ Cell SV kit (GeneAll). This was followed by PCR amplification of the 16SrRNA hypervariable region and sequencing at first Base (Singapore). Sequence homology search was done using the NCBI-nucleotide BLAST tool and alignment was done with ClustalW in MEGA7. Phylogenetic analysis was done using Tamura-Nei algorithm of the neighbour-joining phylogeny method considering 1000 bootstrap replications in MEGA 7 (Fadipe et al. 2019b).

5.2.3 Assessment of Poly-(3)-hydroxyalkanoic Acid Production

This was done following the method described by Fadipe et al. (2019a). Isolate SF4 was grown overnight in a 250 mL flask containing 50 mL of nutrient broth at 37 °C and 120 rpm. Seed culture was harvested by centrifugation at 6000 rpm for 10 min and washed in PHA broth medium (PBM) containing in 100 mL, potassium dihydrogen phosphate—0.33 g, ammonium sulfate—0.2 g, citric acid—0.17 g, magnesium sulfate heptahydrate—0.12 g, and 1 mL trace elements solution. The bacteria pellet was dissolved in 5 mL of PBM and inoculated into a 500 mL conical flask containing 300 mL of PBM (optical density 0.05–0.1 at 600 nm) supplemented with 2% of each carbon source (glucose, sugarcane molasses, glycerol, and starch). The flasks were incubated for 96 h at 37 °C with shaking at 120 rpm. Sample (50 mL) were drawn every 24 h for PHA extraction. Bacteria-PHA biomass were collected by centrifugation at 6500 rpm for 10 min, and the resulting pellet was lyophilized to estimate the dry cell weight (DCW). The growth curve of isolate SF4 over the 96 h incubation period was monitored by measuring the optical density of the media at 600 nm. Each experiment was set up in duplicates conducted under aseptic conditions. Descriptive statistics and Pearson correlation coefficient were done using Microsoft Excel 2016 and results are expressed as arithmetic mean $\pm$ STD.

5.2.4 Extraction of PHA

Poly-(3)-hydroxyalkanoic acid was extracted following modifications to the methods of Ali and Jamil (2016) and Berger et al. (1989). Freshly prepared 5.25% sodium hypochlorite (NaClO), pH 10 was added to the lyophilized bacterial-PHA biomass and vortexed vigorously to obtain an evenly dispersed solution followed by incubation at 37 °C and 120 rpm for 40 min. Labelled conical flasks containing chloroform was prepared and the bacterial-PHA-NaClO mixture was added onto it followed by incubation at 37 °C with shaking at 120 rpm for $1\frac{2}{3}$ h. Using a glass separating funnel, the resulting the mixture was separated to collect the less dense phase containing PHA and chloroform. This was followed by concentrating the PHA-chloroform mixture in a Heidolph 16-G1B diagonal condenser rotary evaporator and drying of the concentrated mixture in a fume cupboard at room temperature. The weight of extracted PHA was estimated and percentage PHA of the DCW was calculated.

5.2.5 Characterization of PHA

Extracted PHA was characterized by Fourier transform infrared spectroscopy (FT-IR) and gas chromatography with mass spectrometry (GC-MS). Ten milligrams (10 mg) of extracted PHA was treated with KBr and subjected to FT-IR in an Agilent Cary 660 FT-IR spectrophotometer. The spectral range of 650–4000 cm^{-1} and resolution of 2 cm^{-1} and 32–64 scans was averaged for PHA functional group detection. The FT-IR spectra were acquired according to the KBr disc protocol (Ali and Jamil 2016; Fadipe et al. 2019a, b).

The method of Ali and Jamil (2014) was used to determine the chemical composition of extracted PHA by GC-MS. Chloroform (1 mL), methanol (0.85 mL), and concentrated H_2SO_4 (0.15 mL) were added to 8 mg of extracted PHA followed by heating at 100 °C for 140 min. One milliliter (1 mL) of the methanolysed PHA sample was injected into a Shimadzu gas chromatograph-mass spectrometer QP2010 in split injection mode (70:30) with nitrogen as the carrier gas at a flow rate of 3 mL/min. The temperature of the oven was set at 60 °C for 2 min at the start, ramped at the rate of 5 °C per minute to 260 °C and held for 15 min. The injector temperature was set at 260 °C and column flow rate was 0.57 cm/s. A GC-MS QP2010 with D1 mass spectrometry system with ion source temperature set at 200 C was used to determine the monomeric composition of methanolysed PHA sample.

5.2.6 Amplification of PHA Synthases C and R Genes in Isolate SF4

The PHA synthase genes, *PhaC* and *PhaR* of isolate SF4 was amplified by end-point polymerase chain reaction (PCR) and amplicons detected on 2% ethidium bromide stained agarose gel in 1× TAE buffer. For the *PhaC* gene, BmphaC015F—

CGTGCAAGAGTGGGAAAAAT and BmphaC931R—TCGCAATATGATCAC GGCTA primer pair with expected amplicon size of 900 bp were used in a gradient PCR reaction (Nayak et al. 2013; Fadipe et al. 2019b). For the detection of *PhaR* gene, PhaR4F—AAACCGAATCTTACTGGGAAG and PhaR4bR—GAAGCTG CTCACCTTGAGTTTTAAT primer pair were used (Fadipe et al. 2019a, b). The PCR cocktail, at a final volume of 50 μL, contained 25 μL 2× Dream Taq Green PCR master mix (Thermo Scientific), 0.5 μL of 100 μM forward and reverse primers, 19 μL nuclease-free H_2O, and 5 μL DNA template.

The PCR gradient cycling conditions for the detection of for *PhaC* gene was 95 °C for 5 min initial denaturation, 95 °C for 60 s denaturation, 55.61, 56.32, 57.28, and 58.32 °C for 60 s annealing, 72 °C for 2 min extension, 72 °C for 5 min final extension, and hold at 4 °C for 40 cycles in a Supercycler Trinity (Kyratec). The PCR cycling conditions for the detection of *PhaR* gene was 95 °C for 5 min initial denaturation, 95 °C for 60 s denaturation, 61.47 °C for 60 s annealing, 72 °C for 2 min extension, 72 °C for 5 min final extension, and hold at 4 °C for 40 cycles in the thermal cycler listed above. The PCR products were purified using the method detailed in the Expin™ Combo GP kit (GeneAll).

The 900 bp (*PhaC*) and 432 bp (*PhaR*) fragments were sequenced using the 3730xl DNA analyser (Applied Biosystems) at Macrogen (Korea). Sequence homology search was done using the NCBI-nucleotide BLAST tool and alignment was done with ClustalW in MEGA7. Phylogenetic relationships with other closely related species were inferred using the predicted amino acids of *PhaC* and *PhaR* according to the Poisson model with the neighbour-joining method in MEGA7 considering 1000 bootstrap replications (Fadipe et al. 2019b).

5.3 Results and Discussion

5.3.1 Detection of PHA Production by Isolate SF4

Poly-(3)-hydroxyalkanoic acid production was observed as orange, blue, and yellow fluorescence with Nile blue A and Nile red stains on PHA detection agar which signifies PHA accumulation (Fig. 5.1). Additionally, orange fluorescence is indicative of the accumulation small chain length PHA, specifically polyhydroxybutyrate (PHB) (Spiekermann et al. 1999; Fadipe et al. 2019a). Sudan Black B stained heat-fixed smears of isolate SF4 showed the presence of blue-black intracellular inclusions (Fig. 5.2) for the four carbohydrates used in this study (Lee and Choi 1999). This further confirms PHA accumulation and capacity of isolate SF4 to utilize these carbohydrates for PHA production (Tan et al. 2014; Ali and Jamil 2016; Salaam et al. 2016).

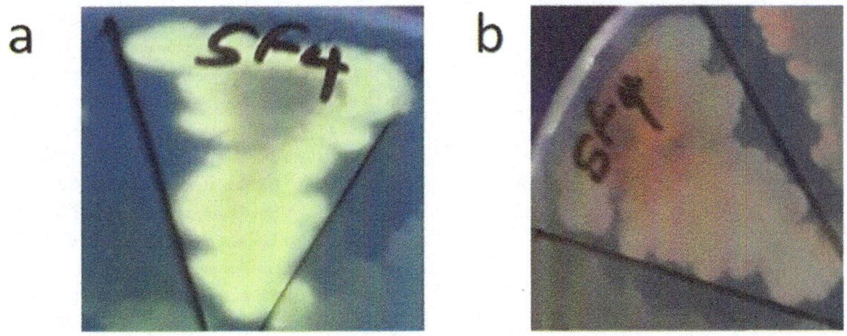

Fig. 5.1 Fluorescence detection of PHA in *B. megaterium* SF4 using sugarcane molasses as carbon source. (**a**)—Nile blue, (**b**)—Nile red

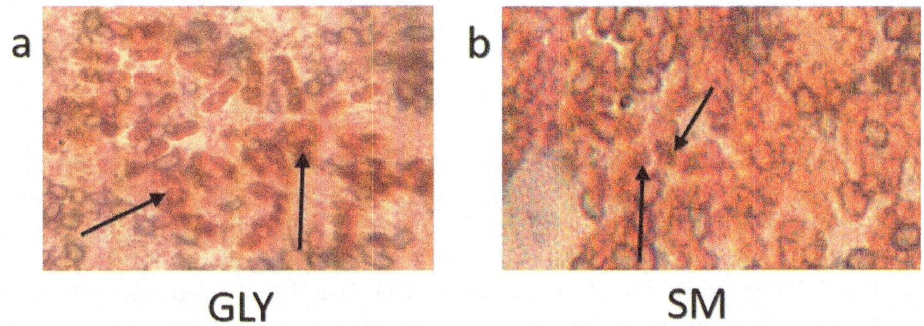

Fig. 5.2 Intracellular detection of PHA in *B. megaterium* SF4 using glycerol and sugarcane molasses as carbon sources. GLY (**a**)—glycerol, SM (**b**)—sugarcane molasses

5.3.2 Colonial, Morphological, Biochemical, and Molecular Characterization of Isolate SF4

Colonial characteristics of isolate SF4 on nutrient agar was observed after 24 h incubation. Isolate SF4 colonies are creamy, unpigmented, circular, motile, and non-slimy with convex elevation and entire edges (Medjeber et al. 2015; Yuli et al. 2017). It is a Gram positive, endospore forming, and rod-shaped bacteria with cells appearing in clusters, chains, and pairs (Medjeber et al. 2015; Yuli et al. 2017). It was positive for glucose, mannitol, and sucrose fermentation and negative for the fermentation of inositol, melibiose, amygdalin, arabinose, rhamnose, sorbitol, and mannose. Furthermore, it was positive for catalase, oxidase, gelatinase tests and negative for urease test. Isolate SF4 was positive citrate utilization and starch hydrolysis and negative for indole production and hydrogen sulfide production (Medjeber et al. 2015; Yuli et al. 2017).

Sequence homology with 16SrRNA nucleotide sequences on NCBI-nucleotide BLAST revealed that isolate SF4 shared closest evolutionary relationship with *Bacillus megaterium* NBRC15308 = ATCC 14581 at 99 % identity score; thus, it is *Bacillus megaterium* SF4. The 16SrRNA sequences have been submitted in the

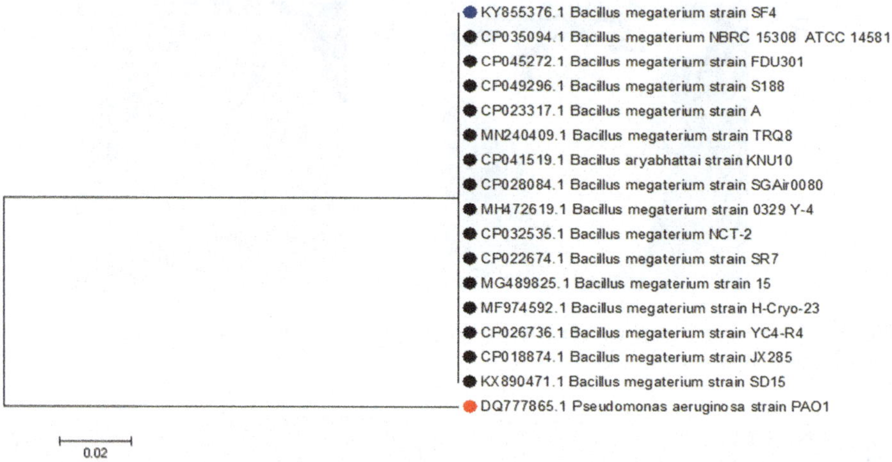

Fig. 5.3 Phylogenetic tree of *B. megaterium* SF4 16SrRNA nucleotide sequences

NCBI-GenBank with accession number KY855376.1. Figure 5.3 shows *Bacillus megaterium* SF4 clustering with *B. megaterium* NBRC15308 = ATCC 14581 (CP035094.1) in the phylogenetic tree below. The colonial, morphological, and biochemical characteristics of the isolate SF4 agrees with previous report by Yuli et al. (2017) expected for *B. megaterium*. The first documented bacterial-PHA producer, a *Bacillus megaterium* strain was identified by Lemogine in 1926 (Muhammadi et al. 2015; Kumar et al. 2020). In addition, several strains of *B. megaterium* have been documented as PHA producers (Nayak et al. 2013; Tan et al. 2014; Medjeber et al. 2015).

The tree was constructed using the neighbour-joining model with *Pseudomonas aeruginosa* (red) as an outgroup considering 1000 bootstrap replications in MEGA7.

5.3.3 Growth Dynamics of Isolate SF4 in Four Different Carbohydrates

Figure 5.4 shows the growth curve of *B. megaterium* SF4 in the four carbohydrates used. In summary, the results show that PHA production in *B. megaterium* SF4 is growth related because PHA accumulation begins with cell growth in the log phase in all carbohydrates used (Aneesh et al. 2017). *B. megaterium* SF4 achieved the best growth in 2% starch supplemented PHA broth medium (PBM). It grew exponentially and peaked at 26 h followed by another sharp peak at 30 h and a sharp decline at 32 h. Thereafter, it grew steadily over the 96-hour incubation period. This observation agrees with reports by Medjeber et al. (2015) where the best growth rate was achieved by *B. megaterium* L9 using 2% starch as carbon source. In 2% glucose supplemented PBM, its growth increased steadily till it peaked at 24 h then achieved another major peak at 48 h and continued to grow gradually over the

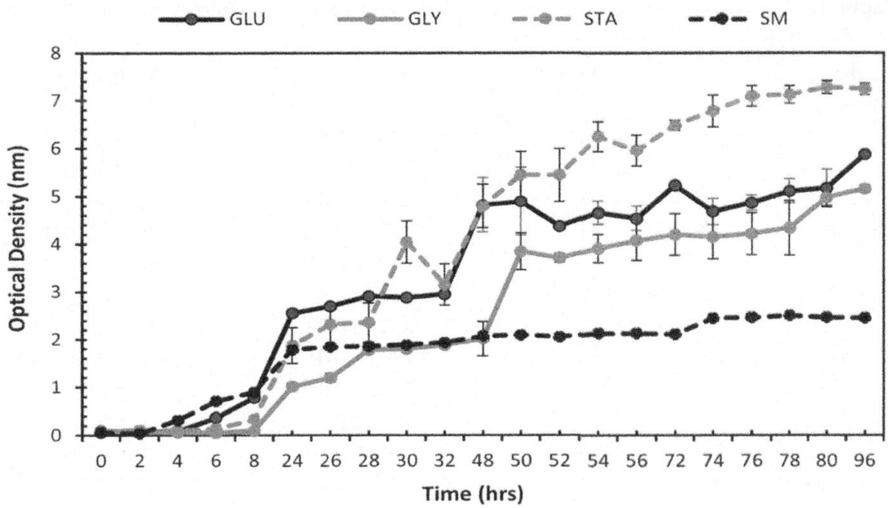

Fig. 5.4 Growth curve of *B. megaterium* SF4 in four carbohydrates as carbon sources. GLU—Glucose, GLY—Glycerol, SM—Sugarcane molasses, STA—Starch

96-hour period. In 2% glycerol supplemented PBM, its growth peaked at 28 h and another major peak at 50 h followed by continued increase throughout the experiment. In 2% sugarcane molasses supplemented PBM, its growth increased exponentially and peaked 24 h then grew steadily and remained constant.

5.3.4 Assessment of Dry Cell Weight and PHA Production in Four Different Carbohydrates

Table 5.1 shows extracted PHA and dry cell weight (DCW) in g/L and Fig. 5.5 shows extracted PHA in % DCW. We report that PHA production (% DCW) shows a weak-positive correlation with biomass accumulation (DCW in g/L) in glucose. However, there is a strong negative correlation between PHA production (% DCW) and biomass accumulation (DCW in g/L) when glycerol, starch, and sugarcane molasses were used as carbon sources.

Figure 5.5 shows the extracted PHA from lyophilized *B. megaterium* SF4 biomass expressed as a percentage of its dry cell weight. *Bacillus megaterium* SF4 produced highest PHA amount of 4.69 ± 0.17% in 2% glucose supplemented PBM at 72 h. We report a lesser yield compared to *B. megaterium* L9 which produced 50% PHA using glucose as carbon source after 48 h incubation at 30 °C (Medjeber et al. 2015). Similarly, another strain of B *megaterium* which produced 39.9% using glucose as sole carbon source (Gouda et al. 2001). In contrast, we report a better PHA yield of 0.28 ± 0.03 g/l and greater dry cell weight (DCW) of 6.13 ± 0.04 g/L at 48 h (Table 5.1) compared to the 0.0143 g/L PHA yield and 0.0203 g/L DCW reported by Okwuobi and Ogunjobi (2013) using a strain of *B. megaterium* and

Table 5.1 Mean dry cell weight and extracted poly-(3)-hydroxyalkanoic acid by *Bacillus megaterium* SF4

Carbon source	Time (h)	Avg. DCW (g/L)	Avg. PHA (g/L)
Glucose	24	3.07 ± 0.16	0.09 ± 0.01
	48	6.13 ± 0.04	0.28 ± 0.03
	72	3.83 ± 0.47	0.18 ± 0.03
	96	4.56 ± 0.45	0.17 ± 0.01
Starch	24	1.77 ± 0.10	0.47 ± 0.04
	48	2.82 ± 0.42	0.33 ± 0.04
	72	5.58 ± 0.40	0.24 ± 0.06
	96	6.12 ± 0.03	0.50 ± 0.06
Glycerol	24	1.77 ± 0.01	0.33 ± 0.01
	48	2.35 ± 0.35	0.31 ± 0.04
	72	3.70 ± 0.34	0.45 ± 0.07
	96	6.55 ± 0.18	0.34 ± 0.06
Sugarcane molasses	24	1.47 ± 0.18	0.15 ± 0.01
	48	3.10 ± 0.00	0.24 ± 0.03
	72	3.38 ± 0.03	0.27 ± 0.01
	96	4.10 ± 0.11	0.20 ± 0.03

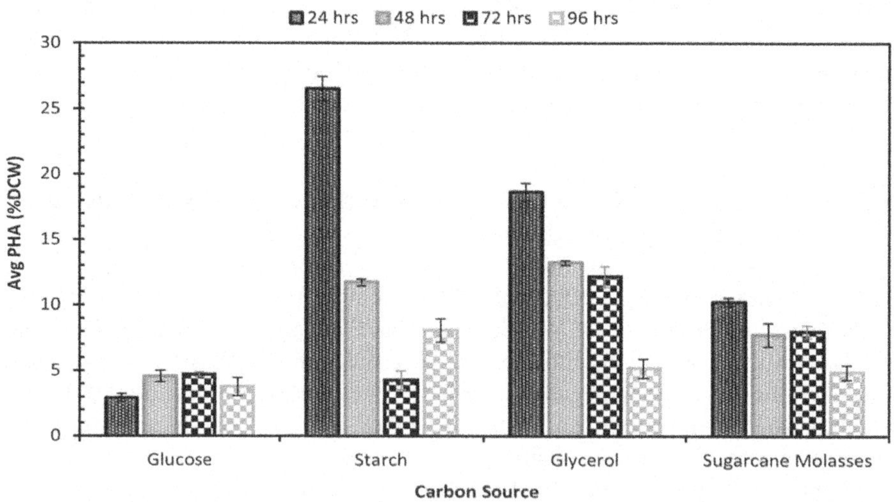

Fig. 5.5 PHA extracted produced by *B. megaterium* SF4 using four carbon sources

glucose as sole carbon source. In their study, the production medium was incubated 37 °C for 48 h. The lesser PHA yield from our work and Okwuobi and Ogunjobi (2013) may be due to partial digestion of the lyophilized bacterial biomass or degradation of PHA during extraction with sodium hypochlorite (NaClO). Nicolas et al. (2008) have suggested that NaClO can serious degradations of PHA and up to 50% decrease in yield. A pretreatment by heating or freezing before NaClO treatment has been shown to improve PHA yields by up to 98% (Nicolas et al. 2008). Another factor which may have contributed to lesser PHA yield may be the higher incubation temperature of 37 °C used in our study compared to 30 °C used by Gouda et al. (2001) and Medjeber et al. (2015). The possible effect of a lower incubation temperature on dry cell weight and PHA yield using *B. megaterium* SF4 are worthy of further investigations.

We report PHA amount of 26.53 ± 0.91% and 1.77 ± 0.10 g/l DCW in starch. This is lower than the yield of 73.46% PHA and 3.07 g/L DCW reported for *B. megaterium* PHB29 in starch at 48 h, pH 7.2, and optimum incubation temperature of 34 °C (Aneesh et al. 2016). In addition, Medjeber et al. (2015) reported a maximum yield of 39.9% PHA and 0.815 g/L DCW in 2% after 48 h of incubation at 30 °C with *B. megaterium* L9. Though we reported a higher amount of DCW compared to Medjeber et al. (2015), these discrepancies in DCW and PHA yield may be due to the higher incubation temperature of 37 °C used in our study. Aneesh et al. (2016) reported a decline in PHA yield and DCW when incubation temperature was increased to 37 °C and 40 °C with *B. megaterium* PHB29 in 2% soluble starch. Krueger et al. (2012) investigated the PHA production potential of four *B. megaterium* strains, namely LAMA073, LAMA095, LAMA262, and LAMA265 using hydrolysed cassava starch by-product at 20% (v/v) and 40% (v/v) at 48 h and 37 °C. *B. megaterium* LAMA095 produced the highest DCW of 4.48 ± 0.5 and 4.97 ± 0.9 g/L which translated to 23.88% and 29.78% PHA yield in 20% (v/v) and 40% (v/v) cassava starch, respectively. Their findings show that an increase in the starch concentration led to an increased DCW and PHA yield. In contrast, we recovered more PHA from a lesser DCW compared to the PHA yields and DCW reported in their work. These differences in yield may be due to different PHA extraction methods used in these studies.

The use of sugarcane molasses as low-cost carbon source for industrial PHA production is been investigated because it is known to contain vitamins and trace elements such as riboflavin thiamine, pyridoxine, and niacin which are important growth factors for microbes (Gouda et al. 2001). *Bacillus megaterium* SF4 produced the maximum amount of 10.22 ± 0.32% PHA (1.47 ± 0.18 g/L) at 24 h and DCW of 4.1 ± 0.11 g/L DCW at 96 h in 2% sugarcane molasses. Gouda et al. 2001 reported a PHA yield of 46.5% in 2% sugarcane molasses using a strain of *B. megaterium*. In addition, *B. megaterium* L9 produced 30% PHA in 2% sugarcane molasses (Medjeber et al. 2015). These yields are significantly higher than that from our study and may also be because of the lower incubation temperatures of 30 °C used for PHA production in these studies.

Bacillus megaterium SF4 produced the maximum amount of 0.33 ± 0.01 g/L PHA and 1.77 ± 0.01 g/L DCW at 24 h. Furthermore, a maximum amount of 6.55 ± 0.18 g/L DCW at 96 h in 2% glycerol. Studies by Cardozo et al. (2016) reported a biomass of 1.23 ± 0.05 g/L and a PHA yield of 2.80 g/L after 48 h for a *Bacillus megaterium* strain in a pH 7.0 production medium incubated at 30 °C. We report a higher biomass of 2.35 ± 0.35 g/L but a lower PHA yield of 0.31 ± 0.04 g/L at 48 h. The discrepancies observed may be due to the lower incubation temperature used by Cardozo et al. (2016), a slightly higher concentration of 5.25% NaClO used in our study compared to 5% used by Cardozo et al. (2016) which may have caused some PHA degradation during extraction and other differences in our method of extraction.

In general, we observed an increase in the number of spores in Sudan Black B stained cells as incubation time increased beyond 48 h for *B. megaterium* SF4. This suggests sporulation as a mechanism of survival under the limiting nutrient growth conditions thus leading to lesser PHA yield. A similar observation was described as a limitation when using spore-forming *Bacilli* for PHA production (Wu et al. 2001). However, sporulation can be prevented by maintaining a low pH in the medium during PHA production or adding up 90 mM α-picolinic acid to the PHA production medium (Akaraonye et al. 2012).

5.3.5 FT-IR Spectra Analysis of Extracted PHA

Figure 5.6 shows the FT-IR spectra of extracted PHA produced by *B. megaterium* SF4 using glucose (a), sugarcane molasses (b), glycerol (c), and starch as carbon sources (d). The four FT-IR spectra validate the presence of poly(3-hydroxybutyric) acid (Krueger et al. 2012; Gumel et al. 2012; Getachew and Woldesenbet 2016) and co-polymer of poly(3-hydroxybutyric) acid and poly3-hydroxyvaleric acid (Kumalaningsih et al. 2011; Shamala et al. 2009). Furthermore, they also reveal conformational changes in amorphous and crystalline phases of small chain length and medium chain length-PHAs in the extracted PHA (Gumel et al. 2012). The spectra show peaks at 2958, 2925, 2923, and 2916 cm^{-1} which indicate irregular CH_2 lateral chains (Gumel et al. 2012). Absorption peaks at 2842 cm^{-1}, 2823 cm^{-1} are symmetrical methyl groups formed because of conformational changes during PHA crystallization (Sharma and Bajaj 2015).

Peaks observed at 1721, 1722, and 1723 cm^{-1} are the PHA characteristic bands corresponding to carbonyl stretching of esters. This indicates the crystallinity and chain length of the PHA (Krueger et al. 2012; Babruwad et al. 2015). Absorption peaks at 1459, 1458, and 1464 cm^{-1} reveal the presence of CH_2 and intracellular bacterial amide (-CO-N) groups (Getachew and Woldesenbet 2016). Further peaks at 1380, 1378, and 1376 cm^{-1} reveal the presence of terminal methyl groups (Kumalaningsih et al. 2011; Krueger et al. 2012).

The peaks observed at 1277 and 1276 cm^{-1} are asymmetric ether groups stretching vibrations (Krueger et al. 2012; Getachew and Woldesenbet 2016). The peaks seen at 1230 and 1229 cm^{-1} indicate asymmetric -C-O- groups (Kumalaningsih et al. 2011; Krueger et al. 2012). Additional peaks as seen at 1181, 1150, 1136, 1133, 1101, 1068, and 1056 cm^{-1} represent C=O stretching vibrations in the amorphous phase (Krueger et al. 2012; Gumel et al. 2012). More visible peaks at 981, 980, 930, 912, 829, 828, 812, 807, 806, and 805 cm^{-1} correspond to alkyl halides in the extracted PHA (Getachew and Woldesenbet 2016).

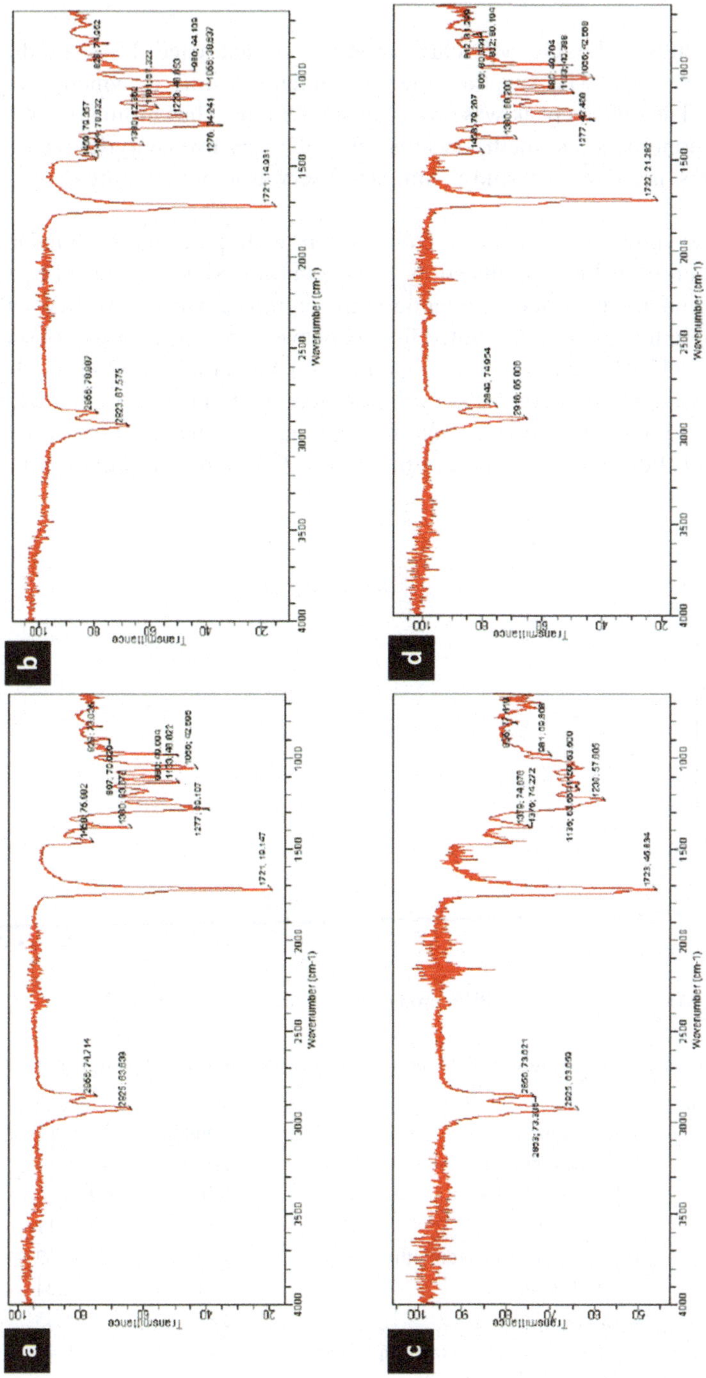

Fig. 5.6 FT-IR spectra of PHA produced by *B. megaterium* SF4 using four carbon sources. (**a**) glucose; (**b**) sugarcane molasses; (**c**) glycerol; (**d**) starch

5.3.6 GC-MS Analysis of Extracted PHA

Figure 5.7 shows the GC-MS chromatogram of extracted PHA produced by *B. megaterium* SF4. The spectra shows the retention time at which compounds were eluted. The highest peak was observed at 4.25 min. The chromatogram showed the main monomers as methyl esters of poly(3-hydroxybutyric) acid for all carbohydrates used in this study. Similar observation was reported by Ali and Jamil (2014).

Table 5.2 shows the chemical composition with percentage abundance and molecular weight of PHA produced by *B. megaterium* SF4 as revealed by GC-MS analysis. In general, the effect of retention time on molecular weight (MW) was also observed as lighter compounds eluted first. However, an exception was observed for nitrobenzene (123.11 g/mol) which eluted after 1-hexanol 2-ethyl (130.23 g/mol). The most abundant compound was cyclohexanone with a percentage of 81.8% and MW of 98.14 g/mol. The least common compound was the oxaspiro cyclic ketone compound (7,9-di-tert-butyl-1-oxaspiro[4,5]deca-6,9-diene-2,8-dione) with 0.41%

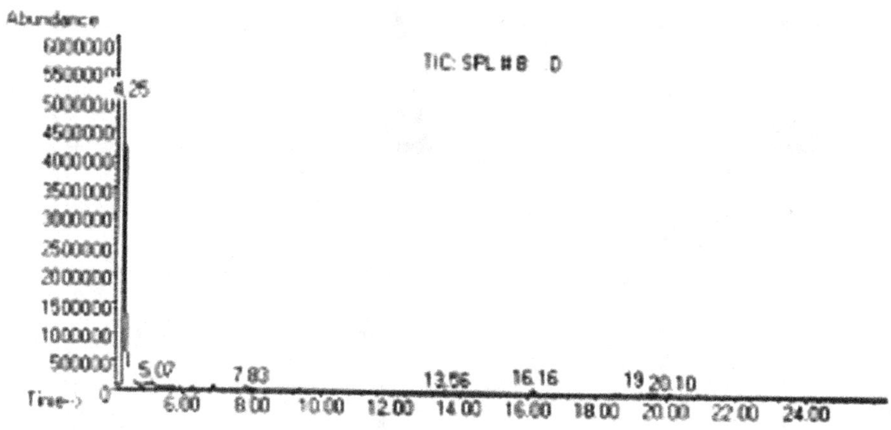

Fig. 5.7 GC-MS chromatogram of PHA produced by *B. megaterium* SF4

Table 5.2 Chemical composition of PHA produced by *B. megaterium* SF4 shown by GC-MS

S/n	Retention time	Compound	% Abundance	Molecular weight (g/mol)
1	4.25	Cyclohexanone	81.80	98.14
2	5.06	1-Hexanol 2-ethyl	9.96	130.23
3	7.83	Nitrobenzene	2.54	123.11
4	13.55	Phenol, 2,4-bis(1,1-dimethylethyl)	0.74	206.32
5	16.15	Diethyl phthalate	2.31	222.24
6	19.38	n-Hexadecanoic acid	2.83	256.43
7	20.10	7,9-Di-tert-butyl-1-oxaspiro[4,5]deca-6,9-diene-2,8-dione	0.41	276.40

abundance and MW of 276.40 g/mol. Our results agree with findings of Okwuobi and Ogunjobi (2013)

5.3.7 Characterization of PHA Synthase Genes

The PHA synthases of class IV PHA producers are encoded by the *PhaC* and *PhaR* genes among members of the *Bacillus* genera and are equally important for polymerization activity during PHA biosynthesis (McCool and Cannon 2001). The amplification of the 900 bp and 481 bp fragments of the *PhaC* (Fig. 5.8a) and *PhaR* (Fig. 5.8b) genes in *B. megaterium* SF4 confirms that it is indeed a class IV PHA producer (McCool and Cannon 2001; Nayak et al. 2013). Nucleotide sequences of *PhaC* and *PhaR* genes shared closest homology with *B. megaterium* NBRC15308 = ATCC 14581 at ≥99% identity.

Figure 5.9 shown below depicts the phylogenetic tree generated using predicted amino acid of the PhaC protein for *B. megaterium* SF4. In the tree, *B. megaterium* SF4 clusters with *B. megaterium* NBRC15308 = ATCC 14581, *Bacillus sp.* ME40, and other *B. megaterium* strains. Figure 5.10 shows the phylogenetic tree generated using predicted amino acid of the PhaR protein for *B. megaterium* SF4. Similarly, the tree shows *B. megaterium* SF4 clustering other *B. megaterium* strains and *Bacillus sp.* Root239 and *Bacillus sp.* Root147. The accession numbers of the partial coding sequences of *PhaC* and *PhaR* genes for *B. megaterium* SF4 in the NCBI database KY855378.1 and MF947449.1, respectively.

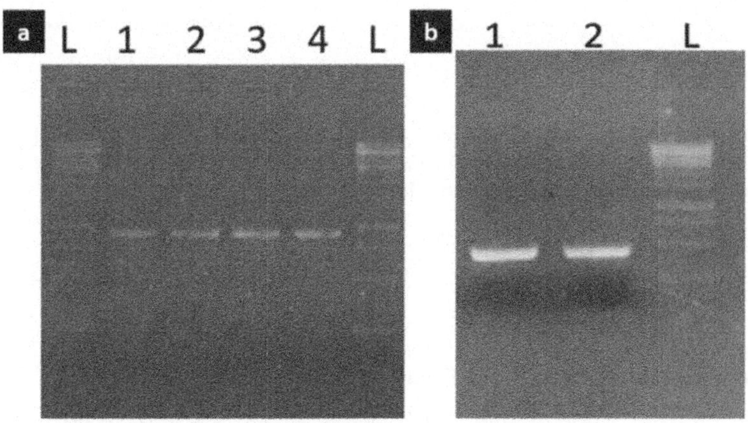

Fig. 5.8 Agarose gel of SF4 PhaC and PhaR genes of *B. megaterium* SF4. L—1 kb DNA ladder, (**a**) 1–4 Gradient annealing temperatures; (**b**) 1 & 2—Gradient annealing temperatures

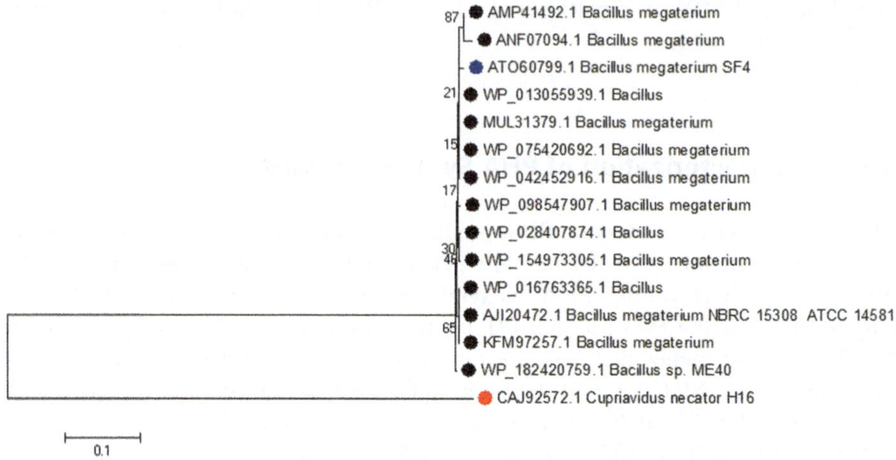

Fig. 5.9 Phylogenetic tree of *B. megaterium* SF4 PhaC predicted amino acids sequences. The tree was constructed using the neighbour-joining model with *Cupriavidus necator* (red) as an outgroup considering 1000 bootstrap replications in MEGA7

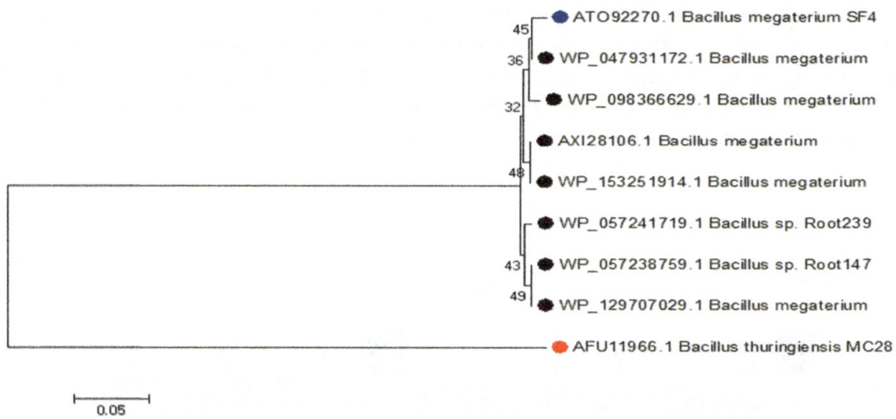

Fig. 5.10 Phylogenetic tree of *B. megaterium* SF4 PhaR predicted amino acids sequences. The tree was constructed using the neighbour-joining model with *Bacillus thuringiensis* (red) as an outgroup considering 1000 bootstrap replications in MEGA7

5.4 Conclusion

Bacillus megaterium SF4 utilized various carbon sources for poly-(3)-hydroxyalkanoic acid production. This potential should be further developed in temperature and pH optimization experiments for industrial application.

Acknowledgments The authors acknowledge the support provided by ICGEB Arturo Falaschi SMART Postdoctoral Fellowship, S/NGA 16-01, and the Department of Microbiology and Molecular Genetics, University of Punjab, Lahore, Pakistan, for research collaboration on this work.

References

Aarthi N, Ramana KV (2011) Identification and characterization of polyhydroxybutyrate producing *Bacillus cereus* and *Bacillus mycoides* strains. Int J Environ Sci 1(5):744–756

Akaraonye E, Moreno C, Knowles JC, Keshavarz T, Roy I (2012) Poly(3-hydroxybutyrate) production by *Bacillus cereus* SPV using sugarcane molasses as the main carbon source. Biotechnol J 7:293–303

Ali I, Jamil N (2014) Enhanced biosynthesis of poly(3-hydroxybutyrate) from potato starch by *Bacillus cereus* strain 64-INS in a laboratory-scale fermenter. Prep Biochem Biotech 44:822–833

Ali I, Jamil N (2016) Biosynthesis and characterization of poly-3-hydroxyalkanote (PHA) from newly isolated bacterium *Bacillus* sp AZR-1. Iran J Sci Technol Trans Sci 42:371. https://doi.org/10.1007/s40995-016-0132-6

Alkotaini B, Sathiyamoorthi E, Kim BS (2015) Potential of Bacillus megaterium for production of polyhydroxyalkanoates using red algae *Gelidium amansii*. Biotechnol Bioprocess Eng 20:865–860

Aneesh BP, Arjun JK, Kavitha T, Harikrishnan K (2016) Production of short chain length Polyhydroxyalkanoates by *Bacillus megaterium* PHB29 from starch feed stock. Int J Curr Microbiol App Sci 5(7):816–823. https://doi.org/10.20546/ijcmas.2016.507.094

Aneesh BP, Arjun JK, Kavitha T, Harikrishnan K (2017) Evaluation of short-chain-length polyhydroxyalkanoate accumulation in *Bacillus aryabhattai*. Braz J Microbiol 48:451–460

Babruwad PR, Prabhu SU, Upadhyaya KP, Hungund BS (2015) Production and characterization of thermostable polyhydroxybutyrate from *Bacillus cereus* PW3A. J Biochem Technol 6 (3):990–995

Berger E, Ramsay BA, Ramsay JA, Chavarie C, Braunegg G (1989) PHB recovery by hypochlorite digestion of non-PHB biomass. Biotechnol Tech 3(4):227–232

Cardozo JRG, Martínez ALM, Pérez MP, Guillermo Antonio Londoño AC (2016) Production and characterization of polyhydroxyalkanoates and native microorganisms synthesized from fatty waste. Int J Polym Sci 2016:1. https://doi.org/10.1155/2016/6541718

Du C, Sabirovab J, Soetaertb W, Carol Lin SK (2012) Polyhydroxyalkanoates production from low-cost sustainable raw materials. Curr Chem Biol 6(1):14–25

Fadipe OT, Jamil N, Lawal KA (2019a) Poly-3-hydroxyalkanoates production potential of *Bacillus cereus* C113 isolated from cassava dumpsite using some carbon sources. Nigerian J Biotechnol 36(1):203–212. https://doi.org/10.4314/njb.v36i1.26

Fadipe TO, Jamil N, Lawal AK (2019b) Polyhydroxyalkanoates synthesis by *Bacillus aryabhattai* C48 isolated from cassava dumpsites in South-Western Nigeria in soil microenvironment for bioremediation and polymer production. Wiley-Scrivener Publishing, Salem, MA, pp 271–281

Getachew A, Woldesenbet F (2016) Production of biodegradable plastic by polyhydroxybutyrate (PHB) accumulating bacteria using low cost agricultural waste material. BMC Res Notes 9 (1):509. https://doi.org/10.1186/s13104-016-2321-y

Gouda MK, Swellam AE, Omar SH (2001) Production of PHB by a *Bacillus megaterium* strain using sugarcane molasses and corn steep liquor as sole carbon and nitrogen sources. Microbiol Res 156:201–207

Gumel AM, Annuar MSM, Heidelberg T (2012) Biosynthesis and characterization of Polyhydroxyalkanoates copolymers produced by *Pseudomonas putida* Bet001 isolated from palm oil mill effluent. PLoS One 7(9):e45214. https://doi.org/10.1371/journal.pone.0045214

Kabilan S, Ayyasamy M, Jayavel S, Paramasamy G (2012) *Pseudomonas* sp. as a source of medium chain length polyhydroxyalkanoates for controlled drug delivery: perspective. Int J Microbiol 2012:317828. https://doi.org/10.1155/2012/317828

Krueger CL, Radetski CM, Bendia AG, Oliveira IM, Castro-Silva MA, Rambo CR, Antonio RV, Lima AOS (2012) Bioconversion of cassava starch by-product into *Bacillus* and related bacteria polyhydroxyalkanoates. Electron J Biotechnol 5(3):13. https://doi.org/10.2225/vol15-issue3-fulltext-6

Kumalaningsih S, Hidayat N, Aini N (2011) Optimization of polyhydroxyalkanoates (PHA) production from liquid bean curd waste by *Alcaligenes latus* Bacteria. J Agric Food Tech 1 (5):63–67

Kumar M, Rathour R, Singh R, Sun Y, Pandey A, Gnansounou E, Lin KA, Tsang DCW, Thakur IS (2020) Bacterial polyhydroxyalkanoates: opportunities, challenges, and prospects. J Clean Prod 263:121500. https://doi.org/10.1016/j.jclepro.2020.121500

Kung SS, Chuang YC, Chen CH, Chien CC (2007) Isolation of polyhydroxyalkanoates-producing bacteria using a combination of phenotypic and genotypic approach. Lett Appl Microbiol 44:364–371. https://doi.org/10.1111/j.1472-765X.2006.02090.x

Lee SY, Choi J (1999) Polyhydroxyalkanoates: a biodegradable polymer. In: Manual of industrial microbiology and biotechnology, 2nd edn. ASM Press, Washington, pp 616–627

Mahmoud MB (2012) Genotypic detection of polyhydroxyalkanoate-producing bacilli and characterization of phaC synthase of *Bacillus sp.* SW1-2. Life Sci J 9(4):518–529

McCool GJ, Cannon MC (2001) PhaC and PhaR are required for polyhydroxyalkanoic acid synthase activity in *Bacillus megaterium.* J Bacteriol 183(14):4235–4243

Medjeber N, Abbouni B, Menasria T, Beddal A, Cherif N (2015) Screening and production of polyhydroxyalkanoates by *Bacillus megaterium* by the using cane and beet molasses as carbon sources. Pharm Lett 7(6):102–109

Muhammadi S, Muhammad A, Shafqat H (2015) Bacterial polyhydroxyalkanoates-eco-friendly next generation plastic: production, biocompatibility, biodegradation, physical properties, and applications. Green Chem Lett Rev 8(3–4):56–77. https://doi.org/10.1080/17518253.2015.1109715

Naheed N, Jamil N (2014) Optimization of biodegradable plastic production on sugar cane molasses in *Enterobacter sp.* SEL2. Braz J Microbiol 45(2):417–426

Nayak PK, Mohanty AK, Gaonkar T, Kumar A, Bhosle SN, Garg S (2013) Rapid identification of polyhydroxyalkanoate accumulating members of bacillales using internal primers for phaC gene of *Bacillus megaterium.* ISRN Bacteriol 2013:562014. https://doi.org/10.1155/2013/562014

Nicolas J, Chi-Wei L, Yu-Hong W, Ho-Shing W, Shaw SW (2008) Isolation and purification of bacterial poly(3-hydroxyalkanoates). Biochem Eng J 39:15–27

Ojumu TV, Yu J, Solomon BO (2004) Production of polyhydroxyalkanoates, a bacterial biodegradable polymer. Afr J Biotechnol 3(1):18–24

Okwuobi PN, Ogunjobi AA (2013) Production and analysis of polyhydroxyalkanoate (PHA) by *Bacillus megaterium* using pure carbon substrates. World Appl Sci J 28(9):1336–1340

Pathak S, Sneha CLR, Mathew BB (2014) Bioplastics: its timeline based scenario & challenges. J Polym Biopol Phys Chem 2(4):84–90

Rehm BHA, Steinbüchel A (1999) Biochemical and genetic analysis of PHA synthases and other proteins required for PHA synthesis. Int J Biol Macromol 25:3–19

Salaam TO, Jamil N, Alebiosu FA, Idowu OO, Khan N, Lawal AK, Elemo GN (2016) Screening of polyhydroxyalkanoates bacteria from cassava and sugar sites in South-Western Nigeria. In: International symposium on biomedical materials Proc, ISBM 2016, Lahore, Pakistan, Interdisciplinary Research Centre in Biomedical Materials (IRCBM) CIIT, Lahore, Pakistan, pp 39–43

Shamala TR, Chandrashekar A, Vijayendra SVN, Kshama L (2003) Identification of polyhydroxyalkanoate (PHA)-producing Bacillus spp. using the polymerase chain reaction (PCR). J Appl Microbiol 94:369–374

Shamala TR, Divyashree MS, Davis R, Kumari KS, Vijayendra SV, Raj B (2009) Production and characterization of bacterial polyhydroxyalkanoate copolymers and evaluation of their blends by Fourier transform infrared spectroscopy and scanning electron microscopy. Indian J Microbiol 49(3):251–258

Sharma P, Bajaj BK (2015) Cost effective substrates for production of polyβhydroxybutyrate by a newly isolated *Bacillus cereus* PS-10. J Environ Biol 36(6):1297–1304

Sheu D, Wang Y, Lee C (2000) Rapid detection of polyhydroxyalkanoates accumulating bacteria isolated from the environment by colony PCR. Microbiology 146:2019–2025

Spiekermann P, Bernd HA, Rehm RK, Dirk B, Alexander S (1999) A sensitive, viable-colony staining method using Nile red for direct screening of bacteria that accumulate polyhydroxyalkanoic acids and other lipid storage compounds. Arch Microbiol 171:73–80

Suchada C (2010) Current trends in biodegradable polyhydroxyalkanoates. J Biosci Bioeng 110 (6):621–632

Tan GY, Chen C, Li L, Ge L, Wang L, Razaad IMN, Li Y, Zhao L, Mo Y, Wang J (2014) Start a research on biopolymer polyhydroxyalkanoate (PHA): a review. Polymers 6:706–754

Valappil SP, Boccaccini AR, Bucke CR (2007) Polyhydroxyalkanoates in Gram positive bacteria: insights from the genera *Bacillus* and *Streptomyces*. Antonie Van Leeuwenhoek 91:1–17

Verlinden RAJ, Hill DJ, Kenward MA, Williams CD, Radecka I (2007) Bacterial synthesis of biodegradable polyhydroxyalkanoates. J Appl Microbiol 102:1437–1449. https://doi.org/10.1111/j.1365-2672.2007.03335.x

Wu Q, Huang H, Hu G, Chen J (2001) Production of poly-3-hydroxybutyrate by *Bacillus* sp. JMa5 cultivated in molasses media. Antonie van Leeuwenhoek 80:111–118

Yuli A, Emma R, Ratu S, Sri RR (2017) Characterization of Bacillus megaterium and Bacillus mycoides Bacteria as probiotic bacteria in fish and shrimp feed. In: 2nd international conference on sustainable agriculture and food security: a comprehensive approach. KnE Life Sciences, pp 127–135. https://doi.org/10.18502/kls.v2i6.1029

6 Mushroom Mycelia-Based Material: An Environmental Friendly Alternative to Synthetic Packaging

Abhik Mojumdar, Himadri Tanaya Behera, and Lopamudra Ray

Abstract

In the circular economy, reduction of the vigorous usages of nonrenewable resources is becoming the leading scenario. Fungal mycelium is the vegetative part of fungus consisting of a number of filamentous fibers that extend out of the fungus and is considered to be natural, fast growing, safe, and renewable. The ability to form self-assembling bonds helps them to grow quickly on biological and agricultural wastes and produce miles of thin fibers which bind to the substrate to form a strong biodegradable material and can easily be shaped for the production of packaging materials, architecture, and various new designed objects. With the benefit of cost-effective raw materials and sustainable substitute to polystyrene like hazardous synthetic materials, this mycelia-based material is becoming the material of choice. This chapter reviews the present scenario of technology-based mushroom cultivation using wastes generated from the agricultural industries and also focuses on a variety of utilizations as an alternative replacement for synthetic polystyrene.

Keywords

Mushroom · Fungal mycelium · Packaging · Architecture

A. Mojumdar · H. T. Behera
School of Biotechnology, KIIT Deemed to be University, Bhubaneswar, Odisha, India

L. Ray (✉)
School of Biotechnology, KIIT Deemed to be University, Bhubaneswar, Odisha, India

School of Law, KIIT Deemed to be University, Bhubaneswar, Odisha, India
e-mail: lopamudra.ray@kls.ac.in; lray@kiitbiotech.ac.in

A. Vaishnav, D. K. Choudhary (eds.), *Microbial Polymers*,
https://doi.org/10.1007/978-981-16-0045-6_6

6.1 Introduction

According to a report by the United Nations, approximately 68% of the total population will live in metropolitan areas by the year 2050, ultimately increasing the construction demands and other domestic uses. However, the natural resources are becoming scarce, demanding a search of newer renewable sources with alternative use of the existing resources. Before the fourth industrial revolution, the sustainability packaging in the industry has never been so vital. More than 350 lakh tons of paperboard packaging and 1600 lakh tons of plastic packaging are annually produced for several uses making their way through the supply chain. The packaging objects are disposed and become a part of the waste stream and again recovered by landfill disposal, recycling, and incineration. Recently, biodesign practices incorporating living organisms like bacteria, algae, or fungi, as essential components in design that can be grown rather than manufacturing are being an important area of research (Ghazvinian et al. 2019).

Mushrooms can be cultivated on agricultural or industrial wastes and are considered as a major source of proteins, amino acids, along with many therapeutic applications. In industries, these can also be used for the biopulping and biobleaching applications. Previously, mushrooms were only considered to be a food, but different other value-added products can be obtained from the wastes such as twig or seed husks containing fungal mycelium. Mycelium can act as a natural glue, able to latch onto whatever around it such as plant stems or cotton shells creating a super-compact web of threads (Miles and Chang 2004). Therefore, the development of a packaging material with the mushroom mycelia can be designed as a cost-effective alternative over traditional plastic and other regular foams. Petroleum or natural gas-derived polystyrene foams are used widely in the prominent packaging, which is not biodegradable or compostable. So the incorporation of mycelium-based materials into the packaging base can help in the reduction of polystyrene consumption leading to eco-friendly packaging with enhanced sustainability and performance (Ashok et al. 2016; Vilaplana et al. 2010; Farmer 2013). The mushroom mycelium or the roots of mushroom can be converted into different shapes by growing them in a mould, and they also rapidly grow into a very compact material. After achieving the desired density with its shape, the dehydration step is done to halt further growth. After the effective life as a packaging material, these can be discarded in the expanse behind the house which can easily be decomposed automatically within a few weeks (Jiang et al. 2017; Abhijith et al. 2018). On the other hand, the use and production of polystyrene results in emission of greenhouse gases and is an energy-consuming process which eventually leads to environmental pollution. When non-recyclable and non-biodegradable polystyrene are used as single-use products such as food containers and food takeout plates, they play as the main element of marine debris and urban litter (Abhijith et al. 2018; Bruscato et al. 2019). The use of polystyrene and resin-based polypropylene is widespread because of many features such as easy recycling processes. However, these are not environment compatible and the processes are not sustainable.

This chapter reviews the present approaches of the technology and also focuses on the mushroom mycelium-based packaging materials and cushioning/padding applications as a promising substitute for the polystyrene foam.

6.1.1 Demand of Ecological Modernization in the Packaging Industry

Since 1950, approximately 91,500 lakh tons of plastics have been manufactured worldwide, out of which 30% are in use, only 9% have been reused, and 60% are in disposal area as the discarded product (Geyer et al. 2017). According to a report, 2,86,000 lakh metric tons of plastic polymers will be produced by 2050 (Geyer et al. 2017). Though the fragments of plastics are known to be weakened by sunlight exposure, plastic packaging is not biodegradable and the impact of millimeter-sized fragments on the environment is still not fully understood (Tudryn et al. 2018). The price of plastic flctuates with the cost of the nonrenwable petroleum resources as theses reused in plastic manufacturing industries for is feedstock. For the transport of plastic manufacturers to downstream wholesalers, approximately 0.49 million British thermal units of energy is consumed with the emission of 0.04 metric tons of carbon dioxide for each ton of plastic wastes (Containers and Good 2016). The type of recycling process and the rate of conversion may differ depending on the type of plastic (Hopewell et al. 2009). Rate of the recycling changes according to the plastic types. For example, plastic water bottles comprising polyethylene terephthalate (PET) can be recurrently recycled with a restoration rate of 19.5% by weight (Containers and Good 2016), whereas the least recycled plastics include polystyrene and its variants such as extruded polystyrene (XPS) and expanded polystyrene (EPS) with a recovery rate of just 0.9% by weight. Moreover, the transportation cost of the polystyrene's polymer structure is also costly and difficult making polystyrene an unattractive material for primary and secondary recycling. The waste-to-energy (WTE) combustion approach is the favored disposal method for plastic waste as compared to any other recycling process. But whatever be the treatment options, like WTE incineration, recycling including the collection of plastic fractions and transport, they harm the human health and environment. It is also apparent that no such promising waste management solutions and also searching for the same is quite challenging (Rigamonti et al. 2014). Due to increasing demands for consumer needs, rapid urbanization with changing consumer life style the expectation of improving the plastic sustainability in anticipation of creating a circular economy seems to be desolated (De los Rios and Charnley 2017). Moreover, only 5% of the retrieved plastics can be held costing around $80 billion and losses of approximately $120 billion annually (MacArthur et al. 2016). Therefore, in response to the highly rigid plastic markets in the 1980s plastic companies used blending options like the introduction of starch-based polymers to the polyethylene blends and also remarketed as "renewably sourced" (Iles and Martin 2013). Still, to meet the highest demand for plastic-based packaging, newer alternatives with null environmental impact and lesser cost needs to be investigated.

6.1.2 Background of Bio-composite Based on Mycelium

6.1.2.1 Mycelium-Based Bio-composite?

The vegetative section of the fungi is known as the mycelium composed of a mass of long, branched filamentous hyphae acting as a growth agent. Mycelium can break down the organic waste to simpler bodies with the enzymes secreted by the hyphae. This process lets the hyphae grow on the waste or any substrate, ultimately setting up a furry or condensed sheet around the substrate used known as "fungal skin" (Appels et al. 2019). The complete degradation process of the substrates is called colonization and during colonization mycelium binds the substrate replacing partially with strong biomass of the fungus. Removal of the fungal skin from the substrate can help in getting pure mycelium (Appels et al. 2019). Drying of the mycelium can stop the fungal growth leading its hibernation, which means at optimum environmental conditions the mycelium can restart growing. Similarly, the heating process permanently stops the growth of mycelium by killing it. However, the end products of either of the two processes are mycelium-based composites, and these can be framed by using non-carbon-based frameworks.

The fungal species used for inoculation, growing conditions, substrates, additives used, processing and foaming techniques define the properties of pure and composite mycelium. Various reports are there comparing the properties of mycelium-based bio-composites with the conventional plastic (Appels et al. 2019; Haneef et al. 2017; Attias et al. 2017; Yang et al. 2017; Jones et al. 2017) such as (1) when compared to plastic, mycelium-based composite shows relatively low density, (2) low compact strength than plastics used for the construction projects, (3) reduced tensile strength, etc.

6.2 Early Uses of Mushroom Packaging

Irrespective of the size of the business and their margins, various strategies could be adopted to lessen the hazardous effects on the environment, emphasizing their sustainability. The packaging is very much necessary for the marketing of a product and almost all of the used packaging materials are discarded as litter once the sale is done. A small amount of consideration on using limited material at the time of packaging might support when it reaches to the shipping part. Mycelia are a natural binder and can be blended with agricultural wastes like oat and corn husks for making an extremely durable material and could be an option for replacement of styrofoam and polystyrene in every practical application. Ecovative pioneers the idea of the use of mushroom-based foam with intact maintenance of the key products. Dell Company has replaced its plastic packaging with mushroom material and announced that the shipping of its products will be by using eco-friendly mycelia foam. Vigorous tests had been carried out by Dell for ensuring the safety of their products similar to the traditional foam. Cushioning by using the mushroom materials proved to be a super green solution, and they are very easily composted after the use (Vendries et al. 2020). After Dell, Ford Company has decided on the application of foam containing mushroom as a major element in their automotive parts such as dashboards, bumpers, and side doors. Nearly about 30 pounds of petroleum-based foams are being used in each car and now the company would prefer for replacement of those with environmentally friendly alternatives. Similarly, Sweden-based flat-pack maker company Ikea has decided to lessen the use of fossil-fuel-based material and mushroom-based foams are being considered as better substitute to polystyrene for their foam packaging (Vendries et al. 2020).

6.3 Potential of Mycelium-Based Material as an Alternative to Synthetic Packaging Materials

Advertising of the products focusing on their environment-friendliness was impressive particularly to a few committed group of buyers, but the gradual increase in the knowledge of the hazardous human effect on the fragile ecosystem has built a few development in viewpoints. The buying option of these environment-friendly packaging materials is mostly approved by younger generations, inspiring many multinational companies and local businesses. Now buyers are much more attracted towards reducing the waste production, and in buying consumable stuffs. The environmentally labeled packaging is recently considered as an important benchmark for the preference that can be professed as rising awareness regarding the significance of righteous and environmental dimensions in choosing the product. Some mushroom-based foam products are depicted in Fig. 6.1. Companies dedicating for ecological and social positivity get a significant rise in the sale. Both consumable and non-consumable goods claiming sustainability on their packaging exhibited a moderate 2% increase in their sell while brands without any sustainability claims showed only 1% of the increase in sell for their brands.

Mushroom aids an extra way of the income to the cultivars as it makes a path for the utilization of their agricultural waste. Instead of remaining through generations, these materials are easily broken down in the landfills and compost. The area of packaging is positively affected by the use of mycelium, eventually obsolescing the use of toxic and persistent oil-based materials. According to the regional availability of the local agricultural by-product materials, the growth substrates of the mushroom may vary making it perfect for manufacturing all over the world. Regional manufacturing units using locally available feedstock can be used to reduce the cost of transportation for raw materials (Jiang et al. 2017; Stamets 2005). The big players in the global economy now remark these new trends. Companies like Unilever and Coca Cola have directed their attention towards making sustainable packaging enhancing their brand's core values. Presently, world consumers are looking for a replacement to the hazardous packaging materials. Sustainable packaging practices will lead the business a path to tap into the attitude of buyers.

Fig. 6.1 Mushroom mycelia based material foam used for packaging

6.4 Production of Mycelium-Based Material for Packaging

Myco-bond, one type of packging material that is resistant to heat and fire, biodegradable, low-energy material and ability to absorb energy has been developed by Ecovative, a start-up at New York by using mushroom mycelia. Compared to traditional foam packing material, only 12.5% of the energy and 10% of the carbon dioxide are required for manufacturing Myco-bond (a packing material). This is produced, allowing the mycelium of the mushrooms to invade the organic material like any agricultural wastes to feed on it (Vendries et al. 2020). There is a series of repeated growth and rupturing, then again greater growth resulting in a mass of mycelia which function almost the same as the polystyrene does in dry conditions, but when exposed to favoring environmental conditions suitable for breakdown, it decomposes easily.

Mushroom roots and any other agricultural by-products such as corn or oat husk, together are kept in varied sizes of trays. These combinations are left for a minimum of 5 days in a dark warehouse to fester and furl. After which a fireproof, water-resistant, and biodegradable new packaging material emerges. Figure 6.2 shows a mushroom-based foam manufacturing facility.

This approach is reframing in many ways which are discussed below:

1. Mycelium possesses unique properties of growing miles of a thread-like root in just 1 day.
2. The organism can grow amazingly very fast and fit into any mould-like dense foam.

Fig. 6.2 Mushroom based foam manufacturing facility

3. These packaging materials can be used for almost everything such as packing for laptops, a wide panel for insulation at homes, and the density of each packaging material can be controlled by ending the growth process.

The source material needs to be disinfected to remove any contesting spores present before starting the process. Disinfection by application of organic cinnamon oil, thyme oil, lemongrass oil, and oregano oil can be used instead of energy-exhaustive steam sterilization process. Packaging foam of the Ecovative is produced by placing the organic wastes such as wood fiber or cotton hulls in a mould and then inoculated with the mushroom spores. Mushroom spores utilize the carbohydrate from waste and obtain energy for growing into the shape of the mould/tray. After reaching sufficient growth, the formation of flowering spores is stopped by subjecting them to heat. Since the mycelium is never allowed to grow for a longer period to form a mushroom, no spores or allergens left in this process (Vendries et al. 2020).

6.5 Mycelium Production and its Environmental Impact

Basal growth media such as sterilized mineral salt medium (MSM) enriched with supplements including carbon, nitrogen, and inorganic compounds was used as the production medium and with a 10% of inoculums (v/v) with a 6 days of incubation at 28 °C with shaking at 150 rpm. Mycelia can be retrieved by centrifuging at 2500 g for 10–15 min. After the repeated washing steps, the recovered mycelia are dried and weighed until an unchangeable weight at 80 °C (Zhu et al. 2015). Effects of pH, auxiliary carbon, and nitrogen source, presence of various inorganic compounds, and organic compounds like $MgSO_4$, $CuSO_4$, Zn $(CH_3COO)_2$ K_2HPO_4, and KH_2PO_4 are important for the growth of microbial mycelia under submerged fermentation conditions. Fermentation temperature, time, speeds of rotation, and inoculums size are some parameters which also affect the fermentation process (Zhu et al. 2015).

Custom shapes of the mycelia are developed by self-assembling bonds formed between the mycelia by growing on the locally sourced agro waste products. Five variables, i.e., (1) active hyphal density, (2) density of the hyphal tip, (3) internal concentration of the substrate, (4) inactive hyphal density, (5) and external substrate concentration are used for defining the model. Branching architecture and filamentous growth habits of the fungi help them in adapting for their growth in the soil. During this process of growth, different types of vesicles providing new cell membrane moves forward, aiding the extension of hyphal tips. This hyphal tip movement with the secretion of lytic enzymes helps the hyphae to penetrate the solid substrate used like chitin and up taking of nutrients in the solute form.

Redistribution of internal metabolites along the fungal mycelium is achieved by the translocation process (Olsson 1995). Hyphae which are involved in the uptake of nutrients, branching, and translocation are known as active hyphae and the tips are denoted as the end of these hyphae. Hyphae no longer involved in the above processes are denoted as inactive hyphae. Whatever may be the environment, a combination of elements and few elements like carbon, hydrogen, nitrogen, sulfur, oxygen, phosphorous, and other metals are required for their growth (Boswell et al. 2003).

6.6 Application of Mushroom Bio-composites

Previously, mycelia were only used in the medical industry to produce small molecules like antibiotics/enzymes/organic acids for more than a century (Wösten 2019). In the 1980s, Japanese scientist, Shigeru Yamanka first investigated the gluing power of the mycelium which can be effectively applied in the paper industry and in manufacture of many building materials. (Girometta et al. 2019). Recently, this mycelium is not only used for the binding and packaging materials but now it is used by many industries like design, fashion, packaging, architectural design, etc.

In building sectors also, mushroom bricks were also tested for building 40 feet tower. Like the rigid board insulation, this mushroom-based foam can be used as an insulation material, providing a tight envelope with thermal bridges (Jones et al. 2017). Without the use of any toxic fire retardants, this can be used as a class A fire rating (Abhijith et al. 2018). MycoBoard is an engineered mushroom-based bio-composite, having the potential to replace wood material. Greensulate, a sustainable insulant and fireproof made from water, perlite, flour, and mushroom spores.

MycoWorks and MOGU Company produced durable, flexible, and sustainable and waterproof synthetic leather (Fig. 6.3a) cultivated using mycelium. Officina Corpus coil started to develop a project "Future of Plastics" focusing on the development of kitchenware and alternatives to single-use plastics (Fig. 6.3b).

Fig. 6.3 Applications of mushroom mycelia-based composites in product designing and architecture

Various furniture was designed from lightweight mycelium composite, regardless of their durability and load-bearing capacity (Fig. 6.3c).

Apart from their use in designing industries, architectural industries also started exploring the applications of mycelium-based composites. Mycelium-based wall and ceiling panels were developed by MOGU (Fig. 6.3d). Over the past years, mycelium is used in building a few architectural projects like Hy-Fi Tower (Fig. 6.3e), the Mycotecture alpha (Fig. 6.3f), MycoTree (Fig. 6.3g). David Benjamine designed the temporary structure of the Hy-Fi tower in the form of three 13-m tall intersecting cylinders built with 10,000 blocks of mycelium (Slavin 2016). This design for the tower was for the experimental purpose to test the durability, biodegradability, and resistance of the mycelium-based masonry unit and the tower remained for 3 months in summer. Phill Ross, the cofounder of MycoWorks designed a small-scale pavilion Mycotecture Alpha. This pavilion consists of 350 mycelium blocks with mycelium floor also (Karana et al. 2018). Recently for the 2017 Seoul Biennale of Architecture and Urbanism, Block Research Group of ETH Zurich in collaboration with Karlsruhe Institute of Technology designed MycoTree by a 3D graphic static method. Besides their use of mushroom-based composite in the masonry architecture, they are recently attempted as a substitute to concrete in a construction project. The *Shell Mycellium* domed building has been constructed in India for Kuchi Muziris Biennale 2016.

Various investigations on the use of mushroom-based material in architecture are moving forward rapidly. The main aim is to find a way for the affordance, enhanced durability, and improved strength against environmental threats and also to augment the chemical and mechanical properties of mycelium-based composites.

6.7 Conclusion

The review explains that fungi can degrade maximum of the organic, inorganic objects. Composites, based upon mycelium are now getting more attention as a sustainable alternative to the synthetic packaging, much new-concept bio-inspired design and insulation panels, etc. The challenges in scaling the rapidly increasing mycelia composite production are still to be looked upon. Overall the same reproducibility, mechanical and thermodynamic parameters could not be expected by these mushroom-based foams when compared to synthetic materials or mono-component natural materials. However, suitable material characterization is still inadequate. Composites made up of mycelium, of which the investigation has only currently initiated, possess immense possibilities in replacing the synthetic hazardous packaging and in many other industries.

References

Abhijith R, Ashok A, Rejeesh C (2018) Sustainable packaging applications from mycelium to substitute polystyrene: a review. Mater Today Proc 5(1):2139–2145

Appels FV, Camere S, Montalti M, Karana E, Jansen KM, Dijksterhuis J, Krijgsheld P, Wösten HA (2019) Fabrication factors influencing mechanical, moisture-and water-related properties of mycelium-based composites. Mater Des 161:64–71

Ashok A, Rejeesh C, Renjith R (2016) Biodegradable polymers for sustainable packaging applications: a review. IJBB 1(11)

Attias N, Danai O, Ezov N, Tarazi E, Grobman YJ (2017) Developing novel applications of mycelium based bio-composite materials for design and architecture. Proceedings of building with biobased materials: best practice and performance specification, 6–7 September, pp 76–77

Boswell GP, Jacobs H, Davidson FA, Gadd GM, Ritz K (2003) Growth and function of fungal mycelia in heterogeneous environments. Bull Math Biol 65(3):447

Bruscato C, Malvessi E, Brandalise RN, Camassola M (2019) High performance of macrofungi in the production of mycelium-based biofoams using sawdust—sustainable technology for waste reduction. J Clean Prod 234:225–232

Containers P, Good ND (2016) Documentation for greenhouse gas emission and energy factors used in the waste reduction model (WARM)

De los Rios IC, Charnley FJ (2017) Skills and capabilities for a sustainable and circular economy: the changing role of design. J Clean Prod 160:109–122

Farmer N (2013) Trends in packaging of food, beverages and other fast-moving consumer goods (FMCG): markets, materials and technologies. Elsevier

Geyer R, Jambeck JR, Law KL (2017) Production, use, and fate of all plastics ever made. Sci Adv 3 (7):e1700782

Ghazvinian A, Farrokhsiar P, Vieira F, Pecchia J, Gursoy B (2019) Mycelium-based bio-composites for architecture: assessing the effects of cultivation factors on compressive strength. In: The eCAADe and SIGraDi Conference, University of Porto, Portugal

Girometta C, Picco AM, Baiguera RM, Dondi D, Babbini S, Cartabia M, Pellegrini M, Savino E (2019) Physico-mechanical and thermodynamic properties of mycelium-based biocomposites: a review. Sustainability 11(1):281
Haneef M, Ceseracciu L, Canale C, Bayer IS, Heredia-Guerrero JA, Athanassiou A (2017) Advanced materials from fungal mycelium: fabrication and tuning of physical properties. Sci Rep 7(1):1–11
Hopewell J, Dvorak R, Kosior E (2009) Plastics recycling: challenges and opportunities. Philos Trans R Soc B Biol Sci 364(1526):2115–2126
Iles A, Martin AN (2013) Expanding bioplastics production: sustainable business innovation in the chemical industry. J Clean Prod 45:38–49
Jiang L, Walczyk D, McIntyre G (2017) A new approach to manufacturing biocomposite sandwich structures: investigation of preform shell behavior. J Manuf Sci Eng 139(2)
Jones M, Huynh T, Dekiwadia C, Daver F, John S (2017) Mycelium composites: a review of engineering characteristics and growth kinetics. J Bionanosci 11(4):241–257
Karana E, Blauwhoff D, Hultink E-J, Camere S (2018) When the material grows: a case study on designing (with) mycelium-based materials. Int J Des 12(2):119–136
MacArthur DE, Waughray D, Stuchtey M (2016) The new plastics economy, rethinking the future of plastics. In: World Economic Forum
Miles PG, Chang S-T (2004) Mushrooms: cultivation, nutritional value, medicinal effect, and environmental impact. CRC Press, Boca Raton
Olsson S (1995) Mycelial density profiles of fungi on heterogeneous media and their interpretation in terms of nutrient reallocation patterns. Mycol Res 99(2):143–153
Rigamonti L, Grosso M, Møller J, Sanchez VM, Magnani S, Christensen TH (2014) Environmental evaluation of plastic waste management scenarios. Resour Conserv Recycl 85:42–53
Slavin K (2016) Design as participation. J Design Sci
Stamets P (2005) Mycelium running: how mushrooms can help save the world. Random House Digital, Inc., New York
Tudryn GJ, Smith LC, Freitag J, Bucinell R, Schadler LS (2018) Processing and morphology impacts on mechanical properties of fungal based biopolymer composites. J Polym Environ 26 (4):1473–1483
Vendries J, Sauer B, Hawkins TR, Allaway D, Canepa P, Rivin J, Mistry M (2020) The significance of environmental attributes as indicators of the life cycle environmental impacts of packaging and food service ware. Environ Sci Technol 54(9):5356–5364
Vilaplana F, Strömberg E, Karlsson S (2010) Environmental and resource aspects of sustainable biocomposites. Polym Degrad Stab 95(11):2147–2161
Wösten HA (2019) Filamentous fungi for the production of enzymes, chemicals and materials. Curr Opin Biotechnol 59:65–70
Yang Z, Zhang F, Still B, White M, Amstislavski P (2017) Physical and mechanical properties of fungal mycelium-based biofoam. J Mater Civ Eng 29(7):04017030
Zhu W, Guo C, Luo F, Zhang C, Wang T, Wei Q (2015) Optimization of Calvatia gigantea mycelia production from distillery wastewater. J Inst Brew 121(1):78–86

An Overview of Microbial Derived Polyhydroxybutyrate (PHB): Production and Characterization

7

Monika Sharma and Harish Kumar Dhingra

Abstract

The glory of synthetic polymer and the production and application can be contributed to their prize, permanence, and endowment for the proficiency in daily living. Although, one time utilization of synthetic plastic, longevity and rebellious characteristics is opened a continuous enchancement in synthetic polymer as a part of garbage. Requirement for the replacement of one-time utilized plastics which is hard to detect has motivated many researchers to work regarding the permanent solution to replace synthetic polymers that are oil-based plastics. Poly-β-hydroxybutyrate is a naturally formed biopolymer that can be used as plastic produced by various microorganisms. It has aquired importance because of its structural variations and similar analogy to synthetic polymer. Additionally, specific physico-chemical, bio-logical, and degradation characterisitics of biodegradable plastics is key factor which make bioplastic important for various uses in different field. This chapter contains classification, biosynthesis, detection techniques, and recovery methods of microbially produced biodegradable polymer. We then discuss the application of biodegradable polymers in various fields to reduce the dependence on synthetic plastic.

Keywords

Biodegradable polymers · poly-β-hydroxybutyrate · biodegradation · bioplastic

M. Sharma (✉) · H. K. Dhingra
School of Liberal Arts and Sciences, Mody University of Science and Technology, Lakshmangarh, Rajasthan, India

A. Vaishnav, D. K. Choudhary (eds.), *Microbial Polymers*,
https://doi.org/10.1007/978-981-16-0045-6_7

7.1 Introduction

7.1.1 Plastics and Problems with Plastics

Macromolecules made up of small molecules present in repeating structure are known as polymers. Polymers are either natural or man-made compounds. Polymers may be classified as homopolymers and heteropolymers on the basis of monomer units. Plastic is oil derived, partially man-made and organic compound which utilized in different kind of product formation. Plastics have many properties like light in weight, strong, and durable. Plastic is also resistant to microbial and chemical decomposition which increases its valuablilty for mankind. Plastic was introduced in 1862 by Alexander Parkes (Lemoigne 1925). Plastic is an important part of our everyday life. Plastic is used to make simple carry bags and complicated medical and surgical implants. Modern lifestyle increases higher use of one-time-use plastics. It is also used for various storage substances like milk, H_2O, coke, extracts, and fruits. It increases consumption of oil-based polymers for storage packets. An upscaled use of man-made polymer increases the resistant waste that is not able to decompose by microbial and chemical decomposition mechanism. Depolymerase enzyme is absent and high molecular weight is high. These characteristics make synthetic plastic sustainable for decomposition. Supervision of oil-based polymers enhances work for garbage managing agency. Waste management authorities either discharges garbage in the earth surce/ocean or burn that reduced the productivity of land/ ocean and cause the air pollution. Different kinds of harmful compounds like HCN produced as a result of burning being the reason for the decline in a person's physical and mental well-being. These synthetic polymer-made packets are dangerous for cattles. The reuse of synthetic polymer is an extremely expensive activity. Very less synthetic polymer carriers get reused. A lot of aquatic creatures and many waves pass on due to synthetic polymer carrier (Heap 2009). The plastic litter ingested by animals clogs their digestive tract which leads to death due to starvation. Birds entangled in synthetic plastic bags cannot fly and as a result they die. In sight of many dangerous effects of synthetic plastics, there is an urgent need for biodegradable plastic, which does not lead to such problems. Bio-based and biodegradable polymers can added to better economical condition by utilization of sustainable assets. In addition, organic polymers give new finish of life as an-aerobic degradation or fertilizing the soil that have lower or no negative impact on the environment (Narancic et al. 2020).

The bio-based plastics have the following advantages over oil-based plastics:

- Bio-based plastics prevent land-filling and various environmental pollutions.
- Bio-based plastics can increase the organic part of the soil; retain water and nutrients in the soil.
- Production of most of the bio-based plastics requires lower energy compared to synthetic plastics.
- It reduces recycling cost and environment cleaning from the contaminated products.

- The renewable source of bio-based polymer will conserve the nonrenewable sources of fossil fuels for a more sustainable enviroment and thereby reducing the emission of greenhouse gas (Pikon and Czop 2014).

7.1.2 Biodegradable Polymers

Biopolymers are also known as natural polymers, and they are produced from the polymerization of monomers from different metabolic process. The biopolymers have been classified into eight classification as follows (Steinbuchel and Hein 2001):

- Nucleic acids
- Polyamides
- Polysaccharides
- Polyhydroxyalkanoic acids, polymalic acid, and cutin
- Inorganic polyesters with polyphosphate
- Poly isoprenoids like gutta-percha
- Poly phenols like lignin

The biodegradable polymers of different classes do not have same properties. Out of these classes, a few biopolymers can be used as plastic. The classes of biodegradable polymers like polythioesters and organic polyoxoesters can act as a source of biodegradable plastics.

7.1.3 Biodegradable Plastics

A biodegradable plastic should be degraded through photo-degradation, oxidation, and hydrolysis without discharging any harmful residue into the environment (Swift 1993). Biopolymers are decomposed by lysis without enzyme activity and hydrolysis with bacterial enzymes (Poirier et al. 1995a, b). Polylactic acids (PLA), polyhydroxyalkanoates (PHA), and poly-propylene are various types of bioplastics (Patwardhan and Srivastava 2004; Datta et al. 1995). Out of them polyhydroxyalkanoate is a highly sustainable bioplastic that will decompose completely.

7.2 Polyhydroxyalkanoates

A number of biopolymers are formed in nature such as nucleic acids, polyamides, polysaccharides, polyoxoesters, polythioesters, polyanhydrides, polyisoprenoids, polyphenols, and polyhydroxyalkanoates (Steinbuchel and Hein 2001). Among the most investigated biodegradable polymers, polyhydroxyalkanoates are promising to other biopolymers because of their close analogy to plastics and structural variations (Kalia et al. 2007). These polyesters are used in a number of applications and have attracted considerable industrial attention (Madison and Huisman 1999). Hence, these polymers are acquired as alternatives to synthetic plastics. A wide variety of Gram positive and Gram negative, aerobic, anaerobic and photosynthetic, lithotrophic, and organotrophic bacteria accumulate PHA intracellularly as carbon and energy source (Senior and Dawes 1973). PHAs are polyesters of various hydroxyalkanoates and hydroxylated at various positions such as 3, 4, 5, and 6. A big amount of different monomer units which can be incorporated in this branch of biopolymer (Anderson and Dawes 1990). Approximately 150 different hydroxyalkanoate monomers have been recognized with various types of out-group structures. All of these are (R)-form chiral molecules that are synthesized by a wide range of microorganisms (Lee 1996a, b; Doi et al. 1992).

7.2.1 Classes of Polyhydroxyalkanoates

Polyhydroxyalkanoates are identified as small chain length PHA, medium chain length PHA, and long chain length PHA based on the size of the hydroxyalkanoic acid units (Steinbuchel and Pieper 1992). The small chain length PHA is made up of three to five units of 3 hydroxy or 4 hydroxy fattyacids. Poly-β-hydroxybutyrate (PHB) is the first of the PHA identified. Mcl and lcl-PHA are composed of C_6 to C_{14} and more than C_{14}, 3-hydroxy fatty acids, respectively. General structural formula is shown in the figure given below (Fig. 7.1 and Table 7.1):

An enzyme for the synthesis of poly-beta-hydroxyalkanoates *A. eutrophus* is able to accept 3-HA with 3–5 C atom rather than enzyme of *P. oleovorans* is able to polymerize 3-HA with 6 to 14 C atom (Khanna and Srivastava 2004).

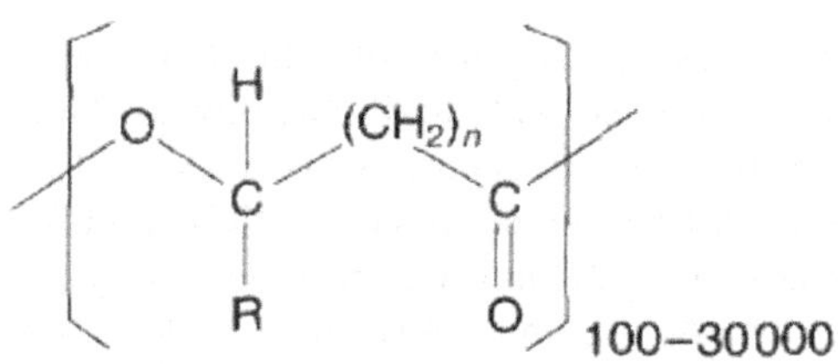

Fig. 7.1 Structure of polyhydroxyalkanoates

Table 7.1 Representative members of polyhydroxyalkanoates (Steinbuchel 1991)

N	R	Types of monomers
1	H_2	Poly 3-hydroxypropionate
1	CH_3-	Poly 3-hydroxybutyrate
1	CH_3CH_2-	Poly 3-hydroxyvalerate
1	CH_3CH_2 CH_2-	Poly 3-hydroxyhexanoate
2	H_2	Poly 4-hydroxybutyrate
2	CH_3-	Poly 4-hydroxyvalerate
3	H_2	Poly 5-hydroxyvalerate
3	CH_3-	Poly 5-hydroxyhexanoate
4	$CH_3(CH_2)_4CH_2$-	Poly 6-hydroxydodecanoate

Microbes can produce PHA with small chain length and medium chain length groups which contain 3 to 14 carbon atoms have been reported. Scl PHA has properties similar to synthetic plastics like poly-propylene, while mcl-PHA shows more elastomeric characteristics (Gopi et al. 2018). These copolymers have characteristics between small chain length and medium chain length PHA on the basis of molecular partition of small chain length and medium chain length units that enhance mechnical and thermal characteristics. Insertion with a small number of medium chain length units, 3 hydroxyhexanoate in biopolymer basic structure, modifies the characterisitics of poly-beta-hydroxyalkanoates The formed poly-beta-hydroxyalkanoates is closest to lower molecular weight poly-ethylene because of this being used at industrial level (Philip et al. 2007; Madison and Huisman 1999; Wang et al. 2007).

7.3 Poly-β-hydroxybutyrate (PHB)

The first and most commonly discovered PHA was poly-β-hydroxybutyrate (PHB). The PHB is nontoxic, biocompatible, and biodegradable plastics that are made up of renewable resources (Reddy et al. 2003). PHB is glassy in nature, rotate the plan polarized light and unsolvable in polar solvants. Properties like this derive PHB very reasonably as a petroleum-based polymer (Lee et al. 2000; Park et al. 2002a, b). Lemoigne (1925) identified homopolymer poly-β-hydroxybutyrate (Jackson and Srienc 1994). Macrae and Wilkinson (1958) observed that PHB was accumulated as a result of imbalance growth produced by nutrient limitation. Senior and Dawes (1973) reviewed on storage polymers and emphasized the role of polyhydroxyalkanoates as carbon storage compounds in microorganisms during nutrient depletion or limiting conditions. Poly-β-hydroxybutyrate synthesized by various Gram-positive and Gram-negative bacteria like *Azotobacter* sp., *Rhizobium* sp., *Pseudomonas* sp., and *Bacillus megaterium*, formed PHB molecules are stored as inclusion body inside the microbial cell (Steinbuchel and Valentin 1995). Plastics made up of PHB have been reported to be completely biodegradable. Microorganisms from various taxa like aerobic, anaerobic, photosynthetic microbes and archae bacteria can produce PHA (Luengo et al. 2003; Merugu et al. 2010). More than 300 species of bacteria are known to biosynthesize PHB (Reddy et al. 2003; Steinbuchel and Schlegel 1991; Suriyamongkol et al. 2007). Within the bacterial domain most of the PHB producers belong to the phylum proteobacteria (Rehm 2007). PHB producers are also found among cyanobacteria (*Synechococcus*), firmicutes (*Bacillus*), etc. PHB-producing strains have been isolated from diverse environments like activated sludge (Byrom 1987; Takeda et al. 1995; Lee et al. 1995; Akar et al. 2006), marine waters (Sabirova et al. 2006; Shrivastav et al. 2010), hypersaline environments, oil-contaminated soils (Dalal et al. 2010; Saharan et al. 2014), rhizosphere (Madison and Huisman 1999; Mohapatra et al. 2015), etc.

7.4 PHB- producing Bacteria

A. vinelandii was not suitable for production at industrial level because it has a low yield of PHB produced. *A. vinelandii* UWD was important because it transformate without outer layer and accumulate 80% PHB (Page 1992). Byrom (1992) worked on the commercial isolation of PHB in the UK. Microbes of interest were from the family Ralstoniaceae at a lab in the UK due to extraction of PHB was easy and had good physical properties. PHB by methylotrophs and *Azotobacter* was also studied. Although that PHB was low in molecular weight and yield was hard to isolate also. *Azotobacter* utilized organic source to produce sugar at the place of PHB; hence, it was not an organism of interest. *Ralstonia eutropha* formed approximately 80% PHB when phosphate is present in restricted amount. The important microbes are those that can produce PHB while cultured on cheap materials.

Lee (1996a, b) discovered that the study of microbial growth pace and PHB production pace by this organism is important for choosing potent microorganism for PHB production. The production of copolymer by 13 microbes from the rRNA superfamily III is discovered (Renner et al. 1996). They suggested that various microbes could synthesize polymer which contains various PHA combinations when cultured on similar materials.

Rawte and Mavinkurve (1998) identified PHA-forming microorganisms from mangroves of Mandovi estuarine. Bacteria were primarily identified at tributyrin medium to check PHB production with the help of Nile blue dye under microscope. From 65 PHA-accumulating strains, some of them could utilize enviroment N_2 then develop nitrogen-less medium. Lokesh et al. (2005) isolated PHB-producing bacterial species and characterized higher PHB producer as *Sphingomonas* sp. The bacteria were developed with various carbon sources and amount of produced polymers were checked. This bacterium might produce PHB when feed on disaccharides, aldo hexose, and sugar alcohols, but failed to accumulate PHB when cultured with ketoses, pentoses, and starch. Aerobic, free-living, and nitrogen-fixing microbes isolated by Pal and Paul (2000) were identified as genus *Azotobacter*; nearly 70% of strains could produce biopolymer. The sustainable strain of A. chroococcum has been shown to store the PHB approximately 70% dried biomass, concurrent cultured on enriched parameters. Mercan et al. (2002) studied production of PHB in different species of *Rhizobium* (*Rhizobium cicer*, *Rhizobium japonicum*, *Rhizobium* sp., and *Bradyrhizobium japonicum*). In the study, they found that *Rhizobium* sp. 2426 strain can produce PHB up to 70% of dry cell weight. In YEM with L-cysteine medium after 48 h, the PHB accumulated in *Rhizobium* sp. was more in YEM when inoculated with cysteine.

Tajima et al. (2003) isolated PHB-producing Gram-positive bacteria from grassland. The PHB productivity from isolate INT005 was more compared to *B. megaterium*. Mukhopadhyay and Paul (2003) studied phototrophic purple non-sulfur strain for the potential PHB accumulation. Thirty bacteria were identified by sludge/water sources taken from various areas of east India. Preliminary screening of these microbes in favor of biopolymer accumulation led to recognition of 5 bacteria by amount of biopolymer between approximately 15% dried biomass, while cultured on acetate-enriched media with 10,000 lux intensity radiance. Carbon sources such as acetate and butyrate were highly suitable for PHB production. Although nitrogen source in the culture medium showed negative effect on PHB production, the microbial growth was increased. Limiting conditions for phosphate and sulfate increase the production of polymer by the bacteria. The effects of factors like pH and light intensity was also studied in PHB production.

Pseudomonas sp. 14–3 (from Antarctic) accumulated poly-β-hydroxybutyrate in high quantity, while cultured upon the media which enriched with octanoate. On the basis of morphological characteristics and rRNA gene sequencing selected bacterium was identified (Ayub et al. 2004). Yilmaz et al. (2005) screened less than 30 *Bacillus* taken by the grass field in the capital of Turkey; those were characterized like *Bacillus licheniformis*, *Bacillus brevis*, *Bacillus circulans*, *Bacillus megaterium*, *Bacillus subtilis*, *Bacillus sphaericus,* and *Bacillus coagulans*. Poly-β-hydroxybutyrate accumulation by the above-mentioned isolates was identified through the spectrophotometer and concluded as bioplastic accumulation between 1.06 and 41.67% (w/v) upon the basis of dried biomass. Maximum bioplastic production (41.67% w/v) was found in *B. brevis* M6.

Singh and Parmar (2011) collected 16 unknown samples, analyzed them, and compared with biopolymer-producing strains like *Ralstonia, Bacillus,* and *Pseudomonas*. Production of bioplastic by these strains was confirmed with microscopic techniques. This was the first report which demonstrates the synthesis of biopolymer experimentally by two novel isolates like *Stenotrophomonas maltophilia* and *Rahnella aquatilis*. *Stenotrophomonas maltophilia* was isolated from the local beach while *Rahnella aquatilis* was isolated from a plastic toy.

Preethi et al. (2012) isolated bacterial isolate SPY-1 which produced high PHB in comparison with different isolates. The strain had the capacity to accumulate 80% of the dry cell weight as PHB.

Hungund et al. (2013) screened soil samples from spent wash and oil mill for the screening of PHB-accumulating strains. One isolate that is *Paenibacillus durus* accumulates PHB in nitrogen-limiting medium in which fructose was present as the main carbon compound. Singh (2014) studied halophilic and halotolerant microbial strains from industrial and agricultural land. These strains were identified as *Bacillus subtilis* and *Pseudomonas* sp. Luria agar medium was identified as the best culture medium for these isolates. Biradar et al. (2015) isolated a total of six bacterial species as PHB producers. Out of them, isolate BBKGBS6 characterized as *Lysinibacillus sphaericus* BBKGBS6 by 16S rRNA sequencing and produced high amount of PHB.

7.5 Importance of PHB to Bacteria

When carbon is present in excess and phosphorous and nitrogen are available in restricted amount, then poly-beta-hydroxybutyrate granules are present as energy-producing bodies. Senior and Dawes (1971) reported that PHB might act like a reservoir of sinking energy; hence, it is known as reducing–oxidaizing reaction regulator material inside microbe. Ensphere of *Azotobacter* in laboratory was associated with PHB production (Stevenson and Socolofsky 1973). However, in natural environment, storage of high quantity of PHB was not required for encystment formation.

Anderson and Dawes (1990) reported microbes with accumulation of biopolymer impeded the debasement of cytosolic parts like protein/RNA during supplement starvation continously, be that as it may, not compulsory always. Poly-beta-hydroxybutyrate increased the survival of few microorganisms and also was used like an organic and vitality hotspot for spore development in *Bacillus* sp. Steinbuchel (1991) published that PHB was accumulated like an intracellular cytoplasmic inclusion body in microbial cell. These inclusion bodies act as a reduced biopolymer, given the bit of leeway and those become inaccessible carbon hotspot in favor of contending bacteria, what's more, that they were osmotically idle, hence influence the osmotic weight of the bacteria. Steinbuchel (1991) announced PHB accumulation in *Ralstonia eutropha*/*Rhodospirillum rubrum* contained pyridine of aceto-acetyl coenzyme-A. In this manner, biopolymer reservoir for diminishing counterparts and viewed as a fermentated product. In Rhizobium/*Bradyrhizobium japonicum*, at some stage in the bacteroid form, the N_2-assimilating enzymes rivaled by PHB production for reducing counterparts (Povolo et al. 1994; Hamieh et al. 2013).

7.6 Physical and Chemical Properties of PHB

Like any other polymer, chains of monomers in PHB form either homopolymer or heteropolymer. Generally, PHB known to be linear are composed of 3-hydroxybutanoic acid monomer units, each unit from an ester bond with the hydroxyl group of the other one.

PHB have an extremely regular structure; all the side gatherings on the polymer chain point a similar way. This makes the chains structure helical structures, with the side gatherings all directing endlessly from the focal point of the helix toward limit steric hindrance. Mass of biopolymer produced generally occurs in between 1×10^4 and 3×10^6 g/mol (Doi et al. 1990). PHB is a very low strength biomolecule that need to improve. High molecular weight PHB is valuable for delivering films and solid strands. The molecular weight of PHBs based on PHB-producing enzymes, that different into substrate necessity (Hazer et al. 2012; Kumar et al. 2013). Carbon sources play a key role in diversification of PHB production. These carbon sources include saccharides, n-alkanes, n-alkanoic acids, and n-alcohols. Waste streams like frying oil, vinegar, fats, food, agriculture, wastewater, and oil mill effluents provide free source of carbon for PHB production (Amy Tan et al. 2014). The PHB composition also depends on the metabolic pathway involved.

The molecular weights have been measured by gel permeation chromatography, light scattering, and sedimentation analysis. Their monomer composition has been analyzed by mass spectroscopy, gas chromatography, and nuclear magnetic resonance (NMR) detection. PHBs crystalline structure, poly dispersities, enthalpy of fusion, melting point, glass transition temperature, and mechanical properties were characterized by the help of various procedures such as differential scanning chromatography.

Poly-beta-hydroxybutyrate, compared with the synthetic plastic like poly-propylene, is identified as having the most similar properties. Likewise, biopolymer not only remains undissolved in aqueous solution or resistant to hydrolytic decomposition but also gets solubilized in organic compounds like chloroform and other chlorinated hydrocarbons. In the case of PHB inside the bacteria, it presents in a cytoplasm, amorphous state. After extraction from the bacteria, PHB becomes crystalline and rigid but a brittle material. The brittleness of PHB during mechanical processes makes it unresistant to stress (Hazer and Steinbüchel 2007). There is another monomer included in the structure of PHB such as 3 HV (3-hydroxyvalerate) which decreases PHB crystallinity and increases its elasticity (Gross et al. 1989).

The glass transition temperature/dissolving temperature of poly-beta-hydroxybutyrate are around 4 °C/180 °C correspondingly. The major problem of PHB is that it decomposes near its melting point. The thickness of formless biopolymer is 1.18 g cm^{-3} and crystalline PHB thickness is 1.26 g cm^{-3}. Augmentation for damage (5%) of poly-beta-hydroxybutyrate is less than expansion for damage of poly-propylene (400%). In general, poly-beta-hydroxybutyrates are hard and weak polymers contrasted with poly-propylene. Weakness is because of the development of enormous translucent spaces as spherulites (Barham and Keller 1986). Super high sub-atomic PHB was set up by modified *E. coli* with qualities from *C. necator*. It has higher mechanical properties than regular PHB (Kusaka et al. 1998). PHB has a few properties like poly-propylene with three exceptional highlights: thermo plastic machinability, protection by aqueous solution like water, and complete biodegradability (Page 1995).

7.7 Genes Involved in PHB Biosynthesis

The first PHB genes were isolated from *Zoogloea ramigera* by Peoples et al. (1987). From that point forward, various qualities associated with the development of PHBs have been described and cloned from different microorganisms.

The partition and quality association of pha (gene-encoding proteins for scl and mcl-PHAs) contrasts with different species (Reddy et al. 2003). In the three-stage scl PHA biosynthetic pathway, the genes are known as *pha-A* (β-ketothiolase), *pha-B* (aceto-acetyl-CoA reductase), and *pha-C* (PHA synthase/PHA polymerase). The gene *pha-A* codes for the compound β-ketoacyl CoA thiolase, which catalyzes the initial phase in the PHB amalgamation. It is an individual from a group of protein associated with the thio-lytic breakage of materials into acyl coenzyme-A and acetyl coenzyme-A. They are found in advanced eukaryotes, yeasts, and prokaryotes and separated in two gatherings dependent upon substrate explicitness. One gathering comprises thiolases expansive explicitness of β-ketoacyl CoA going from 4 to 16 carbon units. These qualities are available in a large portion of the scl

PHA-delivering creatures and in numerous microscopic organisms *pha-B* is related with PHA synthase. In *R. eutropha,* it is situated among *pha-A* and *pha-C* qualities and structures the *pha-CAB* operon (Schubert et al. 1988), while in *Bacillus sp.* it structures *pha-RBC* operon, wherein it is flanked by the *pha-R* and *Pha-C* (McCool and Cannon 1999). The chemical coded by the *pha-B*-NADPH-dependent aceto-acetyl CoA reductase is an (R)-3-hydroxyacyl CoA dehydrogenase, and it catalyzes the next stage in poly-beta-hydroxybutyrate biosynthesis pathway from changing over aceto-acetyl coenzyme-A in 3-hydroxybutyryl coenzyme-A.

Based on the mode of production, subunit arrangement, and material relevence, the PHA synthases are divided into four different classes. Type I PHA synthase is identified from protein of *Ralstonia eutropha* and mol. weight of the subunits is 64.3 kDa. *Pseudomonas* group represents the type II PHA synthase. The molecular weight of the PHA synthase subunit is 62.3 kDa. PHA synthase of type III is identified from protein of *Chromatium vinosum.* It consists of two different subunits coded by two genes *pha-C* and *pha-E* exhibiting the molecular weight of 39.7 and 40.5 kDa, respectively.

McCool and Cannon (1999) studied polyhydroxyalkanoates synthase of *B. megaterium* like a distinctive one and measured as more current sort (type IV). The molecular weight of the PHA synthase is identified as 40 kDa and another protein associated with this pha-R protein (22 kDa), coded by the gene *pha-R,* regulates the PHA synthase activity upon binding. The size of the gene varies from 1 to 2 kb. In *Bacillus megaterium,* the size of *pha-C* is 1089 base pairs (McCool and Cannon 1999). The PHA synthase (encoded by *Pha-C*) uses the coenzyme-A thio-esters of hydroxyalkanoic acids like material; initiate aggregation of various hydroxyalkanoic acids into polyhydroxyalkanoate by discharge of coenzyme-A.

In mcl-PHAs, there are additional genes encoding for the intermediary enzymes, namely *pha-G* (3-hydroxyacyl-acyl carrier protein-coenzyme-A transferase), *pha-J* (enoyl-*CoA* hydratase) besides PHA synthase. The *pha-G* gene was reported like an assimilation connection in between *denovo* fatty acid biosynthesis and PHB biosynthesis by *P. putida* KT2440 (Rehm and Kruger 1998). This enzyme converts 3-hydroxy ACP interceders of fatty acid production equivalent to coenzyme-A derivative, resultant 3-hydroxyacyl coenzyme-A as it may make use of like material in support of PHB synthesis. Two *Pseudomonas aeruginosa* genes homolgous with *pha-JAC,* assigned like the *pha-J1/pha-J2* with different substrate specificities were also cloned, and their role in PHA production was investigated (Tsuge et al. 2000). However, *Pseudomonas* isolates have two kinds of PHB synthases called as *pha-C1/pha-C2* which vary with material relevance; singular unit arrangement of produced polyhydroxyalkanoates (Matsusaki et al. 1998). Other than the synthesis genes, *pha-Z* is likewise needed to prepare the PHB inclusion bodies from delivering carbon and power once the restricted supplement is re-established (Rehm and Steinbüchel 1999).

Umeda et al. (1998) announced that into *Alcaligenes latus, Acinetobacter* species, *Ralstonia eutropha,* and *P. acidophila, phb CAB* genes are paired upon genetic material despite the fact that not really in a similar request. In *Rhizobium meliloti, Paracoccus denitrificans,* and *Z. ramigera, phb-AB* and *phb-C* locus are not related.

A PHB polymerase has two subunits, *Thiocystis violacea, Synechocystis,* and *Chromatium vinosum* determined from *pha-E* and *pha-C* (Hein et al. 1998). The PHA biosynthetic genes *pha-A* (3-ketothiolase), *pha-B* (NADPH-dependent aceto-acetyl coenzyme-A reductase), and *pha-C* (PHA synthase) by acetyl coenzyme-A are bunched and situated in a gene segment *pha-CAB*. The locus determining genes in support of PHB accumulation has been portrayed by 18 variable strains (Madison and Huisman 1999). Six genes determining proteins identified with PHB metabolism are bunched in two different loci, *pha-ZCPR* and *pha-AB*, in a PHB-accumulating bacterium *R. sphaeroides* FJ1 isolated from wastewater. PHB synthase, the key enzyme for PHB production, is continuously produced and its production level is not affected by different growth conditions (Yang et al. 2006).

In *A. caviae,* the PHA polymerase gene is flanked by an additional PHA biosynthetic enzyme enoyl-CoA hydratase encoded by *pha-J* (Fukui and Doi 1997). A few species like *P. denitrificans* had neighboring PHB synthase more genes as *pha-P* (determining phasing)/*pha-R* (determining regulator protein) identified with PHB biosynthesis (Rehm and Steinbüchel 2002).

P. oleovorans and *P. aeruginosa* have two different types of *pha-C* genes that are differenciated from genes for PHB present inside the cell (Rehm and Steinbüchel 2002). Additionally, down stream of synthase gene preparation, the *pha-D* gene is collinearly situated, trailed from *pha-I* and *pha-F*, those are transcribed into reverse track and determine basic and administrative enzymes (Rehm 2003). Isolates, namely *C. vinosum*, *T. violacea*, and *Synechocystis* sp. possess two subunits of PHB synthase, *pha-C* and *pha-E,* which are in one locus yet uniquely situated (Hein et al. 1998). Class IV synthase genes are present in microbes having a place with *Bacillus*; it involves *pha-R* and *pha-C*, those are isolated with *pha-B* (Rehm 2003). The regulation of PHB genes hushes up complex subject to various elements like accessibility of chemicals, elective variables, two-segment administrative frameworks, and auto inciting particles. In *Pseudomonas* species, the PHB synthesis locus other than two PHB synthases, *pha-C1/pha-C2* isolated by *pha-Z* is trailed with *pha-D*, *pha-F,* and *pha-I* (Hoffmann and Rehm 2004). It was discovered that *P. aeruginosa*, *P. putida,* and *P. oleovorans* contain three open reading frames (ORF) such as ORF-1, *pha-I,* and *pha-F* (Prieto et al. 1999). The consensus sequence like sigma 54 (RpoN) of *E. coli* promoter upstream of *pha-C1* was found in *P. aeruginosa* and *P. oleovorans* recommendating which was associated with controlling organization of PHA metabolic reactions into those microorganisms (Timm and Steinbuchel 1992). A *pha-C1* promoter part was not as much dynamic when citric acid/glucose is utilized like a carbon compound yet highly dynamic under occurrence with octanoic acid hence confirming the fact that promoter's initiation relies on the sort of carbon source (Prieto et al. 1999). It was seen into *Pseudomonas oleovorans* that disruption of *pha-F* brought more appearance of *pha-C1* (Prieto et al. 1999).

In *Acinetobacter* sp., PHB production is activated by low phosphate amount as a phosphate inducible promoter was discovered upstream of the initial gene, *pha-B* (Schembri et al. 1995). The production of PHB in *Vibrio harveyi* is directed by a β-hydroxybutyryl homoserine lactone which is a molecule for signaling that collects at high densities (Sun et al. 1994). Mutation reports in *R. eutropha* indicate that expanded transition of acetyl-*CoA* to PHB promotes more production (Park and Lee 1996). Transformation in the *phbHI* modifies the circumstances of PHB production in *R. eutropha* proposing the regulatory function for relating enzymes (Madison and Huisman 1999).

7.8 Biodegradation of PHB

Normally, bioplastic decomposition happens generally through scission of the principle chains or side chains of macromolecules, instigated by warm initiation, oxidation, photolysis, radiolysis, and hydrolysis. The character that recognizes PHA by oil-based synthetic plastics is their biodegradability. The biological decomposing of PHA in common habitats (stream and ocean waters, soil, slope, and manure) is one of the economically appealing highlights (Muhammadi et al. 2015). Biological degradation of PHB in the presence of oxygen brings about H_2O and CO_2, while under the absence of oxygen, the biodegradation compounds are CH_4 and CO_2.

PHB is used by microbes as a vitality source. PHB is biologically degraded in microbial dynamic conditions (Poirier et al. 1995a, b; Lee 1996a, b). Microbes grow upon outer part of a biopolymer and discharge chemicals that decompose PHB into monomers. These monomers are then utilized by the microbes as a carbon source for biomass development. The pace of bioplastic degradation relies upon an assortment of elements containing outside territory, bacterial action of removal condition, acid and base index, heat, dampness, and occurrence of various supplements. PHB is not able to dissolve in water or isn't influenced by dampness. It doesn't decompose under ordinary states of capacity and is steady inconclusively in air (Mergaert et al. 1992). Kim et al. (1997) saw that microbial for the most part surpassed the bacterial biomass and accordingly almost certainly growths may assume an impressive function in debasing polyesters, similarly as they dominatingly play out the deterioration of natural material in the environment.

Biodegradation of PHB was described by the "soil burial experiment." Indicated reduction in the weight of biopolymers covered inside earth surface to 1–2 months was 8.5–57.3% at 28 °C; 11.6–86.7% at 37 °C or afterward diminished up to 7.2–25.9% at 60 °C. PHB decomposed highly at 37 °C than at 28 °C and 60 °C. Practically identical outcomes have been watched already (Kim 2000).

PHB has been accounted in the direction of decomposition into oceanic conditions (Lake Lugano, Switzerland) in 8 months at 6 °C temperature (Johnstone 1990).

7.9 Identification of PHA by Staining Techniques

Whereas confining PHA-producing microorganisms by enviroment, that is important for monitoring quickly large assortment of microbes under brief timeframe. Dyes explicit for PHA are utilized inside location to inclusion bodies. Practical province recoloring strategy has been recommended like technique for quick viewing of PHA-aggregating microorganisms.

Hartman (1940) was pioneer for recommended utilization of Sudan Black-B for microbial lipid staining. Therefore, Burdon et al. (1942) affirmed more noteworthy estimation with stain and adjusted system to show intracellular greasy compound inside microorganisms from getting ready glass slides of microscopic organisms recolored by alcoholic Sudan Black-B stain further counterstained by red colored dye, safranin.

A method to exhibit intracellular greasy compound inside microorganisms with Sudan Black-B stain and the desiccated slide counterstained from red colored dye was given by Lewis (1941). Ostle and Holt (1982) announced that poly-beta-hydroxybutyrate bodies displayed solid orangish fluorescence while recolored from Nileblue-A. Warmth preset isolates were stained by Nileblue-A after that haste was seen at 460 nm.

The culture screening or determination framework was depicted to investigate the accumulation of PHB in *R. capsulatus* (Kranz et al. 1997). Screening by Nile red disintegrated in $(CH_3)_2CO$ recognized PHB-producing and non-producing strains. Spiekerman et al. (1999) suggested the utilization of a delicate feasible colony recoloring strategy utilizing Nile red used to view microorganisms to store PHB by naked eyes. An immediate consideration of stain into media didn't influence the development of microbes yet permitted to locate occurrence of PHB into the feasible cultures whenever throughout development. These PHB-producing microorganisms showed intense fluorescence while seen in ultraviolet rays. Juan et al. (1998) utilized reasonable culture view strategy for quick identification and screening of poly-beta-hydroxybutyrate accumulating exopolysaccharide lacking transformates by parental *R. meliloti*. Alcoholic Sudan Black-B was poured for recoloring microbial culture that was developed upon solid Luria Bertani media and culture was saved untouched for 30 min atleast. Stain was tapped and culture was delicately flushed with absolute alcohol. Bacterial culture unfit for absorbing Sudan Black-B remained uncolored, whereas polyhydroxybutyrate-accumulating strains seemed somewhat dark blue. Phosphine R is a fluorescent stain of PHB identification; in this technique, bacterial colonies are grown on agar plates in the presence of phosphine 3R which is a lipophylic dye. As indicated by this recoloring, the fluorescence of colonies under UV rays shows the presence of the biopolymer. Microbial colonies with high amount of PHB produce bright-green colored fluorescence under UV rays (Bonartseva and Myshkina 1985). Hamieh et al. (2013) detected the PHB in bacterial cell by transmission electron microscopy.

7.10 PHB Extraction and Recovery

PHB is an intracellular compound, which shows up as incorporation bodies in microorganisms (Kapritchkoff et al. 2006). In this manner, if exceptionally clean PHB is expected, cell rupturing ought to be done to break the cell membrame and secrete PHB. As such, extraction of biopolymer from the microbes is an essential phase of the cycle (Anderson and Dawes 1990) and polymer recuperation and cleaning steps are significant commitments to the creation price (Kapritchkoff et al. 2006). Biopolymer recovery steps, notwithstanding the expenses of keeping up unadulterated colonies and significant expenses of natural materials, are one more thing which adds toward significant expense of PHB production. Within a span of previous 20 years, a few recovery measures have been studied and researched so as to locate the monetary method to confine and decontaminate PHB. PHB is an intracellular product, a number of solvents have been studied for PHB extraction with the aim of finding a relatively cheaper method while achieving higher rate of recovery.

Doi et al. (1990) portrayed chloroform recovery strategy. PHB was recovered with heated chloroform into soxhlet to longer than 60 min. At that point, PHB recovered was isolated by fat from hastening through hexane, diethyl ether, and alcohol. At long last, PHB was redisintegrated in chloroform and additional cleansing done by precipitation with hexane.

Ramsay et al. (1994) inspected extraction of PHB by three distinctive chlorine-rich solutions like chloroform, 1,2-dichloroethane, and methylene chloride. They got the best recovery and purity when microbial cells were pretreated with $(CH_3)_2CO$. For all the three solvents, 15 min were ideal digestion time. Extended treatment brought decomposition of PHB weight. A level of extraction after the cells were pretreated with $(CH_3)_2CO$ were 70, 24, and 66%, while refluxed to 15 min by chloroform, methylene chloride, and 1,2-dichloroethane individually. Though degree of cleanliness of those three solutions into ideal situations were 96, 95, and 93%, separately. Ramsay et al. (1990) analyzed PHB extraction measure by *Ralstonia eutropha* utilizing hypochlorite treatment by detergent. Triton X100 and sodium dodecyl sulfate were examined those act as detergent in the reaction. An increase in cleanliness and molecular weight could be gotten from pretreating by detergent before the recovery from sodium hypochlorite. They revealed that detergent separated eliminated from absolute protein and extra enzyme was additionally taken out from sodium hypochlorite treatment.

Fidler and Dennis (1992) revealed framework for extraction of biopolymer bodies from *Escherichia coli* with communicating T7 bacteriophage genes. The lysozyme infiltrated damages the cell membrane being the reason for the release of PHB inclusion bodies. The framework created utilized different extra chromosomal plasmid and communicated at lower point all through the cell division process. Toward finish of collection phase, biomass were reaped then resuspended into sequestering compound. That actuated lysozyme to damage cellular membrane and deliver PHB when aggregation arrived at the most extreme. Triton X100 was additionally supplemented for the help in the cellular membrane disintegration. The productivity of rupturing of membrane was more noteworthy than 99% detailed by them.

Hahn et al. (1994) suggested strategy known as scattering by sodium hypochlorite and chloroform. They guaranteed about technique eliminated the greater part of the non-PHB cell materials during sodium hypochlorite treatment, which encouraged the division of PHB from the microbes. Treatment with sodium hypochlorite diminished the thickness of the chloroform stage. They likewise researched on the ideal state for PHB extraction from *R. eutropha* utilizing scatterings of chloroform and sodium hypochlorite. The ideal parameters from their trails were accounted for to be 90 min assimilation time with 30% solution of sodium hypochlorite and chloroform to watery phase proportion of 1:1 in volume. They acquired level of extraction of 91% and the degree of virtue more compared to 97%. The pace of decomposition expanded like the amount of sodium hypochlorite expanded. Because of the significant expense of dissolvable extraction, the proteolytic assimilation strategy was created via ICI, UK. Parts of that cycle incorporated heat action at 100–150 °C to break the membrane and deform genetic material; proteolytic processing and washing with positively charged detergent to cell particles without PHB. At last, concentrated PHB from centrifugation was treated with H_2O_2.

Now, PHB could be isolated from the liquid with the help of sedimentation technique. A serious damage of PHB during sodium hypochlorite treatment was every now and then announced. A decrease in molecular weight up to 50% of the PHB was reported when the microbial cells were treated with sodium hypochlorite (Lee 1996a, b). As indicated by Steinbuchel (1996), ICI utilized a blend of different enzymatic proteins during the enzymatic processing stage. These catalysts were phospholipase, lysozyme, proteinase, lecithinase, etc. These biological catalysts hydrolyzed a large portion of biomass except PHB but PHB stayed unaffected. Braunegg et al. (1998) detailed that ICI later utilized proteolytic compounds like pepsin, trypsin, and papain and a blend of these catalysts to extract PHB.

PHB recovers from the chloroform phase by non-solvent (aqueous solution of methanol) precipitation and filtration, under ideal parameters, the extraction was up to 91% and the purity of extracted PHB was more than 97%. Shawaphun and Manangan, 2009, reported the extraction of PHB produced by *A. latus* and found that $CHCl_3$ and CH_2Cl_2 gave relatively higher rates of PHB recovery (83.93% and 86.80%) compared to other non-solvents and protic solvents like toluene and hexane.

Bacillus sp. isolates CL1 and *Ralstonia eutropha*, enclosed PHB inside the microbial cytoplasm and phasing gene mutant *R. eutropha* having a solitary big PHB bodies. The optimum PHB bodies are encompassed by phosphor-lipids film and glass-like protein. Concentrated cell mass is dried either by freeze or spraydrying. PHB is recovered from untreated cell mass by dissolving it in a natural dissolvable, generally chloroform, chloro-limited methylene chloride, propylene carbonate, and dichloroethane (Charen et al. 2014; Gomaa 2014). Subsequent to eliminating cell parts by filtration and sedimentation, the PHB is sedimented in solvents that cannot dissolve PHB, for example, cool alcohol (Riedel et al. 2013). Solvents recover the PHB without damaging it by enhancing the cell membrane porousness of the PHB (Kunasundari and Sudesh 2011). Yang et al. (2011) built up a system for PHB utilizing straight alkyl benzene sulfonic acid like an option in contrast for regularly utilizing sodium dodecyl. In this technique, just 21% of the detergent is neccesary, compared to past techniques. Singh et al. (2009) looked into a novel self lyse microbial cell for extraction of PHB accumulated into *Bacillus megaterium*. Here framework, a gene group conveying the membrane rupturing and endo lysin of *Bacillus amyloliquefaciens* phage (Morita et al. 2001) was embedded in *Escherichia coli* and *Bacillus subtilis* transport vector pX. Herein expression scheme, "xyl R-xyl A" aimed genes are incited by xylose however hindered by dextrose (Salzberg and Helmann 2007). This contemporises cycle of unconstrained membrane degradation and material depletion, and this brings about an arrival of amassed PHB.

7.11 Growth Parameters to Increase PHB Production

Suzuki et al. (1986) detailed that *Pseudomonas* sp. developed into alcohol like only carbon and vitality supplement had delivered approximately 65% of PHB upon the basis of dried biomass measurement. So as to get more amount of bioplastic, legitimate media synthesis was used. Here, the amount of phosphate and ammonium were kept up at lower side. Nitrogen inadequacy was discovered as the best method for invigorate production of bioplastic. The restriction of soluble O_2 amount was identified to diminish pace of cell mass development and PHB accumulation. This result was opposing outcomes announced from different researchers.

Sillman and Casida (1986) deliberate vesicle development against PHB production into *Azotobacter vinelandii* ATCC12837. While inoculated into Burks agar with 0.2 mg of glucose per ml that produced bioplastic before vesicle development. Expansion of low degrees of ammonium chloride, urea/adenine caused an addition in vesicle development but PHB development was diminished. Hypo-xanthine and inosine expansion likewise invigorated vesicle development, despite the fact that PHB development was not decreased. It was reasoned that vesicle development by *Azotobacter* didn't need production higher than minimum concentration of PHB.

Poul et al. (1990) contemplated an impact with various carbon compounds at C: N proportion as 15:1. PHB collection was important in extended log phase to fixed period of development with noteworthy amounts of bioplastic produced upon malate like contrasted to acetic acid derivation, pyruvate, lactate, and fructose. More elevated stages were identified to gather while low PO_2 conditions were kept up through development.

As per Doi et al. (1990), two phases of fed-batch culture technique are generally utilized procedure to augment large amount of biomass and inclusion bodies. In the beginning of development stage, ideal wholesome conditions were utilized to build up a high cell number. At that point, a chosen supplement was restricted to invigorate PHB accumulation in the subsequent stage or production stage.

Daniel et al. (1992) detailed so as to *Pseudomonas* strain 135 while developed into ammonium restricted fed-batch culture utilizing alcohol like only carbon and vitality compound; this could accumulate the highest biopolymer that is 55%. At the point when the microbes were cultured on magnesium and phosphate-restricted medium, the accumulation of PHB were approximately 40% and 35%, separately. Kim et al. (1994) detailed as a fed-batch culture of *R. eutropha* strain NCIMB 11599 by controlled sugar amount and nitrogen constraint offers 121 g/L of PHB accumulation. At the point while nitrogen restriction condition was utilized to instigate bioplastic production, sodium hydroxide was added to the media to maintain acid-base equillibrium. Although, because of the cell membrane rupturing brought about by the harmfulness of sodium hydroxide, more cellular aging density was inconceivable. Moreover, this was essential to keep up phosphate and magnesium concentrations more than 0.35 g/L and 10 mg/L individually (Asenjo et al. 1995). Ryu et al. (1997) as a result subsequently embraced phosphate restriction methodology to incite bioplastic production where pH was constrained from expansion of ammonium hydroxide. In these parameters, 232 g/L of bioplastic were acquired.

Shimizu et al. (1994) researched accumulation of PHB by *Ralstonia eutropha* H16 fed in the company of butyric acid and valeric acid. Ideal parameters used for biopolymer accumulation were 3 g/L butyric acid at pH 8.0 by these bacteria. Seventy five percent of PHB was acquired beneath these parameters, whereas lesser biopolymer was produced once pH was set at 8.0. At this point when the amount of butyric acid was set at 0.03, 0.3, and 10 g/l, accumulation of biopolymer was 44%, 55%, and 63%, individually produced.

Bourque et al. (1995) researched the polyhydroxybutyrate accumulation by *Methylobacterium extorquens* ATCC 55366 utilizing methanol like chief carbon and vitality compound into a fed-batch fermentation technique. Polyhydroxybutyrate accumulation somewhere in the range of 40 and 46% on the basis of dried weight was formed by *M. extorquens*. Moreover, biomass synthesis and development pace of *M. extorquens* were influenced from minerals provided in different combinations. Biomass synthesis and the specific development pace of this bacteria was negatively affected in the absence of $FeSO_4$, $CaCl_2$, $(NH_4)_2SO_4$, $MnSO_4$, and $ZnSO_4$.

Parshad et al. (2001) checked for 37 strains and transformates of *Azotobacter chroococcum* in favor of polyhydroxybutyrate accumulation utilizing Sudan Black-B stain. Maximum PHB production was found to be with 2% glucose and 15 mM/L ammonium salt on 36 and 48 h of development into submerged culturing and into immobile culturing individually. They additionally saw that the purpose of polyhydroxybutyrate accumulation was elevated upon sucrose and D-glucose. Among inorganic nitrogen compounds, they discovered 15 mM/L ammonium acetic acid derivative concentration to be the best for polyhydroxybutyrate accumulation.

Ghatnekar et al. (2002) examined polyhydroxybutyrate accumulation by *Methylobacterium* SPV-49. They assessed various carbon compounds. Most extreme production of polyhydroxybutyrate was seen among glucose like the carbon nutrient. Alcohol with saccharides like lactose and sucrose were also additionally stimutaled PHB production. The impact of carbon to nitrogen proportion on PHB production was analyzed. Various procedures for recovery and isolation of the PHB from microbes were studied. The non-dissolvable-based strategy utilizing an elevated stress homogenizer within sight of sodium dodecyle sulfate was discovered to be generally good. Highest isolation of 98% was accomplished by homogenizing the biomass at 400 kg cm^{-2} in 5% sodium dodecyle sulfate.

Mahishi et al. (2003) examined the enhancement of polyhydroxybutyrate accumulation from transgenic *E. coli* holding the polyhydroxybutyrate incorporating genes *Pha*-CSa and *Pha*-BSa from *S. aureofaciens* NRRL 2209. Impacts of various C and N compounds on polymer production from transgenic *E. coli* were examined.

Among the carbon compounds utilized, palm oil, glucose, glycerol, and ethanol upheld PHB production. No polymer production was detected into genetically modified bacteria while sugarcane molasses was utilized like a carbon supply. A mixture of yeast extract, peptone, and corn-steep alcohol was utilized like nitrogen supply. The highest production (60%) of microbial dried biomass was estimated at 48 h of culture development at 37 °C into media containing glycerol as the chief carbon nutrient and yeast extract-peptone mixture like the nitrogen resource.

Khanna and Srivastava (2005) improved cultivation conditions intended for the development of *Ralstonia eutropha* NRRL814690 within sight of supplements to diminish the extraction price of PHB. Along with fructose and $(NH_4)_2SO_4$ like carbon and nitrogen compounds individually, *Ralstonia eutropha* showed highest cell mass of 3.25 g/L with polyhydroxubutyrate amount 1.4 g/L within 2 days. To decide the chances of development capability of *Ralstonia eutropha,* this was developed into various carbon compounds, of which lactic acid, fructose, glucose, and sucrose yielded great biomass formed and polyhydroxubutyrate granules. So as to fuse less expensive N compounds and developmental factors in culture medium, $(NH_4)_2SO_4$ was replaced by $(NH_4)_2NO_3$, urea, and NH_4Cl. The highest PHB production 3.84 g/L after 60 h of incubation obtained with urea as nitrogen source.

Senthil and Prabhakaran (2006) researched an impact of pH on production of polyhydroxubutyrate by *Alcaligenes eutrophus* MTCC1285 in various materials. They examined polyhydroxubutyrate accumulation at pH as 6.9 and 8.0. The concentration of polyhydroxubutyrate accumulation at pH 6.9 was 0.8, 0.5, 0.4, and at pH 8.0, they were 1.1, 0.65, 0.55, and 1.0 g/mL in sago, thippi, glucose, and molasses material containing culture medium, individually. Koller et al. (2007) reported the optimization of polymer accumulation on *Hydrogenophaga pseudoflava* by using hydrolyzed whey and valerate as substrate. Haas et al. (2008) studied the use of potato saccharide, saccharified waste like carbon compound for increased PHB production for *Cupriavidus necator*.

Cavalheiro et al. (2009) evaluated that the molecular weight of PHB accumulated with *C. necator* by glycerol in a range of 7.86×10^2 kDa (used glycerol) and 9.57×10^2 kDa (fresh glycerol) that permits preparing from basic methods of the plastic business.

Chee et al. (2010) screened the effect of lauric acid, myristic acid, oleic acid, palmitic acid, and stearic acid on *Burkholderia* sp. USM.

Nikodinovic Runic et al. (2011) reported that *Pseudomonas putida* produced mcl-PHA with styrene as substrate. Zafar et al. (2012) reported P(3HB) production using *A. lata.* In this investigation, the enhancement of PHB accumulation from *A. lata* MTCC 2311 was completed. From utilizing the hereditary algorithm upon the artificial neural organization, the anticipated most extreme PHB creation of 5.95 g/L originated with 35.2 g/L of sucrose and 1.58 g/L of urea; in any case, the most noteworthy trial PHB 5.25 g/L was accomplished utilizing 36.48 g/L of sucrose.

Povolo et al. (2013) screened lactose and sucrose as carbon source for PHB production on *Hydrogenophaga pseudoflava,* which showed PHB yield of 20.2–62.5% (w/w). Mozumder et al. (2015) investigated impure glycerol for its likely use as a modest feedstock for PHB production. The highest cell mass amount of 104.7 g/L was achieved with a PHB into the liquid cultural medium of 65.6 g/L.

7.12 Factors Affecting PHB Production

A few variables should be considered in the choice of microbes for the PHB production at industrial level for example, the capacity of microorganism to use a reasonable carbon compound, development pace, PHB accumulation intensity and the most extreme degree of PHB production (Ojumu et al. 2004). Different strategies are being developed to enhance the PHB accumulation and to reduce the complete price.

7.12.1 Microorganisms

The main view to consider is the compatibility of the microorganism for systematic accumulation of PHB by inexpensive sources (Keshavarz and Roy 2010) by balancing the limiting conditions required for PHB production. Enhancing the production rates of high yielding PHB cultures is being done. Faster growth rate and developing copolymers blends by using engineering recombinant microbes such as *E. coli* and *P. oleovorans* is also being investigated (Reddy et al. 2003; Suriyamongkol et al. 2007; Preusting et al. 1993).

7.12.2 Medium

The right selection of medium is necessary for high volumetric productivity of PHB. Choice of medium ingredients is responsible for production of homopolymers or copolymers. Several media from cheap sources to reduce the overall cost such as black treacle (Solaiman et al. 2006), corn-steep alcohol (Nikel et al. 2006), whey (Marangoni et al. 2002; Koller et al. 2008), rice and wheat-bran (Van-Thuoc et al. 2008; Huang et al. 2006), starch and starchy wastewaters (Kim and Chang 1998; Quillaguamán et al. 2005; Halami 2008; Haas et al. 2008), outflow from palm and olive oil mill (Pozo et al. 2002; Bhubalan et al. 2008; Ribera et al. 2001), activated sludge (Yan et al. 2006; Jiang et al. 2009), and swine waste (Cho et al. 1997) are being studied extensively. Starting media pH is important for the cell development and PHB production from *C. necator* (CCUG52238T) enhancing the underlying media pH at time spans of 0.5 units influenced both the biomass and PHB development. Both the biomass and PHB amount were enhanced when the starting media pH was enchanced from pH 6.0 to 7.0, that is, from 6.42 to 8.57 g/L for biomass and 20% (w/w) to 34% (w/w) to PHB amount, separately (Zahari et al. 2012).

7.12.3 Fermentation

A few techniques, for example, fed-batch and continuous culture method have been completed to enchance efficiency for PHB production (Doi et al. 1990; Lee 1996a, b; Du et al. 2001; Yu and Wang 2001; Du and Yu 2002). Different carbon sources (Durner et al. 2001), limiting nutrients, different feeding regimes (Sun et al. 2007), and different dilution rates (Durner et al. 2000) have been studied. Introduction of blend cultural medium, for example, activated sludge (Satoh et al. 1999; Chua et al. 2003; Lemos et al. 2006) can add and diminish the expense of bioplastics. Feast/famine sort of fermentation technique has additionally been researched (Dias et al. 2005; Lemos et al. 2006). The best and capable factors integrated shaking rate with confidence 90.04%, trailed by ferrous sulfate, sucrose, and sodium nitrate with certainty stages 81.16, 78.88, and 77.27%, individually (El-Raheem et al. 2013). Solid-state fermentation has been reconsidered like an option to submerged fermentation and could be a potential system in favor of the unexpansive extraction of PHB. The resources venture to solid-state fermentation is generally inferior compared to submerged fermentation and the expense of crude substances for solid-state fermentation would be modest since it utilizes dissipate farming substrates. These positive components make solid-state fermentation an expected procedure for PHB isolation (Sindhu et al. 2015).

7.12.4 Recovery

Since extraction of PHB contributes essentially to the general financial aspects of PHB production, a number of processes have been tried to increase PHB recovery. Pretreatment of cells, cell membrane lysis, enzymatic treatment of non-PHB in the framework, unconstrained freedom of PHB, disintegrated air buoyancy, and extraction utilizing supercritical CO_2 have been tried (Jacquel et al. 2008).

7.13 PHB Quantification and Characterization

Different methods are available for determining the PHB content, the structure of PHB, and the composition of monomer. The first method that was developed by Law and Slepecky (1961) involved transformation of the PHB to unsaturated carboxylic acid by heat treatment with concentrated sulfuric acid and analyzing spectrophotometrically. This method was superseded by gas chromatography (GC) and high-pressure liquid chromatography (HPLC) with a number of variations in sample processing. Braunegg et al. (1978) treated the cells to gentle methano lysis after that checked with gas chromatography. Riis and Mai (1988) improved this method by using propanol and HCl. Chloroform-extracted PHB cells from *B. megaterium* were quantitated by HCl acid ethanolysis and GC-mass spectrometric (MS) (Findlay and White 1983; Tan et al. 2014). Karr et al. (1983) used ion exclusion HPLC. Brandl et al. (1989) identified amount of PHB with gas chromatography after lysis with methanol freeze-dried microbe. After GC analysis, the presence and amount of 3-hydroxybutyric acid were determined. Gel permeation chromatography (GPC), differential scanning calorimeter (DSC), rheology scanning, phase contrast and transmission electron microscopy (TEM) are applied to do the PHB characterization (Yezza et al. 2007; Pereira et al. 2008), GPC for molecular weight measurement, DSC and rheology for melting temperature measurement, and TEM for inclusion body observation.

Nuclear magnetic resonance (NMR) technique has also been applied to characterize PHB. Doi et al. (1986) has analyzed the conformation of PHB with 500-MHz ^{1}H nuclear magnetic resonance spectroscopy and sequence determination of the monomeric units in 3HB-co-3 HV copolymer by analysis of 125-MHz ^{13}C NMR spectra. A number of NMR studies for different PHB to determine the constituent monomers have been done (Rozsa et al. 1996; De Waard et al. 1993; Doi et al. 1995). It has also showed that fourier transform infrared spectroscopy (FTIR) can be used for determining the crystallinity of polymers. Jarute et al. (2004) and Randriamahefa et al. (2003) applied FTIR to check the amount of the PHB in microbial solution (De Rooy et al. 2007).

Thin layer chromatography was used to determine the presence of PHB in production media. Black color bands were observed in TLC which indicated the presence of PHB in the production medium (Sharmila et al. 2011).

7.14 Mutagenesis

Transformation in genes of interest assumes significant role enhancing in bioplastic accumulation. Improvement of bacteria by transformation is extremely financially a savvy technique and deliberately a better methodology for gaining highest yield of polymer. Because of moderate development and less PHB amount, the industrial extraction was moved to *Alcaligenes eutropus*, a table sugar using transformates by first H16 strain (Byrom 1992). ICI side compounds, the UK created PHB valverate on a commercial level from sugar using transformates of *R. eutropha* by a glucose or propionic acid substrate blend like C compound and in phosphate-restricting state (Steinbuchel 1991). Park and Lee (1996) had the option to plan PHB high accumulating from chemical genetic transformation. They utilized N-methyl N′--nitroso guanidine for constraint of transformant bacteria by changes in enzymes associated with tricarboxylic acid cycle. Isocitrate dehydrogenase mutant of *C. necator* was screened after genetic transformation. The biopolymer production was essentially expanded because of the enhanced progression of carbon source to the bioplastic biological production cycle rather than the tricarboxylic acid cycle because of the initiation of biopolymer biosynthesis-associated enzymes promoted by the acetyl coenzyme-A collected inside the cell.

Katircioglu et al. (2003) determined the accepted mutant of *Bacillus megaterium* Y6, *B. sphaericus* X3, *B. subtilis* K8, and *B. firmus* G2 were checked for their poly-β-hydroxybutyrate accumulation limits. *fadA* mutant of *E.coli* strain WA101 showed increased accumulation of 3-hydroxydecanoate monomers up to 93% from sodium decanoate (Park et al. 2002a, b). Aceto-acetyl coenzyme-A reductase and β-ketothiolase were highly expressed in the *CydR* transformate of *Azotobacter vinelandii* prompting toward PHB production through the exponential development stage, dissimilar to the original strain that solitarily produced PHB during stationary stage (Wu et al. 2001).

It has been indicated that transformations that expand the degree of acetyl coenzyme-A act to enhance the production of bipolymer. The presence of a genetic change in strain may help clarify the decline in biopolymer production at 360 h, anyway more noteworthy exploration is important to verify or refute this hypothesis (Verlinden et al. 2007). Adwitiya et al. (2009) in this investigation assessed the poly-beta-hydroxybutyrate synthesis limit selected tranasformation of *Bacillus thuringiensis* IAM 12077. A *Bacillus* sp. distinguished like *Bacillus thuringiensis* IAM 12077 fit to produce 10–15% dry biomass weight poly–beta-hydroxybutyrate when cultured in supplement stock for 48 h was exposed to arbitrary mutagenesis. Within the ultra violate mutant selected, 19 accepted mutants synthesied higher poly–beta-hydroxybutyrate than the original strain, rather 2 bacterial strains synthesized less. B8 mutant showed promising poly-beta-hydroxybutyrate production as 24.68%; 1.54 times with high poly–beta-hydroxybutyrate accumulation that is 1.3 g/L; 5.4 times after that the parental isolate. Chemical transformation produced reputed transformates, of which 6 had a reduction, 3 had no change, and 2 had an enhancement into poly-beta-hydroxybutyrate accumulation. Although the enhancement into production was analogous to enhancement exhibited with the ultraviolet transformate B8, the yields didn't associatively increment. Some mutants of *R. eutropha* such as ATCC 17697 have been reported to use sucrose to accumulate PHB (El-Sayed et al. 2009; Abdelhad et al. 2009; Tanamool et al. 2009), little research has been done to prove that mutants of *R. eutropha* can produce PHB from sucrose as efficiently as with glucose or fructose.

Chen et al. (2011) constructed a series of strains and mutants which were able to produce the PHB in a modified M9 medium supplemented with 20 g/L xylose. Santanu et al. (2011) utilized chemical genetic transformation to make transformate of *C. necator* over producing PHB on cheap carbon compounds. They isolated 5 mutants over producing PHB at minimal-salt media when fructose is used like a carbon compound. Out of them transformates of some were capable of using sugar syrup for more PHB production. Mutations were stimulated with ultraviolet rays, acri-flavin, and 5-bromo urasil. With every bacterium, production of poly-beta-hydroxybutyrate according to dried biomass weight in percentage was identified in a scale of 1.46–63.45%. Acri-flavin created transformation produced on variable strain that is 173A2 that accumulate 200 μg/ml PHB in 24 h of growth utilizing the supplemented cultural media with 1.5% glycerol as carbon compound (Elsayed et al. 2013).

Girdhar et al. (2014) studied the transformation that was done with physical and chemical techniques such as UV light used as physical method and ethyl methyl sulfonate treatment as chemical method on *Bacillus flexus* and *Bacillus megaterium*. The outcomes uncovered that there was a magnificent improvement in biopolymer synthesis with right around two folds, when bacteria treatment performed with ultraviolet light as physical mutagen; however, no such improvement was seen with chemical mutagen like ethyl methyl sulfonate. The biopolymer production enhanced particularly from 15.68 to 18.61% with chemical mutagen while with ultraviolet light treatment the biopolymer production enhanced exceptionally, from 15.68 to 23.70% in case of *Bacillus flexus*.

7.15 Future of PHB

As petroleum cost keeps on raising, the world is being compelled to utilize bioplast for numerous uses. Endeavors have been made to utilize PHB in horticultural, clinical, and in bundling businesses. The headways have been found in the regions of bioplastic like conveyance vehicles for controlled medication delivery and advancement of restorative gadgets including inserts and three-dimensional platforms for tissue building (Lei et al. 2017). In fact, recently their functionality in newer areas like producing intracellular nanostructures has made many companies realize and harness their commercial potential for clinical applications, for example, drug conveyance or antibody advancement. An appealing thought is this utilization of reused bioplastic is a practical alternative for 3D printing, in addition to effectively certain highlights of these innovations, for example, reserve funds on crude materials, low vitality cost, just as low CO_2 emissions (Lanzotti et al. 2019). Right now their creation is costly; however, these plastics are just in their first phase of business advancement (Lee 1996a, b).

An immense body of research has been done in the past 80 years on PHB. Incredible progress have been done into enzyme designing of PHB synthases to change its specificities in order to deliver PHB with wanted monomer arrangement (Suriyamongkol et al. 2007). Wider usage of these composites and blends has allowed overcoming or minimizing the shortcoming in properties that limit their application (Keshavarz and Roy 2010). One of the most promising approaches in overcoming the major drawback of high production cost is the atomic rearing of genetically modified plants communicating practically dynamic PHB biosynthesis pathways and to synthesize PHB legitimately by agribusiness (Steinbuchel and Fuchtenbush 1998); Poirier et al. 1995a, b). Producing bioplastic into crops, in any case, will probably be more costly (3–4 US$/kg) than delivering corn-starch or soyabean-oil, that priced into 0.25–0.50 US$/kg (Suriyamongkol et al. 2007).

Also the extraction procedure and processing from plants still need to be researched. In addition, production of PHB from cheap resources, correct balance of ingredients, potential of different bacterial species needs to be investigated further in setting of expanded PHB production and efficiency, various kinds of PHB and simplicity of bioplastic extraction (Quillaguamán et al. 2010). Artifical synthesis or modification of PHB molecules will without a doubt flourish with the gigantic organic assorted variety of character, where identical enzyme exercises can be gotten by extraordinary spots, whereas genetic modifications turn out to be less and to a lesser extent an innovative obstacle (Madison and Huisman 1999). The change of CO_2 infers a decrease of ozone harming substance outflows. Hydrogen oxidizing microscopic organisms, for example such as *Cupriavidus necator* can store PHB utilizing CO_2 as a carbon source, for example, through an auto-trophic transformation (Mozumder et al. 2015). Notwithstanding, because of its moderate crystallization, thin handling temperature range and inclination to "creep," it isn't appealing for some, applications, requiring advancement so as to defeat these weaknesses (Babu et al. 2013).

7.16 Conclusion

In the twenty-first century, we are living in huge load of pollutions from many sources including polythene wastes. Plastic production also involves a number of harmful chemicals which pose environmental as well as human health risks. Natural mechanisms for self-regeneration cannot manage such harmful pollutants. Thus, this chapter has provided important information regarding the optimum parameters for biopolymer accumulation which may take advantage at commercial stage to biopolymer accumulation of a rapidly rising substitute for oil-based artificial polymer.

References

Abdelhad HM, Abdel Hafez AM, El-sayed AA, Khodair TA (2009) Copolymer [P(HB-CO-HV)] production as affected by strains and fermentation techniques. J Appl Sci Res 5:343–353

Adwitiya P, Ashwini P, Kumar A, Avinash Badri R, Kajal D, Vomsi P, Srividya S (2009) Mutagenesis of *Bacillus thuringiensis* IAM 12077 for increasing poly(-β) hydroxybutyrate (PHB) production. Turk J Biol 33:225–230

Akar A, Akkaya EU, Yesiladali SK, Çelikyilmaz G, Cokgor EU, Tamerler C, Orhon D, Cakar ZP (2006) Accumulation of polyhydroxyalkanoates by *Microlunatus phosphovorus* under various growth conditions. J Ind Microbiol Biotechnol 33:215–220

Amy Tan GY, Chen CL, Ling L, Liya G, Wang L, Razaad IMN (2014) Start a research on biopolymer polyhydroxyalkanoate (PHA): a review. Polymers 6:706–754

Anderson AJ, Dawes EA (1990) Occurrence, metabolism, metabolic role and industrial uses of bacterial polyhydroxyalkanoates. Microbiol Rev 54:450–472

Asenjo JA, Schmidt SA, Anderson PR, Andrews BA (1995) Effect of single nutrient limitation on poly-β-hydroxybutyrate molecular weight distribution in *Alcaligenes eutrophus*. Biotechnol Bioeng 46:497–502

Ayub ND, Pettinari MJ, Ruiz JA, Lopez NJ (2004) A polyhydrobutyrate producing *Pseudomonas* sp. isolated from Antarctic environments with high stress resistance. Curr Microbiol 49:170–174

Babu RP, O'Connor K, Seeram R (2013) Current progress on bio-basedpolymers and their future trends. Prog Biomater 2:1–16

Barham PJ, Keller A (1986) The relationship between microstructure and mode of fracture in polyhydroxybutyrate. J Polym Sci Part B Polym Phys 24:69–77

Bhubalan K, Lee WH, Loo CY, Yamamoto T, Tsuge T, Doi Y, Sudesh K (2008) Controlled biosynthesis and characterization of poly(3-hydroxybutyrateco- 3-hydroxyvalerate-co-3hydroxyhexanoate) from mixtures of palm kernel oil and 3HV-precursors. Polym Degrad Stab 93:17–23

Biradar GG, Shivasharana CT, Kaliwal BB (2015) Isolation and characterization of polyhydroxybutyrate (PHB) producing *Bacillus* species from agricultural soil. Eur J Exp Biol 5:58–65

Bonartseva GA, Myshkina VL (1985) Fluorescence intensity of nodule bacteria (*Rhizobium meliloti, R. phaseoli*) differing in activity, grown in the presence of the lipophilic vital stain phosphine 3R. Microbiology 54:535–541

Bourque D, Pomerleau Y, Groleau D (1995) High cell-density production of poly-β-hydroxybutyrate (PHB) from methanol by *Methylobacterium extorquens*: Production of high-molecular-mass PHB. Appl Microbial Biotechnol 44:367–376

Brandl H, Knee EJ, Fuller RC, Gross RA, Lenz RW (1989) Ability of the phototrophic bacterium *Rhodospirillum rubrum* to produce various poly(β- hydroxyalkanoates): potential sources for biodegradable polyesters. Int J Bio Macromol 11:49–55

Braunegg G, Sonnleitner B, Lafferty RM (1978) A rapid gas chromatographic method for the determination of poly-β-hydroxybutyric acid in microbial biomass. Eur J Appl Microbiol 6:29–37

Braunegg G, Lefebvre G, Genser KF (1998) Polyhydroxyalkanoates, biopolyesters from renewable resources: physiological and engineering aspects. J Biotechnol 65:127–161

Burdon KL, Stokes JC, Kimbrough CE (1942) Studies of the common aerobic spore-forming *Bacilli* staining for fat with Sudan Black B stain. J Bacteriol 43:717–724

Byrom D (1987) Polymer synthesis by microorganisms: technology and economics. Trends Biotechnol 5:246–250

Byrom D (1992) Production of poly-β-hydroxybutyrate: poly-β-hydroxyvalerate copolymers. FEMS Microbiol Rev 103:247–250

Cavalheiro JMBT, De Almeida M, Grandfils C, Da Fonseca M (2009) Poly (3-hydroxybutyrate) production by *Cupriavidus necator* using waste glycerol. Process Biochem 44:509–515

Charen T, Vaishal IP, Kaushalya M, Amutha K, Ponnusami V, Gowdhaman D (2014) Isolation and identification of polyhydroxybutyrate producing bacterial strain (*Bacillus thuringiensis* GVP) from chlorine contaminated soil. Int J Chem Tech Res 6:197–202

Chee JY, Tan Y, Samian MR, Sudesh K (2010) Isolation and characterization of a *Burkholderia* sp. USM (JCM15050) capable of producing polyhydroxyalkanoate (PHA) from triglycerides, fatty acids and glycerols. J Polym Environ 18:584–592

Chen Q, Wang Q, Wei G, Liang Q, Qi Q (2011) Production in *Escherichia coli* of poly (3-hydroxybutyrate-co-3-hydroxyvalerate) with differing monomer compositions from unrelated carbon sources. Appl Environ Microbiol 77:4886–4893

Cho KS, Ryu HW, Park CH, Goodrich PR (1997) Poly(hydroxybutyrate-cohydroxyvalerate)bfrom swine waste liquor by *Azotobacter vinelandii* UWD. Biotechnol Lett 19:7–10

Chua ASM, Takabatake H, Satoh H, Mino T (2003) Production of polyhydroxyalkanoates (PHA) by activated sludge treating municipal wastewater: effect of pH, sludge retention time (SRT) and acetate concentration in influent. Water Res 37:3602–3611

Dalal J, Sarma PM, Lavania M, Mandal AK, Lal B (2010) Evaluation of bacterial strains isolated from oil-contaminated soil for production of polyhydroxyalkanoic acids (PHA). Pedobiologia 54:25–30

Daniel M, Choi JH, Kim JH, Lebeault JM (1992) Effect of nutrient deficiency on accumulation and relative molecular weight of poly-β- hydroxybutyric acid by *Methylotrophic* bacterium, *Pseudomonas* 135. Appl Microbial Biotechnol 37:702–706

Datta R, Tsai SP, Bonsignore P, Moon SH, Frank JR (1995) Technological and economic potential of poly(lactic acid) and lactic acid derivatives. FEMS Microbiol Rev 16:221–231

De Rooy SL, Wahyuni ET, Wiratni Syamsiah S, Ismail J (2007) Purification and characterization of poly-hydroxybutyrate (phb) in *cupriavidus necator*. Indo J Chem 7:243–248

De Waard P, van der Wal H, Huijberts GN, Eggink G (1993) Heteronuclear NMR analysis of unsaturated fatty acids in poly(3-hydroxyalkanoates) Study of betaoxidation in *Pseudomonas putida*. J Biol Chem 268:315–319

Dias JML, Serafim LS, Lemos PC, Reis MAM, Oliveira R (2005) Mathematical modelling of a mixed culture cultivation process for the production of polyhydroxybutyrate. Biotechnol Bioeng 92:209–222

Doi Y, Kunioka M, Nakamura Y, Soga K (1986) 1H and 13C NMR analysis of poly (3-hydroxybutyrate) isolated from *Bacillus megaterium*. Macromolecules 19:1274–1276

Doi Y, Segwa A, Nakaumura S, Junioka MT (1990) Production of biodegradable co-polyesters by *Alcaligenes eutrophus*. In: Dawes EA (ed) New biosynthetic biodegradable polymers of industrial interest from micro-organisms. KAP, London, pp 37–48

Doi Y, Kitamura S, Abe H (1995) Microbial synthesis and characterization of poly(3-hydroxybutyrate-co-3-hydroxyhexanoate). Macromolecules 28:822–828

Doi Y, Kawaguchi Y, Koyama N, Nakamura S, Hiramitsu M, Yoshida Y, Kimura H (1992) Synthesis and degradation of polyhydroxyalkanoates in *Alcaligenes eutrophus*. FEMS Microbiol Rev 103:103–108

Du G, Yu J (2002) Metabolic analysis on fatty acid utilization by *Pseudomonas oleovorans*: mcl-poly(3hydroxyalkanoates) synthesis versus oxidation. Process Biochem 38:325–332

Du G, Chen J, Yu J, Lun S (2001) Continuous production of poly-3- hydroxybutyrate by *Ralstonia eutrophus* in a two-stage culture system. J Biotechnol 88:59–65

Durner R, Witholt B, Egli T (2000) Accumulation of Poly[(R)-3-hydroxyalkanoates] in *Pseudomonas oleovorans* during growth with octanoate in continuous culture at different dilution rates. Appl Environ Microbiol 66:3408–3414

El-Raheem AR, El-Shanshoury E-RK, Amara AA, Salama AF, Kishk SS (2013) Optimization of polyhydroxybutyrate (PHB) production by *Azotobacter vinelandii* using experimental design. Int J Curr Microbiol App Sci 2:227–241

Elsayed NS, Aboshanab KM, Aboulwafa MM, Hassouna NA (2013) Optimization of bioplastic (poly-β-hydroxybutyrate) production by a promising *Azomonas macrocytogenes* bacterial isolates P173. Afr J Microb Research 7:5025–5035

El-Sayed AA, Abdelhady HM, Abdel Hafez AM, Khodair TA (2009) Batch production of polyhydroxybutyrate (PHB) by *Ralstonia eutropha* and *Alcaligenes latus* using bioreactor different culture strategies. J Appl Sci Res 5:556–564

Fidler S, Dennis D (1992) Polyhydroxyalkanoate production in recombinant *E. coli*. FEMS Microbiol Rev 9:231–236

Findlay RH, White DC (1983) Polymeric betahydroxyalkanoates from environmental samples and *Bacillus megaterium*. Appl Environl Microbiol 45:71–78

Fukui T, Doi Y (1997) Cloning and analysis of the poly(3-hydroxybutyrate-co-3-hydroxyhexanoate) biosynthesis genes of *Aeromonas caviae*. J Bacteriol 179:4821–4830

Ghatnekar MS, Pai JS, Ganesh M (2002) Production and recovery of poly-β- hydroxybutyrate from *Methylobacterium* sp V49. J Chem Technol Biotechnol 77:444–448

Girdhar M, Sharma A, Mohan A (2014) Enhancement of commercial production of polymeric PHB material from bacterial strains through mutagenic Strategies. Biosci Biotechnol Res Asia 11:1591–1599

Gomaa EZ (2014) Production of polyhydroxyalkanoates (PHAs) by *Bacillus subtilis* and *Escherichia coli* grown on cane molasses fortified with ethanol. Braz Arch Biol Technol 57:145–154

Gopi S, Kontopoulou M, Ramsay BA, Ramsay JA (2018) Manipulating the structure of medium-chain-length polyhydroxyalkanoate (MCL-PHA) to enhance thermal properties and crystallization kinetics. Int J Biol Macromol 119:1248–1255

Gross RA, DeMello C, Lenz RW, Brandl H, Fuller RC (1989) Biosynthesis and characterization of poly(β-hydroxyalkanoates) produced by *Pseudomonas oleovorans*. Macromolecules 22:1106–1115

Haas R, Jin B, Zepf FT (2008) Production of poly (3-hydroxybutyrate) from waste potato starch. Biosci Biotechnol Biochem 72:253–256

Hahn SK, Chang YK, Lee SY (1994) Recovery and characterization of poly(3hydroxybutyric acid) synthesized in *Alcaligenes eutrophus* and recombinant *Escherichia coli*. Appl Environ Microbiol 61:34–39

Halami PM (2008) Production of polyhydroxyalkanoate from starch by the native isolate *Bacillus cereus* CFR06. World J Microbiol Biotechnol 24:805–812

Hamieh A, Olama Z, Holail H (2013) Microbial production of polyhydroxybutyrate, a biodegradable plastic using agro-industrial waste products. Global Adv Res J Microbiol 2:54–64

Hartman TL (1940) The use of Sudan Black B as a bacterial fat stain. Staining Technol 15:23–28

Hazer B, Steinbüchel A (2007) Increased diversification of polyhydroxyalkanoates by modification reactions for industrial and medical applications. Appl Microbiol Biotechnol 74:1–12

Hazer DB, Kilicay E, Hazer B (2012) Poly (3-hydroxyalkanoate)s: diversification and biomedical applications. A state of the art review. Mat Sci Eng 32:637–647

Heap B (2009) Preface. Philos Trans R Soc Lond B Biol Sci 364:1971–1971

Hein S, Tran H, Steinbuchel A (1998) *Synechocystis* sp. PCC6803 possesses a two component polyhydroxyalkanoic acid synthase gene similar to that of anoxygenic purple sulfur bacteria. Arch Microbiol 170:162–170

Hoffmann N, Rehm BHA (2004) Regulation of polyhydroxyalkanoate biosynthesis in *Pseudomonas putida* and *Pseudomonas aeruginosa*. FEMS Microbiol Lett 237:1–7

Huang TY, Duan KJ, Huang SY, Chen CW (2006) Production of polyhydroxyalkanoates from inexpensive extruded rice bran and starch by *Haloferax mediterranei*. J Ind Microbiol Biotechnol 33:701–706

Hungund B, Shyama VS, Patwardhan P, Saleh AM (2013) Production of polyhydroxyalkanoate from *Paenibacillus durus* BV-1 isolated from oil mill soil. J Microbiol Biochem Technol 5:13–17

Jackson DE, Srienc F (1994) Novel methods to synthesize polyhydroxyalkanoates. Ann N Y Acad Sci 8:134–148

Jacquel N, Lo C-W, Wei Y-H, Wu H-S, Wang SS (2008) Isolation and purification of bacterial poly (3-hydroxyalkanoates). Biochem Eng J 39:15–27

Jarute G, Kainz A, Schroll G, Baena JR, Lendl B (2004) On-line determination of the intracellular poly(β-hydroxybutyric acid) content in transformed *Escherichia coli* and glucose during PHB production using stopped-flow attenuated total reflection FT-IR spectrometry. Anal Chem 76:6353–6358

Jiang Y, Chen Y, Zheng X (2009) Efficient polyhydroxyalkanoates production from a waste-activated sludge alkaline fermentation liquid by activated sludge alkaline fermentation liquid by activated sludge submitted to the aerobic feeding and discharge process. Environ Sci Technol 43:7734–7741

Johnstone B (1990) A throw away answer. Far Eastern Econ Rev 147:62–63

Juan ML, Gonzalez LW, Walker GC (1998) A novel screening method for isolating exopolysaccharide deficient mutants. Appl Environ Microbiol 64:4600–4602

Kalia VC, Lal S, Cheema S (2007) Insight in to the phylogeny of polyhydroxyalkanoate biosynthesis: horizontal gene transfer. Gene 389:19–26

Kapritchkoff FM, Viotti AP, Alli RCP, Zuccolo M, Pradella JGC, Maiorano AE, Miranda BA (2006) Enzymatic recovery and purification of polydroxybutyrate produced by *Ralstonia eutropha*. J Biotechnol 122:453–462

Karr DB, Waters JK, Emerich DW (1983) Analysis of poly-,3-hydroxybutyrate in *Rhizobium japonicum* bacteroids by ion-exclusion high-pressure liquid chromatography and UV detection. Appl Environ Microbiol 46:1339–1344

Katircioglu H, Aslım B, Yuksekdao ZN, Mercan N, Beyatli N (2003) Production of poly-β-hydroxybutyrate (PHB) and differentiation of putative *Bacillus* mutant strains by SDS-PAGE of total cell protein. Afr J Biotech 2:147–149

Keshavarz T, Roy I (2010) Polyhydroxyalkanoates: bioplastics with a green agenda. Curr Opin Microbiol 13:321–326

Khanna S, Srivastava AK (2004) Recent advances in microbial polyhydroxyalkanoates. Process Biochem 40:607–619

Khanna S, Srivastava AK (2005) Statistical media optimization studies for growth and PHB production by *Ralstonia eutropha*. Process Biochem 4:2173–2183

Kim BS (2000) Production of poly(3-hydroxybutyrate) from inexpensive substrates. Enzyme Microb Technol 27:774–777

Kim BS, Chang HN (1998) Production of poly(3-hydroxybutyrate) from starch by *Azotobacter chroococcum*. Biotechnol Lett 20:109–112

Kim BS, Lee SC, Lee SY, Chang HN, Chang YK, Woo SI (1994) Production of poly (3-hydroxybutyric acid) by *Ralstonia eutropha* with glucose concentration control. Biotechnol Bioeng 43:892–898

Kim GJ, Lee IY, Yoon SC, Shin YC, Park YH (1997) Enhanced yield and a high production of medium-chain-length poly(3-hydroxyalkanoates) in a two step fed-batch cultivation of *Pseudomonas putida* by combined use of glucose and octanoate. Enzyme Microb Technol 20:500–505

Koller M, Hesse P, Bona R, Kutschera C, Atlic A, Braunegg G (2007) Potential of various archae and eubacterial strains as industrial polyhydroxyalkanoate producers from whey. Macromol Biosci 7:218–226

Koller M, Bona R, Chiellini E, Fernandes EG, Horvat P, Kutschera C, Hesse P, Braunegg G (2008) Polyhydroxyalkanoate production from whey by *Pseudomonas hydrogenovora*. Bioresour Technol 99:4854–4863

Kranz RG, Gabbert KK, Madigan MT (1997) Positive selection systems for discovery of novel polyester biosynthesis genes based on fatty acid detoxification. Appl Environ Microbiol 63:3010–3013

Kumar P, Patel SKS, Lee JK, Kalia VC (2013) Extending the limits of *Bacillus* for novel biotechnological applications. Biotechnol Adv 31:1543–1561

Kunasundari B, Sudesh K (2011) Isolation and recovery of microbial polyhydroxyalkanoates. Express Polym Lett 5:620–634

Kusaka S, Iwata T, Doi Y (1998) Microbial synthesis and physical properties of ultra-high-molecular-weight Poly[(R)-3-Hydroxybutyrate]. J Macromol Sci Part A 35:319–335

Lanzotti A, Martorelli M, Maietta S, Gerbino S, Penta F, Gloria A (2019) A comparison between mechanical properties of specimens 3D printed with virgin and recycled PLA. Proc Cirp 79:143–146

Law JH, Slepecky RA (1961) Assay of poly-β-hydroxybutyric acid. J Bacteriol 82:33–36

Lee SY (1996a) Bacterial polyhydroxyalkanoates. Biotechnol Bioeng 49:1–14

Lee SY (1996b) Plastic bacteria-progress and prospects for polyhydroxyalkanoate production in bacteria. Trends Biotechnol 14:431–438

Lee SY, Lee YK, Chang HN (1995) Stimulatory effects of amino acids and oleic acid on poly (3-hydroxybutyric acid) synthesis by recombinant *Escherichia coli*. J Ferment Bioeng 79:177–180

Lee SY, Choi J, Lee SH (2000) Production of polyhydroxyalkanoates by fermentation of bacteria. Macromol Symp 159:259–266

Lei T, Guan M, Liu J, Lin HC, Pfattner R, Shaw L, McGuire AF, Huang TC, Shao L, Cheng KT (2017) Biocompatible and totally disintegrable semiconducting polymer for ultrathin and ultralightweight transient electronics. Proc Natl Acad Sci U S A 114:5107–5112

Lemoigne M (1925) Production of β–hydroxybutyric acid by certain bacteria of the B. subtilis group. Ann Inst Pasteur Microbio 39:144–200

Lemos PC, Serafim LS, Reis MAM (2006) Synthesis of polyhydroxyalkanoates from different short-chain fatty acids by mixed cultures submitted to aerobic dynamic feeding. J Biotechnol 122:226–238

Lewis IM (1941) The cytology of bacteria. Bacterial Rev 5:181–236

Lokesh BE, Shamala TR, Chandrashekhar A (2005) Utilization of organic acids and sugars for PHA co-polymer biosynthesis by a native soil isolate *Sphingomonas* spp. In: 46th annual conference of association of microbologists of India, p 48

Luengo JM, Garcia B, Sandoval A, Naharro G, Olivera ER (2003) Bioplastics from microorganisms. Curr Opin Microbiol 6:251–260
Macrae RM, Wilkinson JF (1958) Poly-beta-hyroxybutyrate metabolism in washed suspensions of *Bacillus cereus* and *Bacillus megaterium*. J Gene Microbiol 19:210–222
Madison LL, Huisman GW (1999) Metabolic engineering of poly(3- hydroxyalkanoates):from DNA to plastic. Microbiol Mol Biol Rev 63:21–53
Mahishi LH, Tripathi G, Rawal SK (2003) Poly-β-hydroxybutyrate (PHB) synthesis by recombinant *Escherichia coli* harbouring, *Streptomyces aureoficiens* PHB biosynthesis genes: effects of various carbon and nitrogen sources. Microbiol Res 158:19–27
Marangoni C, Agenor F Jr, Aragão GMF (2002) Production of poly(3- hydroxybutyrate-co-3-hydroxyvalerate) by *Ralstonia eutropha* in whey and inverted sugar with propionic acid feeding. Process Biochem 38:137–141
Matsusaki H, Manji S, Taguchi K, Kato M, Fukui T, Doi Y (1998) Cloning and molecular analysis of the poly(3-hydroxybutyrate) and poly(3- hydroxybutyrate-*co*-3- hydroxyalkanoate) biosynthesis genes in *Pseudomonas* sp. strain 61-3. J Bacteriol 180:6459–6467
McCool GJ, Cannon MC (1999) Polyhydroxyalkanoates inclusion body associated proteins and coding regions in *Bacillus megaterium*. J Bacteriol 181:585–592
Mercan N, Aslim ZN, Yukesekdag BY (2002) Production of PHB by some *Rhizobium*. Turk J Biol 26:215–219
Mergaert J, Anderson C, Wouters A, Swings J, Kersters K (1992) Biodegradation of polyhydroxyalkanoates. FEMS Microbiol Rev 103:317–321
Merugu R, Girisham S, Reddy SM (2010) Production of phb (poly-β-hydroxybutyrate) by *Rhodopseudomonas palustris* ku003 and *Rhodobacter capsulatus* ku002 under phosphate limitation. Int J Appl Biol Pharm Technol 1:847–850
Mohapatra S, Samantaray DP, Samantaray SM (2015) Study on polyhydroxyalkanoates production using rhizospheric soil bacterial isolates of sweet potato. Ind J Sci Technol 8:57–62
Morita M, Tanji Y, Mizoguchi K, Soejima A, Orito Y, Unno HJ (2001) Antibacterial activity of *Bacillus amyloliquefaciens* phage endolysin without holin conjugation. Biosci Bioeng 91:469–473
Mozumder MSI, Gonzalez LG, De Wever H, Eveline IP, Volcke (2015) Poly(3-hydroxybutyrate) (PHB) production from CO2: Model development and process optimization. Biochem Eng J 98:107–116
Muhammadi S, Muhammad A, Hameed S (2015) Bacterial polyhydroxyalkanoates-eco-friendly next generation plastic: production, biocompatibility, biodegradation, physical properties and applications. Green Chem Lett Rev 8:56–77
Mukhopadhyay M, Paul AK (2003) Poly(3-hydroxybutyric acid) accumulation by some indigenous purple non-sulphur bacteria. In: 44th annual conference of association of microbologists of India, p 178
Narancic T, Cerrone F, Beagan N, O'Connor KE (2020) Recent advances in bioplastics: application and biodegradation. Polymers 12:920–958
Nikel PI, Pettinari MJ, Galvagno MA, Mendez BS (2006) Poly(3- hydroxybutyrate) synthesis by recombinant *Escherichia coli* arcA mutants in microaerobiosis. Appl Environ Microbiol 72:2614–2620
Nikodinovic Runic J, Casey E, Duane GF, Mitic D, Hume AR, Kenny ST, O'Connor KE (2011) Process analysis of the conversion of styrene to biomass and medium chain length polyhydroxyalkanoate in a two-phase bioreactor. Biotechnol Bioeng 108:2447–2455
Ojumu TV, Yu J, Solomo BO (2004) Production of polyhydroxyalkanoates, a bacterial biodegradable polymer. Afr J Biotech 3:18–24
Ostle AG, Holt JG (1982) Nile blue a as a fluorescent stain for polyhydroxybutyrate. Appl Environ Microbiol 44:238–241
Page WJ (1992) Production of polyhydroxyalkanoates by *Azotobacter vinelandii* Strain UWD in beet molasses culture. FEMS Microbiol Rev 103:149–158

Page WJ (1995) Bacterial polyhydroxyalkanoates, natural biodegradable plastics with a great future. Can J Microbiol 41:1–3

Pal S, Paul AK (2000) Accumulation of biodegradable plastics by Azotobacter. In: 59th annual conference of association of microbologists of India, p 58

Park J, Lee Y (1996) Metabolic characteristics of isocitrate dehydrogenase leaky mutant of *Alcaligenes eutrophus* and its utilization for poly-β- hydroxybutyrate production. J Ferment Bioeng 81:197–205

Park KW, Kim KJ, Howard AJ, Stark BC, Webster DA (2002a) Vitreoscilla haemoglobin binds to subunit I of cytochrome bo ubiquinol oxidases. J Biol Chem 277:334–337

Park KW, Kim KJ, Howard AJ, Stark BC, Webster DA (2002b) Vitreoscilla haemoglobin binds to subunit I of cytochrome bo ubiquinol oxidases. J Biol Chem 277:33334–33337

Parshad J, Suneja S, Kukeja K, Lakshminarayana K (2001) Poly-β- hydroxybutyrate production by *Azotobacter chroococcum*. Foliar Microbiol 46:315–320

Patwardhan PR, Srivastava AK (2004) Model-based fed-batch cultivation of *R.eutropha* for enhanced biopolymer production. Biochem Eng J 20:21–28

Peoples OP, Masamune S, Walsh CT, Sinskey AJ (1987) Biosynthetic thiolase from *Zoogloea ramigera*. III. Isolation and characterization of the structural gene. J Biol Chem 262:97–102

Pereira SMF, Rodriguez RS, Gomez JGC (2008) Biosynthesis and characterization of biodegradable poly(3-hydroxybutyrate) from renewable sources. Revista Matér 13:1–11

Philip S, Keshaverz T, Roy I (2007) Polyhydroxyalkanoates: biodegradable polymers with a range of applications. J Chem Technol Biotechnol 82:233–247

Pikon K, Czop M (2014) Environmental impact of biodegradable packaging waste utilization. Pol J Environ Stud 23:969–973

Poirier Y, Nawrath C, Somerville C (1995a) Production of polyhydroxyalkanoates, a family of biodegradable plastics and elastomers, in bacteria and plants. Biotechnology 13:142–150

Poirier Y, Somerville C, Schechtman LA, Satkowski MM, Noda I (1995b) Synthesis of high-molecular-weight poly([R]-(β)-3-hydroxybutyrate) in transgenic *Arabidopsis thaliana* plant cells. Int J Biol Macromol 17:7–12

Poul E, Mulard D, Blanc P, Fages J, Homa G, Parelleux A (1990) Effects of partial 'O' pressure and agitation on growth kinetics of *Azospirillum lipoferum* under fermentor conditions. Appl Environ Microbiol 56:3235–3239

Povolo SR, Tombolini A, Morea A, Anderson AJ, Casella S, Nuti MP (1994) Isolation and characterization of mutants of *Rhizobium meliloti* unable to synthesize poly–βhydroxybutyrate. Can J Microbiol 40:823–829

Povolo S, Romanelli MG, Basaglia M, Ilieva VI, Corti A, Morelli A, Chiellini E, Casella S (2013) Polyhydroxyalkanoate biosynthesis by *Hydrogenophaga pseudoflava* DSM1034 from structurally unrelated carbon sources. New Biotechnol 30:629–634

Pozo C, Martinez-Toledo MV, Rodelas B, Gonzalez-Lopez J (2002) Effects of culture conditions on the production of polyhydroxyalkanoates by *Azotobacter chroococcum* H23 in media containing a high concentration of alpechin (waste water from olive oil mills) as primary carbon source. J Biotechnol 97:125–131

Preethi R, Sasikala P, Aravind J (2012) Microbial production of polyhydroxyalkanoate (PHA) utilizing fruit waste as a substrate. Res Biotechnol 3:61–69

Preusting H, Kingma J, Huisman G, Steinbüchel WB (1993) Formation of polyester blends by a recombinant strain of *Pseudomonas oleovorans*: different poly(3-hydroxyalkanoates) are stored in separate granules. J Environ Polym Degrad 1:11–21

Prieto MA, Buhler B, Jung K, Witholt B, Kessler B (1999) *PhaF*, a polyhydroxyalkanoate-granule associated protein of *Pseudomonas oleovorans* GPo1 involved in the regulatory expression system for *pha* genes. J Bacteriol 181:858–868

Quillaguamán J, Hashim S, Bento F, Mattiasson B, Hatti-Kaul R (2005) Poly(beta-hydroxybutyrate) production by a moderate halophile *Halomonas boliviensis* LC1 using starch hydrolysate as substrate. J Appl Microbiol 99:151–157

Quillaguamán J, Guzmán H, Van-Thuoc D, Hatti-Kaul R (2010) Synthesis and production of polyhydroxyalkanoates by halophiles: current potential an future prospects. Appl Microbiol Biotechnol 85:1687–1696

Ramsay JA, Berger E, Ramsay BA, Chavarie C (1990) Recovery of poly-β- hydroxyalkanoic acid granules by a surfactant hypochlorite treatment. Biotechnol Tech 4:221–226

Ramsay JA, Berger E, Voyer R, Charavaric C, Ramasay BT (1994) Extraction of poly-3-hydroxybutyrate using chlorinated solvents. Biotechnol Bioeng 8:589–594

Randriamahefa S, Renard E, Guerin P, Langlois V (2003) Fourier transform infrared spectroscopy for screening and quantifying production of PHAs by *Pseudomonas* grown on sodium octanoate. Biomacromolecules 4:1092–1097

Rawte T, Mavinkurve S (1998) Polyhydroxybutyrates from mangrove flora. Annual Conference of Association of Microbiologists of India, Mangalore, p 241

Reddy CSK, Ghai R, Rashami, Kalia VC (2003) Polyhydroxy alkanotes: an overview. Bioresour Technol 87:137–146

Rehm BHA (2003) Polyester synthases: natural catalysts for plastics. Biochem J 376:15–33

Rehm BHA (2007) Biogenesis of microbial polyhydroxyalkanoate granules: a platform technology for the production of tailor made bioparticles. Curr Issues Mol Biol 9:41–62

Rehm BHA, Kruger SA (1998) A new metabolic link between fatty acid de novo synthesis and polyhydroxyalkanoic acid synthesis. The *PhaG* gene from *Pseudomonas putida* KT2440 encodes a 3-hydroxyacyl-acyl carrier proteincoenzyme a transferase. J Biol Chem 273:24044–24051

Rehm BHA, Steinbüchel A (1999) Biochemical and genetic analysis of PHA synthases and other proteins required for PHA synthesis. Int J Biol Macromol 25:3–19

Rehm BHA, Steinbüchel A (2002) PHA synthases: the key enzymes of PHA synthesis. In: Steinbüchel A, Doi Y (eds) Biopolymers. Wiley-VCH, Weinheim, pp 173–215

Renner G, Haage G, Braunegg G (1996) Production of short-side-chain polyhydroxyalkanoates by various bacteria from the rRNA superfamily III. App Microbiol Biotechnol 46:268–272

Ribera RG, Sanchez MM, Cormenzana AR (2001) Production of polyhydroxyalkanoates by *Pseudomonas putida* KT2442 harboring pSK2665 in wastewater from olive oil mills (alpechin). Electron J Biotechnol 4:116–119

Riedel SL, Brigham CJ, Budde CF, Bader J, Rha C, Stahl U, Sinskey AJ (2013) Recovery of poly (3-hydroxybutyrate-co-3-hydroxyhexanoate) from *Ralstonia eutropha* cultures with non-halogenated solvents. Biotechnol Bioeng 110:461–470

Riis V, Mai W (1988) Gas chromatographic determination of poly-β- hydroxybutyric acid in microbial biomass after hydrochloric acid propanolysis. J Chromatogr 445:285–289

Rozsa C, Gonzalez M, Galego N, Ortiz P, Martinez J, Martinez R, Gomez MR (1996) Biosynthesis and characterization of poly(β-hydroxybutyrate) produced by *Bacillus circulans*. Polym Bull 37:429–435

Ryu HW, Hahn SK, Chang YK, Chang HN (1997) Production of poly(3hydroxybutyrate) by high cell density fed-batch culture of *Alcaligenes eutrophus* with phosphate limitation. Biotechnol Bioeng 55:27–32

Sabirova JS, Ferrer M, Lunsdorf H, Wray V, Kalscheuer R, Steinbuchel A, Timmis KN, Golyshin PN (2006) Mutation in a "tesB-like" hydroxyacyl-coenzyme A-specific thioesterase gene causes hyperproduction of extracellular polyhydroxyalkanoates by *Alcanivorax borkumensis* SK2. J Bacteriol 188:8452–8459

Saharan BS, Grewal A, Kumar P (2014) Biotechnological production of polyhydroxyalkanoates: a review on trends and latest developments. Chin J Biol 1:1–18

Salzberg LI, Helmann JD (2007) An antibiotic-inducible cell wall-associated protein that protects *Bacillus subtilis* from autolysis. J Bacteriol 189:4671–4680

Santanu D, Munmus S, Muralikrishnan V (2011) Random mutagenesis on *Wautersia Eutropha* for improving poly-beta-hydroxybutyrate production from inexpensive carbon sources. Res J Biotechnol 6:5–9

Satoh H, Mino T, Matsuo T (1999) PHA production by activated sludge. Int J Bio Macromol 25:105–109
Schembri MA, Woods AA, Bayly RC, Davies JK (1995) Identification of a 13-kDa protein associated with the polyhydroxyalkanoic acid granules from *Acinetobacter* spp. FEMS Microbiol Lett 133:277–283
Schubert P, Steinbuchel A, Schlegel HG (1988) Cloning of *Alcaligens eutrophus* genes for the synthesis of polyhydroxybutyrate (PHB) and the synthesis of PHB in *E. coli*. J Bacteriol 170:5837–5847
Senior PJ, Dawes EA (1971) Poly-β-hydroxybutyrate biosynthesis and the regulation of glucose metabolism in *Azotobacter beijerinckii*. J Biochem 125:55–66
Senior PJ, Dawes EA (1973) The regulation of poly-β-hydroxybutyrate metabolism in *Azotobacter beijerinckii*. Biochem J 134:225–238
Senthil KB, Prabhakaran G (2006) Production of PHB (bio plastics) using bioeffluents as substrates by *Alcaligens eutropha*. Ind J Biotech 5:76–79
Sharmila T, Meenakshi SA, Kandhymathy A, Bharathidasan R, Mahalingam R, PanneerSelvam A (2011) Screening and characterisation of polyhydroxybutyrate producing bacteria from sugar industry effluents. World J Sci Technol 1:22–27
Shimizu H, Tamura S, Ishihara Y, Shioya S, Suga K (1994) Control of molecular weight distribution and mole fraction in poly(-D(-)- 3hydroxyalkanoates) (PHA) production by *Alcaligenes eutrophus*. In: Doi Y, Fukuda K (eds) Biodegradable plastics and polymers. Elsevier Science B. V., New York, pp 365–372
Shrivastav A, Mishra SK, Shethia B, Pancha I, Jain D, Mishra S (2010) Isolation of promising bacterial strains from soil and marine environment for polyhydroxyalkanoates (PHAs) production utilizing Jatropha biodiesel byproduct. Int J Biol Macromol 47:283–289
Sillman CE, Casida LE (1986) Cyst formation versus poly-β-hydroxybutyric acid accumulation in *Azotobacter*. Soil Biol Biochem 18:23–28
Sindhu R, Pandey A, Binod P (2015) Solid-state fermentation for the production of poly (hydroxyalkanoates). Chem Biochem Eng 29:173–181
Singh R (2014) Isolation and characterization of efficient poly-β-hydroxybutyrate (PHB) synthesizing bacteria from agricultural and industrial land. Int J Curr Microbiol App Sci 3:304–308
Singh P, Parmar N (2011) Isolation and characterization of two novel polyhydroxybutyrate (PHB)-producing bacteria. Afr J Biotechnol 10:4907–4919
Singh M, Patel SKS, Kalia VC (2009) *Bacillus subtilis* as potential producer for polyhydroxyalkanoates. Microb Cell Fact 8:1–11
Solaiman DKY, Ashby RD, Foglia TA, Marmer WN (2006) Conversion of agricultural feedstock and coproducts into poly (hydroxyl alkanoates). Appl Microbiol Biotechnol 71:783–789
Spiekerman P, Rehm BHA, Kalscheuer R, Baumeister D, Qteinbuchel A (1999) A sensitive viable colony staining method using Nile red for direct screening of bacteria that accumulate polyhydroxyalkonic acid and other lipid storage compounds. Arch Microbiol 171:73–80
Steinbuchel A (1991) Polyhydroxyalkanoic acids. In: Macmillan BD (ed) Biomaterials. VCH, New York, pp 123–213
Steinbuchel A (1996) PHB and other polyhydroxyalkanoic acids. In: Rehm HJ, Reed G (eds) Biotechnology. VCH, New York, pp 403–464
Steinbuchel A, Fuchtenbush B (1998) Bacterial and other biological systems for polyester production. Trends Biotechnol 16:419–427
Steinbuchel A, Hein S (2001) Biochemical and molecular basis of microbial synthesis of polyhydroxyalkanoates in microorganisms. Adv Biochem Eng/Biotechnol 71:82–119
Steinbuchel A, Pieper U (1992) Production of a copolyester of 3- hydroxybutyric acid and 3-hydroxyvaleric acid from single unrelated carbon sources by a mutant of *Alcaligenes eutrophus*. Appl Microbial Biotechnol 37:1–6
Steinbuchel A, Schlegel HG (1991) Physiology and molecular genetics of poly(β-hydroxy-alkanoic acid) synthesis in *Alcaligenes eutrophus*. Mol Microbiol 5:535–542

Steinbuchel A, Valentin HE (1995) Diversity of bacterial polyhydroxyalkanoic acids. FEMS Microbiol Lett 128:219–228

Stevenson LK, Socolofsky MD (1973) Role of poly-β-hydroxybutyric acid in cyst formation by *Azotobacter*. Antonie Van Leewen Hoek 39:341–350

Sun W, Cao JG, Teng K, Meighen EA (1994) Biosynthesis of poly-3- hydroxybutyrate in the luminescent bacterium, *Vibrio harveyi* and regulation by the lux autoinducer, N-(3-hydroxybutanoyl) homoserine lactone. J Biol Chem 269:785–790

Sun J, Dai ZW, Chen GQ (2007) Oligomers of polyhydroxyalkanoates stimulated calcium ion channels in mammalian cells. Biomaterials 28:3896–3903

Suriyamongkol P, Weselake R, Narine S, Moloney M, Shah S (2007) Biotechnological approaches for the production of polyhydroxyalkanoates in microorganisms and plants—A review. Biotechnol Adv 25:148–175

Suzuki TY, Amane T, Shimizu S (1986) Mass production of poly-β- hydroxybutyric acid by fully automatic fed-batch culture of methylotroph. Appl Microbial Biotechnol 23:322–329

Swift G (1993) Directions for environmentally biodegradable polymers research. ACC Chem Res 26:105–110

Tajima K, Igari T, Nishimura D, Nakamura M, Satoh Y, Munekata M (2003) Isolation and characterization of *Bacillus* sp. INT005 accumulating polyhydroxyalkanoate (PHA) from gas field soil. J Biosci Bioeng 95:77–81

Takeda MM, Matsuoka H, Ban H, Ohashi Y, Hikuma M, Koizumi JI (1995) Biosynthesis of poly (hydroxybutyrate-co-3-hydroxyvalerate) by a mutant of *Sphaerotilus natans*. Appl Microbiol Biotechnol 44:37–42

Tan GYA, Chen CL, Ge L, Li L, Wang L, Zhao L, Mo Y, Tan SN, Wang JY (2014) Enhanced gas chromatography-mass spectrometry method for bacterial polyhydroxyalkanoates (PHAs) analysis. J Biosci Bioeng 117:379–382

Tanamool V, Danvirutai P, Thanonkeo P, Imai T, Kaewkannetra P (2009) Production of poly-β-hydroxybutyric acid (PHB) from sweet sorghum juice by *Alcaligenes eutrophus* TISTR 1095 and *Alcaligenes latus* ATCC 29714 via batch fermentation. In: The 3th international conference on fermentation technology for value added agroculture products, p 52

Timm A, Steinbuchel A (1992) Cloning and molecular analysis of the poly(3- hydroxyalkanoic acid) gene locus of *Pseudomonas aeruginosa* PAO1. Eur J Biochem 209:15–30

Tsuge T, Fukui T, Matsusaki H, Taguchi S, Kobayashi G, Dio Y (2000) Molecular cloning of two (R)- specific enoyl Coenzyme A hydratase genes from *P. aerogenosa* and their use in polyhydroxyalkanoates synthesis. FEMS Micobiol Lett 184:193–198

Umeda F, Kitano Y, Muraking K, Yogi K, Mivra Y, Mizoguchi T (1998) Cloning and sequence analysis of the poly (3-hydroxyalkanoic acid) synthesis genes of *Pseudomonas acidophila*. Appl Biochem Biotechnol 70:341–352

Van-Thuoc D, Quillaguamán J, Mamo G, Mattiasson B (2008) Utilization of agricultural residues for poly (3-hydroxybutyrate) production by *Halomonas boliviensis* LC1. J Appl Microbiol 104:420–428

Verlinden RA, Hill DJ, Kenward MA, Williams CD, Radecka I (2007) Bacterial synthesis of biodegradable polyhydroxyalkanoates. J Appl Microbiol 102:1437–1449

Wang C, Sheng X, Equi RC, Trainer MA, Charles TC, Sobral BWS (2007) Influence of the poly-β-hydroxybutyrate (PHB) granule-associated proteins (*PhaP1* and *PhaP2*) on PHB accumulation and symbiotic nitrogen fixation in *Sinorhizobium meliloti* Rm1021. J Bacteriol 189:9050–9056

Wu GF, Wu QY, Shen ZY (2001) Accumulation of poly-β-hydroxybutyrate in cyanobacterium *Synechocystis* sp. PCC6803. Bioresour Technol 76:85–90

Yan S, Tyagi RD, Surampalli RY (2006) Polyhydroxyalkanoates (PHA) production using wastewater as carbon source and activated sludge as microorganisms. Water Sci Technol 53:175–180

Yang MK, Lin YC, Shen CH (2006) Identification of two gene loci involved in poly-beta-hydroxybutyrate production in *Rhodobacter sphaeroides* FJ1. J Microbiol Immunol Infect 39:18–27

Yang YH, Brigham C, Willis L, Rha C, Sinskey A (2011) Improved detergent-based recovery of polyhydroxyalkanoates (PHAs). Biotechnol Lett 33:937–942

Yezza A, Halasz A, Levadoux W, Hawari J (2007) Production of poly-β- hydroxybutyrate (PHB) by *Alcaligenes latus* from maple sap. Appl Microbiol Biotechnol 77:269–274

Yilmaz M, Soran HB, Yavuz (2005) Determination of poly-β-hydroxybutyrate (PHB) production by some *Bacillus* spp. World J Microbiol Biotechnol 21:565–566

Yu J, Wang J (2001) Metabolic flux modelling of detoxification of acetic acid by *Ralstonia eutrophus* at slightly alkaline pH levels. Biotechnol Bioeng 73:458–464

Zafar M, Kumar S, Dhiman A (2012) Optimization of poly(3- hydroxybutyrateco-3-hydroxyvalerate) production process by using *Azohydromonas lata* MTCC 2311 by using genetic algorithm on artificial neural network on response surface methodology. Biocatal Agric Biotechnol 1:70–79

Zahari MAKM, Ariffin H, Mokhtar MN, Salihon J, Shirai Y, Hassan MA (2012) Factors affecting poly(3-hydroxybutyrate) production from oil palm frond juice by *Cupriavidus necator* (CCUG52238T). J Biomed Biotechnol 1:1–8

Insight of Biopolymers and Applications of Polyhydroxyalkanoates

8

Rishabh Agrahari, Gargi Sarraf, Naveen Chandra Joshi, Swati Mohapatra, and Ajit Varma

Abstract

In this polluted era while we are talking about eco-friendly environment biological matter comes to our mind. Hence, if we go through the available works and copyrights, there will be no exact authentication on the definitions: "degradable," "biodegradable," "bio-based," "compostable," and "biopolymer"; these words often have overlapping way of explanation and multiple meanings. All these types of biopolymers having different properties and diversified applications. As these biopolymer is the need of the hour to make our environment ecofriendly. The only polymer that carries all the properties of the available properties of the polymer is Polyhydroxyalkanoates. So here in this chapter, we have complied the types and diversities of polymer and highlighted the application of most efficient bio-based, biodegradable, biopolymer "Polyhydroxyalkanoates."

8.1 Introduction

The intention of the environmental science is the polluted material like plastic should be degraded easily. The term "degradable" can be used for a polymer or plastic that disintegrates into other forms and structured by several physicochemical and biological process. Here, the term defines the ability to degrade not necessarily by a biological mechanism. "Biodegradable" is a term used for polymers that have the properties of degradation using microbes or microbial-derived metabolites under certain environmental, physical, and chemical conditions. However, the term

R. Agrahari · G. Sarraf · N. C. Joshi · S. Mohapatra (✉) · A. Varma
Amity Institute of Microbial Technology, Amity University, Uttar Pradesh, Noida, India
e-mail: smohapatra@amity.edu

A. Vaishnav, D. K. Choudhary (eds.), *Microbial Polymers*,
https://doi.org/10.1007/978-981-16-0045-6_8

"biodegradable" is ambiguous on its own. According to the American Society for Testing and Materials standard (ASTM D5488-94de1), the degradation process will be proper if the plastic will be degraded by certain metabolites of the microbes by enzyme action that leads to release of CH_4, CO_2, H_2O, and other inorganic compounds. In other countries, the terms are described differently, for example, the term "Biodegradability" is defined by the Japan BioPlastics Association (JBPA); JBPA explained it as a characteristic of any material capable of undergoing biological disintegration releasing the final product of CO_2 and H_2O, hence recycled in nature. The terms "Biodegradability" and "Disintegration" share an overlapping meaning but have distinguished meaning; disintegration is used for the characteristics of a material to break into separate and small pieces. When it comes to determining the bioplastics as biodegradable, the pre-established criteria of the ISO method are used for evaluation. To be classified as GreenPla®, the biodegradable plastics must meet the criteria like the content of safe intermediate reaction ("Japan BioPlastics Association (JBPA), Green- Pla®" n.d.). There are some other international standards that classify biodegradable polymers, for example, EN 13432:2000, ISO 17088:2012, ASTM D6400–12, EN 14995:2006. The term "BIO-based" focuses on the basis of organic raw material, especially the polymeric material should be generated from regenerated resource via biological processes or natural bases and end up with CO_2, H_2O. The raw material is defined as renewable when it comes from a natural process and has its yield equivalent or more than the consumption rate. Similarly, in this context Farm Security and Rural Investment Act (FSRIA) of 2002 defines bio-based product means "commercial or industrial things" apart from food and feed on biologically derived stuffs such as forest litter, kitchen waste, domestic organic waste, or marine materials.

Carbon-14 can mostly be detected in a bio-based material very rare in case of fossils, this characteristic represents its half-life of 5700 years. To ensure it, the characteristic of carbon from biomass as raw material is exploited to determine the content of a biopolymer; by quantifying carbon atoms released from a short CO_2 cycle (Mohapatra et al. 2017b). Few experiments of 14C suggested, "bio-based materials" can be defined as the degradable organic material derived from non-fossil biological bases. Using liquid scintillation calculating rate of degradation in dpm/g C (disintegrations/minute/ gm carbon) process is measured which can determine the 14C level. A bio-based polyethylene terephthalate (PET) contains nearabout 0.1 dpm/g C of 14C. ASTM D6002-96(2002)e1 proposed the processes of compostable polymer; basically, it is a process of complete decomposition of the plastic (e.g., Cellulose, Hemicellulose) to CO_2 and H_2O (Mohapatra et al. 2017a). This proposed phenomena after getting lots of criticism are modified to meet the international standards as mentioned below:

To meet the above standards, the polymer should have the following characteristics:

1. On composting, it should break down quickly
2. Under the composting, it should quickly biodegrade

3. Compost should retain utility and value and can support plant life
4. It should be nontoxic and biocompatible.

Considering two polymers like biodegradable and compostable plastic, both have their own features such as they need different physicochemical and environmental parameters for their disintegration and the end product of those polymers also make them different from each other. All compostable polymers are by default biodegradable, but not the other way around.

Two criteria define a "biopolymer" (or "bioplastic"): (1) raw material source and (2) bio-deterioration. On this basis, the biopolymer can be differentiated into the following types:

1. Type A: biopolymer is biodegradable and made from renewable sources (bio-based).
2. Type B: biopolymer is not biodegradable but made from renewable sources (bio-based).
3. Type C: biopolymer is biodegradable but it's made from fossil fuels.

Type A:

Type A biopolymer can be produced (with help of microorganisms, plants, and animals) or chemically synthesized using biological materials like corn, sugar, starch, etc. Type A biopolymer includes:

1. Poly (lactic acid) (PLA) considered as synthetic plastic, which can be polymerized in the lab using renewable raw materials.
2. Some microbial-derived lipid easter-like PHAs are the examples of biopolymers.
3. Biopolymer which naturally occurs, for example, starch and protein.

Natural polymers are polymers which are produced in the biosphere by various routes. Starch and PHA are the top-used bio-based biodegradable polymers.

Type B:

The biopolymers which are produced from renewable sources or biomass and lacks biodegradability are Type B biopolymers. Nonbiodegradable bio-based biopolymers (Type B) include various artificial polymers, synthesized by some renewable base materials, for example, certain polyamides derived from castor oil, other examples: biopolyethylene and propylene, biopoly (vinyl chloride) derived from bioethanol, etc.

Type C:

Such type of biopolymer are produced from fossil fuel; though some are derived from synthetic aliphatic polyester, they are biodegradable by composting method. Other polymers such as PCL, PBS, and "aliphatic-aromatic"-derived plastic can be deteriorated through microbial enzyme and metabolites.

The bioplastic material is defined by European bioplastics; in a broad range, it was categorized based on the origin and degradability nature (Ti et al. 2017) (Fig. 8.1). Thus, sugarcane-derived polymer biopolyethylene is classified as a

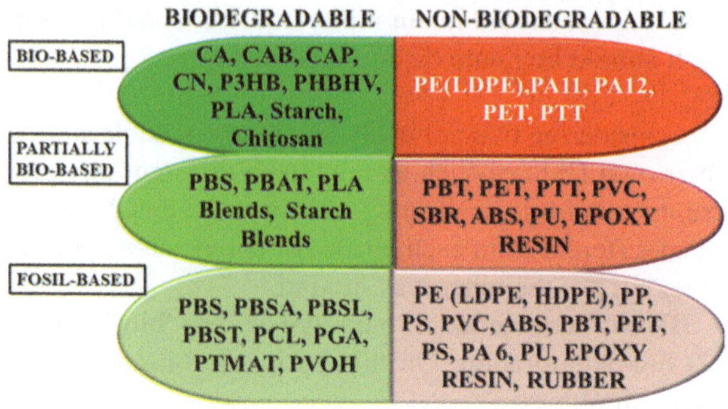

Fig. 8.1 Classification of biopolymers

biopolymer; it is nonbiodegradable but emits a lower amount of greenhouse gases compared to polyethylene; it is nicknamed "green polyethylene."

8.2 Classification of Biopolymers

Biopolymers are divided on the basis of source of origin point, degradation processes, and its end product. Some of the biopolymers are also derived from fossil fuel, due to nontoxicity and easy biodegradability nature that are considered under biopolymer. On the other hand, the bio-based polymer has its sources produced by microorganisms, plants, and animals. Comparing the degradability of bio-based biopolymer, it has a higher number of nonbiodegradable biopolymers compared to biodegradable bio-based biopolymer (Mohapatra et al. 2016; Ravenstijn 2010). Various categories are mentioned under which the three different types of biopolymers are distinguished (Fig. 8.2). Some are bio-based polymers, PLA being able to be synthesized by both renewable sources and fossil fuel; it is largely produced from renewable sources like starch and sugarcane (Dash et al. 2014; Mohapatra et al. 2016; Mohanty et al. 2014). Nowadays, the volume of bio-based thermoset biopolymers surpasses the volume of bio-based thermoplastic biopolymers. Based on the composition of biopolymers, it can be classified as blends, composites, or laminates (Raquez et al. 2010; Ravenstijn 2010).

8.3 Poly(β-, γ-, δ-hydroxyalkanoates)

Polyhydroxyalkanoates are having a vast variety of monomeric polymers. Physical properties of PHA lies in a range of Tm between 45 and 182 °C, Tg between −56 and 15 °C, crystallinity degree between 0 and 80%, and break at an elongation between 5 and 500%. The crystallization rate in a polymer can be controlled (Pati et al. 2020). With reducing molecular weight, the Tm gets reduced; however, Tm remains relatively constant. Elastomers have low crystallinity, an elastomer is PHAs with medium-chain length monomers containing 6–14 C carbon atoms. Poly

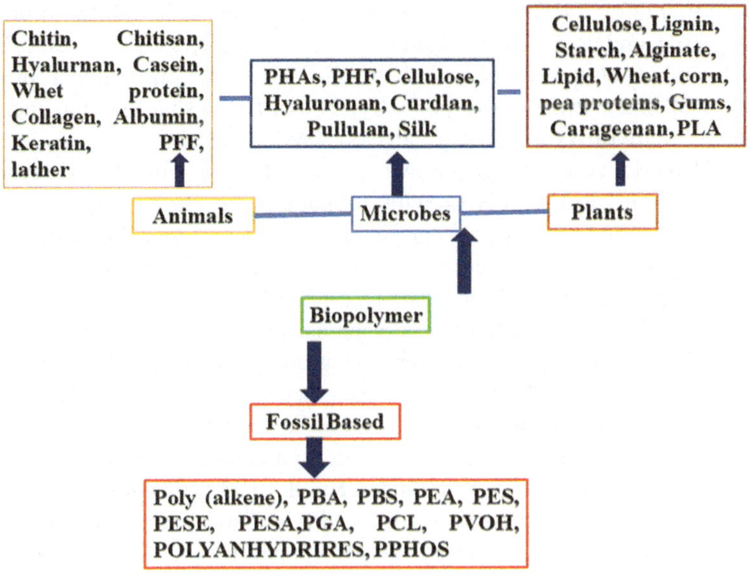

Fig. 8.2 Interrelationship between biopolymers

(3-hydroxybutyrate) (P3HB), poly(4-hydroxybutyrate) (P4HB), and poly (3-hydroxybutyrate-*co*-hydroxyvalerate) (PHBHV) are examples of short-chain PHAs, whereas P3HB is highly crystalline with Tg of 4 °C and a Tm of 180 °C (Mohapatra et al. 2015).

The polymer PHBHV has been characterized by its lower crystallinity and lowers Tm although its character depends upon its composition of 3-hydroxyvalerate (3 HV) no significant change has been observed in crystallinity and elasticity even on increasing 3-hydroxyvalerate (3 HV) content in polymer (Ravenstijn 2010).

Poly(3-hydroxybutyrate-*co*-3-hydroxyhexanoate) (PHBHHx) and poly (3-hydroxybutyrate-*co*-3- ydroxyoctanoate) are some classic examples of medium-chain length PHAs, medium-chain length PHAs characterized with Tg between −25 and 65 °C and Tm between 42 and 65 °C (Rodriguez-Contreras 2019).

When the composition of 3HH is increased in PHBHHx, it showed a significant change in its properties and the crystallinity of PHBHHx rapidly decreases. This is because 3HB and 3HH have structural difference in two methylene groups in the side chain—cf PHBHV (Mohapatra et al. 2017c).

The PHA polymer composed of short-chain length monomers are characterized by little elongation, brittle and easily tear under mechanical constraint, for example, P3HB and PHBHV. On increasing the side-chain of PHA, it will increase the flexibility of polymer, for example, PHBHV is tougher and less brittle compared to P3HB. Likewise, poly(3-hydroxyoctanoate) is an elastomer, and PHAs with longer side-chain has wax-like characteristics (Vroman and Tighzert 2009).

P3HB and polypropylene polymers are quite similar in physical and mechanical properties although their chemical structures are quite different, elongation at break is an exception here. P4HB has a tensile strength comparably similar to polyethylene, it is ductile, and do not exhibit brittle fracture characteristics. However, P4HB is well comparable with poly(trimethylene carbonate) for mechanical properties. PHBHHx and linear low-density polyethylene (LLDPE) share similar mechanical properties. Polyhydroxyalkanoates (PHAs) are polyesters of β-, γ-, δ-, and ε-hydroxyalkanoic acids (also known as 3-, 4-, 5-, and 6-hydoxyacids or -hydroxycarboxylic acids), PHAs with monomers of chain length in between 3 and 5 carbon atoms are semi-crystalline in nature (Rodriguez-Contreras 2019).

PHAs are distinguished based on several factors, mostly by the position of the hydroxyl group concerning the carboxyl group. Moreover, the length of the side chain, a variety of substituents in the side chains, and an additional methyl group on carbon atoms present in between carboxyl group and hydroxyl groups are also considered (Loos 2010). Microorganisms produce a wide range of PHAs based on the chemical structure of the provided carbon substrate, constituents in bacterial PHAs are identified with more than 150 hydroxyalkanoic acids so far (Loos 2010; Steinbüchel and Valentin 1995; Witholt and Kessler 1999). The length of monomers in PHAs affects its properties, most of the PHAs have side-chain monomers in between 3 carbon atoms (3-hydroxypropionate) and 14 carbon atoms (3-hydroxytetradecanoate). Based on the carbon length of monomers, PHAs are classified into short-chain length PHAs (3 to 5 carbon atoms) and medium-chain length PHAs (6 to 14 carbon atoms). On increasing monomer length, it increases the flexibility of polymers. Hence, the PHAs with short-chain length (P3HB) monomers are brittle whereas monomers with medium-chain length are flexible (poly (3-hydroxyoctanoate) (P3HO) is an elastomer). Microorganisms are used for the production of PHAs at a commercial level; in microbes, it is as intercellular carbon and energy storage materials. PHAs are produced naturally by soil bacteria and get degraded on consequent exposure to the same soil bacteria, compost, and marine sediment. PHAs are found as discrete granules inside bacterial cells and can share up to 90% of the bacterial dry cell weight. Biodegradation of PHAs took place when the bacterial cells start growing on their surface and secrete enzymes, as a result, break down the polymer into hydroxy acid and monomeric units. The hydroxy acid obtained from the PHA is utilized by bacteria as a carbon source.

The recent advancement has made it possible to produce monomer and polymer synthetically. PHA has opened the scope for the production of polymer from fossil-fuel to a renewable source, thus attracts researchers around the globe. PHAs have a vast variety of applications such as consumer packaging, disposable diaper linings, garbage bags, food, and medical products. Moreover, it is well-suited with the traditional production process (Niaounakis 2015). PHAs can have a huge biomedical application, thus it is being extensively studied for their properties and their potential uses in controlled release, to use in the formulation of tablets, surgical sutures, wound dressings, lubricating powders, blood vessels, tissue scaffolds, surgical implants to join tubular body parts, bone fracture fixation plates, and other orthopedic uses (1999, WO9932536 A1, METABOLIX INC.). PHAs have a huge diversity of monomeric composition, and hence polymer can be produced in a range of physical properties. PHAs like P3HB, PHBHV, P3HB4HB, PHBHHx, and several others are commercially available and are derived from bacterial fermentations (Alshehrei 2017).

8.4 Applications

Synthetic biodegradable polymers can be used to manufacture numerous products, but its biomedical application dates back to the late 1960s. The first approved biomedical object is bioabsorbable suture. Since then, the use of biopolymers has caused a revolution in biomedical field and manufactured products are not only limited to wound enclosures, body implants, tissue engineering materials but also drug delivery materials for disease treatment and in vivo sensing materials. Biodegradable polymers are suited to be used for medical purposes due to its ability to degrade and convert into nontoxic products and easy elimination from the body without the requirement of alternate specified routes.

There are a few other similar terms which can be categorized under the general definition of biodegradable polymers. Although these have been used interchangeably in the literature, there are slight differences in these terms as per their utilization and metabolism inside the body of the living organism. The terms are enlisted below with exact meanings:

- Bioresorbable: It refers to the biodegradable polymers that undergo bulk erosion and physiological conditions can be degraded into the products and the products reabsorption takes place within the body itself.
- Bioabsorbable: It refers to biodegradable polymers that body fluids can dissolve for their utilization. Furthermore, if the dissolved biopolymer is metabolized and/or eliminated by the body of the consumer as the excretory products, a bioabsorbable polymer can be referred to as bioresorbable.
- Bioerodible: It refers to water-insoluble biodegradable polymers. The biopolymers can undergo surface erosion and under physiological conditions, are converted to water-soluble products, employing physical processes like, dissolution and chemical processes, which are resorbed within the body.

The process of biodegradation of biopolymers can be mediated by the use of enzymes, the degradation can also occur by hydrolysis (breakdown in the presence of water) and/or possible due to the presence of the other chemical compounds inside the body. Hydrolysis of the ester linkage is the most prevalent method for biodegradation. The other labile bonds and sites which are hydrolytically unstable can also be the target for the degradation. This property of biodegradation of biopolymers is due to its instability that is of utmost importance for its application in biomedical sectors.

8.4.1 PHA Application

Polyhydroxyalkanoates (PHAs) naturally occurring biopolyesters, being produced by several microbes in an enormous variety having a range of properties that synthetic polymer didn't have. Properties like biodegradability, biocompatibility, and nontoxicity are crucial in a material having biomedical applications. Additionally, PHAs can be chemically synthesized or can be produced through fermentation of different combinations of carbohydrates, or adding specific inductors can also do the task. PHAs can be produced in a vast variety having diverse material properties through a conventional production process, therefore having large applicability in different fields like biomedical applications, biofuel, bioplastics, agriculture, and the food industry (Rodriguez-Contreras 2019). Moreover, PHAs are a potential alternative to fossil-fuel-based polymers (Zhang et al. 2009).

8.4.2 PHA as Packaging Materials

The first intended use of PHA was to manufacture packaging materials and bottles. Companies like Wella (Germany), Proctor & Gamble, Biomers, Metabolix, and several other companies used PHA for manufacturing articles such as bottles, utensils, feminine hygiene products, cups, medical-surgical garments, containers, and paper coating (Rodriguez-Contreras 2019; Weiner 1997). PHB fibers with high tensile strength were prepared by stretching the fibers after isothermal crystallization near the glass-transition temperature. However, increasing the isothermal crystallization period decreases the strength of PHB fiber. Extending time (more than 24 h) for isothermal crystallization of PHA fiber significantly increases its tensile strength (Tanaka et al. 2007). High molecular weight drawn fiber has a comparatively lower tensile strength compared to lower molecular weight drawn fiber. Vogel et al. (2007) had successfully improved the crystallization of PHA fiber and disabled the inhibition of secondary crystallization in the fiber. They used reactive extrusion with peroxide for improvement of the crystallization of PHB in a melt spinning process. By this method, they created a very strong fiber with a promising application and overcame the brittleness of PHA.

8.4.3 PHA as Biofuels

Zhang et al. (2009) showed that PHA and PHB can be used as biofuel by producing methyl esters of the PHB and mcl PHA. Transesterification process used to produce methyl esters of PHA and PHB which are 3-hydroxybutyrate methyl ester (3HBME) and mcl 3-hydroxyalkanoate methyl ester (3HAME); comparative study for the combustion heat was done and found that 3HBME and 3HAME had combustion heat of 20 and 30 kJ g^{-1}, respectively, hence comparable with ethanol having 27.1 kJ g^{-1}. Furthermore, they studied different fuels like ethanol, n-propanol, n-butanol, diesel, gasoline, blended fuels of 3HAME, and 3HBME.

Ethanol with 10% blend with 3HBME and 3HAME has shown improvement in combustion heat which elevated to 30 and 35 kJ g^{-1}, respectively. However, n-propanol and n-butanol didn't show any improvement in blendings of 3HAME and 3HBME, rather slightly decreased. In a rough estimation, it was found that it would cost US$1200/ton for the PHA-based fuel production (Rodriguez-Contreras 2019; Zhang et al. 2009). Furthermore, esterification of P(3HB-*co*-3 HV) will produce its methyl esters under optimum conditions (reaction temperature 65 °C and time of 60 h) with methanol and 15%(v/v) H_2SO_4. The highest obtained yield of 3HAME was 74.4% with 97% methyl ester content and its properties were able to meet the required standards of ASTM and Thailand's fuel standards (Sangkharak et al. 2020).

8.4.4 Biomedical Applicability of PHAs

8.4.4.1 Tissue Engineering

The area of tissue engineering is multidisciplinary refers to rejuvenation of injured or infected tissues and organs. Mainly it emphasizes on formation tissues using some biocompatible biomaterials, bioactive molecules, and cells (Waghmare et al. 2018). Tissue engineering can be further classified into hard tissue engineering (bones and cartilage) and soft tissue engineering (vascular and skin graft) (Findrik Balogová et al. 2018).

The biomaterial must have the mechanical strength and should serve healthy environment to cell adhesion and proliferation; these features enhance the quality of such biomaterial application in biomedical field (Lizarraga-Valderrama et al. 2019).

8.4.4.2 Hard Tissue

Bone

Bone tissue engineering deals with the recover and re-formation of new bone tissue by inducing cell growth. To compensate for the required mechanical flexibility, stress, and strain; PHA is optimized by blending it with hydroxyapatite (HA), inorganic substances, hydrogels, and nontoxic, and biocompatible materials. Degli Esposti et al. (2019) developed a scaffold having optimum amount of porosity that helps in bone rejuvenation; the scaffold was developed by mixing PHB with HA particles. Incorporation of HA helps to achieve the osteoinductivity and osteoconductivity in bioactive scaffold polymer. Chernozem et al. (2019) have developed biodegradable scaffolds using CaCO3 with PHB and PHBV and demonstrated comparative characterization. Meischel et al. (2016) have studied the response of bone to PHA composites by implanting it in the femora of growing rats.

Cartilage

Tissue engineering has opened the scope in area of rejuvenation of infected or ruptured cartilage. However, there are substantial challenges have been observed in cartilage surgery to carry out the function of cartilage as it was working before. To regain the properties of damaged cartilage, there is an utmost need for an alternative of cartilage that matches the properties, durability, and stability of native hard tissue (Rodriguez-Contreras 2019). Several studies have indicated that the PHA could be a potential for the application. A study was done by Ching et al. (2016) to produce a polymer mimicking the properties of the naturally occurring cartilage. They experimented with the blends of PHB and poly(3-hydroxyoctanoate) (P3HO) and optimized some of its properties like their structures, stiffness biocompatibility, and degradation rate (Bandekar et al. 2017).

8.4.4.3 Soft Tissue

Cardiac Tissue Engineering

Cardiac tissue engineering is among the hot topics in the field because of the utmost need due to massive clinical cases. Among the PHA family, mcl PHA is identified with properties essential for cardiac tissue engineering application. Being medium-chain length PHA, they show elastomeric nature and do meet the other needed properties like increased *Tg*, promising integration of myocardial network, and association of bioactive compounds such as endothelial growth factor enhances cellular viability, adhesion, and growth (Constantinides et al. 2018).

P4HB was identified with the potential applications; it can be used for congenital heart defects heart valves and heart grafts. This pliable material is also found helpful in developing absorbable monofilament sutures and among others like ligament repair, hernia, tendon, etc. (Rodriguez-Contreras 2019). Bagdadi et al. (2018) developed multifunctional cardiac patches based on P3HO having properties bordering that of cardiac muscles. It was considered a potential substitute as they intensify cell count, multiplication, adhesion, etc. When it comes to cardiovascular disease, valvular heart disease stands as the third most leading cardiovascular disease. Heart valve tissue engineering (HVTE) is a potential candidate for the treatment (Mohanty et al. 2015). For HVTE, a polymer is essential which can withstand the mechanical environment, guide tissue regeneration, and able to grow with the patient's heart. Various synthetic biodegradable elastomer which looked for their potential application in HVTE is mentioned in an article by Xue et al. (2017).

Wound Healing

Wound healing requires a more complex biomaterial which has a significant effect on the regeneration of injured skin, and it is a serious concern of reconstructive medicine (Shishatskaya et al. 2016). For the application in wound healing, several natural components such as collagen, fucoidan, hyaluronic acid, chitosan, and synthetic polymers such as polyurethanes, teflon, and methyl methacrylate have been tested previously as dressing materials (Demir and Cevher 2011). For the application of wound healing, a biomaterial has to meet the characteristics of healthy skin, the biomaterial should have effective antimicrobial property, permits gaseous exchange, promotes a moist wound environment, suitably elastic to fit the wound, and provides the mechanical support (Shishatskaya et al. 2016). Chronic bacterial infection is one of the major factors which inhibit the wound healing process. Thus, research has been conducted on the polymer having antimicrobial properties and antibiotic delivery systems. For example, a wound dressing was designed to have anti-biofilm protein-releasing properties; they embedded Dispersin B (DB) in an asymmetrical PHA substrate (Marcano et al. 2017).

8.4.4.4 PHAs for Organ Tissues

Among the PHA family, PHBVHHx is found to have a significant property to support the growth of stem cells. In some studies, PHBVHHx has been used to prepare the supportive scaffolds (three-dimensional) for organ tissue. In several studies, the film and scaffolds were loaded with the stem cells of the human umbilical cord (UC-MSCs) to regenerate the injured liver tissues. In a study, the liver-injured mouse was transplanted with the PHA scaffold biopolymer, and it has shown a positive response in the recovery of the injured liver (Su et al. 2014). Tashjian et al. (n.d.) have found PHA promising for healing tendons. They produced bioresorbable scaffolds to strengthen the suture-tendon interface in the rotator cuff repairs. In a study, a blend of PLA and PHB was developed which was aimed to replace the urethra tubule (Findrik Balogová et al. 2018).

8.4.4.5 PHA for Drug Delivery Systems

The reason for PHA being a potential polymer for drug delivery is because of its biodegradability under different environments. PHA is hydrolyzed into oligomer and monomer by several microbes secreting PHA-hydrolyzing enzymes extracellularly and utilizing these oligomers and monomers as nutrition (Rodríguez-Contreras et al. 2012; Tokiwa and Calabia 2004). Bacterial depolymerase and hydrolytic mechanisms take more than 52 weeks to degrade PHA (Misra et al. 2006).

Besides, several comparative studies compare the degradability of PHA with synthetic and semisynthetic polymers. A comparative study was done by Gil-Castell et al. (2019) to know the durability of different scaffolds which were PHB, polycaprolactone (PCL), PLGA, and polydioxanone (PDO). They found PCL and PHB are more durable than PLGA and PDO; Hence, PDO and PLGA will be suitable for short-term applications and PHB and PCL are the better options for long-term application.

However, several factors govern the biodegradability of PHA; factors like biopolymer composition, molecular mass, stereoregularity, crystallinity, and environmental conditions like temperature, pH, moisture, and nutrition supply plays a major role. Studies have shown that increasing the crystallinity of polymer decreases the biodegradability rate while biopolymers with lower molecular mass biodegrade quickly (Errico et al. 2009). Fabrication of PHA can be done in the desired copolymer composition and molecular mass, which makes it a potential biopolymer for the application of drug delivery. Moreover, several properties like its less toxicity, substantial impact on drug bioavailability, and improved encapsulation have already been demonstrated (Li and Loh 2017).

Masood et al. (2016) discussed the inferences of the encapsulated anticancer agent within different nanoparticles (PHA, cyclodextrin, and poly lactic-co-glycolic acid) are employed to accurately target the tumor sites. Michalak et al. (2017) reviewed the scientific advancement in the production of PHAs functionalized for multifunctional PHA nanoparticles, micelles, PHA for drug delivery, and conjugates of PHA-drug and PHA-proteins. Li and Loh outlined the advancement in therapeutic delivery carriers through the implication of PHA-based nano-vehicles (Li and Loh 2017). Scientists all around the world have studied different structures of the

PHA-based delivery system. For example, a system was developed by Lee et al. (2004), by which a metal stent was coated (through electro spun onto a metal stent) with drug-containing PHA fiber to make a biocompatible stent. For drug delivery, PHA has been used as a matrix to release drug formulation at the desired rate; based on the demand the scaffold casted with antimicrobial and anti-inflammatory agents to nullify the virulence factors. Manero's group studied extensively on PHAs scaffold having diversified cell and tissue development compound coatings (Rodríguez-Contreras et al. 2017, 2019).

8.5 Conclusion

Polyhydroxyalkanoates are renowned bioplastics and the beauty of the polymer is it can be molded in a diversified application based on the monomeric composition, physical and mechanical properties. Basically, these are fatty esters accumulated in the microbial cytosol that serve carbon neutrality. The major challenge of PHAs is the production cost; hence, extensive research is highly essential to reduce the production cost. Moreover, PHAs can be the best biomaterial for a biomedical applications that can compensate for the production cost. Due to the promising characteristics of PHA and its composites, nowadays it is gaining more market in biomedical and research into this area.

References

Alshehrei F (2017) Biodegradation of synthetic and natural plastic by microorganisms. J Appl Environ Microbiol 5(1):8–19

Bagdadi AV, Safari M, Dubey P, Basnett P, Sofokleous P, Humphrey E et al (2018) Poly (3-hydroxyoctanoate), a promising new material for cardiac tissue engineering. J Tissue Eng Regen Med 12(1):e495–e512. https://doi.org/10.1002/term.2318

Bandekar D, Prakash O, Mohapatra S, Hazra M, Hazra S (2017) Putative protein VC0395 _ 0300 from Vibrio cholerae is a diguanylate cyclase with a role in bio fi lm formation. Microbiol Res 202(March):61–70. https://doi.org/10.1016/j.micres.2017.05.003

Chernozem RV, Surmeneva MA, Shkarina SN, Loza K, Epple M, Ulbricht M et al (2019) Piezoelectric 3-D fibrous poly(3-hydroxybutyrate)-based scaffolds ultrasound-mineralized with calcium carbonate for bone tissue engineering: inorganic phase formation, osteoblast cell adhesion, and proliferation. ACS Appl Mater Interfaces 11(21):19522–19533. https://doi.org/10.1021/acsami.9b04936

Ching KY, Andriotis OG, Li S, Basnett P, Su B, Roy I et al (2016) Nanofibrous poly (3-hydroxybutyrate)/poly(3-hydroxyoctanoate) scaffolds provide a functional microenvironment for cartilage repair. J Biomater Appl 31(1):77–91. https://doi.org/10.1177/0885328216639749

Constantinides C, Basnett P, Lukasiewicz B, Carnicer R, Swider E, Majid QA et al (2018) In vivo tracking and 1 H/19 F magnetic resonance imaging of biodegradable polyhydroxyalkanoate/polycaprolactone blend scaffolds seeded with labeled cardiac stem cells. ACS Appl Mater Interfaces 10(30):25056–25068. https://doi.org/10.1021/acsami.8b06096

Dash S, Mohapatra S, Samantaray DP, Sethi AK (2014) Production of polyhydroxyalkanoates by sugar cane Rhizospheric soil bacterial isolates. J Pure Appl Microbiol 8(6):4889–4895

Degli Esposti M, Chiellini F, Bondioli F, Morselli D, Fabbri P (2019) Highly porous PHB-based bioactive scaffolds for bone tissue engineering by in situ synthesis of hydroxyapatite. Mater Sci Eng C 100:286–296. https://doi.org/10.1016/j.msec.2019.03.014

Demir A, Cevher E (2011) Biopolymers as wound healing materials: challenges and new strategies. In: Biomaterials applications for nanomedicine. InTech. https://doi.org/10.5772/25177

Errico C, Bartoli C, Chiellini F, Chiellini E (2009) Poly(hydroxyalkanoates)-based polymeric nanoparticles for drug delivery. J Biomed Biotechnol 2009:1–10. https://doi.org/10.1155/2009/571702

Findrik Balogová A, Hudák R, Tóth T, Schnitzer M, Feranc J, Bakoš D, Živčák J (2018) Determination of geometrical and viscoelastic properties of PLA/PHB samples made by additive manufacturing for urethral substitution. J Biotechnol 284:123–130. https://doi.org/10.1016/j.jbiotec.2018.08.019

Gil-Castell O, Badia JD, Bou J, Ribes-Greus A (2019) Performance of polyester-based electrospun scaffolds under in vitro hydrolytic conditions: from short-term to long-term applications. Nano 9 (5):786. https://doi.org/10.3390/nano9050786

Japan BioPlastics Association (JBPA). Green- Pla®. (n.d.). http://www.jbpaweb.net/english/english.%0Ahtm

Lee IS, Kwon OH, Meng W, Kang I-K, Ito Y (2004) Nanofabrication of microbial polyester by electrospinning promotes cell attachment. Macromol Res 12(4):374–378. https://doi.org/10.1007/BF03218414

Li Z, Loh XJ (2017) Recent advances of using polyhydroxyalkanoate-based nanovehicles as therapeutic delivery carriers. Wiley Interdiscip Rev Nanomed Nanobiotechnol 9(3):e1429. https://doi.org/10.1002/wnan.1429

Lizarraga-Valderrama LR, Taylor CS, Claeyssens F, Haycock JW, Knowles JC, Roy I (2019) Unidirectional neuronal cell growth and differentiation on aligned polyhydroxyalkanoate blend microfibres with varying diameters. J Tissue Eng Regen Med 13(9):1581–1594. https://doi.org/10.1002/term.2911

Loos K (2010) Biocatalysis in polymer chemistry. https://doi.org/10.1002/9783527632534

Marcano A, Bou Haidar N, Marais S, Valleton J-M, Duncan AC (2017) Designing biodegradable PHA-based 3D scaffolds with antibiofilm properties for wound dressings: optimization of the microstructure/nanostructure. ACS Biomater Sci Eng 3(12):3654–3661. https://doi.org/10.1021/acsbiomaterials.7b00552

Masood F (2016) Polymeric nanoparticles for targeted drug delivery system for cancer therapy. Mater Sci Eng C 60:569–578. https://doi.org/10.1016/j.msec.2015.11.067

Meischel M, Eichler J, Martinelli E, Karr U, Weigel J, Schmöller G et al (2016) Adhesive strength of bone-implant interfaces and in-vivo degradation of PHB composites for load-bearing applications. J Mech Behav Biomed Mater 53:104–118. https://doi.org/10.1016/j.jmbbm.2015.08.004

Michalak Michałand Kurcok P, Hakkarainen M (2017) Polyhydroxyalkanoate-based drug delivery systems. Polym Int 66(5):617–622. https://doi.org/10.1002/pi.5282

Misra SK, Valappil SP, Roy I, Boccaccini AR (2006) Polyhydroxyalkanoate (PHA)/inorganic phase composites for tissue engineering applications. Biomacromolecules 7(8):2249–2258. https://doi.org/10.1021/bm060317c

Mohanty S, Samantaray D, Mohapatra S, Rath SN, Mohanty DP, Rath CC (2014) Studies on biodegradation of medium chain length PHAs by soil bacterial isolates. J Adv Microbiol 1 (6):350–357

Mohanty S, Samantaray D, Mohapatra S (2015) Studies on biodegradation of medium chain length PHAs by soil bacterial isolates. J Adv Microbiol 1(6):350–357

Mohapatra S, Samantaray DP, Samantaray SM (2015) Impact of process recovery on PHA production by Bacillus thuringiensis RKD-12. Pollut Res 34(2):395–400

Mohapatra S, Samantaray DP, Samantaray SM, Mishra BB, Das S, Majumdar S et al (2016) Structural and thermal characterization of PHAs produced by Lysinibacillus sp. through submerged fermentation process. Int J Biol Macromol 93:1161–1167. https://doi.org/10.1016/j.ijbiomac.2016.09.077

Mohapatra S, Mohanta PR, Sarkar B, Daware A, Kumar C, Samantaray DP (2017a) Production of polyhydroxyalkanoates (PHAs) by Bacillus strain isolated from waste water and its biochemical characterization. Proc Natl Acad Sci India Sect B Biol Sci 87(2):459–466. https://doi.org/10.1007/s40011-015-0626-6

Mohapatra S, Sarkar B, Samantaray DP, Daware A, Maity S, Pattnaik S, Bhattacharjee S (2017b) Bioconversion of fish solid waste into PHB using Bacillus subtilis based submerged fermentation process. Environ Technol 38(24):3201–3208. https://doi.org/10.1080/09593330.2017.1291759

Mohapatra S, Maity S, Dash HR, Surajit Das SP, Chandi Charan Rath DS (2017c) Bacillus and biopolymer: prospects and challenges. Biochem Biophys Rep 12:206–213. https://doi.org/10.1016/j.bbrep.2017.10.001

Niaounakis M (2015) Definitions of terms and types of biopolymers. In: Biopolymers: applications and trends. Elsevier, pp 1–90. https://doi.org/10.1016/B978-0-323-35399-1.00001-6

Pati S, Maity S, Dash A, Jema S, Mohapatra S, Das S, Samantaray DP (2020) Biocompatible PHB production from Bacillus species under submerged and solid-state fermentation and extraction through different downstream processing. Curr Microbiol 77:1203. https://doi.org/10.1007/s00284-020-01922-7

Raquez JM, Deléglise M, Lacrampe MF, Krawczak P (2010) Thermosetting (bio)materials derived from renewable resources: a critical review. Progr Polym Sci 35:487. https://doi.org/10.1016/j.progpolymsci.2010.01.001

Ravenstijn J (2010) Bioplastics in consumer electronics. Ind Biotechnol 6:252. https://doi.org/10.1089/ind.2010.6.252

Rodriguez-Contreras A (2019) Recent advances in the use of polyhydroyalkanoates in biomedicine. Bioengineering 6(3):82. https://doi.org/10.3390/bioengineering6030082

Rodríguez-Contreras A, Calafell-Monfort M, Marqués-Calvo MS (2012) Enzymatic degradation of poly(3-hydroxybutyrate-co-4-hydroxybutyrate) by commercial lipases. Polym Degrad Stab 97(4):597–604. https://doi.org/10.1016/j.polymdegradstab.2012.01.007

Rodríguez-Contreras A, García Y, Manero JM, Rupérez E (2017) Antibacterial PHAs coating for titanium implants. Eur Polym J 90:66–78. https://doi.org/10.1016/j.eurpolymj.2017.03.004

Rodríguez-Contreras A, Guillem-Marti J, Lopez O, Manero JM, Ruperez E (2019) Antimicrobial PHAs coatings for solid and porous tantalum implants. Colloids Surf B Biointerface 182:110317. https://doi.org/10.1016/j.colsurfb.2019.06.047

Sangkharak K, Khaithongkaeo P, Chuaikhunupakarn T, Choonut A, Prasertsan P (2020) The production of polyhydroxyalkanoate from waste cooking oil and its application in biofuel production. Biomass Conversion Biorefinery. https://doi.org/10.1007/s13399-020-00657-6

Shishatskaya EI, Nikolaeva ED, Vinogradova ON, Volova TG (2016) Experimental wound dressings of degradable PHA for skin defect repair. J Mater Sci Mater Med 27(11):165. https://doi.org/10.1007/s10856-016-5776-4

Steinbüchel A, Valentin HE (1995) Diversity of bacterial polyhydroxyalkanoic acids. FEMS Microbiol Lett 128:219. https://doi.org/10.1016/0378-1097(95)00125-O

Su Z, Li P, Wu B, Ma H, Wang Y, Liu G et al (2014) PHBVHHx scaffolds loaded with umbilical cord-derived mesenchymal stem cells or hepatocyte-like cells differentiated from these cells for liver tissue engineering. Mater Sci Eng C 45:374–382. https://doi.org/10.1016/j.msec.2014.09.022

Tanaka T, Yabe T, Teramachi S, Iwata T (2007) Mechanical properties and enzymatic degradation of poly[(R)-3-hydroxybutyrate] fibers stretched after isothermal crystallization near Tg. Polym Degrad Stab 92(6):1016–1024. https://doi.org/10.1016/j.polymdegradstab.2007.02.017

Tashjian RZ, Kolz CW, Suter T, Henninger HB (n.d.) Biomechanics of polyhydroxyalkanoate mesh-augmented single-row rotator cuff repairs. Am J Orthop 45(7):E527–E533. http://www.ncbi.nlm.nih.gov/pubmed/28005110

Ti P, Maity S, Samantaray DP, Mohapatra S (2017) Bacillus and biopolymer with special reference to downstream processing. Int J Curr Microbiol Appl Sci 6(6):1504–1509. https://doi.org/10.20546/ijcmas.2017.606.177

Tokiwa Y, Calabia BP (2004) Review degradation of microbial polyesters. Biotechnol Lett 26 (15):1181–1189. https://doi.org/10.1023/B:BILE.0000036599.15302.e5

Vogel R, Tändler B, Voigt D, Jehnichen D, Häußler L, Peitzsch L, Brünig H (2007) Melt spinning of bacterial aliphatic polyester using reactive extrusion for improvement of crystallization. Macromol Biosci 7(6):820–828. https://doi.org/10.1002/mabi.200700041

Vroman I, Tighzert L (2009) Biodegradable polymers. Materials 2:307. https://doi.org/10.3390/ma2020307

Waghmare VS, Wadke PR, Dyawanapelly S, Deshpande A, Jain R, Dandekar P (2018) Starch based nanofibrous scaffolds for wound healing applications. Bioactive Materials 3(3):255–266. https://doi.org/10.1016/j.bioactmat.2017.11.006

Weiner R (1997) Biopolymers from marine prokaryotes. Trends Biotechnol 15(10):390–394. https://doi.org/10.1016/S0167-7799(97)01099-8

Witholt B, Kessler B (1999) Perspectives of medium chain length poly(hydroxyalkanoates), a versatile set of bacterial bioplastics. Curr Opin Biotechnol 10:279. https://doi.org/10.1016/S0958-1669(99)80049-4

Xue Y, Sant V, Phillippi J, Sant S (2017) Biodegradable and biomimetic elastomeric scaffolds for tissue-engineered heart valves. Acta Biomater 48:2–19. https://doi.org/10.1016/j.actbio.2016.10.032

Zhang X, Luo R, Wang Z, Deng Y, Chen G-Q (2009) Application of (R)-3-Hydroxyalkanoate methyl esters derived from microbial Polyhydroxyalkanoates as novel biofuels. Biomacromolecules 10(4):707–711. https://doi.org/10.1021/bm801424e

Microbial Pigments and Their Application 9

Selvaraju Vishnupriya, Sundaresan Bhavaniramya,
Dharmar Baskaran, and Arulselvam Karthiayani

Abstract

Microbial pigments are versatile with potential application in food, pharmaceutical, cosmetics, and textile industries as a promising natural source of colorant. The diverse group of microorganisms under stress produces pigments that are biodegradable, non-carcinogenic with evident biological and functional properties. Many new pigment producing organisms are now the most sustainable way to overcome the use of artificial pigments that are harmful. This review mainly focuses on the application of microbial pigments in various industries including the benefits and challenges of using natural pigments with the focus on meeting safety and demand.

Keywords

Microbial pigments · Industrial application · Food industry · Textile industry

9.1 Introduction

Microbial pigments production is a promising field of research implying demand in various industrial application. Biodegradable or bio-based pigments create awareness for advantages such as eco-friendly, has low toxicity compared to synthetic materials. Bio-colorants are in global attraction estimating 45% of total demand in

S. Vishnupriya (✉) · S. Bhavaniramya · A. Karthiayani
College of Food and Dairy Technology, Tamil Nadu Veterinary and Animal Sciences University, Chennai, India

D. Baskaran
Madras Veterinary College, Tamil Nadu Veterinary and Animal Sciences University, Chennai, India

A. Vaishnav, D. K. Choudhary (eds.), *Microbial Polymers*,
https://doi.org/10.1007/978-981-16-0045-6_9

various sectors such as pharmaceutical, cosmetics, and food. Microorganisms are in interest to isolate pigments and develop organic food colorants. Synthetic colors have developed several health deformation and have artificial impact on the outcoming product or design. Hence, moving forward towards natural colors and pigments from microbial sources has showed a wide array of opportunity in various fields. Natural pigments are used in food industry as additives, colorants, and also in other industries such as textiles, cosmetics, and pharmaceuticals for various purpose. Microorganisms and plant sources-based pigments and colorants are studied to replace the carcinogenic and chemically produced colorants. Biopigments play a vital role in global market providing demand for organic pigments, dyes. Furthermore, covering the food industry the focus turns out to be for thickening agents such as xanthan and gellan, flavor producers, essential amino acids, biopigments, viz., carotenoids, flavonoids, melanins, quinones, etc. Microorganisms, especially *Monascus, Rhodotorula, Bacillus,* and *Achromobacter* are capable of producing a large number of biopigments and they are fondly noticed in soil, coast, or deep sea. These pigments also play a supportive part for maintaining the biological activity in food supplements, probiotics, etc. Several bacterial pigments are used for developing colorants such as zeaxanthin (yellow color) extracted from *Flavobacterium sp.* and *Paracoccus sp.*, carotenoids (yellow) isolated from *Streptomyces sp.* Excellent advantages of these pigments include non-toxic, biodegradable, non-carcinogenic, eco-friendly. The reach for natural, sustainable food colors from organic pigments is increasing and the perspective to consume natural products replicates in producing innovative ideas. Microbial pigments through natural fermentation serve a low cost of production and microbes produce various substances such as carotenoids and flavins and they are further utilized in food ingredients. Advanced techniques in biotechnology and metabolic science are providing active production of valuable microbes of interest. Further understanding the biosynthetic pathway of isolated pigments production helps in recognizing conditions and strategies of involved microorganisms suitable for developing the low cost pigment extraction methods. Emerging technology alongside application of natural pigments is effectively utilized using nanotechnology as nano colorants and application in various industry such as the food industry and textile industry are in focus to be developed from microbial pigments. Environmental pollution such as aquatic animal contamination, soil and groundwater remarks are a serious issue with the use of synthetic pigments and therefore more interest on natural products are in high demand.

9.2 Source and Production of Microbial Pigment

The production of pigments by microbes from various sources ranging from soil, marine to agro-industrial and food waste has a prospective demand as they come with the future dimension of having eco-friendly products. These pigments are produced by fermentation of organic waste such as agro-industrial waste otherwise from the nature especially marine source or soil as shown in Fig. 9.1.

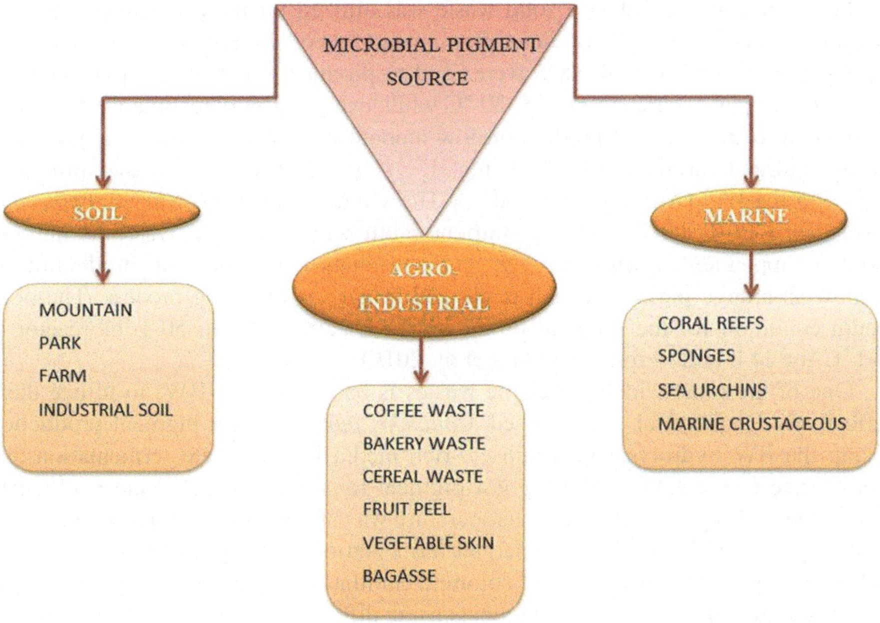

Fig. 9.1 Source of Pigment Producing Organism

9.2.1 Agri-Industrial Waste

Waste generated during the processing of agricultural produce or production is called agro-industrial waste. These waste mostly go underutilized and cause environmental pollution despite the fact it is highly nutritious based on the source (Sadh et al. 2018). However, the waste can be converted to wealth by treating the waste such as seeds, peel, and pulp. Among which microbial pigments are the important value-added product replacing the harmful artificial synthetic colorants and dyes. A wide range of microbes such as bacteria, fungi, yeast, and algae produce a compendium of pigments opening opportunities for various industries (Heer and Sharma 2017).

Pigments such as carotenoids have various advantages such as antioxidant, antibacterial properties. Abdelhafez et al. (2016) used *Serratia marcescens* for the pigment production using agri waste rice bran, sugarcane bagasse, and molasses as substrate. When compared to the other waste residues, molasses served as the best medium with the maximum production of β-carotene (1.1 mg/L) in a rotary shaker at rpm of 150 for 2 days incubated under dark condition with the temperature of 30 °C. Similarly, using *Serratia marcescens* Xd-1prodigiosin was produced by solid-state fermentation of bagasse. Using ultrasonic assisted reflux extraction of higher amount of prodigiosin (40.86 g kg^{-1} dry solid) was obtained with the addition of glycerol and soy peptone (Xia et al. 2016).

From the cannery industry solid waste was utilized for the pigment production, pineapple waste was used for violet pigment at low cost in large scale for industrial application. *C. violaceum*UTM5 were used to produce violet pigment at optimum condition of pH 7, temperature 25–30 °C while using liquid pineapple waste. When compared to the pigment production low amount of violet biocolor was produced using nutrient broth (239 ± 3 mg L^{-1}) while pineapple waste produced 285 ± 5 mg L^{-1} (Aruldass et al. 2015). The same author using *Chryseobacterium artocarpi* CECT 8497 replacing nutrient broth with pineapple waste as nutrient source supplemented with nitrogen and phosphate source for the production of yellowish-orange pigment which is utilized in the soap making process. The optimum condition for the pigment production was carried out in 50 L bioreactor at 30 °C for 24 h at 200 rpm (Aruldass et al. 2016).

One of the most widely available wastes is bakery waste (BW) to utilize them effectively Haque et al. (2016) used *Monascus purpureus* for pigment production using the BW hydrolysates as nutrient-rich media in solid-state fermentation and submerged fermentation. The highest pigment (about 24 AU/g glucose) obtained under the hydrolysis condition of temperature was 50 and 55 °C along with glucose concentration of 5 g L^{-1} in submerged fermentation and also reported to have no adverse effect when used as food colorant. Similarly, Sehrawat et al. (2017) using *Monascus purpureus* produced pigments using different wastes such as sweet potato and potato peel powder, kinnow peel, also rice and wheat bran, as carbon source to compare the maximum yield. After 15 days of fermentation, it was observed to have a maximum amount (3.3 CVU/g) of biopigment was obtained when sweet potato peel powder was used. Followed by this orange waste was utilized by fungal strain *Monascus purpureus* ATCC 16365 and *Penicillium purpurogenum* CBS 113139 for the pigment production using submerged fermentation, solid-state fermentation, and semi-solid-state fermentation for the maximum yield. In comparison *Monascus purpureus* under solid-state fermentation obtained the maximum production of 9 AU/g with high total phenolic content (123.7 mg EAE mL^{-1}) of dry fermented substrate (Kantifedaki et al. 2018).

With sugarcane bagasse (SCB) hydrolysate as a carbon source Hilares et al. (2018) red biocolor was produced using *Monascus ruber Tieghem* IOC 2225. *Monascus ruber* grows well on glucose-based medium however using SCB the yield was 2.5-fold higher under dark conditions after 288 h of fermentation. Different incidences of light were used to compare the yield of red pigment under white light (12.13 AU 490 nm), dark condition (18.71 AU 490 nm) while the biomass under dark condition was (5.78 g L^{-1}). In addition to this, de Paula Bonadio et al. (2018) using the sugarcane juice produced carotenoids with *Rhodotorula rubra* L02. Among the 16 different combinations, it was obtained that the complex sugarcane juice and synthetic media with maltose and sucrose showed best results of about carotenoid production 0.0135 mg/L, 0.0277 mg/L, and 0.029 mg/L, respectively. The coffee waste was utilized for the biopigment using *Rhodotorula mucilaginosa* CCMA 0156 (Collection of Microorganisms of Agricultural Microbiology) with high antimicrobial and antioxidant activity. The maximum yield of pigment using pulp extract was 16.36 ± 0073 mg L^{-1} followed by husk with the yield of

21.35 ± 0067 mg L^{-1}. The antioxidant activity of the pigment from pulp extract (98.26% Protection β-carotene) was comparatively higher than the synthetic antioxidants (BHT—85.92% Protection β-carotene) (Moreira et al. 2018). Similarly, using the corn cob Embaby et al. (2018) using *Monascus purpureus* strain ATCC 16436 through co-solid-state fermentation 10 days at 30 °C orange and red pigments are produced.

In another study, Patthawaro et al. (2019) using *Rhodopseudomonas faecalis* PA2 with different agro-industrial wastes such as cassava meal, soybean meal, and coconut meal for the carotenoid production. As a result, under favorable condition lycopene,1,2-dihydro-3,4-dedihydrolycopene, cis-1,2-dihydrolycopene, 1,2-dihydrolycopene were produced when using soybean meal. These photosynthetic bacteria produced the major carotenoid with only 50% soybean meal in the medium while the pigment production was very low in cassava and coconut meal. The lycopene has good antioxidant effects and bioactive functional properties due to the presence of polyunsaturated hydrocarbons. Followed by the above study Saejung and Puensungnern (2020) also studied using *Rhodopseudomonas faecalis* PA2 for producing carotenoid replacing chemical medium with molasses-based medium while reducing the cost. This photosynthetic bacterium is one among the most preferred commercial sources of carotenoid production that can be used in fortification of food.

Blakeslea trispora (+) MTCC 884 was used along with orange, papaya peels, and carrot waste under solid-state fermentation resulted yield of 0.127 mg/mL of β-carotene a red-orange pigment. For the confirmation of color mass spectroscopy and LCMS was used whose *m/z* value was 537.608 and peak value Rt 13.37. Apart from this, the extracted β-carotene had a scavenging property for about 90 days. At pH range of 6.2–6.5 with temperature between 25 and 32 °C under fermentation of 60–96 h the optimum β-carotene production was obtained (Kaur et al. 2019).

Similarly, using mung bean husk, onion peels, pea pods, and potato skin Sharma and Ghoshal (2020) produced biopigment from *Rhodotorula mucilaginosa* MTCC-1403 in the bioreactor. They obtained the optimum carotenoid production (717.35 μg/g of β-carotene) at temperature of 25.8 °C with the pH 6.1 and agitation 119.6 rpm using flask while under aerobic fermentation the yield was higher 819.23 μg/g which shows carotenogenesis vary with aeration under same condition. Under submerged fermentation using *Monascus ruber* different pigments were produced by varying the pH of maltose syrup residue when used as substrate. At pH 2.0 and 2.5 (highly acidic) yellow pigments were produced, orange pigment when pH was 3.0 and 3.5 and red with pH 4.0 and above while potentially utilizing the waste from the syrup processing industry (De Oliveira et al. 2019).

9.2.2 Marine Source

Compared to soil or terrestrial organisms recently marine-sourced (animals, sponges, sea urchins, marine crustaceans) are gaining a lot of importance with the potential to produce different pigments. These pigments have various features such as scalability, non-toxicity, eco-friendly, and easy gene manipulation (Bălașa et al. 2020).

From the coral reef of the Reunion Island, isolated a *Penicillium* fungi *Talaromyces albobiverticillius 30,548* for pigment production. During the study, they identified 12 compounds among which only 4 (N-glutaryl rubropunctatin, N-threonine-monascorubramine, 6-[(Z)-2-Carboxyvinyl]-NGABA-PP-V, and PP-O) were Monascus-like red pigment with no mycotoxin. Under submerged fermentation for 8 days at temperature 24 °C with the agitation speed of 150 rpm. The red azaphilone biocolorant was confirmed using NMR analysis with the cis configuration at the C10-C11 double bond (Venkatachalam et al. 2018). From the South Yellow Sea in China the researchers isolated *Rhodotorula sp.* RY1801 under fermentation with pH 5.0 and temperature of 28 °C for the production of carotenoids of about 589 to 987 μg/L along with glucose, yeast extract, carbon, and nitrogen sources (Zhao et al. 2019).

Two organisms isolated from the Island of La Reunion for pigment production were *Talaromyces spp.* and *Trichoderma atroviride*. Using submerged culture fermentation both the extracellular and intracellular pigment were extracted using pressurized liquid extraction. N-threonine rubropunctamine and Monascus-like azaphilone were the pigments produced and characterized as non-toxic and non-mycotoxigenic (Lebeau et al. 2017). In addition, many novel pigment producing organisms were isolated and identified from marine environments. Similarly, from the same region various samples such as seawater sediments, living and rubble corals and many others were taken. Among which 31 different species were identified with major genera being *Penicillium, Aspergillus* producing pigments from dark shades to light shade when using PDA medium with wide application in pharmaceutical, fertilizer, and agro industries (Fouillaud et al. 2017).

Marine organisms are the major producers of anthraquinones, *Stemphylium, Fusarium, Eurotium, Microsphaeropsis, and Paecilomyces* play a major role in the production. Apart from these *Xylaria sp.*, a filamentous fungus produced a new compound called xylanthraquinone, *Aspergillus glaucus* produced aspergiolide A, *P. chrysogenum* produced skyrim from the Salt Lake, *Fusarium sp.* produced anhydrofusarubin, *Nigrospora sp.* produced 4-deoxy bostrycin and various other marine-derived fungus has pigment producing property with good antioxidant activity (Fouillaud et al. 2016). In another study in Italy, sea sponge samples were taken from which *Raspaciona aculeata and Dictyonella marsilii* were isolated for the carotenoid production. A typical carotenoid called renieratene, a brown colored pigment, was found in both the species but predominantly in *R. aculeata* 2570 ppm while *D. marsilii* had 277.1 ppm of carotenoid (Salvo et al. 2017).

In the marine ecosystem, astaxanthin is the most common produced by photosynthetic organisms. From the marine water of pacific coast which in the middle Japan the researchers isolated *Brevundimonas* genus a novel strain N-5 producing astaxanthin an important carotenoid pigment on the global market. Under good aerated conditions, these orange-pigmented bacteria (N-5) produces optically pure carotenoid of 364.6 μg g^{-1} dry cells while the total amount was 601.2 μg g^{-1} dry cells. Apart from this other carotenoid such as 2-hydroxy astaxanthin, and

2,2′-dihydroxy astaxanthin were also identified using their retention time in the HPLC (Asker 2017). The same author studied the keto carotenoids such as astaxanthin, adonixanthin, and their hydroxylated derivatives from fresh and marine water, among them *Brevundimonas* and *Erythrobacter* were the highest producers. Different strain from freshwater were isolated FrW-Asx-5, FrW-Asx16 produced astaxanthin content of 46% and 16.6% and the other derivatives 2,2′ dihydroxy astaxanthin, 2-hydroxy-astaxanthin are also produced (Asker 2018). For surface seawater, 88 bacterial strains belonging to *Bacilli, Actinobacteria, Flavobacteriia, α-Proteobacteria, and γ-Proteobacteria* classes were isolated for the production of different carotenoids such as canthaxanthin astaxanthin, lutein, and zeaxanthin. Among the several pigment producing strains Flavobacterium produced 93.1% zeaxanthin of total carotenoids. They have studied both the upper and deep seawater to isolate the pigment producing bacteria; the upper layer had a greater number of bacteria by protecting itself from UV light using pigments (Asker et al. 2018).

Similar to the above study recently, from the scallop pigment producing *Brevundimonas* was isolated. 2,2′-dihydroxy-astaxanthin (82% of total carotenoid) was the major carotenoid obtained from the marine bivalve exhibiting excellent antioxidative activity due to the presence of the extra hydroxyl group. Compared to the other source of astaxanthin producing bacteria, *B. scallop* produced 1303.62 ± 61.06 μg/g dry cells. Low temperature and salinity with high carbon source are the optimum for the carotenoid production (Liu et al. 2020).

9.2.3 Soil

Pigments produced from soil organisms have a huge scope of application in various fields such as cosmetics, medical, food, and textile industry as colorant. Biopigments owing to its non-toxicity have the other advantage of being simple and quick to grow independent of the climatic condition. Various soil samples from different places were taken for the isolation of pigment producing microorganism *Penicillium* spp. from the Brazilian cerrado soil, *P. sclerotiorum* isolated from National Park of Serra Do Cipo produced pencolide, sclerotiorin, and isochromophilone followed by *P. oxalicum* for the production of hydroxyanthraquinone by a Czech company from the soil fungal strains (Akilandeswari and Pradeep 2016).

Bacillus weihenstephanensis was isolated from one of the coldest regions of Northeastern Poland (National park and Farm) for the production of melanin-like pigment with good solubility. During this microbial melanogenesis process, laccase activity contributed to the pigment production, six *B. weihenstephanensis* were melanin-positive isolates among the various samples. However, the *B. weihenstephanensis* from the farm soil exhibited higher pigment production than the park soil sample due to differences in the soil fertilization (Drewnowska et al. 2015). Biopigments from *Streptomyces fradiae* were used in various industries for its hemolytic and antioxidant properties. The organism was isolated from the soil of Sundarbans forest and was tested to replace the synthetic colorants with the eco-friendly biopigments in the textile industry (Chakraborty et al. 2015). Similarly,

Yasmeen et al. (2017) isolated black pigment producing Streptomyces species from the mangrove forest for its potential antimicrobial activities from Andhra Pradesh.

Forest soil from South Korea was used for isolating yellow and flexirubin-type pigment producing *Dyadobacter* genus belonging to family *Cytophagaceae* (Dahal and Kim 2018) and soil bacterium *Flavobacterium humisp* also produced flexirubin-type pigment Kim et al. (2019), while Park et al. (2020) isolated *Chryseobacterium* from mountain soil from Korea reporting the same. He also isolated *Deinococcus* species for deinoxanthin production; this major carotenoid produced by *D. radiodurans* was studied by Kumar et al. (2015) for its antitumor property.

Lebeau et al. (2017) analyzed and isolated *Penicillium purpurogenum rubisclerotium, Fusarium oxysporum* to replace the synthetic red color, the production of water-soluble biopigment was increased either by increasing the fungus growth or the intracellular pigment. The major fungus species isolated from soil are *Penicillium* and *Aspergillus* that is *P. citrinum, P.aurantiogriseum* from forest soil *P. corylophilum, P. fellutanum, A. parasiticus* were isolated from tea plantation while *A. flavus, A. glaucus* were from Rhododendron soil, (Pandey et al. 2018a). From the Indian Himalayan region, orange color pigment producing *Penicillium sp.* was isolated which was produced under stress due to weather conditions. The pigment was insoluble and found to be derivative of carotenoid (Pandey et al. 2018b). Apart from this *Penicillium striatisporum* isolated from Australia and *Penicillium purpurogenum* produced yellow pigments, while *Fusarium verticillioides* produced naphthoquinone pigment isolated from Thailand soil (Kalra et al. 2020).

The mesophilic bacteria *Salinococcus roseus* produced temperature stable orange pigment due to the presence of double bond that are mostly organic compounds when obtained from soil. Hence, it can have various application in textile and food industry (Usman et al. 2018). Fariq et al. (2019) from Punjab, Khewra Salt Range, soil samples were collected for the isolation of halophilic microbes such as *A. elongatus, H. aquamarina,* and *S. sesuvii* for the production of bacterioruberin carotenoids. Resembling this study, soil samples from various places of Gujarat were taken for the pigment producing bacteria among wide group *Staphylococcus xylosus* was identified and also found under stress or harsh condition pigments produced were used in the cosmetic industry (Choksi et al. 2020). *Chaetomium cupreum* was isolated from litter soil for the red colored pigment collected from Karnataka under submerged fermentation and evaluated the extract for its antioxidant activity and application in pharmaceutical (Wani et al. 2020). A blue green phenazine pigment from *Pseudomonas aeruginosa* produce pyocyanin isolated from the soil dumped with agri waste in Dharwad, Karnataka. This pigment has commercial value as it used to resolve multiple environmental and agricultural problems (DeBritto et al. 2020)

9.3 Application of Microbial Pigments in Various Industries

Microbial pigments are the metabolic secretion of microbes that has a wide application in food, textile, and cosmetics industries owing to its beneficial properties. Being eco-friendly these pigments are non-carcinogenic as well as non-toxic in nature acting as a potential replacer for synthetic pigments. These pigments have remarkable therapeutic values with challenges and opportunities in the pharmaceutical industry. Further discussion on extraction methods of the diverse pigments, its scalability from pilot model to industries and to improve the commercial value of any pigments are with a possible road map of pigment extraction to possible applications in various industries as shown in Fig. 9.2 discussed.

9.4 Role of Microbial Pigments in Food Industry

In the food industry, color plays integral part between the acceptance and taste of the food. The more the attractive color of the food the more it has the influence on the selection of food. Synthetic color are generally anticipated for the application in food industry as well in the cosmetic industry as it is cheap and stable meanwhile associated with the harmful effects forced a transition shift from artificial colors. The growing consumer demand for healthy product and increasing awareness on food safety made the bicoloration best alternative to take over the synthetic pigment (Tuli et al. 2015). In spite of having natural pigments from various sources, microbial sources are safe to use without any curtailment in the production due to weather, raw material, cost, and other technical challenges (Panesar et al. 2015). The apparent reason for eschewing artificial colors and additives in foods is due to its allergic effect and hyperactivity in children in the long run when consumed continuously. Furthermore, certain artificial colors are also banned due to the deleterious teratogenic outcome concurrently; most of the biopigments are anticancer, antifungal with antioxidant capacity (Oplatowska-Stachowiak and Elliott 2017).

The essentially increasing need for natural and healthy products in the food industry paves way for researchers to discover new sources of pigment producing organisms and their pigment properties with its varied application to seek demand from the food industry. To improve the color nutrition pigments like carotenoid, riboflavin is used in beverages, snacks, desserts, and also in fresh fruits and meats. Additionally, *Fusarium sporotrichioides,* a fungus and Erwinia uredovora were used in food preparation to increase their visual quality added to the food to increase its appeal (Narsing Rao et al. 2017). In addition to this, FDA has approved pigments as follows: carotene, astaxanthin, and riboflavin from *Monascus, Xanthophyllomyces,* and *Eremothecium* in such a way uplifting the natural source of pigments while meeting the legislation (Numan et al. 2018).

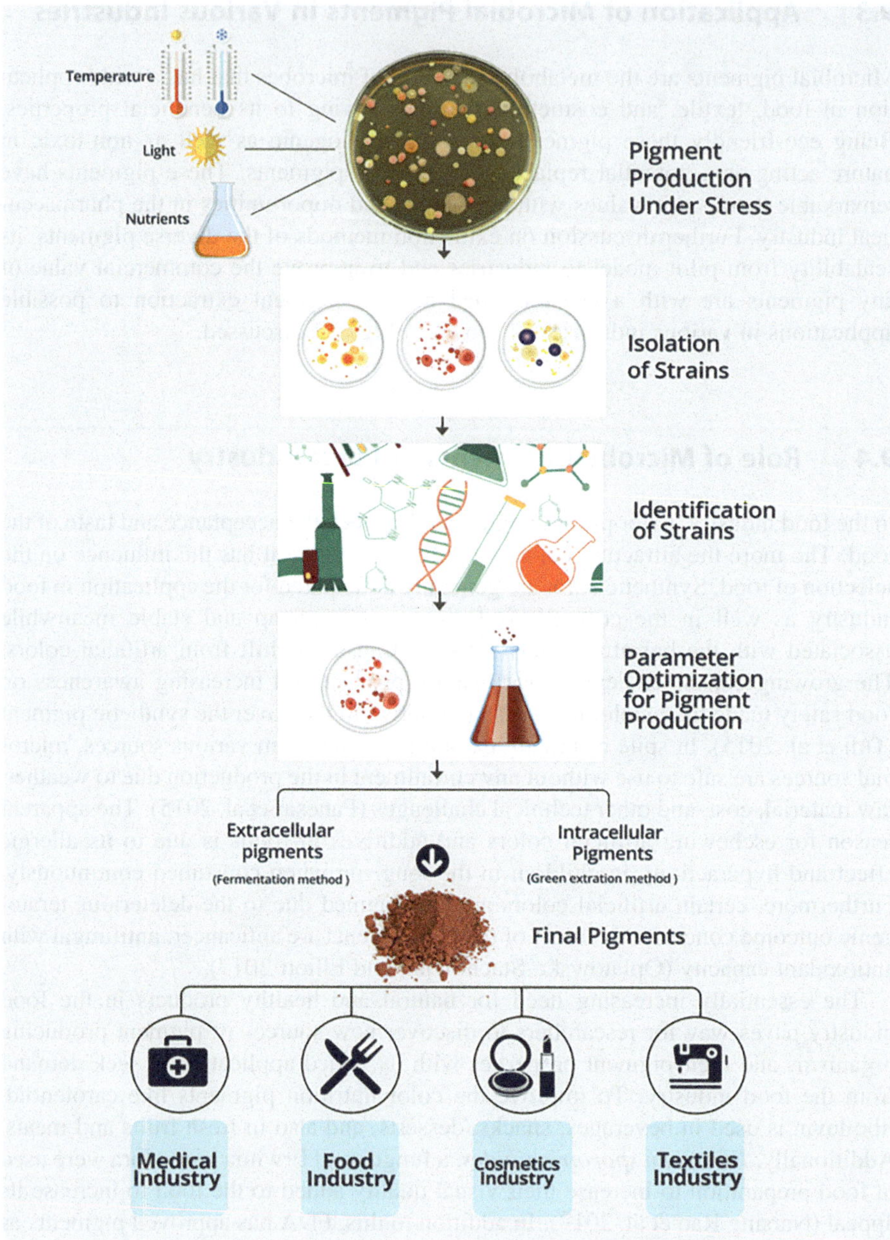

Fig. 9.2 Road map of pigment extraction to possible applications in various industries

9.4.1 Microbial Pigments as Food Color

Considering the organoleptic property, visual impression is the major parameter when it comes to food choices also quality of the food is decided by the color and hence to intensify the color of the meat and egg combination of carotenoid (lutein, astaxanthin, and β-carotene) has been used to feed poultry for the good quality product (Ram et al. 2020a, b). While Venil et al. (2015) used violet biocolor powder in jelly and yogurt produced by *Chromobacterium violaceum UTM 5* as shown in Table 9.1. The extracted pigment was concentrated using a spray dryer with Gum Arabic as encapsulating material for good stability. When analyzing the product quality, the product was stable with no color change or microbial contamination for 1 month in yogurt and jelly. Similarly for the candy preparation carotenoid extracted from *Rhodotorula mucilaginosa* was used when fermented using malt and yeast medium with date syrup as carbon and nitrogen source for the maximum production (Elsanhoty et al. 2017). According to Galaup et al. (2015), coloring of cheese rind is a difficult process but with the presence of *Brevibacterium linens and Brevibacterium aurantiacum* on the surface of vieux-pan French cheese (soft red ripened cheese) resulted in the surface color. Similarly, another Protected Designation of Origin cheese from France called Fourme de Montbrison was investigated for the presence of carotenoid produced by *Brevibacterium linens*. They found chlorobactene and Agelaxanthin A with nine other carotenoids which are responsible for the rind color (Giuffrida et al. 2020). For the red color of the cheese, meat, wine biocolor produced from *Monascus* was used and they are also used to color red koji dates using rice as a fermentation medium (Cardoso et al. 2017).

Among the microbial pigments *Monascus* has an important place to exploit as a colorant, additive in the food industry. The major contributors of pigment production are *Monascus* M3428 *Monascus ruber* CCT 3802, *Monascus purpureus* N11S while showing strong antimicrobial activity. With the hydrolytic enzymes, changes can be made in the morphology of *Monascus* for the pigment production (Kim and Ku 2018). Nitrite is used in the meat industry for the color development but due to its carcinogen production an attempt to replace it with *Monascus* pigment was studied. For the minced meat, *Monascus ruber* MJ-1 pigment was added and after 4 days of fermentation proteolysis and lipid oxidation were analyzed. The result showed inhibition of lipase activity followed by reduction in the thiobarbituric acid reactive value and proved to be effective for the fermented meat application (Yu et al. 2015). Followed by meat to deepen the fish color carotenoid was used along with the feed, in particular to improve the color of the red snapper fish (Pereira da Costa and Campos Miranda-Filho 2019). Red color pigment from *Meiothermus sp.* isolated from the water sample can be used as it showed strong antioxidant activity (Mukherjee et al. 2017). Similarly, blue green algae or the cyanobacteria produced red pigment from *Arthrospira platensis*, it is generally a protein complex which could also be used to replace synthetic color as it is produced using rice husk as substrate medium (Taufiqurrahmi et al. 2017).

Table 9.1 Different colors of pigments and its application

Microbes	Name of micro organism	Pigment name	Application / Uses
Red Pigment			
Fungi	Monascus sp.	Monascorubramine, Rubropunctamine	Antioxidant
	Penicilliumoxalicum	Anthraquinone	Anticancer effect in food and pharmaceuticals
	Penicilliumpurpurogenum	Mitorubrinol	Dyeing of cotton fabrics
	Monascuspurpureus	Azaphilones, Rubropunctamine	Dyeing of cotton fabrics , Wool
	Talaromycesaustralis	Carotenoids, 2, 4-Di-tertbutylphenol	Dyeing of cotton fabrics and Wool
	Talaromycesverruculosus	Polyketide	Cotton Fabric
	Scytalidiumcuboideum	Quinones	Bleached cotton, Spunpolyamide, Spun polyester, polyacrylic, wool
	Fusariumsporotrichioides	Lycopene	Antioxidant
	HaematococcusPluvialis	Astaxanthin	Antioxidant
	Talaromycesatroroseus	Azaphilones	Antioxidant, Anticancer, Antioxidant
	Pseudoalteromonasdenitrificans	Cycloprodigiosin	Antiplasmodial, Anticancer
	FusariumSporotrichioides, Blakesleatrispora	Lycopene	Antioxidant, Anticancer
	Cordycepsunilateralis	Naphtoquinone	Anticancer, Antibacterial, Trypanocidal
	Epicoccum nigrum	Carotenoid	Antioxidants, food coloration, Inhibits HIV-1 replication
	Lecanicilliumaphanocladii	Oosporein	Antifungal activity
	Monascusruber	Azaphilone	Foods
Bacteria	Rugamonasrubra, Streptoverticilliumrubrireticuli, Vibrio, gaogenes, Alteromonasrubra, Serratiamarcescens, Serratiarubidaea	Prodigiosin	Anticancer, immunosuppressant, antifungal, algicidal; dyeing (textile, candles, paper, ink)
	α-Proteobacteria	Heptylprodigiosin	Antiplasmodial
	Pseudoalteromonasrubra	Prodigiosin	Anticancer, DNA Cleavage, Immunosuppressant
	Streptomyces echinoruber	Rubrolone	Antimicrobial
	Streptomyces sp.	Undecylprodigiosin	Antibacterial, Antioxidative, UV-protective, Anticancer
	Proteobacteria	Heptylprodigiosin	Antiplasmodial
Algae	Dunaliellasalina	β-carotene	
	Porphyridiumcruentum and many other microalgae and cyanobacteria	Phycoerythrin	Antioxidant, Antitumor activity, Immunoregulatory
Orange Pigment			
Fungi	Monascus sp.	Monascorubrin, Rubropuntatin, Canthaxanthin	Antibacterial activity, Anticancer activity
	Penicilliumpurpurogenum	Mitorubrin	Food, pharmaceuticals and cosmetics
	Penicilliumsclerotiorum	PencolideSclerotiorin	Antibacterial activity, Antifungal activity
	Neurosporaintermedia	β-Carotene	Various industrial and pharmaceuticals applications
	Neurosporasitophila	Neurosporaxanthin	Antioxidants
	Epicoccumnigrum (CML2971	Orevactaene	Antioxidant

Table 9.1 (continued)

Bacteria	Bradyrhizobium spp. , Lactobacillus pluvalis	Canthaxanthin	Anticancer, Antioxidant
Algae	Dunaliellasalina Microalgae	β- carotene	Anticancer, Antioxidant suppression of cholesterol synthesis
Yellow Pigment			
Fungi	Aspergillussclerotiorum	Neoaspergillic acid	Antibacterial activity
	Aspergillusversicolor	Asperversin	Antifungal activity
	Fusariumverticillioides	Naphthoquinone	Antibacterial activity
	Monascus sp.	Monascin, Ankaflavin	Food colorant, Pharmaceuticals
	Penicilliumherquei	Atronenetin	Food additive, Antioxidant
	Penicilliumpurpurogenum	Purpurogenone	Dyeing of cotton fabrics
	Trichodermaviride	Viridin	Textile dyeing
	Trichodermavirens	Viridol, Virone	Antifungal activity
	Ashbyagossypi	Riboflavin	
	Xanthomonasoryzae	Xanthomonadin	protection against photo damage
	Candida famata	Riboflavin	Baby foods, breakfast cereals, fruit drinks
	Eurotium spp.	Isoquinoline	Antifungal activity
	Penicilliumflavigenum	Anthraquinones	Antioxidants
	Scytalidiumganodermophthorum	Quinones	Textile
	Monascuspilosus	Citrinin	
	Thermomyces sp.	Anthraquinones	Cotton, Silk, Wool
	Trichoderma sp.	Anthraquinones	Cotton, Silk, Cotton Silk
	Aspergillus sp. AN01	Asperyellone	Cotton, Silk, Synthetic and wool fabrics
	Penicilliummurcianum	Carotenoids	Wool
Bacteria	Cytophaga/Flexibacteria AM13,1 Strain	Tryptanthrin	Antioxidant, Anticancer
	Staphylococcus aureus	Zeaxanthin	Photoprotectant, Antioxidant
	Kocuria sp.	Carotenoids	Anti-cancer activity
	Pseudomonas sp.	Phenazine	Biological control
	Bacillus sp.	Riboflavin	Nutritional supplement
	Bacillus subtilis	Riboflavin	Nutritional supplement
	Flavobacterium sp., Paracoccus sp..	Zeaxanthin	Photo protectant, antioxidant
Algae	Chlorella and others, Microalgae	Lutein	Antioxidant
Brown Pigment			
	Aspergillus sp.	Quinones	Cotton, Silk, Silk cotton
	Acrostalagmus (NRC 90)	Quinones	Wool
	Bisporomyces sp.	Quinones (Deep Brown)	Wool
	Penicilliumchrysogenum (NRC 74) , Penicilliumitalicum (NRC E11), Penicilliumregulosum (NRC 50)	Quinones	Wool
Black Pigment			
Fungi	Aspergillus niger	Aspergillin	Antimicrobial activity
	Curvularialunata	Anthraquinones	Silk, Wool
Bacteria	Pseudomonas guinea	Melanin	Antioxidant activity
Yeast	Saccharomyces, Neoformans	Melanin	Antimicrobial, Antibiofilm and antioxidant

Table 9.1 (continued)

		Green Pigment	
	Chlorociboriaaeruginosa	Quinones	Bleached cotton, Spun polyamide,
Bacteria	Pseudomonas Spp.	Phycocyanin	Cytotoxicity, Neutrophil apoptosis, Ciliarydysmotility, Proinflammatory
	Bacillus cereus	Azaphenanthrene	Anticancer, Antibacterial, Textile dyeing
		Blue Pigment	
Fungi	Aspergillusniger	Aspergillin	Antimicrobial activity
Bacteria	Pseudomonas Spp.	Phycocyanin	Cytotoxicity, Neutrophil apoptosis, Ciliarydysmotility, Proinflammatory
	Corynebacteriuminsidiosum	Indigoidine	Protection from oxidative stress, Antioxidant, Antimicrobial
	Erwiniachrysanthemi	Indigoidine	Protection from oxidative stress, Antioxidant, Antimicrobial
Algae	Arthrospira sp. (formerly Spirulina sp.) and many other microalgae and cyanobacteria	Phycocyanin	Antioxidant, Antitumor, Immunoregulatory

9.4.2 Biopigments as Food Additive with its Antioxidant Property

Rhodotorula mucilaginosa AY-01 isolated from the soil collected from Incheon, South Korea was analyzed for the presence of carotenoid as glucose. These glucosidal carotenoids red pigment exhibited strong antibacterial and antioxidant activity (Yoo et al. 2016). Another important pigment from the carotenoid family is zeaxanthin, the International public health organization WHO and FAO recommends about 0–2 mg kg^{-1} zeaxanthin of the body weight (Ram et al. 2020a). From the freshwater zeaxanthin producing *Arthrobacter gandavensis* was isolated for its application in the food and feed industry as additive for its antioxidant property and also in the cosmetic industry due to its prophylactic activity acts as anti-aging agent (Ram et al. 2020b). A novel biopigment isorenieratene, an aromatic carotenoid was isolated from *Rhodococcus sp.* B7740 for the food application. The stability of isorenieratene was compared with dietary carotenoid by mimicking the gastric conditions thus exhibiting good stability with the retention rate after ingestion and hence proved to have application in food (Chen et al. 2018).

Apart from this Mukherjee et al. (2017) isolated *Thermus* strain from the water sample that inhibited protein oxidation activity and the pigments could be recommended for the food preservation application. Similarly, *Kocuria sp.* strain QWT-12 a halotolerant bacterium was isolated from tannery industry wastewater which produced a neurosporene pigment belonging to the carotenoid family with extraordinary antioxidant activity and UV-B radiation protector (Rezaeeyan et al. 2017). *Blakeslea and Phycomyces* are the best producers of antioxidant-rich lycopene which have various applications in dairy, snack food such as chips, soups, sweets to increase the shelf life and also used in the suppression of MSF-7 tumor cells proliferation. It was also used in the surimi and in fish processing industry especially for the salmon fish to intensify the pink color (Numan et al. 2018). The carotenoid from *Pedobacter, Fontibacter flavus* YUAB-SR-25 was reported to contain excellent oxygen scavenging properties with its defense mechanism hence can be used in the fatty-rich products to avoid peroxidation. Followed by violacein, pyomelanin, and carotenoid from *Chromobacterium violaceum, Burkholderiaceno cepacia, Planococcus* including flexirubin from *Chryseobacterium artocarpi* are the other antioxidant-rich pigments which have appreciable use in food industry (Venil et al. 2020a, b). In particular, violacein one of the protective pigments preventing age-related diseases and others such as ulcer, cataract, or degeneration of macular are prevented using microbial pigments (Nigam and Luke 2016). *Bacillus* VITPS12 *and Planococcus maritimus VITP2* isolated from the mangrove forest were the richest source of multiple carotenoids and have therapeutic properties due to the presence of other functional groups (Prathiba and Jayaraman 2018).

9.4.3 Application in Cosmetic and Pharmaceutical Industry

Products produced with natural colors are preferred in spite of the health benefits; they also have essential clinical efficacy in the pharmaceutical industry and potential tools in the cosmetic industry. Numerous secondary metabolites from microorganisms have anticancer, antibiotic, antimalarial, and other beneficial properties. The major colorants from nature are prodigiosin, astaxanthin, and melanin in the cosmetic industry as they are the best source for UV protection.

In general, biopigments are used in cosmetics either as sunscreen or as anti-aging agents. Pigments such as prodigiosin and violacein from *S. marcescens* and *C. violaceum* are used commercially in the cosmetic industry for the sunscreen application as they are naturally antibacterial with antioxidant capacity comparable to ascorbic acid (30% prodigiosin and 20% violacein) (Suryawanshi et al. 2015). For any sunscreen lotion, one of the key parameter is the sun protection factor (SPF). Choksi et al. (2020) isolated multiple organism from soil and water sample among which pigment from two strain (BUY1 and JFY) resulted with the highest SPF value (8.34 and 7.72). Similarly, Silva et al. (2019) *Arthrobacter agilis*, *Arthrobacter psychrochitiniphilus* and *Zobellia laminarie* produced UV-protective pigments that have the ability to withstand the UV-B and C radiation thus protecting the skin. These pigments are also called as photoprotective agents, most of the *trans*-bacterioruberin are used in the sunscreen. Apart from this *Synechocystis pevalekii*, a cyanobacterial pigment can also be used in the sun-protective formula (Tendulkar et al. 2018).

Next most important pigment melanin, one of the major pigments for protecting the skin from disorders, is caused due to ultraviolet rays (Kageyama and Waditee-Sirisattha 2019). Various sources of melanin include *Aspergillus fumigates, Vibrio cholerae, Cryptococcus neoformans, Colletotrichum lagenarium, Alteromonas nigrifaciens,* and most of the *Streptomyces* species (Narsing Rao et al. 2017). One of the significant producers of melanin was *Streptomyces glaucescens NEAE-H* (yield of 31.650 μg/0.1 mL) which was isolated from the soil sample exhibited a significant anticancer activity in skin through the in vitro MTT assay in skin cancer cell line while at increased concentration it showed low cytotoxicity. For this reason, it was used to minimize the tissue destruction in the cosmetics field (El-Naggar and El-Ewasy 2017). Pigment from *Monascus* has been widely used for the multiple fields starting antimicrobial, anti-mutagenic and even one of the wide known applications is its anti-obesity properties. With respect to the cosmetic industry, they were used for hair coloring due to its non-toxic nature but when this pigment is exposed to direct UV rays pigment discoloration is possible (Agboyibor et al. 2018).

9.4.4 Application in the Textile Industry

The concepts of health, organic, natural, and eco-friendly have turned people's perception towards natural colors. Apart from increased marketability, the natural colors have antimicrobial property in spite of the wide range of ecological and medicinal properties. But the required organic substrate for growth of color producing microbes comes with rigorous efforts and is influenced by various process parameters (Tuli et al. 2015). A set of researchers conducted a study to isolate pigment producing fungi *Talaromyces verruculosus* strain from spoiled mangoes, and those red pigments were used in dyeing cotton clothes (Chadni et al. 2017). While in Kerala different soil samples were collected among which researchers isolated 37 Actinomycetes isolate were isolated as test organism for pigment production. The dyeing quality of the pigment were experimented in different cloth material while the uptake of color was good in synthetic fiber (Nair et al. 2017). Similarly, a yellow color pigment with 80% dyeing capacity was produced from *Aspergillus sp.* isolated from the soil sample (Pandiyarajan et al. 2018). Followed by this, Lagashetti et al. (2019) reported different colors such as yellow, green, and red from *S. ganodermophthorum*, *C. aeruginosa*, and *S. cuboideum* with good stability and colorfastness for dying wool, cotton, and spun with no toxic effect. The stringent environmental regulations increase the demand for bio-resources such as fungal pigments over synthetic pigments while commercialization of these pigments becomes tedious process due to the market regulatory affairs. Apart from this, Hernández et al. (2019) also reported the production of yellow and red dyes production from fungi isolated from the wood samples. *Penicillium murcianum* and *Talaromyces australis* pigments are used in the dyeing of woolen clothes at industrial level.

With the greater potential of microbial pigment's staining property, they are exploited through various researches to replace the non-renewable synthetic colorants. Extraordinary ranges of pigments obtained from different classes of fungi are biodegradable and sustainable and are comparatively advantageous to synthetic dyes. Yan et al. (2019) studied the yellow pigment from *Metarhizium anisopliae* for dyeing of silk and wool at optimum temperature between temperatures of 80 and 90 °C under optimum dyeing time of 1 h. Followed by this, red pigment extracted from *Rhodonellum psychrophilum* was isolated from Tso Lake in India which is used in dyeing different fabrics. These fabrics were tested for the antibacterial and antifungal activity, apart from these studies have also reported for the antimicrobial activity of red and yellow color pigment from colored isolated from *Serratia marcescens* and *Micrococcus luteus* (Bisht et al. 2020).

Among important industrial fermentation organisms *Aspergillus* and *Trichoderma sp.*, were studied by Venugopal and Khambhaty (2020) for the pigment production using agro-industrial waste as a replacement for synthetic dyes. Apart from this, pigments from *Penicillium minioluteum, Monascus purpureus,* and *Penicillium chrysogenum* can be used for dyeing of leathers. Different pigment sources are used to dye different fabrics; pink color anthraquinones were used to dye cotton yarn extracted from *Sclerotinia sp.* also the pigments from *Emericella nidulans*, *Monascus purpureus*, *Fusarium verticillioides*, and *Penicillium purpurogenum, Alternaria alternata, Isaria farinosa* were used. Likewise pigment from *Trichoderma sp.* for silk fabric *T. australis* and *P. murcianum* red and yellow pigments was used to dye wool while (Venil et al. 2020a, b). For dyeing of wool, pigments from *Chlorociboria species, Scytalidium cuboideum, Scytalidium ganodermophthorum, and Fusarium oxysporum were* isolated, and they are also used for dyeing of cotton, polyester, nylon, etc. owing to their strong non-degradation property. The pigment quality was tested using different tests such as dipping, submersion, and saturation (Palomino Agurto et al. 2017).

Dyeing properties are important factors for any industrial application of pigments hence sorption kinetics behavior, dyeing rate constant (k), and half-time of dyeing (t1/2) were measured. Apart from this concentration of the pigment, pH and temperature are imporatant in the optimization of dyeing of fabrics (Morales-Oyervides et al. 2017). For example, an orange color pigment isolated from soil Salinococcus roseus under the optimum condition of 40 °C and pH 7, for 96 h had maximum pigment production with better stability (Usman et al. 2018). Multi-facilitated potency of prodigiosin pigment produced by Serratia could be explored not only for antibacterial and antifungal but also as an effective biocontrol agent in agriculture. Inoculation of bacterial isolate in peanut broth at 300 °C for 72 h could yield maximum pigment production and the purity of the pigment extracted was confirmed by the spectrophotometric method and by TLC. When compared with the standard Nistatin, MIC of the pigment was observed at 80 μg for *Alternaria* and for *Fusarium* MIC was observed at 160 μg. Insecticidal activity of 100% mortality was found against cockroaches and tropical ants while 85–71% was observed with termites and pyramid ants (Sagar et al. 2019).

9.5 Conclusion

Exploration of microbial pigments to obtain the natural colorants and its possible application is the need of the day. Novel insights regarding the source of pigmented organisms and the biotechnological tools used to optimize the production to pave way for multifaceted roles creates impact in the current era. From the dietary inclusion, textile dyeing, and pharmaceutical application reveals its scalability in the industrial scale.

Acknowledgments One of the authors, Selvaraju Vishnupriya acknowledges TANUVAS for providing PhD fellowship and for the facilities provided by College of Food and Dairy Technology, TANUVAS.

Conflict of Interest The authors declare no conflict of interest.

Consent for Publication Not applicable.

References

Abdelhafez AA, Husseiny SM, Ali AAA, Sanad HM (2016) Optimization of β-carotene production from agro-industrial by-products by Serratia marcescens ATCC 27117 using Plackett–Burman design and central composite design. Ann Agric Sci 61(1):87–96

Agboyibor C, Kong WB, Chen D, Zhang AM, Niu SQ (2018) Monascus pigments production, composition, bioactivity and its application: a review. Biocatal Agric Biotechnol 16:433–447

Akilandeswari P, Pradeep BV (2016) Exploration of industrially important pigments from soil fungi. Appl Microbiol Biotechnol 100(4):1631–1643

Aruldass CA, Rubiyatno R, Venil CK, Ahmad WA (2015) Violet pigment production from liquid pineapple waste by Chromobacterium violaceum UTM5 and evaluation of its bioactivity. RSC Adv 5(64):51524–51536. https://doi.org/10.1039/c5ra05765e

Aruldass CA, Aziz A, Venil CK, Khasim AR, Ahmad WA (2016) Utilization of agro-industrial waste for the production of yellowish-orange pigment from Chryseobacterium artocarpi CECT 8497. Int Biodeterior Biodegradation 113:342–349. https://doi.org/10.1016/j.ibiod.2016.01.024

Asker D (2017) Isolation and characterization of a novel, highly selective astaxanthin-producing marine bacterium. J Agric Food Chem 65(41):9101–9109

Asker D (2018) High throughput screening and profiling of high-value carotenoids from a wide diversity of bacteria in surface seawater. Food Chem 261:103–111

Asker D, Awad TS, Beppu T, Ueda K (2018) Screening and profiling of natural ketocarotenoids from environmental aquatic bacterial isolates. Food Chem 253:247–254

Bălaşa AF, Chircov C, Grumezescu AM (2020) Marine biocompounds for neuroprotection—A review. Mar Drugs 18(6):290

Bisht G, Srivastava S, Kulshreshtha R, Sourirajan A, Baulmer DJ, Dev K (2020) Applications of red pigments from psychrophilic Rhodonellum psychrophilum GL8 in health, food and antimicrobial finishes on textiles. Process Biochem 94:15

Cardoso LA, Karp SG, Vendruscolo F, Kanno KY, Zoz LI, Carvalho JC (2017) Biotechnological production of carotenoids and their applications in food and pharmaceutical products. In: Carotenoids. InTech Open, p 125

Chadni Z, Rahaman MH, Jerin I, Hoque KMF, Reza MA (2017) Extraction and optimisation of red pigment production as secondary metabolites from Talaromyces verruculosus and its potential use in textile industries. Mycology 8(1):48–57

Chakraborty I, Redkar P, Munjal M, Kumar SS, Rao KB (2015) Isolation and characterization of pigment producing marine actinobacteria from mangrove soil and applications of bio-pigments. Pharm Lett 7(4):93–100

Chen Y, Xie B, Yang J, Chen J, Sun Z (2018) Identification of microbial carotenoids and isoprenoid quinones from Rhodococcus sp. B7740 and its stability in the presence of iron in model gastric conditions. Food Chem 240:204–211

Choksi J, Vora J, Shrivastava N (2020) Bioactive pigments from isolated Bacteria and its antibacterial, antioxidant and sun protective application useful for cosmetic products. Indian J Microbiol 60:379. https://doi.org/10.1007/s12088-020-00870-x

Dahal RH, Kim J (2018) Dyadobacter flavus sp. nov. and Dyadobacterterricola sp. nov., two novel members of the family Cytophagaceae isolated from forest soil. Arch Microbiol. https://doi.org/10.1007/s00203-018-1521-3

De Oliveira CFD, da Costa JPV, Vendruscolo F (2019) Maltose syrup residue as the substrate for Monascus pigments production. Biocatal Agric Biotechnol 18:101101. https://doi.org/10.1016/j.bcab.2019.101101

de Paula Bonadio M, de Freita LA, Mutton MJR (2018) Carotenoid production in sugarcane juice and synthetic media supplemented with nutrients by Rhodotorula rubra l02. Braz J Microbiol 49:872. https://doi.org/10.1016/j.bjm.2018.02.010

DeBritto S, Gajbar TD, Satapute P, Sundaram L, Lakshmikantha RY, Jogaiah S, Ito SI (2020) Isolation and characterization of nutrient dependent pyocyanin from Pseudomonas aeruginosa and its dye and agrochemical properties. Sci Rep 10(1):1–12

Drewnowska JM, Zambrzycka M, Kalska-Szostko B, Fiedoruk K, Swiecicka I (2015) Melanin-like pigment synthesis by soil Bacillus weihenstephanensis isolates from northeastern Poland. PLoS One 10(4):e0125428

El-Naggar NE-A, El-Ewasy SM (2017) Bioproduction, characterization, anticancer and antioxidant activities of extracellular melanin pigment produced by newly isolated microbial cell factories Streptomyces glaucescens NEAE-H. Sci Rep 7(1). https://doi.org/10.1038/srep42129

Elsanhoty RM, Abd El-Razink MM, Al-Turki AI (2017) Production of carotenoids from Rhodotorula mucilaginosa and their applications as colorant agent in sweet candy. WFL Publ Sci Technol 15(2):21–26

Embaby AM, Hussein MN, Hussein A (2018) Monascus orange and red pigments production by Monascus purpureus ATCC16436 through co-solid state fermentation of corn cob and glycerol: an eco-friendly environmental low cost approach. PLoS One 13(12):e0207755

Fariq A, Yasmin A, Jamil M (2019) Production, characterization and antimicrobial activities of bio-pigments by Aquisalibacillus elongatus MB592, Salinicoccussesuvii MB597, and Halomonas aquamarina MB598 isolated from Khewra salt range, Pakistan. Extremophiles 23:435. https://doi.org/10.1007/s00792-019-01095-7

Fouillaud M, Venkatachalam M, Girard-Valenciennes E, Caro Y, Dufossé L (2016) Anthraquinones and derivatives from marine-derived fungi: structural diversity and selected biological activities. Mar Drugs 14(4):64

Fouillaud M, Venkatachalam M, Llorente M, Magalon H, Cuet P, Dufossé L (2017) Biodiversity of pigmented fungi isolated from marine environment in La Réunion island, Indian ocean: new resources for colored metabolites. J Fungi 3(3):36

Galaup P, Sutthiwong N, Leclercq-Perlat MN, Valla A, Caro Y, Fouillaud M, Dufossé L (2015) First isolation of Brevibacterium sp. pigments in the rind of an industrial red-smear-ripened soft cheese. Int J Dairy Technol 68(1):144–147

Giuffrida D, Monnet C, Laurent F, Cacciola F, Oteri M, Le Piver M et al (2020) Carotenoids from the ripening bacterium Brevibacterium linens impart color to the rind of the French cheese, Fourme de Montbrison (PDO). Nat Prod Res 34(1):10–15

Haque MA, Kachrimanidou V, Koutinas A, Lin CSK (2016) Valorization of bakery waste for biocolorant and enzyme production by Monascus purpureus. J Biotechnol 231:55–64. https://doi.org/10.1016/j.jbiotec.2016.05.003

Heer K, Sharma S (2017) Microbial pigments as a natural color: a review. Int J Pharm Sci Res 8 (5):1913–1922

Hernández VA, Galleguillos F, Thibaut R, Müller A (2019) Fungal dyes for textile applications: testing of industrial conditions for wool fabrics dyeing. J Textile Inst 110(1):61–66

Hilares RT, de Souza RA, Marcelino PF, da Silva SS, Dragone G, Mussatto SI, Santos JC (2018) Sugarcane bagasse hydrolysate as a potential feedstock for red pigment production by Monascus ruber. Food Chem 245:786–791

Kageyama H, Waditee-Sirisattha R (2019) Antioxidative, anti-inflammatory, and anti-aging properties of mycosporine-like amino acids: molecular and cellular mechanisms in the protection of skin-aging. Mar Drugs 17(4):222. https://doi.org/10.3390/md17040222

Kalra R, Conlan XA, Goel M (2020) Fungi as a potential source of pigments: harnessing filamentous fungi. Front Chem 8:369

Kantifedaki A, Kachrimanidou V, Mallouchos A, Papanikolaou S, Koutinas AA (2018) Orange processing waste valorisation for the production of bio-based pigments using the fungal strains Monascus purpureus and Penicillium purpurogenum. J Clean Prod 185:882–890

Kaur P, Ghoshal G, Jain A (2019) Bio-utilization of fruits and vegetables waste to produce β-carotene in solid-state fermentation: characterization and antioxidant activity. Process Biochem 76:155–164

Kim D, Ku S (2018) Beneficial effects of Monascus sp. KCCM 10093 pigments and derivatives: a mini review. Molecules 23(1):98

Kim I, Kim J, Chhetri G, Seo T (2019) Flavobacterium humi sp. nov., a flexirubin-type pigment producing bacterium, isolated from soil. J Microbiol 57(12):1079–1085

Kumar BNV, Kampe B, Rösch P, Popp J (2015) Characterization of carotenoids in soil bacteria and investigation of their photodegradation by UVA radiation via resonance Raman spectroscopy. Analyst 140(13):4584–4593. https://doi.org/10.1039/c5an00438a

Lagashetti AC, Dufossé L, Singh SK, Singh PN (2019) Fungal pigments and their prospects in different industries. Microorganisms 7(12):604

Lebeau J, Venkatachalam M, Fouillaud M, Petit T, Vinale F, Dufossé L, Caro Y (2017) Production and new extraction method of polyketide red pigments produced by ascomycetous fungi from terrestrial and marine habitats. J Fungi 3(3):34

Liu H, Zhang C, Zhang X, Tan K, Zhang H, Cheng D et al (2020) A novel carotenoids-producing marine bacterium from noble scallop Chlamys nobilis and antioxidant activities of its carotenoid compositions. Food Chem 320:126629

Morales-Oyervides L, Oliveira J, Sousa-Gallagher M, Méndez-Zavala A, Montañez JC (2017) Assessment of the dyeing properties of the pigments produced by Talaromyces spp. J Fungi 3 (3):38

Moreira MD, Melo MM, Coimbra JM, dos Reis KC, Schwan RF, Silva CF (2018) Solid coffee waste as alternative to produce carotenoids with antioxidant and antimicrobial activities. Waste Manag 82:93–99

Mukherjee T, Bose S, Mukhopadhyay SK (2017). Antioxidant properties of the carotenoid extracts of three Deinococcus–Thermus phylum bacteria, Meiothermus sp. strains RP and TP and Thermus sp. strain YY from Paniphala hot spring, India. Nutrire 42(1). https://doi.org/10.1186/s41110-017-0032-3

Nair AS, Kumar BP, Geo JA (2017) Microbial production of textile grade pigments. Afr J Microbiol Res 11(42):1532–1537

Narsing Rao MP, Xiao M, Li WJ (2017) Fungal and bacterial pigments: secondary metabolites with wide applications. Front Microbiol 8:1113

Nigam PS, Luke JS (2016) Food additives: production of microbial pigments and their antioxidant properties. Curr Opin Food Sci 7:93–100. https://doi.org/10.1016/j.cofs.2016.02.004

Numan M, Bashir S, Mumtaz R, Tayyab S, Rehman NU, Khan AL et al (2018) Therapeutic applications of bacterial pigments: a review of current status and future opportunities. 3 Biotech 8(4):207

Oplatowska-Stachowiak M, Elliott CT (2017) Food colors: existing and emerging food safety concerns. Crit Rev Food Sci Nutr 57(3):524–548

Palomino Agurto ME, Vega Gutierrez SM, Chen HL, Robinson SC (2017) Wood-rotting fungal pigments as colorant coatings on oil-based textile dyes. Coatings 7(10):152

Pandey A, Dhakar K, Jain R, Pandey N, Gupta VK, Kooliyottil R et al (2018a) Cold adapted Fungi from Indian Himalaya: untapped source for bioprospecting. Proc Natl Acad Sci India Sect B Biol Sci 89:1125. https://doi.org/10.1007/s40011-018-1002-0

Pandey N, Jain R, Pandey A, Tamta S (2018b) Optimisation and characterisation of the orange pigment produced by a cold adapted strain of Penicillium sp. (GBPI_P155) isolated from mountain ecosystem. Mycology 9(2):81–92. https://doi.org/10.1080/21501203.2017.1423127

Pandiyarajan S, Premasudha P, Kadirvelu K (2018) Bio-production of novel water-soluble yellow pigment from Aspergillus sp. and exploring its sustainable textile applications. 3 Biotech 8 (9):398

Panesar R, Kaur S, Panesar PS (2015) Production of microbial pigments utilizing agro-industrial waste: a review. Curr Opin Food Sci 1:70–76

Park SA, Ahn SY, Choi KY (2020) Functional microbial pigments isolated from Chryseobacterium and Deinococcus species for bio-paint application. Biotechnol Bioprocess Eng 25(3):394–402

Patthawaro S, Lomthaisong K, Saejung C (2019) Bioconversion of agro-industrial waste to value-added product lycopene by photosynthetic bacterium Rhodopseudomonas faecalis and its carotenoid composition. Waste Biomass Valorization 11:2375. https://doi.org/10.1007/s12649-018-00571-z

Pereira da Costa D, Campos Miranda-Filho K (2019) The use of carotenoid pigments as food additives for aquatic organisms and their functional roles. Rev Aquac. https://doi.org/10.1111/raq.12398

Prathiba S, Jayaraman G (2018) Evaluation of the anti-oxidant property and cytotoxic potential of the metabolites extracted from the bacterial isolates from mangrove Forest and saltern regions of South India. Prep Biochem Biotechnol 48:750–758. https://doi.org/10.1080/10826068.2018.1508038

Ram S, Mitra M, Shah F, Tirkey SR, Mishra S (2020a) Bacteria as an alternate biofactory for carotenoid production: a review of its applications, opportunities and challenges. J Funct Foods 67:103867

Ram S, Tirkey SR, Kumar MA, Mishra S (2020b) Ameliorating process parameters for zeaxanthin yield in Arthrobacter gandavensis MTCC 25325. AMB Express 10:1–13

Rezaeeyan Z, Safarpour A, Amoozegar MA, Babavalian H, Tebyanian H, Shakeri F (2017) High carotenoid production by a halotolerant bacterium, Kocuria sp. strain QWT-12 and anticancer activity of its carotenoid. EXCLI J 16:840

Sadh PK, Duhan S, Duhan JS (2018) Agro-industrial wastes and their utilization using solid state fermentation: a review. Bioresour Bioprocess 5:1–15

Saejung C, Puensungnern L (2020) Evaluation of molasses-based medium as a low cost medium for carotenoids and fatty acid production by photosynthetic bacteria. Waste Biomass Valorization 11(1):143–152

Sagar BSV, Deepak BS, Tejaswini GS, Aparna Y, Sarada J (2019) Evaluation of prodigiosin pigment for antimicrobial and insecticidal activities on selected bacterial pathogens & household pests

Salvo A, Giuffrida D, Rotondo A, Pasquale PD, La Torre GL, Dugo G (2017) Determination and quantification of carotenoids in sea sponges Raspacionaaculeata and Dictyonella marsilii present in the Ganzirri Lake (Messina), Italy. Nat Prod Res 31(20):2397–2404

Sehrawat, R., Panesar, P. S., Panesar, R., & Kumar, A. (2017). Biopigment produced by Monascus purpureus MTCC 369 in submerged and solid-state fermentation: a comparative study. Pigm Resin Technol, 46:425-432

Sharma R, Ghoshal G (2020) Optimization of carotenoids production by Rhodotorula mucilaginosa (MTCC-1403) using agro-industrial waste in bioreactor: a statistical approach. Biotechnol Rep 25:e00407

Silva TR, Tavares RS, Canela-Garayoa R, Eras J, Rodrigues MV, Neri-Numa IA et al (2019) Chemical characterization and biotechnological applicability of pigments isolated from Antarctic bacteria. Mar Biotechnol 21(3):416–429

Suryawanshi RK, Patil CD, Borase HP, Narkhede CP, Stevenson A, Hallsworth JE, Patil SV (2015) Towards an understanding of bacterial metabolites prodigiosin and violacein and their potential for use in commercial sunscreens. Int J Cosmet Sci 37(1):98–107

Taufiqurrahmi N, Religia P, Mulyani G, Suryana D, Ichsan Tanjung FA et al (2017) Phycocyanin extraction in Spirulina produced using agricultural waste. IOP Conf Ser Mater Sci Eng 206:012097

Tendulkar PA, Surve MA, Deodhar MA (2018) Sun protective formulation from carotenoids extracted from thermotolerant genera Synechocystispevalekii. Int J Pharm Sci Res 9 (8):3223–3234

Tuli HS, Chaudhary P, Beniwal V, Sharma AK (2015) Microbial pigments as natural color sources: current trends and future perspectives. J Food Sci Technol 52(8):4669–4678

Usman HM, Farouq AA, Baki AS, Abdulkadir N, Mustapha G (2018) Production and characterization of orange pigment produced by Halophilic bacterium Salinococcus roseus isolated from abattoir soil. J Microbiol Exp 6(6):238–243

Venil CK, Aruldass CA, Abd Halim MH, Khasim AR, Zakaria ZA, Ahmad WA (2015) Spray drying of violet pigment from Chromobacterium violaceum UTM 5 and its application in food model systems. Int Biodeterior Biodegrad 102:324–329

Venil CK, Dufossé L, Renuka Devi P (2020a) Bacterial pigments: sustainable compounds with market potential for pharma and food industry. Front Sustain Food Syst 4:100

Venil CK, Velmurugan P, Dufossé L, Devi PR, Ravi AV (2020b) Fungal pigments: potential coloring compounds for wide ranging applications in textile dyeing. J Fungi 6(2):68

Venkatachalam M, Zelena M, Cacciola F, Ceslova L, Girard-Valenciennes E, Clerc P et al (2018) Partial characterization of the pigments produced by the marine-derived fungus *Talaromyces albobiverticillius* 30548. Towards a new fungal red colorant for the food industry. J Food Compos Anal 67:38–47

Venugopal S, Khambhaty Y (2020) Agro-residues for enhanced production of bio-colorants: a feasible approach for dyeing and surface coating of leather. Clean Technol Environ Policy, 1–14

Wani NA, Khanday WI, Tirumale S (2020) Phytochemical analysis and evaluation of antibacterial activity of different extracts of soil-isolated fungus chaetomiumcupreum. J Nat Sci Biol Med 11 (1):72

Xia Y, Wang G, Lin X, Song X, Ai L (2016) Solid-state fermentation with Serratia marcescens Xd-1 enhanced production of prodigiosin by using bagasse as an inertia matrix. Ann Microbiol 66(3):1239–1247. https://doi.org/10.1007/s13213-016-1208-4

Yan B, Yang M, Zhou Q, Xing T, Chen G, Sheng J (2019) Yellow pigment of Metarhizium anisopliae and its application to the dyeing of fabrics. Color Technol 135(4):267–274

Yasmeen S, Muvva V, Munaganti RK (2017) Isolation and characterization of bioactive Streptomyces from mangrove eco-system of Machilipatnam, Krishna District, Andhra Pradesh. Asian J Pharm Clin Res 9(9):258

Yoo AY, Alnaeeli M, Park JK (2016) Production control and characterization of antibacterial carotenoids from the yeast Rhodotorula mucilaginosa AY-01. Process Biochem 51(4):463–473. https://doi.org/10.1016/j.procbio.2016.01.008

Yu X, Wu H, Zhang J (2015) Effect of Monascus as a nitrite substitute on color, lipid oxidation, and proteolysis of fermented meat mince. Food Sci Biotechnol 24(2):575–581. https://doi.org/10.1007/s10068-015-0075-2

Zhao Y, Guo L, Xia Y, Zhuang X, Chu W (2019) Isolation, identification of carotenoid-producing Rhodotorula sp. from marine environment and optimization for carotenoid production. Mar Drugs 17(3):161

Part II

Microbial Polymers in Agriculture

Extracellular Polymeric Substances from Agriculturally Important Microorganisms

10

Valeria Valenzuela Ruiz, Roel Alejandro Chávez Luzania, Fannie Isela Parra Cota, Gustavo Santoyo, and Sergio de los Santos Villalobos

Abstract

Within microorganisms, there is a group capable of producing extracellular polymeric substances (EPS), which are proteins, polysaccharides, lipids, nucleic acids, and humic substances, produced and excreted outside the cell. Microorganisms have EPS to facilitate their vital metabolic processes, these substances are necessary to carry out chemical reactions within their natural habitat to trap nutrients or protect the plant against abiotic factors such as salinity and drought. Thus, their use in agriculture to promote plant growth and health has been widely studied. EPS have also been a subject of interest in agriculture due to their abilities to improve soil quality, increase soil aggregates, and contribute to improve soil fertility. In addition, use of EPS in soil surrounding plant roots promote benefits in availability of water in contact with plant and in nutrient absorption, these benefits affect both host and surrounding microorganisms, thus creating a more suitable ecosystem for plant developement. These benefits were observed through EPS producer inoculation in wheat, chickpea, and sunflowers, among other crops of agronomic importance.

V. Valenzuela Ruiz · R. A. Chávez Luzania · S. de los Santos Villalobos (✉)
Instituto Tecnológico de Sonora, Ciudad Obregón, Sonora, Mexico
e-mail: sergio.delossantos@itson.edu.mx

F. I. Parra Cota
Campo Experimental Norman E. Borlaug, Instituto Nacional de Investigaciones Forestales, Agrícolas y Pecuarias (INIFAP), Ciudad Obregón, Sonora, Mexico

G. Santoyo
Instituto de Investigaciones Químico Biológicas, Universidad Michoacana de San Nicolás de Hidalgo, Morelia, Michoacán, Mexico

A. Vaishnav, D. K. Choudhary (eds.), *Microbial Polymers*,
https://doi.org/10.1007/978-981-16-0045-6_10

Keywords

Plan growth-promoting microorganisms · Agriculture · Soil health · Beneficial bacteria

10.1 Introduction

Food is one of the most important basic needs for human beings; in turn it has an important implication on society at an economic and personal development level (Etesami and Maheshwari 2018). As technology and knowledge about the food we consume advances, increasing strict parameters have been established for the quality of these products, governmental and competent international organizations have turned to see safe and sustainable food production as a priority (Kenneth et al. 2019). In addition, the population increase around the world also results in an increase in the demand for not only more food production but also at higher quality. It is estimated that there will be around 10 billion people on the planet by the year 2050 (FAO 2017).

To meet this food demand, production must increase by roughly 70% by 2050 and double or triple by 2100 (Crist et al. 2017). Currently, in industrialized agricultural techiques predominate practices that consume fossil fuels and require excessive applications of fertilizers that not only harm soil and crops, but also alter microbial communities that play an important role in plant development; in addition to contaminating the surrounding environment, resulting in *i.e.* saline and compacted soils and water contamination (More et al. 2014). Although some of these practices have immediate advantages in agricultural production, their effects are perpetuated in the environment, causing a problem that ends up affecting the quality of life for humans and other species, as well as long term production (Kenneth et al. 2019). Due to the population increase, the demand for food and the need to generate quality products, research should focus on the development of new methods with sustainable approaches to meet the current food demand (Etesami and Maheshwari 2018).

Plants interact with $\sim 1 \times 10^9$ microbial cells g^{-1} dry soil and 1×10^5 microbial species g^{-1} dry soil (Valenzuela-Aragon et al. 2019). Microorganisms and microbial communities that are found in the phyllosphere [total plant proportion as a habitat for microorganisms, compromising the aerial parts of plants and is dominated by the leaves, where the global bacterial population present could comprise up to 10^{26} cells and fungal population is yet to be estimated (Vorholt 2012)] and the rhizosphere [the soil portion that is influenced by root exudates, where the microbial community residing in this niche is driven by the direct effect of the presence of root exudates, leading to increased microorganisms biomass; and high nutrient availability, which have the ability to change environmental conditions on rhizosphere, creating a microhabitat, whose concentration can be up to 10^{12} cells/g of soil (Brink 2016; Andreote et al. 2014)] in agro-ecosystems, are being targeted by their potential for sustainable agriculture in search of new alternatives to obtain better yields in agriculture and contribute to food security (Valenzuela-Aragon et al. 2019; de los Santos-Villalobos et al. 2018).

The interaction of microorganisms with plants forms a complex network of metabolite exchange, through which biological mechanisms are activated or repressed (Kenneth et al. 2019). Some of these mechanisms, described later, provide microorganisms plant growth promotion and/or biological control, depending on the species, regulated by environmental factors, nutrient availability, and the plant–microorganism interactions (Yadav et al. 2020). Plant growth-promoting microorganisms (PGPM), mainly bacteria (PGPB), promote plant development and overall health, allowing it to grow optimally by reducing stress conditions around it (Rojas-Padilla et al. 2020). On the other hand, microorganisms that have the ability to act as antagonists against pests that affect crops have been classified as biocontrol agents (BCA) (Orozco-Mosqueda et al. 2018).

Within the complex network of metabolites generated by microorganisms in the vicinity of the plant, there is a group whose main differentiator is its polymeric nature, the extracellular polymeric substances (EPS). These substances have diverse functions which are useful for modern agriculture, some PGPB and BCA are EPS producers (Siddiqui and Khan 2017; Valenzuela Ruiz et al. 2020). Thus, in this chapter an overview of the current agronomic researches on EPS are discussed.

10.2 Plant Growth-Promoting Bacteria

This is assigned to those bacteria capable of causing a beneficial effect on plants, they are mainly found in soil, plant root surface up to a distance of 5 mm (and is related to the plant genotype, edaphic-climatic conditions, among others), and within their own tissues (Lakshmanan et al. 2017; Rodriguez et al. 2019).

These microorganisms have stimulating mechanisms (1) indirect mechanisms, plant protecting against adverse factors by the production of antibiotics and lytic enzymes, as well as the synthesis of hydrogen cyanide and volatile compounds (Orozco-Mosqueda et al. 2018); or (2) direct mechanisms, which involve either improving their development through the modulation of hormones or facilitating the acquisition of essential nutrients for the plant by phosphorus, potassium, and zinc solubilization, the biological fixation of nitrogen and the siderophores production, ammonia and secondary metabolites such as EPS with growth-promoting properties (Table 10.1) (de los Santos-Villalobos et al. 2018; Yadav et al. 2020; Valenzuela-Aragon et al. 2019). EPS-producing microorganisms are capable of establishing interactions with plants in a way that promotes their development, stimulating microorganisms present in the surrounding soil community and exudate production, where organic carbon released by roots surfaces favors microbial community proliferation in the rhizosphere, which in turn, promotes soil aggregation, producing mucilaginous EPS, and increasing the amount of soil attached to the roots (Costa et al. 2018). Soil aggregation in the roots by the action of EPS form a favorable plant growth environment with nutrient and water retention properties (Rashid et al. 2016). Likewise, certain microorganisms have biocontrol actions that protect the plant against different microorganisms (Table 10.2) (de los Santos-Villalobos et al. 2018; Yadav et al. 2020;Valenzuela-Aragon et al. 2019).

Table 10.1 Bacteria that promote plant growth in agro-ecosystems

PGPB	Crop	Reference
Rhizobium leguminosarum	Direct growth promotion of canola and lettuce	Noel et al. (1996)
Pseudomonas putida	Early developments of canola seedlings	Glick et al. (1997)
Azospirillum brasilense and *A. irakense*	Growth promotion of wheat and maize plants	Dobbelaere et al. (2002)
P. fluorescens	Growth of pearl millet	Niranjan et al. (2003)
P. putida strain	Tomato plant growth improvement	Gravel et al. (2007)
Azospirillum and Azotobacter	Canola plant growth and productivity improvement	Yasari and Patwardhan (2007)
Pseudomonas aeruginosa and *Bacillus firmus*	HCN production, NH3 production, Siderophore production, antagonist, and phosphate solubilizer	Sandilya et al. (2016, 2017)
Azotobacter, *Azospirillum*, and *Pseudomonas*	Chick pea growth stimulation	Rokhzadi et al. (2008)
Pseudomonas sp. and *R. leguminosmarum*	Improve yield and phosphorus uptake in wheat	Afzal and Asghari (2008)
P. putida	Seed germination, growth parameters of maize seedling in greenhouse and also grain yield of field grown maize	Gholami et al. (2009)
P. putida and *fluorescens*, *A. brasilense,* and *lipoferum*	Improves seed germination, seedling growth, and yield of maize	Nezarat and Gholami et al. (2009)
P. fluorescens	Increase growth, leaf nutrient contents, and yield of banana	Kavino et al. (2010)
Bacillus sp. and *Burkholderia sp.*	Phosphate solubilizer	Oliveira et al. (2009)
Bacillus polymyxa, P. alcaligenes, and *Mycobacterium phlei*	N, P, and K uptake improvement by maize crop	Egamberdiyeva (2007)
Rhodococcus sp., Pseudomonas sp., and *Arthrobacter nicotinovorans*	Phosphate solubilizer	Sofia et al. (2014)
Azospirillum brasilense and *Bradyrhizobium japonicum*	Phytostimulation	Cassan et al. (2009)

The previous tables show success cases where the inoculation of PGPB recognized as EPS producers showed better plant development, *i.e., P. fluorescens* Pf1 and *P. fluorescens* CHA0 (Kavino et al. 2010) have showed growth promotion in banana cultivation and *Bacillus subtilis* strain TE3^{T} (Villa-Rodriguez et al. 2019) has showed biocontrol against *Bipolaris sorokiniana* in wheat.

Table 10.2 Biological control agents used in crop production

BCA	Crop	Reference
Bacillus pumilus, *Kluyvera cryocrescens*, B. *amyloliquefaciens*, and *B. subtilis*	Cucumber mosaic virus (CMV) of tomato	Zehnder et al. (2000)
B. amyloliquefaciens, *B. subtilis*, and *B. pumilus*	Tomato mottle virus	Murphy et al. (2000)
B. pumilus	Bacterial wilt disease in Cucumber	Zehnder et al. (2001)
B. subtilis, *B. pumilus* R7, and *B. pumilus*	Downy mildew in pearl millet	Niranjan et al. (2003)
B. cereus	Foliar diseases of tomato	Silva et al. (2004)
Burkholderia	Maize rot	Hernandez Rodriguez et al. (2008)
Bacillus subtilis	Spot blotch in cereals	Villa-Rodriguez et al. (2019)

The main hypothesis to explain the efficiency of colonization by PGPB or biological control agents is the chemotaxis effect (Yadav et al. 2020). Plant colonization by bacteria and fungi depends on their ability to (1) move toward the root, (2) use nitrogen and carbon sources contained in soil, (3) or those given by exudates of root to shape the root microbiome, (4) resist plant's response reaction, including reactive oxygen species (ROS), and (5) have the ability to form biofilms on the root surface (Kumar and Singh 2020). The study of the behavior of PGPB in agroecosystems has shown a wide range of action mechanisms to achieve their activities, where EPS are an important group of biomolecules produced for these purposes. In addition, these substances promote the production of beneficial biomolecules for plant development or, failing that, the mitigation of stress-causing biomolecules (Parra-Cota et al. 2018).

10.3 Extracellular Polymeric Substances

PGPB are able to produce diverse groups of EPS; these are composed of proteins, nucleic acids, polysaccharides, humic substances, and lipids (Fig. 10.1). Thus, EPS are produced and secreted into the extracellular environment, either because they are essential for life or due to the persistence of microorganisms, because they give an ideal condition to carry out chemical reactions since they trap nutrients or protect against abiotic factors such as drought and salinity (More et al. 2014). The biomolecules included within EPS have a diverse group of compounds that vary their effect on plants according to their nature and chemical composition (Flemming et al. 2016).

EPS are molecules that are produced in exchange for high energy consumption in cells; they are mainly synthesized by environmental signals and provide an advantage comparable to the energy expenditure used (Flemming et al. 2016). The EPS, as

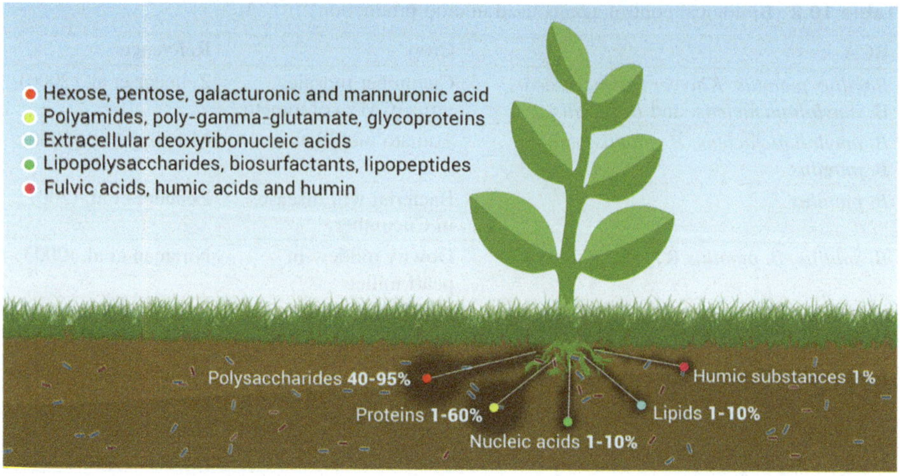

Fig. 10.1 Main EPS produced by bacteria and their composition. Polysaccharides, proteins, nucleic acids, lipids, and humic acids have the ability to improve bioabsorption and soil aggregates although to a greater extent the first two groups

a resource, are important in different ecosystems since they allow the permanence of certain beneficial organisms in the niche. Agro-ecosystems are an important deposit for EPS producers where they provide advantages to crops (Costa et al. 2018). EPS composition grants it important properties for agricultural uses, such as nutrient entrapment; protection against drought and salinity stress; as well as heat and cold stress; antimicrobial protection; better soil texture; and generating symbiosis with plants where these interactions favor soil and plant fertility, resilience and overall plant health (More et al. 2014).

These substances differ from each other by their physical and chemical properties of each type of polymer, i.e., (1) polysaccharides have adhesion, cell aggregation, increased water retention, adsorption properties of organic and inorganic compounds, they form a protective barrier, and are a source of nutrients, (2) proteins possess adhesion, cell aggregation, water retention properties, binding of enzymes, acceptor and electron donor, absorption of organic and inorganic matter, and provides a protective barrier for cells, (3) nucleic acids possess properties of adhesion, aggregation of cells, source of nutrients, exchange of genetic information, export of cellular components, (4) lipids differ by being exporters of cellular components, and (5) humic substances provide adhesion, being electron acceptors and donors (More et al. 2014).

The EPS matrix serves as a reservoir of extracellular enzymes which capture substances from the aqueous phase and favors that these compounds are used as a source of energy, retaining carbon, nitrogen and phosphorus containing compounds (Bach et al. 2010; Flemming and Wingender 2010). During the course of biofilm maturation, soluble EPS forms a filamentous matrix that maximizes nutrient distribution and speeds up the cell aggregation (Berk et al. 2013). EPS have the ability to generate cohesion between bacteria, causing adhesion of biofilms to surfaces, modify spatial organization, permitting the exchange of metabolic information between microorganisms, and allowing adhesion between them; therefore, establishing a stable consortium of mixed population communities, and hence, diverse synergistic relationships (Laspidou and Rittman 2002; Flemming and Wingender 2001).

An example is the nitrification process, which takes place in biofilms and allows the spatial closeness of ammonia oxidizers to nitrite oxidizers (Wan et al. 2013). Also, different protein (adhesins) and extracellular DNA form EPS are related in the process of adhesion of microorganisms to plant surfaces, these being beneficial traits for plant colonization of plant growth-promoting microorganisms (Okshevsky and Meyer 2013; Wan et al. 2013). On the other hand, amino acids present in EPS function as both carbon and nitrogen sources for the root-inhabiting bacteria; for example, agriculturally important bacteria such as *Pseudomonas fluorescens* and *Bacillus subtilis* showed motility toward proteinogenic amino acids which some EPS present (Velmourougane et al. 2017). Nutrient retention and root-inhabiting facility is important for plant growth and health where the retention or availability of nutrients is crucial for plant development (Yadav et al. 2020).

Thus, EPS are considered a viable alternative to replace conventional agrochemicals, standing out for their biodegradability, high efficiency, non-toxicity, and no generation of secondary pollutants (More et al. 2014). Currently, the development of production methods for substances capable of promoting greater plant yield and microorganisms that produce them are of research interest. In recent years, the understanding of the mechanisms involved in plant–microorganisms interaction is determinant to modulate the microbial population in soil for increasing food production.

10.4 Agricultural Important Roles of EPS

10.4.1 Symbiosis

A mechanism widely studied in EPS is its function in symbiosis relation among atmospheric nitrogen-fixing rhizobacteria and plants, where polysaccharides, key in formation of nodules (Costa et al. 2018). Van Workum et al. (1998) stated that EPS produced by *R. leguminosarum* accelerate the root hair curling rate and infection to extent the rhizobial root penetration in *Vicia sativa*. In a recent study, the relation between EPS of *Lotus japonicus* and *Mesorhizobium loti* regulated by a receptor expressed by the plant was investigated, which permits infection of bacteria that produce EPS (Kawaharada et al. 2015). The expression of this receptor shows how the plant can identify EPS structures produced by rhizobial microorganisms. The importance of beneficial relations established between the soil, plant, and microbiota is key for optimal crop production; however, one of the main limitations is crop yield loss due to plant pathogens.

10.4.2 EPS as Pathogenicity/Virulence Factors

Every year around the world, biotic stress in crops caused by diseases, weeds, and insects causes the loss of 31–42% of food production, which represents 500 billion dollars; in addition to an extra loss of 6–20% (120 billion dollars) due to fungi, bacteria, and insects after harvest. These losses tend to be aggravated when dealing with developing countries, where the percentages in comparison with developed countries differ up to 22% (Shameer et al. 2019).

EPS are able to protect plants against pathogens, presenting biologic control activity; however, they may also function as a mechanical wall among phytopathogens and plant defenses (Velmourougane et al. 2017). For instance, *Pseudomonas syringae pv. phaseolicola* and *S. meliloti* polymers are able to protect bacteria against reactive oxygen species that are produced by the plant host during infection, resulting in an oxidative stress decrease (Lehman and Long 2013). Similarly, to *Agrobacterium, Bacillus clavibacter, Erwinia, Pantoea, Pseudomonas, Xanthomonas,* and *Ralstonia* genere (Caro-Astorga et al. 2014;Merritt et al. 2007; Tomlinson et al. 2010; Vlamakis et al. 2013; Ramey et al. 2004).

10.4.3 Drought Stress

In many agriculturally important microorganisms, EPS are really relevant in abiotic and biotic environmental stress combat, e.g., drought, temperature, salt tolerance, and phytopathogens (Bogino et al. 2013). The matrix made of EPS is capable of supporting up to 20 times its weight in water (Flemming and Wingender 2001) and contributes to soil moisture (Roberson et al. 1993). Biofilms are able to hold up water in microenvironments either directly owing to its hygroscopic properties, or for its ability to modify the conformation of the biofilm, an ability that helps to avoid the loss of water by evaporation (Costa et al. 2018). Evaporation losses are not the only way water is retained (Mager 2010) in a study conducted by Cho et al. (2018). They demonstrated that the EPS of *P. chlororaphis* had a positive effect against drought; they caused a limited wilting and improved the water retention capacity, thus increasing the photosynthetic efficiency in *Arabidopsis* plants. Similarly, Timmusk and Zucca (2019) stated that biofilm formation is required in induction of drought tolerance. As a result, the EPS matrix makes microbial cells more tolerant to drought conditions and thereby prolonging their survival rate and residence time in soils a beneficial trait for plant growth-promoting microorganisms (Velmourougane et al. 2017).

10.4.4 Heat Stress

Among the most important stress conditions is stress due to high temperatures, which constitutes one of the conditions with the greatest impact on crops, since heat modifies water relations, the rate of transpiration, the balance of photosynthesis, efficiency of the use of water, protein synthesis and enzymatic activity; this causes burns in branches and leaves, senescence of leaves, less growth, and discoloration of leaves and fruits, which consequently causes a decrease in agricultural yield (Krieger-Liszkay et al. 2018).

In relation to the above, EPS have shown positive effects in the improvement of plant growth in a situation of heat stress, these are a protection factor for microorganisms exposed to places with high temperatures, which allows them to proliferate and develop in such a way that it can continue with its metabolism and carry out the promotion of plant growth, thus the EPS act indirectly on the plants for their benefit (Sengupta and Dey 2019). In addition to the fact that the EPS matrix has important functions as a diffusion barrier to antimicrobial agents, it also provides protection against heat stress; this matrix favors the retention of water in the vicinity of the plant, which allows it to have this resource available for its development, it is in this way that growth promotion occurs in response to abiotic stress (Harimawan and Ting 2016).

10.4.5 Salt Stress

A growing problem in current agriculture are saline soils, where world estimates show that stress from saline conditions is increasing, particularly in irrigated areas, where the augmentation was from 45 million hectares to 62 million hectares between 1990 through 2013 (Qadir et al. 2014). Salinity causes harmful effects on plant growth; this affects the quality of water for agronomic uses, is a factor of soil erosion, and affects sedimentation. Furthermore, EPS have the ability to improve tolerance to saline stress, both for plants and for EPS-producing microorganisms associated with them,(Costa et al. 2018). Bacterial EPS can bind cations including Na+ thereby alleviate salinity stress by decreasing the levels of Na+ in the surrounding environment available to the plant. There are also reports of EPS-producing plant growth-promoting rhizobacteria to significantly enhance the volume of soil macropores and the rhizosphere soil aggregation, resulting in increased water and fertilizer availability to inoculated plants (Upadhyay et al. 2011). Salt-tolerant ACC-deaminase containing PGPB can alleviate soil salinity stress during plant growth by reducing ethylene synthesis while bacterial EPS will ensure its better survival in field and can also help to mitigate salinity stress by reducing the concentration of Na+ in soil (Kumari and Khanna 2015). An example of this mechanism is reported by Kumari and Khanna (2015), in chickpea (*Cicer arietinum* L.) growth promotion by the inoculation of EPS-producing rhizobacteria from genera *Enterobacter Burkholderia, Achromobacter, Bacillus,* and *Pseudomonas.* Similarly, Arora et al. (2010) stated that EPS produced on germination of maize, wheat, and rice by a consortium of cyanobacteria were studied at different concentration of salt. EPS production was found to be stimulated by salts, which in turn had a significant Na+ removal capability from aqueous solution, and the germination of seeds, vigor index, and efficiency of mobilization in all the three crops improved when cyanobacterial EPS were applied.

10.4.6 EPS and Soil Structure

Soil structure plays a fundamental role in ecosystem functions which is crucial for plant development, where nutrient availability, water retention, and the presence of microbial community is dependent of good soil structure, therefore, regulating soil fertility and plant productivity (Vezzani et al. 2018). EPS related to microorganisms present in soils represent 0.1–1.5% of soil organic matter (Chenu 1995). According to Chenu (1993), typical EPS values found in soils (assuming 10^9 cells g^{-1} of soil with total weight of 1.1 mg) ranges between 0.1 and 5.0 mg EPS (or 10–500% of cells' biomass) (Or et al. 2007). Also, EPS produced in situ is expected to improve soil properties, for example, increasing heterogeneity (Davis et al. 2009) and improving soil aggregate stability (Tang et al. 2011). Aggregates in soils are of great importance to improve various characteristics such as erodibility, porosity, agronomic performance, and root elongation. (Bronick and Lal 2005). Evidence of EPS-producing microorganisms involved in soil quality (Table 10.3) are, for example, *Paenibacillus polymyxa* which was implicated in the aggregation of root-adhering soil on wheat (Bezzate et al. 2000); *Pantoea agglomerans* regulates rhizosphere's soil water content by improved soil macroaggregation (Kaci et al. 2005); *Rhizobium* modifies the soil structure around a sunflower root system after inoculation, counteracting the negative effect of a water deficit on plant growth (Alami et al. 2000); *P. putida* is capable of improving soil stability by 150% in salinity, high temperatures, and drought conditions (Vardharajula and Ali 2015), *Bacillus amyloliquefaciens*, *B. licheniformis*, and *B. subtilis* improve aggregate stability under drought stress (Vardharajula and Ali 2014). The production of EPS in cells is given by the internal and external factors that surround them; thus, the factors delimiting production are: genotype, physical, chemical, and biological

Table 10.3 Main producers of EPS and their main benefits to soil aggregation

Strain	Benefit	Reference
P. putida	Improvement of the stability of soil aggregates by more than 50% in conditions of stress due to salinity, drought, and temperature	Vardharajula and Ali (2015)
Paenibacillus	Promote the growth of bacterial heterotrophic community and increase its population, in addition to increasing available N and P	Wu et al. (2014)
Bacillus amyloliquefaciens, *B. licheniformis*, and *B. subtilis*	Aggregate stability improve in stress absence and under drought stress	Vardharajula and Ali (2014)
Microbacterium arborescens	It cause a strong binding in sandy soil better aggregation in a sandy soil than in agricultural	Godinho and Bhosle (2009)
N. muscorum	Improves soil aggravation by 18% and increases total soil carbon by more than 60% and nitrogen by more than double	Rogers and Burns (1994)
Stachybotrys atra	Increase the aggregation of fumigated soil	McCalla et al. (1958)

factors, in addition to the carbon / nitrogen ratio, pH, and temperature (More et al. 2014).

10.5 Inoculation of EPS Producers in Crops

Few of the main problems we are currently facing due to traditional agricultural practices are compacted and saline soils, as well as inefficient crop yields in relation to the amount of agro-chemicals applied, and soil degradation (Lares-Orozco et al. 2016; Herrejón 2014; Miransari 2011; Dotterweich 2013). This last one has impacted 80% of agricultural soils worldwide (Nkonya et al. 2015). EPS inoculated in plants surrounding soils improve water retention, dampen changes in the water potential, promote production of root exudates, and improve nutrient uptake and plant growth. The improvement in soil aggregation and structure is able to promote soil growth because it improves the nutrients absorption and maintains a good humidity degree in soil (Alami et al. 2000; Bezzate et al. 2000; Sandhya et al. 2009). In soil there are different components that promote its aggregation, and microorganisms are essential for soil aggregation. The extent to which microorganisms have in this regard depends on the diversity of species present, the available substrates, and soil management in that region (Costa et al. 2018). *Bacillus*, *Pseudomonas*, *Rhizobium*, and *Pantoea* stand out for being EPS producers and plant growth promoters; these microorganisms can be inoculated both in the soil and in the seedlings (Cipriano et al. 2016). In addition to pure cultures of microorganisms, the combination of these can be an interesting option to achieve a better effect on the soil and plants; however, there are not enough studies on the use of EPS-producing microorganisms consortia.

10.5.1 EPS Application in Agriculture

The inoculation of PGPB is a very widespread and important practice in agriculture due to its multiple benefits; minimal changes in water potential increases nutrient absorption and improves plant growth. Thus, these microorganisms can provide multiple benefits to the plant, and their segregated EPS have the capacity to solve productivity problems and soil aggregation. For example, Tewari and Sharma (2020) carried out field experiments on *Cajanus cajan* using *Rhizobium* and its purified EPS, specifically exopolysaccharides. The experiment was conducted in northern India under natural stress: pH 8.9, 45 °C, and 33 ppm NO_3-N. The test was performed with and without the recommended dose of fertilizer in the form of diammonium phosphate, and in combination with treatments of the microorganism with and without EPS. All treatments significantly improved germination, pod number, seed yield, and protein content by 114, 138, 131, and 137%, respectively, compared to treatments without these two variables (control) (Tewari and Sharma 2020).

In addition, Daud et al. (2019) reported the importance of *Paenibacillus polymyxa* in agriculture and its applicable properties in modern biotechnology focused in crops, for example, tomato (Cheng et al. 2017) maize (Sheela 2013) and rice (de Souza et al. 2015). This bacterium has been noted for its ability to fix atmospheric nitrogen, solubilize phosphates, the production of phytohormones and for the production of important EPS, such as (1) D-glucuronic acid for biofilm formation; (2) production of levanos and β-glucans as antioxidants and antitumor; (3) dehydrogenase and polysaccharide beads for the absorption of heavy metals; (4) polymyxin E1 and E2 as a bactericide; and (5) polymixin B as an antimicrobial (Daud et al. 2019).

Similarly, *Rhizobium* is a wheat growth promoter; it has been shown that the EPS production of this strain improves plant growth, increases the soil attached to the root, and an increase in the amount of soluble aggregates is present in the water due to the decrease in soil water stress. It was found that the polymeric substance produced by this microorganism is composed of glucose, galactose, and mannuronic acid in a molar ratio of 2:1:1 respectively, this was observed when glucose, fructose, mannitol, and sucrose were used as carbon sources; later, it was determined by means of 1D and 2D NMR spectroscopy that consists of a repeating tetrasaccharide unit; this composition also provides polyelectrolytic properties (Kaci et al. 2005).

Another case of success in the use of EPS to improve the productivity of crops is that of inoculation with *Halomonas variabilis* and *Planococcus rifietoensis*, both EPS producers, in chickpea plants. These strains were found to have the ability to form biofilms and accumulate exopolysaccharides by increasing saline stress, this allowed them to tolerate salinity, improved plant growth, and increased soil aggregation by more than 75% under high salt concentrations. These results showed that bacteria could be involved in the development of microbial communities under saline stress; in addition, evidence suggests that these strains are useful in the colonization of bacteria in plant roots and soil particles (Qurashi and Sabri 2012).

10.6 Conclusions and Perspectives

EPS are compounds with functions as diverse and complex as their structures. EPS play a relevant role in agricultural production; however, it is necessary to extend its study in other taxonomic groups that have not been profoundly studied and that are part of the plant microbiome. Likewise, it is required to know the interactions between the different EPS and their possible synergistic actions, either through the co-inoculation of mixtures of pure compounds or microbial strains that produce multiple EPS. Therefore, despite the actual and wide knowledge of EPS, it is also important to continue exploring on the identification of new polymers and potential benefits on plant crops, in order to innovate the bioinoculant formulation for increasing the food production, under current or future scenarios.

References

Afzal A, Asghari B (2008) Rhizobium and phosphate solubilizing bacteria improve the yield and phosphorus uptake in wheat. Int J Agric Biol 10(1):85–88

Alami Y, Achouak W, Marol C, Heulin T (2000) Rhizosphere soil aggregation and plant growth promotion of sunflowers by an exopolysaccharide-producing *Rhizobium sp* strain isolated from sunflower roots. Appl Environ Microbiol 66:3393–3398. https://doi.org/10.1128/Aem.66.8.3393-3398.2000

Andreote F, Gumiere T, Durrer A (2014) Exploring interactions of plant microbiomes. Sci Agric 71 (6):528–539

Arora M, Kaushik A, Rani N, Kaushik CP (2010) Effect of cyanobacterial exopolysaccharides on salt stress alleviation and seed germination. J Environ Biol 31(5):701–704

Bach EM, Baer SG, Meyer CK, Six J (2010) Soil texture affects soil microbial and structural recovery during grassland restoration. Soil Biol Biochem 42(12):2182–2191

Berk V, Fong JC, Dempsey GT, Develioglu ON, Zhuang X, Liphardt J, Yildiz FH, Chu S (2013) Molecular architecture and assembly principles of *Vibrio cholerae* biofilms. Science 337 (6091):236–239. https://doi.org/10.1126/science.1222981

Bezzate S, Aymerich S, Chambert R, Czarnes S, Berge O, Heulin T (2000) Disruption of the *Paenibacillus polymyxa* levansucrase gene impairs its ability to aggregate soil in the wheat rhizosphere. Environ Microbiol 2:333–342. https://doi.org/10.1046/j.1462-2920.2000.00114.x

Bogino PC, Oliva M, Sorroche FG, Giordano W (2013) The role of bacterial biofilms and surface components in plant-bacterial associations. Int J Mol Sci 14(8):15838–15859. https://doi.org/10.3390/ijms140815838

Brink SC (2016) Unlocking the secrets of the rhizosphere. Trends Plant Sci 21:169–170

Bronick CJ, Lal R (2005) Soil structure and management: a review. Geoderma 124:3–22. https://doi.org/10.1016/j.geoderma.2004.03.005

Caro-Astorga J, Pérez-García A, de Vicente A, Romero D (2014) A genomic region involved in the formation of adhesion fibers in *Bacillus cereus* biofilms. Front Microbiol 5:745. https://doi.org/10.3389/fmicb.2014.00745

Cassan F, Maiale S, Masciarellia O, Vidal A, Luna V, Ruiz O (2009) Cadaverine production by Azospirillum brasilense and its possible role in plant growth promotion and osmotic stress mitigation. Eur J Soil Biol 45(1):12–19. https://doi.org/10.1016/j.ejsobi.2008.08.003

Cheng W, Yang J, Nie Q et al (2017) Volatile organic compounds from *Paenibacillus polymyxa* KM2501-1 control Meloidogyne incognita by multiple strategies. Sci Rep 7(1):1–11. https://doi.org/10.1038/s41598-017-16631-8

Chenu C (1993) Clay or sand polysaccharide associations as models for the interface between micro-organisms and soil: water related properties and microstructure. Geoderma 56 (1-4):143–156. https://doi.org/10.1016/0016-7061(93)90106-U

Chenu C (1995) Extracellular polysaccharides: an interface between microorganisms and soil constituents. p. 217–233. In Huang PM, Berthelin J, Bollag JM, McGill WB (ed.) Environmental impact of soil component interactions: land quality, natural and anthropogenic organics. Vol. 1. CRC Press, Boca Raton

Cho SM, Anderson AJ, Kim YC (2018) Extracellular polymeric substances of *Pseudomonas chlororaphis* O6 induce systemic drought tolerance in plants. Res Plant Dis 24(3):242–247. https://doi.org/10.5423/RPD.2018.24.3.242

Cipriano MA, Lupatini M, Lopes-Santos L, Da Silva MJ, Roesch LF, Destéfano SA (2016) Lettuce and rhizosphere microbiome responses to growth promoting *Pseudomonas* species under field conditions. FEMS Microbiol Ecol 92(12):fiw197. https://doi.org/10.1093/femsec/fiw197

Costa OY, Raaijmakers JM, Kuramae EE (2018) Microbial extracellular polymeric substances: ecological function and impact on soil aggregation. Front Microbiol 9:1636. https://doi.org/10.3389/fmicb.2018.01636

Crist E, Mora C, Engelman R (2017) The interaction of human population, food production, and biodiversity protection. Science 356(6335):260–264. https://doi.org/10.1126/science.aal2011

Daud NS, Rosli MA, Azam ZM, Othman NZ, Sarmidi MR (2019) *Paenibacillus polymyxa* bioactive compounds for agricultural and biotechnological applications. Biocatal Agric Biotechnol 18:101092. https://doi.org/10.1016/j.bcab.2019.101092

Davis CA, Pyrak-Nolte LJ, Atekwana EA, Werkema DD, Haugen ME (2009) Microbial-induced heterogeneity in the acoustic properties of porous media. Geophys Res Lett 36 (21):21405–21410

de los Santos-Villalobos S, Parra-Cota FI, Herrera-Sepúlveda A, Valenzuela-Aragon B, Estrada-Mora JC (2018) Colmena: colección de microorganismos edáficos y endófitos nativos, para contribuir a la seguridad alimentaria nacional. Rev Mex Cien Agríc 9(1):191–202

de Souza R, Meyer J, Schoenfeld R, da Costa PB, Passaglia LM (2015) Characterization of plant growth-promoting bacteria associated with rice cropped in iron-stressed soils. Ann Microbiol 65 (2):951–964. https://doi.org/10.1007/s13213-014-0939-3

Dobbelaere S, Croonenborgh A, Thys A, Ptacek D, Okon Y, Vanderleyden J (2002) Effect of inoculation with wild type Azospirillum brasilense and A. irakense strains on development and nitrogen uptake of spring wheat and grain maize. Biol Fertil Soil 36:284–297. https://doi.org/10.1007/s00374-002-0534-9

Dotterweich M (2013) The history of human-induced soil erosion: geomorphic legacies, early descriptions and research, and the development of soil conservation—a global synopsis. Geomorphology 201:1–34. https://doi.org/10.1016/j.geomorph.2013.07.021

Egamberdiyeva D (2007) The effect of plant growth promoting bacteria on growth and nutrient uptake of maize in two different soils. Appl Soil Ecol 36(2–3):184–189. https://doi.org/10.1016/j.apsoil.2007.02.005

Etesami H, Maheshwari DK (2018) Use of plant growth promoting rhizobacteria (PGPRs) with multiple plant growth promoting traits in stress agriculture: action mechanisms and future prospects. Ecotoxicol Environ Saf 156:225–246. https://doi.org/10.1016/j.ecoenv.2018.03.013

Flemming HC, Wingender J (2001) Relevance of microbial extracellular polymeric substances (EPSs)—part I: structural and ecological aspects. Water Sci Technol 43(6):1–8

Flemming HC, Wingender J (2010) The biofilm matrix. Nat Rev Microbiol 8:623–633. https://doi.org/10.1038/nrmicro2415

Flemming HC, Neu TR, Wingender J (2016) The perfect slime: microbial extracellular polymeric substances (EPS). IWA Publishing, London. https://doi.org/10.2166/9781780407425

Food and Agriculture Organization (FAO) (2017) Introduction and overview. In: Bruinsma J (ed) World agriculture: towards 2015/2030: a FAO perspective. Earthscan Publications Ltd, London, pp 1–28. https://www.fao.org/3/a-y4252e.pdf

Gholami A, Shahsavani S, Nezarat S (2009) The effect of plant growth promoting rhizobacteria (PGPR) on germination, seedling growth and yield of maize. World Acad Sci Eng Technol 49:19–24

Glick BR, Chagping L, Sibdas G, Dumbroff EB (1997) Early development of canola seedlings in the presence of plant growth-promoting bacterium Pseudomonas putida GR12-2. Soil Biol Biochem 29:1233–1239. https://doi.org/10.1016/S0038-0717(97)00026-6

Godinho AL, Bhosle S (2009) Sand aggregation by exopolysaccharide producing microbacterium arborescens-AGSB. Curr Microbiol 58:616–621. https://doi.org/10.1007/s00284-009-9400-4

Gravel V, Antoun H, Tweddell RJ (2007) Growth stimulation and fruit yield improvement of greenhouse tomato plants by inoculation with Psedomonas putida or Trichoderma atroviride: possible roe of indole acetic acid (IAA). Soil Biol Biochem 39:1968–1977. https://doi.org/10.1016/j.soilbio.2007.02.015

Harimawan A, Ting YP (2016) Investigation of extracellular polymeric substances (EPS) properties of *P. aeruginosa* and *B. subtilis* and their role in bacterial adhesion. Colloids Surf B Biointerfaces 146:459–467. https://doi.org/10.1016/j.colsurfb.2016.06.039

Hernandez Rodriguez A, Heydrich Perez M, Acebo Guerrero Y, Velazquez del Valle MG, HernandezLauzardo AN (2008) Antagonistic activity of Cuban native rhizobacteria against Fusarium verticillioides (Sacc.) Nirenb. in maize (Zea mays L.). Appl Soil Ecol 39:180–186. https://doi.org/10.1016/j.apsoil.2007.12.008

Herrejón MA (2014) Dinámica estacional de comunidades microbianas en huertas de aguacate con diferente uso de suelo. Biol Rev DES Cien Biol Agropecuarias Universidad Michoacana de San Nicolás de Hidalgo 16(1):19–24

Kaci Y, Heyraud A, Barakat M, Heulin T (2005) Isolation and identification of an EPS-producing *Rhizobium* strain from arid soil (Algeria): characterization of its EPS and the effect of inoculation on wheat rhizosphere soil structure. Res Microbiol 156(4):522–531. https://doi.org/10.1016/j.resmic.2005.01.012003000046

Kavino M, Harish S, Kumar N, Saravanakumar D, Samiyappan R (2010) Effect of chitinolytic PGPR on growth, yield and physiological attributes of banana (*Musa spp.)* under field conditions. Appl Soil Ecol 45:71–77. https://doi.org/10.1016/j.apsoil.2010.02.003

Kawaharada Y, Kelly S, Nielsen MW, Hjuler CT, Gysel K, Muszynski A et al (2015) Receptor-mediated exopolysaccharide perception controls bacterial infection. Nature 523:308–312. https://doi.org/10.1038/nature14611

Kenneth OC, Nwadibe EC, Kalu AU, Unah UV (2019) Plant growth promoting rhizobacteria (PGPR): a novel agent for sustainable food production. Am J Agric Biol Sci 14:35–54

Krieger-Liszkay A, Krupinska K, Shimakawa G (2018) The impact of photosynthesis on initiation of leaf senescence. Physiologia Plantarum, Wiley, 2018. ffhal-02397915f

Kumar A, Singh J (2020) Biofilms forming microbes: diversity and potential application in plant–microbe interaction and plant growth. In: Yadav AN, Singh J, Rastegari AA, Yadav N (eds) Plant microbiomes for sustainable agriculture. Sustainable development and biodiversity, vol. 25, pp 193–197. https://doi.org/10.1007/978-3-030-38453-1

Kumari P, Khanna V (2015) ACC-deaminase and EPS production by salt tolerant rhizobacteria augment growth in chickpea under salinity stress. Int J Bioresour Stress Manag 6(5):558–565. https://doi.org/10.5958/0976-4038.2015.00084.6

Lakshmanan V, Ray P, Craven KD (2017) Toward a resilient, functional microbiome: drought tolerance-alleviating microbes for sustainable agriculture. In: Plant stress tolerance. Humana Press, Totowa, pp 69–84. https://doi.org/10.1007/978-1-4939-7136-7_4

Lares-Orozco MF, Robles-Morúa A, Yepez EA, Handler RM (2016) Global warming potential of intensive wheat production in the Yaqui Valley, Mexico: a resource for the design of localized mitigation strategies. J Clean Prod 127:522–532. https://doi.org/10.1016/j.jclepro.2016.03.128

Laspidou C, Rittman BE (2002) A unified theory for extracellular polymeric substances, soluble microbial products, and active and inert biomass. Water Res 36(11):2711–2720. https://doi.org/10.1016/s0043-1354(01)00413-4

Lehman AP, Long SR (2013) Exopolysaccharides from *Sinorhizobiummeliloti* can protect against H_2O_2-dependent damage. J Bacteriol 195:5362–5369. https://doi.org/10.1128/jb.00681-13

Mager DM (2010) Cyanobacterial soil crusts: analysing resilience in Kalahari sand soils. In: Degenovine KM (ed) Semi-arid environments: agriculture, water supply and vegetation. NOVA Scientific Publishers, Hauppauge

McCalla TM, Haskins FA, Curley RD (1958) Soil aggregation by microorganisms following soil fumigation. Agron Hortic 22:311–314

Merritt PM, Danhorn T, Fuqua C (2007) Motility and chemotaxis in *Agrobacterium tumefaciens* surface attachment and biofilm formation. J Bacteriol 189:8005–8014. https://doi.org/10.1128/JB.00566-07

Miransari M (2011) Soil microbes and plant fertilization. Appl Microbiol Biotechnol 92 (5):875–885. https://doi.org/10.1007/s00253-011-3521-y

More TT, Yadav JSS, Yan S, Tyagi RD, Surampalli RY (2014) Extracellular polymeric substances of bacteria and their potential environmental applications. J Environ Manag 144:1–25. https://doi.org/10.1016/j.jenvman.2014.05.010

Murphy JF, Zehnder GW, Schuster DJ, Sikora EJ, Polston JE, Kloepper JW (2000) Plant growth-promoting rhizobacterial mediated protection in tomato against tomato mottle virus. Plant Dis 84(7):779–784. https://doi.org/10.1094/PDIS.2000.84.7.779

Niranjan SR, Deepak SA, Basavaraju P, Shetty HS, Reddy MS, Kloepper JW (2003) Comparative performance of formulations of plant promoting rhizobacteria in growth promotion and

suppression of downy mildew in pearl millet. Crop Prot 22:579–588. https://doi.org/10.1016/S0261-2194(02)00222-3
Nkonya E, Mirzabaev A, Von Braun J (2015) Economics of land degradation and improvement: a global assessment for sustainable development. Springer, New York
Noel TC, Sheng C, Yost CK, Pharis RP, Hynes MF (1996) Rhizobium leguminosarum as a plant growth-promoting rhizobacterium: direct growth promotion of canola and lettuce. Can J Microbiol 42(3):279–283. https://doi.org/10.1139/m96-040
Okshevsky M, Meyer RL (2013) The role of extracellular DNA in the establishment, maintenance and perpetuation of bacterial biofilms. Crit Rev Microbiol 41:341–352. https://doi.org/10.3109/1040841X.2013.841639
Oliveira CA, Alves VMC, Marriel IE, Gomes EA, Scotti MR, Carneiro NP, Guimaraes CT, Schaffert RE, So NMH (2009) Phosphate solubilizing microorganisms isolated from rhizosphere of maize cultivated in an oxisol of the Brazilian Cerrado biome. Soil Biol Biochem 41:1782–1787. https://doi.org/10.4067/S0718-95162016005000043
Or D, Phutane S, Dechesne A (2007) Extracellular polymeric substances affecting pore-scale hydrologic conditions for bacterial activity in unsaturated soils. Vadose Zone J 6(2):298–305. https://doi.org/10.2136/vzj2006.0080
Orozco-Mosqueda M, Rocha-Granados M, Glick BR, Santoyo G (2018) Microbiome engineering to improve biocontrol and plant growth-promoting mechanisms. Microbiol Res 208:25–31. https://doi.org/10.1016/j.micres.2018.01.005
Parra-Cota FI, Coronel-Acosta CB, Amézquita-Avilés C, de los Santos-Villalobos S, Escalante-Martínez D (2018) Diversidad metabólica de microorganismos edáficos asociados al cultivo de maíz en el Valle del Yaqui, Sonora. Rev Mex Cien Agríc 9(2):431–442. https://doi.org/10.29312/remexca.v9i2.1083
Qadir M, Quillérou E, Nangia V, Murtaza G, Singh M, Thomas RJ, Drechsel P, Noble AD (2014) Economics of salt-induced land degradation and restoration. Nat Res Forum 38:282–295. https://doi.org/10.1111/1477-8947.12054
Qurashi AW, Sabri AN (2012) Bacterial exopolysaccharide and biofilm formation stimulate chickpea growth and soil aggregation under salt stress. Braz J Microbiol 43:1183–1191. https://doi.org/10.1590/S1517-838220120003000046
Ramey BE, Koutsoudis M, von Bodman SB, Fuqua C (2004) Biofilm formation in plant-microbe associations. Curr Opin Microbiol 7(6):602–609. https://doi.org/10.1016/j.mib.2004.10.014
Rashid MI, Mujawar LH, Shahzad T, Almeelbi T, Ismail IMI, Oves M (2016) Bacteria and fungi can contribute to nutrients bioavailability and aggregate formation in degraded soils. Microbiol Res 183:26–41. https://doi.org/10.1016/j.micres.2015.11.007
Roberson EB, Chenu C, Eirestone MK (1993) Microstructural changes in bacterial exopolysaccharides during desiccation. Soil Biol Biochem 25:1299–1301. https://doi.org/10.1016/0038-0717(93)902.30-9
Rodriguez PA, Rothballer M, Chowdhury SP, Nussbaumer T, Gutjahr C, Falter-Braun P (2019) Systems biology of plant microbiome interactions. Mol Plant 12:804–821. https://doi.org/10.1016/j.molp.2019.05.006
Rogers SL, Burns RG (1994) Changes in aggregate stability, nutrient status, indigenous microbial populations, and seedling emergence, following inoculation of soil with Nostocmuscorum. Biol Fertil Soils 18:209–215. https://doi.org/10.1007/BF00647668
Rojas-Padilla J, Chaparro Encinas LA, Montoya RIR, de los Santos Villalobos S (2020) Growth promotion on wheat (Triticum turgidum L. subsp. durum) by co-inoculation of native Bacillus strains isolated from the Yaqui Valley, Mexico. Nova Sci 12(24). https://doi.org/10.21640/ns.v12i24.2136
Rokhzadi A, Ahmad A, Farrokh D, Ghorban NM, Eslam M (2008) Influence of plant growth-promoting rhizobacteria on dry matter accumulation and yield of Chickpea (Cicer arietinum L.) under field conditions. Am Eur J Agric Environ Sci 3:253–257. https://doi.org/10.1007/978-981-10-5589-8_2

Sandhya V, SK ZA, Grover M, Reddy G, Venkateswarlu B (2009) Alleviation of drought stress effects in sunflower seedlings by the exopolysaccharides producing *Pseudomonas putida* strain GAP-P45. Biol Fertil Soils 46(1):17–26. https://doi.org/10.1007/s00374-009-0401-z

Sandilya SP, Bhuyan PM, Gogoi DK, Kardong D (2016) Phosphorus solubilization and plant growth promotion ability of rhizobacteria of R. communis L. growing in Assam, India. Proc Natl Acad Sci India Sect B Biol Sci 88(3):959–966. https://doi.org/10.1007/s40011-016-0833-9

Sandilya SP, Bhuyan PM, Vijay N, Gogoi DK, Kardong D (2017) Impact of Pseudomonas aeruginosa strain MAJ PIA03 affecting the growth and phytonutrient production of castor, a primary host-plant of Samiaricini. J Soil Sci Plant Nutr 17(2):499–515. https://doi.org/10.4067/S0718-95162017005000036

Sengupta S, Dey S (2019) Microbial exo-polysaccharides (EPS): role in agriculture and environment. Agric Food 1(12):4–8

Shameer K, Naika MB, Shafi KM, Sowdhamini R (2019) Decoding systems biology of plant stress for sustainable agriculture development and optimized food production. Prog Biophys Mol Biol 145:19–39. https://doi.org/10.1016/j.pbiomolbio.2018.12.002

Sheela T (2013) Influence of plant growth promoting Rhizobacteria (PGPR) on the growth of maize (*Zea mays L.*). Golden Res Thoughts 3(6).: GRT-3093 ref.21

Siddiqui ZA, Khan M (2017) Biofilm formation by Pseudomonas spp and their significance as a biocontrol agent. Biofilms Plant Soil Health 14:69. https://doi.org/10.1002/9781119246329.ch5

Silva HSA, da Silva RR, Macagnan M, de Almeida HVB, Pereira MCB, Mounteer A (2004) Rhizobacterial induction of systemic resistance in tomato plants: non-specific protection and increase in enzyme activity. Biol Control 29:288–295. https://doi.org/10.1111/j.1439-0434.2004.00853.x

Sofia IA, Paula P, Castro ML (2014) Phosphate solubilizing rhizobacteria enhance Zea mays growth in agricultural P-deficient soils. Ecol Eng 73:526–535. https://doi.org/10.1016/j.ecoleng.2014.09.060

Tang J, Mo Y, Zhang J, Zhang R (2011) Influence of biological aggregating agents associated with microbial population on soil aggregate stability. Appl Soil Ecol 47(3):153–159. https://doi.org/10.1016/j.apsoil.2011.01.001

Tewari S, Sharma S (2020) Rhizobial exopolysaccharides as supplement for enhancing nodulation and growth attributes of Cajanus cajan under multi-stress conditions: a study from lab to field. Soil Tillage Res 198:104545. https://doi.org/10.1016/j.still.2019.104545

Timmusk S, Zucca C (2019) The plant microbiome as a resource to increase crop productivity and soil resilience: a systems approach. J Cameroon Acad Sci 14(3):181–187. https://doi.org/10.4314/jcas.v14i3.2

Tomlinson AD, Ramey-Hartung B, Day TW, Merritt PM (2010) Agrobacterium tumefaciens ExoR represses succinoglycan biosynthesis and is required for biofilm formation and motility. Microbiology 156:2670–2681. https://doi.org/10.1099/mic.0.039032-0

Upadhyay SK, Singh JS, Singh DP (2011) Exopolysaccharide-producing plant growth-promoting rhizobacteria under salinity condition. Pedosphere 21:214–222. https://doi.org/10.1016/s1002-0160(11)60120-3

Valenzuela Ruiz V, Gálvez Gamboa G, Villa Rodríguez E, Parra Cota F, Santoyo G, de los Santos Villalobos S (2020) Lipopéptidos producidos por agentes de control biológico del género *Bacillus*: revisión de herramientas analíticas utilizadas para su estudio. Rev Mex De Cienc Agríc 11(2):419–432. https://doi.org/10.29312/remexca.v11i2.2191

Valenzuela-Aragon B, Parra-Cota FI, Santoyo G, Arellano-Wattenbarger GL, de los Santos-Villalobos S (2019) Plant-assisted selection: a promising alternative for in vivo identification of wheat (Triticum turgidum L. subsp. Durum) growth promoting bacteria. Plant Soil 435:367–384. https://doi.org/10.1007/s11104-018-03901-1

Van Workum WAT, Wijffelman CA, Spaink HP, Van Brussel AAN, Pollock TJ, Kijne JW (1998) Biosynthesis and function of extracellular polysaccharides of rhizobium leguminosarum. In: Elmerich C, Kondorosi A, Newton WE (eds) Biological nitrogen fixation for the 21st Century.

Current plant science and biotechnology in agriculture, vol 31. Springer, Dordrecht. https://doi.org/10.1007/978-94-011-5159-7_93

Vardharajula S, Ali SZ (2014) Exopolysaccharide production by drought tolerant *Bacillus spp.* and effect on soil aggregation under drought stress. J Microbiol Biotechnol Food Sci 4:51–57. https://doi.org/10.15414/jmbfs.2014.4.1.51-57

Vardharajula S, Ali SZ (2015) The production of exopolysaccharide by *Pseudomonas putida* GAP-P45 under various abiotic stress conditions and its role in soil aggregation. Microbiology 84:512–519. https://doi.org/10.1134/S0026261715040153

Velmourougane K, Prasanna R, Saxena AK (2017) Agriculturally important microbial biofilms: present status and future prospects. J Basic Microbiol 57(7):548–573. https://doi.org/10.1002/jobm.201700046

Vezzani FM, Anderson C, Meenken E, Gillespie R, Peterson M, Beare MH (2018) The importance of plants to development and maintenance of soil structure, microbial communities and ecosystem functions. Soil Tillage Res 175:139–149

Villa-Rodríguez E, Parra-Cota F, Castro-Longoria E, López-Cervantes J, de los Santos-Villalobos S (2019) Bacillus subtilis TE3: a promising biological control agent against Bipolaris sorokiniana, the causal agent of spot blotch in wheat (Triticum turgidum L. subsp. Durum). Biol Control 132. https://doi.org/10.1016/j.biocontrol.2019.02.012

Vlamakis H, Chai Y, Beauregard P, Losick R (2013) Sticking together: building a biofilm the *Bacillus subtilis* way. Nat Rev Microbiol 11:157–168. https://doi.org/10.1038/nrmicro2960

Vorholt JA (2012) Microbial life in the phyllosphere. Nat Rev Microbiol 10(12):828–840. https://doi.org/10.1038/nrmicro2910

Wan Z, Brown PJB, Elliott EN, Brun YV (2013) The adhesive and cohesive properties of a bacterial polysaccharide adhesin are modulated by a deacetylase. Mol Microbiol 88:486–500. https://doi.org/10.1111/mmi.12199

Wu N, Pan H-X, Qiu D, Zhang Y-M (2014) Feasibility of EPS-producing bacterial inoculation to speed up the sand aggregation in the Gurbantünggüt Desert, Northwestern China. J Basic Microbiol 54:1378–1386. https://doi.org/10.1002/jobm.201400355

Yadav AN, Singh J, Rastegari AA, Yadav N (2020).Plant microbiomes for sustainable agriculture. In: Sustainable development and biodiversity. Springer International Publishing. https://doi.org/10.1007/978-3-030-38453-1

Yasari E, Patwardhan AM (2007) Effects of Azotobacter and Azospirilum inoculations and chemical fertilizers on growth and productivity of Canola. Asian J Plant Sci 6:77–82. https://doi.org/10.3923/ajps.2007.77.82

Zehnder GW, Yao C, Murphy JF, Sikora EJ, Kloepper JW (2000) Induction of resistance in tomato against cucumber mosaic cucumovirus by plant growth-promoting rhizobacteria. Biocontrol 45:127–137. https://doi.org/10.1023/A:1009923702103

Zehnder G, Murphy JF, Sikora EJ, Kloepper JW (2001) Application of rhizobacteria for induced resistance. Eur J Plant Pathol 107:39–50. https://doi.org/10.1023/A:1008732400383

Significance of Bacterial Polyhydroxyalkanoates in Rhizosphere

11

Sundaresan Bhavaniramya, Selvaraju Vishnupriya, and Dharmar Baskaran

Abstract

Polyhydroxyalkanoates PHAs are groups of polyesters synthesized from different types of microorganisms utilized for production of biopolymers. Rhizospheres from soil produce PHA developing microbes thereby improving the ecological balance. Significance of producing bacterial PHA in rhizospheres has developed interest over the years. Global demand for replacing synthetic plastics through bio-based plastics and biodegradable polymers are widely in focus. Hence, the least expensive, universally available method of producing bacterial-based PHA for biopolymer manufacture can be optimized and established through soil rhizobium. The knowledge on bacterial PHA synthesis from soil rhizosphere has been advantageous and provides a wide range of applications including textile, packaging and polymer, food products packaging designs industries, etc. Recent approaches in developing bacterial PHA from different sources have been discussed in this chapter.

Keywords

Biopolymer · Bacterial PHA · Rhizosphere · Soil

S. Bhavaniramya (✉) · S. Vishnupriya
College of Food and Dairy Technology, Tamil Nadu Veterinary and Animal Sciences University, Chennai, India

D. Baskaran
Madras Veterinary College, Tamil Nadu Veterinary and Animal Sciences University, Chennai, India

A. Vaishnav, D. K. Choudhary (eds.), *Microbial Polymers*,
https://doi.org/10.1007/978-981-16-0045-6_11

11.1 Introduction

PHAs are groups of bio-polyesters synthesized from various range microorganisms, viz., Gram-positive, Gram-negative bacteria. Commonly, PHAs are classified based on short chain length PHA containing C4-C5 monomers and medium chain length with C6-C24 monomers. The functions of PHA play a strategic role in the survival characteristics of microbes under varied environmental conditions. They are produced under limited conditions with excess carbon for the utilization in bioplastic production. Rhizospheres are defined as distinct soil compartments with varied organic acids, sugars representing the biology of soil. The rhizosphere microbiome enhances plant nutrition by nitrogen fixation and aiding uptake of plant nutrients. The interest for bacterial PHA from the rhizosphere focuses sustainable development of an eco-friendly environment. The biological origin of the polymers makes it biocompatible, biodegradable describing under green plastics. PHAs are accumulated by several bacteria in the form of intercellular granules that serve as a reservoir of energy. PHA was first discovers in 1925 by French Scientist Lemoigne in *Bacillus megaterium* in the form of (3-polyhydroxybutyrate) PHB (Chee et al. 2010). Generally, bacteria store PHA in the cytoplasm to require nutrients. Bacteria vary accordingly with PHA production, nutrient limiting bacteria accumulate PHA and those that do not accumulate during growth phase. For example, *Pseudomonas putida, Pseudomonas oleovorans* belong to the former group and recombinant *E. coli* to the latter (Muhammadi et al. 2015). Subsequent changes in soil organic matter actively monitored and that are associated with certain variations in the soil conditions. The bacterial PHA produced from rhizosphere soil bacteria involves biodegradable, economically effective polymers for the environment. Abundance and global dependence on synthetic plastics is a serious problem, further, waste management concerns on petroleum-derived plastics, petrochemicals cause threat to the environment. Generally, the production process is tedious, and recycling of synthetic plastics remains a challenging, time-consuming method. In such cases, biodegradable polymers, plastics from bacterial sources are the best solution to environmental hazards. The interest towards bacterially derived PHA has been settled for many years, and the cost of production along with slow growth of microorganisms possess a remarkable concern. Several industrial wastes are focused on candidates for PHA production, and PHA remains an obvious alternative to synthetic plastic wastes. PHAs are being produced industrially and cover broad spectrum end products from food packaging to medical applications. Regardless of the quality characteristics and advantages to the environment, PHAs are impaired for high cost of production (Amaro et al. 2019). The importance of high cost of production implies the use of costly fermenters for carbon source production. Thus, soil bacteria are ideal sources for carbon development and storehouse of nutrients. Rhizospheres from soil organic matter are significant producers of PHA solving environmental problems. The function of PHAs is storage reservoir of carbon utilized by microorganisms whenever required and that are generally present in the form of granules. Currently, PHAs are used as a blend form or in pure form covering a variety of applications: PHAs (short chain length) are often used in food packaging and disposable items, and PHAs (medium chain length) are applicable for high value-added applications, viz., materials for suture, surgical implants, etc. Industrialization and global usage of PHAs is limited due to high costs of production that improves the prices of petrol-based polymers, from polypropylene (PP) and polyethylene (PE). Further, industrialized the production of PHAs, have problems with the purity of substrates, and concerns of some carbon sources identified from human food supplies, and hence huge amounts of solvents required for the extraction in large-scale synthesis and makes the production process complicated. Biosynthesis of PHA occurs by diverse microbial species in the environment. This chapter details the wide knowledge of bacterial PHA and their significance in soil rhizosphere.

11.2 Origin of Rhizosphere

Plant roots are surrounded by regions of rhizospheres that are centered around the root and their habitat includes plant living with roots and are associated with soil where exudates deposited stimulate several microbial activities that are significantly implied in plant health and disease, and further focusing on the ecosystems terrestrial and agriculture. Microbes whose abundance and metabolic activity promotes the growth of plant, whereas other harmful microbes develop on root exudates by growing in the rhizosphere (Amaro et al. 2019). Hiltner's findings summarize the beginning of research in rhizosphere microbiology and also provides basic and applied scientific inventions that are useful for future. Indicators such as microbial density, enzymatic activity, and root mapping are various gradients of chemicals present in spatial limits of the rhizosphere (Pernicova et al. 2019). The composition the biotic population is mostly contingent on plant species, their texture of root, location of carbon source in plants, physico-chemical properties of soil, and a variety of microbial population, amongst the other factors. Rhizospheres are extraordinarily nutrient dense regions compared to dense soil regions and the root exudates energy, molted root cells, decaying root tissue, mucilage secretion proteinaceous substances. Interactions amongst plants and microorganisms remains symbiotic and improves carbon source to the surrounding soil from root. Soil with nutrient-rich compounds support the growth of dense and diversely populated microorganisms that mainly include bacteria and fungi.

Microorganisms extract nutrients from small metabolites, which are released actively or passively into the surrounding soil from the plant. These root exudates contain various metabolites, viz., amino acids, organic acids secondary plant compounds including flavonoids and terpenoids (Pakalapati et al. 2018). Microbial population utilizes these compounds for some specificity, and further individual microbial species are utilized more effectively than another species. Invertebrates such as insects, arthropods, and nematodes are considered as primary consumers and feed on live or dead materials of plants. The digestive system of these herbivorous functions as a tool for nutrients in planta and used for transporting between the roots and the soil via evacuation (Mostafa et al. 2020). They feed commonly on microbes present in rhizosphere, in a category that comes under primary or secondary consumers.

Plant roots can contribute carbon to the rhizosphere and deliver organic metabolites from the rhizosphere (Koller et al. 2005). 200–2000% of secondary microbial metabolites efflux increased in plant-derived amino acids through root to the rhizosphere, and this approach are thought to play a representative role in the capability of plants to modify the rhizosphere (Johnston et al. 2018). Rhizospheres are nutrient-rich regions of soil that are dynamic in nature supporting dense and diverse fauna (Ciesielski et al. 2015). Despite the challenges correlated with ecological interactions present in the soil, researchers have started to understand the complex behavioral interactions taking place in the rhizosphere. Rhizosphere biology knowledge learnt by agricultural researchers have focused soil microbes on their positive and negative relationship of plants. The detailed descriptions have been represented for many rhizosphere interactions, and the ambiguous nature of the rhizosphere still remains a challenge to scientists interested in the ecology of the plants present in soil (Reddy et al. 2003).

11.3 Sources of Bacterial PHA

11.3.1 Biosynthesis of PHA

Numerous microorganisms are associated with biosynthetic pathways and belong to a diverse class of polymers that have approximately hundred different types of monomers. The alternatives for petroleum-based plastic are now unearthed to be bio-based PHAs for biopolymer production as they are from renewable resources (Adeleye et al. 2020). Bacteria producing acetyl CoA converted to PHB through several biosynthetic enzymes. Several central metabolic pathways represented including glycolytic, pentose pathway, and Krebs cycle, of which PHA biosynthetic pathway reported acetyl CoA remains key element for biosynthesis of short chain length (SCL) and medium chain length (MCL) PHAs forming PHB (3-polyhydroxy butyrate). 3-ketothiolase with two molecules of acetyl CoA forms aceto-acetyl CoA (Mitra et al. 2020) (Fig. 11.1).

11.3.2 Reduction of 3-Ketothiolase by PhaB

Second step of PHB biosynthesis, acetoacetyl-CoA with NADH as a cofactor by PhaB which is reduced to (R)-3-HB-CoA. Carbon of acetoacetyl-CoA, resulting in the formation of (R)-3-HB-CoA further leveling up to polymerization (Yu 2010).

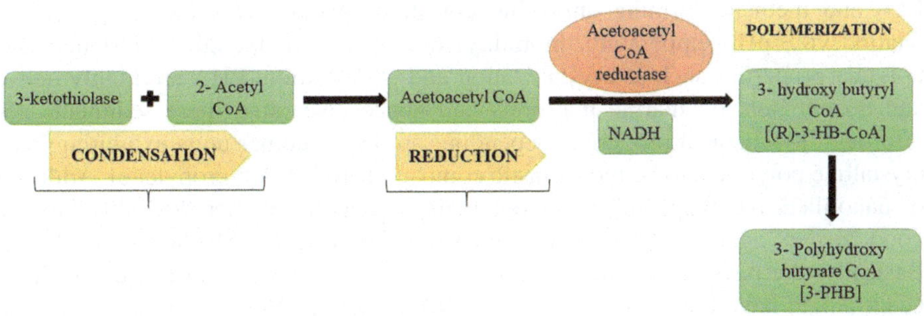

Fig. 11.1 Biosynthetic pathway of PHA

11.3.3 PHA Polymerization of PhaC

PhaC is the important enzyme in PHA biosynthesis representing CoA thioesters of various HAs and polymerization. More than 150 HA CoA are required to characterize PhaC and recently only 60 different phaC genes have been denoted which defines the important role of PhaC that produces PHAs. The crystal structure of PhaC has been determined through various attempts and elucidated its catalytic mechanism over the past 30 years. However, crystallization of PhaC has so far proven unsuccessful (Kumar et al. 2019). Hydroxy butyryl-CoA polymerizes to form PHB and the propagation proceeds by additional bonding to develop PHAs representing 80% of bacterial cells. Solvent extracted PHAs bacterial cell mass grown in the presence of sugars, carbohydrates, and are focused on gene transfer pertaining to significant research activity (Mostafa et al. 2020).

11.4 Properties of PHA

PHB coming under PHA develops as good resistance materials to moisture and aroma barrier properties. High biodegradability and biocompatibility of PHAs makes it eco-friendly polymers for various applications in different fields. The most important physical and thermal property of PHA refers to molecular mass and crystallinity index (Anjum et al. 2016). Due to the chemical diversity, PHAs properties are considered similar to conventional plastics. The rate of degree of polymerization maximizes up to 30,000 that validates high molecular masses. Further consistent molecular mass and molecular weight distribution of a particular polymer represents characteristics for its commercial suitability and stability for manufacture (Costa et al. 2019). According to Chee et al. (2010), the crystallinity range of PHA between 60 and 80% referred as rigid and remains more elastic, flexible crystallinity with low degree of PHA makes it applicable for various industrial processing. Polyhydroxy butyrates are brittle and fragile with the ability to elongate less and a breaking point below 15%. PHA are generally degraded by various microbes in different environmental conditions. This results in production of

CO_2 and methane forming anaerobic digestion several years including various factors, viz., pH, temperature, humidity, microbial activity, other nutrients, and polymer properties such as crystallization and composition. PHA are highly dense and do not float on marine platforms and hence discarding these sediments are effortless. Mechanical properties depending on the monomer units exhibiting hard crystalline polyesters to be more elastic comparable to PP(Polypropylene). Addition of nanofillers blending with other polymers decreases polymer flexibility but are good barriers against organic solvents (Anjum et al. 2016; Mostafa et al. 2020). Basically, the different monomer units of PHA, viz., short chain length, medium chain length have high crystallinity and with respect to Young's modulus, tensile strength, and oxygen permeability of PHB like isotactic PP(Polypropylene) representing excellent packaging material. Replacing synthetic plastics with these bioplastics are important for degradation problems prevailing in the environment.

11.5 Types of PHA

PHAs differ in R group and consist of 3-hydroxy fatty acids and common types are in Table 11.1. Linear chain polyesters of PHAs containing monomers of hydroxy acid (HA) that are linked to an ester bond. The ester bond is recognized by a carboxyl group that is connected to the hydroxyl group (Singh et al. 2017). Mcl-PHA are certain side chain length HA showing elastomeric properties. Different temperature conditions indicate various polymers and make them suitable for packaging films. Examples of Scl PHA are PHO (3-Polyhydroxy octanoate) and PHN (3-Polyhydroxy nonanoate). Typical examples of Mcl-PHA are 3-hydroxyhexanoate (HHx), 3-hydroxyheptanoate (HH), and 3-hydroxydecanoate (HD) are identified in the largest group of natural polyesters (Table 11.2).

PHA Polyhydroxyalkanoates (PHAs) polymers are produced by several microorganisms and stores as intracellular organelles. For example, Chemolithotrophic bacteria *Cupriavidus necator* (*Ralstonia eutropha*), *Cupriavidus metallidurans,* and *Alcaligenes latus* (Suriyamongkol et al. 2007). Several

Table 11.1 Classification of PHA

S. No	Name of PHA	R group	Short form
1	Poly(3-hydroxybutyrate)	CH_3	PHB
2	Poly(3-hydroxyvalerate)	CH_2CH_3	PHV
3	Poly(3-hydroxyhexanoate)	$CH_2CH_2CH_3$	PHHx

Table 11.2 Types of PHA on chain length

S. No	Scl PHA (short chain length)	Mcl (PHA medium chain length)
1	3-Polyhydroxyoctanoate (PHO)	3-hydroxyhexanoate (HHx)
2	3-Polyhydroxynanonoate (PHN)	3- hydroxyheptanoate (HH)
3		3- hydroxydecanoate (HD)

Pseudomonas sp. such as *P. oleovorans* and *P.aeruginosa* produce PHA under Gram-negative bacteria. Gram-positive bacteria producing PHAs are Bacillus sp. *B. subtilis, B. thuringiensis, B. cereus,* and increase PHA synthesis. Several Halophilic and Halomonas sp. based on various factors molecular weight and type of feed produce PHA.

11.6 Ecological Niche of Bacterial PHA Production

11.6.1 Hydrocarbons

Hydrocarbons form effective production of PHAs approximately 30% yield of the bacterial cell weight. PHAs from hydrocarbons have propyl, dodecyl alkyl groups, and aromatic hydrocarbons develop substrates from terephthalic acid in PET. They emulsify the effective production of hydrocarbons by a number of Pseudomonas strains. Recently, PHA production from hydrocarbons are low compared to others and further improvements in production, processing has to be established for industrial scale. PHA derived from hydrocarbon in waste material showed high impact with synthetic plastics. Hydrocarbons utilized from bacterial strains such as Pseudomonas sp., *R. eutropha* are decomposed rapidly compared to plastic particles (Samorì et al. 2019). Other hydrocarbons in interest for PHA production are organic matter in oil fields and substrates of methane. Furthermore, soil bacteria rhizosphere shows favorable production hydrocarbons and provides a sustainable approach for the synthesis of PHA (Wongsirichot et al. 2020).

11.6.2 Halophiles

Halophiles or Halophilic bacteria generally reported in oceans, salt lakes, marshy land areas, are Archaea bacteria that produce PHA. Bhardwaj et al. (2014) reported that the microbes require salt to grow and optimally 5–10% stating PHA accumulation by archaea in the Dead Sea. PHA from halophiles grow in salt conditions and about 50–60% production in its cell weight, representing an excellent carbon source. Halophiles at 3–5% salt concentration accumulate 55% PHB. Several studies show halophilic archaea bacteria such as *Halococcus* sp. and *Halorubrum* sp., independent of nutrient conditions develop PHB. Halophiles produce PHA for synthesis of polymers at industrial level. Halotolerance bacteria are from moderate to extreme from 20 to 30% salt concentration. Use of halophiles in processing of PHA makes it a low-cost substrate and more efficiently produced from the bacterial cells and further develops biocatalysts for industrial PHA production (Sanhueza et al. 2019). Researches described show excellent potential of halophiles for production of PHA via biotechnological techniques for various industrial applications.

11.6.3 Photosynthetic Bacteria

Cyanobacteria are represented as photosynthetic, and they produce PHA during oxygenic photosynthesis that have natural sites to store PHA. Species-specific bacteria producing PHB reports the presence of PHA in cyanobacteria (Kai and Loh 2013). Studies reported production of PHA by cyanobacteria and their capability to produce PHB sunlight energy which reduces the production of greenhouse gases (Gasser et al. 2009).

11.6.4 Antibiotic Factors

Rhizosphere are untapped reservoirs of PHA, and some microbes are *Burkholderia terricola, Lysobacter gummosus, and Pseudomonas extremaustralis* produce PHA. *Azospirillum brasilense is a growth*-promoting rhizobacterium that has been reported to produce PHA. Antibiotic factors are reserved in the roots of plants that impulse hydroxyalkanoates production, and these microbes include the ability to adapt environmental conditions to promote growth of rhizosphere habitat (Obruca et al. 2018). *Azospirillum brasilense* synthesizes nitric oxide, carotenoids, and other nutrients further utilizing PHA polymers for shelf life of bacteria. Mostly, production of soil inoculants, biofertilizers containing beneficial bacteria for agricultural purposes. Molecular techniques and cultivation patterns characterize PHA production from rhizopheres. PHA producers by the fact the rhizosphere roots contain ample sources of inorganic and carbon nutrients are in limited quantity. PHB production is found in different rhizobia established in *Sinorhizobium meliloti* within *alfalfa*. PHA stored in roots helps in rapid colonization, efficient nodulation, and nitrogen fixation (Volova et al. 2019). PHB production plays an important role in antibiotic factors thereby imposing nitrogen supplementation.

11.7 Bacterial PHA in Rhizosphere

11.7.1 Screening of PHA from Bacterial Sources

PHA produced from microbes provide unbalanced growth conditions, and this happens when nutrients such as nitrogen, phosphorus, or sulfate are in limited quantity. PHA and PHB are present in numerous microorganisms. Bacteria such as Acetobacter, Bacillus, Archaebacteria, Methyl bacteria, and Pseudomonas have been found to synthesize PHA. To isolate PHA-accumulating bacteria from diverse sources, several researches have been carried out and different strains were selected efficiently. Industrial effluents, dairy waste, domestic sewage, and activated sludges are found to be the prime sources of active PHA. Higher level PHA producers from tannery effluent and sewage sludge and further isolating PHA-accumulating bacteria from nature, it is necessary to screen collection of bacteria through molecular techniques. Stains specific to PHA are detected viable colony staining technique and the ability of the isolates was further studied by Sudan Black staining method (Sujatha et al. 2005). Food waste, paper mill waste, molasses, cane sugar waste, and oil plant effluents are sources of PHA, and other agricultural wastes also provide PHA-producing microorganisms.

11.7.2 Characterization and Identification of PHA from Bacterial Sources

Isolates of PHA from bacterial sources were screened under Sudan Black staining for 48 h incubation on E2 medium and analyzed for Gram type of isolated bacterial colonies. Identification of the bacterial colonies are further characterized by biochemical tests. PHA are a range of Gram-positive and Gram-negative bacteria, mostly bacillus species. Several researchers reported PHA production from different Bacillus species. Singh Saharan et al. (2014) reported PHA production from B. subtilis, B. megaterium, and characterization from industrial wastes. Pseudomonas species have also been studied for formation of PHA by *Pseudomonas oleovorans*. Pseudomonas sp. isolated from large quantities industrial waste showed polyhydroxy butyrate (PHB) (Ayub et al. 2009). Fermented molasses mash from sugar industry waste represents potential carbon sources and corn oil also showed high PHA content. The bacterial colonies are developed around 48–72 h to screen PHA development with axenic strains nature (Ayub et al. 2007). DNA extraction and characterization from cultures developed using manufacturer's instructions manual of the DNeasy Blood and Tissue kit (QIAGEN GmbH, Hilden, Germany) that showed uniform colonies and cellular morphology which were assessed spectrophotometer (Suriyamongkol et al. 2007). Further characterization of these bacterial PHA were processed through PCR amplifications of the 16S rRNA and *phaC* genes with general procedures and effective bacteria-producing PHA are identified. The identification and characterization of the bacterial PHA in the rhizosphere was reported by Gasser et al. (2009) depicting Pha genes stress resistance associated with PHA and PHB accumulation. Changing environmental conditions and other features are responsible for production and biosynthesis.

11.7.3 PHA Production from Bacterial Sources

High quantities of PHA-producing bacteria present in marine, soil, and effluents has been done screening using culture-specific approaches. Gram-positive and Gram-negative bacteria synthesize various PHA. PHA synthases are key enzymes degraded by *phaC* gene that polymerize PHA monomers (Tarrahi et al. 2020). PHAs of short chain length contain enzymes of only one type (PhaC), and medium chain length PHAs (mcl-PHA) contain class II and polymerize showing different rheological properties. Desirable biotechnological developments, culture-independent methods for the detection potential of PHA producers are showcased and some bacteria occupy more than one *phaC Cupriavidus necator*. Evidence for

PHA production in Alpha proteobacteria and Gamma proteobacteria in hypersaline environments characterized by metabolic diversity (Roja et al. 2019). In situ analysis of PHA and its monomer's are studied through biotechnology that can be implemented on an industrial scale for various applications in food, pharmaceutical, and cosmetic agents. Within the last few decades, the main contributors of high cost production in PHA are highly pure substrates that account for 45% in total cost of production. Furthermore, renewable sources are explored by researchers and they develop bioprocesses for by-product utilization. By-product are important for PHA production.

11.8 Factors Affecting PHA Production

Nutrient limitation is the primary factor affecting PHA production that is directed towards increased biomass growth. Phosphorus and nitrogen in low concentration provide limited conditions for PHB synthesis. Bacteria removing substances like phosphorus was obtained from waste and are able to produce PHB. Optimizing the conditions for PHA production focuses factors like oxygen or nitrate as electron acceptor, nitrogen concentration (NH_4Cl), phosphorus (KH_2PO_4), and carbon concentration (acetate). The carbon/nitrogen (C/N) ratio for PHA production is 28.3 that reflects 1.8 folds increase during PHB production.

Activated sludge also a source of PHA was found to be effective. Box-Behnken design requires optimization of the cultural conditions that maximizes PHB production and parameters for cultivation implies ammonium sulfate as nitrogen source, glucose as carbon source, KH_2PO_4, and Na_2HPO_4 as phosphorus source that provides maximum production.

PHA production is prominent by the addition of various acids including citrate (0.5%) and acetate (0.5%) production from *Aulosira fertilissima,* resulting to form PHB. Further, mixture of acetic acid and butyric acid represented high productivity of PHB (Lee et al. 2019).

PHA polymer composition acidogenic effluents (AE) varied and PHA derived from synthetic acids were found to be 91%, and AE were found to be 84%. *Pseudomonas putida* and *Pseudomonas aeruginosa* were studied and reported to specify the route of PHA metabolic biosynthesis (Singh Saharan et al. 2014).

Composition of monomers producing PHA demands challenging in research field for its production polymers and high substrate availability of PHB results for faster PHB production (Wong et al. 2020).

pH plays a vital role in PHA productivity and its composition of monomer. Study on effect of pH on fermentation with an alkaline pH of 9 showed high production level of PHAs. pH range also affected composition of polymer and PHA accumulated pH 8.5–9.5 increases PHA production. Microorganisms include cytoplasmic pH compatible with functioning of cell that responds to change in pH of the external environment. PHA production increased at pH of 5.5–9.5 on an account of 10–30% mol (Velázquez-Sánchez et al. 2020).

11.9 Applications of PHA

Microbial PHAs have considerable potential towards various fields including pharmaceuticals, agriculture, regulated for pesticides, manufacturing biofuel, and medicine in surgical substitutes.

PHA as nanocomposites are utilized in bioengineering. The biopolymer plays a vital role with biocompatibility, bioavailability, and biodegradability properties. PHA-based nanocomposites are utilized for cell proliferation in tissue engineering. Biodegradable, eco-friendly agricultural products, viz., fertilizers, insecticides, and pesticides for protecting the plants and saplings. PHAs are biodegradable polyesters offering greater solution in waste management and present as substitution for conventional plastic. Nanocomposites from PHA are also used in developing antibacterial applications. For example, ZnO-based nanocomposites are good disinfectants preventing bacterial proliferation in food packaging and medical instruments.

PHA in enzyme immobilization highly acknowledged application to provide impact on biomaterials. Enzyme immobilization is a familiar field for purification and nanoencapsulation of active materials from preventing degradation through various factors. Nanoprecipitation technique to synthesize PHA from dairy waste developed and performed active production PHA from microorganisms (Zhang et al. 2018).

Drug delivery of nanoparticles is ideal to implement delivery of bioactive compounds reducing drug degradation and deactivation. PHA-based nanoparticles are biodegradable and finally optimize the patient compatibility as drug delivery systems should be removed in the body to prevent potent toxicity. These polymers are nonspecific esterase and lipases that are widely incorporated for hydrophobic rugs and proteins (Chanprateep 2010). PHAs are less toxic when compared with synthetic polymers and are designated as promising sources for therapeutic applications.

Protein purification has been a crucial process. PHA nanoparticles play a major role in protein purification. PHAs are used in the food industry for various applications, viz., antimicrobial films, thermal stability, provide strength and toughness for food packaging material (Liu et al. 2020).

Polyhydroxyalkanoates (PHAs) are highly biodegradable and processing versatility to replace fossil fuel-based plastics. They generally originated from microbial cultures possessing main properties as water insolubility and resistance to ultraviolet rays. Further, biocompatibility and degradability are anaerobically compared to other polymers. PHB degradation takes place with production of polymers with short chain length and PHA made by extrusion for developing flexible plastics molded to various shapes (Tarrahi et al. 2020).

PHAs can be applied in a variety of applications including wall paint coatings, textile materials, films for food packaging, medicinal products, farming agriculture application, for commercial availability.

PHA can be used as bioplastics in cables, connectors, and in electronics manufacturing companies, viz., Nokia, Samsung, Sony, Fujitsu, for their goods.

The premium cost of production develops support for green products and consumer interests. PHA biopolymers have a flexible nature that provides extension to produce materials such as caps, bottles, blister packs, and other containers for various customer goods industries. Represents as a promising polymer with properties implying barrier and mechanical strength compared to other bioplastics such as polylactic acid (Adeleye et al. 2020).

PHA polymer provides advanced mechanical strength making the packaging foils suitable for packaging food products. Further, possess basic property such as excellent oxygen transmission rate (OTR) and water vapor transmission rate (WVTR) representing the primary role of any packaging material that makes it a potential choice for food packaging (Costa et al. 2019).

Flexibility of PHAs shows promising property for various applications in agriculture, and biodegradable plastic requires release mechanism showed control characteristics and can be implemented in fertilizer. PHA pellets are developed technologies that are placed on fields of the soil showing gradual degradation over a long period further reduce the use of fertilizer and cost of labor.

Industrial scale production of PHA is in focus by engineering biocatalysts via synthetic biology tools. Industrial biotechnology focuses the effort to minimize PHA production cost mainly by engineering strains of higher PHA production and thereby provide less energy consumption, for high value-added applications during PHA production,

11.10 Future Prospects and Challenges

PHA represents a high potency of low-cost biodegradable plastics, and there are still some disadvantages for the consumption focusing on production. Nonbiodegradable plastics and other problems influencing production are PHA properties. Economic use of waste feedstock from microbial cultures leads to the production of PHAs and provides greater sustainability. Ongoing research on different methods for the improvement of the yield of PHA are done using genetic modification of the bacteria that are isolated from waste materials. Diverse polymers available in industrial applications with high value functionalities should be focused for improved PHA production. Further investigations were made to reduce the cost of production of PHA polymers and have an impact on several researchers for developing environmentally friendly polymers. Despite the significant advances in PHA production and accumulation research towards molecular advances enhance the survivability of bacteria. Advances in this area in future could benefit the industry of bacterial inoculants and the capabilities of the microorganisms to establish the target niche is of utmost importance. Advanced technology requires upgradation to develop an integrated system for microbially synthesized polymers from soil bacteria rhizosphere. PHAs produced from various sources, effluents, and wastes developed using similar microorganisms would help to reduce environmental pollution and hence recommended for commercialization.

Acknowledgments One of the authors SB, thankfully acknowledges Tamil Nadu Veterinary and Animal Sciences University for providing the necessary facilities to carry out the work.

References

Adeleye AT, Odoh CK, Enudi OC, Banjoko OO, Osiboye OO, Toluwalope Odediran E, Louis H (2020) Sustainable synthesis and applications of polyhydroxyalkanoates (PHAs) from biomass. Process Biochem 96:174–193. https://doi.org/10.1016/j.procbio.2020.05.032

Amaro TMMM, Rosa D, Comi G, Iacumin L (2019) Prospects for the use of whey for polyhydroxyalkanoate (PHA) production. Front Microbiol 10(May):1–12. https://doi.org/10.3389/fmicb.2019.00992

Anjum A, Zuber M, Zia KM, Noreen A, Anjum MN, Tabasum S (2016) Microbial production of polyhydroxyalkanoates (PHAs) and its copolymers: A review of recent advancements. Int J Biol Macromol 89:161–174. https://doi.org/10.1016/j.ijbiomac.2016.04.069

Ayub ND, Pettinari MJ, Méndez BS, López NI (2007) The polyhydroxyalkanoate genes of a stress resistant Antarctic Pseudomonas are situated within a genomic island. Plasmid 58(3):240–248. https://doi.org/10.1016/j.plasmid.2007.05.003

Ayub ND, Tribelli PM, López NI (2009) Polyhydroxyalkanoates are essential for maintenance of redox state in the Antarctic bacterium Pseudomonas sp. 14-3 during low temperature adaptation. Extremophiles 13(1):59–66. https://doi.org/10.1007/s00792-008-0197-z

Bhardwaj U, Dhar P, Kumar A, Katiyar V (2014) Polyhydroxyalkanoates (PHA)-cellulose based nanobiocomposites for food packaging applications. ACS Symp Ser 1162:275–314. https://doi.org/10.1021/bk-2014-1162.ch019

Chanprateep S (2010) Current trends in biodegradable polyhydroxyalkanoates. J Biosci Bioeng 110 (6):621–632. https://doi.org/10.1016/j.jbiosc.2010.07.014

Chee J, Yoga S, Lau N, Ling S, Abed RMM (2010) Bacterially produced polyhydroxyalkanoate (PHA): converting renewable resources into bioplastics, pp 1395–1404

Ciesielski S, Mozejko J, Pisutpaisal N (2015) Plant oils as promising substrates for polyhydroxyalkanoates production. J Clean Prod 106:408–421. https://doi.org/10.1016/j.jclepro.2014.09.040

Costa SS, Miranda AL, de Morais MG, Costa JAV, Druzian JI (2019) Microalgae as source of polyhydroxyalkanoates (PHAs)—A review. Int J Biol Macromol 131:536–547. https://doi.org/10.1016/j.ijbiomac.2019.03.099

Gasser I, Müller H, Berg G (2009) Ecology and characterization of polyhydroxyalkanoate-producing microorganisms on and in plants. FEMS Microbiol Ecol 70(1):142–150. https://doi.org/10.1111/j.1574-6941.2009.00734.x

Johnston B, Radecka I, Hill D, Chiellini E, Ilieva VI, Sikorska W, Musioł M, Zięba M, Marek AA, Keddie D, Mendrek B, Darbar S, Adamus G, Kowalczuk M (2018) The microbial production of polyhydroxyalkanoates from waste polystyrene fragments attained using oxidative degradation. Polymers 10(9):957. https://doi.org/10.3390/polym10090957

Kai D, Loh XJ (2013) Polyhydroxyalkanoates: chemical modi fi cations toward biomedical applications

Koller M, Bona R, Braunegg G, Hermann C, Horvat P, Kroutil M, Martinz J, Neto J, Pereira L, Varila P (2005) Production of polyhydroxyalkanoates from agricultural waste and surplus materials. Biomacromolecules 6(2):561–565. https://doi.org/10.1021/bm049478b

Kumar G, Ponnusamy VK, Bhosale RR, Shobana S, Yoon JJ, Bhatia SK, Rajesh Banu J, Kim SH (2019) A review on the conversion of volatile fatty acids to polyhydroxyalkanoates using dark fermentative effluents from hydrogen production. Bioresour Technol 287(May):121427. https://doi.org/10.1016/j.biortech.2019.121427

Lee S, Kang M, Bae JH, Sohn JH, Sung BH (2019) Bacterial valorization of lignin: strains, enzymes, conversion pathways, biosensors, and perspectives. Front Bioeng Biotechnol 7 (September):1–18. https://doi.org/10.3389/fbioe.2019.00209

Liu L-Y, Xie G-J, Xing D-F, Liu B-F, Ding J, Ren N-Q (2020) Biological conversion of methane to polyhydroxyalkanoates: current advances, challenges, and perspectives. Environ Sci Ecotechnol 2:100029. https://doi.org/10.1016/j.ese.2020.100029

Mitra R, Xu T, Xiang H, Han J (2020) Current developments on polyhydroxyalkanoates synthesis by using halophiles as a promising cell factory. Microb Cell Factor 19(1):1–30. https://doi.org/10.1186/s12934-020-01342-z

Mostafa YS, Alrumman SA, Otaif KA, Alamri SA, Mostafa MS, Sahlabji T (2020) Production and characterization of bioplastic by polyhydroxybutyrate accumulating Erythrobacter aquimaris isolated from mangrove rhizosphere. Molecules 25(1):179. https://doi.org/10.3390/molecules25010179

Muhammadi S, Afzal M, Hameed S (2015) Bacterial polyhydroxyalkanoates-eco-friendly next generation plastic: production, biocompatibility, biodegradation, physical properties and applications. Green Chem Lett Rev 8(3–4):56–77. https://doi.org/10.1080/17518253.2015.1109715

Obruca S, Sedlacek P, Koller M, Kucera D, Pernicova I (2018) Involvement of polyhydroxyalkanoates in stress resistance of microbial cells: biotechnological consequences and applications. Biotechnol Adv 36(3):856–870. https://doi.org/10.1016/j.biotechadv.2017.12.006

Pakalapati H, Chang CK, Show PL, Arumugasamy SK, Lan JCW (2018) Development of polyhydroxyalkanoates production from waste feedstocks and applications. J Biosci Bioeng 126(3):282–292. https://doi.org/10.1016/j.jbiosc.2018.03.016

Pernicova I, Kucera D, Nebesarova J, Kalina M, Novackova I, Koller M, Obruca S (2019) Production of polyhydroxyalkanoates on waste frying oil employing selected Halomonas strains. Bioresour Technol 292(August):122028. https://doi.org/10.1016/j.biortech.2019.122028

Reddy CSK, Ghai R, Rashmi, Kalia VC (2003) Polyhydroxyalkanoates: an overview. Bioresour Technol 87(2):137–146. https://doi.org/10.1016/S0960-8524(02)00212-2

Roja K, Ruben Sudhakar D, Anto S, Mathimani T (2019) Extraction and characterization of polyhydroxyalkanoates from marine green alga and cyanobacteria. Biocatalysis Agric Biotechnol 22:101358. https://doi.org/10.1016/j.bcab.2019.101358

Samorì C, Kiwan A, Torri C, Conti R, Galletti P, Tagliavini E (2019) Polyhydroxyalkanoates and crotonic acid from anaerobically digested sewage sludge. ACS Sustain Chem Eng 7(12):10266–10273. https://doi.org/10.1021/acssuschemeng.8b06615

Sanhueza C, Acevedo F, Rocha S, Villegas P, Seeger M, Navia R (2019) Polyhydroxyalkanoates as biomaterial for electrospun scaffolds. Int J Biol Macromol 124:102–110. https://doi.org/10.1016/j.ijbiomac.2018.11.068

Singh Saharan B, Grewal A, Kumar P (2014) Biotechnological production of polyhydroxyalkanoates: a review on trends and latest developments. Chin J Biol 2014:1–18. https://doi.org/10.1155/2014/802984

Singh AK, Sharma L, Mallick N, Mala J (2017) Progress and challenges in producing polyhydroxyalkanoate biopolymers from cyanobacteria. J Appl Phycol 29(3):1213–1232. https://doi.org/10.1007/s10811-016-1006-1

Sujatha K, Mahalakshmi A, Shenbagarathai R (2005) A study on accumulation of PHB in native Pseudomonas isolates LDC-5 and LDC-25. Indian J Biotechnol 4(2):216–221

Suriyamongkol P, Weselake R, Narine S, Moloney M, Shah S (2007) Biotechnological approaches for the production of polyhydroxyalkanoates in microorganisms and plants—A review. Biotechnol Adv 25(2):148–175. https://doi.org/10.1016/j.biotechadv.2006.11.007

Tarrahi R, Fathi Z, Seydibeyoğlu MÖ, Doustkhah E, Khataee A (2020) Polyhydroxyalkanoates (PHA): from production to nanoarchitecture. Int J Biol Macromol 146:596–619. https://doi.org/10.1016/j.ijbiomac.2019.12.181

Velázquez-Sánchez C, Espín G, Peña C, Segura D (2020) The modification of regulatory circuits involved in the control of polyhydroxyalkanoates metabolism to improve their production. Front Bioeng Biotechnol 8(April). https://doi.org/10.3389/fbioe.2020.00386

Volova T, Kiselev E, Zhila N, Shishatskaya E (2019) Synthesis of polyhydroxyalkanoates by hydrogen-oxidizing bacteria in a pilot production process. Biomacromolecules 20 (9):3261–3270. https://doi.org/10.1021/acs.biomac.9b00295

Wong JX, Ogura K, Chen S, Rehm BHA (2020) Bioengineered polyhydroxyalkanoates as immobilized enzyme scaffolds for industrial applications. Front Bioeng Biotechnol 8(March). https://doi.org/10.3389/fbioe.2020.00156

Wongsirichot P, Gonzalez-Miquel M, Winterburn J (2020) Integrated biorefining approach for the production of polyhydroxyalkanoates from enzymatically hydrolyzed rapeseed meal under nitrogen-limited conditions. ACS Sustain Chem Eng 8(22):8362–8372. https://doi.org/10.1021/acssuschemeng.0c02406

Yu J (2010) Biosynthesis of polyhydroxyalkanoates from 4-ketovaleric acid in bacterial cells. ACS Symp Ser 1043:161–173. https://doi.org/10.1021/bk-2010-1043.ch012

Zhang J, Shishatskaya EI, Volova TG, da Silva LF, Chen GQ (2018) Polyhydroxyalkanoates (PHA) for therapeutic applications. Mater Sci Eng C 86:144–150. https://doi.org/10.1016/j.msec.2017.12.035

[illegible]

Role of Microbial Biofilms in Agriculture: Perspectives on Plant and Soil Health

12

Anupam Gogoi, Mandeep Poudel, Jagajjit Sahu, and Geetanjali Baruah

Abstract

Microbial communities, especially beneficial microbes are a boon to plant health as well as mankind. Regardless of the works that have been done in this field, still, a long way is ahead of the researchers to employ notable microbes in the benefit of plants to fight against various stresses and diseases. Biofilms are the multicellular assemblies or aggregates of differentiated microbial communities embedded in an extracellular matrix which have proven to be an efficient tool for the sustainable management of crops against multiple stresses. This chapter aims to provide deep insight into the biofilm-producing microbiota and their roles in crop protection. The knowledge garnered from the published work led us to information regarding the functional attributes of various biofilm-producing microbiota. As the literature suggested maximum numbers of biofilms used are from the bacterial community such as *Azospirillum brasilense*, *Bacillus amyloliquefaciens*, *and Pseudomonas fluorescens*. The biofilms are beneficial in enhanced nutrient solubilization, nitrogen fixation, sequestering, enhancement of plant biomass, and many other important physiological processes. Besides, we also attempted a basic scientometrics approach to provide the most frequent keywords for the published articles for various types of microbes forming biofilms. The prospects of biofilm

A. Gogoi · M. Poudel
Department of Plant Sciences, Faculty of Biosciences (BIOVIT), Norwegian University of Life Sciences (NMBU), Ås, Norway

J. Sahu
National Center for Cell Sciences (NCCS), University of Pune Campus, University Road, Ganeshkhind, Pune, Maharashtra, India

G. Baruah (✉)
Environment Division, Assam Science Technology and Environment Council, Guwahati, Assam, India

A. Vaishnav, D. K. Choudhary (eds.), *Microbial Polymers*,
https://doi.org/10.1007/978-981-16-0045-6_12

in crop protection depend upon a better understanding and may be combined with multi-omics approaches. This chapter establishes the efficiency of the use of biofilms disclosing sustainable ways of fighting against stress conditions and maintaining plant health.

Keywords

Beneficial microbe · Sustainable agriculture · Microbiota · EPS · Soil restoration · Quorum sensing

12.1 Introduction

Plants encounter diverse groups of microbial communities such as beneficial, pathogenic and commensals, and abiotic stresses such as drought, salinity, heat-stress, pH, and temperature in naturally growing environments (Knief et al. 2011; Narsai et al. 2013; Ahmad et al. 2019). Due to the ever-increasing world population and constant global climatic changes (Cleland 2013; Ekwurzel et al. 2017), threats to food security and sustainability have become a growing concern. Meanwhile, a combination of environmental stresses and deterioration of soil health due to excessive use of synthetic fertilizers and pesticides hampered the whole agricultural scenario. It is estimated that disease caused by biotic factors account for 20–40% of annual crop losses worldwide (Oerke and Dehne 2004; Savary et al. 2012). The abiotic stresses also contribute significantly to the reduction of crop yield and productivity (Ciais et al. 2005; Francini and Sebastiani 2019). Several management tactics including the use of pesticides, synthetic fertilizers in combination with resistance breeding, and agronomic practices have been employed in recent decades to overcome the environmental stresses. However, the continuous application of synthetic chemicals not only deteriorates soil health but also imparts residual toxicity in foods (Aktar et al. 2009; Bourguet and Guillemaud 2016). The rapid evolution of pathogens to overcome plant resistance genes and pesticides limit their long-term use in the field (McDonald and Stukenbrock 2016). Thus, an alternative, efficient, and sustainable approach is needed to tackle the present problems associated with plant and soil health.

Most soil-borne pathogenic microbes in the cropping environment are often counterbalanced by the metabolic activities of associated beneficial microbiome (Wei et al. 2019). These beneficial microbes have been a subject of research and have drawn attention towards sustainable management of crops against environmental stresses. Microbial cells physically interact with plant surfaces via the formation of multicellular assemblies known as biofilms (Danhorn and Fuqua 2007). Biofilms are self-assembled, aggregates of microbial cells or a coherent mixture of differentiated microbial community embedded in an extracellular matrix of polymeric substances that are associated with an inert or biotic surface (Stoodley et al. 2002; Danhorn and Fuqua 2007; Vlamakis et al. 2013; Velmourougane et al. 2017). Cells adhere to surfaces via extracellular polymeric substances (EPS), which is a complex mixture of exopolysaccharides, lipopolysaccharides, proteins, and DNA (Whitchurch et al. 2002). Biofilm formation and maturation is mediated by EPS production and triggered by environmental and quorum sensing signals (Flemming et al. 2016). Since the production of these biomolecules is energy driven, their functions should carry advantage to the producer microbial community. Most of the benefits of biofilms are attributed to environmental protection such as shielding from antimicrobial compounds, drought, and heavy metals (Wolfaardt et al. 1999; Sandhya and Ali 2015; Flemming et al. 2016; Zainab et al. 2020). Slime formed by EPS helps in sequestering carbon, water, nutrients, and facilitated cell–cell communication within the microbial community and with plant cells (Moons et al. 2006; Islam et al. 2016). Studies on EPS extracellular DNA revealed that the presence of DNA molecules can be a template for genetic exchange, signaling, and provide structural integrity (Flemming et al. 2016). Biofilm formation is an inherent mechanism adapted by many microbes including bacteria, fungi, algae, and archaea, irrespective of their beneficial or pathogenic traits to their hosts (Velmourougane et al. 2017). Beneficial biofilm-producing microbes are a subject of interest for agricultural and industrial applications. Several microbial species including bacteria such as *Bacillus*, *Pseudomonas*, *Rhizobium*, *Azospirillum*, *Azotobacter*, and *Burkholderia*; cyanobacteria—*Anabaena* and *Nostoc*; filamentous fungi—*Trichoderma* promotes plant growth and soil health. These beneficial plant microbiomes help solubilizing and up taking of macro- and micronutrients such as Phosphorus (P), Nitrogen (N), Potassium (K), Magnesium (Mg), Zinc (Z), Copper (Cu), and Iron (Fe) by their hosts (Al-Nahidh and Gomah 1991; Ma et al. 2009; Wang et al. 2020). They also protect crops from various biotic and abiotic stress including drought, low/high temperature, change in soil pH, and salt tolerance. The slime layer produced by the EPS matrix act like a glue that helps in soil aggregation and thus improves soil structure, health, and fertility (Chenu and Cosentino 2011; Costa et al. 2018). In this chapter, we addressed the roles of microbial biofilms in various aspects of stress management in plants and how they promote soil health and sustainability in agriculture. We have also tried to shed light on the publication patterns and research outputs in this field via a basic scientometrics analysis to help young researchers get clarity in this regard.

12.2 Biofilm-Producing Microbes Categorically with Special Emphasis on Agriculturally Important Microbes (AIMs)

Plant and soil-associated microbial communities impart several traits either beneficial or pathogenic to their hosts and their growing environments. Most microbes form biofilms to interact with their ancestral population via a series of developmental stages. Stoodley et al. described five biofilm developmental stages as (1) primary attachment of cells to a physical surface, (2) formation of EPS to firmly adhere cells together, (3) cell differentiation and biofilm development, (4) maturation of biofilm architecture and the slime-layer formation, and (5) disassembly and dispersal of individual cells from the differentiated biofilm (Stoodley et al. 2002). Cell–cell communication during the differentiation process involves signaling mechanisms that control the phenotypes of microbial communities. In bacteria, these mechanisms include the exchange of organic molecules, proteins, or transmission of electrical signals by ion channels (Prindle et al. 2015). Studies on the role of chemical signaling (also known as quorum sensing, QS) that involve several diffusible sensing molecules have been a subject of intensive research. These include acyl-homoserine lactones (AHLs), autoinducing peptides (AIPs), γ -butyrolactones. 3,4-dihydroxy-2-heptylquinoline (PQS), 3OH palmitic acid methyl ester, and cyclic dipeptides, commonly known as autoinducer (reviewed by Waters and Bassler 2005). QS is a signaling process initiated by specific diffusible sensing molecules that carry inter-cellular chemical signals which result in an inducible gene expression at a population level to adapt to the local environmental changes. The knowledge of QS is not only limited to bacteria but has been extended for closely related cyanobacteria (Sharif et al. 2008) and other organisms including filamentous fungi (Padder et al. 2018), algae (Zhou et al. 2016), oomycetes (Kong et al. 2010a, b), and archaea (Rajput and Kumar 2017; Kaur et al. 2018). Some bacteriophage can also bind to bacterial QS molecules via their genome encoded LuxR-type receptor (Silpe and Bassler 2019) or can sense the metabolic state of host cyclic-3′,5′-AMP (cAMP) receptor protein (CRP) to determine host population density (Laganenka et al. 2019). An increase in bacterial population modulates the transition of virus particles from lysogenic to lytic state (Laganenka et al. 2019).

12.3 Roles of Microbial Biofilm in Crop Protection

12.3.1 Disease and Pest Resistance

M Biofilm-producing beneficial microbes from different taxa including bacteria, fungi, algae, and archaea play important roles in shaping plant growth and development parameters and protect crops against various biotic and abiotic stresses (Ariosa et al. 2004; Ribaudo et al. 2006; Sun et al. 2010; Fontenelle et al. 2011; Song et al. 2019). These microbes colonize in the rhizoplane (Hansen et al. 1997; Benizri et al. 2001), or in the outermost zone of the rhizosphere, i.e., in the soil compartments called as ectorhizosphere (Whipps 2001; Lopes et al. 2016), or inside plant tissues such as root hair, cortex, and vascular bundles to establish a mutualist interaction with their host and its environment (Hallmann 2001; Compant et al. 2010; Jia et al. 2016; Liu et al. 2017). Colonization of beneficial microbes within the plant intercellular spaces are known as endophytes. Endophytic cohabitation and rhizosphere inhabiting microbes enhance plant immunity mechanisms, thereby primed the host to combat subsequent attack by pathogenic microbes and pests (Van Wees et al. 2008). Plant tissues colonized by a biofilm often generate mobile signals that trigger induced systemic resistance (ISR) in the whole plant. For instance, root inoculation of *Paenibacillus polymyxa* MAS100 in potato (*Solanum tuberosum*) suppressed growth of the bacterial wilt pathogen *Ralstonia solanacearum* and reduced disease incidence up to 20% after inoculation of the pathogen (Soliman 2020). Besides, the rhizobacteria *Paenibacillus* spp. could potentially be used as a biocontrol agent due to its dual role as an antifungal and nematicide against *Fusarium* wilt and root gall disease caused by *Meloidogyne incognita,* respectively (Son et al. 2009). Seed inoculation of two endophytic bacteria *Bacillus pumilus* and *B. amyloliquefaciens* in tomato reduced disease severity of up to 62% against bacterial speck disease (Lanna-Filho et al. 2017). Several studies have reported that these bacterial species induced expression of several pathogenesis-related protein-coding genes including *PR1*, *PR2* and *PR3,* and other defense-related enzymes to reduce viral load and multiplication of many viruses including *Bean common mosaic virus*, *Cucumber mosaic virus*, *Papaya ringspot virus*, *Pepper mild mottle virus*, *Tobacco streak virus*, *Tomato chlorotic spot virus*, *Tomato mottle virus*, and *Tomato yellow leaf curl virus* in many crop species (Murphy et al. 2000; Ahn et al. 2002; Jetiyanon and Kloepper 2002; Udaya Shankar et al. 2009; Abdalla et al. 2017; Vinodkumar et al. 2018; Guo et al. 2019). Many other bacterial species including *P. fluorescens*, *P. putida*, *B. subtilis, B. cereus, B. sphaericus, and related species induce ISR to protect crops against many viral diseases* (Beris and Vassilakos 2020). Some fungal species including *Aspergillus* sp. strain NBP-08, *Penicillium chrysogenum, Penicillium oxalicum*, *Penicillium* sp. strain NBP-45, *Talaromyces* sp. (isolates NBP-61, NBP-66 *and* NBP-67), *Trichoderma harzianum,* and many others can also contribute in plant immunity by modulating hormonal signaling pathways and thereby the expression of defense-related genes and gene products such as β- 1,3-glucanase, peroxidase, chitinase, lipoxygenase, polyphenol oxidase, and superoxide dismutase (Chowdappa et al. 2013; Murali et al. 2013; Murali and Amruthesh 2015). These antimicrobial proteins in plant functions as inhibitory enzymes by blocking pathogen growth via disintegration of cell wall through hydrolytic cleavage of β glycosidic bonds, polysaccharide cross-linking and lignification in plant tissues, facilitate the accumulation of reactive oxygen species (ROS) such as hydrogen peroxide (H_2O_2) and superoxide ($O^-{}_2$) that are toxic to the invading pathogen (Sudisha et al. 2012).

The role of hormonal pathways including salicylic acid (SA), jasmonic acid (JA), ethylene (ET), and auxin (Aux) signaling in plant defense against pathogens, and pests have been well documented and reviewed (Pieterse et al. 2014). Regulation and crosstalk between phytohormonal signaling pathways are key mechanisms of ISR to biotrophic and necrotrophic pathogens (Glazebrook 2005). Increased expression of phenylalanine ammonia lyase (PAL), a central regulator of the SA biosynthetic pathway has a clear link to resistance against most biotrophs, whereas JA/ET hormonal pathways are more active against necrotrophic pathogens. Nevertheless, five rhizospheric plant growth-promoting fungi (PGPF) antagonistic to the necrotrophic fungi *Colletotrichum capsici* showed induced PAL enzyme activity in Chilli (*Capsicum annum* L.) (Naziya et al. 2020). Plant hormones also play protective roles against insect herbivores and are triggered by beneficial microbes. For instance, ISR triggered by root-colonizing microbial biofilms increased expression of JA-dependent gene—*Lipoxygenase 2* (*LOX2*) and JA/ET-dependent genes—*Plant Defensin 1.2* (PDF1.2) and *Hevein-like protein* (*HEL*) in *Arabidopsis* and thereby primed host against leaf chewing and phloem-feeding insect pests (Van Oosten et al. 2008; Pineda et al. 2012; Pangesti et al. 2015). Although, most microbially induced resistance against plant-feeding insects are JA-dependent, while JA-independent mechanisms via phenylpropanoid and terpenoid biosynthetic pathways are no exception (Valenzuela-Soto et al. 2010; Zebelo et al. 2016; Sharma et al. 2017). Thus, different microbial species induce ISR via various phytohormone signaling pathways, and a conserved link to describe its role against biotrophy, necrotrophy, and herbivorous pests are poorly understood.

Cell wall lignification and callose deposition are two important signature mechanisms of defending priming in plants. The facilitated accumulation of these metabolites by microbial biofilms helps in strengthening of the cell wall and acts as a physical barrier for pathogen invasion and insect herbivory (Pieterse et al. 2014; Harun-Or-Rashid and Chung 2017). Several bacterial species produce volatile organic compounds or facilitate the release of herbivore-induced plant volatiles that collectively act as insect repellents, toxic to herbivores, or attract predators to feed parasitize herbivores (Ryu et al. 2004; Zebelo et al. 2016; Aartsma et al. 2017). These secondary metabolites including 2,3-butanediol, acetoin, gossypol, and other volatile compounds are air diffusible that systematically primed other parts of the same plant so that they become prepared to resist subsequent attack by pests. Thus, biofilm-producing microbes trigger ISR via various mechanisms at different levels of resistance to protect plants against various biotic stresses caused by bacteria, fungi, viruses, and insect pests.

12.3.2 Protection from Abiotic Stress

In agriculture, apart from the biotic stresses, increasing exposure to various abiotic stresses including soil salinity, temperature fluctuations (low and high temperature), drought and waterlogging, mineral deficiency, and heavy metal toxicity are hostile to plant growth and development processes (Khan et al. 2014; He et al. 2018). These multilevel abiotic stress factors negatively affecting the global food production and thereby, socio-economic outcomes (Ghassemi et al. 1995; Ashraf et al. 2012). It has been estimated that abiotic factors globally contribute more to yield loss (kg/hectare) than biotic factors in major crops including corn, wheat, soybean, millet, oats, and barley (Bayer Crop Science 2008). For instance, saline stress, one of the most predominate, that cover 10% of the land surface (950 million ha; Mha) and 50% of arable land (230 Mha) globally (Carrow and Duncan 1998), is reducing the use of agricultural land by 1–2% every year (Islam et al. 2016). High soil salinity impinges various stages of plant physiological processes, viz. seed germination, seedling establishment, growth, and fertility, resulting from induced osmotic imbalance across the plasma membrane, ionic toxicity, the release of ROS, and nutrients unavailability (He et al. 2018). An interlinked of this is drought stress resulting from low rainfall that can be aggravated by a change in global temperature, high intensity of light, soil structure, and reduced water retention capacity (Salehi-Lisar and Bakhshayeshan-Agdam 2016). Drought stress is multidimensional as it affects the acquisition of plant nutrients and water, and the rate of transcription that result in reduced biomass production and crop productivity. Also, metal contamination in soils resulting from metal mining, sewage sludge, waste disposal, and irregular agricultural practices (application of fertilizers and pesticides) has become a serious concern as it affects about 235 Mha of agricultural land. The heavy metal contaminants (e.g., Al, Cd, Cu, Cr, Ni, Pb, and Hg) not only affect crop yields but may enhance residual toxicity in foods as it can translocate to the edible plant parts (Hart et al. 2006; Khan et al. 2014). Thus, abiotic stresses influence various parameters of plant health and affect agro-ecosystems and may pose serious threats to food security, human and animal health.

The employment of microbial biostimulants is an eco-friendly and sustainable approach for effective management of environmental stresses and in promoting plant health. These microbial biofilms perform multiple beneficial roles that include enhanced tolerance to salinity and drought stress (Alami et al. 2000; Gusain et al. 2014; Islam et al. 2016), nutrient availability (e.g., N, P, K, Mg) (Chowdappa et al. 2013; Islam et al. 2016), heavy metal tolerance and ROS scavenging (Faisal and Hasnain 2005; Islam et al. 2016; Crossay et al. 2020). The mechanisms of alleviating salt stress by the plant-associated microbes include nullifying hormonal imbalance by secreting indole acetic acid (IAA) to improve plant growth parameters and counterbalance the release of stress hormone abscisic acid (ABA). During saline stress, excessive Na^{+} ions can induce blockage or reduce the efficiency of transporters such as high-affinity K^{+} transporter 1 (HKT1) and proton pump ATPase that control Na^{+} uptake. Beneficial microbes such as *Pseudomonas* induce expression of these ion transporters, thereby help plants to accumulate more K^{+}, Mg^{2+}, and Ca^{2+} to induce plant growth and to avoid ionic imbalance caused during salt stress (Yang et al. 2009; Yao et al. 2010). Previous studies have revealed that plant-associated rhizobacteria *Paenibacillus polymyxa* enhance transcript abundance of *EARLY RESPONSIVE TO DEHYDRATION 15* (*ERD15*), a drought stress-responsive gene in Arabidopsis and *Achromobacter piechaudii* ARV8 produce 1-aminocyclopropane-1-carboxylate (ACC) deaminase to lower the drought stress-elevated ethylene levels and restored normal plant growth (Timmusk and Wagner 1999; Mayak et al. 2004).

12.4 Role in Soil Health

Microbial biofilms can not only promote plant health but also maintain soil architecture and improve water retention capacity by forming soil aggregates (Alami et al. 2000; Sandhya et al. 2009; Deka et al. 2019), nutrient composition, nutrient solubilization, and availability (Lee et al. 2008; Wani and Khan 2010; Malviya et al. 2011; Wang et al. 2014; Pérez-De-Luque et al. 2017), carbon sequestration, nitrogen fixation, and soil fertility enhancement (Hafeez et al. 2000; Rinaudi and Giordano 2010; Prasanna et al. 2016). With increasing concern of heavy metal contamination in arable soils, bioaugmentation—"an application of cultured microbial biomass on the soil surface to remove or degrade specific contaminants from the soil" could serve as an effective strategy for improving metal toxicity. For instance, co-inoculation of sunflower (*Helianthus annuus*) with bacteria *Pseudomonas libanensis* strain TR1 and arbuscular mycorrhizal fungus (AMF) *Claroideoglomus claroideum* strain BEG210 increased plant biomass and promote phytoremediation of nickel (Ni) from contaminated saline soil. Increased production of ACC deaminase, IAA, and siderophore by these microbes might help in efficient uptake of Ni^{2+} ions from the soil (Ma et al. 2019).

An important attribute of soil health is soil structure while influencing both the nature and content of organic matter. The basic units of soil structure are soil aggregates, which determine the physical and mechanical properties of soil such as porosity, hydraulic conductivity, water retention capacity, aeration, and temperature. Plant mucilage and microbial EPS play important role in binding soil units together and increase organic matter content (Watt et al. 1993). Several studies have shown that the application of microbial EPS resulted in stable soil aggregation and substantially increased nutrient uptake by plants. For instance, root adhering and EPS-producing *Rhizobium* spp. strain YAS34 significantly increased soil macropore volume of up to 12–60 μm in diameter after inoculation in soil and increased sunflower shoot and root dry mass up to 50 and 70%, respectively (Alami et al. 2000). A similar effect was observed in wheat inoculated with *Paenibacillus polymyxa* CF43 that increased the ratio of root-adhering soil dry mass to root tissue dry mass by 30–100% in treated soil (Bezzate et al. 2000). Microbial biofilms could also play an important role in maintaining carbon (C) cycling via sequestration of atmospheric CO_2 in soil, providing the potential for large-scale C sinks. Thus, it can aid in the possible management of leading global problems such as global warming and soil conservation.

12.5 Impact on Plant Growth

Plant growth parameters are essential features for monitoring crop yield and productivity. With diminishing natural resources, it is important to focus on different growth parameters for greater yield with fewer inputs. On the other hand, increasing focus on plant biomass and intensifying food production by sustainable means are among the top 12 areas of the 2030 agenda of the Food and Agricultural Organization (FAO) to end poverty and hunger. It has been estimated that food production must rise to 60% to feed the global population that is projected to reach 10 billion by 2050 (FAO 2009). An overall increment for food production by sustainable means is a challenging goal for the current agricultural sector. In this case, a balance employment of microbial communities to enhance plant growth parameters could potentially create a new avenue for sustainable agriculture.

The use of microbes to enhance plant growth and biomass has been well documented in many crop species (Ribaudo et al. 2006; Lee et al. 2008; Canellas et al. 2013). These include Amaranth (*Amaranthus gangeticus*), Arugula (*Eruca sativa*), Capsicum (*Capsicum annuum*), Chickpea (*Cicer arietinum*), Chinese cabbage (*Brassica rapa* subsp. *chinensis*), Mustard (*Brassica campestris*), Chrysanthemum (*Chrysanthemum morifolium*), Cotton (*Gossypium barbadense* and *Gossypium hirsutum*), Cucumber (*Cucumis sativus*), *Groundnut (Arachis hypogaea)*, Lettuce

(*Lactuca sativa*), Maize (*Zea mays*), Mungbean (*Vigna radiata*), Red sage (*Salvia miltiorrhiza*), *Ribbed bog moss (Aulacomnium palustre*), Rice (*Oryza sativa*), Sesban (*Sesbania sesban*), Soybean (*Glycine max*), Strawberry (*Fragaria. spp.* and *F. ananassa*), Sunflower (*Helianthus annuus*), Tea (*Camellia sinensis*), Tomato (*Lycopersicon esculentum* and *Solanum lycopersicum*), and Wheat (*Triticum aestivum*). A detailed list of microbial communities involved in plant growth and development, and their role in defense against biotic and abiotic stresses, and promotion of soil health is documented in Table 12.1.

In the interest to access, the connections between different species producing biofilms and the crop plants that have been benefited, a chord diagram was created using Circos (Krzywinski et al. 2009) (Fig. 12.1). The input for the Circos tool was created with the help of in-house R scripts from Table 12.1. As the figure mentions, the *Azospirillum brasilense*, *Bacillus amyloliquefaciens,* and *Pseudomonas fluorescens* were the three most connected biofilm-producing microbiota. All the three species were bacteria with every six connections to six different crop species. In the case of crop species, both Tomato and Rice were found to be connected to a maximum of ten species producing biofilms.

Most of the functional attributes influenced by beneficial microbes include enhanced nutrient solubilization and uptake, fixation of atmospheric nitrogen by nodules formation, sequestering iron via the production of siderophores. These actions positively affect plant growth parameters such as (1) increase in root and shoot length, (2) seed germination, and seedling vigor, (3) increase in grain production, and overall plant biomass. Change in the level of two phytohormones—IAA and ethylene both related to plant growth is influenced by microbial biofilms. Most studied *Rhizobium* and *Azospirillum* species produce IAA that facilitate nodule formation in the host plant (Ribaudo et al. 2006; Leggett et al. 2017). Similarly, mutants of soybean *Bradyrhizobium elkanii* deficient in IAA production induced fewer root nodules than the wildtype strain (Fukuhara et al. 1994). In contrary to IAA, ethylene can affect plant growth parameters in different ways such as inhibiting root elongation, promoting fruit ripening and senescence, and inhibiting nodule formation. Many plant growth-promoting bacteria such as *Rhizobium leguminosarum*, *Pseudomonas* spp., *Azospirillum* spp., and *Burkholderia unamae* synthesize the enzyme ACC deaminase, which converts ACC (aprecursor of ethylene biosynthesis) into ammonia and α-ketobutyrate (Glick et al. 2007). As a result, the level of ethylene decreases, and therefore the plant growth inhibitory role of the hormone could be overcome (Onofre-Lemus et al. 2009; Glick 2012). Thus, these microbes modulate phytohormone pathways to influence plant growth parameters and enhance plant biomass.

12.6 Factors Affecting Biofilm Formation

There is a lot of information is yet to unveil contributing to the current knowledge on how plants regulate the microbial association. The process of biofilm formation is dependent on multiple factors. Some of such key factors are mentioned below.

Table 12.1 Examples of biofilm-producing microbiota from different taxa involve in promoting plant and soil health

Microbes	Functional attributes	Crop plant	References
Bacteria			
Azospirillum brasilense *Azospirillum lipoferum*	Enhanced production of phytohormones, enhanced root and shoot length, increased root hairs and side roots	*Capsicum annuum* *Gossypium barbadense* *Gossypium hirsutum* *Lycopersicon esculentum* *Triticum aestivum*	Ribaudo et al. (2006), Bashan (1998)
Azospirillum brasilense	Increased production of wetland rice	*Oryza sativa*	Hahn et al. (2016)
Bacillus xiamenensis	Boost leaf relative water content, fresh weight and dry weight, decrease in electrolyte leakage, reduced proline content, malondialdehyde and lipid peroxidation, induced heavy metal stress tolerance (Chromium)	*Sesbania sesban*	Din et al. (2020)
Bacillus cereus	Increased antioxidant enzymes (superoxide dismutase (SOD), catalase (CAT), peroxidase (POD) and proline, decreased Na^+ accumulation under saline stress and improved mineral content; N (33%), P (48%), K (50%) in saline stressed plants Increased in dehydrogenase and phosphatase	*Vigna radiata*	Islam et al. (2016)
Bacillus subtilis *Bacillus amyloliquefaciens*	Activation of induced systemic resistance (ISR) and plant growth promotion	*Arabidopsis thaliana*	Ryu et al. (2004)
Bacillus subtilis strain QST71	Increased plant growth, increased phosphorus mobilization and uptake (40%), high P adsorption capacity	*Cucumis sativus*	Garcia-Lopez and Delgado (2016)

(continued)

Table 12.1 (continued)

Microbes	Functional attributes	Crop plant	References
Bacillus subtilis OTPB1	Enhanced systemic resistance and induction of defense-related enzymes against *Alternaria solani* and *Phytophthora infestans*	*Lycopersicon esculentum*	Chowdappa et al. (2013)
Bacillus sp. PSB10	Enhanced growth of plant under chromium stressed soil, ability to tolerate and detoxify chromium metal ion, solubilization of inorganic phosphate	*Cicer arietinum*	Wani and Khan (2010)
Bacillus velezensis LM2303	Reduced incidence and severity against *Fusarium graminearum* (Fusarium head blight)	*Triticum aestivum*	Chen et al. (2018)
Bacillus velezensis YC7010	Induced systemic resistance against green peach aphid (*Myzus persicae*) resulted in callose deposition and hypersensitive response	*Arabidopsis* sp.	Harun-Or-Rashid et al. (2017)
Bacillus amyloliquefaciens p16	Exopolysaccharide produced by bacteria aided in improvement of soil physical structure and enhance water retention capacity	–	Deka et al. (2019)
Bacillus amyloliquefaciens *Bacillus pumilus*	Reduced disease severity against *Cucumber mosaic virus, Bean common mosaic virus, Pepper mild mottle virus, Tobacco Streak virus, Tomato chlorotic spot virus, Tomato mottle virus, Tomato yellow leaf curl virus*	*Cucumber, Cowpea, Tobacco, Cotton, Solanum lycopersicum L. (Tomato)*	Jetiyanon and Kloepper (2002), Udaya Shankar et al. (2009), Ahn et al. (2002), Vinodkumar et al. (2018), Abdalla et al. (2017), Murphy et al. (2000), Guo et al. (2019)
Bacillus subtilis BEB-DN	Induced systemic resistance against whitefly *Bemisia tabaci*	*Solanum lycopersicum*	Valenzuela-Soto et al. (2010)

(continued)

Table 12.1 (continued)

Microbes	Functional attributes	Crop plant	References
Ochrobactrum intermedium	Promotion of plant growth and reduced uptake and toxicity of chromium from contaminated soil	*Helianthus annuus*	Faisal and Hasnain (2005)
Rhodopseudomonas sp.	Enhanced shoot growth, increased fruit/flower ratio, increase in average fruit weight and lycopene content Solubilization of insoluble phosphate	*Lycopersicon esculentum*	Lee et al. (2008), Koh and Song (2007)
Rhizobium sp. (strain YAS34)	Exopolysaccharides produced by the bacteria modified the soil structure around root system, relieved plant from water stress, increased nitrogen uptake and promoted plant growth	*Helianthus annuus*	Alami et al. (2000)
Bradyrhizobium japonicum	Increased in average crop yield and improved nutrient (N, P, Mg) uptake	*Glycine max*	Leggett et al. (2017), Egamberdieva et al. (2018)
Gluconacetobacter diazotrophicus Pal5	Induced systemic resistance to drought and abiotic stress, increase in productivity parameters, antioxidant defense induction, increase in plant biomass and higher gas exchange and osmoprotectant solutes	*Oryza sativa*	Filgueiras et al. (2019)
Gluconacetobacter diazotrophicus Pal5	Increase phosphorus content, higher biomass production, root area and growth index	Strawberry (*Fragaria spp.*)	Delaporte-Quintana et al. (2017)
Herbaspirillum seropedicae	Early lateral root emergence, increased grain production	*Zea mays*	Canellas et al. (2013)
Herbaspirillum seropedicae DX35	Increase nitrogenase activity and phosphorus solubilization	*Oryza sativa*	Wang et al. (2014)
Paenibacillus spp.	Promoted plant growth, reduced	*Lycopersicon esculentum*	Son et al. (2009)

(continued)

Table 12.1 (continued)

Microbes	Functional attributes	Crop plant	References
	disease impact and symptom development, antifungal and nematicidal activity		
Paenibacillus polymyxa	Ameliorated soil structure, improved water retention and nutrient transfer	*Triticum aestivum*	Bezzate et al. (2000), Gouzou et al. (1993)
Pseudomonas fluorescens Pf1	Disease control and growth promotion	*Oryza sativa Camellia sinensis*	Vidhyasekaran et al. (1997), Saravanakumar et al. (2007)
Pseudomonas fluorescens SS101	Enhanced resistance against *Pseudomonas syringae pv tomato* and *Spodoptera exigua*	*Arabidopsis thaliana*	van de Mortel et al. (2012)
Pseudomonas putida	Stable soil aggregate, better soil structure, increase plant biomass, increase survivability under drought stress	*Helianthus annuus*	Sandhya et al. (2009)
Pseudomonas putida Rs-198	Increase plant stand and germination, production of phytohormones, i.e., increased IAA and reduced ABA, protection against salt stress	*Gossypium hirsutum*	Yao et al. (2010)
Pseudomonas simiae WCS417r	Enhanced synthesis of camalexin and aliphatic glucosinolates, induced systemic resistance against leaf chewing herbivore *Mamestra brassicae*	*Arabidopsis thaliana*	Pangesti et al. (2016)
Pseudomonas fluorescens WCS417r	Enhanced attraction of parasitoid wasp *Microplitis mediator* to *Mamestra brassicae* (Cabbage moth; generalist insect herbivore) infested plant	*Arabidopsis thaliana*	Pangesti et al. (2015)
Pseudomonas fluorescens	Elicited priming and induced systemic resistance against herbivorous insects	*Arabidopsis thaliana*	Pineda et al. (2012)

(continued)

Table 12.1 (continued)

Microbes	Functional attributes	Crop plant	References
Burkholderia megapolitana	Antifungal activity, plant growth promotion	*Aulacomnium palustre*	Vandamme et al. (2007)
Burkholderia unamae	Expression of ACC (1-aminocyclopropane-1-carboxylate) deaminase thereby enhancing plant growth	*Lycopersicon esculentum*	Onofre-Lemus et al. (2009)
Azospirillum amazonense	Increased nitrogen accumulation, grain yield and panicle number	*Oryza sativa*	Rodrigues et al. (2008)
Paenibacillus polymyxa MAS100	Suppression of disease symptoms against *Ralstonia solanacearum*	*Solanum tuberosum*	Soliman (2020)
Rhizobium leguminosarum	Nitrogen fixation, soil fertility enhancement	*Lens culinaris* *Oryza sativa*	Hafeez et al. (2000), Kecskés et al. (2016), Rinaudi and Giordano (2010), Russo et al. (2015)
Sinorhizobium meliloti	Nitrogen fixation	Alfalfa (Medicago sativa), *Medicago truncatula*	Marx et al. (2016), Mitsui et al. (2004)
Bacillus licheniformis Rt4M10, *Pseudomonas fluorescens* Rt6M10	Acted as stress alleviators against drought by producing abscisic acid (ABA), indole3-acetic acid (IAA)	*Vitis vinifera*	Salomon et al. (2014)
Algae			
Spirulina platensis	Enhanced plant growth and seed germination	*Eruca sativa* *Ameranthus gangeticus* *Brassica rapa ssp. chinensis*	Wuang et al. (2016)
Spirulina platensis	Promoted growth and yield	*Vigna radiata*	Aung (2011)
Chlorella vulgaris	Increased overall plant growth and decreased soluble carbohydrates	*Lactuca sativa*	Faheed and Fattah (2008)
Chlorella vulgaris, Spirulina platensis	Higher amino acid content and mineral composition, increased crop yield	*Vigna radiata*	Dineshkumar et al. (2020)

(continued)

Table 12.1 (continued)

Microbes	Functional attributes	Crop plant	References
Fungi			
Trichoderma harzianum, Fusarium pallidoroseum	Increased biomass production, enhanced water stress resistance by induction of antioxidant enzymes	*Oryza sativa*	Gusain et al. (2014)
Aspergillus niger, Penicillium notatum	Significant increase in soil available phosphorus, increase in phosphorus, nitrogen, protein, oil content and yield of plant	*Arachis hypogaea* (Ground nut)	Jitendra et al. (2011)
Gigaspora rosea	Enhanced weight and nitrogen fixation of nodules	*Glycine max*	Sakamoto et al. (2013)
Penicillium oxalicum	Promoted plant growth and induced systemic resistance against *Sclerospora graminicola* (Downy mildew)	*Pennisetum glaucum (Pearl millet)*	Murali and Amruthesh (2015)
Penicillium chrysogenum	Increase in seed germination and seedling vigor, activated defense-related genes against *Sclerospora graminicola* (Downy mildew)	*Pennisetum glaucum*	Murali et al. (2013)
Penicillium sp. EU0013	Increased plant biomass, reduced disease severity against *Fusarium oxysporum* f. sp. *lycopersici*	*Solanum lycopersicum*	Sartaj et al. (2011)
Penicillium spp. GP15-1	Increased plant biomass, increased root and shoot growth, reduced disease severity against damping-off	*Cucumis sativus* (Cucumber)	Hossain et al. (2014)
Trichoderma harzianum OTPB3	Increased root and shoot growth, increased nutrient uptake, enhanced activity of defense-related enzymes like peroxidase, polyphenol oxidase and superoxide dismutase	*Lycopersicon esculentum*	Chowdappa et al. (2013)

(continued)

Table 12.1 (continued)

Microbes	Functional attributes	Crop plant	References
Trichoderma harzianum	Increased macronutrient uptake, increased seed germination and vigor, induced disease resistance to *Plasmopara halstedii* (downy mildew)	*Helianthus annuus*	Nagaraju et al. (2012)
Trichoderma spp.	Promoted plant growth and activation of defense mechanism against *Xanthomonas euvesicatoria* (bacterial spot) and *Alternaria solani* (early blight)	*Solanum lycopersicum*	Fontenelle et al. (2011)
Cladosporium sphaerospermum	Promoted plant growth, higher amount of Gibberellin production	*Oryza sativa* *Glycine max*	Hamayun et al. (2009)
Alternaria sp. A13	Significantly enhanced dry and fresh plant biomass, increased secondary metabolites accumulation related to medicinal importance	*Salvia miltiorrhiza*	Zhou et al. (2018)
Aspergillus tubingensis, Penicillium sp., Talaromyces funiculosus, Trichoderma harzianum, Trichoderma asperellum	Enhanced seed and plant growth along with induced resistance to *Colletotrichum capsici* causal agent of anthracnose disease	*Capsicum annuum*	Naziya et al. (2020)
Hypovirulent Binucleate *Rhizoctonia spp.* (HBNR)	Induced systemic resistance against anthracnose, enhanced lignin deposition and peroxidase activity	*Cucumis sativus*	Muslim et al. (2019)
Trichoderma viride, Chaetomium spp., T. harzianum, T. asperellum	Induced production of defense-related enzymes, accumulated higher phenolics	*Curcuma longa*	Vinayarani et al. (2019)
Aspergillus niger, Trichoderma koningiopsis	Activated induced systemic resistance and systemic acquired resistance against wilt disease caused by *Rhizoctonia solani*	*Triticum aestivum*	El-Maraghy et al. (2020)

(continued)

Table 12.1 (continued)

Microbes	Functional attributes	Crop plant	References
Acaulospora saccata—A. fragilissima—Scutellospora ovalis—Rhizophagus neocaledonicus—Claroideoglomus etunicatum—Pervetustus simplex	Improved fitness and tolerance to metals, reduced translocation of toxic metals to aerial plant part	*Sorghum vulgare*	Crossay et al. (2020)
Piriformospora indica	Promoted root and shoot growth, expressed drought-related genes and conferred resistance against biotic and abiotic stress	*Brassica campestris* L. ssp. Chinensis	Sun et al. (2010)
Piriformospora indica	Promoted plant growth, seed production and increased oil content	*Helianthus annuus*	Bagde et al. (2011)
Rhizophagus irregularis	Enhanced soil nutrient solubilisation and uptake, induced systemic resistance against pathogen	*Triticum aestivum*	Pérez-de-Luque et al. (2017)
Colletotrichum tropicale	Increased lignin and cellulose content leading to reduced pathogen and herbivore damage	*Theobroma cacao*	Mejía et al. (2014)
Archaea			
Nitrosocosmicus oleophilus MY3	Promoted plant growth and induced systemic resistance against necrotrophic bacterium *Pectobacterium carotovorum* and biotrophic bacterium *Pseudomonas syringae*	*Arabidopsis thaliana*	Song et al. (2019)
Mixed species			
Chara vulgaris—Epiphytic cyanobacteria	Nitrogen fixation	*Oryza sativa*	Ariosa et al. (2004)
Claroideoglomus claroideum—Pseudomonas libanensis	Phytoremediation of metal-contaminated soil, enhanced plant growth	*Helianthus annuus*	Ma et al. (2019)
Anabaena sp.—Trichoderma sp.	Increment in available soil nitrogen	*Chrysanthemum morifolium*	Prasanna et al. (2016)

(continued)

Table 12.1 (continued)

Microbes	Functional attributes	Crop plant	References
	concentration, enhancement in photosynthetic pigments and soil organic carbon concentration		
Anabaena sp.—Azotobacter sp.	Increase in plant fresh and dry weight, higher nitrification potential	*Chrysanthemum morifolium*	Prasanna et al. (2016)
Mesorhizobium ciceri—Trichoderma viride	Accumulation of defense-related enzymes—L-phenylalanine ammonia lyase (PAL), peroxidase (POX) and polyphenol oxidase (PPO). Reduced disease incidence	Chickpea (*Cicer arietinum*)	Das et al. (2017)
Aspergillus sp.—*Enterobacter* sp.	Increased yield and plant growth	*Fragaria ananassa*	Singhalage et al. (2019)

12.6.1 pH

Multiple factors whether biotic or abiotic influence the soil microbial community structure among which the role of pH is significant in the biofilm production. Microbes can tolerate a wide range of pH be it acidic, neutral, or alkaline and therefore, their efficiency in biofilm formation also depends on the optimal pH of the medium they grow on. The microbes with a broad range of pH tolerance soil have a better endurance rate or chance of flourishing compared to microbes having a limited range of pH tolerance. Several microbes belonging to the rhizosphere and rhizoplane of diverse groups of the plant denoted as plant growth-promoting bacteria (PGPB) has been found to be of broader range of pH tolerance (3.0–13.0) and can be utilized for agricultural improvement to enhance plant growth and yield (Kumar et al. 2018). Recent reports showed extreme pH can decrease the growth of certain populations in biofilms (Ansari et al. 2017). Ansari et al. (2017) showed extreme pH negatively affects rhizobia biofilm formation. Similarly, in the case of *B. subtilis*, growth was not limited in extreme pH and temperature that did not result in pellicle formation. Hostacká et al. (2010) reported that biofilm production of *P. aeruginosa* increased to 139–244% when subjected to pH 8.5 from 136 to 164% when pH applied was 7.5 in comparison with pH 5.5. Similarly, in the case of *K. pneumonia*, the production rate increased to 151–319% when pH was 8.5 and production dropped to 113–177% when pH was set to 7.5 compared with pH 5.5. This study also showed that production of *V. cholerae* non-O1 and O1 biofilm had increased to 204–329% upon pH 8.5 from 123 to 316% while at pH 7.5 as compared with the production upon application of pH 5.5. Hostacká et al. (2010) showed the biofilm

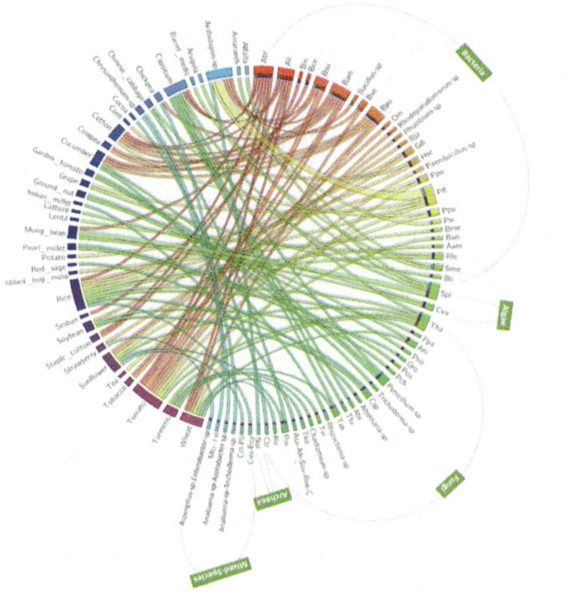

ABBREVIATIONS	SCIENTIFIC NAMES
Abr	Azospirillum brasilense
Ali	Azospirillum lipoferum
Bxi	Bacillus xiamenensis
Bce	Bacillus cereus
Bsu	Bacillus subtilis
Bam	Bacillus amyloliquefaciens
Bacillus-sp	Bacillus sp.
Bve	Bacillus velezensis
Bpu	Bacillus pumilus
Oin	Ochrobactrum intermedium
Rhodopseudomonas-sp	Rhodopseudomonas sp.
Rhizobium-sp	Rhizobium sp.
Bja	Bradyrhizobium japonicum
Gdi	Gluconacetobacter diazotrophicus
Hse	Herbaspirillum seropedicae
Paenibacillus-sp	Paenibacillus sp.
Ppo	Paenibacillus polymyxa
Pfl	Pseudomonas fluorescens
Ppu	Pseudomonas putida
Psi	Pseudomonas simiae
Bme	Burkholderia megapolitana
Bun	Burkholderia unamae
Aam	Azospirillum amazonense
Rle	Rhizobium leguminosarum
Sme	Sinorhizobium meliloti
Bli	Bacillus licheniformis
Spl	Spirulina platensis
Cvu	Chlorella vulgaris
Tha	Trichoderma harzianum
Fpa	Fusarium pallidoroseum
Ani	Aspergillus niger
Pno	Penicillium notatum
Gro	Gigaspora rosea
Pox	Penicillium oxalicum
Pch	Penicillium chrysogenum
Penicillium-sp	Penicillium sp.
Trichoderma-sp	Trichoderma sp.
Csp	Cladosporium sphaerospermum
Alternaria sp	Alternaria sp.
Atu	Aspergillus tubingensis
Tfu	Talaromyces funiculosus
Tas	Trichoderma asperellum
Rhizoctonia-sp	Hypovirulent Binucleate Rhizoctonia sp.
Tvi	Trichoderma viride
Chaetomium-sp	Chaetomium sp.
Tha	T. harzianum
Tas	T. asperellum
Tko	Trichoderma koningiopsis
Asa-Afr-Sov-Rne-Cet-Psi	Acaulospora saccata - A. fragilissima - Scutellospora ovalis - Rhizophagus neocaledonicus - Claroideoglomus etunicatum - Pervetustus simplex
Pin	Piriformospora indica
Rir	Rhizophagus irregularis
Ctr	Colletotrichum tropicale
Nol	Nitrosocosmicus oleophilus
Cvu-Ecy	Chara vulgaris – Epiphytic cyanobacteria
Ccl-Pli	Claroideoglomus claroideum - Pseudomonas libanensis
Anabaena-sp-Trichoderma-sp	Anabaena sp. – Trichoderma sp.
Anabaena-sp-Azotobacter sp	Anabaena sp. – Azotobacter sp.
Mci-Tvi	Mesorhizobium ciceri -Trichoderma viride
Aspergillus-sp-Enterobactor-sp	Aspergillus sp. - Enterobactor sp.

Fig. 12.1 Chord diagram representing the association between the microbiota and crop plants. The different types of microbes have been clustered and labeled in the figure for easy illustration; the table in the figure represents the abbreviations and their full forms. The common names of the crop plants mentioned in the table has been used to avoid graph layout issue

production increased to about 169% when pH was changed to 7.5 from 5.5, the rise of pH to 8.5 from 5.5 caused an typical difference of 229%. Kumar et al. (2019) revealed differential growth in response to pH were observed in the case of 10 bacterial isolates, which contributed to in vitro plant growth-promoting traits such as phosphate solubilization, nitrogen fixation, production of indole acetic acid (IAA) and siderophore. Some bacterial isolates having plant growth-promoting properties showed positive response under a broad range of pH from acidic to alkaline (about 3.5–12.5) which has the potential use for application as a bacterial consortium in agricultural lands to augment crop productivity.

12.6.2 Temperature and Light Intensity

Temperature and light intensity play a major role in several biofilm productions. The fluctuation of temperature and light intensity affect the photosynthetic biofilm production critically. It was reported that a large quantities of specific algal assemblies in biofilms were affected due to fluctuations in seasonal light and temperature observed mainly in wastewater treatments (Congestri et al. 2006). Kebede-Westhead and team in 2004 showed that diatoms were more prevalent in low light (270 μmol/m^2/s) unlike to high light conditions (390 μmol/m^2/s). Similarly, in a work reported by Villanueva and team in, 2011 showed that diatoms contributed to a major biomass fraction under lower temperatures, i.e., 7–11 °C in contrast to higher temperatures, i.e., 11–15 °C. An increase in temperature from 12, 18, and 24 °C increased cyanobacteria proportions in relation to the green algae and diatom populations as cyanobacteria (Schnurr and Allen 2015). This study also showed that application of increased phosphorus and nitrogen loadings, carbon concentrations (inorganic), and also light intensities tend to increase the accumulation of photosynthetic biomass and proportions in comparison to non-photosynthetic biomass.

12.6.3 Oxygen

Biofilm formation gets affected under hypoxic conditions in contrast to normoxia (Ghotaslou and Salahi 2013). Among the other key factors, high levels of oxygen activate the production of polysaccharides (Bayer et al. 1990). It has been concluded that polysaccharide concentration follows the similar trend as concentration of dissolved oxygen (DO) within biofilms (Ahimou et al. 2007; Tay et al. 2001). With the increase in the aeration rates, the polysaccharide content has been seen to increase (Tay et al. 2001).

12.6.4 EPS

The quality and character of the EPS content of a biofilm may differ depending on the environmental factors (Ahimou et al. 2007). The content of EPS is crucial to cohesion and adhesion of biofilms to the surfaces (Ahimou et al. 2007). The material properties of biofilm such as the strength of the biofilm were largely determined by the EPS (Klapper et al. 2002). Multiple environmental factors also contribute positively to EPS production process. Some of such factors are high availability of oxygen (Bayer et al. 1990), nitrogen limitation (Jarman et al. 1978; Mian et al. 1978), low pH (Ryu et al. 2004), nutrient deficiency (Mao et al. 2001), dehydration (Ophir and Gutnick 1994), and low temperature (Junkins and Doyle 1992).

Other than these various other factors such as QS, messenger molecules, salinity and presence of soil enzymes and biocides, also contribute either negatively or positively to the process of biofilm formation.

12.7 Biosafety Concern, Regulatory Mechanisms, and Use-Associated Issues

With the increasing food production demand, the rise in the adoption of approaches to improve crop productivity is evident more than ever before. Application of beneficial microbes although still at a commencing stage, its global commercial market is predicted to reach USD 11.6 billion by 2025 (Marketsandmarkets 2020; Singh 2017) because of its deluging potential to complement chemical-based products (Trivedi et al. 2017). This surge for alternative plant growth promotion products has led to a need for risk assessment from the early stage of product development, i.e., isolation of an organism from its natural habitat, to its release and colonization in the targeted regions.

Despite its rising popularity, the use of microbes in agriculture has multiple challenges regarding its commercialization, quality control, and standardization of products ((Arjjumend and Koutouki 2018). For example, in the European Union context, there are three strict registration procedures (Regulation (EC) No 1107/2009; Commission Regulation (EU) No 283/2013; Commission Regulation (EU) No 284/2013) for plant protection products to reach the commercial market (Lugtenberg 2018). A registration dossier in the EU must submit all the requested information regarding the identification of active microorganisms at its strain level including its biological properties as well as the physical and chemical properties of the product, its safety efficacy, and its safety for humans and the environment. Genotoxicity tests are also to be conducted in vitro and in vivo for mutagenic and carcinogenic studies, respectively. Furthermore, the characterization of the impact on non-targeted organisms especially birds, bees, aquatic organisms, and soil dwellers are observed in laboratory conditions (Pliego et al. 2011). Similarly, in the United States, the federal agencies; EPA (Environmental Protection Agency), and USDA (United States Department of Agriculture) regulate the registration of biostimulants with state-specific mandates and registration compliance (Backer et al. 2018; Du Jardin 2015).

Some other countries regard biocontrol agents and biopesticides as pesticides under pesticide acts (Wilson 2003). An example of which is India where the Insecticides Act, 1968, and Insecticides Rules 1971 to a large extent regulate biopesticides and biocontrol agents (Arjjumend and Koutouki 2018). However, the treatment of biofertilizers in India is different. An amendment in 2006 and 2009 on The Fertilizer (Control) Order 1985 has made provisions for addressing biofertilizers (Arjjumend and Koutouki 2017). Apart from this, India also has the guidelines for regulating the export, import, and release of biological control agents and other beneficial organisms approved by the National Plant Protection Organization (NPPO) in 2006. Besides, The Directorate of Plant Protection, Quarantine & Storage on behalf of NPPO is responsible to carry necessary pest risk analysis of the intended import of biological control agent and beneficial organism, before its release and must maintain appropriate records relevant to the organism. Some other country like Canada, for example, groups products based on their intended benefits. Approval of the product is more straightforward if the claimed label on the product relates to fertility and nutrient aspects whereas if the product is a biocontrol agent then additional verification and scrutiny will be involved in the process (Backer et al. 2018).

Besides these registration and commercialization hurdles, some growth-promoting microbes can never be used as inoculum due to their pathogenic and other detrimental traits over humans, animals, and other plants. *Burkholderia cepacia,* for example, has several beneficial attributes on plant growth and defense. *B. cepacia* can induce nodules in crops (Van et al. 2000), has bioremediation properties (Schlömann et al. 1990), and also hold antagonistic properties against several phytopathogens like *Meloidogyne incognita* (Meyer et al. 2001), *Pythium* sp., *Fusarium* sp. (Meyer et al. 2001). Apart from these aforementioned properties, *B. cepacia* is also involved in infection of cystic fibrosis patients and immunocompromised individuals (Rojas-Rojas et al. 2019; Speert 2001; Tabacchioni et al. 2002). Similarly, several species of bacteria under the genus *Stenotrophomonas* have plant growth-promoting attributes, but *S. maltophilia* a plant-associated bacterium can also be responsible for human infection (Coenye et al. 2004; Ryan et al. 2009). For these reasons and to avoid other associated impacts, proper taxonomic position of the microorganism has to be identified before continuing the studies on plant's performance (Santos et al. 2019).

Microbes in general do not exist independently in their natural environmental conditions. But when it comes to evaluating the efficacy of inoculum for plant-promoting attributes, it is based on laboratory screening of pure culture isolate (Ahmad et al. 2008). Only those with better performance are then tested in pot experiments and finally in field conditions (Barriuso et al. 2008). Sometimes those with negligible in vitro performance with alternative promoting mechanisms could be discarded based on classical screening results (Cardinale et al. 2015). Moreover, the effectiveness of microbial inoculation depends on several factors; its colonization efficiency and biocompatibility with the target cultivar, climatic factors, soil properties, root exudates, spatial location, and other living components (Brimecombe et al. 2007; Burns et al. 2015; Rilling et al. 2019), a major aspect of which is studied negligibly (Rilling et al. 2019; Thijs et al. 2016). For developing an inoculum with effective colonization potential in the rhizospheres, it needs to hold a longer self-life and with the appropriate carrier. In fact, for large-scale adoption of microbial products, they need to prove their consistency over a range of climatic and edaphic conditions.

"*Everything is everywhere, but, the environment selects*" (De Wit and Bouvier 2006) a statement by Baas-Becking (1934). The principle behind the old microbiological tenet can set a base to understand how the sole release of a microbe into the environment does not determine the consequences of its release but, the interaction between the released microbe and the environment does (Yoshikura 2015). This consideration is important to study the aftermath of the microbial release. Although the complexity behind the study of such interaction among several ecological components seems exacting, probing plant–microbiome–host interactions seems to be possible by the use of microfluidics-based techniques (Trivedi et al. 2017). This technique can be used to quantify chemotactic processes and responses in high spatiotemporal resolutions by the use of a range of optical microscopies ((Massalha et al. 2017; Stanley et al. 2014; Stanley and van der Heijden 2017). Besides this technique, another novel promising method for tracking microbes includes the use of piezoelectric quartz microbalances, where the mass of crystal proportionating to the baseline oscillation can detect target strain (Bunroddith et al. 2018). Such techniques along with omics approach (proteomics, metagenomics, and metatranscriptomics) and next-generation sequencing (NGS) can be useful not only to screen the antagonist strain before its mass production but also to monitor the impact on indigenous microbial communities under field conditions after release ((Lombard et al. 2006; Rilling et al. 2019; Yi and Chun 2015).

The rise in the application of biotechnology in agriculture has put light on the use of genetically modified microorganisms (GMM). Although preferentially the use of genetically well-studied microorganisms is chosen for modifications the procedure of GMM use is entangled with several issues, issues that relate to the environment, economics, regulation, and ownership ethics of the product (Nelson 2001). GMM application, no doubt, would raise questions regarding the risk of horizontal gene transfer, the possibility of changing the abiotic environment, and its potential to outcompete/antagonize local microbiota ((van Elsas et al. 2015). The release of a GMM, unlike genetically modified plants, is almost an irreversible process. GMM release is expected to have a functional and structural shift where the indigenous microbes' community has low resiliency, but in a robust community, the release of GMM would have a negligible impact on the dominant microbial community or microbiological soil functions (Tebbe and Miethling-Graff 2006; Tebbe 2015). Therefore, while carrying environmental risk assessment considerations on the indigenous microbial community's resiliency is crucial.

12.8 Keyword Mining

With the exponential increase in the literature data, the requirement to extract meaningful information automatically has been indispensable as handling these massive data can be really cumbersome. Though the data analytic approaches are in the developing stage, there are still a good number of tools and methods to mine text data to understand the basic nature such as topic analysis, keyword frequencies, and much other similar information. The mining of published articles to evaluate the research productivity using different bibliometric indicators as well as understanding the structure with the help of topology of tokens (which are basically pre-processed important words in the articles) has obtained an upsurge interest in the last decade. The keyword-based text-mining frameworks are very commonly used ones with simple analysis methods which can be easily applied to most type of articles and also proven to be of great help to develop systematic reviews (Li et al. 2016).

To have an idea about the publication patterns and research outputs, we performed a basic scientometrics analysis. A search on PubMed with the query term < "biofilm" AND "agriculture" > retrieved a total of 1018 records. In order to get a clearer view of the subject of the PubMed reports, a simple Venn diagram was

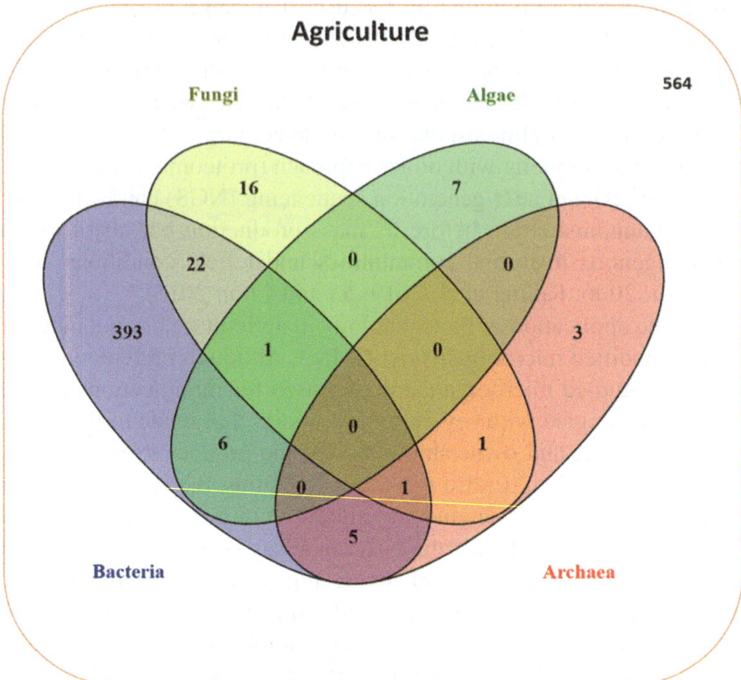

Fig. 12.2 Venn diagram based on keyword mining

created with the help of Venny (Oliveros 2015) based on the output of this scientometrics analysis (Fig. 12.2). The Venn diagram is based on the mention of the biofilm-producing microbial taxa that have been found in the PubMed reports based on the simple keyword searches. It can be seen that out of the 1018 reports, the highest mentioned category was bacteria followed by fungi, algae, and archaea.

The idea was to obtain keywords by mining the titles and abstracts of the related articles on PubMed and infer into the specific areas in agriculture where biofilms are being used. The PubMed identifiers (PMID) were downloaded and parsed through an R script to create a word cloud for the frequent keywords. The PMIDs were used as input and with the help of the RISmed package, the titles and abstracts were retrieved (Kovalchik 2017). The widely used text-mining package tm was used to create a corpus out of the titles and abstracts of all the records (Feinerer et al. 2008). The corpus was pre-processed to get rid of noises such as stop-words, punctuations, and extra spaces. A term "document matrix" was created using the same package following a record of the words and its corresponding frequencies. Finally, a word "cloud" was created keeping the frequency of the words as a metric reflected with the font size directly proportional to the frequencies (Fig. 12.3).

The word "cloud"was restricted to the most frequent 500 words to get a clear picture of the over-represented words in the articles. There are still some unnecessary words which are trivial English words which can be ignored while trying to note the

Fig. 12.3 Word Cloud based on keyword mining

research areas. Although some of the words such as "colonization," "surface," and "extracellular" are frequency wise less in number but can be seen due to the relevance to the subject of the study conducted. As expected, the most frequent word was "biofilm" with a total frequency of 2774. A redundant word which is "biofilms" again came in the third position with 1044 occurring. The second frequent word was "formation" which must be due to its frequent association with the word "biofilm." The places from fourth to eighth contained words "bacterial," "strains," "study," "bacteria," "strain," respectively, suggesting that the bacterial biofilms are the most frequent ones that have been used in the field of agriculture. This purely coincides with the fact that the bacterial biofilm reports are abundant as mentioned above in the Venn diagram.

12.9 Conclusion

The current scenario with respect to concurrent events of the gradual loss of agricultural lands and increasing food demand to fulfill the need for the population outburst is paving the way to look for sustainable alternatives. Biofilms produced by agriculturally important microbes hold a very important space in promoting sustainable agricultural practices. The current research scenes in this old yet budding concept in terms of applicability shows much more focus on bacterial biofilms whereas other biofilm-producing microbial taxa should also get equal attention. It surely will help in the extraction of novel information and thus strengthen the knowledge base. There is a need to increase the funding by the government in this sector and more efforts to popularize the applicability of the beneficial biofilm-producing microbes to uplift the crop and soil health keeping the probable harmful effects in check.

References

Aartsma Y, Bianchi FJJA, van der Werf W, Poelman EH, Dicke M (2017) Herbivore-induced plant volatiles and tritrophic interactions across spatial scales. New Phytol 216:1054–1063. https://doi.org/10.1111/nph.14475

Abdalla OA, Bibi S, Zhang S (2017) Application of plant growth-promoting rhizobacteria to control *Papaya ringspot virus* and tomato chlorotic spot virus. Arch Phytopathol Pflanzenschutz 50:584–597. https://doi.org/10.1080/03235408.2017.1352248

Ahimou F, Semmens MJ, Haugstad G, Novak PJ (2007) Effect of protein, polysaccharide, and oxygen concentration profiles on biofilm cohesiveness. Appl Env Microbiol 73(9):2905–2910. https://doi.org/10.1128/AEM.02420-06

Ahmad F, Ahmad I, Khan M (2008) Screening of free-living rhizospheric bacteria for their multiple plant growth promoting activities. Microbiol Res 163(2):173–181

Ahmad B, Raina A, Khan S (2019) Impact of biotic and abiotic stresses on plants, and their responses. In: Disease resistance in crop plants, pp 1–19. https://doi.org/10.1007/978-3-030-20728-1_1

Ahn IP, Park K, Kim CH (2002) Rhizobacteria-induced resistance perturbs viral disease progress and triggers defense-related gene expression. Mol Cells 13:302–308

Aktar W, Sengupta D, Chowdhury A (2009) Impact of pesticides use in agriculture: their benefits and hazards. Interdiscip Toxicol 2:1–12. https://doi.org/10.2478/v10102-009-0001-7

Alami Y, Achouak W, Marol C, Heulin T (2000) Rhizosphere soil aggregation and plant growth promotion of sunflowers by an exopolysaccharide-producing *Rhizobium* sp. strain isolated from sunflower roots. Appl Environ Microbiol 66:3393–3398. https://doi.org/10.1128/AEM.66.8.3393-3398.2000

Al-Nahidh SI, Gomah AHM (1991) Response of wheat to dual inoculation with VA-mycorrhiza and azospirillum, fertilized with NPK and irrigated with sewage effluent. Arid Soil Res Rehabil 5:83–96. https://doi.org/10.1080/15324989109381269

Ansari FA, Jafri H, Ahmad I, Abulreesh H (2017) Factors affecting biofilm formation in in vitro and in the rhizosphere. In: Biofilms in plant and soil health, pp 275–290. https://doi.org/10.1002/9781119246329

Ariosa Y, Quesada A, Aburto J, Carrasco D, Carreres R, Leganés F, Valiente EF (2004) Epiphytic cyanobacteria on Chara vulgaris are the main contributors to N2 fixation in rice fields. Appl Environ Microbiol 70:5391–5397. https://doi.org/10.1128/AEM.70.9.5391-5397.2004

Arjjumend H, Koutouki K (2017) Ukrainian legislation for safeguarding the agroecosystems and environmental health: the challenges ahead. In Kharkiv international legal forum on law and problems of sustainable development in globalized world, Kharkiv, Ukraine, pp 3–17
Arjjumend H, Koutouki K (2018) Science of biopesticides and critical analysis of Indian legal frameworks regulating biocontrol agents. Int J Environ 11(3):579–587
Ashraf M, Ahmad MSA, Öztürk M, Aksoy A (2012) Crop improvement through different means: challenges and prospects. In: Ashraf M, Öztürk M, Ahmad M, Aksoy A (eds) Crop production for agricultural improvement. Springer, Dordrecht, pp 1–15
Aung KLN (2011) Effect of *Spirulina* biofertilizer suspension on growth and yield of *Vigna radiata* (L.) Wilczek. Univ Res J 4(1):351–363
Baas-Becking L (1934) Geobiologie of inleiding tot de milieukunde (Geobiology as introduction for environmental research), vol. 18/19. WP van Stockum & Zoon: The Hague, p 263
Backer R, Rokem JS, Ilangumaran G, Lamont J, Praslickova D, Ricci E, Subramanian S, Smith DL (2018) Plant growth-promoting rhizobacteria: context, mechanisms of action, and roadmap to commercialization of biostimulants for sustainable agriculture. Front Plant Sci 9:1473
Bagde US, Prasad R, Varma A (2011) Influence of culture filtrate of *Piriformospora indica* on growth and yield of seed oil in *Helianthus annus*. Symbiosis 53:83. https://doi.org/10.1007/s13199-011-0114-6
Barriuso J, Solano BR, Lucas JA, Lobo AP, García-Villaraco A, Mañero FJG (2008) Ecology, genetic diversity and screening strategies of plant growth promoting rhizobacteria (PGPR). J Plant Nutr 4:1–17
Bashan Y (1998) Azospirillum plant growth-promoting strains are nonpathogenic on tomato, pepper, cotton, and wheat. Can J Microbiol 44(2):168–174. https://doi.org/10.1139/w97-136
Bayer Crop Science (2008) “We need a second green revolution”—Safeguarding food for a growing world population
Bayer ASF, Eftekhar JT, Nast CC, Speert DP (1990) Oxygen-dependent up-regulation of mucoid exopolysaccharide (alginate) production in *Pseudomonas aeruginosa*. Infect Immunol 58:1344–1349
Benizri E, Baudoin E, Guckert A (2001) Root colonization by inoculated plant growth-promoting rhizobacteria. Biocontrol Sci Tech 11:557–574. https://doi.org/10.1080/09583150120076120
Beris D, Vassilakos N (2020) Plant beneficial microbes: do they have a role as antiviral agents in agriculture? INC
Bezzate S, Aymerich S, Chambert R, Czarnes S, Berge O, Heulin T (2000) Disruption of the *Paenibacillus polymyxa* levansucrase gene impairs its ability to aggregate soil in the wheat rhizosphere. Environ Microbiol 2:333–342. https://doi.org/10.1046/j.1462-2920.2000.00114.x
Bourguet D, Guillemaud T (2016) Impact of fertilizers and pesticides on soil microflora in agriculture
Brimecombe MJ, De Leij F, Lynch JM (2007) Rhizodeposition and microbial populations. The rhizosphere: biochemistry and organic substances at the soil-plant interface. CRC Press, Taylor & Francis Group, Boca Raton, pp 73–109
Bunroddith K, Viseshakul N, Chansiri K, Lieberzeit P (2018) QCM-based rapid detection of PCR amplification products of Ehrlichia canis. Anal Chim Acta 1001:106–111
Burns JH, Anacker BL, Strauss SY, Burke DJ (2015) Soil microbial community variation correlates most strongly with plant species identity, followed by soil chemistry, spatial location and plant genus. AoB Plants 7:plv030
Canellas LP, Balmori DM, Médici LO, Aguiar NO, Campostrini E, Rosa RCC, Façanha AR, Olivares FL (2013) A combination of humic substances and *Herbaspirillum seropedicae* inoculation enhances the growth of maize (Zea mays L.). Plant Soil 366:119–132. https://doi.org/10.1007/s11104-012-1382-5
Cardinale M, Ratering S, Suarez C, Montoya AMZ, Geissler-Plaum R, Schnell S (2015) Paradox of plant growth promotion potential of rhizobacteria and their actual promotion effect on growth of barley (Hordeum vulgare L.) under salt stress. Microbiol Res 181:22–32

Carrow RN, Duncan RR (1998) Salt-affected Turfgrass sites: assessment and management. Ann Arbor Press, Chelsea

Chen L, Heng J, Qin S, Bian K (2018) A comprehensive understanding of the biocontrol potential of *Bacillus velezensis* LM2303 against Fusarium head blight. PLoS One 13(6):e0198560. https://doi.org/10.1371/journal.pone.0198560

Chenu C, Cosentino D (2011) Microbial regulation of soil structural dynamics. In: Ritz K, Young I (eds) The architecture and biology of soils: life in inner space, 1st edn. CABI, Wallingford, pp 37–70

Chowdappa P, Mohan Kumar SP, Jyothi Lakshmi M, Upreti KK (2013) Growth stimulation and induction of systemic resistance in tomato against early and late blight by *Bacillus subtilis* OTPB1 or Trichoderma harzianum OTPB3. Biol Control 65:109–117. https://doi.org/10.1016/j.biocontrol.2012.11.009

Ciais P, Reichstein M, Viovy N, Granier A, Ogée J, Allard V, Aubinet M, Buchmann N, Bernhofer C, Carrara A, Chevallier F, De Noblet N, Friend AD, Friedlingstein P, Grünwald T, Heinesch B, Keronen P, Knohl A, Krinner G, Loustau D, Manca G, Matteucci G, Miglietta F, Ourcival JM, Papale D, Pilegaard K, Rambal S, Seufert G, Soussana JF, Sanz MJ, Schulze ED, Vesala T, Valentini R (2005) Europe-wide reduction in primary productivity caused by the heat and drought in 2003. Nature 437:529–533. https://doi.org/10.1038/nature03972

Cleland J (2013) World population growth; past, present and future. Environ Resour Econ 55:543–554. https://doi.org/10.1007/s10640-013-9675-6

Coenye T, Vanlaere E, Falsen E, Vandamme P (2004) Stenotrophomonas africana Drancourt et al. 1997 is a later synonym of Stenotrophomonas maltophilia (Hugh 1981) Palleroni and Bradbury 1993. Int J Syst Evol 54(4):1235–1237

Compant S, Clément C, Sessitsch A (2010) Plant growth-promoting bacteria in the rhizo- and endosphere of plants: their role, colonization, mechanisms involved and prospects for utilization. Soil Biol Biochem 42:669–678. https://doi.org/10.1016/j.soilbio.2009.11.024

Congestri R, Francesca DP, Philippis RD, Buttino I, Paradossi G, Albertano P (2006) Seasonal succession of phototrophic biofilms in an Italian wastewater treatment plant: biovolume, spatial structure and exopolysaccharides. Aquat Microb Ecol 45:301–312. https://doi.org/10.3354/ame045301

Costa OYA, Raaijmakers JM, Kuramae EE (2018) Microbial extracellular polymeric substances: ecological function and impact on soil aggregation. Front Microbiol 9:1–14. https://doi.org/10.3389/fmicb.2018.01636

Crossay T, Cavaloc Y, Majorel C, Redecker D, Medevielle V, Amir H (2020) Combinations of different arbuscular mycorrhizal fungi improve fitness and metal tolerance of sorghum in ultramafic soil. Rhizosphere 14:1002042. https://doi.org/10.1016/j.rhisph.2020.100204

Danhorn T, Fuqua C (2007) Biofilm formation by plant-associated bacteria. Annu Rev Microbiol 61:401–422. https://doi.org/10.1146/annurev.micro.61.080706.093316

Das K, Rajawat MVS, Saxena AK, Prasanna R (2017) Development of *Mesorhizobium ciceri*-based biofilms and analyses of their antifungal and plant growth promoting activity in chickpea challenged by Fusarium Wilt. Indian J Microbiol 57:48–59. https://doi.org/10.1007/s12088-016-0610-8

De Wit R, Bouvier T (2006) Everything is everywhere, but, the environment selects'; what did Baas Becking and Beijerinck really say? Environ Microbiol 8(4):755–758

Deka P, Goswami G, Das P, Gautom T, Chowdhury N, Boro RC, Barooah M (2019) Bacterial exopolysaccharide promotes acid tolerance in *Bacillus amyloliquefaciens* and improves soil aggregation. Mol Biol Rep 46:1079–1091. https://doi.org/10.1007/s11033-018-4566-0

Delaporte-Quintana P, Grillo-Puertas M, Lovaisa NC, Teixeira KR, Rapisarda VA, Pedraza RO (2017) Contribution of *Gluconacetobacter diazotrophicus* to phosphorus nutrition in strawberry plants. Plant Soil 419:335–347. https://doi.org/10.1007/s11104-017-3349-z

Din BU, Rafique M, Javed MT, Kamran MA, Mehmood S, Khan M, Sultan T, Munis FH, Chaudhary HJ (2020) Assisted phytoremediation of chromium spiked soils by *Sesbania Sesban* in association with *Bacillus xiamenensis* PM14: a biochemical analysis. Plant Physiol Biochem 146:249–258. https://doi.org/10.1016/j.plaphy.2019.11.010

Dineshkumar R, Duraimurugan M, Sharmiladevi N, Lakshmi LP, Rasheeq AA, Arumugam A, Sampathkumar P (2020) Microalgal liquid biofertilizer and biostimulant effect on green gram (*Vigna radiata* L) an experimental cultivation. Biomass Conv Biorefin. https://doi.org/10.1007/s13399-020-00857-0

Du Jardin P (2015) Plant biostimulants: definition, concept, main categories and regulation. Sci Hortic 196:3–14

Egamberdieva D, Jabborova D, Wirth SJ, Alam P, Alyemeni MN, Ahmad P (2018) Interactive effects of nutrients and *Bradyrhizobium japonicum* on the growth and root architecture of soybean (*Glycine max* L.). Front Microbiol 9:1000. https://doi.org/10.3389/fmicb.2018.01000

Ekwurzel B, Boneham J, Dalton MW, Heede R, Mera RJ, Allen MR, Frumhoff PC (2017) The rise in global atmospheric CO2, surface temperature, and sea level from emissions traced to major carbon producers. Clim Chang 144:579–590. https://doi.org/10.1007/s10584-017-1978-0

El-Maraghy SS, Tohamy TA, Hussein KA (2020) Role of plant-growth promoting fungi (PGPF) in defensive genes expression of *Triticum aestivum* against wilt disease. Rhizosphere 15:100223. https://doi.org/10.1016/j.rhisph.2020.100223

Faheed FA, Fattah ZA (2008) Effect of *Chlorella vulgaris* as bio-fertilizer on growth parameters and metabolic aspects of lettuce plant. J Agri Soc Sci 4:165–169

Faisal M, Hasnain S (2005) Bacterial Cr(VI) reduction concurrently improves sunflower (*Helianthus annuus* L.) growth. Biotechnol Lett 27:943–947. https://doi.org/10.1007/s10529-005-7188-2

FAO (2009) How to feed the world in 2050. http://www.fao.org/fileadmin/templates/wsfs/docs/expert_paper/How_to_Feed_the_World_in_2050.pdf

Feinerer I, Hornik K, Meyer D (2008) Text mining infrastructure in R. J Stat Softw 25(5):1–54

Filgueiras L, Silva R, Almeida I, Vidal M, Baldani JI, Meneses CHSG (2019) *Gluconacetobacter diazotrophicus* mitigates drought stress in *Oryza sativa* L. Plant Soil 451:57–73. https://doi.org/10.1007/s11104-019-04163-1

Flemming HC, Wingender J, Szewzyk U, Steinberg P, Rice SA, Kjelleberg S (2016) Biofilms: an emergent form of bacterial life. Nat Rev Microbiol 14:563–575. https://doi.org/10.1038/nrmicro.2016.94

Fontenelle A, Guzzo S, Lucon C, Harakava R (2011) Growth promotion and induction of resistance in tomato plant against *Xanthomonas euvesicatoria* and Alternaria solani by Trichoderma spp. Crop Prot 30:1492–1500. https://doi.org/10.1016/j.cropro.2011.07.019

Francini A, Sebastiani L (2019) Abiotic stress effects on performance of horticultural crops. Horticulturae 5. https://doi.org/10.3390/horticulturae5040067

Fukuhara H, Minakawa Y, Akao S, Minamisawa K (1994) The involvement of indole-3-acetic acid produced by *Bradyrhizobium elkanii* in nodule formation. Plant Cell Physiol 35:1261–1265. https://doi.org/10.1093/oxfordjournals.pcp.a078722

Garcia-Lopez AM, Delgado A (2016) Effect of *Bacillus subtilis* on phosphorus uptake by cucumber as affected by iron oxides and the solubility of the phosphorus source. Agric Food Sci 25 (3):216–224. https://doi.org/10.23986/afsci.56862

Ghassemi F, Jakeman AJ, Nix HA (1995) Salinisation of land and water resources: human causes, extent, management and case studies. In: Salinisation of land and water resources: human causes, extent, management and case studies. CAB International, Wallingford

Ghotaslou R, Salahi B (2013) Effects of oxygen on in-vitro biofilm formation and antimicrobial resistance of Pseudomonas aeruginosae. Pharm Sci 19(3):96–99

Glazebrook J (2005) Contrasting mechanisms of defense against biotrophic and necrotrophic pathogens. Annu Rev Phytopathol 43:205–227. https://doi.org/10.1146/annurev.phyto.43.040204.135923

Glick BR (2012) Plant growth-promoting bacteria: mechanisms and applications. Scientifica (Cairo) 2012:963401. https://doi.org/10.6064/2012/963401

Glick BR, Todorovic B, Czarny J, Cheng Z, Duan J, McConkey B (2007) Promotion of plant growth by bacterial ACC deaminase. CRC Crit Rev Plant Sci 26:227–242. https://doi.org/10.1080/07352680701572966

Gouzou L, Burtin G, Philippy R, Bartoli F, Heulin T (1993) Effect of inoculation with *Bacillus polymyxa* on soil aggregation in the wheat rhizosphere: preliminary examination. Geoderma 56:479–491. https://doi.org/10.1016/0016-7061(93)90128-8

Guo Q, Li Y, Lou Y, Shi M, Jiang Y, Zhou J, Sun Y, Xue Q, Lai H (2019) *Bacillus amyloliquefaciens* Ba13 induces plant systemic resistance and improves rhizosphere microecology against tomato yellow leaf curl virus disease. Appl Soil Ecol 137:154–166. https://doi.org/10.1016/j.apsoil.2019.01.015

Gusain YS, Singh US, Sharma AK (2014) Enhance activity of stress related enzymes in rice (*Oryza sativa* L.) induced by plant growth promoting fungi under drought stress. Afr J Agric Res 9:1430–1434. https://doi.org/10.5897/AJAR2014.8575

Hafeez FY, Shah NH, Malik KA (2000) Field evaluation of lentil cultivars inoculated with *Rhizobium leguminosarum* bv. Viciae strains for nitrogen fixation using nitrogen-15 isotope dilution. Biol Fertil Soils 31:65–69. https://doi.org/10.1007/s003740050625

Hahn L, de Sá ELS, Filho BDO, Machado RG, Damasceno RG, Giongo A (2016) Rhizobial inoculation, alone or coinoculated with *Azospirillum brasilense*, promotes growth of wetland rice. Rev Bras Ciênc Solo 40:e0160006. https://doi.org/10.1590/18069657rbcs20160006

Hallmann J (2001) Plant interactions with endophytic bacteria. In: Jeger MJ, Spence NJ (eds) Biotic interactions in plant-pathogen associations. CAB International, Wallingford, pp 87–119

Hamayun M, Khan SA, Ahmad N, Tang DS, Kang SM, Na CI, Sohn EY, Hwang YH, Shin DH, Lee BH, Kim JG, Lee IJ (2009) *Cladosporium sphaerospermum* as a new plant growthpromoting endophyte from the roots of *Glycine max* (L.) Merr. World J Microbiol Biotechnol 25(4):627–632. https://doi.org/10.1007/s11274-009-9982-9

Hansen M, Kregelund L, Nybroe O, Sørensen J (1997) Early colonization of barley roots by *Pseudomonas fluorescens* studied by immunofluorescence technique and confocal laser scanning microscopy. FEMS Microbiol Ecol 23:353–360. https://doi.org/10.1016/S0168-6496(97)00037-8

Hart JJ, Welch RM, Norvell WA, Kochian LV (2006) Characterization of cadmium uptake, translocation and storage in near-isogenic lines of durum wheat that differ in grain cadmium concentration. New Phytol 172:261–271. https://doi.org/10.1111/j.1469-8137.2006.01832.x

Harun-Or-Rashid M, Chung YR (2017) Induction of systemic resistance against insect herbivores in plants by beneficial soil microbes. Front Plant Sci 8:1816. https://doi.org/10.3389/fpls.2017.01816

Harun-Or-Rashid M, Khan A, Hossain MT, Chung YR (2017) Induction of systemic resistance against aphids by endophytic *Bacillus velezensis* YC7010 via expressing *PHYTOALEXIN DEFICIENT4* in *Arabidopsis*. Front Plant Sci 8:211. https://doi.org/10.3389/fpls.2017.00211

He M, He CQ, Ding NZ (2018) Abiotic stresses: general defenses of land plants and chances for engineering multistress tolerance. Front Plant Sci 871:1–18. https://doi.org/10.3389/fpls.2018.01771

Hossain MM, Sultana F, Miyazawa M, Hyakumachi M (2014) Plant growth-promoting fungus *Penicillium* spp. GP15-1 enhances growth and confers protection against damping-off and anthracnose in the cucumber. J Oleo Sci 63(4):391–400. https://doi.org/10.5650/jos.ess13143

Hostacká A, Ciznár I, Stefkovicová M (2010) Temperature and pH affect the production of bacterial biofilm. Folia Microbiol (Praha) 55(1):75–78. https://doi.org/10.1007/s12223-010-0012-y

Islam F, Yasmeen T, Arif MS, Ali S, Ali B, Hameed S, Zhou W (2016) Plant growth promoting bacteria confer salt tolerance in *Vigna radiata* by up-regulating antioxidant defense and biological soil fertility. Plant Growth Regul 80:23–36. https://doi.org/10.1007/s10725-015-0142-y

Jarman TR, Deavin L, Slocombe S, Righelato RC (1978) Investigation of the effect of environmental conditions on the rate of exopolysaccharide synthesis in Azotobacter vinelandii. J Gen Microbiol 107:59–64

Jetiyanon K, Kloepper JW (2002) Mixtures of plant growth-promoting rhizobacteria for induction of systemic resistance against multiple plant diseases. Biol Control 24:285–291. https://doi.org/10.1016/S1049-9644(02)00022-1

Jia M, Chen L, Xin HL, Zheng CJ, Rahman K, Han T, Qin LP (2016) A friendly relationship between endophytic fungi and medicinal plants: a systematic review. Front Microbiol 7:906. https://doi.org/10.3389/fmicb.2016.00906

Jitendra M, Kiran S, Vaibhavi J (2011) Effect of phosphate solubilizing fungi on growth and nutrient uptake of ground nut (*Arachis hypogaea*) plants. Adv Biores 2(2):110–113

Junkins AD, Doyle MP (1992) Demonstration of exopolysaccharide production by enterohemorrhagic Escherichia coli. Curr Microbiol 25:9–17

Kaur A, Capalash N, Sharma P (2018) Quorum sensing in thermophiles: prevalence of autoinducer-2 system. BMC Microbiol 18:1–16. https://doi.org/10.1186/s12866-018-1204-x

Kebede-Westhead E, Pizarro C, Mulbry WW (2004) Treatment of dairy manure effluent using freshwater algae: elemental composition of algal biomass at different manure loading rates. J Agric Food Chem 52:7293–7296

Kecskés M, Choudhury A, Casteriano A, Deaker R, Roughley R, Lewin L, Ford R, Kennedy I (2016) Effects of bacterial inoculant biofertilizers on growth, yield and nutrition of rice in Australia. J Plant Nutr 39(3):377–388. https://doi.org/10.1080/01904167.2015.1016172

Khan PSSV, Nagamallaiah G V, Rao MD, Sergeant K, Hausman JF (2014) Abiotic stress tolerance in plants: insights from proteomics, 2nd edn. Elsevier

Klapper I, Rupp CJ, Cargo R, Purevdorj B, Stoodley P (2002) Viscoelastic fluid description of bacterial biofilm material properties

Knief C, Delmotte N, Vorholt JA (2011) Bacterial adaptation to life in association with plants - a proteomic perspective from culture to in situ conditions. Proteomics 11:3086–3105. https://doi.org/10.1002/pmic.201000818

Koh RH, Song HG (2007) Effects of application of *Rhodopseudomonas* sp. on seed germination and growth of tomato under axenic conditions. J Microbiol Biotechnol 17(11):1805–1810

Kong P, Lee BWK, Zhou ZS, Hong C (2010a) Zoosporic plant pathogens produce bacterial autoinducer-2 that affects *Vibrio harveyi* quorum sensing. FEMS Microbiol Lett 303:55–60. https://doi.org/10.1111/j.1574-6968.2009.01861.x

Kong P, Tyler BM, Richardson PA, Lee BW, Zhou ZS, Hong C (2010b) Zoospore interspecific signaling promotes plant infection by *Phytophthora*. BMC Microbiol 10:313. https://doi.org/10.1186/1471-2180-10-313

Kovalchik S (2017) RISmed: download content from NCBI databases. R package version 2.1.7. https://CRAN.R-project.org/package=RISmed

Krzywinski M, Schein J, Birol I, Connors J, Gascoyne R, Horsman D, Jones SJ, Marra MA (2009) Circos: an information aesthetic for comparative genomics. Genome Genome Res 19 (9):1639–1645. https://doi.org/10.1101/gr.092759.109

Kumar A, Singh VK, Tripathi V, Singh PP, Singh AK (2018) Plant growth-promoting rhizobacteria (PGPR): perspective in agriculture under biotic and abiotic stress. In: Prasad R, Gill SS, Tuteja N (eds) New and future developments in microbial biotechnology and bioengineering. Elsevier, Amsterdam, pp 333–342

Kumar A, Kumari M, Swarupa P, Shireen S (2019) Characterization of pH dependent growth response of agriculturally important microbes for development of plant growth promoting bacterial consortium. J Pure Appl Microbiol 13(2):1053–1061. https://doi.org/10.22207/JPAM.13.2.43

Laganenka L, Sander T, Lagonenko A, Chen Y, Link H, Sourjik V (2019) Quorum sensing and metabolic state of the host control lysogeny-lysis switch of bacteriophage T1. MBio 10:3–8. https://doi.org/10.1128/mBio.01884-19

Lanna-Filho R, Souza RM, Alves E (2017) Induced resistance in tomato plants promoted by two endophytic bacilli against bacterial speck. Trop Plant Pathol 42:96–108. https://doi.org/10.1007/s40858-017-0141-9

Lee KH, Koh RH, Song HG (2008) Enhancement of growth and yield of tomato by Rhodopseudomonas sp. under greenhouse conditions. J Microbiol 46:641–646. https://doi.org/10.1007/s12275-008-0159-2

Leggett M, Diaz-Zorita M, Koivunen M, Bowman R, Pesek R, Stevenson C, Leister T (2017) Soybean response to inoculation with *Bradyrhizobium japonicum* in the United States and Argentina. Agron J 109:1031–1038. https://doi.org/10.2134/agronj2016.04.0214

Li D, Wang Z, Wang L et al (2016) A text-mining framework for supporting systematic reviews. Am J Inf Manag 1(1):1–9

Liu H, Carvalhais LC, Crawford M, Singh E, Dennis PG, Pieterse CMJ, Schenk PM (2017) Inner plant values: diversity, colonization and benefits from endophytic bacteria. Front Microbiol 8:2552. https://doi.org/10.3389/fmicb.2017.02552

Lombard N, Bailly J, Marmeisse R, Simonet P (2006) La métagénomique des communautés microbiennes

Lopes LD, de Cássia Pereira e Silva M, Andreote FD (2016) Bacterial abilities and adaptation toward the rhizosphere colonization. Front Microbiol 7:1341. https://doi.org/10.3389/fmicb.2016.01341

Lugtenberg B (2018) Putting concerns for caution into perspective: microbial plant protection products are safe to use in agriculture. J Plant Dis Protect 125(2):127–129

Ma Y, Rajkumar M, Freitas H (2009) Inoculation of plant growth promoting bacterium *Achromobacter xylosoxidans* strain Ax10 for the improvement of copper phytoextraction by Brassica juncea. J Environ Manag 90:831–837. https://doi.org/10.1016/j.jenvman.2008.01.014

Ma Y, Rajkumar M, Oliveira RS, Zhang C, Freitas H (2019) Potential of plant beneficial bacteria and arbuscular mycorrhizal fungi in phytoremediation of metal-contaminated saline soils. J Hazard Mater 379:120813. https://doi.org/10.1016/j.jhazmat.2019.120813

Malviya J, Singh K, Joshi V (2011) Effect of phosphate solubilizing fungi on growth and nutrient uptake of ground nut (Arachis hypogaea) plants. J Appl Sci Res 2:110–113

Mao Y, Doyle MP, Chen J (2001) Insertion mutagenesis of wca reduces acid and heat tolerance of enterohemorrhagic *Escherichia coli* O157:H7. J Bacteriol 183:3811–3815

Marketsandmarkets (2020) Agricultural microbials market worth $11.6 billion by 2025. https://www.marketsandmarkets.com/PressReleases/agricultural-microbial.asp

Marx H, Minogue CE, Jayaraman D, Richards AL, Kwiecien NW, Siahpirani AF, Rajasekar S, Maeda J, Garcia K, Valle-Echevarria ARD (2016) A proteomic atlas of the legume *Medicago truncatula* and its nitrogen-fixing endosymbiont *Sinorhizobium meliloti*. Nat Biotechnol 34 (11):1198–1205. https://doi.org/10.1038/nbt.3681

Massalha H, Korenblum E, Malitsky S, Shapiro OH, Aharoni A (2017) Live imaging of root–bacteria interactions in a microfluidics setup. PNAS 114(17):4549–4554

Mayak S, Tirosh T, Glick BR (2004) Plant growth-promoting bacteria confer resistance in tomato plants to salt stress. Plant Physiol Biochem 42(6):565–572. https://doi.org/10.1016/j.plaphy.2004.05.009

McDonald BA, Stukenbrock EH (2016) Rapid emergence of pathogens in agro-ecosystems: global threats to agricultural sustainability and food security. Philos Trans R Soc B Biol Sci 371:20160026. https://doi.org/10.1098/rstb.2016.0026

Mejía LC, Herre EA, Sparks JP, Winter K, García MN, Van Bael SA, Stitt J, Shi Z, Zhang Y, Guiltinan MJ, Maximova SN (2014) Pervasive effects of a dominant foliar endophytic fungus on host genetic and phenotypic expression in a tropical tree. Front Microbiol 5:479. https://doi.org/10.3389/fmicb.2014.00479

Meyer SL, Roberts DP, Chitwood DJ, Carta LK, Lumsden RD, Mao W (2001) Application of Burkholderia cepacia and Trichoderma virens, alone and in combinations, against Meloidogyne incognita on bell pepper. Nematropica 31:75–86

Mian FA, Jarman RT, Righelato RC (1978) Biosynthesis of exopolysaccharide by Pseudomonas aeruginosa. J Bacteriol 134:418–422

Mitsui H, Sato T, Sato Y, Ito N, Minamisawa K (2004) *Sinorhizobium meliloti* RpoH$_1$ is required for effective nitrogen-fixing symbiosis with alfalfa. Mol Gen Genomics 271(4):416–425. https://doi.org/10.1007/s00438-004-0992-x

Moons P, Van Houdt R, Aertsen A, Vanoirbeek K, Engelborghs Y, Michiels CW (2006) Role of quorum sensing and antimicrobial component production by *Serratia plymuthica* in formation of biofilms, including mixed biofilms with Escherichia coli. Appl Environ Microbiol 72:7294–7300. https://doi.org/10.1128/AEM.01708-06

Murali M, Amruthesh KN (2015) Plant growth-promoting fungus *Penicillium oxalicum* enhances plant growth and induces resistance in pearl millet against downy mildew disease. J Phytopathol 163:743–754. https://doi.org/10.1111/jph.12371

Murali M, Sudisha J, Amruthesh KN, Ito SI, Shetty HS (2013) Rhizosphere fungus *Penicillium chrysogenum* promotes growth and induces defence-related genes and downy mildew disease resistance in pearl millet. Plant Biol 15:111–118. https://doi.org/10.1111/j.1438-8677.2012.00617.x

Murphy JF, Zehnder GW, Schuster DJ, Sikora EJ, Polston JE, Kloepper JW (2000) Plant growth-promoting rhizobacterial mediated protection in tomato against tomato mottle virus. Plant Dis 84:779–784. https://doi.org/10.1094/PDIS.2000.84.7.779

Muslim A, Hyakumachi M, Kageyama K, Suwandi S (2019) Induction of systemic resistance in cucumber by hypovirulent binucleate *Rhizoctonia* against Anthracnose caused by *Colletotrichum orbiculare*. Trop Life Sci Res 30(1):109–122. https://doi.org/10.21315/tlsr2019.30.1.7

Nagaraju A, Sudisha J, Murthy SM, Ito S (2012) Seed priming with *Trichoderma harzianum* isolates enhances plant growth and induces resistance against *Plasmopara halstedii*, an incitant of sunflower downy mildew disease. Australasian Plant Pathol 41(6):609–620. https://doi.org/10.1007/s13313-012-0165-z

Narsai R, Wang C, Chen J, Wu J, Shou H, Whelan J (2013) Antagonistic, overlapping and distinct responses to biotic stress in rice (*Oryza sativa*) and interactions with abiotic stress. BMC Genomics 14:93. https://doi.org/10.1186/1471-2164-14-93

Naziya B, Murali M, Amruthesh KN (2020) Plant growth-promoting fungi (PGPF) instigate plant growth and induce disease resistance in Capsicum annuum L. upon infection with Colletotrichum capsici (Syd.) Butler & Bisby. Biomol Ther 10:4–6. https://doi.org/10.3390/biom10010041

Nelson GC (2001) Genetically modified organisms in agriculture: economics and politics. Elsevier

Oerke EC, Dehne HW (2004) Safeguarding production—losses in major crops and the role of crop protection. Crop Prot 23:275–285. https://doi.org/10.1016/j.cropro.2003.10.001

Oliveros JC (2007–2015) Venny. An interactive tool for comparing lists with Venn's diagrams. https://bioinfogp.cnb.csic.es/tools/venny/index.html

Onofre-Lemus J, Hernández-Lucas I, Girard L, Caballero-Mellado J (2009) ACC (1-aminocyclopropane-1-carboxylate) deaminase activity, a widespread trait in *Burkholderia* species, and its growth-promoting effect on tomato plants. Appl Environ Microbiol 75:6581–6590. https://doi.org/10.1128/AEM.01240-09

Ophir T, Gutnick DL (1994) A role for exopolysaccharide in the protection of microorganisms from desiccation. Appl Environ Microbiol 60:740–745

Padder SA, Prasad R, Shah AH (2018) Quorum sensing: a less known mode of communication among fungi. Microbiol Res 210:51–58. https://doi.org/10.1016/j.micres.2018.03.007

Pangesti N, Pineda A, Dicke M, van Loon JJA (2015) Variation in plant-mediated interactions between rhizobacteria and caterpillars: potential role of soil composition. Plant Biol 17:474–483. https://doi.org/10.1111/plb.12265

Pangesti N, Reichelt M, van de Mortel JE, Kapsomenou E, Gershenzon J, van Loon JJ, Dicke M, Pineda A (2016) Jasmonic acid and ethylene signaling pathways regulate glucosinolate levels in plants during rhizobacteria-induced systemic resistance against a leaf-chewing herbivore. J Chem Ecol 42(12):1212–1225. https://doi.org/10.1007/s10886-016-0787-7

Pérez-De-Luque A, Tille S, Johnson I, Pascual-Pardo D, Ton J, Cameron DD (2017) The interactive effects of arbuscular mycorrhiza and plant growth-promoting rhizobacteria synergistically enhance host plant defences against pathogen. Sci Rep 7:1–10. https://doi.org/10.1038/s41598-017-16697-4

Pieterse CMJ, Zamioudis C, Berendsen RL, Weller DM, Van Wees SCM, Bakker PAHM (2014) Induced systemic resistance by beneficial microbes. Annu Rev Phytopathol 52:347–375. https://doi.org/10.1146/annurev-phyto-082712-102340

Pineda A, Zheng SJ, van Loon JJA, Dicke M (2012) Rhizobacteria modify plant-aphid interactions: a case of induced systemic susceptibility. Plant Biol 14:83–90. https://doi.org/10.1111/j.1438-8677.2011.00549.x

Pliego C, Kamilova F, Lugtenberg B (2011) Plant growth-promoting bacteria: fundamentals and exploitation. In: Bacteria in agrobiology: crop ecosystems. Springer, pp 295–343

Prasanna R, Kanchan A, Kaur S, Ramakrishnan B, Ranjan K, Singh MC, Hasan M, Saxena AK, Shivay YS (2016) Chrysanthemum growth gains from beneficial microbial interactions and fertility improvements in soil under protected cultivation. Hortic Plant J 2:229–239. https://doi.org/10.1016/j.hpj.2016.08.008

Prindle A, Liu J, Asally M, Ly S, Garcia-Ojalvo J, Süel GM (2015) Ion channels enable electrical communication in bacterial communities. Nature 527:59–63. https://doi.org/10.1038/nature15709

Rajput A, Kumar M (2017) Computational exploration of putative LuxR solos in archaea and their functional implications in quorum sensing. Front Microbiol 8:1–16. https://doi.org/10.3389/fmicb.2017.00798

Ribaudo CM, Krumpholz EM, Cassán FD, Bottini R, Cantore ML, Curá JA (2006) Azospirillum sp. promotes root hair development in tomato plants through a mechanism that involves ethylene. J Plant Growth Regul 25:175–185. https://doi.org/10.1007/s00344-005-0128-5

Rilling J, Acuña J, Nannipieri P, Cassan F, Maruyam F, Jorquera M (2019) Current opinion and perspectives on the methods for tracking and monitoring plant growth–promoting bacteria. Soil Biol Biochem 130:205–219

Rinaudi LV, Giordano W (2010) An integrated view of biofilm formation in rhizobia. FEMS Microbiol Lett 304:1–11. https://doi.org/10.1111/j.1574-6968.2009.01840.x

Rodrigues EP, Rodrigues LS, de Oliveira ALM, Baldani VLD, dos Santos Teixeira KR, Urquiaga S, Reis VM (2008) *Azospirillum amazonense* inoculation: effects on growth, yield and N_2 fixation of rice (*Oryza sativa* L.). Plant Soil 302:249–261. https://doi.org/10.1007/s11104-007-9476-1

Rojas-Rojas FU, López-Sánchez D, Meza-Radilla G, Méndez-Canarios A, Ibarra JA, Estrada-de Los Santos P (2019) El controvertido complejo Burkholderia cepacia, un grupo de especies promotoras del crecimiento vegetal y patógenas de plantas, animales y humanos. Rev Argent Microbiol 51(1):84–92

Russo DM, Abdian PL, Posadas DM, Williams A, Vozza N, Giordano W, Kannenberg E, Downie JA, Zorreguieta A (2015) Lipopolysaccharide O-chain core region required for cellular cohesion and compaction of in vitro and root biofilms developed by *Rhizobium leguminosarum*. Appl Environ Microbiol 81(3):1013–1023. https://doi.org/10.1128/AEM.03175-14

Ryan RP, Monchy S, Cardinale M, Taghavi S, Crossman L, Avison MB, Berg G, Van Der Lelie D, Dow JM (2009) The versatility and adaptation of bacteria from the genus Stenotrophomonas. Nat Rev Microbiol 7(7):514–525

Ryu C-M, Farag MA, Hu C-H, Reddy MS, Kloepper JW, Paré PW (2004) Bacterial volatiles induce systemic resistance in Arabidopsis. Plant Physiol 134:1017–1026. https://doi.org/10.1104/pp.103.026583

Sakamoto K, Ogiwara N, Kaji T (2013) Involvement of autoregulation in the interaction between rhizobial nodulation and AM fungal colonization in soybean roots. Biol Fertil Soils 49(8):1141–1152. https://doi.org/10.1007/s00374-013-0804-8

Salehi-Lisar SY, Bakhshayeshan-Agdam H (2016) Drought stress in plants: causes, consequences, and tolerance. Drought Stress Toler Plants 1–16:1–526. https://doi.org/10.1007/978-3-319-28899-4_1

Salomon MV, Bottini R, de Souza Filho GA, Cohen AC, Moreno D, Gil M, Piccoli P (2014) Bacteria isolated from roots and rhizosphere of *Vitis vinifera* retard water losses, induce abscisic acid accumulation and synthesis of defense-related terpenes in in vitro cultured grapevine. Physiol Plant 151(4):359–374. https://doi.org/10.1111/ppl.12117

Sandhya V, Ali SZ (2015) The production of exopolysaccharide by *Pseudomonas putida* GAP-P45 under various abiotic stress conditions and its role in soil aggregation. Microbiology (Russian Fed) 84:512–519. https://doi.org/10.1134/S0026261715040153

Sandhya V, Ali SK Z, Grover M, Reddy G, Venkateswarlu B (2009) Alleviation of drought stress effects in sunflower seedlings by the exopolysaccharides producing *Pseudomonas putida* strain GAP-p45. Biol Fertil Soils 46:17–26. https://doi.org/10.1007/s00374-009-0401-z

Santos MS, Nogueira MA, Hungria M (2019) Microbial inoculants: reviewing the past, discussing the present and previewing an outstanding future for the use of beneficial bacteria in agriculture. AMB Express 9(1):205

Saravanakumar D, Vijayakumar C, Kumar N, Samiyappan R (2007) PGPR-induced defense responses in the tea plant against blister blight disease. Crop Prot 26(4):556–565. https://doi.org/10.1016/j.cropro.2006.05.007

Sartaj AS, Sakamoto K, Inubushi K (2011) Effect of *Penicillium* sp. EU0013 inoculation on tomato growth and Fusarium wilt. Hortic Res 65:69–73

Savary S, Ficke A, Aubertot JN, Hollier C (2012) Crop losses due to diseases and their implications for global food production losses and food security. Food Security 4:519–537. https://doi.org/10.1007/s12571-012-0200-5

Schlömann M, Schmidt E, Knackmuss H (1990) Different types of dienelactone hydrolase in 4-fluorobenzoate-utilizing bacteria. J Bacteriol 172(9):5112–5118

Schnurr P, Allen D (2015) Factors affecting algae biofilm growth and lipid production: a review. Renew Sustain Energy Rev 52:418–429. https://doi.org/10.1016/j.rser.2015.07.090

Sharif DI, Gallon J, Smith CJ, Dudley E (2008) Quorum sensing in cyanobacteria: N-octanoyl-homoserine lactone release and response, by the epilithic colonial cyanobacterium *Gloeothece* PCC6909. ISME J 2:1171–1182. https://doi.org/10.1038/ismej.2008.68

Sharma E, Anand G, Kapoor R (2017) Terpenoids in plant and arbuscular mycorrhiza-reinforced defence against herbivorous insects. Ann Bot 119:791–801. https://doi.org/10.1093/aob/mcw263

Silpe JE, Bassler BL (2019) Phage-encoded LuxR-type receptors responsive to host- produced bacterial quorum-sensing autoinducers. mBio 10:e00638–e00619. https://doi.org/10.1128/mBio.00638-19

Singh BK (2017) Creating new business, economic growth and regional prosperity through microbiome-based products in the agriculture industry. Microb Biotechnol 10(2):224–227

Singhalage ID, Seneviratne G, Madawala HMSP, Wijepala PC (2019) Profitability of strawberry (*Fragaria ananassa*) production with biofilmed biofertilizer application. Sci Hortic 243:411–413. https://doi.org/10.1016/j.scienta.2018.08.033

Soliman M (2020) *Paenibacillus polymyxa* and Bacillus aryabhattai as biocontrol agents against Ralstonia solanacearum in vitro and in planta. J Plant Prot Pathol 11:197–203. https://doi.org/10.21608/jppp.2020.87024

Son SH, Khan Z, Kim SG, Kim YH (2009) Plant growth-promoting rhizobacteria, *Paenibacillus polymyxa* and Paenibacillus lentimorbus suppress disease complex caused by root-knot nematode and fusarium wilt fungus. J Appl Microbiol 107:524–532. https://doi.org/10.1111/j.1365-2672.2009.04238.x

Song GC, Im H, Jung J, Lee S, Jung MY, Rhee SK, Ryu CM (2019) Plant growth-promoting archaea trigger induced systemic resistance in *Arabidopsis thaliana* against Pectobacterium carotovorum and Pseudomonas syringae. Environ Microbiol 21:940–948. https://doi.org/10.1111/1462-2920.14486

Speert DP (2001) Understanding Burkholderia cepacia: epidemiology, genomovars, and virulence. Inf Med 18(1):49–56

Stanley CE, van der Heijden MG (2017) Microbiome-on-a-chip: new frontiers in plant–microbiota research. Trends Microbiol 25(8):610–613

Stanley CE, Stöckli M, van Swaay D, Sabotič J, Kallio PT, Künzler M, deMello AJ, Aebi M (2014) Probing bacterial–fungal interactions at the single cell level. Integr Biol 6(10):935–945

Stoodley P, Sauer K, Davies DG, Costerton JW (2002) Biofilms as complex differentiated communities. Annu Rev Microbiol 56:187–209. https://doi.org/10.1146/annurev.micro.56.012302.160705

Sudisha J, Sharathchandra RG, Amruthesh KN, Kumar A, Shetty HS (2012) Pathogenesis related proteins in plant defense response. In: Mérillon JM, Ramawat KG (eds) Plant defence: biological control, progress in biological control, vol 12. Springer, Dordrecht, pp 379–403

Sun C, Johnson JM, Cai D, Sherameti I, Oelmüller R, Lou B (2010) *Piriformospora indica* confers drought tolerance in Chinese cabbage leaves by stimulating antioxidant enzymes, the expression of drought-related genes and the plastid-localized CAS protein. J Plant Physiol 167:1009–1017. https://doi.org/10.1016/j.jplph.2010.02.013

Tabacchioni S, Bevivino A, Dalmastri C, Chiarini L (2002) Bulkholderia cepacia complex in the rhizosphere: a mini review. Ann Microbiol 52(2):103–118

Tay JH, Liu QS, Liu Y (2001) The role of cellular polysaccharides in the formation and stability of aerobic granules. Lett Appl Microbiol 33:222–226

Tebbe CC(2015) Risk assessment considerations of genetically modified micro-organisms for releases. Biosafety and the environmental used of micro-organisms

Tebbe CC, Miethling-Graff R (2006) Marker genes as tools to study deliberately released soil bacteria. In: Nucleic acids and proteins in soil. Springer, pp 421–447

Thijs S, Sillen W, Rineau F, Weyens N, Vangronsveld J (2016) Towards an enhanced understanding of plant–microbiome interactions to improve phytoremediation: engineering the metaorganism. Front Microbiol 7:341

Timmusk S, Wagner EGH (1999) The plant-growth-promoting rhizobacterium *Paenibacillus polymyxa* induces changes in *Arabidopsis thaliana* gene expression: a possible connection between biotic and abiotic stress responses. Mol Plant-Microbe Interact 12:951–959. https://doi.org/10.1094/MPMI.1999.12.11.951

Trivedi P, Schenk PM, Wallenstein MD, Singh BK (2017) Tiny microbes, big yields: enhancing food crop production with biological solutions. Microb Biotechnol 10(5):999–1003

Udaya Shankar AC, Chandra Nayaka S, Niranjan-Raj S, Bhuvanendra Kumar H, Reddy MS, Niranjana SR, Prakash HS (2009) Rhizobacteria-mediated resistance against the blackeye cowpea mosaic strain of bean common mosaic virus in cowpea (Vigna unguiculata). Pest Manag Sci 65:1059–1064. https://doi.org/10.1002/ps.1791

Vandamme P, Opelt K, Knöchel N, Berg C, Schönmann S, De Brandt E, Eberl L, Falsen E, Berg G (2007) Burkholderia bryophila sp. nov. and Burkholderia megapolitana sp. nov., moss-associated species with antifungal and plant-growth-promoting properties. Int J Syst Evol Microbiol 57(Pt 10):2228–2235

Valenzuela-Soto JH, Estrada-Hernández MG, Ibarra-Laclette E, Délano-Frier JP (2010) Inoculation of tomato plants (*Solanum lycopersicum*) with growth-promoting Bacillus subtilis retards whitefly Bemisia tabaci development. Planta 231:397–410. https://doi.org/10.1007/s00425-009-1061-9

van de Mortel JE, de Vos RC, Dekkers E, Pineda A, Guillod L, Bouwmeester K, van Loon JJ, Dicke M, Raaijmakers JM (2012) Metabolic and transcriptomic changes induced in Arabidopsis by the rhizobacterium *Pseudomonas fluorescens* SS101. Plant Physiol 160(4):2173–2188. https://doi.org/10.1104/pp.112.207324

van Elsas JD, Semenov AV, Pereira e Silva M, Salles JF (2015) Lessons of the impact of genetically engineered micro-organisms on natural ecosystems like soil. In: Biosafety and the environmental uses of micro-organisms: conference proceedings. OECD Publishing, Paris

Van Oosten VR, Bodenhausen N, Reymond P, Van Pelt JA, Van Loon LC, Dicke M, Pieterse CMJ (2008) Differential effectiveness of microbially induced resistance against herbivorous insects in *Arabidopsis*. Mol Plant Microbe Interact 21:919–930. https://doi.org/10.1094/MPMI-21-7-0919

Van Wees SC, Van der Ent S, Pieterse CM (2008) Plant immune responses triggered by beneficial microbes. Curr Opin Plant Biol 11:443–448. https://doi.org/10.1016/j.pbi.2008.05.005
Van VT, Berge O, Ke SN, Balandreau J, Heulin T (2000) Repeated beneficial effects of rice inoculation with a strain of Burkholderia vietnamiensison early and late yield components in low fertility sulphate acid soils of Vietnam. Plant Soil 218(1–2):273–284
Velmourougane K, Prasanna R, Saxena AK (2017) Agriculturally important microbial biofilms: present status and future prospects. J Basic Microbiol 57:548–573. https://doi.org/10.1002/jobm.201700046
Vidhyasekaran P, Rabindran R, Muthamilan M, Nayar K, Rajappan K, Subramanian N, Vasumathi K (1997) Development of a powder formulation of *Pseudomonas fluorescens* for control of rice blast. Plant Pathol 46:291–297. https://doi.org/10.1046/j.1365-3059.1997.d01-27.x
Villanueva VD, Font J, Thomas S, Romaní AM (2011) Biofilm formation at warming temperature: acceleration of microbial colonization and microbial interactive effects. Biofouling 27:59–71. https://doi.org/10.1080/08927014.2010.538841
Vinayarani G, Madhusudhan KN, Prakash HS (2019) Induction of systemic resistance in turmeric by rhizospheric isolate *Trichoderma asperellum* against rhizome rot disease. J Plant Pathol 101:965–980. https://doi.org/10.1007/s42161-019-00303-9
Vinodkumar S, Nakkeeran S, Renukadevi P, Mohankumar S (2018) Diversity and antiviral potential of rhizospheric and endophytic *Bacillus* species and phyto-antiviral principles against tobacco streak virus in cotton. Agric Ecosyst Environ 267:42–51. https://doi.org/10.1016/j.agee.2018.08.008
Vlamakis H, Chai Y, Beauregard P, Losick R, Kolter R (2013) Sticking together: building a biofilm the *Bacillus subtilis* way. Nat Rev Microbiol 11:157–168. https://doi.org/10.1038/nrmicro2960
Wang X, Cao Y, Tang X, Ma X, Gao J, Zhang X (2014) Rice endogenous nitrogen fixing and growth promoting bacterium Herbaspirillum seropedicae DX35. Acta Microbiol Sin 54:292–298
Wang J, Li R, Zhang H, Wei G, Li Z (2020) Beneficial bacteria activate nutrients and promote wheat growth under conditions of reduced fertilizer application. BMC Microbiol 20:1–12. https://doi.org/10.1186/s12866-020-1708-z
Wani PA, Khan MS (2010) Bacillus species enhance growth parameters of chickpea (Cicer arietinum L.) in chromium stressed soils. Food Chem Toxicol 48:3262–3267. https://doi.org/10.1016/j.fct.2010.08.035
Waters CM, Bassler BL (2005) Quorum sensing: cell-to-cell communication in bacteria. Annu Rev Cell Dev Biol 21:319–346. https://doi.org/10.1146/annurev.cellbio.21.012704.131001
Watt M, McCully ME, Jeffree CE (1993) Plant and bacterial mucilages of the maize rhizosphere: comparison of their soil binding properties and histochemistry in a model system. Plant Soil 151:151–165. https://doi.org/10.1007/BF00016280
Wei Z, Gu Y, Friman VP, Kowalchuk GA, Xu Y, Shen Q, Jousset A (2019) Initial soil microbiome composition and functioning predetermine future plant health. Sci Adv 5:1–12. https://doi.org/10.1126/sciadv.aaw0759
Whipps JM (2001) Microbial interactions and biocontrol in the rhizosphere. J Exp Bot 52:487–511. https://doi.org/10.1093/jexbot/52.suppl_1.487
Whitchurch CB, Tolker-Nielsen T, Ragas PC, Mattick JS (2002) Extracellular DNA required for bacterial biofilm formation. Science 295:1487. https://doi.org/10.1126/science.295.5559.1487
Wilson MJ (2003) Biological control agents: safety and regulatory policy. Biocontrol 48:477–484
Wolfaardt GM, Lawrence JR, Korber DR (1999) Function of EPS. In: Wingender J, Neu R, H.C.F. (eds) Microbial extracellular polymeric substances. Berlin: Springer, p 171–200
Wuang SC, Khin MC, Chua PQD, Luo YD (2016) Use of *Spirulin*a biomass produced from treatment of aquaculture wastewater as agricultural fertilizers. Algal Res 15:59–64. https://doi.org/10.1016/j.algal.2016.02.009
Yang J, Kloepper JW, Ryu C-M (2009) Rhizosphere bacteria help plants tolerate abiotic stress. Trends Plant Sci 14:1–4. https://doi.org/10.1016/j.tplants.2008.10.004

Yao L, Wu Z, Zheng Y, Kaleem I, Li C (2010) Growth promotion and protection against salt stress by *Pseudomonas putida* Rs-198 on cotton. Eur J Soil Biol 46:49–54. https://doi.org/10.1016/j.ejsobi.2009.11.002

Yi H, Chun J (2015) Next generation sequencing-based metagenomics for monitoring soil microbiota. In: Biosafety and the environmental uses of microorganisms. OECD, Paris

Yoshikura H (2015) Reflection on environmental risk assessment of micro-organisms. In: Biosafety and the environmental uses of micro-organisms: conference. OECD, Paris

Zainab N, Amna DBU, Javed MT, Afridi MS, Mukhtar T, Kamran MA, Qurat ul ain, Khan AA, Ali J, Jatoi WN, Hussain Munis MF, Chaudhary HJ (2020) Deciphering metal toxicity responses of flax (*Linum usitatissimum* L.) with exopolysaccharide and ACC-deaminase producing bacteria in industrially contaminated soils. Plant Physiol Biochem 152:90–99. https://doi.org/10.1016/j.plaphy.2020.04.039

Zebelo S, Song Y, Kloepper JW, Fadamiro H (2016) Rhizobacteria activates (+)-δ-cadinene synthase genes and induces systemic resistance in cotton against beet armyworm (*Spodoptera exigua*). Plant Cell Environ 39:935–943. https://doi.org/10.1111/pce.12704

Zhou J, Lyu Y, Richlen ML, Anderson DM, Cai Z (2016) Quorum sensing is a lanuange of chemical signals and plays an ecological role in algal-bacterial interactions. CRC Crit Rev Plant Sci 35:81–105. https://doi.org/10.1080/07352689.2016.1172461

Zhou LS, Tang K, Guo SX (2018) The plant growth-promoting fungus (PGPF) *Alternaria* sp. A13 markedly enhances *Salvia miltiorrhiza* root growth and active ingredient accumulation under greenhouse and field conditions. Int J Mol Sci 19(1):270. https://doi.org/10.3390/ijms19010270

13 Biological Soil Crusts to Keep Soil Alive, Rehabilitate Degraded Soil, and Develop Soil Habitats

Atoosa Gholamhosseinian, Adel Sepehr, Behnam Asgari Lajayer, Nasser Delangiz, and Tess Astatkie

Abstract

Soil is a mixture of liquids, gases, minerals, organic matters, and microorganisms that support life together and play a vital role in the ecosystem cycles. Soil is an important part of agriculture, which is a major source of food. Healthy soils produce plenty of healthy nutrients. Adding chemical fertilizers has been considered as a good solution to increase agricultural production and secure human needs for food for many years. As a result, excessive use of these fertilizers has led to soil contamination and reduced fertility. The use of chemicals has also endangered the environment, and to mitigate this, using environment friendly biofertilizers have been considered. Soil biocrusts, which include cyanobacteria, lichens, mosses, and fungi, live on the surface and inside the soil. Biocrusts play an important role in ecological remediation to increase soil water, dust capture, interaction with vascular plants, increase soil stability, reduce water and wind erosion, fixing N and P, etc. In this chapter, what biological crusts are and their

A. Gholamhosseinian · A. Sepehr
Department of Desert and Arid Zones Management, Ferdowsi University of Mashhad, Mashhad, Iran

B. Asgari Lajayer (✉)
Health and Environment Research Center, Tabriz University of Medical Sciences, Tabriz, Iran
e-mail: h-asgari@tabrizu.ac.ir

N. Delangiz
Department of Plant Breeding and Biotechnology, Faculty of Agriculture, University of Tabriz, Tabriz, Iran

T. Astatkie (✉)
Faculty of Agriculture, Dalhousie University, Truro, NS, Canada
e-mail: astatkie@dal.ca

A. Vaishnav, D. K. Choudhary (eds.), *Microbial Polymers*,
https://doi.org/10.1007/978-981-16-0045-6_13

role in improving soil conditions, reducing erosion, and their use in plant germination, and as biofertilizer is reviewed.

Keywords

Aggregate stability · Biocrust · Biofertilizer · Seed germination

13.1 Introduction

The current population of the world is 7.6 billion, and at the current rate of population growth, the population of the world is forecasted to reach 9.8 billion in 2050, and 11.2 billion in 2100. To meet the needs of this growing population, the production of agricultural products including cereals needs to be increased, but this is becoming unattainable due to land and soil constraints, which means that there is a strong need for solutions to feed the growing population of the world. This goal of achieving food security is primarily the responsibility of the agricultural sector. This target can only be achieved by converting more land for cultivation or increasing the productivity of available cultivated land. Since the first option is not feasible due to land constraints, increasing the productivity of agricultural products is the only viable option. The best ways of increasing agricultural productivity are improving soil fertility and using environment friendly tools.

At present, agricultural production is dependent on chemical fertilizers, pesticides, over-irrigation, etc. These practices have caused environmental and health problems including loss of soil fertility, overuse of water and land resources, reduced plant diversity, and environmental pollution (Singh and Strong 2016). Sustainable agricultural practices can solve all problems related to food production and issues related to quality and environmental health (Mason 2003). Among the sustainable agricultural practices, the use of biofertilizers is one of the best. Biofertilizers are living microorganisms that can secrete certain organic compounds into the soil that increase seed growth, plant growth, and soil fertility.

Early living entities on an evolutionary path of indefinite duration led to the creation of the last common ancestor that had the basic characteristics of all cells. Early life evolution was accompanied by metabolic diversification, and as soon as life appeared, it started to evolve and diversify (López-García and Moreira 2006). Biological soil crusts (BSCs/biocrust) are the pioneers of life on Earth and include cyanobacteria, green algae, moss, and lichens (Belnap et al. 2006), and they occupy the surface and the soil in the top few millimeters (Warren et al. 2019).

In recent years, considering the importance of replacing chemical fertilizers by biofertilizers for the creation and development of sustainable agriculture, studies on the effects of biocrust have increased. Furthermore, biocrust microbes are the most dominant community types (Zaady et al. 2013), and one of the first forms of earthly organisms whose life probably dates back to 2.6 billion years. Biocrust microbes can be found in different ecological, succession, and agronomic conditions where infertility resulted in colonization opportunities of bare soil surfaces (Elumeeva et al. 2011). Biocrusts are more common in arid and semi-arid ecosystems, where diversity and vegetation cover are low.

Communities dominated by mosses, lichens, green algae, and cyanobacteria are biological keys to polar, temperate, and arid ecosystems worldwide, where they have key roles in controlling ecosystem structure and function (Belnap and Lange 2003). However, there is very limited information on the potential effects of different species on soil performance and microbial communities.

13.2 Cyanobacteria and Green Algae

Algae are found all over the world in a variety of environments such as freshwater, seawater, snow, rock, and on the surface of plants and animals. Many algae, such as *Oscillatoria Brevis, Scytonema elongates, Heterhormogonium,* can tolerate temperatures between 50 and70 °C, which is making them common in tropical soils (Fig. 13.1). Although many algae can tolerate a lot of sunlight, their growth rate is low under light conditions. Many green algae can be found in a wide range of pH (Heidarpour et al. 2019); however, blue-green algae prefer neutral or alkaline pH (Bilgrami and Saha 2004). After the death and decomposition of algae, organic matter is added to the soil, and after some years, the soil becomes suitable for cultivation.

Cyanobacteria species that are found all over the world in fresh and saline waters, in temperate, arid, and tropical regions, on tree trunks, and in soils as microbial communities (Thajuddin and Subramanian 2005) play a vital role in soils' physico-chemical properties. They are also able to occur in all terrestrial and aquatic habitats,

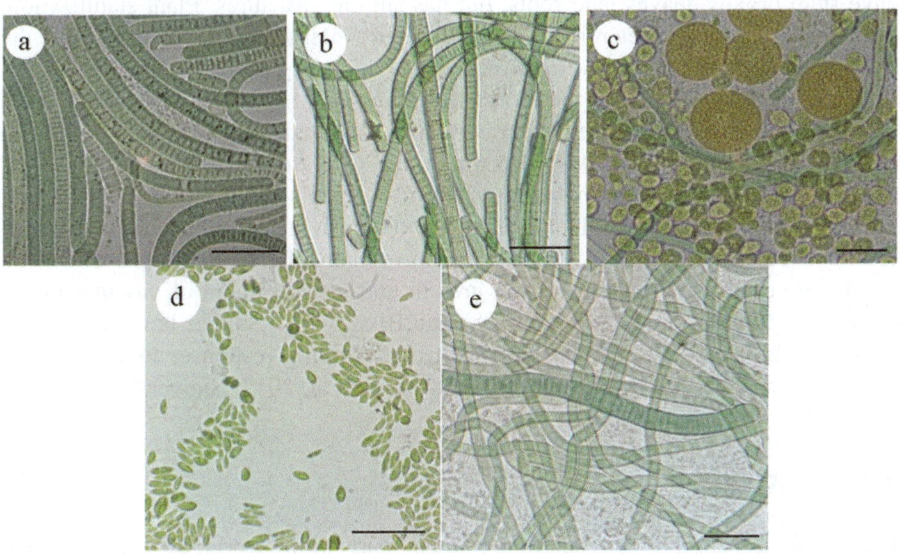

Fig. 13.1 Cyanobacteria and algae cultivated in the semi-arid region of northeastern Iran. (**a**) *Osillatoria sp.*, (**b**) *Phormidium sp.*, (**c**) cyanobacteria and green algae, (**d**) *Coccomyxa sp.*, (**e**) *Phormidium sp*

such as oceans, rivers, freshwater, wet rocks, and bare soil, or in contact with other plants and animals such as lichens, sponges, and Protestants (Vaughan 2011).

Cyanobacteria are photosynthetic and can manufacture their own food. They are more than 3.5 billion years old fossils (Soudzilovskaia et al. 2011). It is understood that cyanobacteria played an important role in producing oxygen for the atmosphere and oceans (Demoulin et al. 2019). Cyanobacteria are prokaryotic phototrophs that have been considered in many industries, including food supplementation and the production of some valuable essential polymers (Arias et al. 2020). It is also used for various purposes, including dietary supplements and food, due to its high content of vitamins and proteins, as well as its high digestibility (Abed et al. 2009; Rosgaard et al. 2012; Colica et al. 2014).

Cyanobacteria have played an important role in early evolution and life on Earth. Moreover, cyanobacteria can grow and survive in unfavorable conditions for a long period of time when environmental conditions are not conducive to growth (Whitton 2000). The high compatibility of cyanobacteria with any environment indicates that their culture, like other microorganisms, does not require energy-rich compounds (Lau et al. 2015).

13.3 Mosses

Mosses are groups of the simplest early plants. These plants have an intermediate position between green thallophytes (algae) and vascular plants (Jackson et al. 2011). This slow-growing plant is one of the first terrestrial plants and does not have stem organs, leaves, and roots, but has similar structures. Plant stabilization, soil uptake, aggregate retention, and uptake by a filamentous structure called the rhizoid, which consists of several brownish-red cells (Fig. 13.2c).

The impact of mosses on the inorganic environment is substantial (Cornelissen et al. 2007) as they control the temperature and moisture fluctuations in soils (Bueno et al. 2016; Gornall et al. 2007; Jackson et al. 2011; Soudzilovskaia et al. 2011). Mosses can also affect the availability of soil nutrients (De Long et al. 2016) by affecting soil temperature, reducing albedo, and improving substrate quality (Gornall et al. 2007; Lang et al. 2009). Moss species have many effects on soil properties that can improve the environment for the growth and establishment of vascular plants (During and van Tooren 1990; Sohlberg and Bliss 1987). Studies have shown that moss species have different establishment and seedling responses to warming or rainfall when other factors are kept constant (Lett et al. 2017; Stuiver et al. 2014).

13.4 Lichens

Lichens are a section of the plant flora in many parts of the world; they were the first multicellular creatures to settle on land (Clauzade and Ozenda 1970). Lichens are complex and highly developed plants. Lichens have high endurance due to photosynthesis and stabilization of solar energy by algal part and resistance of fungal

Fig. 13.2 (**a**) Lichen (*Xanthoria parietina*) with mosses; (**b**) *Syntrichia sp.*; (**c**) The morphology of moss consisting two parts: sporophyte (1,2) and gametophyte (3), 1. Capsule 2. Seta 3. leaf 4. Rhizoid; (**d**) Moss grown on rock in the semi-arid region of northeastern Iran

filaments in harsh environmental conditions and can survive at minus 60 °C. Lichens are living organisms composed of mainly mycobionts and photosynthetic algae and/or cyanobacteria (photobionts; one species of mycobiont (fungi) and one species of photobiont (algae)) form a symbiotic relationship in a thallus.

The body of lichens is composed of several layers: upper hard layer of fungal filaments, algae layer that is mostly made of algal or cyanobacterial cells, called the medulla, the central part, which consists of fungi filaments and some of the materials made by lichen, is stored in this part, and the lowest and thickest layer is made of fungal threads with a thickness of 0.5 mm. Soil living lichen subspecies (e.g., *Lepraria crassissima* and *Diploschistes diacapsis*) have higher respiration rate, enzyme activity, and concentration ability of the inorganic soil in comparison with cyanobacteria living soils (Kroken and Taylor 2000; Muggia et al. 2010; Miralles et al. 2012, 2013) (Fig. 13.3).

Fig. 13.3 (**a**, **b**) View of lichens in the northeast of Khorasan Razavi province; (**c**) view of terrestrial lichens (*psora sp.*); (**d**) the red circles show the soil lichens and the blue circles show the mosses; (**e**) *xanthoria parietina;* and (**f**) an example of lichens on a tree trunk

13.5 Ecological Roles of Biocrusts

Studies have shown that biocrusts play a significant role in ecological remediation. There are many ecological roles of biocrusts: (1) the redistribution of precipitated rainwater; (2) the capture, collection, and use of airborne and soil nutrients; (3) interaction with vascular plants; and (4) enhancing the soil physical structure and stability; (5) reducing the wind and water erosion; (6) increasing nutrients accessibility; (7) increasing the roughness of the surface, so enhancing dust trapping capacity (Fearnehough et al. 1998; Belnap and Lange 2003; Warren et al. 2019; Weber et al. 2016); (8) improving the essential nutrient concentration in plants (Belnap and Harper 1995); (9) enhancing nitrogen and carbon fixation rate (Evans and Lange 2001; Yang et al. 2014); so, increasing the mineralization and bioavailability rate of nitrogen (Delgado-Baquerizo et al. 2013; Hu et al. 2015); thus, biocrusts nutrient enrichment has an important role in soil fertility and nutrient bioavailability. Also, it is important in the biogeochemical cycles.

13.5.1 The Role of Soil Microorganisms in Inhibiting Runoff

Runoff is one of the important research topics in hydrology (Martinez-Hernández et al. 2017). Erosion determines the rate of sedimentation as well as plant establishment (Le Bissonnais 2016), and it is considered as a valuable tool for studying the structure, distribution of plants, and the function of ecosystems (Neilson 1995).

Microorganisms such as earthworms, insects, bacteria, mosses, cyanobacteria, algae, fungi, and lichens are accumulated on the surface of soils with several millimeters of thickness. As ecosystem engineers, they form the biological surface of the soil thus providing the ground for the growth and development of their space by changing soil surface conditions that has an impact on other animal, plant, and soil habitats, as well as combating soil erosion. One of the most important functions of the biological crust is soil stabilization, which leads to a reduction in water and wind erosion (Williams et al. 1993). Soil microorganisms are useful and efficient agents to improve the quality and fertility of soils (Saghafi et al. 2019a, b, 2020). They also decrease soil degradation (Bowker 2007; Rossi et al. 2017; Maestre et al. 2017). It was suggested that inoculating biological crusts that manipulate soil communities could be a key action to restore damaged ecosystems (Wubs et al. 2016). A study on the Colorado Plateau showed that under rainfall simulation conditions, in samples inoculated with Cyanobacterium, sediment erosion of 400 g/m^2 and samples containing lichens and mosses were almost zero (Belnap et al. 2012). Also, a 100% decrease in water erosion was observed in the Loess Plateau Region, in the presence of moss crust relative to bare soil (Zhao and Xu 2013).

13.5.2 Hydrology and Available Soil Water

Soil moisture plays a key role in the rate of residue decomposition and mineralization of organic phosphorus. There are obvious contradictions about the role of biocrusts in relation to their effects on soil hydrology (effect on runoff infiltration). Several researchers concluded that biocrusts increases penetration, while some believe that the presence of the biocrust reduces it. Climate change influences the way in which biocrust affects hydrology, and how it increases runoff (and not sediment loading process). It may also increase infiltration (Chamizo et al. 2016).

An extensive review by Warren et al. (2019) revealed that much of the variation can be attributed to soil texture. When the content of sand in the soil exceeds about 80%, and the soil is not frost-heaved, biocrusts generally reduce water infiltration compared to soil without crust. This seems to be attributable to soil porosity. Porosity in sandy soils is possible with the accumulation of biocrust in finer soil particles when polysaccharides exudates, root structures, fungal hyphae, or cyanobacterial filaments block the pores of the soil, preventing water from moving and penetrating (Rodriguez-Caballero et al. 2018). When sand content is less than about 80%, biocrust organisms contribute to soil aggregation, and, as a result, it leads to higher rates of porosity and more penetration than soil without biocrust. Biocrusts increase the surface roughness, which increases the retention time of water on the surface (Faist et al. 2020), it also affects the movement and retention of water in the soil. The process of runoff production and re-infiltration plays an important role in ecosystem efficiency and large-scale hydrological response (Kidron 2016). When the biological crusts get wet, their volume increases up to ten times, which results in the absorption and retention of water in the soil.

13.5.3 Application of EPS on Improving Soil Properties

Exopolysaccharides (EPS) are the most vital chemicals that are synthesized by cyanobacteria. They play an important and vital role in improving soil parameters (Weber et al. 2016). These biopolymers form a layer around cells called capsule (De Philippis and Vincenzini 1998). It protects cells that are under physical stress such as drought (Costa et al. 2018). Exopolysaccharides regulate the function of cell loss or water absorption (Mazor et al. 1996; Potts 1994; Adessi et al. 2018) to protect cells from harmful effects of dehydration-hydration cycles and environmental stresses (Liu et al. 2016). One of the most important studied roles of the EPS matrix is the accumulation capacity of soil particles that is important for improving soil structure and fertility (Boonchai et al. 2014). EPS acts like a glue bond to clay and ions and holds soil particles together (Chenu 1995).

These exo-polymers that are studied thoroughly can improve the quality, fertility, and structure of soils (Hwang et al. 2004; Elisashvili et al. 2009). EPSs are produced in response to environmental stresses such as salinity, pH changes, drought, temperature changes, etc., to protect the cell (Wingender et al. 1999). EPSs also enhance water retention and maintain moisture for a longer time (Rossi et al. 2012; Colica et al. 2014; Adessi et al. 2018) and defend microorganisms from dehydration and nutrient constraint to help them to survive (Zhang 2005; Parikh and Madamwar 2006).

EPS produces a layer around microbes that protects them from the negative effects of heavy metals and antimicrobial chemicals. EPS matrix can also protect cyanobacteria and the environment against drought by retaining water. EPSs are widely used in the food and agricultural industries. Their glycosidic linkage is a kind of gelling or thickening agent (Kraan 2012). Also, EPS can play the role of water binding, stabilizer, or emulsifier agent in food industries (Whistler and Daniels 1990; Singha and Hemachandran 2012). Moreover, EPSs are kind of antibacterial agents, and anticancer (polysaccharide from Spirulina sp., inhibits tumor incursion), anti-inflammatory, and immunomodulatory agents (Kraan 2012; Orsavova et al. 2015; De Jesus Raposo et al. 2015). Also, EPS secreted by Cyanobacterium *aponinum sp.* (Gudmundsdottir et al. 2015) and *Lyngbya sp.* are known to serve as an immune system booster and used in pharmaceuticals (Pignolet et al. 2013; Liu et al. 2016; Raposo and Stoorvogel 2013). *Chlorella sp.* and *Spirulina sp.* are rich in vitamins and proteins, so they are important in the nutraceutical industry (Choi et al. 2013).

13.5.4 Biocrust and Soil Nutrition

The activity of microorganisms in poor soils depends on conditions such as temperature, humidity, and the presence of organic carbon compounds. During the process of photosynthesis, the autotrophic activity of biocrust causes carbon dioxide to be absorbed by the air and stabilized in the soil as carbohydrates and polysaccharides. EPS is essential for microbial life because it traps nutrients, provides a suitable environment for chemical reaction, and protects microbes from stresses such as drought and saline soils. As microbial EPS improves the aggregation ability of soil particles, it can increase the water retaining and nutrients trapping capacity. Whereas Cyanobacteria have the ability of N_2 and CO_2 stabilization, these biocrusts can stabilize a wide range of compounds such as P and vitamins (Priya et al. 2017). Thus, they can improve the fertility of soils in arid and semi-arid regions, where water and N are the most limiting factors (Noy-Meir 1973).

Studies in the Sonoran Desert lowland and the Colorado Plateau highland indicated coincident trends of biogenic elements –N and C- enrichment and the non-biogenic elements, such as Zr, AS, Zn, Cu, Ca, Mn, and Cr depletion (Beraldi-Campesi et al. 2009). The biocrusts physiological and survival activities depend on many elements. Also, the composition of macro- (Ca, Mg, N, P, and K) and micro-elements (Fe, Zn, Cu, and Mn) is related to the composition and distribution of biocrusts in soil (Bowker et al. 2016). For instance, *D. diacapsis* can increase the concentration of micro- and macro-elements in comparison with the bare areas without lichens (Concostrina-Zubiri et al. 2013). Other studies showed that lichens and mosses have positive correlations with manganese, zinc, potassium, and magnesium, but negative correlation with phosphorus (Bowker et al. 2005; Ochoa-Hueso et al. 2011). Housman et al. (2006, 2007) and Li et al. (2010) reported that microbial biomass and nutrient concentration of the moss-covered regions is higher than regions with cyanobacterial crust. Soils with higher carbonate content are suitable for the growth and establishment of lichens and mosses. All these functions help to improve the dust capture capacity and the accumulation of essential elements and organic matters. Thus, the development of the soil surface and biocrusts formation will occur (Li et al. 2006).

13.5.5 Soil Texture and Aggregate Stability

Soil texture is the coarseness or fineness of the particles, which is mainly obtained from the relative proportions of different particles such as silt, sand, and clay. Soils are the product and result of the performance of various processes, which are formed over time genetic horizons. The location of soil formation, intensity, and weakness of influential factors such as time, determine the speed of soil formation and evolution. Other factors that affect soil formation are bedrock material, weathering processes, mineral composition, and type and composition of organic matter. Soil texture has a great impact on soil physical and chemical performance (Bowman and Hutka 2002) such as water and nutrient mobility, air availability, and soil water content.

Cyanobacterial crusts with the same species are concentrated in fine-textured soils closer to the surface than in coarse soils because light and moisture are rapidly limited in depth (Garcia-Pichel and Belnap 1996). Polysaccharides can bond with soluble minerals when soils with a weak structure are wet and after drying. In fine-textured soils, the recovery rate of cyanobacteria is faster because they require less energy for colonization and accumulation due to smaller soil pores (Zheng et al. 2011; Rosenstein et al. 2014); as a result, they are easily attached to soil aggregates

Fig. 13.4 Electron microscope images are taken from the cross-section of biocrust soils, showing hyphae and polysaccharide filaments of mosses and lichens (red arrow) that have kept the soil cohesive (LEO 1450VP, Germany-central lab in Ferdowsi University of Mashhad)

through the secretion of exopolysaccharides (Chock et al. 2019). The decaying material mixes with the soil and acts as a glue to bind the particles. Increasing humus in the soil is important to maintain moisture.

Algae and cyanobacteria synthesize extracellular polymeric materials (EPS), which contain various biopolymers and help EPS to bind to the surface of minerals (Flemming and Wingender 2010). In most soils, especially in arid and desert areas, cyanobacteria are the predominant species in the biological crust and provide an environment where other biological crust components can be located (Abed et al. 2009). So, the filaments of the cyanobacteria can facilitate the formation of a stable soil surface through binding to the soil particles. These stable surfaces make a suitable environment for organisms such as lichens and mosses to colonize and develop (Deng et al. 2020).

The Scanning electron micrographs (SEM) shown in Fig. 13.4 has details of the internal structure of soil crusts and the role that cyanobacteria play in soils as well as the infiltration of fungal hyphae in the soils. According to research conducted by Wang et al. (2017), in the Atacama Desert, biocrusts are able to protect soil from wind erosion as well as to increase soil cohesion by trapping dust and atmospheric salts.

Sepehr et al. (2018) examined the relationship between cyanobacteria and soil parameters and found that cyanobacteria increase organic carbon and nitrogen and their EPS secretions cause more stability in aggregates. The effect of biocrusts on the amount of water in a region of southwestern Spain was studied using a pressure plate device at two pressures of −33 and −1500 kPa and in two layers of 0–1 and 1–5 cm covered by crusts. They observed a significant increase in the amount of water at a pressure of −33 kPa in the order of the least developed crust, in the direction of the crust with the most developed in all layers, and at a pressure of −1500 kPa the types of crusts had a similar amount of water (Chamizo et al. 2018).

13.6 Sustainable Agriculture

13.6.1 Wastewater Treatment

Domestic, agricultural, and industrial wastewaters contain high concentrations of phosphorus and nitrogen, which may lead to eutrophication and toxicity (Pouliot et al. 1989; Magro et al. 2012). Over the past few decades, efforts have been made to use microalgae for secondary effluents' biological treatment (Oswald and Gotaas 1957; de la Noüe et al. 1992; Heidarpour et al. 2019). Cyanobacteria were successfully used in the treatment of municipal, livestock, and industrial wastewater and has met with considerable success (Muñoz and Guieysse 2006; Lee et al. 2015).

Cyanobacteria can be widely used in wastewater treatment by removing nutrients, reducing biological oxygen demand (BOD), and chemical oxygen demand (COD), as well as having a high ability to remove heavy metals (Abdel-Raouf et al. 2012; Honda et al. 2012). Cyanobacteria are one of the largest groups of Gram-negative prokaryotes that have great diversity in morphology, cell division patterns, physiology, cell differentiation, and habitats (Rippka et al. 1979). In general, cyanobacteria have different properties that make them a good option for wastewater treatment. The capacity and ability of cyanobacteria to grow in various effluents such as municipal and industrial wastewater has already been studied (Parmar et al. 2011; Vijayakumar 2012).

The use of cyanobacteria in wastewater treatment, in addition to economically valuable by-products, has also been considered to produce clean- or at least cleaner-water (Markou et al. 2014). Cyanobacteria can accumulate phosphorus and mineral nitrogen store it in the form of polyphosphate and cyanophycean and use it for growth (Fay 1983; Abdel-Raouf et al. 2012). Cyanobacteria can survive and thrive in highly variable conditions with contaminated effluents.

Another advantage of filamentous cyanobacteria (epilithic or benthic) is their natural tendency to aggregate and precipitate easily in the absence of stirring, which makes them excellent candidates for mass cultivation for wastewater treatment (de la Noüe and Proulx 1988; Mespoulède 1997). During numerous studies, the ability of cyanobacteria to biodegrade and bio-sorb persistent heavy metals have been demonstrated (Kirkwood et al. 2001; El-Bestawy 2008). Their minimal nutrient needs have made them suitable for using them as low-cost, environment friendly technologies for wastewater treatment as well as to produce biological products (Lau et al. 2015).

Blue-green algae are known to store polymeric compounds such as polyhydroxyalkanoates (PHA) and carbohydrates under environmental stress and nutrient deficiency conditions (Samantaray et al. 2011). PHA has received a lot of attention due to its potential use as a substrate for bioplastics and biofuels. This type of wastewater treatment is an economical and environment friendly process without pollution (Honda et al. 2012; Rawat et al. 2011).

13.6.2 Seed Germination and Establishment of Vegetation

Biocrust can lead to stunted or optimal plant growth (Condon and Pyke 2018a; Havrilla et al. 2019); the biocrusts may act as armor coatings to prevent seed penetration, but if the seeds can be successfully established, the presence of biocrusts can increase seedling growth rate (Ferrenberg et al. 2018; Slate et al. 2019). The positive effect of cyanobacteria on the germination and growth of wheat and rice seeds has been reported (Muñoz-Rojas et al. 2018). Furthermore, the use of beneficial microorganisms such as cyanobacteria on the seeds can be a useful tool for transferring inoculation to the soil after germination (O'Callaghan and van Sinderen 2016). Soaking the seeds in cyanobacterial suspension allows them to be impregnated into the seed to create desirable conditions for inoculation (Mahmood et al. 2016).

Research has shown that in paddy fields, cyanobacteria help to accelerate seed germination and to promote seedling growth (Gupta and Lata 1964). This research showed that cyanobacteria are richer in protein and therefore have a positive effect on grain yield and quality. The potential of cyanobacteria as a biological primer to restore soil in an arid region of Western Australia was investigated, and the results showed positive effects on the growth of native plants and on the improvement of soil substrate characteristics (Flemming and Wingender 2010).

Cyanobacteria have more than 800 secondary metabolites including linear, cyclic peptides, fatty acid, ribosomal peptides, and amides (Costa et al. 2008). In addition, they produce various compounds in soils that can improve soil fertility. They are also known as biocontrol against fungi and harmful bacteria for plants. Some of these include polypeptides, proteins, vitamins, biotins, amino acids, and growth-promoting factors (Singh et al. 2016). This secondary metabolite shows different clinical activities such as antimicrobial, anticancer, antitumor, and anti-inflammatory (Nagarajan et al. 2012).

Cyanobacteria are able to exude plant growth regulators (PGRs) (Venkataraman 1981); Cytokinin-like (Strick et al. 1997) and gibberellins-like compounds (Shen-Rui and Shen 1997) are some of them. Furthermore, cyanobacteria exude pharmaceutically active compounds (Metting and Pyne 1986; Chamizo et al. 2020), and organic acids, toxins, antibiotics, and algicides (Hellebust 1974). The Chemicals produced by cyanobacteria have essential roles in regulating metabolism, plant growth, development, and improving germination (Hashtroudi et al. 2013).

Cyanobacteria can lead to favorable microhabitats for soil and other plants, and also improve the establishment and growth of annual plants (Xu et al. 2013; Samantaray et al. 2011). It was also reported that the use of cyanobacterial extract increased the growth of plants as well as their dry weight (Offer et al. 1992). The use of dilute aqueous extracts of cyanobacteria was successful in the control of a damping-off disease caused by fungal strain (Caire et al. 1976). The use of cyanobacteria to increase seed germination is recommended as a biological fertilizer and to increase the growth of many plants (Strick et al. 1997).

After regeneration and establishment of vegetation, experimental findings showed that the concentrations of essential elements in the soil with cyanobacterial-lichen crust, moss crust increased significantly. Therefore, plant revegetation also has a mutual effect on the development of cyanobacteria-lichen and moss crusts. Due to these characteristics, their high dispersion and establishment in any environment, cyanobacteria can be known as ecosystem engineers and can be used as a potential tool for soil repair in arid areas (Rossi et al. 2017).

13.6.3 Biofertilizer

Biofertilizers are good replacements of chemical fertilizers due to their environmental compatibility, because of which they have received a lot of attention in recent decades. The widespread use of chemical fertilizers and pesticides for a long time led to the contamination of soil (Khoshru et al. 2020; Asgari Lajayer et al. 2019). Recently, agriculture has relied on chemical fertilizers to increase yield, which also reduces soil microbiota and thus reduces soil fertility (Singh et al. 2016). It has been found that biological soil crusts contribute to soil fertility resulting in increased crop yield and reduced pollution of the environment, water, and soil (Rossi et al. 2017). Soil fungi are one of the main pathogens in crop fields. Kulik (1995) stated that cyanobacteria have a high yield as a biological fertilizer for plant pathogenic bacteria and fungi. Anabaena and Nostoc species are two of them that are considered as biofertilizers (Li et al. 2002). *Nostoc sp.*, *Anabaena sp.*, and *Oscillatoria sp.* have a high abundance, especially in arid areas and can produce a lot of EPS and survive during environmental stresses.

13.6.4 Biocrusts Functions and Utility in Restoration

Arid and semi-arid lands are the most extensive biomes on Earth, covering about 45% of the surface of Earth (Reynolds et al. 2007; Pravalie 2016). Biocrusts are able to live in areas that are difficult for vascular plants to survive, and they work on basic functions of ecosystems such as protecting soil from erosion, conserving moisture adjustment hydrology, and providing nutrient cycling of carbon and nitrogen fixation (Belnap et al. 2016). Restoration of biocrusts in arid and degraded areas should be a top priority because the regeneration of biocrusts leads to the recovering of many desired environmental functions (Bowker 2007).

During the biological crust development process, the transfer of dominant functional groups from cyanobacteria to mosses and lichens changes the structure of bacterial communities because of the different effects of nursing on soil bacteria (De Velasco et al. 2017). However, knowledge about the role and diversity of the bacterial community in successive biological sequences in environmental regeneration remains limited (Condon and Pyke 2018b; Liu et al. 2016). Recently, studies showed the role of biocrust functions in the ecological restoration of the ecosystem of deserts (Bowker 2007; Winkler et al. 2018), and the critical role they play in predicting desert ecosystem responses in different climatic conditions (Weber et al. 2016). Cyanobacteria EPS properties are important factors for selecting the type of crust for soil restoration. Specific properties of cyanobacterial EPS in terms of monosaccharide and macromolecular composition can provide useful understandings in this area.

13.7 Conclusion

Biological soil crusts have an important role in physical and chemical properties of soil, particularly in arid environments, although most of these effects are magnified on the surface layers (0–5 cm). Biological soil crusts are ecological engineers in nature and their positive activities in the soil introduced them as an indicator of soil quality. Biological soil crusts conserve soil and protect it against erosion. This important role led researchers to utilize them for erosion control and actions related to soil restoration and combating desertification. They have an essential role in soil stability, regulation of salinity, water balance, and fertility enhancement particularly in arid ecosystems and environments with hyper conditions regarding climate and erosion.

References

Abdel-Raouf N, Al-Homaidan AA, Ibraheem IBM (2012) Microalgae and wastewater treatment. Saudi J Biol Sci 19:257–275

Abed RMM, Dobretsov S, Sudesh K (2009) Applications of cyanobacteria in biotechnology. J Appl Microbiol 106:1–12

Adessi A, Cruz de Carvalho R, De Philippis R, Branquinho C, Marques da Silva J (2018) Microbial extracellular polymeric substances improve water retention in dryland biological soil crusts. Soil Biol Biochem 116:67–69

Arias DM, García J, Uggetti E (2020) Production of polymers by cyanobacteria grown in wastewater: current status, challenges and future perspectives. New Biotechnol 55:46–57

Asgari Lajayer B, Najafi N, Moghiseh E, Mosaferi M, Hadian J (2019) Effects of gamma irradiated and non-irradiated sewage sludge on growth characteristics, leaf area index and macronutrients concentrations in basil. J Soil Sci Plant Nutr 19:580–591

Belnap J, Harper KT (1995) Influence of cryptobiotic soil crusts on elemental content of tissue of two desert seed plants. Arid Land Res Manag 9(2):107–115

Belnap J, Lange OL (eds) (2003) Biological soil crusts: structure, function, and management. Springer, Berlin, pp 3–30

Belnap J, Phillips SL, Troxler T (2006) Soil lichen and moss cover and species richness can be highly dynamic: the effects of invasion by the annual exotic grass Bromus tectorum, precipitation, and temperature on biological soil crusts in SE Utah. Appl Soil Ecol 32:63–76

Belnap J, Wilcox BP, Van Scoyoc MW, Phillips SL (2012) Successional stage of biological soil crusts: an accurate indicator of ecohydrological condition. Ecohydrology 6:474–482

Belnap J, Weber B, Büdel B (2016) Biological soil crusts as an organizing principle in drylands. In: Weber B, Büdel B, Belnap J (eds) Biological soil crusts: an organizing principle in drylands, Ecological studies series. Springer, Berlin, pp 3–16

Beraldi-Campesi H, Hartnett HE, Anbar A, Gordon GW, Garcia-Pichel F (2009) Effect of biological soil crusts on soil elemental concentrations: implications for biogeochemistry and as traceable biosignatures of ancient life on land. Geobiology 7:348–359

Bilgrami KS, Saha L (2004) A textbook of algae. CBS Publishers & Distributors, p 263

Boonchai R, Kaewsuk J, Seo G (2014) Effect of nutrient starvation on nutrient uptake and extracellular polymeric substance for microalgae cultivation and separation. Desal Water Treat 55:360–367

Bowker MA (2007) Biological soil crust rehabilitation in theory and practice: an underexploited opportunity. Restor Ecol 15(1):13–23

Bowker MA, Belnap J, Davidson DW, Phillips SL (2005) Evidence from micronutrient limitation of biological soil crusts: importance to arid-lands restoration. Ecol Appl 15:1941–1951

Bowker MA, Belnap J, Budel B, Sannier C, Pietrasiak N, Eldridge DJ, Rivera-Aguilar V (2016) Controls on distribution patterns of biological soil crusts at micro- to global scales. In: Weber B, Budel B, Belnap J (eds) Biological soil crusts: an organizing principle in drylands. Springer International Publishing, Cham

Bowman GM, Hutka J (2002) Particle size analysis. In: McKenzie N, Coughlan K, Cresswell H (eds) Soil physical measurement and interpretation for land evaluation. CSIRO Publishing, Victoria, pp 224–239

Bueno CG, Williamson SN, Barrio IC, Helgadóttir Á, Hik DS (2016) Moss mediates the influence of shrub species on soil properties and processes in alpine tundra. PLoS One 11(10):e0164143

Caire G, Zaccaro de Mulé MC, Doallo S, Halperin DR, Halperin L (1976) Accio'n de extractosalgalesacuosos y ete'reos de Nostoc muscorum Ag. (n_79a). I Efectosobrepla'ntulas de mijo (Panicum miliaceum L.) mediantetratamiento de sussemillas. Bol Soc Arg Bot 17 (3/4):289–300

Chamizo S, Cantón Y, Rodríguez-Caballero E, Domingo F (2016) Biocrusts positively affect the soil water balance in semiarid ecosystems. Ecohydrology 9(7):1208–1221

Chamizo S, Mugnai G, Rossi F, Certini G, De Philippis R (2018) Cyanobacteria inoculation improves soil stability and fertility on different textured soils: gaining insights for applicability in soil restoration. Front Environ Sci 6:49

Chamizo S, Adessi A, Certini G, De Philippis R (2020) Cyanobacteria inoculation as a potential tool for stabilization of burned soils. Restor Ecol 28:S106–S114

Chenu C (1995) Extracellular polysaccharides: an interface between microorganisms and soil constituents natural and anthropogenic organics. In: Huang PM, Berthelin J, Bollag JM, McGill WB, Page AL (eds) Environmental impact of soil component interactions. Natural and anthropogenic organics. CRC Press, Boca Raton, pp 217–233

Chock T, Antoninka A, Faist AM, Bowker MA, Belnap J, Barger NN (2019) Responses of biological soil crusts to rehabilitation strategies. J Arid Environ 163:77–85

Choi AM, Ryter SW, Levine B (2013) Autophagy in human health and disease. N Engl J Med 368 (7):651–662

Clauzade G, Ozenda P (1970) Les Lichens. Masson, Paris, p 801

Colica G, Li H, Rossi F, Li D, Liu Y, De Philippis R (2014) Microbial secreted exopolysaccharides affect the hydrological behavior of induced biological soil crusts in desert sandy soils. Soil Biol Biochem 68:62–70

Concostrina-Zubiri L, Huber-Sannwald E, Martínez I, Flores Flores JL, Escudero A (2013) Biological soil crusts greatly contribute to small-scale soil heterogeneity along a grazing gradient. Soil Biol Biochem 64:28–36

Condon LA, Pyke DA (2018a) Fire and grazing influence site resistance to Bromus tectorum through their effects on shrub, bunchgrass and biocrust communities in the Great Basin (USA). Ecosystems 21(7):1416–1431

Condon LA, Pyke DA (2018b) Resiliency of biological soil crusts and vascular plants varies among morphogroups with disturbance intensity. Plant Soil 433:271–287

Cornelissen JC, Lang SI, Soudzilovskaia NA, During HJ (2007) Comparative cryptogam ecology: a review of bryophyte and lichen traits that drive biogeochemistry. Ann Bot 99(5):987–1001

Costa JAV, Santana FB, Andrade MR, Lima MB, Frank DT (2008) Microalgae biomass and biomethane production in the south of Brazil. Biotechnol Lett 136:402–403

Costa OYA, Raaijmakers JM, Kuramae EE (2018) Microbial extracellular polymeric substances: ecological function and impact on soil aggregation. Front Microbiol 9:1636 de Jesus Raposo MF, De Morais AMB, De Morais RMSC (2015) Marine polysaccharides from algae with potential biomedical applications. Mar Drugs 13(5):2967–3028

de la Noüe J, Proulx D (1988) Biological tertiary treatment of urban wastewaters with chitosan-*immobilized Phormidium* sp. Appl Microbiol Biotechnol 29:292–297

de la Noüe J, Laliberté G, Proulx D (1992) Algae and wastewater. J Appl Phycol 4:247–254

De Long JR, Dorrepaal E, Kardol P, Nilsson MC, Teuber LM, Wardle DA (2016) Contrasting responses of soil microbial and nematode communities to warming and plant functional group removal across a post-fire boreal Forest successional gradient. Ecosystems 19:339–355

De Philippis R, Vincenzini M (1998) Exocellular polysaccharides from cyanobacteria and their possible applications. FEMS Microbiol Rev 22:151–175

De Velasco G, Je Y, Bossé D, Awad MM, Ott PA, Moreira RB, Schutz F, Bellmunt J, Sonpavde GP, Hodi S, Choueiri TK (2017) Comprehensive meta-analysis of key immune-related adverse events from CTLA-4 and PD-1/PD-L1 inhibitors in cancer patients. Cancer Immunol Res 5 (4):312–318

Delgado-Baquerizo M, Maestre FT, Gallardo A, Bowker MA, Wallenstein MD, Quero JL, Ochoa V, Gozalo B, García-Gómez M, Soliveres S, García-Palacios P (2013) Decoupling of soil nutrient cycles as a function of aridity in global drylands. Nature 502(7473):672–676

Demoulin CF, Lara YJ, Cornet L, François C, Baurain D, Wilmotte A, Javaux EJ (2019) Cyanobacteria evolution: insight from the fossil record. Free Radic Biol Med 140:206–223

Deng S, Zhang D, Wang G, Zhou X, Ye C, Fu T (2020) Biological soil crust succession in deserts through a 59-year-long case study in China: how induced biological soil crust strategy accelerates desertification reversal from decades to years. Soil Biol Biochem 141:107665

During HJ, van Tooren BF (1990) Bryophyte interactions with other plants. Bot J Linn Soc 104 (1–3):79–98

El-Bestawy E (2008) Treatment of mixed domestic–industrial wastewater using cyanobacteria. J Ind Microbiol Biotechnol 35:1503–1516

Elisashvili V, Kachlishvili ET, Wasser SP (2009) Carbon and nitrogen source effects on basidiomycetes exopolysaccharide production. Appl Biochem Microbiol 45:531–535

Elumeeva TG, Soudzilovskaia NA, During HJ, Cornelissen JHC (2011) The importance of colony structure versus shoot morphology for the water balance of 22 subarctic bryophyte species. J Veg Sci 22(1):152–164

Evans RD, Lange OL (2001) Biological soil crusts and ecosystem nitrogen and carbon dynamics. In: Biological soil crusts: structure, function, and management. Springer, Berlin, pp 263–279

Faist A, Antoninka A, Belnap J, Bowker M, Duniway M, Garcia-Pichel F, Nelson C, Reed S, Giraldo-Silva A, Velasco-Ayuso S, Barger N (2020) Inoculum and habitat amelioration efforts in biological soil crust recovery vary by desert and soil texture. Restor Ecol 28. https://doi.org/10.1111/rec.13087

Fay P (1983) The blue-greens. Edward Arnold Publishers Ltd, London, 87 pp

Fearnehough W, Fullen MA, Mitchell J, Trueman IC, Zhang J (1998) Aeolian deposition and its effect on soil and vegetation changes on stabilised desert dunes in northern China. Geomorphology 23:171–182

Ferrenberg S, Faist A, Howell A, Reed S (2018) Biocrusts enhance soil fertility and Bromus tectorum growth and interact with warming to influence germination. Plant Soil 429:77–90

Flemming HC, Wingender J (2010) The biofilm matrix. Nat Rev Microbiol 8(9):623–633

Garcia-Pichel F, Belnap J (1996) Microenvironments and microscale productivity of cyanobacterial desert crusts. J Phycol 32:774–778

Gornall JL, Jónsdóttir IS, Woodin SJ, van der Wal R (2007) Arctic mosses govern below-ground environment and ecosystem processes. Oecologia 153(4):931–941

Gudmundsdottir T, Winther J, de Fine LS, Bonnesen T, Asdahl P, Tryggvadottir L, Anderson H, Wesenberg F, Malila N, Hasle H, Olsen J (2015) Cardiovascular disease in adult life after childhood cancer in Scandinavia: a population-based cohort study of 32,308 one-year survivors. Int J Cancer 137(5):1176–1186

Gupta AB, Lata K (1964) Effect of algal growth hormones on the germination of paddy seeds. Hydrobiologia 24(1–2):430–434

Hashtroudi MS, Ghassempour A, Riahi H, Shariatmadari Z, Khanjir M (2013) Endogenous auxins in plant growth-promoting cyanobacteria—Anabaena vaginicola and Nostoc calcicola. J Appl Phycol 25(2):379–386

Havrilla CA, Chaudhary VB, Ferrenberg S, Antoninka AJ, Belnap J, Bowker MA, Eldridge DJ, Faist AM, Huber-Sannwald E, Leslie AD, Rodriguez-Caballero E (2019) Towards a predictive framework for biocrust mediation of plant performance: a meta-analysis. J Ecol 107 (6):2789–2807

Heidarpour A, Aliasgharzad N, Khoshmanzar E, Khoshru B, Asgari Lajayer B (2019) Bio-removal of Zn from contaminated water by using green algae isolates. Environ Technol Innov 16:100464. https://doi.org/10.1016/j.eti.2019.100464

Hellebust JA (1974) Extracellular products. In: Stewart WSP (ed) Algal physiology and biochemistry, 10. Botanical monographs, Berkeley, pp 838–865

Honda R, Boonnorat J, Chiemchaisri C, Chiemchaisri W, Yamamoto K (2012) Carbon dioxide capture and nutrients removal utilizing treated sewage by concentrated microalgae cultivation in a membrane photobioreactor. Bioresour Technol 125:59–64

Housman DC, Powers HH, Collins AD, Belnap J (2006) Carbon and nitrogen fixation differ between successional stages of biological soil crusts in the Colorado plateau and Chihuahuan desert. J Arid Environ 66(4):620–634

Housman DC, Yeager CM, Darby BJ, Sanford RL, Kuske CR, Neher DA, Belnap J (2007) Heterogeneity of soil nutrients and subsurface biota in a dryland ecosystem. Soil Biol Biochem 39(8):2138–2149

Hu R, Wang XP, Pan YX, Zhang YF, Zhang H, Chen N (2015) Seasonal variation of net N mineralization under different biological soil crusts in Tengger desert, North China. Catena 127:9–16

Hwang HJ, Kim SW, Xu CP, Choi JW, Yun JW (2004) Morphological and rheological properties of the three different species of basidiomycetes Phellinus in submerged cultures. J Appl Microbiol 96:1296–1305

Jackson BG, Martin P, Nilsson MC, Wardle DA (2011) Response of feather moss associated N2 fixation and litter decomposition to variations in simulated rainfall intensity and frequency. Oikos 120(4):570–581

Khoshru B, Moharramnejad S, Hosseinzadeh Gharajeh N, Asgari Lajayer B, Ghorbanpour M (2020) Plant microbiome and its important in stressful agriculture. In: Varma A, Tripathi S, Prasad R (eds) Plant microbiome paradigm. Springer Nature, Basel. https://doi.org/10.1007/978-3-030-50395-6_2

Kidron GJ (2016) Linking surface and subsurface properties of biocrusted and non-biocrusted habitats of fine-grained fluvial sediments (playas) from the Negev Desert. J Hydrol Hydromechan 64(2):141–149

Kirkwood AE, Nalewajko C, Fulthorpe RR (2001) The occurrence of cyanobacteria in pulp and paper waste-treatment systems. Can J Microbiol 47:761–766

Kraan S (2012) Algal polysaccharides, novel applications and outlook. In: Chang C-F (ed) Carbohydrates: comprehensive studies on glycobiology and glycotechnology. InTech, Maastricht, pp 489–532

Kroken S, Taylor JW (2000) Phylogenetic species, reproductive mode, and specificity of the green alga Trebouxia forming lichens with the fungal genus Letharia. Bryologist 103:645–660

Kulik MM (1995) The potential for using cyanobacteria (blue-green algae) and algae in the biological control of plant pathogenic bacteria and fungi. Eur J Plant Pathol 101:585–599

Lang SI, Cornelissen JHC, Klahn T, Van Logtestijn RSP, Broekman R, Schweikert W, Aerts R (2009) An experimental comparison of chemical traits and litter decomposition rates in a diverse range of subarctic bryophyte, lichen and vascular plant species. J Ecol 97(5):886–900

Lau N-S, Matsui M, Abdullah AA-A (2015) Cyanobacteria: photoautotrophic microbial factories for the sustainable synthesis of industrial products. Biomed Res Int 2015:9

Le Bissonnais Y (2016) Aggregate stability and assessment of soil crustability and erodibility: I. Theory and methodology. Eur J Soil Sci 67:1–21

Lee CS, Lee S-A, Ko SR, Oh HM, Ahn CY (2015) Effects of photoperiod on nutrient removal, biomass production, and algal-bacterial population dynamics in lab-scale photobioreactors treating municipal wastewater. Water Res 68:680–691

Lett S, Nilsson MC, Wardle DA, Dorrepaal E (2017) Bryophyte traits explain climate-warming effects on tree seedling establishment. J Ecol 105(2):496–506

Li XR, Wang XP, Li T, Zhang JG (2002) Microbiotic soil crust and its effect on vegetation and habitat on artificially stabilized desert dunes in Tengger Desert, North China. Biol Fertil Soil 35:147–154

Li XR, Xiao HL, He MZ, Zhang JG (2006) Sand barriers of straw checkerboards for habitat restoration in extremely arid desert regions. Ecol Eng 28:149–157

Li Y, Han D, Hu G, Sommerfeld M, Hu Q (2010) Inhibition of starch synthesis results in overproduction of lipids in Chlamydomonas reinhardtii. Biotechnol Bioeng 107(2):258–268

Liu YY, Wang Y, Walsh TR, Yi LX, Zhang R, Spencer J, Doi Y, Tian G, Dong B, Huang X, Yu LF (2016) Emergence of plasmid-mediated colistin resistance mechanism MCR-1 in animals and human beings in China: a microbiological and molecular biological study. Lancet Infect Dis 16 (2):161–168

López-García P, Moreira D (2006) Selective forces for the origin of the eukaryotic nucleus. Bioessays 28(5):525–533

Maestre FT, Solé R, Singh BK (2017) Microbial biotechnology as a tool to restore degraded drylands. Microb Biotechnol 10:1250–1253

Magro CD, Deon MC, Rossi AD, Reinehr CO, Hemkemeier M, Colla LM (2012) Chromium (VI) biosorption and removal of chemical oxygen demand by Spirulina platensis from wastewater-supplemented culture medium. J Environ Sci Health A 47(12):1818–1824

Mahmood K, Xu Z, El-Kereamy A, Casaretto JA, Rothstein SJ (2016) The Arabidopsis transcription factor ANAC032 represses anthocyanin biosynthesis in response to high sucrose and oxidative and abiotic stresses. Front Plant Sci 7:1548

Markou G, Vandamme D, Muylaert K (2014) Microalgal and cyanobacterial cultivation: the supply of nutrients. Water Res 65:186–202

Martinez-Hernández C, Rodrigo-Comino J, Romer-Díaz A (2017) Impact of lithology and soil properties on abandoned dryland terraces during the early stages of soil erosion by water in south-east Spain. Hydrol Proc 31:3095–3109

Mason J (2003) Sustainable agriculture, 2nd edn. Landlinks Press, Collingwood

Mazor G, Kidron GJ, Vonshak A, Abeliovich A (1996) The role of cyanobacterial exopolysaccharides in structuring desert microbial crusts. FEMS Microbiol Ecol 21:121–130

Mespoulède V (1997) Sélection et étude d'une souche de cyanobactéries polaires pour l'épuration d'un effluent eutrophisant. MSc Thesis, Université Laval, Québec, 56 pp

Metting B, Pyne JW (1986) Biologically active compounds from microalgae. Enzyme Microbiol Technol 8:386–394

Miralles I, Domingo F, García-Campos E, Trasar-Cepeda C (2012) Biological and microbial activity in biological soil crusts from the Tabernas desert, a sub-arid zone in SE Spain. Soil Biol Biochem 55:113–121

Miralles I, Trasar-Cepeda C, Leirós CM, Gil-Sotres F (2013) Labile carbon in biological soil crusts in the Tabernas desert, SE Spain. Soil Biol Biochem 58:1–8

Muggia L, Zellnig G, Rabensteiner J, Grube M (2010) Morphological and phylogenetic study of algal partners associated with the lichen-forming fungus Tephromela atra from the Mediterranean region. Symbiosis 51:149–160

Muñoz R, Guieysse B (2006) Algal–bacterial processes for the treatment of hazardous contaminants: a review. Water Res 40:2799–2815

Muñoz-Rojas M, Román JR, Roncero-Ramos B, Erickson TE, Merritt DJ, Aguila-Carricondo P, Cantón Y (2018) Cyanobacteria inoculation enhances carbon sequestration in soil substrates used in dryland restoration. Sci Total Environ 636:1149–1154

Nagarajan M, Maruthanayagam V, Sundararaman M (2012) A review of pharmacological and toxicological potentials of marine cyanobacterial metabolites. J Appl Toxicol 32:153–185
Neilson RP (1995) A model for predicting continental-scale vegetation distribution and water balance. Ecol Appl 5:362–385
Noy-Meir I (1973) Desert ecosystems: environment and producers. Annu Rev Ecol Syst 4(1):25–51
O'Callaghan A, van Sinderen D (2016) Bifidobacteria and their role as members of the human gut microbiota. Front Microbiol 7:925
Ochoa-Hueso R, Hernandez RR, Pueyo JJ, Manrique E (2011) Spatial distribution and physiology of biological soil crusts from semi-arid central Spain are related to soil chemistry and shrub cover. Soil Biol Biochem 43:1894–1901
Offer ZY, Goossens D, Shachak M (1992) Aeolian deposition of nitrogen to sandy and loessial ecosystems in the Negev Desert. J Arid Environ 23:355–363
Orsavova J, Misurcova L, Ambrozova JV, Vicha R, Mlcek J (2015) Fatty acids composition of vegetable oils and its contribution to dietary energy intake and dependence of cardiovascular mortality on dietary intake of fatty acids. Int J Mol Sci 16(6):12871–12890
Oswald W, Gotaas HB (1957) Photosynthesis in sewage treatment. Trans Am Soc Civ Eng 122:73–105
Parikh A, Madamwar D (2006) Partial characterization of extracellular polysaccharides from cyanobacteria. Bioresour Technol 97(15):1822–1827
Parmar A, Singh NK, Pandey A, Gnansounou E, Madamwar D (2011) Cyanobacteria and microalgae: a positive prospect for biofuels. Bioresour Technol 102:10163–10172
Pignolet O, Jubeau S, Vaca-Garcia C, Michaud P (2013) Highly valuable microalgae: biochemical and topological aspects. J Ind Microbiol Biotechnol 40(8):781–796
Potts M (1994) Desiccation tolerance of prokaryotes. Microbiol Rev 58:755–805
Pouliot Y, Buelna G, Racine C, de la Noüe J (1989) Culture of cyanobacteria for tertiary wastewater treatment and biomass production. Biol Wastes 29:81–91
Pravalie R (2016) Drylands extent and environmental issues. A global approach. Earth Sci Rev 161:259–278
Priya P, Krishnan R, Mujumdar M, Houze RA (2017) Changing monsoon and midlatitude circulation interactions over the Western Himalayas and possible links to occurrences of extreme precipitation. Clim Dyn 49(7–8):2351–2364
Raposo G, Stoorvogel W (2013) Extracellular vesicles: exosomes, microvesicles, and friends. J Cell Biol 200(4):373–383
Rawat I, Kumar RR, Mutanda T, Bux F (2011) Dual role of microalgae: phycoremediation of domestic wastewater and biomass production for sustainable biofuels production. Appl Energ 88(10):3411–3424
Reynolds JF, Stafford-Smith DM, Lambin E, Turner BL III, Mortimore M, Batterbury SPJ (2007) Global desertification: building a science for dryland development. Science 316:847–851
Rippka R, Deruelles J, Waterbury JB, Herdman M, Stanier RY (1979) Generic assignments, strain histories and properties of pure cultures of cyanobacteria. J Gen Microbiol 111:1–61
Rodriguez-Caballero E, Belnap J, Büdel B, Crutzen PJ, Andreae MO, Pöschl U, Weber B (2018) Dryland photoautotrophic soil surface communities endangered by global change. Nat Geosci 11(3):185–189
Rosenstein O, Zaady E, Katra I, Karnieli A, Adamowski J, Yizhaq H (2014) The effect of sand grain size on the development of cyanobacterial biocrusts. Aeolian Res 15:217–226
Rosgaard L, de Porcellinis AJ, Jacobsen JH, Frigaard NU, Sakuragi Y (2012) Bioengineering of carbon fixation, biofuels, and biochemicals in cyanobacteria and plants. J Biotechnol 162:134–147
Rossi F, Potrafka RM, Pichel FG, De Philippis R (2012) The role of the exopolysaccharides in enhancing hydraulic conductivity of biological soil crusts. Soil Biol Biochem 46:33–40
Rossi F, Li H, Liu Y, De Philippis R (2017) Cyanobacterial inoculation (cyanobacterisation): perspectives for the development of a standardized multifunctional technology for soil fertilization and desertification reversal. Earth Sci Rev 171:28–43

Saghafi D, Delangiz N, Asgari Lajayer B, Ghorbanpour M (2019a) An overview on improvement of crop productivity in saline soils by halotolerant and halophilic PGPRs. 3 Biotech 9:261. https://doi.org/10.1007/s13205-019-1799-0
Saghafi D, Ghorbanpour M, Shirafkan Ajirloo H, Asgari Lajayer B (2019b) Enhancement of growth and salt tolerance of canola (Brassica napus L.) seedlings by halotolerant rhizobium strains containing ACC-deaminase activity. Plant Physiol Rep 24:225–235
Saghafi D, Asgari Lajayer B, Ghorbanpour M (2020) Engineering bacterial ACC deaminase for improving plant productivity under stressful conditions. In: Sharma V, Salwan R, Al-Ani LKT (eds) Molecular aspects of plant beneficial microbes in agriculture. Elsevier Inc. https://doi.org/10.1016/B978-0-12-818469-1.00022-5
Samantaray S, Nayak JK, Mallick N (2011) Wastewater utilization for poly-β-hydroxybutyrate production by the cyanobacterium Aulosira fertilissima in a recirculatory aquaculture system. Appl Environ Microbiol 77:8735–8743
Sepehr A, Hasanzadeh M, Rodriguez-Caballero C (2018) The protective role of cyanobacteria on soil stability in two Aridisols in northeastern Iran. Geoderma Reg 16:e00201
Shen-Rui Z, Shen RZ (1997) A broad spectrum biopesticide type biofertilizer anaerobic fermentation effluent and plants adverse resistance. Acta Agric Shanghai 13(2):89–96
Singh JS, Strong PJ (2016) Biologically derived fertilizer: a multifaceted bio-tool in methane mitigation. Ecotoxicol Environ Safety 124:267–276
Singh JS, Kumar A, Rai AN, Singh DP (2016) Cyanobacteria: a precious bio-resource in agriculture, ecosystem, and environmental sustainability. Front Microbiol 7:529
Singha M, Hemachandran K (2012) Content based image retrieval using color and texture. Signal Image Process 3(1):39–57
Slate ML, Callaway RM, Pearson DE (2019) Life in interstitial space: biocrusts inhibit exotic but not native plant establishment in semi-arid grasslands. J Ecol 107:1317–1327
Sohlberg EH, Bliss LC (1987) Responses of Ranunculus sabinei and Papaver radicatum to removal of the moss layer in a high-arctic meadow. Can J Bot 65(6):1224–1228
Soudzilovskaia NA, Graae BJ, Douma JC, Grau O, Milbau A, Shevtsova A, Cornelissen JHC (2011) How do bryophytes govern generative recruitment of vascular plants? New Phytol 190 (4):1019–1031
Strick WA, Staden JV, Van-Staden J (1997) Screening of some south African seaweeds for cytokinin-like activity. S Afr J Bot 63(3):161–164
Stuiver BM, Wardle DA, Gundale MJ, Nilsson MC (2014) The impact of moss species and biomass on the growth of Pinus sylvestris tree seedlings at different precipitation frequencies. Forests 5 (8):1931–1951
Thajuddin N, Subramanian G (2005) Cyanobacterial biodiversity and potential applications in biotechnology. Curr Sci 89(1):47–57
Vaughan T (2011) Mammalogy. Jones and Barlett, Boston, p 21. isbn:9780763762995
Venkataraman GS (1981) Blue-green algae for rice production—a manual for its promotion. FAO Soils Bull 46:1–52
Vijayakumar S (2012) Potential applications of cyanobacteria in industrial effluents—a review. J Bioremed Biodegr 3:1–6
Wang F, Michalski G, Luo H, Caffee M (2017) Role of biological soil crusts in affecting soil evolution and salt geochemistry in hyper-arid Atacama Desert, Chile. Geoderma 307:54–64
Warren SD, Clair LL, Stark LR (2019) Reproduction and dispersal of biological soil crust organisms. Front Ecol Evol 7:e344
Weber B, Bowker M, Zhang Y, Belnap J (2016) Natural recovery of biological soil crusts after disturbance. In: Weber B, Büdel B, Belnap J (eds) Biological soil crusts: an organizing principle in drylands, Ecological studies series. Springer, Berlin, pp 479–498
Whistler R, Daniels R (1990) Function of polysaccharides. In: Food additives. Marcel Dekker, New York, pp 399–406
Whitton BA (2000) Soils and rice-fields. In: Whitton BA, Potts M (eds) Ecology of cyanobacteria. Their diversity in time and space. Kluwer Academic Publishers, Dordrecht, pp 235–255

Williams J, Dobrowolski J, West N (1993) Microphytic crust influence on interrill erosion and infiltration capacity. Trans ASAE 38(1):139–146

Wingender J, Neu TR, Flemming HC (1999) What are bacterial extracellular polymeric substances? In: Wingender J, Neu TR, Flemming HC (eds) Microbial extracellular polymeric substances. Springer, Berlin, pp 1–20

Winkler T, Sass FA, Duda GN, Schmidt-Bleek K (2018) A review of biomaterials in bone defect healing, remaining shortcomings and future opportunities for bone tissue engineering: the unsolved challenge. Bone Joint Res 7(3):232–243

Wubs ER, Van Der Putten WH, Bosch M, Bezemer TM (2016) Soil inoculation steers restoration of terrestrial ecosystems. Nat Plants 2:16107

Xu X, Cao X, Zhao L (2013) Comparison of rice husk-and dairy manure-derived biochars for simultaneously removing heavy metals from aqueous solutions: role of mineral components in biochars. Chemosphere 92(8):955–961

Yang Y, Guan J, Yin J, Shao B, Li H (2014) Urinary levels of bisphenol analogues in residents living near a manufacturing plant in South China. Chemosphere 112:481–486

Zaady E, Arbel S, Barkai D, Sarig S (2013) Long-term impact of agricultural practices on biological soil crusts and their hydrological processes in a semiarid landscape. J Arid Environ 90:5–11

Zhang Y (2005) The microstructure and formation of biological soil crusts in their early developmental stage. Chin Sci Bull 50:117–121

Zhao YG, Xu MX (2013) Runoff and soil loss from revegetated grasslands in the hilly Loess Plateau region, China: influence of biocrust patches and plant canopies. J Hydrol Eng 18:387–393

Zheng Y, Xu M, Zhao J, Bei S, Hao L (2011) Effects of inoculated Microcoleus vaginatus on structure and function of biological soil crusts of desert. Biol Fertil Soil 47:473–480

[illegible]

Fungal Chitosan: The Importance and Beneficiation of this Biopolymer in Industrial and Agricultural Process

14

Allwin Ebinesar, Veena S. More, D. L. Ramya, G. R. Amrutha, and Sunil S. More

Abstract

Chitosan is used as an alternative effective bio-pesticide to manage the crop diseases and to increase the fertility of soil surface. Chitosan is a chitin derivative composed of poly(β-(1–4)-D-glucosamine) isolated from fungal cell walls, crustacean, and shrimp shell. It is a polymer that is non-toxic, biocompatible, and conveniently biodegradable. Chitosan is capable of persuading callose formation in plants, by synthesising phytoalexins to inhibit the growth of parasites in plant cells. In addition, it enhances stomatal conductance and decreases the phase of transpiration without affecting the plants' physical nature. Furthermore, to alter the permeability of the membrane, Chitosan is used as a coating agent for nuts, fruits, and vegetables. It raises sugar and proline concentrations and increases the activity of peroxidase, catalase, phenyl aniline ammonia lyase, and tyrosine-ammonia lyase enzymes. It also has the potential to release the fertiliser in a continuous manner, and it increases the soil water retention property. In addition, chitosan has a high potential to resist pathogenic fungus formation, sporulation, viability of spores, and fungal germination. For four decades, chitosan has been used in a variety of applications for various applications, such as enzyme immobilisation, enhancement of healing activity, heavy metal removal, industrial residue treatment, food preservation, cosmetics, and medical applications. In addition, it is used in the manufacture of vaccines for the adsorption of protein on the surface of the mucosa, value-added product in carbohydrates and applied in different industrial products such as nutrition, biomedical, and prebiotics. In

A. Ebinesar · V. S. More (✉) · D. L. Ramya · G. R. Amrutha
Department of Biotechnology, Sapthagiri College of Engineering, Bangalore, India

S. S. More
School of Basic and Applied Sciences, Dayananda Sagar University, Bangalore, India

A. Vaishnav, D. K. Choudhary (eds.), *Microbial Polymers*,
https://doi.org/10.1007/978-981-16-0045-6_14

this book chapter, we mainly focus on discussion about the application of chitosan in agriculture and industrial processes.

Keywords

Fungi · Chitin · Chitosan · Bioremediation · Soil fertility

14.1 Introduction

Biopolymers are natural polymers that, along with their immunogenic properties, have the characteristic of being easily degraded by enzymes (Pokhrel et al. 2015). Different numbers of biological products, such as silk, alginate, elastin, chitin, and collagen, are known to be natural biopolymers. The advancement of natural biopolymers offers a hopeful direction for the development of research in a variety of fields such as agriculture, food, and treatment of medicinal and wastewater. (Correlo et al. 2005). Chitin is one of the commonly available biopolymers with a 1–4 glycosodic bond molecular complex and N-acetyl—D-glucosamine units that can be derived from many living organisms, such as insects, microbial cell walls, molluscs, and crustacean outer layers (Muzzarelli 1977). Chitosan is another biopolymer derived from the deacetylation of chitin biomolecule end up with formation of 1–4 glycosodic bond and glucosamine poly-carbohydrates molecular combination. These biopolymers are used in several industrial processes, such as the treatment of contaminated water: the remediation of metal ions, the flocculation and coagulation of protein and amino acid biomolecules and coloured dyes; in the food industry: the removal of suspended solids, the stabilisation agent for colour and food conservation and protection, the preservation agent, the gelling agent and the thickener; in the field of medicine: cholesterol management, bone fracture wound healing, inhibition of dental plaque, blood clotting agents and drug delivery; in the field of agriculture: seed coating, continuous release of agrochemicals and fertilisers, and also in the field of cosmetics: body lotions and creams, moisturisers (Davis 2011). Knorr (1991) reported that chitosan was isolated from crab shells and prawns over the decades by using alkalis under deacetylation at high temperatures. However, due to variations in the availability and cost of the above-mentioned raw materials, the output rate will not be constant (Crestini et al. 1996). In addition, the heterogeneous physical properties of the chitosan extracted from the crabs and prawns are shown. To solve these problems, through the fermentation process, the chitosan is effectively isolated from the farming of the respective fungi. Chitin has persisted with the cell surface of various fungi such as Ascomycetes septa, Deuteromycetes, Zygomycetes, Basidiomycetes to preserve their shape, the veracity of the cell structure and the strength of the cell wall (Ruiz-Herrera et al. 1992). The first-person isolated chitin from various forms of fungal species such as *Agaricus acris, Agaricus cantharellus, Agaricus volvaceus, Hydnum hybridum, Hydnum repandum,* and *Boletus viscidus* was Braconnot during the year 1811. In addition to that Odier (1823) isolated the chitin and chitosan successfully isolated from the

Table 14.1 Comparison between different characteristic of chitin and chitosan biopolymers between fungal mycelium species and crustacean species

Basic parameters	In fungal mycelia	In crustacean sources
Accessibility	Easily available in all seasons and no topographical limitations	Availability according to the seasons
Inorganic constituents	Lower	Higher
Method of extraction process	Very simple process	Huge solvents are needed
Effectiveness of cost for waste management	Less	More
Pre-treatment	No need	Required
Physicochemical properties	Remains same	Varied
Orientation of the plant immune activity	Very effective	Less effective

insects crutile and from mushrooms. Novak et al. (2003) introduced the role and function of the group of amines and also explained the characteristics of the formation of chitosan. In the year 1894, Hoppe-Seyler (1894) was the first person proposed the name chitosan. White et al. (1979) is the first-person isolated chitosan from the fungal mycelium. Araki and Ito (1974) found catalytic mechanism involved for the bioconversion of chitin residue present in the cell wall into chitosan by the chitin deacetylase enzyme using Mucor rouxii fungal species. Very recently, the industrial processing of deacetylated chitin (chitosan) has been introduced, gaining more interest and being adopted by many countries. Compared with chitosan isolated from crustacean sources, the notable advantages of chitosan isolated from fungal mycelium were described (Knorr and Imeri 1989) by comparing different physiochemical parameters such as effect or degree of acetylation, dispersal of charged groups, molecular weight of chitin, and chitosan present on fungal chitosan, as shown in Table 14.1 (Adams 2004).

In this book chapter, we discussed about the various physiological function of chitin and chitosan, synthesis from fungal species, application of chitosan and its derivatives in food industry, biomedical, cosmetics and agriculture.

14.2 Physiological Function of Fungal Chitin and Chitosan

The key constituents of neutral polysaccharides, chitin, glycoproteins and chitosan are the cell wall of the fungus mycelium, and the macro amounts of galactosamine polymers, melanin, fatty acids, and polyuronides (Wu 2004). In the cell wall of the hyphae and reproductive structure (spores), chitin biopolymers were observed conjugating with glucan molecules and forming microfibrils. In addition, to provide the rigidity and morphology of the cell wall, the microfibrils are enshrined as an unstructured matrix. The presence of chitin in certain fungal mycelium, such as *Mucor, Rhizopus, and absidia*, has been observed (Ruiz-Herrera et al. 1992), which contributes to the integrity of the cell wall and enhances the defence against foreign materials and also increases stability at high temperature (Adams 2004). Specht et al. (1996) depicted that the other functions of chitin and chitosan biopolymers compute by the effect of mutants bearing defect mechanisms in synthesis of chitin, accumulation of polysaccharides on cell walls and intracellular transferring in chitin synthases.

14.2.1 Biosynthesis of Chitin and Chitosan Biopolymers

The synthesis of chitin is initiated by glycogen, the source material. The phosphorylase enzyme catalyses glycogen in this synthesis process and contributes to glucose-1 phosphate conversion. Glucose 6 phosphate was further transformed by the enzyme phosphomutase and then by the obtained product into fructose 6 phosphate in the presence of hexokinase. And then, the process of amination and acetylation results in the transformation of Fructose 6 phosphate into N-acetyl glucosamine. In addition, isomerisation was performed by the addition of a 1, 6 carbon phosphate group catalysed by the enzyme phospho-N-acetyl glucosamine mutase. Additionally, the inter-conversion reaction followed to form uridine diphosphate N-acetyl glucosamine by the interaction of uridine triphosphate. Eventually, the converted UDP N-acetyl glucosamine formed chitin in the presence of chitin acetylase enzyme (EC 3.5.1.41) shown in Fig. 14.1 (Batista et al. 2018).

14.2.2 Fungi Used for the Isolation of Chitin and Chitosan

In order to produce chitosan, Hu and colleagues studied various forms of fungal species. The ability to isolate chitosan is present in these isolated fungal strains. They found that the fungi, namely Basidiomycetes, Deuteromycetes, and Ascomycetes are not having native chitosan but the fungal zygomycetes comprised with chitosan fungi (Kirk et al. 2008) (Table 14.2).

The quantity of chitin in fungal cell wall is precise to environmental factors, species, and age of the microbial sources. The percentage of chitin content specifically vary from 2 to 42% in the Euascomycetes, fungal, and yeast.

14.2.3 Factors Distressing Fungal Chitin and Chitosan Conversion

Campos-Takaki et al. 1983 suggested that due to environmental factors, substrate conditions and intrinsic characteristics of fungal organisms, the amount of chitin and chitosan biopolymer found on the cell of the fungi differs. According to the mechanisms involved in the biosynthesis process in fungal cells, they found the impact of different factor affects chitosan formation by using submerged fermentation process. These variables are primarily concerned with enhancing the synthesis of chitin and also optimising the physiochemical characteristics (Tanaka 2001). Additionally, these features improved the metabolic activity of fungi followed by

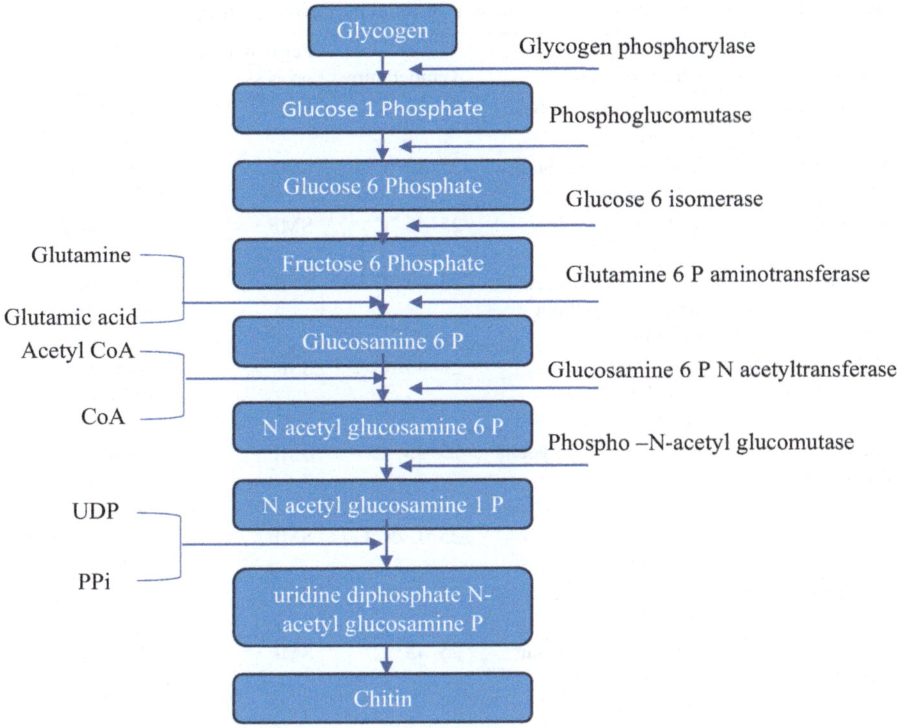

Fig. 14.1 Schematic representation of synthesis of chitin

catabolic repression (Denuziere et al. 2001). Researchers have found that, by adding different nutrients, the growth of fungal mycelium and the formation of biopolymers can be enhanced. Nwe and Stevens (2004) observed a substantial uplift of total chitosan generation in the *Gongronella butleri* fungal species by the addition of urea. The added urea serves as a source of nitrogen and helps increase the substantial elevation of biopolymer molecular weight in the two quantities of chitosan. Mishra and Kumar (2007) investigated the efficacy of chitosan production by using different sources of nitrogen such as yeast extract, dry cyanobacterial biomass, urea, *Pleurotus ostreatus,* and ammonium sulphate and they found the huge amount of biopolymer collected in the occurrence of cyanobacterial biomass because of its nutritional value. An experiment was performed by Benjamin and Pandey (1997) and his colleagues to investigate the production of biopolymers using coconut oil cake as a substrate for the production of chitosan under well-optimised conditions in candida rugosa. In addition, the influence of different sources of carbon such as sucrose, pectin, glucose, and galacturonic acid on the growth of *Aspergillus niger* fungal biomass was examined by submerged fermentation processes (Solís-Pereira et al. 1993). Finally, they observed that the complementary carbon nutrients induce to increase the biomass amount. Xie and West (2009) showed that the high acidic pH action of the solid substrate helps to increase the biomass content by a large amount.

Table 14.2 List of fungi used in the production of fungi by various methods

Fungal strain	Chitosan yield	Temperature	Fermentation process	References
Rhizopus oryzae	4.3 g/kg of soya bean residue	25–45 °C	SSF	Suntornsuk et al. (2002)
Mucor rouxii/ SMF	0.61 g/L for molasses salt medium	25–45 °C	SMF	Streit et al. (2009)
A. glauca	7:4% of dry cell weight	25–45 °C	SMF	Hu et al. (2004)
M. rouxii	7.3% of dry mycelial weight—Synthetic medium	25–45 °C	SMF	Synowiecki et al. (1997)
G. butleri USDB 0201	4–6 g/100 g mycelia—Sweet potato	25–45 °C	SSF	Nwe et al. (2002)
Mucor racemosus	35:1 mg/g dry mycelia	25–45 °C	SMF	Knorr and Teutonico (1986)
Cunninghamella elegans	20:5 mg/ g dry mycelia	25 °C45 °C	SMF	Da Silva et al. (2001)
P. blakesleeanus	10:1 6 2:9% dry mycelia	25–45 °C	SMF	Da Silva et al. (2001)
Gongronella butleri	1.19 g/L of chitosan per litre of apple pomace extract	25–45 °C	SMF	Streit et al. (2009)
R. oryzae	1.13 6 0.10 g/L of whey medium (in the presence of gibberellic acid)	25–45 °C	SMF	Chatterjee et al. (2008)
R. oryzae TISTR 3189/SMF	138 mg/g or 14% of dry mycelial weight—PDB medium	25 °C45 °C	SMF	Pochanavanich and Suntornsuk (2002)
G. butleri IF08081	730 mg/L of Shochu distillery effluent	25–45 °C	SMF	Yokoi et al. (1998)
A. coerulea/SMF	4.11 6 0.08 g/L of synthetic medium	25–45 °C	SMF	Mélida et al. (2015)

Chatterjee et al. (2008) optimised the production of chitosan biopolymer in *M. rouxii and R. oryzae* in the presence of salt treated molasses and whey growth medium and the effect of auxins, gibberellic acid, and kinetin. He found that, in the presence of low plant hormone concentrations, the chitosan content increased in both fungal species and, in the case of higher plant hormone concentrations, the growth of both microbial species was decreased. Some research studies depicted that the amount of biopolymer separated from the fungal species which harvest at their log phase.

14.3 Application of Chitosan in Food Industries

Because of its physiochemical parameters such as molecular weight (2–200 kDa), chitosan is used in various disciplines such as food science, medicine, agriculture, and cosmetics to break down the acetyl compound from chitin based on the solubility of the polymer chain in the presence of weak or strong acid. The solubility of chitin was improved by the protonation of the D-glucosamine amine group and eventually converted under acidic conditions to polysaccharides. Due to its biological activity and its protection for human beings, the comprehensive application of chitosan would be more favoured. In food industries, the application of chitosan has more fascinated consideration as a natural preservative because its antimicrobial activity in contradiction of different foodborne microbes like fungi, bacteria, and yeast. Due to the electrostatic interaction between the cationic charged biological ligand and the anionic charged microbial cell surface ligand, this biopolymer attempts to adjust cell diffusivity. The degradation of protein and other cellular compounds (Young et al. 1982) is accomplished by this interaction. In addition, by avoiding the diffusion of hydrolysed mRNA products and the synthesis of protein compounds, metal complexation, spore forming elements and nutrients, this biopolymer contributes to the antimicrobial mechanism (Hadwiger and Loschke 1981). Uchida et al. (1989) reported that chitosan has an important antimicrobial activity to inhibit fungal and bacterial growth in contrast to chitosan oligomers. In addition, the biopolymer's antimicrobial effect also depends on its molecular weight, degree of deacetylation, and type of bacterium (Tsai et al. 2005). The loss and gain of microbial growth inhibition by mixing chitosan into various food compounds containing carbohydrates, fat, protein, salts, and vitamins was studied by Roller and Covill (1999). In addition, Devlieghere et al. (2004) found that there was a negative effect on the addition of salt, starch, oil, and proteins to prevent the growth of microbial flora.

14.3.1 Effect of Chitosan in Bread

Bread, the most disbursed food mainly made from wheat flour used worldwide. Martins et al. (2017) suggested that the synergistic impact on gluten molecules such as intolerance, allergy, and celiac disease is a major challenge for researchers to explain the alternative source of bread output. Starch is a raw material component that can be found in various sources, such as maize, potatoes, rice, and cassava (Ferreira 2019). In addition, the pigment rice has higher levels of starch, protein, vitamin B, and minerals and consists of high nutritional values of pro-anthocyanides, carotenoids, and anthocyanins. Garry Kerch (2015) studied staling phenomena in order to verify the consistency of the bread's flavour and texture with regard to time. By hindering the initial toughness of a starch gel and limiting microbial growth, this blending of chitosan increases the shelf-life span. The research to increase the shelf life of baguette by impregnating chitosan with a molecular weight of 493 kDa was conducted by Park and his co-workers. The efficacy of the addition of chitosan was tested by coating various chitosan concentrations on the surface of the dough in the range of 0.5–1.5%. He noted that due to the moisture-resistant property of chitosan, 1% of the chitosan-coated sample showed many characteristics such as hardness, less weight loss, and retrogradation than the uncoated baguette at 26 °C for 36 h (Butler et al. 1996). The chitosan creates a barrier layer on the dough surface and resists the presence of moisture on the bread surface, thereby reducing weight loss, retaining hardness, retro deterioration, and relative to the regulation, the shelf life also increased one more day. Similarly, Park and Chong (2002) detected that the shelf life of bread can also extend to 24 h by coating the chitosan that has the molecular weight of kDa. Not only does the chitosan biopolymer boost retrogradation, but it can also increase the shelf life and consistency of the bread by inhibiting the growth of the microbes and preserving the property of antioxidation. Similarly, with a molecular weight of 120 kDa combined with 0.3% lactic acid, the chitosan concentration increased from 1 to 2%, decreasing the bacterial count and thio-complexed barbituric acid reactive compounds and increasing the water content after 8 days at room temperature compared to uncoated chitosan. The presence of mould was found after 4 days in the control. Lee et al. (2002) revealed that the carboxymethyl modified chitosan also increases the shelf-life period of fermented bread by inhibiting the microbial growth and delaying interaction between amylose and amylopectin in starch molecule. Similarly, the same research group examined the antimicrobial activity of chitosan by varying the molecular weight from 1, 5, 30, and 120 kDa with a degree of deacetylation of lower (Mw 1 and 5) and higher molecular weight chitosan was (30 and 120 kDa) 92% and 85%, respectively, and also in the presence of lactic acid and reacted with varying dough concentrations 0.01%, 0.1%, 0.3%, and 0.5%, respectively. They found that, after 8 days of incubation at room temperature, the lower concentration of chitosan and control showed similar results in bacterial growth of about 106 CFU/mL viable cells. Colour, water content, pH, and titratable acidity analysed by (Silva et al. 2020), basic volume and textural properties of gluten off bread by reaction between red rice flour starch and cassava flour and transglutaminase enzyme with a chitosan concentration of 0, 1, and 2%. The presence of chitosan and the activity of the transglutaminase enzyme alter the bread structure and the enzyme and chitosan increase the stabilised network. Although the presence of chitosan may also have some unfavourable consequences, it decreases the lag time of yeast fermentation follow-on in bread with less precise volume and greater inflexibility and chewiness.

14.3.2 Effect of Chitosan in Fruits and Vegetables

In general, due to the lack of proper storage devices, fruits and vegetables are contaminated by yeast, fungi, bacteria, and mould after harvesting (El Ghaouth et al. 1991). By edible coating and storage of fruits and vegetables in cold temperatures, Park et al. (2005) attempted to minimise microbial infection. The non-toxic coating serves as a protective shield, decreasing the rate of fruit respiration and transpiration, not allowing microbial species to develop and change in colour, increasing the quality of texture (Kester and Fennema 1986). By altering the levels of ethylene, oxygen, and carbon di oxide, the semipermeable membrane shown by the coating keeps back the ripening of the fruit. In addition, the chitosan coating avoids the diffusion of levels of carbon di oxide as compared to oxygen in order to avoid the growth of anaerobic microbes in the inner atmosphere of fruit and vegetable tissue and to extend the storage time and to regulate the fruit deterioration. El Ghaouth et al. (1991) analysed the antimicrobial activity on the surface of strawberry fruit by coating chitosan biopolymer in the presence of *Botrytis cinerea and Rhizopus sp*. As a result, the decay of the fruit and the consistency of the fruit are decreased by the subsequent dip of the fruit into chitosan solution (1 and 1.5%) at 13 °C. The chitosan forms a powerful barrier, prolongs the time for anthocyanin synthesis and high acidity, decreases the rate of growth and breathing of fungicides (Reddy et al. 2000).

El Ghaouth et al. (1991, 1992) delineated that the biopolymer has the capability to inhibit the growth of fungal by persuading the chitinase enzyme to chitosan has the ability to inhibit growth of several fungi to induce chitinase, a defence enzyme, and to provoke phytoalexin in pea shells. Enzyme-induced defence mechanism by the combination of chitinase and β-1,3-glucanase. Notably, the enzymatic defence mechanism is not so effective in the presence of chitosan in unbroken fruit. This is mainly because of the intact fruit surface that does not allow the surface of chitosan and tissue to interact. In strawberry cuticles that have no porus on the surface, a similar form of observation was made. The edible coatings have recently been used as a method for combining different functional components such as flavours, chromophores, nutraceuticals, and antimicrobials agents (Han et al. 2004; Hernández-Muñoz et al. 2006). Calcium, potassium, vitamin E, and oleic acid are combined with chitosan and used as an edible coating to improve the shelf life and nutritional value of fruits and vegetables. The capacity of antimicrobial activity and moisture barrier, weight loss, colour on strawberries and red raspberries were demarcated by Han et al. (2004), by the three distinct chitosans such as native chitosan, 5% glucanol modified chitosan and 0.2% DL-alpha-tocopheryl acetate modified chitosan. This three distinct chitosan has a deacetylation degree of about 87% and is treated as 1 and 2% with separate acetic acid concentrations and incubated at lower temperatures. The results obtained have shown that the chitosan coatings selected are capable of increasing antimicrobial activity, reducing the rate of colour changes and decreasing pH changes. The addition of calcium and vitamin E, however, results in substantial improvements in weight loss and regulation of moisture. Due to the hydrophilic aspect of the chitosan films, the calcium and vitamin E modified chitosan decreases the penetration of water molecules through the matrix (Amarante and Banks 2001).

14.3.3 Effect of Chitosan in Kimchi

Kimchi, a mixture of Chinese cabbage and various ingredients, such as red pepper, green onion, garlic, and fermented fish sauce (Choi et al. 2016), is a fermented vegetarian food. No et al. (1995) elucidated the edible period of the kimchi food can extensively increase by the addition of chitosan oligosaccharides with the concentrations of 0.05–0.02% in 0.2% of water. They reported that, relative to the control at 20 °C, the addition of chitosan increased the shelf-life span from 2 to 6 times (Yoo et al. 1998). No et al. (1995) evidenced that the period of deterioration can be extended for 10 days by salt-modified chitosan (0.1–0.15%) comparatively to control. Chitosan biopolymer decreases the softening of cabbage tissues and the interaction of chemical compounds. In addition, chitosan blended in preservatives, medical herb extracts, sodium benzoate, liquid calcium form which helps to improve quality and minimise degrading effects (Son et al. 1996).

14.3.4 Effect of Chitosan in Meat

In general, meat has the property of being quickly affected by lipid oxidation, contributing to rancidity on the surface of the meat. Chitosan serves as an antioxidant and antimicrobial ability to keep back lipid oxidation and hinder the growth of microbes that spoil food on the meat surface, as shown by several researchers. Darmadji and Izumimoto (1994a) showed that the addition of 1% of chitosan decreases the value of thio-barbituric acid (TBA) on the meat surface by 70% compared to the control stored at 4 °C for 3 days. The value of TBA remains same for 10 days in the presence of 0.5–1.0% of chitosan. Likewise, chitosan has the property to enhance the colour of beef at the time of storage. Darmadji and Izumimoto (1994b) analysed the stability of the storage of minced beef blended with 100 ppm of nitrite and *Lactobacillus plantarum* with chitosan cumulated with nitrite and *Lactobacillus plantarum*. He noticed that bacterial growth decreased to double the log cycle, TBA decreased to 36%, and the remaining nitrite degenerated by 63%, creating a better colour on the fermented surface. The growth of bacterial cells inhibited on chilled pork by the addition of 0.3–0.6% of chitosan glutamate to immature minced pork was investigated by Sagoo et al. (2002). He found that the chitosan reduces the count of viable cells, lactic acid bacteria by 3 log of CFU/g yeast, and moulds for 18 days at 4 °C. Juneja et al. (2006) depicted that the additional 3% of chitosan glutamate which has deacetylation of around 86% significant effect on reducing *Clostridium perfringens* spore at 7.2 °C in 12 and 15 h. In addition, the shelf life of the pork can be increased from 0.1 to 1.0% by the different chitosan concentration, which has a molecular weight of 30 and 120 kDa. The dipping of the fork into the chitosan biopolymer was observed, and 5% trisodium phosphate decreased the number of aerobic bacteria on the surface of the pork. Wu et al. (2000) disclosed that soy protein did not have a major effect on reducing or regulating lipid oxidation when packing pork with chitosan (DD-85%: in the presence of 1 and 2% formic acid). Because of that, on the beef surface, the existence of TBA increased, leading to increased oxygen permeability.

14.3.5 Effect of Chitosan Added with Seafood and Seafood Products

Due to lipid oxidation of fatty acids, the mechanism catalysed by the presence of hematin compounds and metal ions in the fish muscle (Decker and Hultin 1992) can possibly destroy seafood. In addition, the consistency of seafood deteriorated by autolysis, protein loss, and contamination demonstrated by microbe development (Jeon et al. 2002). Kamil et al. (2002) investigated the antioxidant activity of seafood with different viscosities in the range of 14–360 cP, molecular weight of chitosan 660–1800 kDa in cooked, comminuted flesh. Chitosan concentrations of 50–200 ppm and traditional antioxidants such as tert-butylhydroquinone, butylated hydroxy anisole, and butylated hydroxytoluene were evaluated for the effect of oxidative stability and maintained at 4 °C. From the obtained result, they observed that optimised parameters for reducing lipid oxidation on surface at viscosity 14 cP, chitosan concentration at 200 ppm for 8 days stored at 4 °C. It clearly reported that the antioxidant potential on the fish surface was obtained at the molecular weight of chitosan and at its concentration of 200 ppm and 14 cP viscosity. Peng et al. (1998) showed that chitosan modified by ferrous ions decreased lipid oxidation on the surface of the fish. Amino groups have been active in metal ion chelation.

14.4 Applications of Chitosan in Pharmaceutical and Biomedical Field

Chitosan is used in pharmaceutical formulation and drug delivery in order to increase sustained release of various drugs–vaccines, antibiotics, proteins, anti-inflammatory drugs, peptides, factor of growth, antimicrobial therapies, gene therapy, and delivery of genes. In addition to that, this polymer used in the area of tissue engineering for the regeneration of wounds and burns, tissue, bones, neurons, muscles, cartilage, tendons, liver, ligaments, and also applicable in different diagnostic approaches in skin treatment, cancer therapy, ophthalmology, and dental care. Moreover, the chitosan has the property of sensing heavy metals, bioimaging, matrix for holding the enzymes and in the field of veterinary medicines.

14.4.1 Oral Sources of Dosage

Sawayanagi, Y investigated the extent of disintegrating of oral dosage by the blending of chitin, chitosan or MCC and mixture of lactose or. From the result he observed that the chitosan has more potential on sustained drug release with respect to time function which is shown in Fig. 14.2 (Sawayanagi et al. 1982; Ofori-Kwakye and Fell 2001). The lag time of disintegration potential of the three biopolymers additives are arranged in the increasing order (chitosan > chitin > MCC) with respect to their respective concentrations. It is mainly due to the angle of rest and it was considerably decreased by enhancing the concentration of chitin or chitosan biopolymers mixed with lactose or potato starch comparatively to the MCC. Chitin

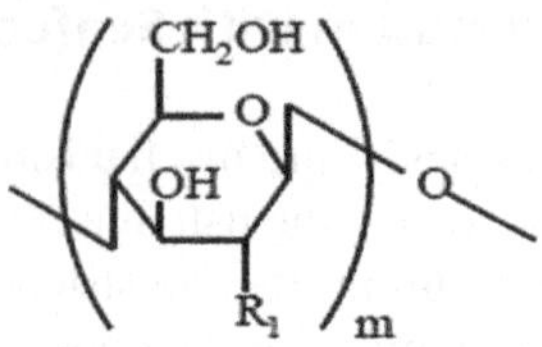

Materials	R_1
Chitin	$NHCOCH_3$
Chitosan	NH_2
N-Succinyl-chitosan	$NHCO(CH_2)_2COOH$
N-Trimethyl-chitosan	$N^+(CH_3)_3I$
Mono-N-carboxymethyl-chitosan	$NHCH_2COOH$
Chitosan-thioglycolic acid	$NHCOCH_2SH$

Fig. 14.2 Chemical constituents of chitin and chitosan biopolymers (Kato et al. 2003)

and chitosan are extensively applied in the place of MCC due to their higher lubricating property. Likewise, the complexation of chitosan and lactose increases the tablet hardness with respect to rise in chitosan concentration, suggesting that chitosan also has a strong combinative property than MCC (Sawayanagi et al. 1982). The important additives for direct compression have been reported to be chitin and its derivative, Chitosan.

14.4.2 Dressing of Wounds

To accelerate the early recovery, lesion dressing is essential for burn wound healing. The purpose of bandaging is to reduce the development of hematoma and inflammation, to limit dead space, to defend against further infection or trauma, to retain drainage, to create sufficient oxygen stress, to maintain a humid atmosphere, and to minimise movement (Choate 1994). A moist condition that facilitates angiogenesis is important for the distribution of wound healing cellular components (Liptak 1997). The effective wound dressing will have the following characteristics: (1) avoidance of growth of bacteria on the injured skin, (2) exudate captivation, and (3) epidermal formation and granulation. Currently, the Beschitin® -W (Unitika Ltd., Osaka, Japan) introduced the chitin in the bandage as a protective coverage material for the wound healings. The therapeutic effects were studied using a rat thermal damage model by the four different forms of wound dressings which include three films and one spongy membrane, i.e. chitin/chitosan mixture spongy membrane, chitin/chitosan mixture film, chitin film, and chitosan film (Tachihara et al. 1997a, b). Directly after drenched in the buffer, the chitin/chitosan spongy membrane fascinated some solvent, and then incremental water absorption was observed, indicating that higher absorption was observed in the spongy membrane compared to any other films examined. Similarly, in comparison with the other film forms, the spongy membrane also showed high silver sulfadiazine (SSD) content. The spongy membrane believes to have more microscopic structure in turn which has the ability for high drug concentration.

14.4.3 Muco-Adhesive Oral

Mucus, viscoelastic gel adhered to tissue of the mucosal region in various organs such as respiratory, reproductive tracts, eyes and gastrointestinal in human body. It comprised with H_2O (90–98%), 0.25–5% w/w of mucins, salinity level 0.5–1%, percentage of protein 0.5% w/v, DNA, bacteria, lipids, and cellular debris (Khutoryanskiy 2011). Mucins is the major component of the mucus; it has gel-like property due to the presence of glycoprotein. The glycoprotein has protein core, and it is covalently bonded with the side chain of its carbohydrate molecule through *O*-glycosidic linkages (Peppas and Buri 1985). Several research articles stated that the chitosan derivative has the muco-adhesive properties. The hydroxyl and amine functional group of chitosan has the characteristic of increasing the solubility of chitosan at lower pH conditions. In acidic condition, the pronated amine group electrostatically interact with the sialic acid and the epithelial surface. Importantly, the muco-adhesion characteristic formed by the hydrogen bond and apolar interactions. Recently, chitosan derivatives used as a tool for transmucosal drug delivery process in the discipline of nasal vaccination and eye drops. The commercial oral muco-adhesive film was initially introduced by WaplonP® (Kowa Co., Ltd., Aichi, Japan) for the treatment of stomatitis. In this product, the muco-adhesive property brought by the combination of chitosan and sodium alginate. Kawasaki stated that modified chitosan helps for the sustained release of and diltiazem which shows the efficiency more than two times than oral administration (Kawase et al. 1997).

14.4.4 Adhesive for Water Resistance

Naturally, the marine animals used water-resistant protein to attach on the submerged surfaces. The protein biomolecules built with repetitive groups of hydroxyproline, dihydroxyphenyl aniline, and lysine. The adhesion process carried out by conversion of o-diphenolic residues to O-quinone mediated by catechol oxidase enzyme (Waite and Tanzer 1981). Then the O-quinone followed a non-enzymatic reaction to make complex with lysine protein molecules to form protein gels which give more adhesive strengths. But there is no well-defined literature proof for the interactions so far. Yu and Hwang (1999) reported that the oxidation of dihydroxyphenyl aniline has the capability of gel formation but the non-oxidative form of DOPA can be used as a water-resistant adhesion material. Waite stated that the mechanism involved in the water-resistant adhesion by the quinone chemistry is very complex. In addition to that Conner and Hemingway (1989) narrated three different roles in the adhesion process by the phenols and quinones that are very remarkable. The three steps are (1) the phenolic residues present in phenol formaldehyde is having similar structure with the polyphenolic adhesive proteins, (2) the synthetic quinone comprised polymeric compounds has adhesive properties, (3) dihydroxyphenyl aniline converted to O-quinones by auto-oxidation process. Domard (1987) used to tyrosinase enzyme oxidise the low molecular weight compound dopamine and then the quinone reactions facilitated by the biopolymer (Chitosan) comprised with polysaccharides, and it is not a protein. The amine group present in the chitosan deprotonated at dissociation constant 6–6.5. Muzzarelli et al. (2001) reported that the adaptation of phenolic on chitosan mediated by tyrosinase converted into quinone-tan modified chitosan, formed crosslinked gel. In earlier studies, the same researchers revealed that the quinone-tanned chitosan behave like ag gels which has water-resistant adhesive properties.

14.5 Carriers for Drugs

14.5.1 Microparticles/Nanoparticles

Studies performed, gadolinium neutron capture for cancer treatment mediated by chitosan nanoparticles acts like a chitosan nanoparticle act like drug carrier for gadolinium (Yamada et al. 2000). In this study, the drug gadolinium was administered concentration in tumour tissue on mice around 74% and delivered with the help of gadopentetic acid-adapted chitosan nanoparticles which is approximately 430 nm. The result revealed that the tumour growth was effectively repressed by chitosan-modified gadolinium. Chitosan colloidal particles can be successfully used as an oral vaccine due to its quick absorption on lymphoid tissue (Tokumitsu et al. 2000). Similarly, the interaction of (IgG) immunoglobin and local (IgA) with DT-adapted chitosan shows a noteworthy effect comparatively than intragastrical alimentation with diphtheria toxoid (DT) in PBS (Van der Lubben et al. 2000). From the safety point of view, non-viral gene transmission has gained a lot of attention. In addition to that, several studies have reported that the different types of cell lines could be transfected with the use of DNA-nanospheres (200–750 nm) mediated by gelatin and chitosan (Van der Lubben et al. 2001). But the DNA-chitosan nanospheres could not be crosslinked during the preparation. While DNA-chitosan nanospherical particles had relatively low in vitro transfection efficiency and were close to DNA-gelatin nanospheres, many non-viral vectors suffer from the problem. During the fabrication of nanoparticles, different types of water loving and water hating drugs loaded with chitosan nanoparticles. The loading efficacy of the drug mainly depends upon the method of separation and physiochemical properties. In cancer therapy process, the polar 5-flurouracil completely adhered with chitosan nanoparticles (200–300 nm) chemically crosslinked in the presence of glutaraldehyde in water in oil emulsion process. Mitra (2001) described about the suppression of tumour by conjugating the doxorubicin drug with chitosan nanoparticle using reverse micelle process (Van der Lubben et al. 2003). Janes et al. (2001) reported that tripolyphosphate and crosslinked chitosan nanoparticles is used to deliver lots of small molecular drugs and bio-macromolecular treatment. For example, through ionotropic gelation of the chitosan with tripolyphosphate, researchers effectively encapsulated DOX with the chitosan nanoparticles. The cytotoxicity of DOX-binded nanoparticles in A375 cells of human lump and C26 cells of murine colorectal carcinoma showed that nanoparticle dextran sulphate-containing formulations were able to slow cytostatic action relative to unmodified DOX.

14.5.2 Conjugates

The functional groups such as amine and carboxylic group present in N-succinyl complexed chitosan (Suc-Chi) derivatives can easily react with several compounds shown in Fig. 14.2. The conjugation of (Suc-Chi) with water-soluble carbodiimide (EDC) was mediated by direct link of mitomycin C (MMC) and changed in water-insoluble compound and water-soluble conjugates (Leong et al. 1998). After administration, the distribution of (Suc-Chi) in different organs such as liver, kidney, and spleen is less than 5% of dose per gram of tissue (Mitra et al. 2001). Moreover, these conjugates with less than 10% of the dose/g tissue administered and tested in preputial gland, small intestine, backbone, femoral muscle, and peritoneum organs. But more than 10% of N-succinyl-modified chitosan was transferred to the prostate and lymph nodes for treatments (Janes et al. 2001). The conjugates can increase the molecular weight, and it lowers the biodegradability when it is complexed with succinyl groups. The anionic charges help extend the half-life of the polymers in the systematic circulation process due to minimum interaction with tissues and blood vessels (Song et al. 1992) and supermolecules are described to stay longer in the body related with low molecular weight molecules (Kato et al. 2000a, b).

14.5.3 Antitumor Activity

Hydrophilic Suc-Chi-mitomycin C conjugate, antitumor activity was examined in 180-, M5076-, and MH134-bearing sarcoma mice (Kato et al. 2003). Such conjugates show strong antitumor activity. Suc-Chi-MMC conjugates (Suc-Chi-MMC conjugates) were also shown to have strong antitumor activity contrary to sarcoma 180, melanoma of B16, and murine leukaemia (L1210 and P388) (Takakura et al. 1987; Sezaki and Hashida 1984). Although the conjugate compounds are much better, with a slight reduce in body weight but no lethality, but the excess quantity of free drug subsidised to lethal side effects. The anticancer activity of MMC decreased by the chemical modification of the aziridine ring, one of the active groups when the substituents is increased (Kato et al. 2002a, b).

14.5.4 Enhancers for Intestinal Absorption

N-trimethyl-chitosan acts as an ampholyte ion consisting of quaternary amines by methylation of amino groups and mono-N-carboxymethyl-chitosan (mCM-Chi) shown in Fig. 14.2. As intestinal absorption enhancers, these chitosan derivatives are productive because they open tight intracellular junctions through interaction with mucosal epithelial cells.

14.5.5 Tissue Engineering

Researchers have stated that conjugates of chitosan and thioglycolic acid used as an innovative material for the scaffold process. In its structure, Chi-TGA has immobilised HS^- groups (Fig. 14.3), and the formation of disulphide bonds with the polymeric system was revealed by the conjugate process. The e application of sulphide groups with chitosan forms a new polymer showing dissimilar properties (water solubility, biodegradability, and muco-adhesion, in situ gelling). At a physiological pH and temperature (at 37 °C), the gel-like scaffolds organised using Chi-TGA conjugates can form a regular shape parallel to the native tissue because in situ gelling depends on the quantity of thiol groups encapsulated to the polymer at the concerned pH.

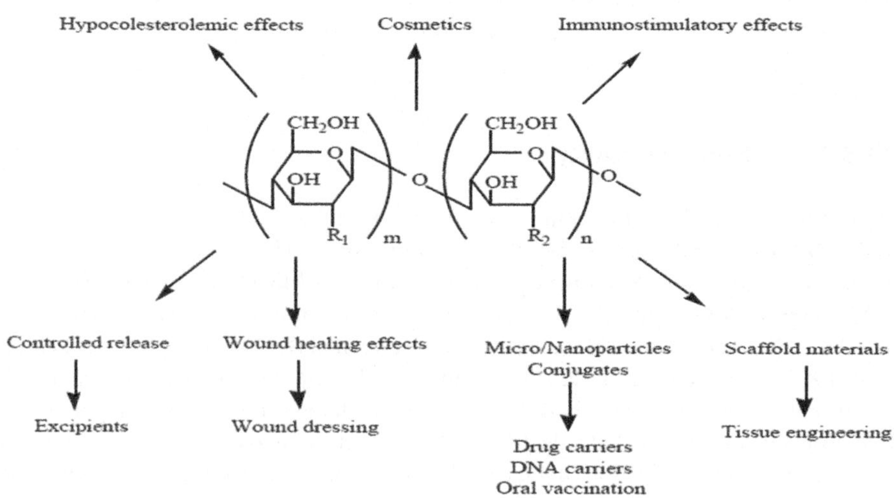

Fig. 14.3 Application of chitin and chitosan derivatives

14.5.6 Dentistry

The application of chitosan in dental medicine has the properties of bioactivity, wound healing, anti-inflammatory, bone repair, and hemostatic actions (Shirahata 1984; Hamman et al. 2003). The different forms of chitosan are used as a solution form (acetic or methane sulphonic acid used as solvents) in the form of microspheres, hydrogel and toothpastes, and its combination with additives such as amorphous calcium phosphate, amelogenin, and quinic acid has enhanced the ability of these chitosan preparations to prevent tooth decay and enamel degradation. The gel/hydrogel form of chitosan refers to the treatment of chronic periodontitis and canker sores. Chitosan can play an important role in preventive dentistry because of its inherent flexibility, efficacy, and the ability to act as a protective shield to the intrusion of acids into enamel and its mineral loss. Antimicrobial effect on oral biofilm and decrease in the number of Streptococcus mutans in the oral cavity are functions of toothpastes, mouthwashes, and chewing gums dependent on chitosan and herbal fullness. Chitosan complexes and microparticles of fluoride improve the absorption and safety of fluoride cavities. Chitosan-based endodontic cements minimise inflammation and help the rebuilding of bone.

14.5.7 Veterinary Medicine

In veterinary medicine, researchers investigated the applications of chitosan curing of wound, regeneration of bone, antimicrobial and analgesic effects (Queiroz et al. 2015). The application of chitosan for drug and vaccine delivery in veterinary animals and as a dietary element was also addressed. The chemotherapeutics such as painkillers, antiparasitics, anaesthetics, antibiotics, and growth promoters to mucosal epithelium for absorption for local or systemic operation are the veterinary drug delivery, areas most likely to benefit from chitosan and the delivery of immunomodulatory agents to the lymphoid tissue associated with the mucosa for local immune response activation or modulation.

14.5.8 Cosmetics

Generally, most of the chitosan products has huge molecular weight which is very difficult to enter into the skin and makes a significant effect in skin care. The major chitosan derivatives used for the skin care products are chitosan complexed hydrochloride, acetate, lactate, carboxymethyl, oligosaccharides, as well as chitin sulphate, carboxymethyl chitin, and quaternised derivatives. It can be used on both dissolved and solid forms in aqueous solutions. The chitosan-blended cationic cosmetics are bacteriostatic, fungistatic, film-forming, antistatic moist-retaining and retains the moisture content very less and preserve the hair style with high humidity and also it releases the bioactive compounds very slowly. The chitosan biopolymer is more companionable with the components such as starch, glucose,

Table 14.3 Applications of chitosan and its oligosaccharides in dentistry (Queiroz et al. 2015)

Topic	Form	Application
Surgery in dental	Solution	Periodontitis, erosive lesion, dental plaque inhibitor, agent to prevent dental caries
Dental treatment	Hydrogel	Constituent in dentifrices
Dental constituents	Toothpastes	Dental care: Toothpaste, chewing gum, Oral care
Transplants	Bio-adhesive	Increase salivary secretion, antibacterial effects
Drug delivery systems	Powder	Fluoride delivery
Oral hygiene	Granule	Acts as adhesives for dental
Antibacterial effect	Nanoparticle	Nano-biomaterials
Healing dentistry	Sponge	Used to enhance wound therapeutic in bone tissue
Dental composites	Composites	Use for scaffolds and transporters for molecular treatment
Wound healing		Cell-defensive action
Nano-dentistry		Cell shielding action

polyols, oils, fats, saccharose, acids, nonionic emulsifiers, waxes, and nonionic water-soluble gums. Additionally, this biopolymer used in hydrating agents, solar filters, and other bioactive materials. Some of the applications for chitosan (Table 14.3) are: hair care, e.g. shampoos, colouring products, hair spray, and setting lotion, creams, and lotions (products for the face, hand, and body), colour cosmetics (make-up, nail polish, eye shadow, and lipstick), deodorising products, active agent microencapsulation, and dental care, e.g. toothpaste, tooth gel, and mouthwash (Queiroz et al. 2015). Chitin is also beneficial in cosmetology since the skin tolerates it well. It is an efficient moisturising agent and a film-forming tensor that often has two advantages: it provides water and prevents dehydration. Chitosan and chitin also have metal-chelating properties that are responsible for a very large number of exposure allergies. There are several commercially available products based on chitosan for cosmetic use: Curasan™, Hydamer™, Zenvivo™, Ritachitosan®, Chitosan MM222, etc (Tables 14.4 and 14.5).

14.5.9 Antimicrobial Agent

The antimicrobial effect of chitosan leads by the electrostatic interaction between chitosan and the functional groups present on the surface of the bacteria and damage the physiological membrane, e.g. by facilitating the leakage of intracellular components and also by suppressing the flow of nutrients into the cells. Chitosan's

Table 14.4 Applications of chitosan and its oligosaccharides in veterinary medicine (Queiroz et al. 2015)

Topic	Form	Application
Types of drug delivery systems	Solution	Sustained release of drugs in animals
Delivery of vaccines	Powder	Mucosal formations
Accessory	Microsphere	Mucosal delivery of antigens
Properties of biological effect	Microcapsule	Enhance the immune response
Immunisation of mucosal	Nanoparticle	Haemostatic
Healing of wound	Gel, hydrogel	Wound healing action
Regeneration of tissues	Sponge	Regenerative medicine in the field of tissue engineering
Nutritional supplement	Film	Vaccines
Nanomedicine	Filament	Surgical threads

Table 14.5 Applications of chitosan and its oligosaccharides in cosmetics (Queiroz et al. 2015)

Topic	Form	Application
Toiletry	Solution	Functional additives
Hair care	Powder	Moisturisers: Maintain skin moisture, tone skin
Personal care	Film	Thickening agent
Cosmeceuticals		Hydrating and film-forming agent
Oral care		Role in surfactant stability; stabilise emulsion
Dental care		Antistatic effect
Hygiene		Bacteriostatic
Skin care		Delivery systems
Essential oils		Products: Shampoos, creams, skin creams, creams for acne treatment, lotions, bath lotions, nail polish, fixtures, make-up powder, lacquers, nail lacquers, nail enamel, varnishes, hair sprays, hair colourants, wave agents

antimicrobial effect against a spectrum of microbial and fungi arising from its polycationic existence is well known.

14.5.10 Anticholesterolaemic Effect

It is known that the entrapment instigated by a viscid polysaccharide solution decreases the solubility and absorption of dietary fatty acids. Conversely, the electrostatic force of interaction between chitosan and anion substances such as bile acids and fatty acids is measured by the involvement of the amino group in its structure. The interaction of chitosan with anionic surface-active materials depends on its three reactive functional groups: the secondary amino group, C-3 and C-6 primary and secondary hydroxyl groups, respectively. While considerable efforts have been made to find a connection between the physicochemical characteristics of chitosan and its ability to bind fat, only a few important associations have been shown (Morin-Crini et al. 2019).

14.5.11 Antioxidative Activity

Chitosan has substantial scavenging ability against various free radical species, which gives similar effect that caused with commercial antioxidants. The existence of toxic-free radical will be scavenged by the chitosan derivatives such as 1.1-diphenyl-2-picrylhydrazyl (DPPH) radical, superoxide, hydroxyl, and alkyl radical samples isolated from crab shell with DD (degree of deacetylation) of about 90%, 75%, and 50%, respectively. The obtained results were showed that the higher DD chitosan biopolymer exhibited the highest scavenging activity against the free radicals. Chitosan Mw was found to have a negative association with activity. The chitosan sulphate derivatives had a higher scavenging effect on peroxide radicals, but the lowest Mw chitosan demonstrated a higher ferrous ion-chelating potential versus others (Chen and Chou 2005). The major reason for selecting chitosan as a natural antioxidant for stabilising the lipid containing foods and also to extend their shelf life by the chelation of metal ions. By chelating ferrous ions existing in the system, chitosans can retard lipid oxidation, thereby eliminating their pro-oxidant activity or their convert to ferric ions.

14.6 Applications of Chitosan in Agriculture

Generally, the plant pathogens are most economical species used as agricultural microbial source around the world. The pathogens are mainly controlled by chemical pesticides. But the usage of chemical pesticides leads two major issues (1) propagation of confrontation in pest inhabitants and (2) increasing the contamination of unpreserved agricultural crops. Several environmentally friendly approaches have been used to reduce the harmful effects of synthetic pesticides in order to solve these problems. These methods are capable of reducing the concentration of pesticides but these processes are not economically feasible (Chen and Chou 2005). Subsequently, to minimise the effect of pesticides, certain traditional approaches are used, but these techniques are not economically viable and are very difficult to apply in practical circumstances. So, an alternative that is very healthy and eliminates the harmful effects on human health, and the environment is introduced by natural polymer chitosan (Lim and Hudson 2004).

14.6.1 Chitosan in Managing Plant Diseases

The chitosan biopolymer can be extensively used to monitor the pathogens occurring in the plants and the successive of the application of the polymer depends upon the

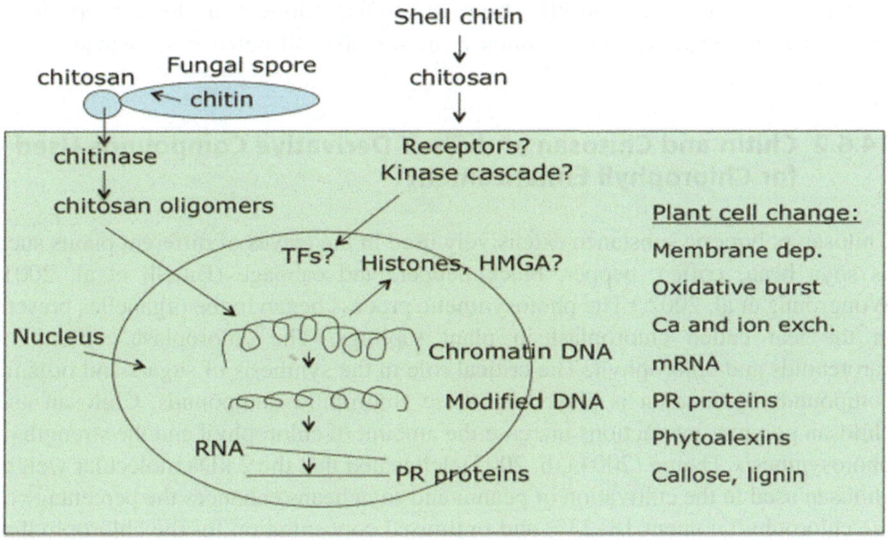

Fig. 14.4 Some projected properties of chitosan act as elicitor of plant defence mechanism (Orzali et al. 2017)

pathosystem, chitosan derivatives, concentration of the biopolymer, degree of deacetylation of chitosan, viscosity, and effect of treatment (Jeon and Kim 2002). Clearly, chemically modified chitosans such as N-carboxymethyl chitosan, methyl pyrrolidinone chitosan, N-phosphono methyl chitosan, and N-carboxymethyl chitosan limit fungal growth in plants (Lafontaine and Benhamou 1996). Chitosan has the properties of enhancing the plant's growth and reducing the diseases borne by the pathogenic soil. He noted that fungal species such as *Fusarium oxysporum* and *Radicis-lycopersici*, bottled up by the modification of chitosan, made the root rotten (Fig. 14.4).

The low molecular weight chitosan acts like a highly potential biotic elicitor, recognising plants (pathogen signal metabolites) that help activate plant defences. These mechanisms try to activate multiple pathways that increase resistance to crop diseases. Falcón-Rodríguez A.B observed that, by synthesising new enzymatic molecules, the defence process produces a chemical and mechanical barrier for agricultural plants. Furthermore, the biopolymer induces the hypertensive effect of destroying the microbes around the infection region. They recently found that, accompanied by the synthesis and accumulation of secondary metabolites, the defence mechanism. Callose, phytoalexins, lignin, PR proteins, and the enzymes involved in the process are PAL, peroxidases, and chitinase, which are the main compounds used in the self-defence process. The plant cell receptors recognise the elicitors through the transmembrane cell, but the apt mechanism was not yet elucidated properly. Figure 14.2 shows the interaction between DNA and chitosan for the direct evocation on activity of gene activity in cell defence. The predictive model shows that by altering the structure of DNA along with the transcriptional factor of a mobile group HMG, chitosan induces the self-defence mechanism.

14.6.2 Chitin and Chitosan and Their Derivative Compounds Used for Chlorophyll Enhancement

Chitosan polymeric substance extensively used in the leaves of different plants such as soya bean, coffee, pepper, black pepper, and cabbage (Bittelli et al. 2001, Wongroung et al. 2002,) The photosynthetic process began in the organelles present in the leaf called chloroplast in plant saplings. The chloroplast consists of carotenoids and chlorophyll. The critical role in the synthesis of sugars and organic compounds by fixation is played by these chloroplast compounds. Chitosan and chitosan polymer interactions increase the amount of chlorophyll and the strength of photosynthesis. Dzung (2004a, b, 2005) delineated that the 2 kDa molecular weight chitosan used in the cultivation of peanut and soya beans enhances the percentage of the chlorophyll content 18–23% and optimised concentration for the chlorophyll at 30 ppm. With the addition of chitosan around 40–50 ppm, many plants such as black pepper and coffee increases their antimicrobial activity. Table 14.6 shows the amount and effect and on concentration of chitosan oligomers present in induces the chlorophyll in some plants. Nitar et al. (2004) investigated property of the leaves of chlorophyll improved by the addition of chitosan around 20 ppm for 1 day. It is observed that the colour of the leaves is greater than control.

14.6.3 Stimulating Seed Germination

It is one of the most effective methods for stimulating the germination process and is by increasing the bioactivity of chitosan and its derivatives. This phase of bioactivity was started in 1996. Hadwiger et al. (1986), Yano and Tsugita (1988) observed the chitosan derivative such as depolymerised chitin, and hydroxyethyl-chitin, CM-chitin enhanced the germination of soya bean around 6%. The efficacy of seed germination was investigated by Li and Wu (1998) using various parameters such as the degree of acetylation, the percentage of chitosan used, and the organic acid used. Different plants, such as cabbage, mung bean, radish, lettuce, green pea, alfalfa, and spoon cabbage, were involved in this research. He observed that

Table 14.6 Effect of chitosan concentration to enhance the photosynthesis process

Conc. of Chitooligomer	0 ppm	20 ppm	30 ppm	40 ppm	80	100
Peanut	4.35	5.42	5.75	5.18	5.08	–
Soya bean	2.76	3.16	3.23	3.10	2.92	
Coffee	2.24	2.27	2.59	2.31	2.31	2.14
	2.21 2.18	2.24	2.13	2.453	2.18	2.52
Coffee seedling	1.05	1.15	1.43	1.70	1.38	

germination is found in the system which was dispersed in chitosan, DD, and acid 1% acetic acids. But the researchers discovered that they were not mediated in the absence of acetic acid. Similarly, the same research was conducted to improve wheat seed germination by coating chitosan for 15 min with a molecular concentration of 2–8 mg/L. Finally, he found that the untreated wheat has reached 50–57% and the treated gains 70–90%. Sarathchandra and Jaj (2004) did the experiment to observe the germination of treat pearl millet by fixed concentration of chitosan around 4%. The millet seeds were immersed in different chitosan solution which was prepared by different concentrations such 1:5, 1:10, 1:15, 1:19, and 1:25 respectively, soak time is around 3–9 h. Eventually, the dilute sample at 1: 19, soaking time: is around 6 h shows a significant increase in seed germination around 91% compared with control is around 83%.

14.6.4 Improve Mineral Nutrient Uptake of Plants

Dzung (2005) recorded that chitosan and its derivatives were squirted with a 25-ppm concentration on the cabbage leave, and then about 11% of the amount of nitrogen diffused into the leaves of the cabbage. Li and Wu (1998) defined that the components present in spoon cabbage would never be altered, even if the seed was coated with chitosan or in the case of successful yield treatment. Likewise, a similar kind of studies was carried out by Li and Wu (1998). He researched this about 20–80 ppm of the chitosan oligomer sprayed on leaves. This was delineated by the results obtained; the amount of nitrogen content increased by the accumulation of minerals in the plants about 3.3–13.6%, respectively. Chitosan oligomers have shown an effective capacity of about 50 ppm for enhancing mineral assimilation in coffee. In the coffee leaves, the addition of 50 ppm chitosan helps to accumulate about 9.49%, 11.76%, 37.7%, and 18.75%, respectively, of nitrogen, potassium, calcium, and manganese.

14.6.5 Chitosan as Soil Amendment

From the previous discussion, we demonstrated that the biopolymer of chitosan has the potential to minimise microbial activity and helps to retain bioproducts without altering their characteristics and features. In addition, the mixing of chitosan on the surface of the soil often has a direct impact on reducing microbial levels and improving physiochemical conditions.

Several researchers observed that, after adding the biopolymer, the growth of fungi such as *Cylindrocladium floridanum, Aspergillus flavus* infections, and Alternaria solani in the soil is effectively reduced. The assimilation of the soil's chitosan biopolymer serves as an elicitor, which provides a signal to trigger the plant's DNA self-defence mechanism. Similarly, the same kind of mechanism was observed between the enzymes, namely G-POD, PPO, PAL, and APX with the plants. Chitosan's alteration of soil characteristics is an environmentally responsible process before the chitosan of the soil is deteriorated. Owing to the presence of bacterial flora present in the soil, the chitosan was blended into the soil. Several studies confirm that the existence of chitinase enzymes is primarily due to the diversity of bacterial communities in a significant fraction of the population. The bacterial and fungal species can degrade chitin molecules widely known as chitinolytic process. Chitin promotes the development of the bacteria as compared to the fungi in this process. Chitin is effectively degraded by the key bacteria, namely *Bacillus species, Streptomyces,* and *Stenotrophomonas*, but the mechanisms are not clearly described properly. The chitin changes the equilibrium conditions with the fungal species in the realistic situation and demonstrates an antimicrobial impact on some pathogens and is even more advantageous for many other bacterial species such as Bacillus, Pseudomonas fluorescens, actinomycetes, mycorrhiza, and rhizobacteria. There are two hypotheses formulated by the researchers based on the mechanism found between the chitosan and fungi (a) Chitinolytic microorganisms have the ability to break down pathogenic bacteria's chitinous hyphae through hydrolysis. (b) Chitosan secondary products have a negative effect against pathogens (Dzung 2005; Li and Wu 1998).

14.6.6 Methods of Application of Chitin, Chitosan, and the Derivatives for Agriculture

In water and even in diluted organic acids, such as lactic acids and acetic acids, the chitosan biopolymer and its derivatives typically have greater solubility. That is why chitosan is easily used for many purposes, such as soil enrichment, foliar spraying, agricultural process supplementation, seed coatings, and even material supplements used in plant culture media. It is very important to know the difficulties in the application of agriculture and standardisation of chitosan due to its growth elevations, enhancement in yield of crops, degree of deacetylation, molecular weight, and concentration of chitosan. The preferable concentration of chitosan used in coating on seed surface is about 0.2–2%, for the soil improvement 0.1–0.5%, percentage of chitosan (Mw 20–50 ppm) used in foliar spraying 0.1–0.2%, concentrations of oligomers of the chitosan (molecular weight 2–3) for spraying process, concentration of chitosan oligomer used in plant tissue culture around 15 ppm and higher Mw is preferred in high concentrations.

14.7 Conclusion

Chitosan is a biopolymer derived from chitin deacetylation found in different organisms, such as fungi, prawns, and bacteria. Chitosan can be used in various sectors, such as the food, pharmaceutical, agricultural, cosmetic and wastewater treatment industries. The ability or effectiveness of chitosan depends on the molecular weight, the solubility, the barrier of moisture, and the degree of parameters of deacetylation. Chitosan has a greater potential for antimicrobial activity and increases the shelf life of various products, such as bread, seafood, meat, vegetables, and pharmaceuticals. The chitosan derivatives also show a promise effect on antimicrobial activity and helps in increasing shelf life of various products. Chitosan helps communicate the self-defence mechanism for plants in the agricultural sector, increases the lifetime with coating, increases the accumulation of nitrogen sources, decreases the creation of anaerobic conditions within vegetables by reducing the surface permeability. Finally, via the enzymatic process, the chitosan can be easily degraded.

References

Adams DJ (2004) Fungal cell wall chitinases and glucanases. Microbiology 150(1):2029–2035. https://doi.org/10.1099/mic.0.26980-0

Amarante C, Banks NH (2001) Postharvest physiology and quality of coated fruit and vegetables. Hortic Rev 26(4):161–238

Araki Y, Ito E (1974) A pathway of chitosan formation in Mucor rouxii. Enzymatic deacetylation of chitin. Eur J Biochem 55(1):71–78. https://doi.org/10.1111/j.1432-1033.1975.tb02139.x

Batista ACL, Souza Neto FE, Paiva WS (2018) Review of fungal chitosan: past, present and perspectives in Brazil. Polímeros 28(3):275–283. https://doi.org/10.1590/0104-1428.08316

Benjamin S, Pandey A (1997) Coconut cake—a potent substrate for the production of lipase by Candida rugosa in solid-state fermentation. Acta Biotechnol 17(3):241–251. https://doi.org/10.1002/abio.370170308

Bittelli M, Flury M, Campbell GS (2001) Reduction of transpiration through foliar application of chitosan. Agric For Meteorol 107(3):167–175. https://doi.org/10.1016/S0168-1923(00)00242-2

Butler BL, Vergano PJ, Testin RF (1996) Mechanical and barrier properties of edible chitosan films as affected by composition and storage. J Food Sci 61(5):953–955, 961. https://doi.org/10.1111/j.1365-2621.1996.tb10909.x

Chatterjee S, Adhya M, Guha AK, Chatterjee BP (2008) Chitosan from Mucor rouxii: production and physico-chemical characterization. Process Biochem 40(1):395–400. https://doi.org/10.1016/j.procbio.2004.01.025

Chen Y, Chou C (2005) Factors affecting the susceptibility of *Staphylococcus aureus* CCRC 12657 to water soluble lactose chitosan derivative. Food Microbiol 22(1):29–35. https://doi.org/10.1016/j.fm.2004.05.005

Choate CS (1994) Wound dressings. A comparison of classes and their principles of use. J Am Pediatr Med Assoc 84:463–469. https://doi.org/10.7547/87507315-84-9-463

Choi C, Nam JP, Nah JW (2016) Application of chitosan and chitosan derivatives as biomaterials. J Ind Eng Chem 33:1–10. https://doi.org/10.1016/j.jiec.2015.10.028

Conner AH, Hemingway RW (1989) In: Conner HA, Branham SJ (eds) Adhesives from Renewable Resources, ACS Symp. Ser. 385. American Chemical Society, New Orleans

Correlo VM, Bhattacharya L, Mano JF (2005) Properties of melt processed chitosan. England, pp 5–37

Crestini C, Kovac B, Giovannozzi-Sermanni G (1996) Production and isolation of chitosan by submerged and solid-state fermentation from Lentinus edodes. Biotechnol Bioeng 50(2):207–210. https://doi.org/10.1002/bit.260500202

Da Silva ARV, De Souza W, Fukushima K, De Campos-Takaki GM (2001) Faster chitosan production by Mucoralean strains in submerged culture. Braz J Microbiol 32(1):20–23. https://doi.org/10.1590/S1517-83822001000100005

Darmadji P, Izumimoto M (1994a) Effect of chitosan in meat preservation. Meat Sci 38(2):243–254. https://doi.org/10.1016/0309-1740(94)90114-7

Darmadji P, Izumimoto M (1994b) Effects of chitosan and nitrite on the properties of fermented meat. Anim Sci Technol 65(7):639–646

Davis SP (2011) Chitosan manufacture, properties, and usage. Nova Science Publishers, Hauppauge

Decker EA, Hultin HO (1992) Lipid oxidation in muscle foods via redox iron. In: Angelo AJ (ed) Lipid oxidation in food. American Chemical Society, Washington, DC, pp 33–54. https://doi.org/10.1021/bk-1992-0500.ch003

Denuziere A, Sorlier P, Viton C (2001) Relation between the degree of acetylation and the electrostatic properties of chitin and chitosan. Biomacromolecules 2(3):765–772. https://doi.org/10.1021/bm015531

Devlieghere F, Vermeulen A, Debevere J (2004) Chitosan: antimicrobial activity, interactions with food components and applicability as a coating on fruit and vegetables. Food Microbiol 21 (6):703–714. https://doi.org/10.1016/j.fm.2004.02.008

Domard A (1987) pH and c.d. measurements on a fully deacetylated chitosan: Application to CuII-polymer interactions. Int J Biol Macromol 9:98–103

Dzung NA (2004a) Study on effect of chitosan oligomer on the growth and development of some short-term crops in Dak nong province. Final report of project of Science & Technology. Department of Dak Nong (Vietnamese)

Dzung NA (2004b) Study on effect of bioactive substances on the growth and yield production of kohlrabi. In: Scientific works for developing agriculture and rural of Central Highland, Agricultural Publisher, Hanoi, Vietnam, pp 178–184 (Vietnamese)

Dzung NA (2005) Application of chitin, chitosan and their derivatives for agriculture in Vietnam. J Chitin Chitosan 10(3):109–113

El Ghaouth A, Arul J, Ponnambalam R, Boulet M (1991) Chitosan coating effect on storability and quality of fresh strawberries. J Food Sci 56(6):1618–1620. https://doi.org/10.1111/j.1365-2621.1991.tb08655.x

El Ghaouth A, Ponnampalam R, Castaigne F (1992) Chitosan coating to extend the storage life of tomatoes. Hortscience 27(9):1016–1018. https://doi.org/10.21273/hortsci.27.9.1016

Ferreira S (2019) Physicochemical, morphological and antioxidant properties of spray-dried mango Kernel starch. J Agric Food Res 25(1):1000–1012. https://doi.org/10.1016/j.jafr.2019.100012

Hadwiger LA, Loschke DC (1981) Molecular communication in host-parasite interactions: hexosamine polymers (chitosan) as regulator compounds in race-specific and other interactions. Phytopathology 71(7):756–762

Hadwiger LA, Kendra DF, Fristensky BW et al (1986) Chitosan both activates genes in plants and inhibits RNA synthesis in fungi. In: Muzzarelli RAA, Plenum C (eds) Chitin in nature and technology. Plenum Press, New York, p 215. https://doi.org/10.1007/978-1-4613-2167-5_28

Hamman JH, Schultz CM, Kotzé AF (2003) N-trimethyl chitosan chloride: optimum degree of quaternization for drug absorption enhancement across epithelial cells. Drug Dev Ind Pharm 29 (2):161–172. https://doi.org/10.1081/DDC-120016724

Han C, Zhao Y, Leonard SW, Traber MG (2004) Edible coatings to improve storability and enhance nutritional value of fresh and frozen strawberries *(Fragaria ananassa)* and raspberries *(Rubus idaeus)*. Postharvest Biol Technol 33(1):67–78. https://doi.org/10.1016/j.postharvbio.2004.01.008

Hernández-Muñoz P, Almenar E, Ocio MJ (2006) Effect of calcium dips and chitosan coatings on postharvest life of strawberries *(Fragaria ananassa)*. Postharvest Biol Technol 39(3):247–253

Hoppe-Seyler F (1894) Ueber chitosan und zellulose. Ber Deutsche Chemische Gesellschaft 27:3329. https://doi.org/10.1002/cber.189402703135

Hu Y, Jiang X, Ding Y, Ge H, Yuan Y, Yang C (2004) Synthesis and characterization of chitosan poly (acrylic acid) nanoparticles. Biomaterials 23(15):3193–3201. https://doi.org/10.1016/S0142-9612(02)00071-6

Janes K, Fresneau M, Marazuela A, Fabra A (2001) Chitosan nanoparticles as delivery systems for doxorubicin. J Control Release 73(1–3):255–267. https://doi.org/10.1016/s0168-3659(01)00294-2

Jeon YJ, Kim SK (2002) Antitumor activity of chitosan oligosaccharides produced in ultrafiltration membrane reactor system. J Microbiol Biotechnol 12(3):503–507
Jeon YJ, Park PJ, Kim SK (2002) Antimicrobial effect of chito-oligosaccharides produced by bioreactor. Carbohydr Polym 44:71–76. https://doi.org/10.1016/S0144-8617(00)00200-9
Juneja VK, Thippareddi H, Bari L (2006) Chitosan protects cooked ground beef and Turkey against Clostridium perfringens spores during chilling. J Food Sci 71(6):M236–M240. https://doi.org/10.1111/j.1750-3841.2006.00109.x
Kamil JYVA, Jeon YJ, Shahidi F (2002) Antioxidative activity of chitosans of different viscosity in cooked comminuted flesh of herring *(Clupea harengus)*. Food Chem 79(1):69–77. https://doi.org/10.1016/s0308-8146(02)00180-2
Kato Y, Onishi H, Machida M (2000a) A novel water-soluble N-succinyl chitosan-mitomycin C conjugate prepared by direct carbodiimide coupling: physicochemical properties, antitumor characteristics and systemic retention. STP Pharm Sci 10(2):133–142
Kato Y, Onishi H, Machida Y (2000b) Evaluation of N-succinyl chitosan as a systemic long-circulating polymer. Biomaterials 21(15):1579–1585. https://doi.org/10.1016/s0142-9612(00)00044-2
Kato Y, Onishi H, Machida Y (2002a) In vitro and in vivo evaluation of microparticulate drug delivery systems composed of macromolecular prodrugs. J Pharm Pharmacol 13(9):2136–2155. https://doi.org/10.3390/molecules13092136
Kato Y, Onishi H, Machida Y (2002b) Lactose aminated and intact N-succinyl-chitosans as drug carriers in liver metastasis. Anticancer Res 22(1–2):2771–2776
Kato Y, Onishi H, Machida Y (2003) Curr Pharm Biotechnol 4(5):303–309. https://doi.org/10.2174/1389201033489757
Kawase M, Michibayashi N, Nakashima Y, Kurikawa N (1997) Application of glutaraldehyde-crosslinked chitosan as a scaffold for hepatocyte attachment. Biol Pharm Bull 20:708–710. https://doi.org/10.1248/bpb.20.708
Kerch G (2015) The potential of chitosan and its derivatives in prevention and treatment of age-related diseases. Mar Drugs 13:2118–2182. https://doi.org/10.3390/md13042158
Kester JJ, Fennema OR (1986) Edible films and coatings: a review. Food Technol 40(12):47–59
Khutoryanskiy VV (2011) Advances in cohesion and mucoadhesive polymers. Macromol Biosci 11:748–764. https://doi.org/10.1002/mabi.201000388
Kirk PM, Cannon PF, Minter DW, Stalpers JA (2008) Dictionary of the fungi, 10th edn. CAB International, Wallingford fermentations. Appl Microbiol Biotechnol 39:36–41
Knorr D (1991) Recovery and utilization of chitin and chitosan in food processing waste management. Food Technol 26:114–122
Knorr D, Imeri AG (1989) Effects of chitosan treatments on the clarity and color of apple juice. J Food Sci 54(2):495. https://doi.org/10.1111/j.1365-2621.1989.tb03119.x
Knorr D, Teutonico RA (1986) Chitosan immobilization and permeabilization of Amomnthus tricolor cells. J Agric Food Chem 34(1):96–97. https://doi.org/10.1021/jf00067a026
Lafontaine JP, Benhamou N (1996) Chitosan treatment: an emerging strategy for enhancing resistance of greenhouse tomato plants to infection by *Fusarium oxysporum* f. sp. radicislycopersici. Biocontrol Sci Technol 6:111–124. https://doi.org/10.1080/09583159650039575
Lee HY, Kim SM, Kim JY (2002) Effect of addition of chitosan on improvement for shelf-life of bread. J Korean Soc Food Sci Nutr 31(3):445–450
Leong KW, Mao HQ, Truong-Le VL (1998) DNA-polycation nanospheres as non-viral gene delivery vehicles. Control Release 53(1–3):183–193. https://doi.org/10.1016/s0168-3659(97)00252-6
Li CF, Wu JC (1998) Chitosan benefits cultivation of vegetables. In: Chen RH, Chen HC (eds) Advances in chitin science, vol III. Rita Advertising Co.Ltd, Taipei, pp 448–451
Lim S, Hudson S (2004) Synthesis and antimicrobial activity of a water-soluble chitosan derivative with a fiber-reactive group. Carbohydr Res 339(9):313–319. https://doi.org/10.1016/j.carres.2003.10.024

Liptak JM (1997) An overview of the topical management of wounds. Aust Vet J 75:408–413. https://doi.org/10.1111/j.1751-0813.1997.tb14342.x

Martins ZE, Pinho O, Ferreira IMPLVO (2017) Food industry by-products used as functional ingredients of bakery products. Trends Food Sci Technol 67(2):106–128. https://doi.org/10.1016/j.tifs.2017.07.003

Mélida H, Sain D, Stajich JE et al (2015) Deciphering the uniqueness of Mucoromycotina cell walls by combining biochemical and phylogenomic approaches. Environ Microbiol 17(5):1649–1662

Mishra A, Kumar S (2007) *Cyanobacterial* biomass as N-supplement to agro-waste for hyper-production of laccase from *Pleurotus ostreatus* in solid state fermentation. Process Biochem 42 (4):681–685. https://doi.org/10.1016/j.procbio.2006.09.022

Mitra S, Gaur U, Ghosh P, Maitra A (2001) Tumor targeted delivery of encapsulated dextran-doxorubicin conjugate using chitosan nanoparticles as carrier. J Control Release 74 (1–3):317–323. https://doi.org/10.1016/s0168-3659(01)00342-x

Mitra S, Gaur U, Ghosh PC, Maitra AN (2001) Tumour targeted delivery of encapsulated dextran–doxorubicin conjugate using chitosan nanoparticles as carrier. J Control Release 74 (1–3):314–323. https://doi.org/10.1016/S0168-3659(01)00342-X

Morin-Crini N, Lichtfouse E, Torri G (2019) Applications of chitosan in food, pharmaceuticals, medicine, cosmetics, agriculture, textiles, pulp and paper, biotechnology, and environmental chemistry. Environ Chem Lett 17(2):83–92. https://doi.org/10.1007/s10311-019-00904-x

Muzzarelli RAA (1977) Chitin. Pergamon Press Ltd, Headington Hill Hall

Muzzarelli RAA, Muzzarelli C, Tarsi R (2001) Fungistatic activity of modified chitosan's against Saprolegnia parasitica. Biomacromolecules 2(1):165–169. https://doi.org/10.1021/bm000091s

Nitar N, Chandrkrachang S, Stevens WF (2004) Application of chitosan in Myanmar's agricultural sector. In: Khor E, Hutmacher D, Yong LL (eds), Proceedings of sixth Asia-Pacific chitin and chitosan symposium, Singapore, isbn: 981-05-0904-9

No HK, Park IK, Kim SD (1995) Extension of shelf-life of kimchi by addition of chitosan during salting. J Korean Soc Food Nutr 24(6):932–936

Novak K, Cupp MJ, Tracy TS (2003) Chitosan. In: Cupp MJ, Tracy TS (eds) Dietary supplements: toxicology and clinical pharmacology. Humana Press, Totowa

Nwe N, Stevens WF (2004) Effect of urea on fungal chitosan production in solid substrate fermentation. Process Biochem 39(11):1639–1642. https://doi.org/10.1016/S0032-9592(03)00301-7

Nwe N, Stevens WF, Tokura S, Tamura H (2002) Characterization of chitosan and chitosan–glucan complex extracted from the cell wall of fungus Gongronella butleri USDB 0201 by enzymatic method. Enzym Microb Technol 42(3):242–251

Odier A (1823) Mémoir sur la composition chimique des parties cornées des insectes. Mémoirs de la Societé d'Histoire. Naturelle 1:29–42

Ofori-Kwakye K, Fell JT (2001) Biphasic drug release: the permeability of films containing pectin, chitosan and HPMC. Int J Pharm 226(1–2):139–145. https://doi.org/10.1016/s0378-5173(01)00802-x

Orzali L, Corsi B, Forni C (2017) Chitosan in agriculture: a new challenge for managing plant disease. Intech. https://doi.org/10.5772/66840

Park CS, Chong HS (2002) Effect of soluble chitosan on the quality of paeksulgis. Korean J Food Preserv 9(3):321–326

Park BC, Park YH, Park HM (2005) Activation of chsC transcription by AbaA during asexual development of Aspergillus nidulans. FEMS Microbiol Lett 220(2):241–246

Peng C, Wang Y, Tang Y (1998) Synthesis of crosslinked chitosan-crown ethers and evaluation of these products as adsorbents for metal ions. J Appl Polym Sci 70(3):501–506. https://doi.org/10.1002/(SICI)1097-4628(19981017)70:3<501::AID-APP11>3.0.CO;2-3

Peppas NA, Buri PA (1985) Surface, interfacial and molecular aspects of polymer bio-adhesion on soft tissues. J Control Release 2(1):257–275. https://doi.org/10.1016/0168-3659(85)90050-1

Pochanavanich P, Suntornsuk W (2002) Fungal chitosan production and its characterization. Lett Appl Microbiol 35:17–21. https://doi.org/10.1046/j.1472-765X.2002.01118.x

Pokhrel S, Yadav PN, Adhikari R (2015) Applications of and aliphatic polyester blends. Mater Sci Eng 403(1–2):57–68. https://doi.org/10.3126/njst.v16i1.14363
Queiroz J, Fernandes SKSC, Azevedo EP, Barbosa AA, Fook MVL (2015) Chitosan: applicability in preventive dentistry. Dent Mater 31(1):58–59. https://doi.org/10.1016/j.dental.2015.08.130
Reddy BMV, Belkacemi K, Corcuff R (2000) Effect of pre-harvest chitosan sprays on post-harvest infection by Botrytis cinerea and quality of strawberry fruit. Postharvest Biol Technol 20 (1):39–51. https://doi.org/10.1.1.1073.799
Roller S, Covill N (1999) The antifungal properties of chitosan in laboratory media and apple juice. Int J Food Microbiol 47(1–2):67–77. https://doi.org/10.1016/s0168-1605(99)00006-9
Ruiz-Herrera J, Sentandreu RR, Martinez JP (1992) Chitin biosynthesis in fungi. In: Handbook of applied mycology, vol 4, pp 281–312
Sagoo S, Board R, Roller S (2002) Chitosan inhibits growth of spoilage micro-organisms in chilled pork products. Food Microbiol 19(2):175–182. https://doi.org/10.1006/fmic.2001.0474
Sarathchandra RG, Jaj SN (2004) A chitosan formulation Alexa induce Downy Mildew disease resistance and growth promotion in pearl millet. Crop Prot 23(10):881–888. https://doi.org/10.1016/j.cropro.2003.12.008
Sawayanagi Y, Nambu N, Nagai T (1982) Directly compressed tablets containing chitin or chitosan in addition to lactose or potato starch. Chem Pharm Bull 30(8):2935–2940. https://doi.org/10.1248/cpb.30.2935
Sezaki H, Hashida M (1984) Macromolecular carrier systems for targeted drug delivery: pharmacokinetic considerations on biodistribution. Crit Rev Ther Drug Carrier Syst 1(13):1–38. https://doi.org/10.1023/A:1016084508097
Shirahata K (1984) Basic studies-Chapter 1: Chemical; Section VI: Derivatives and activities. In: Taguchi T (ed) Mitomycin. Kyowa Kikaku Tsushin, Tokyo, pp 25–43
Silva HA, Paiva EG, Lisboa HM (2020) Role of chitosan and transglutaminase on the elaboration of gluten-free bread. J Food Sci Technol 57(5):1877–1886. https://doi.org/10.1007/s13197-019-04223-5
Solís-Pereira S, Favela-Torres E, Viniegra-González G (1993) Effects of different carbon sources on the synthesis of pectinase by *Aspergillus niger* in submerged and solid-state. https://doi.org/10.1007/BF00166845
Son YM, Kim KO, Jeon DW, Kyung KH (1996) The effect of low molecular weight chitosan with and without other preservatives on the characteristics of kimchi during fermentation. Korean J Food Sci Technol 28(5):888–896
Song Y, Onishi H, Machida Y (1992) Biological properties of conjugates of mitomycin C with estradiol benzoate and estradiol: their stability characteristics in biological media and their binding abilities to estrogen receptor. Chem Pharm Bull 40(20):2822–2825. https://doi.org/10.1248/bpb.20.1096
Specht CA, Liu Y, Robbins PW, Bulawa CE et al (1996) The chsD and chsE genes of Aspergillus nidulans and their roles in chitin synthesis. Fungal Genet Biol 20(2):153–167. https://doi.org/10.1006/fgbi.1996.0030
Streit F, Koch F, Laranjeira MCM et al (2009) Production of fungal chitosan in liquid cultivation using apple pomace as substrate. Braz J Microbiol 40:20–25
Suntornsuk W, Pochanavanich P, Suntornsuk L (2002) Fungal chitosan production on food processing by-products. Process Biochem 37(2):727–729. https://doi.org/10.1016/S0032-9592(01)00265-5
Synowiecki J, Al-Khateeb NA (1997) Mycelia of Mucor rouxii as a source of chitin and chitosan. Food Chem 60:605–610
Tachihara K, Onishi H, Machida Y (1997a) J Pharm Sci Technol Jpn 57:40–49
Tachihara K, Onishi H, Machida Y (1997b) J Pharm Sci Technol Jpn 57:159–167
Takakura Y, Takagi A, Hashida M, Sezaki H (1987) Pharmacokinetics and targeted delivery of proteins and genes. Pharm Res 41(1–2):293–300. https://doi.org/10.1016/0168-3659(96)01360-0
Tanaka H (2001) Scale-up for fermentative production. J Biosci Bioeng 109(2):138–144

Tokumitsu H, Hiratsuka J, Sakurai Y (2000) Gadolinium neutron-capture therapy using novel gadopentetic acid-chitosan complex nanoparticles: in vivo growth suppression of experimental melanoma solid tumor. Cancer Lett 150(2):177–182. https://doi.org/10.1016/s0304-3835(99)00388-2

Tsai GJ, Zhang SL, Shieh PL (2005) Antimicrobial activity of a low-molecular-weight chitosan obtained from cellulase digestion of chitosan. J Food Prot 67(2):396–398. https://doi.org/10.4315/0362-028X-67.2.396

Uchida Y, Izume M, Ohtakara A (1989) Preparation of chitosan oligomers with purified chitosanase and its application. In: Skjak-Brack G, Anthonsen T, Sandford P (eds) Chitin and chitosan. Elsevier, London, pp 373–382

Van der Lubben IM, Konings FAJ, Borchard G (2000) Chitosan for mucosal vaccination. Adv Drug Deliv Rev 52(2):139–144. https://doi.org/10.1016/s0169-409x(01)00197-1

Van der Lubben IM, Konings FAJ, Borchard G (2001) Chitosan and its derivatives in mucosal drug and vaccine delivery. Adv Chitin Sci 4:250–253. https://doi.org/10.1016/s0928-0987(01)00172-5

Van der Lubben IM, Kersten G, Fretz MM (2003) Chitosan microparticles for mucosal vaccination against diphtheria: oral and nasal efficacy studies in mice. Vaccine 21(13–14):1400–1408. https://doi.org/10.1016/s0264-410x(02)00686-2

Waite JH, Tanzer ML (1981) Polyphenolic substance of Mytilus edulis: novel adhesive containing L-DOPA and hydroxyproline. Science 212(1):1038–1040. https://doi.org/10.1126/science.212.4498.1038

White SA, Farina PR, Fultong L (1979) Production and isolation of chitosan from Mucor rouxii. Appl Environ Microbiol 38(2):323–328. https://aem.asm.org/

Wongroung, S, Seesuriyachun, P, Ruangpradith, P (2002) Effect of chitosan sprays as a crop protection on Pe-tsai (*Brassica campestris spp.pekinensis*). In: Suchiva K, Chandrkrachang S, Methacanon P, Peter MG (eds) Advances in chitin science, vol. V, Bangkok, pp 479–481

Wu T (2004) Production and characterization of fungal chitin and chitosan. Master's thesis. Tennessee: University of Tennessee

Wu Y, Rhim JW, Weller CL, Hamouz F (2000) Moisture loss and lipid oxidation for precooked beef patties stored in edible coatings and films. J Food Sci 65(2):300–304. https://doi.org/10.1111/j.1365-2621.2000.tb15997.x

Xie G, West TP (2009) Citric acid production by *Aspergillus Niger* ATCC 9142 from a treated ethanol fermentation co-production using solid-state fermentation. Lett Appl Microbiol 48(5):639–644. https://doi.org/10.1111/j.1472-765X.2009.02586.x

Yamada K, Chen T, Kumar G, Vesnovsky O (2000) Chitosan based water-resistant adhesive. Analogy to mussel glue. Biomacromolecules 1(2):252–258. https://doi.org/10.1021/bm0003009

Yano S, Tsugita T (1988) Chitosan-containing seed coatings for yield enhancement. Japanese Kokai Tokyo Koho JP 63(139):102–107

Yokoi H, Aratake T, Nishio S, Hirose J et al (1998) Chitosan production from shochu distillery wastewater by funguses. J Ferment Bioeng 85:246–249. https://doi.org/10.1016/S0922-338X(97)86777-3

Yoo EJ, Lim HS, Kim JM, Song SH, Choi MR (1998) The investigation of chitosan oligosaccharide for prolongating fermentation period of kimchi. J Korean Soc Food Sci Nutr 27(5):869–874

Young DH, Kohle H, Kauss H (1982) Effect of chitosan on membrane permeability of suspension cultured Glycine max and *Phaseolus vulgaris* cells. Plant Physiol 70(5):1449–1454. https://doi.org/10.1104/pp.70.5.1449

Yu M, Hwang JD (1999) Role of L-3,4-dihydroxyphenylalanine in mussel adhesive proteins. J Am Chem Soc 121(24):5825–5826. https://doi.org/10.1021/ja990469y

Role of Microbial Extracellular Polymeric Substances in Soil Fertility

15

Alok Bharadwaj

Abstract

In the present scenario, conventional chemical polymers are substituted by biopolymers having various characteristics features such as high effectiveness, rapidly biodegradable, nontoxic, and possess no pollution. Presently, microorganisms used for producing extracellular polymeric substances along with biopolymers have been studied by many research workers as an effective flocculent having its application in relevance in different types of water, polluted water, and water treatment practices. These polymeric substances are extremely hydrated polymers which are secreted by a number of microbes that are chiefly made up of DNA, proteins, and polysaccharides. Tocarryout different microbial activities, extracellular polymeric substances play an important role as it provides ideal conditions for performing chemical changes, nutrient availability, and shield them from environmental stresses like drought and salinity. Moreover, it (microbial extracellular polymeric substances) increases the soil aggregation ability and assists plants by trapping moisture and nutrients. In addition to this, it also possesses some distinctive features like biocompatibility, potential to have gelling and thickening ability along with its industrial applications. Besides this, the ability of extracellular polymeric substances has been analyzed and it has been observed that only a small amount of them are being utilized in various areas, mainly in agriculture. In this chapter, an effort has been made to impart recent information associated with the various characteristics of microbial extracellular polymeric substances in the betterment of soil health by improving soil fertility and its physicochemical properties.

A. Bharadwaj (✉)
Department of Biotechnology, GLA University, Mathura, Uttar Pradesh, India
e-mail: alok.bhardwaj@gla.ac.in

A. Vaishnav, D. K. Choudhary (eds.), *Microbial Polymers*,
https://doi.org/10.1007/978-981-16-0045-6_15

Keywords

Biopolymer · Exopolysaccharide · Soil aggregation · Soil fertility

15.1 Introduction

Nowadays chemical flocculants have been replaced by certain microbial metabolites like extracellular polymeric substances because microbial metabolites are cheap, eco-friendly, and environment friendly. These are the biopolymers (Wingender et al. 1999). It has been clear from the previous literature that the extracellular polymeric substance matrix is produced as microbial secretions, which contains various biochemicals, discharge of different cell components as well as organic compounds present in the culture medium. In addition to it, these microorganisms, whether prokaryotic or eukaryotic, naturally secretes these extracellular polymeric substances (Flemming and Wingender 2001). Normally, an extracellular polymeric substance is found as a viscous biofilm matrix that contains 50–90% of the total organic matter. On the basis of its organization with the cell or the procedure applied for its extraction from the cell, extracellular polymeric substances classified as capsular extracellular polymeric substance (CEPS), slime extracellular polymeric substance (SEPS), loosely bound extracellular polymeric substance (LBEPS), and tightly bound extracellular polymeric substance (TBEPS) (Table 15.1).

These extracellular polymeric substances are secreted by certain microbial strains act as polymers, i.e., made up of mainly DNA, proteins, and polysaccharides and it has been noticed that the secretion of these substances are elicited by the environmental signals. The biosynthesis of these extracellular polymeric substances required a high amount of energy. So, these substances must provide some benefits to the microbe that produced it (Flemming and Wingender 2010). Henceforth, the

Table 15.1 Extracellular polymeric substances: types and functions

S. No.	Type of extracellular polymeric substance	Specific function	References
1	Capsular extracellular polymeric substance	Capsular extracellular polymeric substances are an essential portion of the cell membrane and attached to the bacterial cell	–
2	Slime extracellular polymeric substance	When biomass is centrifuged, the supernatant contains slime polymers. These are present in soluble form	Tian (2008), Wingender et al. (1999)
3	Loosely bound extracellular polymeric substance	These are also called soluble extracellular polymeric substances that are produced in the surrounding environment in dissolved form	Tian (2008), Wingender et al. (1999)
4	Tightly bound extracellular polymeric substance	These extracellular polymeric substances are firmly attached to the cell. e.g., sheaths, capsular polymers, condensed gel	–

manufacturing and utility of these extracellular polymeric substances are of immense importance for decades.

In previous studies, some researchers have defined the functions of extracellular polymeric substances like these substances helps in attachment with any surface, the formation of bacterial aggregates, promote biofilm formation along with cell-cell recognition, proactive barrier for cells, minimizing water loss, reducing cell desiccation, and promote sorption of inorganic and organic components and ions (Tian 2008). Extracellular polymeric substances are mainly composed of proteins, carbohydrates, and humic materials. Due to the presence of such components, the extracellular polymeric substance matrix exhibit adsorption capability, biodegradability along with hydrophilicity/hydrophobicity. Moreover, extracellular polymeric substances displayed significant function in biofilm development, mass transfer through biofilm, surface deposition of various metals, organic and inorganic compounds along with providing structural support to the biofilm (Czaczyk and Myszk 2007; Flemming and Leis 2003; Neyens et al. 2004; Flemming et al. 2005).

As explained earlier, one of the major functions of extracellular polymeric substances is the collection of soil particles (aggregation), which enhances soil fertility by improving the soil physicochemical characteristics. The slimy texture of extracellular polymeric substance act as a glue that helps in attachment with ions and clay that holds the solid particles together (Chenu 1995). Likewise, the structure of extracellular polymeric substances are quite variable, and henceforth their effectiveness also varies accordingly in the soil. These extracellular polymeric substances are formulated in vitro conditions for improving soil health, structure, fertility, and yield.

In the present review, an effort has been made to explain the various functions of extracellular polymeric substances and their utility in the aggregation of soil particles. Many research workers have already described the ingredients, biosynthesis, and other aspects of extracellular polymeric substances in detail (Nouha et al. 2017; Flemming et al. 2016; More et al. 2014; Sheng et al. 2010; Flemming and Wingender 2010; Wingender et al. 1999b).

15.2 Ecological Characteristics

The term EPS is an acronym for "extracellular polysaccharides," "exopolysaccharides," or "exopolymers." The extracellular polymeric substances are mainly produced by bacteria, cyanobacteria, algae (Boonchai et al. 2014; Parikh and Madamwar 2006), fungi (Elisashvili et al. 2009; Hwang et al. 2004), yeasts (Pavlova and Grigorova 1999), and protists (Lee Chang et al. 2014; Jain et al. 2005). Extracellular polymeric substances are basically constituted from carbohydrates, enzymes, proteins, lipids, humic acid, etc. (Flemming and Wingender 2010; Wingender et al. 1999a, b).

As discussed earlier that the biosynthesis of extracellular polymeric substance is energy-based process, and henceforth the producing microbial strain must get some benefit in the environment. It has been observed in laboratory cultures that there is no

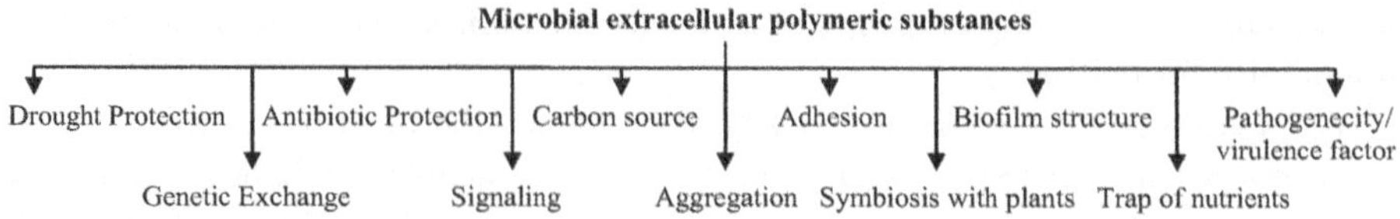

Fig. 15.1 Microbial extracellular polymeric substances (EPS): Various advantages in soil

impact of extracellular polymeric substances on cell viability and cell growth, which means they are not necessary for existence. Moreover, it has been found that under natural conditions mostly microorganisms exist in aggregates like flocs and biofilms. Such structures are formed by extracellular polymeric substances that are structurally and functionally necessary (Wingender et al. 1999b). The most common efficiency of extracellular polymeric substances is to provide protection to the producing microorganism. The synthesis of extracellular polymeric substances can be triggered up by variation in abiotic environmental factors like pH, temperature, drought, and salinity (Vardharajula and Ali 2015; Kumar et al. 2007; Wingender et al. 1999b). The functions of extracellular polymeric substances are precise in Fig. 15.1.

15.3 Impact of Extracellular Polymeric Substances on Soil Aggregation

15.3.1 Role of Microbial Population on Soil Aggregation

Soil is the uppermost layer of earth-crust and soil aggregates are the fundamental element of soil structure that is made up of pores and the solid material formed by the rearrangement of small elements, flocculation, and cementation. These are the major components that express the physicochemical properties of the soil like soil temperature, soil moisture, water holding capacity, aeration and influence the physical, chemical, and biological processes (Tang et al. 2011; Alami et al. 2000). Soil aggregates play a vital role in the enhancement of the soil porosity, productivity, and crop yield via increasing the growth of plant root and shoot (Bronick and Lal 2005; Dinel et al. 1992).

The process of development of these soil aggregates depends on various factors like soil microorganisms, soil vegetation, and cation exchange between organic matter and soil particles (Kumar et al. 2013). The consistency of these aggregates is based on their porosity, internal organization, pore wall hydrophobicity, and tortuosity (Chenu and Cosentino 2011). Improved aggregation along with superior soil organization is the essential requirement of sustainable agriculture and environment (Amézketa 1999). On the basis of the hierarchical model, it can be advocated that huge (large) aggregates are constituted from the combination of smaller aggregates (Tisdall and Oades 1982).

These smaller aggregates (size 2–20 mm) are constituted by the amalgamation of clay component with inorganic amorphous binding element like various oxides, aluminosilicates, humic materials, soil polysaccharides along with metal ions. Such long-lasting smaller aggregates combined with each other leads to the formation of larger (big) aggregates (size 20–250 mm) with the help of fungal hyphae and plant root hairs. It has been observed that these smaller aggregates stick with each other with the help of transient binding agents like polyuronides and polysaccharides to constitute large aggregates (size >250 mm diameter). There are several factors that influenced the process of aggregate formation like microbial population, organic and mineral components, plant variety, and soil type (Tisdall and Oades 1982). The major component that engaged in the formation of small aggregates are found to be clay minerals, $CaCO_3$ and Fe- and Al-(hydr)oxides (Totsche et al. 2018). Extracellular polymeric substances are found to be associated with the development of organo-mineral complex in the soil that can affect the amount of organic substance along with the reactivity of minerals (Liu et al. 2013). The mineral aggregate of extracellular polymeric substances develop an association between various minerals and increasing their capability to hold more water by adsorbing onto the mineral surfaces (Henao and Mazeau 2009). In numerous studies, the adsorption of extracellular polymeric substances was performed and visualized through microscopy.

It has been found in the previous research that electrostatic bonding is necessary between the extracellular polymeric substance of *Pseudomonas putida X4* and goethite, minerals kaolinite and montmorillonite (Lin et al. 2016). At low pH, due to the protonation mechanism, the extracellular polymeric substance of C, N, and P portion showed maximum adsorption capacity. Moreover, to study the division of proteins, polysaccharides, and nucleic acid in mineral–extracellular polymeric substance complexes, the confocal laser scanning microscopy (CLSM) technique is used by the authors. Liu et al. (2013) observed that adsorption of extracellular polymeric substance from *Bacillius subtilis168* to goethite and found that goethite-adsorbed extracellular polymeric substance complex becomes enriched with extracellular polymeric substance portion that mainly contains proteins and lipids but lacking in polysaccharides. The adsorbed proteins were found to be sulfur-rich that can be obtained from sulfur enriched amino acids. Whereas, the pure extracellular polymeric substances were found to be rich in protein and polysaccharides having very less quantity of lipids and nucleic acids. Research has been carried out for many years to know various microbial populations present in different aggregate categories through various experimental setup and techniques (Blaud et al. 2012, 2017).

Some literature has been available on the type of microbial population that reside in aggregates of various size in different agriculture management systems (Kravchenko et al. 2014; Mummey and Stahl 2004; Sessitsch et al. 2001); on the other hand, there is no study available that describe the type of microbial population accountable for aggregate formation. It has also been reported that the properties of aggregation of soil is dependent on the isolated microbial species and such studies like impact of microbes and their extracellular polymeric substance on soil aggregation have been studied for decades. In one study, bacterial strains were isolated from two agriculture ecosystems (9 years no tillage and 40 years tillage) by using smaller aggregates (250–50 mm) and check their aggregation capability by using fatty acid methyl ester (FAME) technique (Caesar-TonThat et al. 2007). They found that isolated microbes like *Stenotrophomonas, Sphingobacterium, Bacillus,* and *Pseudomonas species* have the potential to stabilize and enhance the capacity of aggregate in artificial aggregates and these bacterial strains were found freely in the normal soils. Another study has revealed that there were several factors like tillage, irrigation, and cropping system that influenced the bacterial strains that reside inside the soil aggregate (Caesar-Tonthat et al. 2014). They have isolated more than 50 soil aggregating bacteria out of 1296 isolates, out of which dominating bacterial strains isolated belonged to *Bacillus* spp. and *Pseudomonas* spp. In other studies, it has also been found that *Bacillus species* and *Pseudomonas species* have the potential to produce extracellular polymeric substances and biofilms, i.e., responsible for improving soil structure and soil health.

15.3.2 Inoculation of Extracellular Polymeric Substance Producers in Soils

From previous research, it has been proved that in the process of soil aggregation and stabilization, microorganisms play a very important role. Moreover, there are many factors like type of microbial species, available soil nutrients, and their management that influence the microorganisms that are responsible for soil structure management and stabilization (Beare et al. 1994; Umer and Rajab 2012). Among the microbial community, mainly bacteria and fungi secrete extracellular polymeric substances and carry out breakdown of aromatic humic substances that produces clay–metal–organic matter complexes which are ultimately required for the stabilization of soil structure (Umer and Rajab 2012). In addition, fungi contribute with the help of their hyphae although with less perseverance. In previous studies, a number of bacterial and fungal species have been utilized for representing the impact of pure microbial cultures on soil aggregation and it has been found that the whole process is dependent on the type of microbial strain. Henceforth, for improving the soil productivity and soil quality, various microbial strains and extracellular polymeric substances have to be searched out for the components that increase the aggregation capability in various types of soils.

From previous research, it has become evident that the bacterial strains, that produce extracellular polymeric substances and have the capability to produce soil aggregates belong to the genus *Bacillus, Pseudomonas,* and *Paenibacillus.* These bacterial strains can be easily grown at in vitro conditions and secrete a large quantity of extracellular polymeric substances. Moreover, some fungal species like *Penicillium* and *Streptomyces* also exert a noteworthy positive impact on erodibility and soil loss after rainfall simulation (Gasperi-Mago and Troeh 1979). It has been found that inoculation of *Pseudomonas putida strain GAP-P45* in soil enhances the soil aggregation capacity and its stability in more than 50% cases and it depends on certain factors like salt, temperature, and drought stress. Moreover, during the situation of stress microbial strains secretes large amounts of extracellular polymeric substances that help the bacteria in defending it from water stress and improved soil structure (Vardharajula and Ali 2015).

Another study was performed from the Gurbantünggüt Desert, where an unknown bacterial strain has been isolated from biological soil crusts (BSCs). This bacterial strain not only stabilizes the sand surface but also increases the aggregation process along with decreasing the soil water evaporation only after 8 days of inoculation. Moreover, the extracellular polymeric substances secreted by the bacterium possess a glue-like substance that helps in binding the soil and sand particles that can be examined by scanning electron microscopy (HuiXia et al. 2007). Another study was performed on *Paenibacillus KLBB0001* that secretes a large quantity of extracellular polymeric substances at Gurbantünggüt Desert. In this study this bacterial strain when inoculated in desert soil, it improves the recovery of biological soil crusts (BSC). It has been observed after 1 year of experimentation that soil becomes enriched with a higher number of bacteria, phosphorus, and available nitrogen. Moreover, at the inoculation area, some sticky substance aggregating sand and soil particles showed the presence of extracellular polymeric substance confirmed from the microscopic studies (Wu et al. 2014).

Several studies revealed that in place of pure culture, a mixture of microorganisms may be an alternative approach for the same. Swaby (1949) found in his study that when mixed microbial culture was applied in soil, the rate of aggregation of soil particles maximized may be due to interaction between various microbial strains. Different microbial species have different types of extracellular polymeric substances produced. When the extracellular polymeric substance of one microbial species come in contact with the extracellular polymeric substance of other microbial species like extracellular polymeric substance-coated fungal hyphae, the resulting product has a greater aggregation capability of binding the soil and sand particles together in comparison to physical involvement by hyphae (Aspiras et al. 1971). In addition, the amalgamation of microbial inoculants with organic fertilizers enhances the microbial aggregation capability, resulting in increased production of extracellular polymeric substances, ultimately ensuring improved soil fertility, structure, and function (Rashid et al. 2016).

15.3.3 Inoculation of Extracellular Polymeric Substance Producers in Plants

Another important and latest available technology for sustainable agriculture is the inoculation of plant with arbuscular mycorrhizal fungi and plant growth promoter rhizobacteria. In this technology, microbial strains established themselves within plant roots increase their population along with the secretion of exudates. The population of microbial species enhanced in the rhizosphere region, as a result of the liberation of organic carbon by the plant roots. In response to it, a gelatinous extracellular polymeric substance is produced that encourage soil aggregation capacity and increasing root-adhering soil (RAS). An environment has been established by the RAS aggregation, through which plants utilize water and nutrients for their growth and expansion. The availability of these advantageous microorganisms helps the plants in the accessibility of major nutrients like N, P, K, and iron for their growth and expansion (Rashid et al. 2016).

The data available on the recent research clearly revealed that the most promising bacterial strains, that secretes large quantity of extracellular polymeric substances and increase plant growth belong to the genus *Rhizobium, Pseudomonas, Bacillus,* and *Pantoea.* These bacterial strains when added either directly into the soil or in seedling exerts the optimistic approach on the plant development and growth also (Cipriano et al. 2016). The extracellular polymeric substances secreted in the rhizospheric region protect the plant from desiccation, variation in the water potential, rising nutrient uptake as well as supports plant growth. With the enhancement in the soil aggregation capability, the soil structure improves that ultimately enhances the growth and yield of seedlings because it encourages the uptake of water and nutrients (Sandhya et al. 2009; Bezzate et al. 2000; Alami et al. 2000). A lot of research has been carried out on PGPR and AMF but at that time the major emphasis was on plant growth but not on soil aggregation.

Rhizobium strain KYGT207, i.e., a wheat growth promoter and extracellular polymeric substance secreting bacterium was picked from arid soil in Algeria. This bacterium is capable of improving the soil structure and its productivity. With the inoculation of *Rhizobium strain KYGT207* strain in wheat plant, the root-adhering soil dry mass/root dry mass ratio was increased by 137%. In addition, it also increased the percentage of water stable aggregates by reducing the soil water stress through the secretion of extracellular polymeric substances (Kaci et al. 2005). These inoculated microbial species interact with the microorganisms that are naturally present in the rhizosphere region and showed a considerable impact on soil characteristics and quality by enhancing plant productivity. Unwanted germination and root growth must be inhibited for getting improved soil structure and increased aggregate formation. The use of these microbial inoculants has been continued for the last several years but more efforts must be required for providing optimum conditions for the growth of a particular microorganism to obtain more efficient inoculants, better biomass, and quality extracellular polymeric substances. Henceforth, these strains gave a maximum performance in soil with an increase in their population and become dominant on the natural population of microorganisms.

15.3.4 Inoculation of Pure Extracellular Polymeric Substance into Soil

Previous research also connects the products of microbial secretion with soil aggregation capability. For improving the soil structure and productivity, polysaccharides also play an important role, although they are not the major soil aggregating molecules. Moreover, humic acid has the ability to enhance the soil aggregation capacity as well as improving soil structure. When natural and synthetic soil aggregates were treated with certain chemicals like tetraborate and periodate, they do not show a stable pattern as polysaccharides are necessary components required for a constant soil aggregation process (Sparling and Cheshire 1985; Mehta et al. 1960).

To prove this statement, an experiment was performed for observing the correlation between carbohydrate content and mean weight of soil aggregate, and the results obtained were clearly depicted that for enhancing the capability of soil aggregation, carbohydrate content was essentially required in soils with various crops (Angers and Mehuys 1989). Moreover, the addition of sodium periodates in soil before wet sieving also mentioned the contribution of carbohydrates in maintaining the stable soil aggregates.

Biopolymers, that are resistant to breakdown mechanism may contribute to the improved soil structure. In other words, we can say that the greater is the resistance, the extended is the longevity of extracellular polymeric substances in soils. Colloids and metal ions present in the soil may affect the degradation of biopolymers. Now investigation was initiated for the inoculation of polymers in the soil, it has been found that there is variation in the binding capacity of microbial polysaccharides and plant itself. Nevertheless, certain parameters of the soil like pH also affect the mechanism of polysaccharides' action. It has occurred due to the charge on the molecules that is mandatory for binding of the particles (Martin 1971). There are certain features of polysaccharides that affect the binding capacity like flexible nature, linear structure, and length, that ultimately permit the Van der Waals forces, hydrogen bonding, and acyl groups, allowing the ionic binding to clays (Martin 1971).

There are many bacteria and fungi that produced a variety of extracellular polymeric substances and their performance has been checked as soil aggregating agents. For this, in place of the addition of microbial strains, an alternative approach has been used by directly applying the polymers to the soil. The extracellular polymeric substances secreted by *L. mesenteroides, B. subtilis,* and *Leuconostoc dextranicum* have been evaluated for their aggregating potential by Geoghegan and Brian (1948). It has been found that various extracellular polymeric substances produced have a significant role in soil aggregation capability and were examined by wet sieving process and small quantity of levans (0.125–0.05%), that have a significant role in stabilizing the aggregates.

Chromobacterium violaceum is a Gram-negative coccobacilli that secrete a specific extracellular polymeric substance which exerts more resistance to degradation than other plant polysaccharides. The extracellular polymeric substance secreted by this coccobacilli displayed the highest aggregation capacity among all polysaccharides and maintains the soil structure along with its pH (Martin and Richards 1963).

There are certain extracellular polymeric substances that exert very high water holding capacity. Chenu and Roberson (1996) have demonstrated the water holding capacity of xanthan and found that water holding capacity of 15 times its weight. On the other hand, when dextran was analyzed for the same activity exerts less water holding capacity may be due to variation in structure. It has been observed from the previous research that the extracellular polymeric substances that possess higher water holding capacity can able to protect plants, soil, and microorganisms from drought stress, encouraging hydrating conditions and bridging between clay and soil particles.

It has also been observed from the previous research that in place of direct inoculation of extracellular polymeric substances in soil, its production can be improved by N management. The impact of the incorporation of N in extracellular polymeric substances and soil aggregation has been observed by Roberson et al. (1995) by using indirect measurements and monosaccharide composition. He depicted that when intermediate and high amount of N fertilization provide the same crop yield, the data obtained for soil characteristics showed variation in the results. That is because the intermediate N fertilization encouraged improved aggregation and saturated hydraulic conductivity. Henceforth, the production of extracellular polymeric substances could be enhanced by inoculation of nutrients directly into soil ultimately improving the soil structure and soil aggregation capacity.

15.4 Conclusion

For survival in soil, microorganisms undergo various mechanisms to tolerate certain environmental stresses. Extracellular polymeric substances provide favorable conditions for the growth of microorganisms like wet atmosphere, ease of trapping nutrients, assisting chemical reactions, and provide a shield against environmental stress and predators. These extracellular polymeric substances when released by the microorganism contain a variety of compounds and perform several functions based on their organization and composition. A lot of previous research has been demonstrated that due to the production of extracellular polymeric substances, the soil aggregation capacity, soil structure, and soil quality along with soil fertility improved drastically. Furthermore, besides improving soil structure, these extracellular polymeric substances also improve the uptake of nutrients and availability of water to both microbes and plants, and henceforth not only adventitious for the plant but also the environment as a whole. As we all know that microorganisms possess huge capabilities and can be increased by improving the structure and composition of extracellular polymeric substances along with using a consortium of microorganisms for huge quantity production of extracellular polymeric substances.

Due to the valuable characteristics like biocompatibility, biodegradability, emulsion, gelling, and aggregating capabilities, these extracellular polymeric substances are the choice of interest since years back. For obtaining a higher yield, manipulation was carried out in the polymers, i.e., based on description and knowledge of physiological characterization of extracellular polymeric substance producing microorganisms. A number of microorganisms produced extracellular polymeric substances but because of the change in the polymer composition, many prospects remain open for investigation and experimentation.

References

Alami Y, Achouak W, Marol C, Heulin T (2000) Rhizosphere soil aggregation and plant growth promotion of sunflowers by an exopolysaccharide-producing Rhizobium sp strain isolated from sunflower roots. Appl Environ Microbiol 66:3393–3398. https://doi.org/10.1128/Aem.66.8.3393-3398.2000

Amézketa E (1999) Soil aggregate stability: a review. J Sustain Agric 14:83–151. https://doi.org/10.1300/J064v14n02_08

Angers DA, Mehuys GR (1989) Effects of cropping on carbohydrate content and water-stable aggregation of a clay soil. Can J Soil Sci 69:373–380. https://doi.org/10.4141/cjss89-037

Aspiras RB, Allen ON, Harris RF, Chesters G (1971) Aggregate stabilization by filamentous microorganisms. Soil Sci 112:282–284. https://doi.org/10.1097/00010694-197110000-00012

Beare MH, Hendrix PF, Cabrera ML, Coleman DC (1994) Aggregate-protected and unprotected organic matter pools in conventional and no-tillage soils. Soil Sci Soc Am J 58:787. https://doi.org/10.2136/sssaj1994.03615995005800030021x

Bezzate S, Aymerich S, Chambert R, Czarnes S, Berge O, Heulin T (2000) Disruption of the Paenibacillus polymyxa levansucrase gene impairs its ability to aggregate soil in the wheat rhizosphere. Environ Microbiol 2:333–342. https://doi.org/10.1046/j.1462-2920.2000.00114.x

Blaud A, Lerch TZ, Chevallier T, Nunan N, Chenu C, Brauman A (2012) Dynamics of bacterial communities in relation to soil aggregate formation during the decomposition of 13C-labelled rice straw. Appl Soil Ecol 53:1–9. https://doi.org/10.1016/j.apsoil.2011.11.005

Blaud A, Van Der Zaan B, Menon M, Lair GJ, Zhang D, Huber P et al (2017) The abundance of nitrogen cycle genes and potential greenhouse gas fluxes depends on land use type and little on soil aggregate size. Appl Soil Ecol 125:1–11. https://doi.org/10.1016/j.apsoil.2017.11.026

Boonchai R, Kaewsuk J, Seo G (2014) Effect of nutrient starvation on nutrient uptake and extracellular polymeric substance for microalgae cultivation and separation. Desal Water Treat 55:360–367. https://doi.org/10.1080/19443994.2014.939501

Bronick CJ, Lal R (2005) Soil structure and management: a review. Geoderma 124:3–22. https://doi.org/10.1016/j.geoderma.2004.03.005

Caesar-TonThat TC, Caesar AJ, Gaskin JF, Sainju UM, Busscher WJ (2007) Taxonomic diversity of predominant culturable bacteria associated with microaggregates from two different agroecosystems and their ability to aggregate soil in vitro. Appl Soil Ecol 36:10–21. https://doi.org/10.1016/j.apsoil.2006.11.007

Caesar-Tonthat TC, Stevens WB, Sainju UM, Caesar AJ, West M, Gaskin JF (2014) Soil-aggregating bacterial community as affected by irrigation, tillage, and cropping system in the northern great plains. Soil Sci. 179:11–20. https://doi.org/10.1097/ss.0000000000000036

Chenu C (1995) Extracellular polysaccharides: an interface between microorganisms and soil constituents. In: Huang PM, Berthelin J, Bollag JM, Mcgill WB, Page AL (eds) Environmental impact of soil component interactions. Natural and anthropogenic organics. CRC Lewis Publishers, Boca Raton

Chenu C, Cosentino D (2011) Microbial regulation of soil structural dynamics. In: Ritz K, Young I (eds) The architecture and biology of soils: life in inner space. CABI, Wallingford, pp 37–70

Chenu C, Roberson EB (1996) Diffusion of glucose in microbial extracellular polysaccharide as affected by water potential. Soil Biol Biochem 28:877–884. https://doi.org/10.1016/0038-0717(96)00070-3

Cipriano MA, Lupatini M, Lopes-Santos L, Da Silva MJ, Roesch LF, Deståfano SA et al (2016) Lettuce and rhizosphere microbiome responses to growth promoting Pseudomonas species under field conditions. FEMS Microbiol Ecol 92:fiw197. https://doi.org/10.1093/femsec/fiw197

Czaczyk K, Myszk K (2007) Biosynthesis of extracellular polymeric substances (EPS) and its role in microbial biofilm formation. Pol J Environ Stud 16:799–806

Dinel H, Lévesque PEM, Jambu P, Righi D (1992) Microbial activity and long-chain aliphatics in the formation of stable soil aggregates. Soil Sci Soc Am J 56:1455–1463. https://doi.org/10.2136/sssaj1992.03615995005600050020x

Elisashvili VI, Kachlishvili ET, Wasser SP (2009) Carbon and nitrogen source effects on basidiomycetes exopolysaccharide production. Appl Biochem Microbiol 45:531–535. https://doi.org/10.1134/s0003683809050135

Flemming HC, Leis A (2003) Sorption properties of biofilms. In: Encyclopedia of environmental microbiology. Wiley, Hoboken

Flemming HC, Wingender J (2001) Relevance of microbial extracellular polymeric substances (EPSs) Part I: structural and ecological aspects. Water Sci Tech 43:1–8

Flemming HC, Wingender J (2010) The biofilm matrix. Nat Rev Microbiol 8:623–633

Flemming HC, Wingender J, Griebe T, Mayer C (2005) Chapter 2: physicochemical properties of biofilms. In: Evans LV (ed) Biofilms: recent advances in their study and control. Harwood Academic Publishers, Reading

Flemming H-C, Wingender J, Szewzyk U, Steinberg P, Rice SA, Kjelleberg S (2016) Biofilms: an emergent form of bacterial life. Nat Rev Microbiol 14:563–575. https://doi.org/10.1038/nrmicro.2016.94

Gasperi-Mago RR, Troeh FR (1979) Microbial effects on soil erodibility. Soil Sci Soc Am J 43:765–768. https://doi.org/10.2136/sssaj1979.03615995004300040029x

Geoghegan MJ, Brian RC (1948) Aggregate formation in soil. 1. Influence of some bacterial polysaccharides on the binding of soil particles. Biochem J 43:5–13. https://doi.org/10.1042/bj0430005

Henao LJ, Mazeau K (2009) Molecular modelling studies of clay–exopolysaccharide complexes: soil aggregation and water retention phenomena. Mater Sci Eng C 29:2326–2332. https://doi.org/10.1016/j.msec.2009.06.001

HuiXia P, Zhengming C, Xuemei Z, Shuyong M, Xiaoling Q, Fang W (2007) A study on an oligotrophic bacteria and its ecological characteristics in an arid desert area. Sci China Ser D Earth Sci 50:128–134. https://doi.org/10.1007/s11430-007-5015-4

Hwang HJ, Kim SW, Xu CP, Choi JW, Yun JW (2004) Morphological and rheological properties of the three different species of basidiomycetes Phellinus in submerged cultures. J Appl Microbiol 96:1296–1305. https://doi.org/10.1111/j.1365-2672.2004.02271.x

Jain R, Raghukumar S, Tharanathan R, Bhosle NB (2005) Extracellular polysaccharide production by Thraustochytrid protists. Mar Biotechnol 7:184–192. https://doi.org/10.1007/s10126-004-4025-x

Kaci Y, Heyraud A, Barakat M, Heulin T (2005) Isolation and identification of an EPS-producing Rhizobium strain from arid soil (Algeria): characterization of its EPS and the effect of inoculation on wheat rhizosphere soil structure. Res Microbiol 156:522–531. https://doi.org/10.1016/j.resmic.2005.01.012

Kravchenko AN, Negassa WC, Guber AK, Hildebrandt B, Marsh TL, Rivers ML (2014) Intra-aggregate pore structure influences phylogenetic composition of bacterial community in macroaggregates. Soil Sci Soc Am J 78:1924–1939. https://doi.org/10.2136/sssaj2014.07.0308

Kumar AS, Mody K, Jha B (2007) Bacterial exopolysaccharides—a perception. J Basic Microbiol 47:103–117. https://doi.org/10.1002/jobm.200610203

Kumar R, Rawat KS, Singh J, Singh A, Rai A (2013) Soil aggregation dynamics and carbon sequestration. J Appl Nat Sci 5:250–267. https://doi.org/10.31018/jans.v5i1.314

Lee Chang KJ, Nichols CM, Blackburn SI, Dunstan GA, Koutoulis A, Nichols PD (2014) Comparison of thraustochytrids Aurantiochytrium sp., Schizochytrium sp., Thraustochytrium sp., and Ulkenia sp. for production of biodiesel, long-chain omega-3 oils, and exopolysaccharide. Mar Biotechnol 16:396–411. https://doi.org/10.1007/s10126-014-9560-5

Lin D, Ma W, Jin Z, Wang Y, Huang Q, Cai P (2016) Interactions of EPS with soil minerals: a combination study by ITC and CLSM. Colloids Surf B 138:10–16. https://doi.org/10.1016/j.colsurfb.2015.11.02

Liu X, Eusterhues K, Thieme J, Ciobota V, Höschen C, Mueller CW et al (2013) STXM and nanoSIMS investigations on EPS fractions before and after adsorption to goethite. Environ Sci Technol 47:3158–3166. https://doi.org/10.1021/es3039505

Martin JP (1971) Decomposition and binding action of polysaccharides in soil. Soil Biol Biochem 3:33–41. https://doi.org/10.1016/0038-0717(71)90029-0
Martin JP, Richards SJ (1963) Decomposition and binding action of a polysaccharide from Chromobacterium violacium in soil. J Bacteriol 85:1288–1294
Mehta NC, Streuli H, Müller M, Deuel H (1960) Role of polysaccharides in soil aggregation. J Sci Food Agric 11:40–47. https://doi.org/10.1002/jsfa.2740110107
More TT, Yadav JSS, Yan S, Tyagi RD, Surampalli RY (2014) Extracellular polymeric substances of bacteria and their potential environmental applications. J Environ Manage 144:1–25. https://doi.org/10.1016/j.jenvman.2014.05.010
Mummey DL, Stahl PD (2004) Analysis of soil whole- and inner microaggregate bacterial communities. Microb Ecol 48:41–50. https://doi.org/10.1007/s00248-003-1000-4
Neyens E, Baeyens J, Dewil R, De heyder B (2004) Advanced sludge treatment affects extracellular polymeric substances to improve activated sludge dewatering. J Hazard Mater 106:83–92
Nouha K, Kumar RS, Balasubramanian S, Tyagi RD (2017) Critical review of EPS production, synthesis and composition for sludge flocculation. J Environ Sci 66:225–245. https://doi.org/10.1016/j.jes.2017.05.020
Parikh A, Madamwar D (2006) Partial characterization of extracellular polysaccharides from cyanobacteria. Bioresour Technol 97:1822–1827. https://doi.org/10.1016/j.biortech.2005.09.008
Pavlova K, Grigorova D (1999) Production and properties of exopolysaccharide by Rhodotorula acheniorum MC. Food Res Int 32:473–477. https://doi.org/10.1016/s0963-9969(99)00110-6
Rashid MI, Mujawar LH, Shahzad T, Almeelbi T, Ismail IMI, Oves M (2016) Bacteria and fungi can contribute to nutrients bioavailability and aggregate formation in degraded soils. Microbiol Res 183:26–41. https://doi.org/10.1016/j.micres.2015.11.007
Roberson EB, Shennan C, Firestone MK, Sarig S (1995) Nutritional management of microbial polysaccharide production and aggregation in an agricultural soil. Soil Sci Soc Am J 59:1587–1594. https://doi.org/10.2136/sssaj1995.03615995005900060012x
Sandhya V, Ali SKZ, Minakshi G, Reddy G, Venkateswarlu B (2009) Alleviation of drought stress effects in sunflower seedlings by the exopolysaccharides producing Pseudomonas putida strain GAP-P45. Biol Fertil Soils 46:17–26. https://doi.org/10.1007/s00374-009-0401-z
Sessitsch A, Weilharter A, Gerzabek MH, Kirchmann H, Kandeler E (2001) Microbial population structures in soil particle size fractions of a long-term fertilizer field experiment. Appl Environ Microbiol 67:4215–4224. https://doi.org/10.1128/aem.67.9.4215-4224.2001
Sheng G-P, Yu H-Q, Li X-Y (2010) Extracellular polymeric substances (EPS) of microbial aggregates in biological wastewater treatment systems: a review. Biotechnol Adv 28:882–894. https://doi.org/10.1016/j.biotechadv.2010.08.001
Sparling GP, Cheshire MV (1985) Effect of periodate oxidation on the polysaccharide content and microaggregate stability of rhizosphere and non-rhizosphere soils. Plant Soil 88:113–122. https://doi.org/10.1007/bf02140670
Swaby RJ (1949) The relationship between micro-organisms and soil aggregation. J Gen Microbiol 3:236–254. https://doi.org/10.1099/00221287-3-2-236
Tang J, Mo Y, Zhang J, Zhang R (2011) Influence of biological aggregating agents associated with microbial population on soil aggregate stability. Appl Soil Ecol 47:153–159. https://doi.org/10.1016/j.apsoil.2011.01.001
Tian Y (2008) Behaviour of bacterial extracellular polymeric substances from activated sludge: a review. Int J Environ Pollut 32:78–89
Tisdall JM, Oades JM (1982) Organic matter and water-stable aggregates in soils. J Soil Sci 33:141–163. https://doi.org/10.1111/j.1365-2389.1982.tb01755.x
Totsche KU, Amelung W, Gerzabek MH, Guggenberger G, Klumpp E, Knief C et al (2018) Microaggregates in soils. J Plant Nutr Soil Sci 181:104–136. https://doi.org/10.1002/jpln.201600451
Umer MI, Rajab SM (2012) Correlation between aggregate stability and microbiological activity in two Russian soil types. Eur J Soil Sci 1:45–50

Vardharajula S, Ali SZ (2015) The production of exopolysaccharide by Pseudomonas putida GAP-P45 under various abiotic stress conditions and its role in soil aggregation. Microbiology 84:512–519. https://doi.org/10.1134/s0026261715040153

Wingender J, Neu TR, Flemming HC (1999) Microbial extracellular polymeric substances: characterization, structure and function. Springer, Berlin

Wingender J, Jaeger K-E, Flemming H-C (1999a) Interaction between extracellular polysaccharides and enzymes. In: Wingender J, Neu TR, Flemming HC (eds) Microbial extracellular polymeric substances. Springer, Berlin, pp 231–251

Wingender J, Neu TR, Flemming H-C (1999b) What are bacterial extracellular polymeric substances? In: Wingender J, Neu TR, Flemming H-C (eds) Microbial extracellular polymeric substances: characterization, structure and function. Springer, Berlin, pp 1–19

Wu N, Pan H-X, Qiu D, Zhang Y-M (2014) Feasibility of EPS-producing bacterial inoculation to speed up the sand aggregation in the Gurbantünggüt Desert, Northwestern China. J Basic Microbiol 54:1378–1386. https://doi.org/10.1002/jobm.201400355

Microbes Derived Exopolysaccharides Play Role in Salt Stress Alleviation in Plants

16

Purnima Singh, Vibha Pandey, and Prerana Parihar

Abstract

Several factors are responsible for the annihilation of agriculture production. Contribution of salt stress as an abiotic factor is gaining a striking position among all. The reason assigned for it is pressure on land for higher production to fetch the ever-increasing gigantic world population coupled with ever-changing climatic conditions. Erratic rainfall and higher temperature in arid and steppe climatic regions are further aggravating this problem. It is estimated that more than 50% of all arable land will suffer from salinization till 2050 and every year 1.5 million ha irrigated land is falling out of agricultural production. Salinity affects plant growth by disrupting cell ionic balance that ultimately causes deviation in cell division, photosynthesis, transpiration, plant signaling system, etc. Coping with salinity is a global issue to ensure sustainable crop production. Since the major upshots of soil salinization are loss of biological activity and deterioration of physiochemical fabric. Coevolution of plant and microbes always provide an opportunity to get a sustainable answer from rich microbiota of plant rhizosphere. Halophilic and halotolerant microbes residing in salt-stressed soil are capable of improving soil health and crop productivity through secretion of several polysaccharides, viz., intracellular storage polysaccharides, extracellular polysaccharides (also called Exopolysaccharide, EPS), and capsular polysaccharides. These polysaccharides contribute to biofilm formation and pathogenicity. Artificial tailoring of polymers having unique attributes such as

P. Singh · P. Parihar
Department of Plant Pathology, College of Agriculture, Rajmata Vijayaraje Scindia KrishiVishwa Vidyalaya, Gwalior, Madhya Pradesh, India

V. Pandey (✉)
Department of Plant Pathology, College of Agriculture, Jawaharlal Nehru Krishi Vishwa Vidhaylaya, Jabalpur, Madhya Pradesh, India

A. Vaishnav, D. K. Choudhary (eds.), *Microbial Polymers*,
https://doi.org/10.1007/978-981-16-0045-6_16

biocompatibility, gelling, and thickening as of ESP can be made superior through industrial application. Hence, comprehensive knowledge about EPS biosynthesis mechanism and its mode of action will further expand its role in salinity amelioration to address salinity problems. Therefore, microbial-derived EPS widens the scope of its use as a potential sustainable alternative for enhancement and protection of agricultural crops in salt stress arena.

Keywords

EPS · Biofilm · Salt stress · Biosynthesis of EPS · Salinity amelioration

16.1 Introduction

Perpetual agriculture is indispensable for today's world as it provides the possibility to fulfill our farming exigency. In developing nations, crop production is chiefly curtailed by the bleaker ramification of biotic and abiotic stress. Plants are sessile, so they are different kinds of biotic and abiotic stresses such as flood, dry spell, salt stress, heat and cold, exposure to heavy metals, nutrient deficiency as well as plant pathogens, and pests. Among various environmental factors, the presence of salt in the soil is one of the main issues responsible for lowering plant growth and crop productivity across the globe. As per salt-affected soil portal of FAO, the land affected by salinity throughout the world accounts for more than 1000 million hectares. The enhanced stress (high conc. of NaCl) not only led to high ethylene production but also induced ion toxicity, oxidative stress, which disturbed the osmotic potential in plants. Soil stress affects all the physiological functions like photosynthesis, nitrogen fixation, respiration, etc., which are essential for plants, resulting in a decrease in farm production and productivity. Due to the irrational application of agrochemicals, the trouble of soil salinity is getting more pronounced in arid and steppe climate. The use of PGPR is an assured biological approach to extenuate the salt stress. They can cope up with salt and other abiotic stresses by using a mechanism like induced systemic resistance, which helps them to thrive easily in harsh conditions. However, fungal diversification increases as the salt imposed stress increases upto 24.1 g NaCl/L concentration, but it was remarkably reduced to 44.1 g NaCl/L. However, club fungi (Basidiomycota) progressively replaced by Ascomycota as salinity increased while club fungi community flourished in the lack of NaCl (Cortés-Lorenzo et al. 2016).

In the semiarid areas, the main restricting feature to crop productivity is the use of water which has high soluble salt content. So, the identification, exploitation, and use of halotolerant microorganisms may lower salt stress and will provide a new eco-friendly substitute for stress tolerance in crops. In the mesophilic prokaryotic *Escherichia coli* bacteria first halotolerant gene was described. This gene *proB-74* was responsible for determining the deposition of osmolytes like proline in (Serrano and Gaxiola 1994). Similarly, Khan and coworkers (2009) have proposed long-term plan for revamping the diversity of mangroves through mechanistic application of

EPS-producing halotolerant microbes for mitigating derelict soil and enhancing yield.

EPS (exopolysaccharides) is known to produce by many algae, yeast, bacteria, and fungus. The production of EPS by many microorganisms is an immediate reaction to selective pressure exerted by the natural environment. Microbial EPS are biodegradable in nature; hence, causes no damage to the environment in comparison to synthetic polymers. This makes them environment friendly. The use of microbial polymer has begun in 1960 and thereafter there has been an exceptional enhancement in their remunerative use. Microbial EPS play a key role in micro and macro environment as well as, in cellular associations and in nutrition. EPS-producing microbes can live below the required threshold nutrient amount and better survive under an oligotrophic environment. This chapter aims to present overall information on microbial EPSs, their biosynthesis, and how they can be used to combat salt stress.

16.2 Salinity and Crop Production

Most grown crops are vulnerable to salinity stress. The term salinity remits soluble salt concentration present in soil and water. According to Mahmuduzzaman et al. (2014), in the coastal region, salt stress is a key problem decreasing the production of standing crops. Based on origin, salinization is of three types according to Zhou et al. (2013):

(a) **Primary salinization**:	natural salinity
(b) **Secondary salinization**:	main cause is the rise in groundwater level
(c) **Tertiary salinization**:	this happens due to irrigation

Based on adaptive evolution, plants are roughly categorized into two major types: plants which can withstand salinity stress known as halophytes and second plants which cannot withstand salinity stress and die are known as glycophytes. Unfortunately, most of the agricultural crops belong to the second category. Planting soil tolerant crop like Barley and Canola is a usual practice (Fita et al. 2015). However, these crops are capable of tolerating normal salt stress, so they have restricted reach and cannot be used in region which has moderate or high soil electrical conductivity (Morton et al. 2019). According to Mass and Hoffman model the crop yield is not affected until the salinity threshold surpasses up to a certain limit. Since optimum amount of salt concentration is required by plant for growth. High amounts of salt (Na^+ and Cl^-) are toxic to plants. Moreover, salt concentration beyond tolerable limit (Gupta and Huang 2014) lead to crop failure as plant fails to absorb water and wilting occurs. Salt stress also influences the plant growth by disrupting cell ionic balance that ultimately causes deviation in cell division, photosynthesis, transpiration, plant signaling system, etc. Thus, salt stress is a major vicious abiotic stress that hamper crop yield around the world (Flowers 2004).

As reflected by the growth responses of plants, they greatly show differences in their salinity tolerance. For example, cereals rice (*O. sativa*), foxtail barley (*Hordeaum vulgare*), bread wheat (*Triticum aestivum*), and durum wheat (*T. turgidum* ssp. *durum*) are most sensitive, most tolerant, moderately, and less tolerant to salt stress, respectively (Munns and Tester 2008). There is a greater variation in salinity tolerance in dicot plants in comparison to monocot plants. Some legumes are even more sensitive than rice (Lauchli 1984); *Medicago sativa* (Alfa-alfa or Lucern) is very tolerant and continues to grow at higher salinity.

16.3 EPS and Biofilm Producing Microorganism

Salt stress is critical abiotic stress which negatively influences the biochemical and physiological mechanism of plants and severely hampers crop productivity worldwide (Per et al. 2017). There are several groups of microorganism (Table 16.1) like bacteria, microalgae, cyanobacteria, and fungi that can live even in grating conditions. Production of EPPs and biofilm by halotolerant or halophilic microorganism are some added mechanisms which provide salinity resistance (Steele et al. 2014) in microbes. EPSs are a natural class of biopolymer which have great ecological importance. EPS are a general attribute of biofilm, where they have structural and protective functions (Rossi and De Philippis 2015). Microbial EPS represents a sustainable and alternative for the protection and enhancement of crops from salinity stress. The EPS has capability to attenuate the repercussions of salinity stress on the growth of agricultural crops grown in different salinity levels. EPSs alleviate salt stress by decreasing the amount of available Na^+ absorption in plant.

A crop when employed with EPSs exhibit enhanced soil structure and improved resistance to both salinity and water stress (Sandhya et al. 2009). Acidic EPSs have the capacity to cohere with Na^+ ions and making them unobtainable to plants under stress condition (Shrivastava and Kumar 2015), these bacteria mostly produce succinoglycan-type acidic EPSs. A study on root-zone inhabiting bacteria *Aeromonas hydrophila*, *B. insolitus*, and *Bacillus* sp. isolated from wheat seedling growing under usara soils revealed that they considerably increased dry matter of roots and shoots. Due to enhanced assimilation of calcium, sodium, and potassium ions from salt, microbial EPSs resulted in increased crop productivity (Ashraf et al. 2004).

Salt tolerant PGPR are key microbes in extenuating salt stress by inducing physical and chemical modifications. Indirectly, under stress conditions they regulate the conc. of ethylene in plants through the action of ACC deaminase (Glick 2005) and they produce siderophores to fight against disease-causing entities as well as accelerate nutrient absorption through the formation of various phytohormones (Goswami et al. 2016). A conglomerate of bacteria does so by producing bacterial EPSs which are involved in the adherence of bacteria in environmental surfaces and this association is known as biofilm (Mah and O'Toole 2001). To recoup the stress foisted by salt, salt-tolerant bacteria aid to metabolism and form biofilm and produce EPS. Halophilic bacteria can form biofilm and produce EPS at higher salt stress. EPS

Table 16.1 EPS-producing microorganism

S. no.	Microbes	Crop	Remark	Reference
1.	*Azotobacter chroococcum*	*Vicia faba*	Decreased Na^+ and Cl^- conc. significantly in salt stressed Faba beans	Abd El-Ghany and Attia (2020)
2.	*Dunaliella salina* (microalgae)	*Solanum lycopersicum*	3 and 6 g L^{-1} NaCl	
3.	*Halomonas* sp.	*Oryza sativa*	Under both salt and arsenic stress enhanced growth	Mukherjee et al. (2019)
4.	*Halomonas variabilis*	*Cicer arietinum*	Tendency to form biofilm at higher salinity	Qurashi and Sabri (2012)
5.	*Planococcus rifietoensis*	*Cicer arietinum*	Tendency to form biofilm at higher salinity	Qurashi and Sabri (2012)
6.	*Dietzia natronolimnaea STR1*	*Triticum aestivum* L.	Enhanced wheat tolerance to salinity stress	Bharti et al. (2016)
7.	*Bacillus alcalophilus*, *B. thuringiensis*, and *Gracilibacillus saliphilus*	*A. macrostachyum*	Mitigated the effects of high salt Physiological performance of the plant	Navarro-Torre et al. (2017)
8.	*M. yunnanensis*, *P. rifietoensis*, and *V. paradoxus*	*Beta vulgaris* L.	Enhanced salinity tolerance	Zhou et al. (2017)
9.	*S. brachiata*	Transgenic Tobacco	SbASR-1 gene transcript showed upregulation under salt stress conditions	Jha et al. (2012)
10.	*Pseudomonas putida* (GAP-P45)	*Helianthus annus*	Release EPS and alleviates salt stress	Sandhya et al. (2009)
11.	*Pseudomonas extremorientalis* (TSAU20) and *Pseudomonas chlororaphis* (TSAU13)	*Solanum lycopersicum* L.	Reduced oxidative stress	Egamberdieva et al. (2017)
12.	*P. alcaligenes* (PsA15), *B. polymyxa* (BcP26), and *M. phlei* (MbP18)	–	Tolerate high temperatures and salt conc. in aridisol	Shrivastava and Kumar (2015)
13.	*Aneurinibacillus aneurinilyticus* and *Paenibacillus sp.*	*Phaseolus vulgaris*	Decreased stress stimulated ethylene production	Gupta and Pandey (2019)

encourages bacterial colonization near plants root, and they can finally be added in soil. Till now, the rhizobial EPS synthesis has been studied in two specie *S. meliloti* and *R. leguminosarum* (Janczarek 2011). It is well documented that nutrient and other stress conditions are responsible for biofilm formation in *S. meliloti* and *R. Leguminosarum.*

In recent decades, many microalgal extracellular polysaccharides have been reported and their functional properties were documented. A number of microalgae like *Arthrospira* sp. (*Spirulina*), *Botryococcus braunii*, *Chlamydomonas reinhardtii*, *Dunaliella salina*, *Dunaliella tertiolecta*, *Isochrysis galbana,* and *Porphyridium cruentum* can be a fount of EPSs (Rossi and De Philippis 2016).

Ethylene is produced in multiple stresses and is also associated with the management of many physiological activities of crops, but the ethylene production in plants is influenced by climate change, which imposed a notable depletion in crop growth and development (Iqbal et al. 2017; Dubois et al. 2018). ACC is the precursor of ethylene synthesis, which is secreted as root exudates by plants, growing under stressful conditions (Wang et al. 2013; Liu et al. 2015; Abiri et al. 2017). ACC deaminase has the capacity to enhance plant growth and development which are surviving under adverse conditions (Gupta and Pandey 2019). The presence of more amount of salt in soil stimulates the biosynthesis of ethylene by precursor ACC; hence, it is called salt stress-induced ethylene (Chookietwattana and Maneewan 2012). PGPR bacteria can lower salt stress by reducing the concentration of ethylene. Ammonia and α-ketobutyrate are formed after the breakdown of ACC by an enzyme ACC deaminase that is present in certain PGPR; therefore, lowering the conc. of ethylene within the plant. Therefore, they have the capability to suppress the abiotic stress induced by ethylene production and its associated harmful effect on plants. There are many research papers suggesting that plants when inoculated with PGPR having ACC deaminase make the plant more resistant to different abiotic stresses.

Cyanobacteria are prokaryotic, unicellular or filamentous, oxygen-producing microorganisms among which some of them can fix atmospheric nitrogen. Green algae are eukaryotic lower plants. Numerous cyanobacteria and green algae are surrounded by a special mucilaginous covering around their cells or filaments. Filaments are mainly made up of carbohydrates which are synthesized and secreted extracellularly. Many studies have analyzed the composition of EPS using various chromatographic and mass spectrometric techniques. The principal sugars present in EPS of cyanobacteria and microalgae are glucose, galacturonic acid, fucose, galactose, rhamnose, arabinose, xylose, mannose, orthomethyl sugar, and acidic residues of glucuronic acid. Depending upon species EPS form different types of layers around the cells and called as slime, sheath, or capsules (Drews and Weckesser 1982; De Philippis and Vincenzini 2003).

In fields of science and industry, for fungal PSs there are many research avenues open (Osin'ska-Jaroszuk et al. 2015). The information on the diversity of fungal EPS, their mode of action and mode of interaction with higher organisms are still finite; therefore, there is a need to explore the efficacy of the EPSs produced fungi.

16.4 Chemical Structure of EPS

A wide range of bacteria secrete EPS which are carbohydrate polymers, which have an important role in symbiosis and pathogenesis. The EPSs can be homopolymer or heteropolymer and may have a diversity of non-carbohydrate substituents. The most common homopolymers found in plant interactive rhizobium bacteria family are the periplasmic {3-1,2 glucans (Hisamatsu et al. 1987), cellulose in *Agrobacterium* (Matthysse 1983) and *Rhizobium* species, levan in *Erwinia* (Bennett and Billing 1980) and *Pseudomonas* (Fett et al. 1989) species, and some alginates in fluorescent *Pseudomonas* sp. (Fett et al. 1989).

Based on location in bacterial cell EPSs may be assembled into two groups: extracellular or intracellular. The intracellular biopolymers are less and have limited utilization in comparison to extracellular biopolymers. The extracellular biopolymers can be grouped into four major classes; (i) polyamides (ii) polyesters (iii) inorganic polyanhydrides, and (iv) polysaccharides (Cerning 1995; Rehm 2010) and are commonly known as extracellular polymerhic substances (Rehm 2010), slime, and microcapsular polysaccharides (Sutherland 1972; Ruas-Madiedo et al. 2002) by others. Among the four components of the extracellular biopolymer, polysaccharide component is the most abundant that forms the basis of its classification. Their presence on the cell wall serve as protective and structural functions and is found as one of the constituents of teichoic acid. Outside the cell, they may take a shape of the morphological entity known as capsule (Ruas-Madiedo et al. 2002) or completely release in the environment as a slime layer (Cerning 1995).

Fungal Polysaccharides are classified based on their systemic propinquity, structure, sugar composition (homo and heteropolysaccharides), bonds present [β(1-3), β(1-6) and α-(1-3)], and their location in cell (cell wall polysaccharides, exopolysaccharides, and endopolysaccharides). Exopolysaccharides (EPSs) are the most studied fungal polysaccharides but their definition, origin, and classification are unclear and need more explanation (Osinka-Jaroszuk et al. 2015). The different ecological functions are performed by Ascomycota and Basidiomycota. Therefore, EPSs play a vital role in different biological functions with other organisms and host plants. The biochemical and biological properties of synthesized EPSs by Basidiomycota and Ascomycota manifest a lofty of variation (Donot et al. 2012; Mahapatra and Banerjee 2013).

16.5 Biosynthesis of EPS

EPSs are synthesized intracellularly throughout the growth phase. Microbial EPSs are generally synthesized in four (Stanford 1979) steps, involving four different types of enzymes (Kumar et al. 2007);

1. Uptake of a specific substrate (e.g., glucose).
2. Metabolism of sugar proceeds and the substrate is phosphorylated.
3. Polymerization.
4. Modification and release.

The rate of EPS production depends upon the stress and microbes produce different types of EPS by using different biosynthetic pathways. The genes which organize biosynthesis of EPS are generally clustered inside the genome of the particular producing organism. Based on biological functions polysaccharides produced by microbes can be classified into capsular polysaccharides, intracellular storage polysaccharides (glycogen), extracellular bacterial polysaccharides which are also called as exopolysaccharides (alginate, cellulose, xanthan, sphingan, etc.) which are necessary for biofilm formation and pathogenicity. Mannose, galactose, and glucose are some common monomers found in polysaccharides of various species. Amino sugars, neutral sugar (galacturonic), uronic (fucose and rhamnose), ester-linked substituents, and pyruvate ketals are EPS components (Abbas and Naqqash 2019). Till now four general mechanisms are known for the synthesis of carbohydrate polymers: (i) ATP-binding cassette (ABC) transporter-dependent pathway (ii) Wzx/Wzy-dependent pathway (iii) extracellular synthesis by use of a single sucrase protein, and (iv) synthase-dependent pathway.

Since it is known that acidic EPS binds to ions present in saline soil, this chapter will only focus on the biosynthesis of acidic EPS. The acidic heteropolysaccharides produced by rhizobia has a complex structure as well as the synthesis process. Wzx/Wzy-dependent pathway produces acidic EPS succinoglycan in several bacterial strains like *Pseudomonas*, *Alcaligens*, *Agrobacterium,* and *Rhizobium* (Harada and Yoshimura 1964; Zevenhuizen 1997). For the production of succinoglycan, the model organism discussed in this chapter is *R. meliloti.*

Succinoglycan (Fig. 16.1) is a branched heteropolysaccharide consists of repeating units of an oligosaccharide and acetate, succinate and pyruvate are present as many alternative decorations. The monomers present in the repeating unit are β-linked glucose and galactose and they are present in the 7:1 ratio. According to Stredansky (2005) pyruvate present in a stoichiometric ratio, while depending up on cultural conditions and strain, the presence of acetate and succinate decoration depends. Succinoglycan is essential in plants as it is necessary for plant symbiosis and other biological functions (Leigh and Walker 1994).

The biosynthesis of these acidic EPS involves four major steps (Boulnois and Jann 1989; Jann and Jaon 1991; Long et al. 1988) which are as under:

1. Nucleotide sugar diphosphate intermediates synthesis.
2. Assembly of the repeating oligosaccharide subunit stepwise.
3. Addition of decorations.
4. Transfer of the formed polysaccharide chain from its carrier lipid to the new subunit.

Very less is known about the polymerization step, but it is assumed that polymerization occurs in cytoplasmic membrane inner face, after which Bayer adhesion zones present between the inner and outer membrane transports the EPS through it.

A total of 19 sets of genes are required in *R. meliloti* SU47 for the biosynthesis of succinogylcan, two clusters of genes are recognized, one (Long et al. 1988; Reuber et al. 1991) for succinoglycan synthesis and second (Glazebrook and Walker 1989)

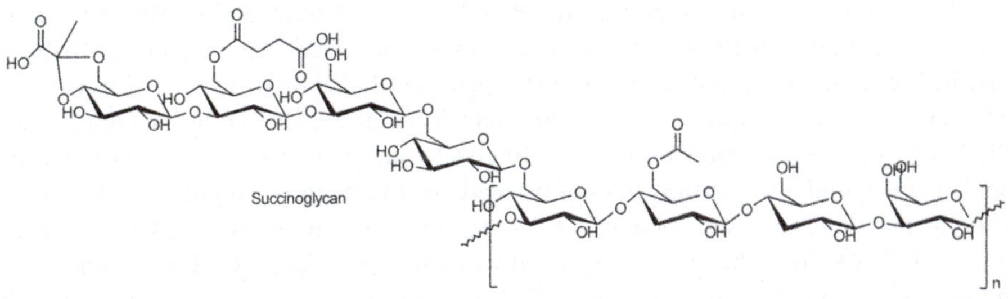

Fig. 16.1 Chemical structure of Succinoglycan

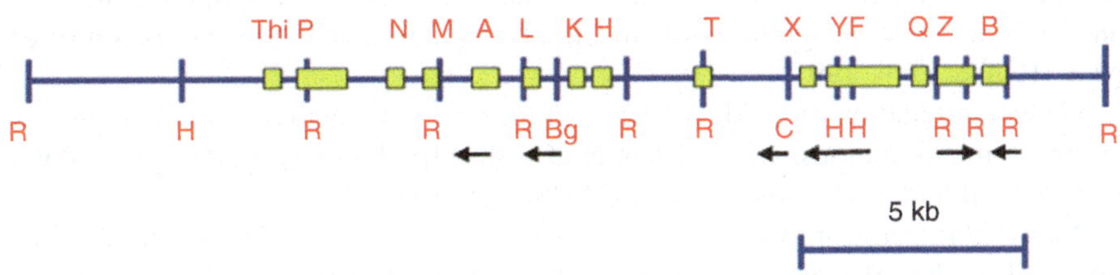

Fig. 16.2 Genetic map of the *R. meliloti* SU47 *exo* gene cluster (The map is derived from Long et al. (1988)). Yellow boxes show the minimum boundaries of the *exo* loci designated above the line; Thi- thiamine biosynthetic locus. Vertical line -cutting sites by restriction enzymes designated below the line. Abbreviations are: R-*Eco*RI; H-*Hind*III; Bg-*Bgl*II; and C-*Cia*I. Not all of the sites for the latter two enzymes have been determined. Horizontal arrows indicate the direction of transcription)

for EPSb (EPS II) synthesis are found on the indigenous megaplasmid pRmeSU47b (pSymb) (Hynes et al. 1986; Finan et al. 1986). The succinoglycan synthesis genes are called as *exo* genes, while EPSb synthesis genes are called as *exp* or *muc*. The *exo* loci are necessary for the biosynthesis of acidic EPS in *R. Meliloti* (Reuber et al. 1991). The *exo* cluster has been distinguished by genetic complementation and sequencing and contains at least 12 genetic complementation groups within a 22 kb region (Fig. 16.2). Also, several loci located outside the *exo* gene clusters affect the biosynthesis of succinoglycan. While *exp* clusters contain seven complementation groups (Glazebrook and Walker 1989).

Nucleotide sugar precursors are first synthesized in the biosynthesis of succinoglycan. Gene *exoC* codes for an enzyme phosphoglucomutase which is responsible for transferring phosphate group from the 6′ to the 1′ position in reverse direction (glucose-6-phosphate → glucose-1-phosphate) (Uttaro et al. 1990). UDP-glucose-4-epimerase is encoded by gene *exoB,* which converts UDP-glucose to UDP-galactose (Buendia et al. 1991). Mutation in *exoB* and *exoC* causes the lack of EPS I production but also affects the synthesis of other polymers like EPS II, LPS, and β-glucans. Protein which displays UDP-glucose pyrophosphorylase activity is encoded by *exoN* gene. Mutation in *exoN* gene leads to a reduction in EPS I production (Glucksmann et al. 1993; Becker et al. 1993).

EPS synthesis in rhizobium is regulated transcriptional and posttranslational levels. The finest distinguished transcriptional regulator acts negatively. Many researchers concluded that *R. melitoti* (Doherty et al. 1988; Reed et al. 1991a) has a negative regulatory locus, *exoR* which affects succinoglycan synthesis. On pSymb, the *exoR* gene is chromosomal and unlinked to the *exo* cluster. The existence of several independent insertion mutations leads to the negative regulatory nature of *exoR* gene, which causes enhanced EPS synthesis and they are recessive natured mutations. Alkaline phosphatase activities of *exo-phoA* fusions indicate that *exoR* regulates the transcription or translation of most of the genes of the *exo* cluster with the exception of the *galE* equivalent *exoB* (Doherty et al. 1988; Reuber et al. 1991). Another succinoglycan regulatory gene, *exo*S which resembles like *exoR*. This gene (*exoS*) also lack genetic linkage to the *exo* cluster, mutation of both the genes *exoR* and *exoS* affected the synthesis of succinoglycan and also on the expression of *exo* gene (Reuber et al. 1991).

Genetic regulation of EPSb synthesis in *R .melitoti* is relatively less known. Two chromosomal mutations, *mucR* (Zhan et al. 1989) [probably similar to *rexA* (Pühler et al. 1991)] and *expR* cause increased EPSb synthesis.

Many *Rhizobium* sp. has a new regulatory mechanism which is conserved and imparted by the *exoX* and *exoY* genes. The *exoX* (psiA)-exoY (pssA) are posttranslational regulatory system of *Rhizobium* sp. the *exoX* gene is found in *R. meliloti* (Reed et al. 1991b; Zhan and Leigh 1990) that encodes a small protein with hydrophilic as well as hydrophobic domains. In *R. meliloti*, *exoX* affects succinoglycan synthesis but not EPS synthesis (Reed et al. 1991b; Zhan et al. 1991).

16.6 Inoculation of EPS Producers in Soil and Plant

16.6.1 Inoculation in Soil

Microbial polysaccharides are the most active organic constituents in soil aggregate formation and stabilization. According to several studies, microbes are elementary for stabilization and aggregation of soil, the effect of microbes on soil structure and stability depends upon the type of microbial species present, soil management and available substrate (Umer and Rajab 2012). The microbial polysaccharide binds the soil particle through polymer bridges.

Pseudomonas, *Bacillus*, and *Paenibacillus* are the most explored EPS-producing bacterial species for soil aggregation, these genera are easily cultured in laboratory conditions and produce a decent quantity of EPS. When GAP-P 45 strain of *P. putida* is inoculated under stress conditions, the strain produced an increased amount of EPS in the soil as a result, it enhanced aggregate stability by more than 50% in soil (Vardharajula and Ali 2015).

Under salt stress condition many cyanobateria survives by accumulating increased amount of saccharides in cells as well as outside the cell (Padhi et al. 1997). Soil aggregation rose at higher salinity up to 100 mM in inoculated soil (Qurashi and Sabri 2012) and aggregate formation was more pronounced near roots.

This indicates biofilm formation near roots in response to salt stress leads to bacterial attachment and soil particles adhere together with roots in response to increased biofilm formation at higher EPS (Ashraf et al. 2005). This binding property of EPS would act like cementing material and provide strength to soil aggregate formation (Ashraf and Hafeez 2004; Ashraf and Foolad 2007) and favor development of plants under harsh conditions.

In addition, inoculation of cyanobacteria in soil enhances the structure of the soil. Due to EPS production, these bacteria stabilize the soil surface. Moreover, the inoculation of cyanobacteria is extended up to subcrust, improves soil fertility, structure, and nutrient bioavailability. Rossi and De Philippis (2015) explored that when degraded and desertified land was treated with cyanobaceria; it improved the quality of arable land. They are the best candidate, as they possess the benefits like oligotrophy, stress tolerance, and drought resistance, which improve soil stability and moisture content (Guo et al. 2007). Cyanobacterial EPS produce binding and gluing mesh on soil particles, which apply mechanical impact on soil particles. Due to water-loving nature of EPS they retain available moisture (Mugnai et al. 2017).

Similarly, inoculation of filamentous fungi enhances soil aggregation with variable potency in comparison to bacteria. The best known fungal species belong to genera of *Absidia*, *Aspergillus*, *Chaetomium*, *Mucor*, *Rhizopus*, *Fusarium*, and *Aspergillus*.

16.6.2 Inoculation in Plant

Beneficial microbes elevate the uptake of nutrients like N, P, and K upon inoculation in crop plants. PGPR and AMF inoculation in plant is an important practice in agriculture. The crop plants secrete low molecular weight organic substances from their root passively known as root exudates, which leads to microbial community development in the root zone. The microbe in turn establishes interaction with the host plant and thereafter enhancing soil aggregation, plant growth, and escalating RAS (root adhering soil). Rashid et al. (2016) inferred that plant can absorb nutrients and water by RAS easily.

Cipriano et al. (2016) summarized that *B. pantoea*, *Rhizobium* sp., and *Pseudomonas* sp. are well-known EPS and plant growth promoters, so they are the most investigated and best candidate for inoculation. They are directly inoculated into the seedlings. The production of EPS by microbes in the crop root zone protects the plant from several environmental stresses, also enhances the nutrient uptake in crop plants as well improves soil structure and aggregation (Alami et al. 2000; Bezzate et al. 2000).

In chickpea inoculation of *H. variabilis* (HT-1) and *Planococcus rifietoensis* (RT-4) ameliorate salt stress, enhanced soil aggregation under salinity more than 75%, promoted plant growth, improved soil fertility, and under stress enhance the growth of the plant. Similarly, *Rhizobium* strain (YAS-34) reasonably affected water, nitrogen, and soil aggregation and RAS up to 100% under normal and water-stressed conditions. The strain enhanced the volume of soil macropore, increased plumule and radical biomass, and performed as a plant growth promoter. These repercussions of EPS production turned down water loss and top up water holding capacity (WHC) of soil (Alami et al. 2000). Aeromonas and Bacillus strain aggregates around the root of wheat as assessed by Ashraf et al. (2004), they observed restricted Na^+ uptake in moderate saline soil. Similarly, Siddikee et al. (2010) in canola reported *P. rifietoensis* (RS 18) promotes plant growth, which is a halotolerant bacterium.

16.7 Amelioration of Salt Stress by Microbes

Microbial polymers produced by microorganisms are involved in salt stress tolerance, which are not only beneficial for producing microorganism but also for the associated crop. To counterbalance the harmful impact of salt stress, halotolerant bacteria form biofilm and produce EPS which have a strong affinity for water and aid in their metabolism. So, halophilic bacteria when exposed to high salt stress, it uses both strategies to vanquish the salinity stress. The Na^+ ion uptake can be decreased by $NaCl^-$tolerant strains, as they produce the polymers which can decrease the availability of Na^+ ion (Upadhyay et al. 2011). Consequently, promotes survival of microorganism in saline environment by preventing osmotic stress and nutrient disparity. For example, *S. meliloti* (EPBI), which thrives in salty marsh at 0.3 M salt concentration, shows reduced growth when *S. meliloti* culture is inoculated at low concentration, for bacteria this reduced levels of salt concentration is a stressful condition (Lloret et al. 1998).

Biofilm coordinates between EPS matrix and bacterial cells, i.e., bacterial biofilm has an interdependent lifestyle. It has been reported that three bacteria *Halomonas aqamariana*, *H. meridiana,* and *Kushneria indalinina* produce vigorous EPS and biofilm at 1 molar salinity levels. They also concluded that under saline conditions salt-loving bacteria (halophiles) colonizes crop plants root zones more effectively than normal conditions and helps plants to thrive under such stressful condition.

In usara soils, halotolerant bacteria can be used as a bioremediation tool. Shah et al. (2017) quoted the potency of two halophilic bacteria *Ocenobacillus kapalis* and *T. devorans*; they produce EPS molecule which alleviates salt stress by decreasing the assimilation of Na^+ ion by roots. Also, these strains ameliorate the saline soil of *M. sativa* (alfa-alfa) plant, as well as favored soil aggregation under salt stress and water stress equally. The glucuronic acid, glucose, mannose, and galactose are the main composition as well as the metabolic products of muran type of EPS, so they could also function as plant nutrient (Arias et al. 2003). In other legumes also a similar pattern of productivity has been reported, when chickpea (*C. arientum*) plant co-inoculated with the six halotolerant strains of *Bacillus* sp., they enhanced symbiosis effectiveness. In addition dual inoculation indeed raises the growth of plant thus suppressing the repressive effect of salt stress.

16.8 Conclusion

Adaptation to varying environmental conditions by microorganisms especially residing in the soil through different mechanisms has well been documented after decades of studies by researchers. However, extracellular polymers secreted by microorganisms enable to maintain enough moisture around them, absorb nutrient; and also protect them from harmful organisms and stress conditions through facilitating signal-based chemical reactions. Moreover, these extracellular polymers are highly diverse owing to the difference in their composition and structure; therefore, their action is very selective.

Presently, the world is moving towards the principal of sustainability where feasibility to subside the non-biodegradable and bio-incompatible compounds can be possible through the use of EPS. Gelling and emulsion capacities of EPS are contributory factors in improving soil fertility and enhancing nutrient uptake by plants. However, plant–microbial interactions are governed by the consistency and biocompatibility of extracellular polymeric substances. Characterization of EPS-producing microorganisms having different physiological potential is essential to make need-based manipulation at genetic level in these microbes to obtain improved polymers. Although, several researchers have documented that EPS ameliorates soil quality and plant growth but complex mechanism involved in the production of these compounds need to be explored extensively. Moreover, there is also a need to develop a consortium of genetically modified microorganism that can fulfill ever-increasing demand of agriculture.

16.9 Future Prospectus

Moving from green revolution to green economy in ever-increasing world population, characterization of several new EPS from microorganism is seen as the only viable solution. Hence, the use of several highly specific modern techniques to characterize new EPS and their mechanism of action need to be prioritized. Improving soil structure and fertility along with plant health management are well-established facts. However, implementation of techniques like next generation sequencing, CLSM, scanning electronic microscopy, nuclear magnetic resonance for characterization of EPS could open a window of its applicability in other areas too like garbage management, wastewater treatment, soil reclamation, postharvest management, etc. However, understanding of EPS structure, composition, genetic regulations under different conditions are inter-related aspects that need an in-depth knowledge base. Exploration and characterization of new polymer from microorganisms could lead to its application in other fields of scientific discoveries. Besides, studies on environment-driven signal-based EPS production and its interactions with soil and plant system are yet unexplored. Hence, it is time to integrate different fundamental knowledge together to widen the scope of better understanding and use of EPS in ever-changing climatic conditions.

References

Abbas R, Naqqash T (2019) Halotolerant PGPR: a hope for cultivation of saline soil. J King Saud Univ-Sci 31(4):1195–1201

Abd El-Ghany MF, Attia M (2020) Effect of exopolysaccharide-producing bacteria and melatonin on faba bean production in saline and non-saline soil. Agro 10(3):316

Abiri R, Shaharuddin NA, Maziah M, Yusof ZNB, Atabaki N, Sahebi M (2017) Role of ethylene and the APETALA 2/ethylene response factor superfamily in rice under various abiotic and biotic stress conditions. Environ Exp Bot 134:33–44

Alami Y, Achouak W, Marol C, Heulin T (2000) Rhizosphere soil aggregation and plant growth promotion of sunflowers by an exopolysaccharide-producing Rhizobium sp. strain isolated from sunflower roots. Appl Environ Microbiol 66:3393–3398

Arias S, Del Moral A, Ferrer MR, Tallon R, Quesada E, Bejar V (2003) Mauran, an exopolysaccharide produced by the halophilic bacterium *Halomonas maura*, with a novel composition and interesting properties for biotechnology. Extremophiles 7(4):319–326

Ashraf M, Foolad MR (2007) Roles of glycine betaine and proline in improving plant abiotic stress resistance. Environ Exp Bot 59:206–216

Ashraf M, Hafeez M (2004) Thermotolerance of pearl millet and maize at early growth stages: growth and nutrient relations. Biol Plant 48:81–86

Ashraf M, Hasnain S, Berge O, Mahmood T (2004) Inoculating wheat seedlings with exopolysaccharide-producing bacteria restricts sodium uptake and stimulates plant growth under salt stress. Biol Fertil Soils 40:157–162

Ashraf M, Hasnain S, Hussain F (2005) Exo-polysaccharides (EPS) producing biofilm bacteria in improving physicochemical characteristics of the salt affected soils. In: Iftikhar AR, Pervez A, Irshad M, Siddique M, Shahzadi I (eds) Proc. International Conference on Environmentally Sustainable Development (ESDev-2005). COMSATS Institutes of Information and Technology, Abbottabad, pp 1527–1536

Becker A, Kleickmann A, Keller M, Arnold W, Pühler A (1993) Identification and analysis of the *Rhizobium meliloti exoAMONP* genes involved in exopolysaccharide biosynthesis and mapping of promoters located on the *exoHKLAMONP* fragment. Mol Gen Genet 241:367–379

Bennett RA, Billing E (1980) Origin of the polysaccharide ooze from plants infected with *Erwinia amylovora*. J Gen Microbiol 116:341–349

Bezzate S, Aymerich S, Chambert R, Czarnes S, Berge O, Heulin T (2000) Disruption of the *Paenibacillus polymyxa* levansucrase gene impairs its ability to aggregate soil in the wheat rhizosphere. Environ Microbiol 2:333–342

Bharti N, Pandey SS, Barnawal D, Patel VK, Kalra A (2016) Plant growth promoting rhizobacteria *Dietzia natronolimnaea* modulates the expression of stress responsive genes providing protection of wheat from salinity stress. Sci Rep 6:347–368

Boulnois GJ, Jann K (1989) Bacterial polysaccharide capsule synthesis, export and evolution of structural diversity. Mol Microbiol 3:1819–1823

Buendia AM, Enenkel B, Koplin R, Niehaus K, Arnold W, Pühler A (1991) The *Rhizobium meliloti exoZl exoB* fragment of megaplasmid 2: ExoB functions as a UDP-glucose 4-epimerase and ExoZ shows homology to NodX of *Rhizobium leguminosarum* biovar *viciae* strain TOM. Mol Microbiol 5:1519–1530

Cerning J (1995) Production of exopolysaccharides by lactic acid bacteria and dairy propionibacteria. Lait 75:463–472

Chookietwattana K, Maneewan K (2012) Selection of efficient salt-tolerant bacteria containing ACC deaminase for promotion of tomato growth under salinity stress. Soil Environ 31:30–36

Cipriano MA, Lupatini M, Lopes-Santos L, Da Silva MJ, Roesch LF, Destéfano SA (2016) Lettuce and rhizosphere microbiome responses to growth promoting Pseudomonas species under field conditions. FEMS Microbiol 92(12). https://doi.org/10.1093/femsec/fiw197

Cortés-Lorenzo C, González-Martínez A, Smidt H, González-López J, Rodelas B (2016) Influence of salinity on fungal communities in a submerged fixed bed bioreactor for wastewater treatment. Chem Eng J 285:562–572

De Philippis R, Vincenzini M (2003) Outermost polysaccharidic investments of cyanobacteria: nature, significance and possible applications. Recent Res Dev Microbiol 7:13–22

Doherty D, Leigh JA, Glazebrook J, Walker GC (1988) *Rhizobium meliloti* mutants that overproduce the *R. meliloti* acidic Calcofluor-binding exopolysaccharide. J Bacteriol 170:4249–4256

Donot F, Fontana A, Baccou JC, Schorr-Galindo S (2012) Microbial exopolysaccharides: main examples of synthesis, excretion, genetics and extraction. Carbohydr Polym 87:951–962

Drews G, Weckesser J (1982) Function, structure and composition of cell walls and external layers. In: Carr NG, Whitton BW (eds) The biology of cyanobacteria. Blackwell, Oxford, pp 333–357

Dubois M, Van Den Broeck L, Inzé D (2018) The pivotal role of ethylene in plant growth. Trends Plant Sci 23:311–323

Egamberdieva D, Davranov K, Wirth S, Hashem A, Abd Allah EF (2017) Impact of soil salinity on the plant-growth–promoting and biological control abilities of root associated bacteria. Saudi J Biol Sci 24:1601–1608

Fett WF, Osman SF, Dunn MF (1989) Characterization of exopolysaccharides produced by plant-associated fluorescent *pseudomonads*. Appl Environ Microbiol 55:579–583

Finan TM, Kunkel B, DeVos GF, Signer ER (1986) Second symbiotic megaplasmid in *Rhizobium meliloti* carrying exopolysaccharide and thiamine synthesis genes. J Bacteriol 167:66–72

Fita A, Rodriiguez-Burrenzo A, Boscaiu M, Prohens J, Vincente O (2015) Breeding and adapting crops tolerant to drought and salinity: a new paradigm for increasing food production. Front Plant Sci 6:978

Flowers TJ (2004) Improving crop salt tolerance. J Exp Bot 55(396):307–319

Glazebrook J, Walker GC (1989) A novel exopolysaccharide can function in place of the Calcofluor-binding exopolysaccharide in nodulation of alfalfa by *Rhizobium meliloti*. Cell 56:661–672

Glick BR (2005) Modulation of plant ethylene levels by the bacterial enzyme ACC deaminase. FEMS Microbiol Lett 251:1–7

Glucksmann MA, Reuber TL, Walker GC (1993) Genes needed for the modification, polymerization, export, and processing of succinoglycan by *Rhizobium meliloti*: a model for succinoglycan biosynthesis. J Bacteriol 175:7045–7055

Goswami D, Thakker JN, Dhandhukia PC (2016) Portraying mechanics of plant growth promoting rhizobacteria (PGPR): a review. Cogent Food Agric 2(1):1127500

Guo Y, Zhao H, Zuo X, Drake S, Zhao X (2007) Biological soil crust development and its topsoil properties in the process of dune stabilization, Inner Mongolia. China Environ Geol 54:653–662

Gupta B, Huang B (2014) Mechanism of salinity tolerance in plants: physiological, biochemical and molecular characterization. Int J Genomics 701:596

Gupta S, Pandey S (2019) ACC Deaminase producing bacteria with multifarious plant growth promoting traits alleviates salinity stress in french bean (*Phaseolus vulgaris*) plants. Front Microbiol 10:1506

Harada T, Yoshimura T (1964) Production of a new acidic polysaccharide containing succinic acid by a soil bacterium. Biochim Biophys Acta 83:374–376

Hisamatsu M, Yamada T, Higashiura T, Ikeda M (1987) The production of acidic,O-acetylated cyclosophoraos [cyclic(l-2)f3-D-glucans] by *Agrobacterium* and *Rhizobium* species. Carbohydr Res 163(1):15–22

Hynes MF, Simon R, Muller P, Niehaus K, Labes M, Piihler A (1986) The two megaplasmids of Rhizobium meliloli are involved in the effective nodulation of alfalfa. Mol Gen Genet 202:356–362

Iqbal N, Khan NA, Ferrante A, Trivellini A, Francini A, Khan M (2017) Ethylene role in plant growth, development and senescence: interaction with other phytohormones. Front Plant Sci 8:475

Janczarek A (2011) Environmental signals and regulatory pathways that influence exopolysaccharide production in rhizobia. Int J Mol Sci 12:7898–7933

Jann K, Jaon B (1991) Genetic approaches to oligosaccharide metabolism. Biochem Soc Trans 19:623–628

Jha B, Lal S, Tiwari V, Yadav SK, Agarwal PK (2012) The *SbASR-1* gene cloned from an extreme halophyte *Salicornia brachiata* enhances salt tolerance in transgenic tobacco. Mar Biotechnol 14:782–792. https://doi.org/10.1007/s10126-012-9442-7

Khan MS, Zaidi A, Musarrat J (2009) Microbial strategies for crop improvement. Springer, Berlin

Kumar AS, Mody KH, Jha B (2007) Bacterial exopolysaccharides—a perception. J Basic Microbiol 47:103–117

Lauchli A (1984) Salt exclusion: an adaptation of legumes for crops and pastures under saline conditions. In: Staples RC (ed) Salinity tolerance in plants: strategies for crop improvement. Wiley, New York, pp 171–187

Leigh JA, Walker GC (1994) Exopolysaccharides of *Rhizobium*: synthesis, regulation and symbiotic function. Trends Genet 10:63–67

Liu J, Xie B, Shi X, Ma J, Guo C (2015) Effects of two plant growth-promoting rhizobacteria containing 1-aminocyclopropane-1-carboxylate deaminase on oat growth in petroleum-contaminated soil. Int J Environ Sci Technol 12:3887–3894

Lloret J, Wulff BB, Rubio JM, Downie JA, Bonilla I, Rivilla R (1998) Exopolysaccharide II production is regulated by salt in the halotolerant strain *Rhizobium meliloti* EFB1. Appl Environ Microbiol 64:1024–1028

Long S, Reed JW, Himawan J, Walker GC (1988) Genetic analysis of a cluster of genes required for synthesis of the Calcofluor-binding exopolysaccharide of *Rhizobium meliloti*. Bacteriol 170:4239–4248

Mah TFC, O'Toole GA (2001) Mechanisms of biofilm resistance to antimicrobial agents. Trends Microbiol 9:34–39

Mahapatra S, Banerjee D (2013) Fungal exopolysaccharide: production, composition and applications. Microbiol Insights 6:1–16

Mahmuduzzaman M, Ahmed ZU, Nuruzzaman AKM, Ahmed FRS (2014) Causes of salinity intrusion in coastal belt of Bangladesh. Int J Plant Res 4:8–13

Matthysse AG (1983) Role of bacterial cellulose fibrils in *Agrobacterium tumefaciens* infections. Bacteriol 154:906–915

Morton MJ, Awalia M, Al-Tamimi N, Saades S, Pailles Y, Negrao S (2019) Salt stress under the scalper-dissecting the genetics of salt tolerance. Plant J 97:144–163

Mugnai G, Rossi F, Felde VJMNL, Colesie C, Büdel B, Peth S et al (2017) Development of the polysaccharidic matrix in biocrusts induced by a cyanobacterium inoculated in sand microcosms. Biol Fertil Soils 54:27–40

Mukherjee P, Mitra A, Roy M (2019) Halomonas rhizobacteria of *Avicennia marina* of indian sundarbans promote rice growth under saline and heavy metal stresses through exopolysaccharide production. Front Microbiol 10:1207

Munns R, Tester M (2008) Mechanisms of salinity tolerance. Annu Rev Plant Biol 59:651–681

Navarro-Torre S, Barcia-Piedras JM, Mateos-Naranjo E, Redondo-Gómez S, Camacho M, Caviedes MA (2017) Assessing the role of endophytic bacteria in the halophyte *Arthrocnemum macrostachyum* salt tolerance. Plant Biol 19:249–256

Osinka-Jaroszuk M, Jarosz-Wilkołazka A, Jaroszuk-Ściseł J, Szałapata K, Nowak A, Jaszek M, Ozimek M, Majewska M (2015) Extracellular polysaccharides from Ascomycota and Basidiomycota: production conditions, biochemical characteristics, and biological properties. World J Microbiol Biotechnol 31:1823–1844

Padhi H, Rath B, Adhikary SP (1997) Tolerance of nitrogen-fixing cyanobacteria to NaCl. Biol Plant 40:261–268

Per TS, Khan NA, Reddy PS, Masood A, Hasanuzzaman M, Khan MIR, Anjum NA (2017) Approaches inmodulating proline metabolism in plants for salt and drought stress tolerance: phytohormones, mineral nutrients and transgenics. Plant Physiol Biochem 115:126–140

Pühler A, Arnold W, Buendia-Claveria A, Kapp D, Keller M, Niehaus K, Quandt J, Roxlau A, Weng WM (1991) The role of the *Rhizobium meliloti* exopolysaccharides EPSI and EPSII in the infection process of alfalfa nodules. In: Hennecke H, Verma DPS (eds) Advances in molecular genetics of plant–microbe interaction, vol 1. Kluwer Academic Publishers, Dordrecht, The Netherlands, pp 189–194
Qurashi AW, Sabri AN (2012) Bacterial exopolysaccharide and biofilm formation stimulate chickpea growth and soil aggregation under salt stress. Braz J Microbiol 43:1183–1191
Rashid MI, Mujawar LH, Shahzad T, Almeelbi T, Ismail IMI, Oves M (2016) Bacteria and fungi can contribute to nutrients bioavailability and aggregate formation in degraded soils. Microbiol Res 183:26–41
Reed JW, Glazebro Ok J, Walker GC (1991a) The *exoR* gene of *Rhizobium meliloti* affects RNA levels of other exo genes but lacks homology to known transcriptional regulators. J Bacteriol 173:3789–3794
Reed JW, Capage M, Walker GC (1991b) *Rhizobium meliloti* exoG and exoJ mutations affect the ExoX-ExoY system for modulation of exopolysaccharide production. J Bacteriol 173:3776–3788
Rehm BHA (2010) Bacterial polymers: Biosynthesis, modifications and applications. Nat Rev Microbiol 8:578–592
Reuber TL, Long S, Walker GC (1991) Regulation of *Rhizobium meliloti* exo genes in free-living cells and in planta examined by using Tnpho A fusions. J Bacteriol 173:426–434
Rossi F, De Philippis R (2015) Role of cyanobacterial exopolysaccharides in phototrophic biofilms and in complex microbial mats. J Life 5:1218–1238
Rossi F, De Philippis R (2016) Exocellular polysaccharides in microalgae and cyanobacteria: chemical features, role and enzymes and genes involved in their biosynthesis. In: Borowitzka MA, Beardall J, Raven JA (eds) The physiology of microalgae. Springer, Dordrecht, pp 565–590
Ruas-Madiedo P, Hugenholtz J, Zoon P (2002) An overview of the functionality of exopolysaccharides produced by lactic acid bacteria. Int Dairy J 12:163–171
Sandhya VZAS, Grover M, Reddy G, Venkateswarlu B (2009) Alleviation of drought stress effects in sunflower seedlings by the exopolysaccharides producing *Pseudomonas putida* strain GAP-P45. Biol Fertil Soils 46:17–26
Serrano R, Gaxiola R (1994) Microbial models and salt stress tolerance in plants. CRC Crit Rev Plant Sci 13:121–138
Shah G, Jan M, Afreen M, Anees M, Rehman S, Daud MK, Malook I, Jamil M (2017) Halophilic bacteria mediated phytoremediation of salt-affected soils cultivated with rice. J Geochem Explor 174:59–65
Shrivastava P, Kumar R (2015) Soil salinity: a serious environmental issue and plant growth-promoting bacteria as one of the tools for its alleviation. Saudi J Biol Sci 22:123–131
Siddikee MA, Chauhan PS, Anandham R, Han G, Sa T (2010) Isolation, characterization, and use for plant growth promotion under salt stress, of ACC deaminase-producing halotolerant bacteria derived from coastal soil. J Microbiol Biotechnol 20(11):1577–1584
Stanford PA (1979) Exocellular, microbial polysaccharides. In: Tipson RS, Horton D (eds) Advances in carbohydrate chemistry and biochemistry, vol 36. Academic, London/New York, pp 266–303
Steele DJ, Franklin DJ, Underwood GJ (2014) Protection of cells from salinity stress by extracellular polymeric substances in diatom biofilms. Biofouling 30:987–998
Stredansky M (2005) Succinoglycan. In: Biopolymers online. Wiley-VCH Verlag GmbH & Co. KGaA, Weinheim. application.wiley-vch.de
Sutherland IW (1972) Bacterial exopolysaccharides. In: Rose AH (ed) Advances in microbial physiology. Academic Press, New York, NY
Umer MI, Rajab SM (2012) Correlation between aggregate stability and microbiological activity in two Russian soil types. Eur J Soil Sci 1:45–50

Upadhyay SK, Singh JS, Singh DP (2011) Exopolysaccharide-producing plant growth-promoting rhizobacteria under salinity condition. Pedosphere 21:214–222

Uttaro AD, Cangelosi GA, Geremia RA, Nester EW, Ugalde RA (1990) Biochemical characterization of avirulent *exoC* mutants of *Agrobacterium tumefaciens*. J Bacteriol 172:1640–1646

Vardharajula S, Ali SZ (2015) The production of exopolysaccharide by *Pseudomonas putida* GAP-P45 under various abiotic stress conditions and its role in soil aggregation. Microbiology 84:512–519

Wang F, Cui X, Sun Y, Dong C-H (2013) Ethylene signalling and regulation in plant growth and stress responses. Plant Cell Rep 32:1099–1109

Zevenhuizen LPTM (1997) Succinoglycan and galactoglucan. Carbohydr Polym 33:139–144

Zhan H, Leigh JA (1990) Two genes that regulate exopolysaccharide production in *Rhizobium meliloti*. J Bacteriol 172:5254–5259

Zhan H, Levery SB, Lee CC, Leigh JA (1989) A second exopolysaccharide of *Rhizobium meliloti* strain SU47 that can function in root nodule invasion. Proc Natl Acad Sci U S A 86:3055–3059

Zhan H, Lee CC, Leigh JA (1991) Low phosphate regulation of the second exopolysaccharide (EPSb) in *Rhizobium meliloti* strain SU47. J Bacteriol 173:7391–7394

Zhou D, Lin Z, Liu L, Zimmermann D (2013) Assessing secondary soil salinization risk based on the PSR sustainability framework. J Environ Manage 128:642–654

Zhou N, Zhao S, Tian C-Y (2017) Effect of halotolerant rhizobacteria isolated from halophytes on the growth of sugar beet (*Beta vulgaris* L.) under salt stress. FEMS Microbiol 364(11). https://doi.org/10.1093/femsle/fnx091

Part III

Microbial Polymers in Industrial Sectors

Microbial Exopolysaccharides: Structure and Therapeutic Properties

17

Hiran Kanti Santra and Debdulal Banerjee

Abstract

Microbial exopolysaccharides are the metabolic products of the bacterial or fungal cells that are composed of sugar, proteins, organic acid in various ratios and serve multiple protecting functions to the host cell against drought, pathogenic attack, antibiotics treatment, etc. They were first investigated during the 1880s and day by day have gained popularity in the sectors of food, pharmaceutics for anthropogenic benefits. They possess huge structural diversity according to their different types of sugar compositions and bond formations. Structural diversity and wide functional property have made EPSs as potent antibacterial, antioxidant, anticancer, and antitumor agents with acceptability in nano drug delivery, bioactive compound encapsulation, as prebiotics (fermented milk products like kefir, villi, taetea), etc. EPSs from LAB (lactic acid bacteria), extremophiles like marine halophiles, thermophiles, and psychrophiles are major contributors to the modern pharmaceutical industry with evidenced pharmacological applications. Their unique composition of sugars with different types of linkages is the prime cause of their multidimensional bioactivity. These exopolymers are not only useful in direct disease prevention but Also helps in therapeutic diagnosis, drug delivery, wound healing, etc. In this chapter, some recent discoveries of exopolysaccharides from a variety of microbial sources (archaebacteria, mesophiles, algal members, endophytic fungi, and bacteria) are mentioned with their utilization as antioxidative agents, antimicrobial compounds, antiproliferative compound, etc., along with their structural information.

H. K. Santra · D. Banerjee (✉)
Microbiology and Microbial Biotechnology Laboratory, Vidyasagar University, Midnapore, West Bengal, India
e-mail: db@mail.vidyasagar.ac.in

A. Vaishnav, D. K. Choudhary (eds.), *Microbial Polymers*,
https://doi.org/10.1007/978-981-16-0045-6_17

Keywords

EPS · LAB · Antioxidative agents · Antimicrobial activity · Antiproliferative agents

17.1 Introduction

Exopolysaccharides from microbial sources are generally considered as polymers of carbohydrates having high molecular weight. Polysaccharides are water-soluble jelly-like thick high molecular weight long-chain polymers of sugars which have a wide range of utility in the food industry as an emulsifier, stabilizer, encapsulator, bio-thickeners and controls crystallization, syneresis. Actually, they are grouped as exopolysaccharides, capsular polysaccharides, or sometimes lipopolysaccharides. It is very common that capsular polysaccharides are responsible for the pathogenicity of the virulent strains whereas lipopolysaccharides are generally stored in the outer membrane of the bacterial cell. EPSs are composed of carbohydrate and non-carbohydrate parts (like-acetate, phosphate, succinate, pyruvate, etc.). According to the monosaccharide compositions they are of two types—homo and heteropolysaccharides which are made up of the same (pullulan, cellulose, dextran, etc.) or different (hyaluronic acid, xanthan) type of repeating monosaccharide units. Due to their viscous pseudoplastic behavior, they are popular emulsifying, stabilizing, and viscosifying agents that retain moisture in food, improves blood flow, acts as a microcarrier in tissue culture and matrix immobilizing agents. Their intense and diverse biological function is obtained due to their composition, structure, and molecular weight. Polysaccharides are known to be present in different type of life forms ranging from prokaryotic to eukaryotic ones and display multidimensional biochemical structures and physiochemical qualities. They are necessary biological natural products reported from microorganisms, animals, and plants with popular biological potentials, e.g., rheology modifiers in food and pharmaceutical industries (as antimicrobial—antibacterial, antifungal, antiviral, antitumor, antioxidant, anticancer, cholesterol lowering, prebiotics, etc.), cosmetics, and textile industries. They offer the host cell adhesion and protection against harsh environmental situations. Microbial EPSs are characterized by short production times and simple extraction procedures. Microbial EPS are of high demand than plant or based on their physiochemical (e.g., freezing, boiling, and also melting) characteristics and rheological (characters related to flow of the viscous liquid) characteristics. They are manufactured at logarithmic (exponential or log) or stationary (stable, i.e., birth of new microorgnisms = death of old organisms) are found in insoluble or soluble way in the culture extracts as slime components. Their monomeric (simplest and smallest) units, glycosidic linkages (covalent bond that joins sugar) and diverse branching pattern has made them unique. Microscopic fungi derived EPS are of higher demand and easy to handle than the fruiting body derived EPS due to their easy isolation, simple purification, minimum cost, and gigantic production. The optimum conditions include the utilization of submerged culture where the dispersed mycelial filaments produce interwoven hyphal pellets excreting polymeric substances.

Plant (alginate, carrageenan, guar gum, locust bean gum, pectin, and starch) and animal (casein, gelatin) polysaccharides after certain modifications are widely used as hydrocolloids but their use is restricted due to limitations of their rheological properties and stability. So the focus is shifted towards the microbial source. Microbial (microalgal, bacterial, fungal) extracellular polysaccharides or exopolysaccharides remains associated with cellular surfaces forming a capsular or slimy appearance and are sometimes involved in immune response (capsular EPS and O-antigen in case of *Streptococcus pneumoniae* and *S. agalactiae*). Microbial cells use their EPS as a shield of protection against antibiotics treatment, disinfecting chemicals like ethanol, sulfur dioxide, metal ions, desiccation, phagocytosis, predation by protozoan, osmotic stress, and macrophage attack. It promotes biofilm formation, adhesion to solid substratum, and facilitates cellular recognition. Popular microbial biopolymers include bacterial alginates, pullulan, dextrans, yeast glycans, xanthan, and gellan. The prime hindrance with microbial polysaccharide is the detailed knowledge of the biosynthetic pathway, genetic regulations, expensive recovery costs, and critical bioprocess formulations. But once the issues are resolved microbial polysaccharides positively influence the food and pharmaceutical sectors. Another major drawback is the safety issues related to no GRAS (generally recognized as safe) organisms like Xanthan and gellan from phytopathogenic bacterial strains of *Xanthomonas campestris* and *Sphingomonas elodea,* respectively. In this respect, principle sources of exopolysaccharides are food-grade microorganisms, GRAS-LAB, etc.

17.2 LAB Polysaccharides

Microbial polysaccharides not only function as a direct therapeutic agent but also acts as a prebiotic that promote the growth of beneficial microbiomes in intestinal tracts and suppress the flora of putrefying or gas-forming bacteria. Lactic acid bacteria also known as LAB mainly isolated from dairy products like yogurts, kefirs, fermented milks, cheese, etc., are potent producers of exopolysaccharides with multidimensional application (Fig. 17.1). Other than LAB Propionibacteria, Bifidobacteria are also EPS producers. The popular fermented milks of Scandinavian origin like villi, taetea, skyr are of high demand due to their slimy to thick consistency because of the souring effect of *Lactobacillus lactis* subsp. lactis and *L. lactis* subsp. cremoris and texture of heteropolysaccharide. Other than these drinks kefir, a type of alcoholic to acidic effervescent drink fermented from grains is also acceptable due to its exopolysaccharide contents of the LAB strains (Vuyst and Degeest 1999). LAB, when fall under GRAS category, can provide a better natural end product in terms of enhanced and improved texture with rheological property, low viscosity, increased moisture content, and higher water retention capacity. Major drawbacks of LAB exopolysaccharides are in vivo biofilm formation and bio-fouling leading to loss of palatability of milk products, spoilage of wines, beers, ciders,

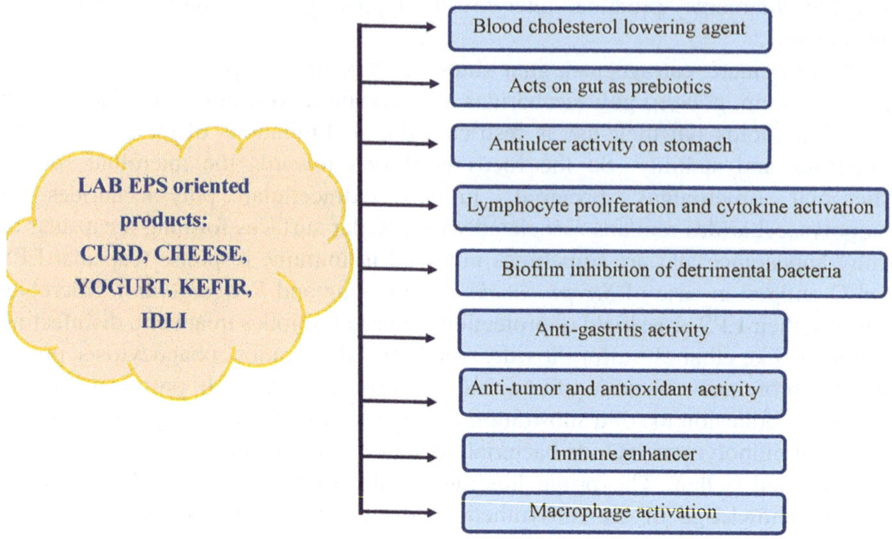

Fig. 17.1 Common fermented food with LAB exopolysaccharides and their potential multidiscipline therapeutic applications

fermented sausage and airtight cooked meats, etc. Dextran derivatives, levan, and alternan from LAB are used in the paper and metal plating industry, develop products that improve blood flow and extend blood volume, stabilize food syrup as a thickener, acts as a bulking agent with low viscosity, acts as an extender in cosmetics and food industry, etc.

In the year of 1969 US FDA (United States FOOD and Drug administration authority) first legally approved the commercial use and application of bacterial exopolysaccharide (xanthan from *Xanthomonas campestris*) as a food additive for thickening, gelling of sugar beet and cane syrups. Hydrated polymers of xanthan are popularly used in soups, sauces, yogurts, dairy desserts, salad dressings, and fermented milk products due to its thixotropic and shear thinning (pseudoplastic) properties. Its rheology falls rapidly upon constant pouring, shaking, and stirring but regains its nature immediately after the shaking is withdrawn. Its wide range of adaptability over pH changes, different ionic strengths, and temperature has made it more compatible with other food products. The negative relationship between shearing force and retention of viscosity is highly recommended for palatability, i.e., mouth fullness texture. Slime mold producing bacterial cultures are coupled with the starter LAB cultures as they improve product viscosity, lower syneresis problems, and avoid graininess issues to some extent. These ropy strains maintain the regularity of stirred type yogurt as the viscous EPS is less susceptible to mechanical damage done by blending and pumping procedures making the coagulum more thermal and physical shock resistant. Another positive side of this type of approach is that it does not require any artificial stabilizers making the product much more natural and acceptable. The higher acceptability of fermented milk products are due to their viscosity that is enhanced by the inoculating bacterial strains changing the rheology making the products more popular. The viscous nature depends on the pH, fermentation time, culture medium, culture strains, and also total solids of the medium. Not only the use of ropy, moderately ropy, or non-ropy strains affects the texture of the product but also the aggregated proteins influence the taste parameters. The ropy (*Lactobacillus delbrueckii* subsp. *bulgaricus* strain) and non-ropy both types of starter cultures are potent enough to form a uniform layer of EPS strands in between the microbial cells and casein protein networks modifying the protein texture and aggregation properties. Yogurts are made up of stabilizers; either commercially available chemicals or derived from the ropy starter culture of some particular LAB. Yogurts with natural stabilizers from microorganisms' source have distinct rheology and palatability than the artificial ones.

Another popular milk product with increased adhesiveness due to the action of polysaccharide producing LAB strains (*L. lactis* subsp. *cremoris*) is known as Villi. It prepares a thick slimy matrix where the proteins are trapped forming the casein micelle cluster increasing the consistency of the product.

The main two constituents of the kefir grains are LAB (*Lactobacillus kefiranofaciens*) and yeasts (*Candida kefir*) which prepare the polysaccharide gel-like matrix (kefiran). The homo and heterofermentative strains are *Lactobacillus acidophilus*, *Lactobacillus kefirgranum,* and *L. kefir* and *L. parakefir*.

Panthavee et al. (2017) isolated two EPS producing strains of thermophilic LAB (LY45-*Pediococcus pentosaceus*, PY45-*Lactobacillus amylovorus*) from Thai tropical fruits. The strains (LY45 and PY45) produced neutral to acidic polysaccharides and manufactured glucose ($C_6H_{12}O_6$), mannose (D-mannose), galactose, respectively. The EPS shows anti-hyaluronidase activity in a similar resolution with sodium cromoglicate and dipotassium glycr-rhizinate. So these types of strain are very much useful in pharmaceutical product development that can act as both probiotics and anti-allergic or anti-inflammatory.

Luang-In et al. (2018) determined the bioactivities of EPS isolated from *Bacillus tequilensis* PS21 obtained from kefir of Kampaeng Petch, Thailand. FTIR, HPLC, and SEM data confirmed the heteropolysaccharide (monosaccharide components are xylose, glucose, ribose, rhamnose, galactose with 17.65, 2.54, 1.83, 1.23, 1%, respectively) nature of the protein-bound polysaccharide with a grainy appearance and uneven porous surface. The antioxidant nature of the EPS was confirmed by DPPH, ABTS, and FRAP methods. It is also characterized by anti-tyrosinase activity with 34.9% inhibition. This type of study emphasizes the multidimensional role of EPS not only as prebiotics but also as a source of antioxidants.

Yu et al. (2018) screened a new LAB (*Weissella cibaria* 27) from kimchi, a Korean dish and documented the antibacterial activity of the EPS produced by this strain under optimum conditions (22 °C, pH 62, 24 h). EPS is made up of dextran with α-1,6 glycosidic linkages having a molecular weight of 1.2×10^7Da. The use of sucrose has changed the surface property of EPS and made it more hydrophobic for better inhibition of bacterial pathogens. Thus, these types of EPS possess potent application in the sector of food, pharmaceuticals, cosmetics industry, and development of chemotherapeutic agents with antibacterial activity.

17.3 EPSs from Marine Microbial Sources

In most of the cases, EPSs are of low poisonous activity and this property has made them being widely used as a therapeutic agent. EPSs suppress the tumor development by their immunomodulatory activity in vivo. They exhibit toxicity to malignant tumor cells by boosting up the host's defense and resistance.

Marine microorganisms also play an important role in the production of exopolysaccharides with therapeutic properties. Marinactan is an exopolysaccharide isolated from marine *Flavobacterium uliginosum* (an isolate of surface seaweed) a heteroglycan made up of glucose, mannose, and fucose monomer shows anticancer activity towards solid and ascites form of sarcoma 180 form (Umezawa et al. 1983).

In another study conducted by Ruiz-Ruiz et al. (2011) they screened the two EPS producing strain for antitumor activity and reported that the heteropolysaccharides restrict the proliferation of tumor infected cell lines and oversulfated ones are effective against T-cell lines (obtained from acute lymphoblastic (blood and bone marrow cancer) leukemia-intrinsic pathway). Chen et al. 2011 isolated *Aspergillus* sp. Y16 from the leaves of mangrove plant *Ipomoea pescaprae* (Linn.) and extracted, purified a homogeneous exopolysaccharide (As1-1) by AEC (anion-exchange) and GPC (gel permeation chromatography). 1D and 2D NMR spectroscopical data suggested As1-1 was made up of mannose (molecular weight-15 kDa). Backbone of the polysaccharide is made up of (1-2)-linked α-D-mannopyranose units, substituted at C-6 by the (1-6)-linked α-D-mannopyranose, (1-)-linked β-D-galactofuranose and (1-)-linked β-D-mannopyranose units. The polysaccharide was characterized by in vitro antioxidative potency tested using DPPH (1,1-diphenyl-2-picrylhydrazyl), superoxide radicals, and in vivo affectivity was confirmed using lipid peroxidation and superoxide radicals.

Ye et al. (2012) isolated and purified an acidic exopolysaccharide from a marine *Pseudomonas* sp. PF-6 which was water-soluble but insoluble or immiscible in organic (acetone and benzene, etc.) solvents and odorless with a (MW) molecular weight of 8.83×10^5 Da. After oxidization, it turns to uronic acid and IR spectroscopic analysis confirmed the absorption bands at 927.49 cm^{-1} and 811.76 cm^{-1} which elucidate the occurrence of the pyran group. The EPS possess antioxidative potentiality confirmed through DPPH, OH radical scavenging activity. The antioxidative nature was directly proportional to the concentration of EPS.

Ye et al. (2016) isolated a selenium tolerant strain of *Pseudomonas* sp. PT-8 from Antarctic Ocean mud and characterized the selenium-rich polysaccharide (Se-EPS) with high uronic acid content. EPS exhibited a yield of maximum 256.7 mg/kg of selenium upon optimization. EPS was purified into two fractions; one being neutral and another being acidic where the second acidic fraction (EPS2) being a homogenous polysaccharide with rhamnose ($C_6H_{12}O_5$)19.58:arabinose ($C_5H_{10}O_5$) 19.28: xylose ($C_5H_{10}O_5$) 5.97: mannose ($C_6H_{12}O_6$) 18.99:glucose ($C_6H_{12}O_6$) 23.70:galactose ($C_6H_{12}O_6$) 12.48 with 7.3 kDa MW (molecular weight) having a molecular weight of 7.3 kDa. The amount of rhamnose and the molecular weight of the EPS is involved in a negative relationship. This selenium-rich EPS is a good agent of antioxidative activity confirmed through DPPH and OH radical scavenging assay.

Another example includes YCP which is a (1,4)-α-D-glucan obtained *Phoma herbarum* S4108 (a filamentous fungi found in sediments) can influence the activation of macrophage (2). Two different types of neutral exopolysaccharides (GX1-1, GX2-1) were reported from an indigenous fungus *Penicillium* sp. gxwz446 isolated from a coral *Echinogorgia flora* occurring in South China region (Sun et al. 2016). EPS were purified by AE (anion-exchange), GPC (Gel Permeation chromatography). Spectroscopic analysis coupled with chemical testing revealed the nature of the EPS as glucan with a molecular weight of 5.0 kDa. GX1-1 has (1 → 4)-linked α-D-glucopyranose units as the backbone with a substitution at C-2 by a single α-D-glucopyranose on every sixth sugar residues. GX2-1 was made up of galactofuranose with mannogalactoglucan with a molecular weight of 9.5 kDa. The prime linkages consists of (1 → 4)-β-D-Glc*p*, (1 → 5)-β-D-Gal*f*, (1 → 3,5)-β-D-Gal*f*, (1 → 6)-α-D-Man*p* and (1 → 2, 6)-α-D-Man*p*. GX1-1 can cause macrophage activation. GX1-1 was subjected to sulfated modification and there was one sulfate substitution on every sugar ring basically at O-6. The sulfated derivatives were more efficient to enhance the pinocytic ability of RAW264.7 cells and also promote NO production.

It is not always that exopolysaccharides derived from microorganisms directly contribute to therapeutic purposes but sometimes in a passive way they modify the production of some bioactive component in another eukaryotic host such as plant cells (Chen et al. 2016). One such example includes the higher production of volatile oil components of *Atractylodes lancea* due to the elicitation of exopolysaccharide components produced by the endophytic fungus *Gilmaniella* sp. AL12 (a stem isolate of *A. lancea*). Researchers identified that the active component is an extracellular mannan having a polymerization degree of 26–42. Results of 2D electrophoresis membrane proteomics study revealed that 83 different proteins are expressed and probably mannan maintains an antagonistic between endophyte AL12 and host *A. lancea* interactions. Mannose was produced from mannan that mediates activation of enzyme hexokinase and also promotes anabolic biochemical reaction photosynthesis promoting photosynthesis coupled with the metabolism of energy (breakdown of product for the generation of energy-ATP) that mediates important metabolic fluxes (turnover of components into something new) for terpenoid biosynthesis. The remnant part of the mannan enhances autoimmunity of the host plant by Growth protein-mediated signal transfer coupled with pathways that bind mannose and lectin. As a result of the immune-stimulating activity, the higher accumulation of volatile components occur. The prime constituents of the *A. lancea* volatile oil are sesquiterpenes and polyacetylene that includes atractylone, atractylodin, β–eudesmol, b-caryophyllene, caryophyllene oxide, hinesol, b-sesquiphellandrene, and zingiberene (Yuan et al. 2016). The plant constituents were used as basic raw or crude drug for the treatment of rheumatic diseases (problems of tendon and ligament, etc.), gastric problems, and also as diuretics in Asian countries like China, Japan, and Korea (Ouyang et al. 2012). Other pharmacological utilities of this herb include anti-inflammatory (prime component for mast cell-mediated inflammatory diseases), antimicrobial, anticancer, antihypertensive, lipase inhibition, anti-obesity, antiplatelet, anti-ulcer, antipyretic, etc. (Jiao et al. 2014).

Yan et al. (2016) evaluated the antioxidative potentiality of an EPS obtained from *Aspergillus versicolor* N2bc which was mannoglucogalactan (molecular weight-20.5 kDa) in nature. The main chain of EPS consists of →2)-D-Glcp-(1→, →2)-D-Glcp-(1→ and →6)-D-Manp-(1 → units, substituted at C-6 position of →2)-D-Glcp-(1 → units. Branches were made up of galactofuranose (five-membered ring form of galactose)-oligosaccharides (made up of a small number of monosaccharide units) built up of →5)-D-Galf-(1→, →6)-D-Galf-(1 → and terminal -D-Galf units).

El-Newary et al. (2017) evaluated the antitumor, antioxidant, and anti-inflammatory (COX-1 and COX-2) activity of the EPS which is acidic in nature and composed of monosaccharide units of glucose, galactose, glucuronic acid (molar ratio of 1.6:1.0:0.9) having a molecular mass of 3.76 × 104 g/mol. Except reducing power all the other common measurement units of antioxidative nature like free radical scavenging, antiproliferative activity towards cancer cells (breast cancer) MCF7 exhibited IC50 and IC90 values of 70 mg/mL and 127.40 mg/mL, respectively. The antitumor activity of the EPS was due to uronic acid contents that may find potential application in the food and pharmaceutical industry.

Hassan and Ibrahim (2017) investigated the antitumor (effective against MCF-7, HCT-116, and HepG2), antibacterial against Aeromonas hydrophila (Gram-negative), Pseudomonas aeruginosa (Gram-negative), Streptococcus faecalis (Gram-positive) effect of heteropolysaccharide obtained from marine bacterial species *Bacillus subtilis* SH1 which was composed of monosaccharide units of maltose and rhamnose having a compact film-like structure. The production parameters were optimized 2.5% peptone-protein source, 4.5% yeast extract-nitrogen source, and malt extract-sugar source to enhance the yield up to 1.4 times. It was also characterized by antiviral activity along with antioxidant potency at higher concentrations. Due to its film-like appearance, it could be beneficial for the preparation of plasticized films.

Abdrabo et al. (2018) isolated 13 marine bacterial species from algal and rock surfaces of Suez Gulf, Egypt and detected the antimicrobial and antitumor effect of the biofilm-associated exopolysaccharide. *Halomonas saccharevitans* AB2 was the most potent isolate with antimicrobial (antibacterial and antifungal against *Vibrio fluvialis* and *Aspergillus niger* ATCC 16404, respectively) and antitumor activity (against eight cell lines) with an IC50 value of 20.3 μg mL against A-549 (human lung carcinoma) making this strain a therapeutically valuable EPS reserve.

Mohamed et al. (2018a, b) screened 29 bacterial strains or the production of exopolysaccharide from soil sediments of mangrove area of Ras Mohamed area, Read Sea Cost of Sinai Peninsula, Egypt. One strain (identified as *Bacillus altitudinis* MSH2014) provided a satisfactory yield of EPS and monosaccharide components were mannuronic acid, glucose, sulfate, and molecular weight is 4.23×10^5. EPS shows potent anticancer activity against two cancer cell lines EACC and lung cancer A-549.

Exopolysaccharides from marine microorganisms are known to be effective against liver cancer (malignant) cell lines also (Yang et al. 2013, 2014; Peng et al. 2014). Eleven bacterial strains were obtained from saltwater marine sediments of the Mediterranean and the Red Sea and they were tested for their cytotoxic nature against HepG2 cells. It was reported that five strains (*Bacillus amyloliquifaciens* MGA2, *Bacillus thuringiensis* E4, *Brevundimonas subvibrioides* MSA1, Pseudomonas fluorescens SGA3 (Gram-negative), Advanella kashmirensis NRC-7 (Gram-negative)) were the most effective against HepG2. These are acidic heteropolysaccharides (made up of different units of monosaccharides) with hexose sugar glucose, galactose, mannuronic acid, mannose, glucuronic acid as monosaccharide units. The occurrence of sulfate and carboxylic groups in different amounts (percentages) were obtained from FTIR data. The molecular weight ranged from 1.94×10^4 to 7.95×10^5 g/mol in an average. (Asker et al. 2018).

Abinaya et al. (2018) documented the antibiofilm (biofilm inhibitory) and antilarval functions of exopolysaccharide obtained from a marine *Bacillus licheniformis* Dahb1 which was composed of aromatic compounds and various functional groups. EPS exhibited strong antioxidative property based on DPPH (2, 3, 5 diphenyl picryl hydrazine) free radical scavenging ability. The antibiofilm activity was against Gram-negative bacteria—*Pseudomonas aeruginosa* and *Proteus vulgaris*; Gram-positive bacteria—*Bacillus subtilis* and *Bacillus pumilus*; *Candida albicans* at 75 μg/mL concentrations. Antilarval activity was against *Anopheles stephensi* and *Aedes aegypti* with an LC_{50} value of <80 μg/mL. This multipotent EPS exhibited zero toxicity when applied upto 5 mg/mL which emphasizes its environmental application and entomological acceptability.

Antibacterial activity against the wide range of human pathogenic microorganisms *Acinobacter* sp., *Bacillus* sp., *Campylobacter* sp., *Enterobacter* sp., *Enterococcus* sp., *Escherichia* sp., *Proteus* sp., *Pseudomonas* sp., *Salmonella* sp., *Streptococcus* sp., *Staphylococcus* sp. was exhibited by EPS produced from the marine isolates of Mauritius coastal water. The EPS producing strains were known as *Bacillus* sp., *Halomonas* sp., *Psychrobacter* sp., *Alcaligenes* sp., by morphological (external characters), biochemical, and 16 s rRNA (based on protein, DNA, and RNA). Eight EPS were effective in a broad spectrum way against almost half of the pathogenic bacterial strains. The best two EPS have a MIC value ranging from 62.5 μg/mL to 500 μg/mL. FTIR (Fourier Transform Infrared Spectroscopy) coupled with TLC study demonstrated the occurrence of carboxyl (CHO), hydroxyl (OH), amide, and sulfate groups with the two main sugar components glucose and fructose in the EPS (Aullybax et al. 2019).

A recent study on *Deinococcus radiodurans* R1 (*dra0033*) revealed that it is resistant to a wide range of environmental stresses. Generally, the exopolysaccharide is a cell wall component but its antioxidant ability has also been established. Data from bioinformatics suggests that dra0033 encodes an ExoP-like protein that mediates the synthesis of DeinoPol and mutated dra0033 strains rapidly decreases the survival rate according to increasing stress. The molecular weight of the monosaccharide unit composition ranged from 80 to 100 kDa. This is the first report of bioactive exopolysaccharide (from Deinococcal source) with antioxidative potency (ROS scavenging and protective effect on human keratinocytes) and useful in cosmetics, pharmaceutical sectors as a safe agent (Lin et al. 2020).

EPS are useful antioxidants, antitumor, antimicrobial agents, also beneficial for wound healing, minimize bacterial or viral infections, promote host immune response, and enhance chemotherapeutic drug activities.

17.4 Extremophilic Microbes as Exopolysaccharide Producers

In recent times, the focus has always been shifted to microbiomes of extreme environments (cold habitat, desert areas, high salt seawater, volcanic lavas, hydrothermal vents, etc.) as they are potent enough to survive in odd situations where the normal life force ceases to proliferate. Due to their unique metabolic pathways coupled with resistant enzymes they are wonders of nature.

They are highly attractive agents for utilization in the food and pharmaceutical sectors due to their high biodegradability, nontoxic nature, etc. They are easy to manipulate as they are influenced by the fermentation parameters and involves simple purification, characterization. Some examples of exopolysaccharides from extremophilic microbes with their possible application and structure are summarized in Table 17.1.

Manca et al. (1996) worked on *Bacillus thermoantarticus* (a thermophilic strain) and isolated two EPS (1 and 2) by cold ethanol precipitation from liquid cultures with almost a yield of 400 mg/liter when mannose is used as the preferred carbon source. The EPS and dry or wet biomass of the organism were involved in a proportional relationship. The two EPS represents 27% and 71% of carbohydrate content from their overall biomass. Mannose and glucose were present in a 1.0 and 0.7 molar ratio in case of EPS1 which is proved to be a sulfated polysaccharide. Mannose was also the major component in case of EPS2 sulfated homopolysaccharide. Hexose was being dextrorotatory for both the EPS. NMR confirms the occurrence of α-D-mannose and β-D-glucose. H^1 NMR indicated the occurrence of pyruvic acid.

Moriello et al. (2003) isolated *Geobacillus* sp. 4004 from sea sand of Maronti, near Sant' Angelo (Ischia) and extracted three different types of polysaccharides (in a medium optimum situation of 60 °C and pH 7.0) each with a monosaccharide composition of EPS-1- glucose/mannose/galactose (0.3/1.0/minimum amount- relative ratio of monosaccharides), EPS-2- glucose/mannose/galactose (03/1.0/trace amount), and EPS-3-galactose (1.0):mannose (0.8):glucosammine (0.4):arabinose (0.2) with a pentasaccharide unit.

Llamas et al. (2011) conducted a polyphasic taxonomic (the data are obtained from different types of factors like biochemical, anatomical, phylogenetic, etc.) study on two halophilic (salt-loving species) bacterial strains (Gram-negative) N12T and B-100. The isolates were obtained from two separate saline (soil with high salt concentration) soil sample (San Pedro del Pinatar and Sabinar saline (high salt concentration) wetland saline wetland respectively in Murcia, south-east Spain) and produced sulfated exopolysaccharides with biotic potentials. Their optimum growth conditions were in NaCl concentrations (5–10%-w/v), temperature (20–32 °C), medium pH (7–8). The isolates' 16S rRNA analysis revealed that they

Table 17.1 EPS from extremophiles and their biological applications

Sl. No.	Extremophilic microorganism	Origin of isolates	Structure of EPS	Biological role	Reference
1	*Vibrio diabolicus* (strain-HE800)	Deep-sea hydrothermal area situated at East Pacific Rise (from a Pompei worm tube)	Tetrasaccharide linear unit Uronic acid:GlcN:GalN 1:0.5:0.5 → 3)-β-D-GlcpNAc-(1 → 4)- β-D-GlcpA-(1 → 4)-β-D-GlcpA-(1 → 4)-α-D-GalpNAc-(1→)	Helps in bone (calcite cells) rehabilitation bone and acts as a cicatrizing (cementing) material	Raguénès et al. (1997a)
2	*Alteromonas infernus* sp. nov	Obtained from deep-sea hydrothermal vent	Monosaccharide units are made up of glucose (6 carbon), galactose (6 carbon), galacturonic acid (oxidized form of D-galactose), glucuronic acids	Plays role in NHS (normal human serum) activation, regenerates bone tissues and skin tissues, useful in drug localization (delivery to the target site) and tissue engineering	Raguénès et al. (1997b)
3	*Alteromonas infernus* (strain-785)	Hydrothermal vent of Guaymas basin near Gulf of California from a fluid sample of *Riftia pachyptila*	Glc:Gal:GlcA:GalA (1:1:07:0.4)	Acts as an anticoagulant agent	Colliec et al. (2001)
4	*Alteromonas macleodii* subsp. *fijiensis*	Deep-sea hydrothermal vent (cracks on sea floor) of North Fijian Basin	Heteropolysaccharides (made up of repeating units of diverse monosaccharides) was sulfated and of high uronic acid contents along with pyruvate. It contains a repeating unit of branched (side chain) hexasaccharide (with six carbon atom) with Glc, Man, Gal, GlcA, GalA, pyruvated mannose	Heals bone problems and treats cardiovascular diseases	Colliec et al. (2001)
5	*Alteromonas macleodii* subsp. *fijiensis* biovar	Obtained from epidermal cells of a polychaete annelid (*Alvinella pompejana*)	Gal:Glc:Rha:Fuc:Man:GlcA: GAlA:3–0-(1-carboxylethyl)-D-	Useful component in cosmetics (make-up) industry	Cambon et al. (2002)

(continued)

Table 17.1 (continued)

Sl. No.	Extremophilic microorganism	Origin of isolates	Structure of EPS	Biological role	Reference
	deepsane HYD657 (strain)		GlcA (1:0.42:0.85:0.5:0.42:0.5:0.5:0.5)		
6	*Halomonas maura.*	Collected from saline soil of a solar saltern in Asilah, Morocco	Polyanionic sulfated exopolysaccharide with high uronic acid content and occurrence of glucose, mannose, galactose	Antioxidative ability, antihemolytic and antithrombogenic activity, agent of drug delivery, used for cancer chemotherapy, bioimaging, nanotechnology	Raveendran et al. (2013), Arias et al. (2003)
7	*Vibrio diabolicus*	Isolated from deep-sea hydrothermal vents.	Structure is 3)-β-D-GlcpNAc-(1 → 4)-β-D-GlcpA-(1 → 4)-β-D-GlcpA- (1 → 4)-α-D-GalpNAc-(1 → 3)-β-D-GlcpNAc-(1 → 4)-β-D-GlcpA-(1 → 4)-β-D-GlcpA- (1 → 4)-α-D-GalpNAc-(1→)	Bone healing property with no side effects	Zanchetta et al. (2003)
8	*Bacillus licheniformis* B3-15	Shallow or facile hydrothermal vent hot spring of Vulcani Island, Italy	Prime monosaccharide is mannose. Contains the tetrasaccharide (made up of four monosaccharides) repeating (same combination of sugar again and again) with manno-pyaranosidic arrangement	Exhibits antiviral activity	Arena et al. (2006)
9	*Geobacillus tepidamans* (strain V-264)	Hot springs of Velingrad region of Central Bulgaria	Glu:Gal:Fuc:Fru-0.07:0.04:0.02-molar ratio of sugars, α-gluco-galacto configuration	Antitoxic activity	Kambourova et al. (2009)

10	*Bacillus thermodenitrificans* (strain B3-72)	Shallow or facile hydrothermal vent hydrothermal vent of Vulcano Island, Italy	Persisting unit of trisaccharide with a manno-pyranosidic ordering (Man:Glc-1:0.2)	Possess immunomodulatory (enhances immune function) property and acts as antiviral agent.	Arena et al. (2009)
11	*Thermus aquaticus*	–	Tetrasaccharide repeating units of galactofuranose, galactopyranose, and N-acetylgalactosamine (1:1:2)	Induces immune response through Toll like receptors. Induces macrophage cells for production of cytokines TNF-α and IL-6	Lin et al. (2011)
12	*Brevibacillus themoruber* (strain-423)	Hot springs of Kaynarca (82 °C, pH 6.8) and Karamustafapasa (58 °C, pH 6.6) located in Bursa (Turkey)	Heteropolymer structure and glucose being the prime monosaccharide unit	Tissue engineering and drug delivery	Yildiz et al. (2014)
13	*Pseudomonas* sp. ID1	Collected from marine sediments of South Shetland Islands (Antarctica)	Made up of fucose, galactose, and glucose	Pharmaceutically important due to its emulsifying ability	Carrión et al. (2015)
14	*Geobacillus thermodenitrificans*, B3-15	Isolated from submarine shallow vents of the Eolian Islands (Italy)	Made up of trisaccharide repeating units with manno-pyranosidic anomeric configuration. Molecular weight 400 KDa. Man/Glu-1:0.2 ratio	Immunomodulatory in nature with antiviral property	Gugliandolo et al. (2015)
15	*Bacillus licheniformis* B3-72	Isolated from submarine shallow vents of the Eolian Islands (Italy)	Made up of tetrasaccharide repeating units with manno-pyranosidic anomeric configuration. Molecular weight 600 KDa	Immunomodulatory in nature with antiviral property	Gugliandolo et al. (2015)
16	*Haloterrigena turkmenica*	Isolated from sulfate saline soils of Turkmenistan	A type of sulfated heteropolysaccharide made up of glucose, galactose, glucosamine, galactosamine, glucuronic acid	Pharmaceutically valuable for its antioxidative nature	Squillaci et al. (2016)

(continued)

Table 17.1 (continued)

Sl. No.	Extremophilic microorganism	Origin of isolates	Structure of EPS	Biological role	Reference
17	*Bacillus velezensis* MHM3	Isolated from the seashore of Red Sea Coast (El-Ein El Sokhina), Suez governorate of Egypt	Uronic acid (24%) and sulfate (18.19%), monosugars are glucuronic acid, glucose, fructose, and rhamnose having a molar ratio of 4.00: 2.00: 1.00: 0.13, average molecular weight = 1.145 × 104 g/mol and the number average of molecular weights of = 5.155 × 103 g/mol. FTIR analysis and periodate oxidation revealed the existence of α-(1–4) linkage in acidic polysaccharide	Anticancer activity by inducing apoptosis of Human Breast Cancer MCF-7 Cell lines through interfering a mitochondrial Pathway	Mahgoub et al. (2018)

belong to Halomonas genus under the Gammaproteobacteria class. They exhibited a closest affinity with *Halomonas nitroreducens* (98.7% and 98.3% similarity with N12T and B-100, respectively). G + C contents were 61.9 and 63.8%, respectively. There was 43.9%, 30.5%, 39.2%, 46.3%, 52.9%, 51.3%, 100% relatedness among the strains N12T and *H. nitroreducens* CECT 7281 T, *H. ventosae* CECT 5797 T, H. *fontilapidosi* CECT 7341 T, *H. maura*, *H. saccharevitans* LMG 23976 T, *H. koreensis* JCM 12237 T, B-100, respectively (according to DNA-DNA hybridization). The prime fatty acid components of the strain N12T (*Halomonas stenophila* sp. nov strain N12T) are C12:0 3-OH (5.42%), C15:0 iso 2-OH/C16:1v7c (17.37%, C16:0 (21.62%), C18:1v7c (49.19%).

Jiang et al. (2011) demonstrated the antibiofilm activity of the marine bacterium *Vibrio* sp. QY101 against *Pseudomonas aeruginosa*, *Staphylococcus aureus*. The polysaccharide was composed of galacturonic acid, glucuronic acid, glucosamine, and rhamnose (confirmed by IEC-ion exchange chromatography, GFC-gel filtration chromatography) with an average MW (molecular weight) of 546 KDa. EPS enhanced the aminoglycosides antibiotics activity against *P. aeruginosa* but has no direct antibacterial activity. Both the cellular aggregates and attachments of *P. aeruginosa* and *S. aureus* were hampered by EPS action but the inhibition of cell surface interaction was only found in case of *S. aureus* and dispersion of preformed cell aggregates occurred in case of P. *aeruginosa*. This type of investigation emphasizes the fact that EPS are useful for ameliorating the infections related to biofilm formation on medical devices like catheter, pacemaker, etc.

Bai et al. (2012) investigated the role of heteropolysaccharide of Antarctic bacteria *Pseudoaltermonas* sp. S5 as an immunomodulation agent (immunological effect on peritoneal macrophage) which was composed of galactose, glucose, and mannose in 44.3:50.9:4.8 ratio with a molecular weight of 397 kDa. EPS is reported to cause higher occurrence of macrophage activation in a dose-dependent manner (2.5–200 μg/mL) after 24 h of treatment along with increase in NO production and also secretion of TNF-α (tumor necrosis factor) and IL-1β (interleukin). So this type of heteropolysaccharide can act as a natural biological response surface modifier with multiple biological activities.

Sarilmiser et al. (2014) reported that exopolysaccharide produced from *Halomonas smyrnensis* is effective for its bio flocculating ability, anti-cytotoxicity, bio-compatibility, and finally important for its antitumor activity. They aimed at the development of polysaccharide-based nanocarrier system for anticancer drugs. It was reported that through periodate oxidation levan is modified chemically and contains a higher amount of aldehyde groups. Their compatibility with mouse fibroblast cell (L929) was checked and satisfactory results were obtained. The anticancer potency was dependent on cellular type in a dose-dependent manner when checked on four different type of cell lines A549 (human lung adenocarcinoma), AGS (human gastric adenocarcinoma), HepG2/C3A (human liver hepatocellular carcinoma), MCF- 7 (human breast adenocarcinoma) due to caspase-3/7-based apoptosis. This type of findings supports future concepts of polysaccharide-based nanocarrier systems developing anticancer drugs.

Radchenkova et al. (2013, 2014, 2015) worked on *Aeribacillus pallidus* 418 and characterized two EPS; EPS1 with 3200 kDa molecular weight and mannose, trehalose, galactosamine, glucosamine, galactose, glucose, ribose being the monosaccharide composition with a molar ratio of 69.3/7.8/6.3/5.4/4.7/3.4/2.9 and EPS2 with 700 kDa molecular weight and mannose, galactose, glucose, galactosamine, glucosamine, ribose, and arabinose being the prime sugar components with 33.9, 17.9, 15.5, 11.7, 8.1, 5.3, 4.9 molar ratio. The EPSs were of pseudoplastic rheological behavior, foaming ability, and emulsifying property.

Sun et al. (2015) isolated Polaribacter (a strain of Flavobacter-SM1127) from an arctic Pheophycean brown algae *Laminaria* sp. The EPS was composed of N-acetyl glucosamine-NAG, mannose, and glucuronic residues with a mass of 220 kDa and were bound by heterogeneous linkages. It was characterized by high viscosity followed by good shear thinning property with high tolerance (resistance ability) to salinity (high salt concentration) and pH change. It was perfect for oral administration and external use on skins for its fibroblast protection ability, moisture retention, and antioxidative potency. So it is a potent agent for food, pharmaceutical, cosmetics, and biomedical application.

Sardari and his coworkers (2016) isolated EPS from *Rhodothermus marinus* (a thermophilic bacteria) and characterized the EPS from two different strains (DSM4252T and MAT493). The EPS were produced in both the logarithmic (exponential) and stationary (birth and death of microorganisms cell) phase supplementation with glucose, sucrose, lactose, and maltose in optimum conditions. Both the EPS were heteropolymeric in nature with a molecular weight of 8×10^4 Da and composed of monosaccharide units of Glucose/arabinose/xylose with a relative ratio of 1/1.57/3.72 (for MAT493 strain) and Glucose/arabinose/xylose/mannose with a relative ratio of 1/3.75/3.02/1.87 (DSM4252T). The occurrence of hydroxyl (OH), carboxyl (CHO), N-acetyl, amine (NH_2), and sulfate ester groups is confirmed by FTIR spectrum and elucidates the fact that this species produces sulfated EPS with high arabinose (aldopentose) and xylose (aldopentose).

Brian et al. (2016) investigated the biofilm inhibitory activity of soluble, loosely and tightly bound polymeric substances from the planktonic and biofilm cultures of *Pseudoalteromonas ulvae* TC14. EPS1 and EPS2 have a molecular weight of 1000 kDa and 4000 kDa. The main constituents were glucan and poly (glutamyl-glutamate) fractions. The EPS compositions varied qualitatively and quantitatively in biofilm and planktonic cultures. Antibiofilm nature contributes to the antifouling ability of the strain.

Gugliandolo et al. (2013) demonstrated the anti-cytotoxicity, antiviral activity, antibiofilm activity, and immunomodulatory ability (against herpes simplex virus type 2; HSV-2) of exopolysaccharide isolated from *Bacillus licheniformis* T14. The EPS produced at optimum growth conditions was made up of monosaccharide unit of fructose (1.0): fucose (0.75): glucose (0.28): galactosamine (trace): mannose (trace). MW (molecular weight) was known to be 1000 kDa and occurrence of trisaccharide repeating units with β-mano-pyranosidic configuration.

Arun et al. (2017) extracted an EPS from *Halobacillus miurensis* which contains 56.1%, 12.1%, and 1.5% of carbohydrate, ash, and protein contents. The prime monosaccharides were galactose and glucose with functional groups like hydroxyl, alkenes, amide, carboxyl, etc., and also aliphatic amine and alkynes at the fingerprint region. EPS exhibited potent antioxidant activity in the forms of hydroxyl and superoxide radical scavenging assay along with DPPH free radical scavenging.

Wang et al. (2017) isolated a strain of *Geobacillus* sp. (strain TS3-9) from a radioactive radon-Rn hot spring and characterized the EPS as a heteropolysaccharide made up of D-mannose, D-glucose, rhamnose with a ratio of 1:0.14:0.06. The uronic and acetyl contents were measured as 18.6% and 6.07%, respectively. They were potent agents with antioxidative properties confirmed by hydroxyl radical scavenging, superoxide radical scavenging, and DPPH radical scavenging assay. The EPS was able to check the growth of hepatoma carcinoma cells also.

Caruso et al. (2018) investigated four species of sponge (Porifera) associated Antarctic bacteria (Cold region) named as *Winogradskyella* sp. (CAL384 and CAL396), *Colwellia* sp. (GW185), *Shewanella* sp. (CAL606). Particularly, these species were selected due to their high mucous content found in the culture petri plates. Production parameters were optimized by varying the carbon sources and other cultural factors like pH, temperature, and NaCl concentrations. EPSs are reported to be of high sugar (15–28%), protein (3–24%), and uronic acid content (3.2–11.9%). The EPS is composed of galactose, glucose, mannose, and galactosamine as the prime constituents. The high protein content provides the emulsifying ability of the EPS. The higher freeze thaw survival ratio is responsible for its cryoprotective activity. The tolerance to mercury and cadmium is due to the occurrence of sulfate groups and uronic acids acting as a cation ligand. The EPS possess multidimensional application in the sectors of cosmetics, food, pharmaceuticals, heavy metal chelation, and environmental clearance due to its biodegradability, cryoprotecting ability, emulsifying, and biofilm forming activity.

Panosyan et al. (2018) worked on extremophilic microbial EPS isolated from Arzakan geothermal spring, Armenia. Out of 16 isolates, the two best EPS producers were identified (by 16S rRNA analysis) as *Geobacillus thermodenitrificans* ArzA-6-archaebacteria-thermotolerant and *Geobacillus toebii* ArzA-8-archaebacteria. EPS productions from these two species of bacilli were evaluated in varying culture conditions to check the highest production yield. Molecular mass of the two purified EPS were known as 5×10^5 Da and 6×10^5 for *G. thermodenitrifcans* ArzA-6 and *G. toebii* ArzA-8, respectively. Mannose was the prime monosaccharide composition in both the heteropolysaccharide as confirmed by GC-MS, HPAE-PAD, and NMR studies. Values of optical rotation ($[\alpha]_D^{25°C}$) for the two EPSs (2 mg mL − 1 H_2O) were determined as −142,135 and −128,645 for *G. thermodenitrifcans* ArzA-6 and *G. toebii* ArzA-8, respectively. The molar ratio of monosaccharide units were mannose-1, galactose-0.5, glucose-0.2, arabinose-0.05 with a molecular weight of 600 kDa for *G. tobeii*. Mannose/galactose/arabinose/fructose/glucose were the monosaccharides for *G. thermodenitrificans* with 1/0.13/0.1/0.06/0.05 relative ratio with a molecular weight of 500 kDa.

Radchenkova et al. (2018) isolated halophilic bacterium *Chromohalobacter canadensis* 28 from Pomorie salterns. The optimum culture conditions (lactose as the prime carbon source, salt concentration-NaCl-15%w/v) provide a two field increase of EPS production. Purified polymer is made up of two major fractions of protein (72%w/w) and carbohydrate (14.3%w/w). HPLC study reveals the occurrence of polyglutamic acid (75.7%w/w) as a protein component. The prime sugar components are as follows glucosamine %ww-36.7, glucose %ww-32.3, rhamnose %ww-25.4, xylose%ww-1.7 3.9%w/w of unknown sugar. It was the first report of the occurrence of PGA (polyglutamic acid) in an EPS fraction and it bears good emulsifying, stabilizing, water retention behavior and potential application in the cosmetics industry.

Chikkanna et al. (2018) isolated a halophilic bacterial strain *Halomonas nitroreducens* (strain-WB1) from a hydrothermal vent of Bakreeswar Sakti Peeth, West Bengal, India and the EPS obtained from this strain was characterized by the presence of monomer of glucose, galactose, mannose, rhamnose, arabinose, xylose, fructose, galacturonic acid which were arranged in different configurations forming three different EPSs. EPS1, EPS2 and EPS3 contains glucose-%ww-28, mannose-%ww-64, galactose-%ww-6, xylose-%ww traces; glucose-%ww-18.5, mannose-%ww-44, rhamnose-%ww-2, arabinose-%ww-1.5, xylose-traces; and glucose-%ww-19, mannose-%ww-56.5, galactose-%ww-14.2, galacturonic acid-%ww-1.5, fructose-%ww traces. The EPS was found at higher concentrations during the logarithmic and early stationary phases. The unique composition of this EPS has made it less susceptible to environmental diagenesis. It has an antioxidative, heavy metal binding, and emulsifying functions. This type of investigation enlightens the fact that thermal hot springs of the temperate region are even storehouse of extremophilic microbes with industrial or medicinal importance.

17.5 Endophytic Fungi as EPS Producers

Endophytic fungi are all-square in respect to bioactive compounds production ranging from antimicrobial to anticancer in the field of pharmaceuticals for the development of the new drug to agriculture in order to control post-harvest decay but they are quite unexplored regarding EPS production (Santra and Banerjee 2020). The attempts that had been made by the researchers in order to screen bioactive exopolysaccharides from common or uncommon medicinal or indigenous plants have opened up millions of opportunities in the field of pharmaceutical sectors as a source of potent bioactive novel agents with medicinal importance (Figs. 17.2 and 17.3). A few examples are elucidated here chronologically (Table 17.2). Chen et al. (2013) isolated an endophytic bacterium (endophyte-MD-b) *Bacillus amyloliquefaciens* from the medicinal plant of Asparagaceae family *Ophiopogon japonicas* from the medicinal plant *Ophiopogon japonicas*. The exopolysaccharides produced by this endophytic bacterial strain was effective towards gastric carcinoma (cancer of gastric cells) cell (MC-4 and SGC-7901) as an antitumor agent. MTT (3-(4, 5-dimethylthiazol2-yl)-2,5-diphenyltetrazolium bromide) assays and microscopy were conducted for accurate estimation of cell viability (MC-4 and SGC-7901) using different concentrations (14, 22, and 30 μg/μL) of exopolysaccharide. There was a concentration-dependent response of exopolysaccharides on cell viability with

Fig. 17.2 Various types of endophytic exopolysaccharides

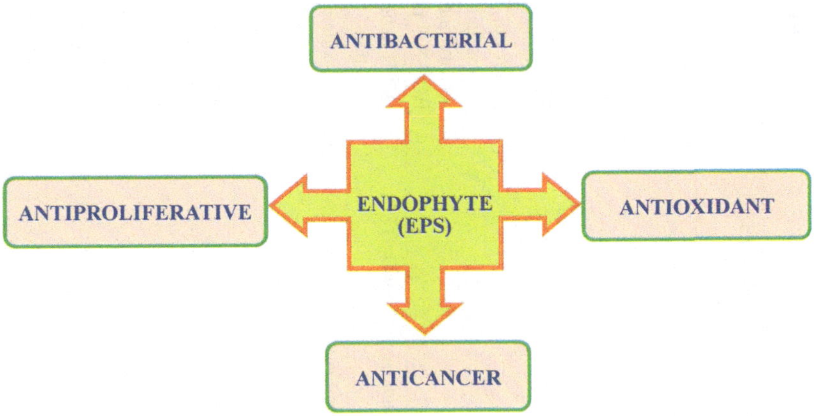

Fig. 17.3 Role of endophytic EPS

Table 17.2 A short profile of endophytic exopolysaccharides with potent therapeutic activity

Sl. No.	Microbe	Biological activity	Structural composition	Reference
1	*Paenibacillus polymyxa* EJS-3	Antioxidative potency (scavenge DPPH (2,3,5 diphenyl picryl hydrazine), SOD (superoxide radical), OH (hydroxyl) radical, modifies thymus (T-cell) and spleen indices. The functionalities of superoxide dismutase (SOD), catalase enzyme action, GPx and TAC (Total antioxidant capacity), but decreases MDA (malondialdehyde) levels in both blood serum and liver tissues level in both serums and livers, exhibit anticancer activity (inhibits human gastric cancer cell BGC-823 proliferation)	Molecular weights of EPS1 and EPS2 is $1.22{*}10^6$ and 8.69×10^5 Da, respectively. They were composed of mannose, fructose and glucose in a molar ratio of 2.59:29.83:1 and 4.23:36.59:1	Liu et al. (2009)
2	*Fusarium oxysporum* Dzf17	DPPH radical scavenging and ferrous ion chelation	–	Li et al. (2011)
3	*Bkleasmium* sp. Dzf12 from rhizome of *Dioscorea zingiberensis*	Antioxidative potential checked by DPPH radical scavenging assay, Fe^{3+} reducing and ferrous ion ($Fe^{2+)}$ chelating activity, hydroxyl radical scavenging assay	–	Li et al. (2012)
4	*Fusarium solani* SD5	Anti-inflammatory (erythrocyte membrane protection from hemolysis), anti-allergic (inhibits 48/80 induced mast cells degranulation), prebiotic activity (selectively stimulates *Lactobacillus* sp. growth and inhibits *Clostridium perfringens*) and antioxidative activity (scavenge DPPH radicals, reduces MDA, oxidized glutathione contents, increases SOD, GPx, and GSH levels in macrophage cells)	Terminal α-L-rhamnopyranosyl, (1 → 2)-α-L-rhamnopyranosyl, (1 → 4)-β-D-galactopyranosyl, (1 → 4,6)-β-D-galactopyranosyl moieties in a molar ratio of nearly 1:1:3:1. Rhamnogalactan with $M_w \sim 1.87 \times 10^5$ Da. EPS have a diameter of 1 nm and melting temp of 172–178 °C	Mahapatra and Banerjee (2012 and 2013b)

5	*Bacillus Cereus* SZ1 from *Artemisia annua* L.	Antioxidative and DNA damage protective activity (superoxide radical, hydroxyl radical), increases glutathione (GSH) level and catalase (CAT) activities, decreases the level of malondialdehyde (MDA) and lactate dehydrogenase (LDH) activity	FTIR spectra confirm the occurrence of characteristic hydroxyl group, C-H bonds, it is an acidic polysaccharide, and 77.75% and 15.56% is the total sugar and protein content, respectively.	Zheng et al. (2016)
6	*Bacillus cereus* RCR08 from *Ricinus communis*	Antioxidative activity confirmed by DPPH radical scavenging assay, cytotoxicity towards hepatocellular carcinoma (Huh 7.5) cell line	FTIR spectrum confirms occurrence of 3404, 2933, 1655 and 1042 cm^{-1}. Polymer is composed of carbohydrate (88.8%), protein (3.18%), and RNA (6.0%), DNA (3.2%)	Mukherjee et al. (2017)
7	*Bacillus* sp. (14 DS-1) from *Codonopsis pilosula* root	Induces macrophage activation and infiltration, TNF-α and nitric oxide production. Suppresses cancer cell migration, arrest S phase, hampers spindle orientation and positioning	Arabinose, mannose, rhamnose, fucose, glucose, galactose	Chen et al. (2018)
8	*Fusarium equiseti* ANP2 from Krishna estuarine mangrove sediments	Scavenge hydroxyl radicals and shows promising antioxidative activity	Mannose (72.6%) and glucose (27.3%)	Prathyusha et al. (2018)
9	*Fusarium* sp. A14 isolated from *Fritillaria unibracteata* Hsiao et KC Hsia	In vitro antioxidative activity and antiproliferative activity on HepG2 cells	Mannose 0.31: rhamnose 0.55: glucose 10: galactose 0.34: xylose 0.3: arabinose 0.6 where pyranose is connected by a glycosidic linkage with side chains of rhamnose, arabinose, glucose, and galactose. The backbone was of rhamnose, xylose, arabinose, and glucose. A14EPS-2 was composed of mannose, rhamnose, glucose, galactose, xylose, and arabinose in a ratio of 16:0.88:10.00:0.39:0.06:0.06.	Pan et al. (2019)

(continued)

Table 17.2 (continued)

Sl. No.	Microbe	Biological activity	Structural composition	Reference
			Pyranose being dominant was visualized in α- and β-organizations in A14EPS-2	
10	*Fusarium solani* DO7 from *Dendrobium officinale*	It uplifts the TNF-α, IL-6, and iNOs mRNA (messenger RNA) expression levels of RAW264.7 cells, and induces the formation of TNF-α (Tumor necrosis factor a) and IL6 (interleukin 6) and NO. As it enhances macrophage activity it could be exploited as a therapeutic agent for the treatment of hypo-immunity patients	EPS (DG2) had a molecular weight of 152.8 kDa and was made up of Galp, Manp, Glcp, and Araf at a molar ratio of 4.3:1.9:2.8: 1.0. The prime linkage of DG2 were (1→)-β-D-Glc, (1 → 4)-β-D-Glc, (1 → 3,6)-β-D-Gal, and (1 → 6)-β-D-Gal, with no harmful effect on embryonic kidney cells (of human origin)	Zeng et al. (2019)
11	*Penicillium citreonigrum* from corm of saffron (*Crocus sativus* L.)	Protects cochlear hair cells from ototoxicity exposure, i.e., otoprotective in nature	EPS has a main backbone of (1-2)-linked α-D-Manp, (1-2, 4)-linked _-D-Manp, (1-4)-linked α-D-Xylp, (1-2, 3, 5)-linked β-D-Araf, (1-6)- linked α-D-Glcp with α-D-Glcp-(1- and α-D-Galp-(1- as sidegroups	Li et al. (2019)
12	*Alternaria tenuissima* F1 from *Angelica sinensis*	Possess in vitro antioxidative property confirmed hydroxy (OH) radical scavenging activity, SO (superoxide anion) scavenging effect along with reducing power ability	D-galacturonic acid 0.45: rhamnose 3.02: D-mannose 3.25: glucose 1.0: D-galactose 0.95	Wang et al. (2019)
13	*Glutamicibacter halophytocola KLBMP 5180* from *Ligustrum sinense* (Chinese Privet)	A potent in vitro antioxidant	Two EPS fragments were found 5180 EPS1, 5180 EPS2 made up of monosaccharide units of rhamnose, galacturonic acid, glucose, glucuronic acid, xylose, arabinose	Xiong et al. (2020)

an IC50 value of 19.7 and 26.8 μg/μL for MC-4 and SGC-7901, respectively. The EPS exhibited cellular abnormalities like mitochondrial dysfunction leading to antitumoral activity. This type of investigation leads to the discovery of novel anticancer drugs from natural products.

Orlandelli et al. (2016) isolated endophytic fungi (*Diaporthe* sp., *Marasmius* sp., *Phlebia* sp., *Phoma* sp., *Phyllosticta* sp., and *Schizophyllum* sp.) from *Piper hispidum* Sw and screened their EPS producing ability in submerged culture (96 h of fermentation). *Diaporthe* sp. JF766998 harbors a higher amount of carbohydrate content (100%) after 168 h of fermentation than the other strains and other fermentation conditions of the same strain. The optimum parameters were marked as 100 mL of medium and 168 h of fermentation. Glucose and galactose were the prime monosaccharide component after 96 and 168 h of fermentation respectively whereas glucose was the prime component of 96 h fermentation product with a molecular weight of 46.6 kDa. The EPS is of higher medicinal importance as glucan rich fractions are characterized by antitumor activity.

Mahapatra and Banerjee (2013a, b) worked on exopolysaccharide of endophytic *Fusarium solani* (strain SD5) isolated from an Indian tree species *Alstonia scholaris* revealed that it has numerous health benefits when tested in mice model and could further be exploited for pharmaceutical application as dietary fiber and immune enhancer or as an edible antioxidant component. Its nontoxic nature and ability to modify gut microbial flora has made it more suitable for medicinal application. The determined molecular weight was Mw $\sim 1.87 \times 10^5$ Da. Structural data revealed the occurrence of terminal α-l-rhamnopyranosyl, (1 → 2)-α-l-rhamnopyranosyl, (1 → 4)-β-D-galactopyranosyl, (1 → 4,6)-β-D-galactopyranosyl moieties in a molar ratio of almost 1:1:3:1. The fibrillar structures were elucidated by TEM studies. The fiber length was almost 1 nm and 172–178 °C was the melting point detected. It has broad-spectrum application as anti-inflammatory, anti-allergic, erythrocyte protection, protection of mast cell from degranulation, etc. The EPS increased the population of total aerobes, anaerobes, *Lactobacillus* sp. and other enterobacteria but significantly reduced the population of harmful bacteria *Clostridium perfringens*. So it is evident that this type of unique EPS components enhances intestinal health by the elimination of the detrimental bacterial population from gut and colon. The intestinal and colonic microflora largely influences individual health by fermenting non-digestible dietary residues and generating short-chain fatty acid molecules that stimulated Calcium (Ca^{2+}), magnesium (Mg^{2+}), Iron (Fe^{2+}) absorption finally regulating bowel movements and proliferation of epithelial cells. So, this type of microbial polysaccharides opens up new frontiers of edible antioxidants.

In another study, Mahapatra and Banerjee (2016) demonstrated and optimized the production parameters of an exopolysaccharide from an endophytic fungus *Pestalotiopsis* sp. (strain BC55) isolated from an Indian medicinal herb *Andrographis paniculata*. It was reported that it contains (1 → 3)-linked β-D-glucopyranosyl moiety and the frame of randomly occurring unit was established as (– → 3)-β-D-Glcp-(1 →).

Hu et al. (2019) isolated a novel type of exopolysaccharide from a Bacillus bacteria *Bacillus* sp. S-1 from a popular Chinese Sincuhan pickle. The monosaccharides were glucose, galactose, mannose with an average molecular weight of 1.765 × 10^4 Da. EPS exhibited potent antioxidant activity by DPPH, superoxide, and hydroxyl radical scavenging assay. EPS was reported to protect oxidative injuries by restricting H_2O_2 induced RAW264.7 cell apoptosis. So these types of EPS are useful nutraceuticals with potent application as antioxidants. NMR data suggest that EPS is composed of β configuration glycosidic bond and a minute quantity of α-glycosidic linkage. Backbone of the chain consists of →4-Glc-1 → and →4-Gal-1→. FTIR data suggests the occurrence of hydroxyl groups, C-O-C glycosidic bond, C-O-H bond with pyranose unit, α-type arrangement of the mannose units, and β-type glycosidic linkage. GC-MS data revealed the occurrence of five major peaks of methylated sugars; one unit of mannose, and two units of galactose and glucose with Manp-(1→, Galp-(1→, 4)-Galp-(1→, →4)-Glcp-(1 → and →4,6)-Glcp-(1 → with a molar ratio of 1.1, 4.0, 25.7, 19.4, 3.2, respectively.

Zhao et al. (2019) characterized an EPS from *Leuconostoc lactis* (strain-L2) from a popular fermented food hami melon which is made up of *Cucumis melo* var. *saccharinus*. Glucose was known to be the monosaccharide with a molecular weight of 3.7 × 10^3 kDa (determined by High performance size exclusion chromatography). EPS is composed of a compact, branched, tubular structure with a smooth, consistent matrix having α-(1 → 6)-linked polymer with D-pyranose residues. It was characterized by a high water solubility index and water retention capacity. This type of character is useful to have an emulsifying property and holds potent application in the field of food, medicine, and pharmaceuticals.

Zheng et al. (2019) isolated an endophytic fungus *Fusarium solani* DO7 from an orchid *Dendrobium officinale* and extracted exopolysaccharides from both the host and endophytic fungus. It was revealed that both the structures are same and similar products are achieved under optimum fermentation conditions. The main fungal polysaccharide chain consists of (→3,6)-β-L-Man-(1→, α-D-Glc-(1→, →4)-α-D-Glc-(1→, →3,6)-β-D-Gal-(1→, and →6)-β-D-Gal-(1→, while the side chain was α-D-Glc-(1→). Both the FP (fungal polysaccharide) and DOP (*D. officinale* polysaccharide) exhibited safe use on cell lines for in vivo application and were effective antioxidant, antiglycation, and immunomodulatory activities. Another part of the study confirms that the membrane receptors of the RAW264.7 cells for fungal polysaccharides are TLR2 and TLR4. This type of work not only illuminates the pharmaceutical application of EPS but also reestablishes the host–endophyte relationship and opens up new biotechnological scopes for microbial product utilization from the untapped source of plants.

Fooladi et al. (2019) isolated a fungus *Neopestalotiopsis* sp. SKE15 in PDA (potato dextrose agar) as mold which was present together with a wood decomposing fungus belonging to Agaromycetes of Basidiomycota. The production parameters were optimized by Plackett-Burman test with (RSM) Response surface methodology and yield was enhanced. The monomers of the EPS are D-glucose, sorbitol, D-galactose with a molecular weight of 1.5–2.1 × 10^6 Da. EPS was an efficient antibacterial and antioxidative agent confirmed by anti *S. aureus* (ATCC 25923)—Gram-positive, *B. subtilis* (ATCC 6633)—Gram-positive, *P. aeruginosa* (ATCC 278533)—Gram-negative activity and DPPH, hydroxyl radical scavenging assay, respectively. Moreover, this exhibited cytotoxic behavior against three cancer cell lines at a final concentration of 100 and 200 μg/mL. This type of investigation shows promising aspect for the medicinal application of EPS.

Zhang et al. (2020) documented the antioxidative potency of the EPS produced by the endophytic *Pilidiella guizhouensis* ZJSRU-M1 isolated from *Eupatorium chinese* L. It was a homogenous polysaccharide made up of glucose with a molecular weight of 13.4 kDa. The antioxidative potency was confirmed by DPPH (DPPH (2,3,5 diphenyl picryl hydrazine), OH (hydroxyl), and superoxide radical scavenging ability which describes its potential future applications in the food and pharmaceutical industry as health products.

17.6 Some Recent Investigation on Structure and Function of EPS

Exopolysaccharides are structurally and functionally vivid as well as of diverse nature. Bioactivities of some popular exopolysaccharides with their structure are recorded here (Table 17.3 and Figs. 17.4 and 17.5).

Guo et al. (2013) reported the antioxidative potency (confirmed by in vitro tests like DPPH radical scavenging, superoxide radical scavenging, and lipid peroxidation inhibition) of a water-soluble exopolysaccharide (Os-2) isolated from *Oidiodendron truncatum* GW (isolated from under lichen soils) which was composed of glucose and minute amounts of glucosamine with a molecular weight of 9.6 kDa. NMR spectroscopy confirms the presence of (1 → 6)-linked-D-glucopyranose residues with minor amounts of (1 → 2)-linked-D-glucopyranose residues.

Pei et al. (2012) examined the antiviral activity of the EPS produced by *Aspergillus versicolor* PH1008 from the mangrove forest of Hainan province. EPS exhibited antiviral activity against H_3N_2 influenza virus with an IC_{50} value of 34.32 μg/mL and CoxB3 (Coxsackievirus) with 7.81 μg/mL-IC_{50} value confirmed by observing drug cytotoxicity and cytopathic effect.

Li et al. (2015) reported the antibacterial and antibiofilm activity of a novel exopolysaccharide obtained from *Arthrobacter* sp. B4 and the production parameters were optimized (glucose, yeast extract, and KH_2PO_4 in an amount of 16.94 g, 2.33 g, and 5.32 g/L) by RSM (Response surface methodology) coupled with PBD (Plackett-Burman design).

Another example is of Lasiodiplodan which is a commercially valuable molecule obtained from fungus *Lasiodiplodia theobromae* that has utility in the pharmaceutical industry for the manufacture of antiaging creams, ointments, self-healing hydrogels, in the food industry as probiotics, etc. (Cunha et al. 2019).

Kumar et al. (2018) characterized an interesting exopolysaccharide Lasionan from *Lasiodiplodia* sp. (MTCC 6000-strain number) as glucomannan with glucose and mannose in a 1:1 ratio having a molecular mass of 29.3 kDa. Lasionan was efficient antibacterial (broad spectrum—*Staphylococcus aureus*, *Bacillus subtilis*, *Micrococcus luteus*, *Klebsiella planticola*, *Escherichia coli*, *Pseudomonas aeruginosa*, *Chromobacterium violaceum* and even against MDR strains and

Table 17.3 Structure and function of some popular microbial exopolysaccharides

Sl. no.	Name of the EPS	Structure of EPS	Applications	Reference
1	Cyclosophorans from *Agrobacterium* sp., *Rhizobium* sp., *Xanthomonas* sp.	Cyclic (1 → 2)-β-D-glucans	Useful in drug encapsulation and development of food items	Vandamme and Soetaert (1995)
2	Emulsan from *Acinetobacter calcoaceticus*	Heteropolysaccharide with repeating trisaccharide [GalNAc → GalNAcUA → an unidentified N-acetyl amino sugar]	As an emulsifying and mobilizing agent	Sutherland (1998), Kumar et al. (2007), Vu et al. (2009)
3	Acetan from *Acetobacter xylinum*	Branched homopolysaccharide (tetrasaccharide) of glucose with (1 → 4 β) linkage and α-(1,3)- side chain [β–D–Glc–(1 → 4)–β–D–Glc–]$_n$ ↓[α(1 → 3)] Rha–α(1 → 6)Glc–β(1 → 6)Glc–α (1 → 4)GlcA β(1 → 2)Man	Acts as a viscosifier and gelling agent	Sutherland (1998), Kumar et al. (2007), Vu et al. (2009)
4	Succinoglucan from *Alcaligenes faecalis* var. myxogenes	It is a glucose-galactose heteropolymer with succinate and pyruvate. [β–D–Glc–(1 → 4)–β–D–Gal–(1 → 3)]n	It is used for the preparation of low-calorie jams and jellies, also for the formation of films that promote immobilization (fixation) of enzymes	Sutherland (1998)
5	Cellulose from *Acetobacter xylinum Agrobacterium* sp., *Pseudomonas* sp., *Rhizobium* sp.	It is a linear homopolymer of glucose with (1 → 4 β) linkage [β–D–Glc–(1 → 4)–β–D–Glc–]n	It is useful for the recovery of burns (fire wounds) and also surgical (during the time of surgery) wounds	Sutherland (1998)
6	Gellan from *Sphingomonas paucimobilis*, *Aureomonas elodea*, *S. elodea*	It is a linear heteropolysaccharide of glucose, glucuronic acid, and rhamnose [–D–Glc–(1 → 4)β–D-GlcUA–(1 → 4)βD–Glc(1 → 4) β –L–Rha(1 → 3)α–]$_n$	It is used as a vector for ophthalmic drug delivery and in food industry, it is useful for the manufacturing of low calories jam jellies, acts as a thinning agent for the preparation of low calories alcoholic beverages like wines	Sutherland (2002)

7	Pullulan obtained from *Aureobasidium pullulans*	It is a linear homopolysaccharide (similar monosaccharides in repeating units), made up of maltotriose (glucose trisaccharide), and maltotetraose (glucose tetrasaccharide) units with both α-(1 → 6) and α(1 → 4) linkages in regular alternation	Possess pharmaceutical applications, clinical, and health care applications. Acts as a dietary fiber and prebiotics promoting the growth of *Bifidobacteria* sp. used as ingredients in wound healing products. As a dietary supplement, breathe freshener. Used in pharmaceutical coatings, microencapsulation like sustained-release formulations and dental care products. Used in low-calorie food as an additive. Used as nontoxic conjugates for the development of vaccines and interferon, facilitates, delivery of liposome. Pullulan-based hydrogels coupled with nanoparticles have a variety of uses	Leathers (2003)
8	Alternan from *Leuconostoc mesenteroides* NRRL B1355	It is a branched homopolymer with alternating α-(1 → 3) and α-(1 → 6)-linked D-glucose residues	Used as an ingredient of prebiotic food. It can replace polydextrose (synthetic polymer of glucose), maltodextrins, gum arabic...bulking (high amount) agents in food products. Used as a component of cosmetic creams and ointments	Leathers et al. (2003)
9	Kefiran from *Lactobacillus hilgardii*, L. *rhamnosus*, L. *kefiranofasciens*	[→6)β–DGlc–(1 → 2{6})β–DGal–(1 → 4)α–DGal–(1 → 3)β–DGal–(1 → 4) β--DGlc–(1]$_{21}$	It is traditionally consumed as self-carbonated and slightly alcoholic fermented milk drink and other dairy products (brewed milk) as prebiotic. It has a cytotoxic effect towards human cancer cell lines	Matsuda et al. (2003)

(continued)

Table 17.3 (continued)

Sl. no.	Name of the EPS	Structure of EPS	Applications	Reference
10	Elsinan from *Elsinoe leucospila*, *E. fawcetti*, *E. ampelina*	It is a linear polymer made up of (1 → 4)- and (1 → 3)-α-D-glucosidic linkages in a ratio of 2.6:1	Useful as an oxidation barrier in vitamin, enzyme, and hormone preparations. Useful in biomedical sealing, tableting, or encapsulating the drugs, acts as a hypercholesterolemic drug, possess antitumor activities	Misaki (2004)
11	Hyaluronan *Streptococcus equi*, *Streptococcus zooepidemicus*	It is a linear Heteropolysaccharide of glucuronic acid and N-acetylglucosamine (1 → 4 β) and (1 → 3 β) [–DGlcUA–(1 → 4)β–DGalNAc–(1 → 3)β–]	It has roles in the prevention of angiogenesis and cancer, arrest cell motility, induce wound healing, promotes cell adhesion. So useful for skin moisturizers, osteoarthritis treatment, useful replacement of eye fluid in ophthalmic surgery, wound healing, acts as a lubricant for joints, prevention of adhesion in case of abdominal surgery	Widner et al. (2005)
12	Dextran from *Leuconostoc mesenteroides*, species of Lactobacillus, Leuconostoc, Streptococcus.	It is a linear polysaccharide of D-glycopyranose containing α-(1 → 4) and α-(1 → 6)	Acts as a blood plasma extender, useful in microsurgery with antithrombotic effect, antiviral activity against Human Immunodeficiency syndrome, used for the treatment of arteriosclerosis, used in surgical sutures	Khan et al. (2007)
13	Scleroglucan from *Sclerotium glucanicum*, *S. rolfsii*	It is a homopolysaccharide made up of a linear chain of D-(1–3)-glucopyranosyl and -D-(1–6)-glucopyranosyl groups	It has antitumor, antiviral, immune stimulatory (enhances immunity), anti-neoplastic (inhibits cancerous cell proliferation), and antimicrobial (antibacterial, antifungal, and antiviral) useful in accurate drug delivery and reduces blood cholesterol level excreting the lipid from the blood.	Survase et al. (2007)

			Used in the preparation of outer layer or cover of the tablets, opthalmic solution (eye soothing medicine) injectable antibiotic suspensions, and calamine lotions (other skin preparations like creams, protective lotions, emollients, demulcents, and antisoilants)	
14	Levan obtained from *Bacillus subtilis*, *B. polymyxa*, *Aerobacter levanicum*, *Streptococcus* sp., *Pseudomonas* sp., *Corynebacterium laevaniformans*	It is a homopolymer of fructose which is made up of repeating five-member fructofuranosyl rings which were connected by β-(2 → 6) links. The main chain is branched through β-(2 → 1) linkage of the fructofuranosyl rings	Useful in the pharmaceutical industry as a blood plasma extender, acts as a hypocholesterolemic agent, modifies cell membrane (in case of tumor cells), acts as an immunomodulatory agent (affects macrophages and humoral or cell-mediated immunity), and also useful as a tablet binder. It has anti-inflammatory effect against inflammatory reactions towards skin irritants (creates an odd feelings in the upper epidermal portion of the skin), cell-proliferative effect in bioartificial skin	Rairakhwada et al. (2007)
15	Curdlan from *Alcaligenes faecalis* var. myxogenes	It is a linear homopolysaccharide (made up of similar types of sugar) made up of glucose (1 → 3β) linkage [β–D–Glc–(1 → 3)–β–D–Glc–]n	It is useful in the release of drugs so used in the pharmaceutical industry	Kumar and Mody (2009)
16	Alginate from *Azotobacter vinelandii*, *Pseudomonas aeruginosa*	It is made up of a variety of proportions of 1,4-linked β-D-mannuronic acid (M) and its C-5 epimer α-L-guluronic (G)	It is used as an emulsion stabilizer (balancer), gelling agent (increases the concentration of the liquid) and thickener (provides a dense appearance), foam stabilizer (balances)	Remminghorst and Rehm (2009)

(continued)

Table 17.3 (continued)

Sl. no.	Name of the EPS	Structure of EPS	Applications	Reference
			and suspending agent, water-binding component. Useful for pharmaceutical applications like cellular and enzyme encapsulation for delayed release, and in the dressings of wounds and dental materials	
17	Xanthan gum from *Xanthomonas campestris*	It is a homopolysaccharide of glucose (1 → 4 β) linkage with side chains α-(1,3)- trisaccharides [β–D–Glc–(1 → 4)–β–D–Glc–]$_n$	It is used in the packaging of foods, manufacturing of cottage cheese and creaming emulsions, important stabilizing ingredients for toothpaste, replaces calf milk, and used as a cattle feed supplement	Kumar and Mody (2009)
18	Welan from *Alcaligenes* sp.	It is a Heteropolysaccharide of glucose, glucuronic acid, rhamnose with a backbone of [D–Glc–(1 → 4)β–D-GlcUA–(1 → 4)βD–Glc(1 → 4) α–L–Rha/Man(1 → 3) β–]$_n$	Acts as a stabilizer (sustainer) and viscosifier (thickener)	Kumar and Mody (2009)
19	Galactopol from *Pseudomonas* sp.	It is made up of galactose, mannose, glucose, rhamnose, acetate, succinate, pyruvate	Used as a hydrocolloid in food, feed, pharmaceutical industry. Also useful in coating and packages	Freitas et al. (2009)
20	Fucopol from *Enterobacter* strain A47 DSM 23,139	It is made up of fucose, galactose, glucose, acetate, succinate, pyruvate	Important hydrocolloid in food, feed, pharmaceuticals, and medicine industry. Also useful in coating and packages.	Freitas et al. (2011)
21	Colanic acid obtained from *E. coli*, *Enterobacter* sp. and *Salmonella* sp.	Made up of fucose, glucose, glucuronate, and galactose	Useful ingredients in food and cosmetics industry	Nwodo et al. (2012)

22	Glucuronan *Sinorhizobium meliloti* M5N1C5	Made up of glucuronic acid	Useful ingredients in food and cosmetics industry	Nwodo et al. (2012)
23	Hetero-EPS Enterobacter bacteria strains	Made up of fucose (hexose deoxy sugar), glucose (reducing sugar), galactose (similar to glucose), glucuronic acid (hexose sugar), pyruvate ($CH_3COCOOH$), succinate (salt ester of succinic acid), acetate (acetic acid with base)	Hypoglycemic (reduces sugar level in blood) and hypolipidemic (reduces cholesterol levels in the blood) action for type 2 diabetic mice	Huang et al. (2015)
24	Xanthouronan	produced by C-6 oxidation of xanthan using NaOCl/NaBr/TEMPO catalyst	Possess in vitro (no involvement of living cell) antioxidative property	Delattre et al. (2015)
25	Sphingans from *Sphingomonas* sp.	It is made up of 13% glucuronic acid, neutral sugars (without any charge) like mannose 1:glucose 2.28: rhamnose 2.12 A-L-rha-(1 → 4)β-L-Rha-(1 → 4)-β-D-Glc(1 → 3)-α-D-Glc	Useful in food, oil, and ink industry	Li et al. (2016)

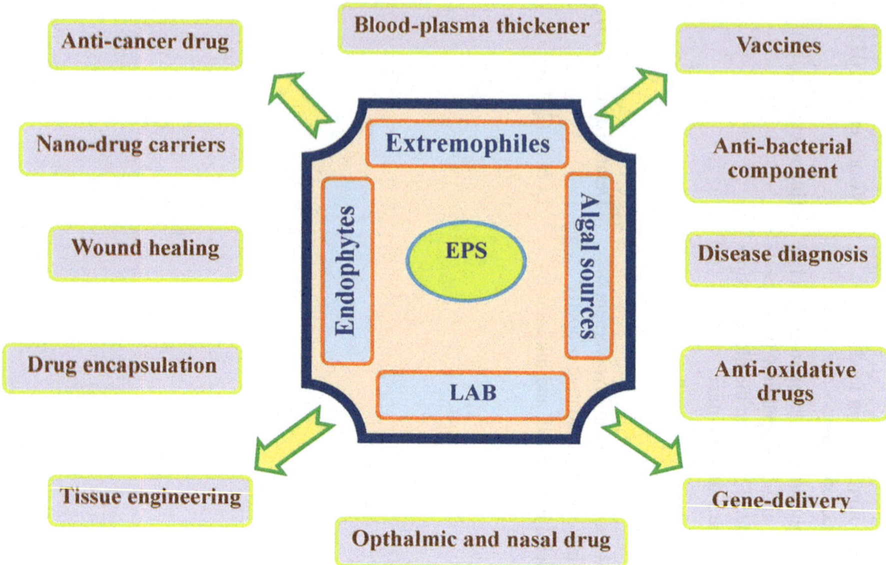

Fig. 17.4 Medicinal applications of microbial exopolysaccharides

candida pathogens—*Candida albicans* MTCC strains), antioxidant (DPPH free radical, superoxide anion scavengability along with lipid peroxyl radical scavenging ability), antibiofilm agent accompanied with immunomodulation qualities (in case of RAW 264.7 macrophage cells). Lasionan directly inhibited the LPS mediated ROS and NO generation. It also downregulates the production of LPS mediated pro-inflammatory mediators like TNF-α, IL-6. So this basidiomycota originated EPS possess qualitative characters for their therapeutic and biomedical applications.

Costa et al. (2019) worked on the exopolysaccharide produced by a fungus *Aspergillus terreus* and optimized the culture conditions as 28 °C temp, 8.79 pH, 98 agitation rpm, carbon source-glucose 2.39%, and nitrogen source-ammonium nitrate 0.957%. Monosaccharide analysis revealed the occurrence of mannose, galactose, glucose in 24.3, 40.5, and 35.2 mol%. NMR coupled with IR data suggests the occurrence of galactomannans (mannose with galactose backbone), β-1,3-glucans, and glycogen-like glucans. This novel EPS are able to activate murine macrophages, increasing the production of NO (nitric oxide), superoxide anions, TNF-α and IL-6 (interleukin-6). This EPS could further be exploited as an immunomodulating agent in the biomedical sectors.

Asgher et al. (2020) characterized a mesophilic bacterial EPS and determined its functional properties. The organism is known to produce a heteropolysaccharide with glucose, mannose, and fructose with a percentage of 46.80, 20.60, and 32.58, respectively linked by glycosidic linkages and the surface being smooth and less porous. The production parameters were optimized using RSM-CCD (Response surface methodology and Central composite design) and there was a 3.2 fold

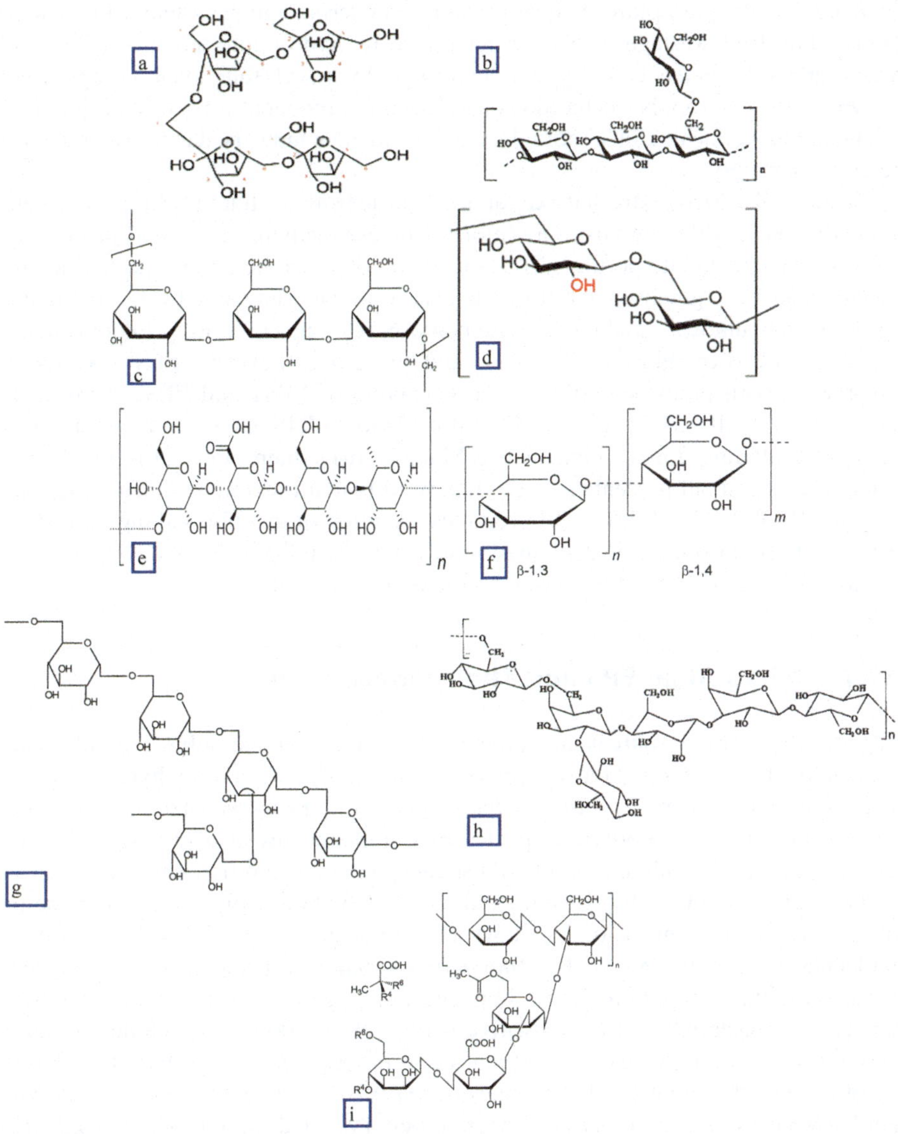

Fig. 17.5 Chemical structures of exopolysaccharides (**a**) levan, (**b**) scleroglucan, (**c**) pullulan, (**d**) lasiodiplodan, (**e**) gellan, (**f**) 1,6-β-D glucan, (**g**) dextran, (**h**) kefiran, (**i**) xanthan

increase than the pre-optimized one in a solid state fermentation system using mango peels, daily fruit wastage. It was characterized by hydrophilic moieties with good water and oil uptake capacity and carboxyl, hydroxyl functional groups were present. This heteropolysaccharide is an efficient component for the food, pharmaceutical, biochemical, biotechnological, and cosmetics industry due to water and oil retention capacity.

Wang et al. (2020) extracted exopolysaccharide from the liquid cultures of a well-known popular edible fungus of traditional Chinese medicine *Monascus purpureus*. EPS was separated into portions EPS1- made up of six carbon monosaccharide unit sugars mannose, glucose, galactose in 8:1:11 molar ratio and EPS2- rhamnose (6-deoxy-L-mannose), arabinose (aldopentose-5C sugar), xylose (aldopentose), mannose (hexose), glucose (hexose), and galactose (hexose). EPS was found complexed with pigment portions. The backbone of EPS1 and EPS2 were made up of 5-β-D-Galf (31.74 wt%) and 2-β-D-Manp (17.48 wt%) with other sugar residues including 2,6-β-D-Galf, T-α-D-Manp, 6-α-D-Manp, and 3,5-β-D-Galf was remaining at a small percentage. EPS-1 showed potent immunomodulatory activity towards RAW2647 cell lines and promotes secretion of cytokines along with IL-6, TNF-α, IL-10, also improved expression levels of the mRNA. So, the EPS could be exploited for immunomodulatory and antioxidative activity.

17.7 Micro Algal EPS and Their Bioactivities

Algae from marine sources are potent synthesizers of exopolysaccharides and especially the members of cyanophyceae, chlorophyceae, rhodophyceae are the major contributors here. *Gyrodinium impudicum* strain KG_{03} (isolated from seawater sample in Kang-Gu, Korea), a species of dinoflagellates also produces sulfated exopolysaccharide with antiviral (against encephalomyocarditis virus with an EC_{50} value of 26.9 μg/mL) activity (Yim et al. 2004). The homopolysaccharide made up of galactose with uronic acid (2.96% wt/wt), sulfate group (10.32% wt/wt) possess a molecular weight of 1.87×10^7. *Anabaena spiroides*, a member of cyanophyceae possess multidisciplinary activities like antithrombogenic, anti-atherogenic, anticoagulant, antibacterial, and antioxidative (Freire et al. 2005). Hypocholesterolemic activity was seen by the exopolysaccharide of *Porphyridium* sp. (Dvir et al. 2009). Another cyanobacterial species *Arthrospira platensis* is known to exhibit antilarval, antibacterial (*Staphylococcus epidermis* (Gram +ve) and *Salmonella typhimurium* (Gram −ve)) and antioxidative activity. *Rhodella reticulate* is a member of rhodophyte which was reported to produce exopolysaccharide that was known to exhibit antioxidative and free radical scavenging ability (Chen et al. 2010). Sun et al. (2012) reported that exopolysaccharides from *Porphyridium cruentum* are known to have antiviral, antibacterial, immunomodulatory, and antitumor activity. Zhang et al. (2019) extracted two exopolysaccharides from two different species of green microalgae *Chlorella zofingiensis*, *C. vulgaris,* and characterized them as heteropolysaccharides of D-manno, ribose, L-rhamnose, glucosamine, glucuronic acid, galacturonic acid, D-glucose, galactosamine, D-galactose, D-arabinose, L-fucose in a ratio of 43.33, 0.28, 0.99, 4.12, 1.75, 0.48, 17.42, 0.55, 30.08, 0.85, 0.04 and 10.60, 0.57, 4.79, 28.00, 0, 0.24, 7.31, 0.71, 45.43, 2.30, 0.06, respectively. They reported average production of 208.4 and 364.3 mg/L and the molecular weight was reported to be 2.66×10^4, 1.88×10^4 Da, respectively with antioxidant and antitumor (against human colon cancer cell lines-HCT8-28.3% and 18.0%, respectively) activities. Moreover, the antitumor effects were studied using human colon cancer cell lines HCT8. Further investigations may end up discovering potent agents with antitumor activity.

Yu et al. (2019) extracted six different types of exopolysaccharides from green algae *Chlorella vulgaris* according to six different types of treatments like (i) repeated freezing and thawing (RFP) method, (ii) Microwave-assisted polysaccharide (MAP), (iii) Ultrasonic wave (UWP)-mediated polysaccharide, (iv) alkali-based extraction (AEP), (v) Hot water-based (HWP) extraction, and (vi) Cellulose-based extraction (CEP). The exopolysaccharides are potent antioxidative agents with in vitro (DPPH radical scavenging assay) and in vivo (using *Caenorhabditis elegans* as a model organism) effectivity. The best antioxidative activity was exhibited by the exopolysaccharide which was extracted according to UWP and the DPPH, superoxide, hydroxyl radical scavenging rates were reported to be 65.1% ± 2.4%, 61.2% ± 2.7%, and 56.2% ± 2.2%, respectively along with impressive metal chelating activity (63.6% ± 2.5% at 0.4 mg/mL concentration). Ultrasonic wave obtained exopolysaccharide enhanced the amount of stress-related enzymes in the *C. elegans* cells; catalase and superoxide dismutase were enhanced by 7.29% ± 1.8% and 24.41 ± 4.8%, respectively. This type of study focuses on the antioxidative potency of the exopolysaccharides produced by green algae. Also, the extraction processes vary from one method to another and the process of extraction is directly related to their antioxidative ability.

Anticancer drugs that are widely used most of the times possess a harmful side effect which is needed to be excluded. To overcome this hindrance search of bioactive compounds from natural sources with no side effects is the prime target. They either can replace the toxic chemotherapeutic agents or can be jointly used with the existing popular drug to increase the drug efficacy. A recent study conducted by Nikolave and his coworkers (2019) demonstrated that a red algae Porphyridium sordidum produced EPS with a composition of xylose-1: glucose-0.52: galactose-0.44: rhamnose-0.31. This heteropolysaccharide was potent enough in restricting the growth of MCF7, MDA-MB231 (breast cancer) cell lines. Transport of exopolysaccharide was increased to the targeted cells by reverse electroporation. A mixture of 75 μg/mL polysaccharide with 200 V/cm electroporation induces the decrease of MDA-MB231 cell viability by 40% and drastic changes in cell morphology. So hopes of chemotherapeutic agents with zero or less toxicity are from microalgal source to combat the cancer disease.

17.8 Medicinal/Therapeutic Applications of Exopolysaccharides

Exopolysaccharides possess huge medicinal and therapeutic applications starting from nanoparticle-based bioactivity to drug delivery. Some such examples are represented here.

In case of medical imaging exopolysaccharides also play an important role in the diagnosis of cancer cells or tumors. After attaching a specific ligand group to a biopolymer, EPS gets functionally much more active like pullulan when modified adjusted by amino functional groups provide a better resolution of fluorescence in tumor cells better than traditional quantum dots-based liposomes (Moscovici 2015). Nanoparticles when found to be crosslinked with dextran exopolysaccharide provide greater opportunity for a broad spectrum of imaging like DMR (diagonostic magnetic resonance), MRI (Magnetic resonance imaging), PET (Positron electron tomography), and FMT (fluorescence molecular tomography). Some examples include killing of atheroma cells in carotid arteries by irradiation and pullulan overlayed nanoparticles with magnetic nature kills almost 80% of KB tumor cells (Tassa et al. 2011; Bataille et al. 2011). The antitumor function towards cancer cell lines is quite a slow process in case of EPS treatment in comparison to the conventional chemotherapeutic and radiation therapies but a combination of both can surely improve the treatment conditions where EPS acts as a synergistic or prophylactic agent. A popular exopolysaccharide levan produced from a variety of microorganisms is known to exhibit antiviral (influenza, type40 Adeno-virus, AIDS, hepatitis B), anticancer activity (MCF-7, AGS, HepG2/C3A, BGC-823), hypoglycemic, antioxidative potential (Sarilmiser and Öner 2014; Liu et al. 2012; Zong et al. 2012; Srikanth et al. 2015). EPS are useful agents of drug delivery and are called as the intelligent drug delivery system that mediates the absorption of the active constituents of drug or hold the drug suspension at the same time retains water and also possess an insoluble matrix, so easily drug delivery is completed. This type of association increases drug penetration to the targeted cells. Pullulan is a popular exopolysaccharide with thermal, pH, and cation sensitive drug delivery options. Being neutral in nature it is an easy target of derivatization and development of thermo and pH-sensitive microspheres. Silver nanoparticles of pullulan are known to show higher antibacterial activities (Kumar et al. 2012). Carboxymethyl pullulan when grafted with polyether-amine provides sustained drug delivery options and provides a platform for retention and delivery of antioxidant, anticancer, antiproliferative, antibiotic, anticoagulant, NSAID, anti-hypertensive molecules like lutein, α-tocopherol, 5-fluorouracil, indomethacin, gatifloxacin, heparin, sulindac, pindolol, respectively (Mocanu et al. 2012; Matricardi et al. 2013). Another example of gellan includes the delivery of nasal drug solving the problem of slow absorption of oral administration of drugs (Hagerstrom, 2003). Gellan beads and dextran nanoparticles mediate the appropriate controlled release of ketoprofen (an inflammatory drug) and antibiotic amoxicillin (Prezotti et al. 2014; Saldias et al. 2015). Polymeric substances can increase the bioavailability of small drug molecules or proteins peptide chains, siRNA (small interfering RNA), hormones, and vaccines to our body by preventing them from getting deformed in the liver or other body parts thus increasing their ability to cross the plasma membrane and mucosal surfaces (Ganguly et al. 2014). Pullan nanoparticles protect insulin from enzymatic degradation (Akiyoshi et al. 1998) Dextran coated components mediate the delivery of insulin and also growth hormones maintaining their bioactivities (Chalasani et al. 2007; Yuan et al. 2012).

Bacterial exopolysaccharides are efficient agents of wound healing or skin repair also. Bacterial cellulose takes a lead role in this respect due to their high water absorption capacity and also a higher degree of resistance towards enzymatic degradation. Chemical derivatization may increase its efficiency to be used as a tissue engineering component. Components like alginate, gellan, hyaluronan, chitosan are used for this purpose. They are efficient healers in case of diabetic foot ulcers, psoriasis, dryness-induced skin disease, etc. (Fu et al. 2013).

Not only in external wound regeneration or drug delivery microbial EPS (alginate, dextran, pullulan, etc.) possess a role in the maintenance of heart health by being a prime component in designing of structures that can efficiently support (dextran + poly L-lysine = bioadhesive in surgery), replace the tissues (also promotes cardiomyoblast growth) that have been damaged due to myocardial infarction (Silva et al. 2015; Thankam and Muthu, 2014; Matsumura et al. 2014). In case of osteoarthritis or arthroplasty cartilage regeneration is to some extend mediated by alginate mediated stem cell cultures. In addition to chitosan, they demonstrate higher antimicrobial (antibacterial) and wound protecting or healing properties (Ivanova et al. 2014). Alginate components are also used for the development of bioartificial bones, pancreas, liver cell cultures, and also vasculatures (Aravamudhan et al. 2014).

EPS can act as antigen carriers and known as polysaccharide vaccines. One such example includes the tetanus toxoid which remains entrapped in carboxymethyl pullulan matrix and gets delivered to its target place (Mocanu et al. 2011). Some examples on mice model yields satisfactory results where the alginate, curdlan, and hyaluronic acid mediates the transport of vaccine and induce strong immunoresponse against tetanus and hepatitis B virus (Sharma et al. 2009).

Pullulans are also valuable for gene delivery over viral vectors to solve the problem of lysosomal degradation which is made up of plasmid incorporation into pullulan nanoparticles (Gupta and Gupta 2004). Dextran and curdlan derivatives can efficiently transfect human lung, epithelial carcinoma, leukemia, and colon cancer cell lines proving it a good agent for gene delivery (Raemdonck et al. 2010; Naeye et al. 2010; Han et al. 2015).

EPS helps in the encapsulation of drugs of importance like diabetes, liver disorders, etc. This process provides safe transportation of drug to the target organ and also protection from the hostile immune reaction. Bacterial cellulose is used for these purposes (Rokstad et al. 2014).

17.9 Conclusion

Microbial exopolysaccharides are one of the miracles of nature with multidiscplinary utility in almost one and all sectors of medicinal-pharmaceutical and industrial arenas. They are reported from a wide range of microbial populations from extremophilic archaebacteria to eukaryotic microscopic fungi. Their efficient utilization and skilled exploitation can solve many burning issues of pharmaceutical and allied sectors. Scientists are engaged in developing new ways for their enhanced production and purification, development of tools for drug delivery, polysaccharide vaccine, encapsulation agents, etc. Their broad-spectrum utility as antibacterial, antioxidant, agents for drug delivery has retained the interests of several schools of researchers. This chapter incorporates some recent and age-old landmark discoveries of microbial exopolysaccharides with special emphasis on their bioactivity in vivo, in vitro, and throwing some lights on their structures also. A lot of paths is still to complete for appropriate commercialization of several exopolysaccharide-based products for the ultimate benefits of mankind.

References

Abdrabo MAA, Ibrahim HAH, Hassan SW, Abdul-Raouf UM (2018) Antimicrobial and anti-tumor activities of exopolysaccharides produced by the biofilm of marine Halomonas saccharevitans AB2 isolated from Suez Gulf, Egypt. Egypt J Aquat Biol Fish 22(5 (Special Issue)):99–119

Abinaya M, Vaseeharan B, Divya M, Vijayakumar S, Govindarajan M, Alharbi NS, Khaled JM, Al-anbr MN, Benelli G (2018) Structural characterization of *Bacillus licheniformis* Dahb1 exopolysaccharide—antimicrobial potential and larvicidal activity on malaria and Zika virus mosquito vectors. Environ Sci Pollut Res 25(19):18604–18619

Akiyoshi K, Kobayashi S, Shichibe S, Mix D, Baudys M, Kim SW (1998) Self-assembled hydrogel nanoparticle of cholesterol-bearing pullulan as a carrier of protein drugs: complexation and stabilization of insulin. J Control Rel 54:313–320

Aravamudhan A, Ramos DM, Nada AA, Kumbar SG (2014) Natural polymers: polysaccharides and their derivatives for biomedical applications. In: Kumbar S, Laurencin C, Deng M (eds) Natural and synthetic biomedical polymers. Elsevier, Brazil, pp 67–89

Arena A, Maugeri TL, Pavone B, Iannello D, Gugliandolo C, Bisignano G (2006) Antiviral and immunoregulatory effect of a novel exopolysaccharide from a marine thermotolerant *Bacillus licheniformis*. Int Immunopharmacol 6:8–13

Arena A, Gugliandolo C, Stassi G, Pavone B, Iannello D, Bisignano G, Maugeri TL (2009) An exopolysaccharide produced by *Geobacillus thermodenitrificans* strain B3-72: antiviral activity on immunocompetent cells. Immunol Lett 123:132–137

Arias S, Moral AD, Ferrer MR, Tallon R, Quesada E, Bejar V (2003) Mauran, an exopolysaccharide produced by the halophilic bacterium *Halomonas maura*, with a novel composition and interesting properties for biotechnology. Extremophiles 7:319–326

Arun J, Selvakumar S, Sathishkumar R, Moovendhan M, Ananthan G, Maruthiah T, Palavesam A (2017) In vitro antioxidant activities of an exopolysaccharide from a salt pan bacterium *Halolactibacillus miurensis*. Carbohydr Polym 155:400–406

Asgher M, Urooj Y, Qamar SA, Khalid N (2020) Improved exopolysaccharide production from *Bacillus licheniformis* MS3: optimization and structural/functional characterization. Int J Biol Macromol 151:984–992

Asker MS, El Sayed OH, Mahmoud MG, Yahya SM, Mohamed SS, Selim MS, El Awady MS, Abdelnasser SM, Abo Elsoud MM (2018) Production of exopolysaccharides from novel marine bacteria and anticancer activity against hepatocellular carcinoma cells (HepG2). Bull Natl Res Centre 42(1):30

Aullybux AA, Daneshwar P, Theeshan B, Jeewon R (2019) Phylogenetics and antibacterial properties of exopolysaccharides from marine bacteria isolated from Mauritius seawater. Ann Microbiol 69(9):957–972

Bai Y, Zhang P, Chen G, Cao J, Huang T, Chen K (2012) Macrophage immunomodulatory activity of extracellular polysaccharide (PEP) of Antarctic bacterium *Pseudoaltermonas* sp. S-5. Int Immunopharmacol 12(4):611–617

Bataille I, Meddahi-Pellé A, LeVisage C, Letourneur D, Chaubet F (2011) Pullulan for biomedical uses. In: Popa V (ed) Polysaccharides in medicinal and pharmaceutical applications. Smithers Publication, Shawbury, pp 145–182

Brian-Jaisson F, Molmeret M, Fahs A, Guentas-Dombrowsky L, Culioli G, Blache Y, Cérantola S, Ortalo-Magné A (2016) Characterization and anti-biofilm activity of extracellular polymeric substances produced by the marine biofilm-forming bacterium *Pseudoalteromonas ulvae* strain TC14. Biofouling 32(5):547–560

Cambon-Bonavita MA, Raguénès G, Jean J, Vincent P, Guezennec J (2002) A novel polymer produced by a bacterium isolated from a deep-sea hydrothermal vent polychaete annelid. J Appl Microbiol 93:310–315

Carrión O, Delgado L, Mercade E (2015) New emulsifying and cryoprotective exopolysaccharide from Antarctic *Pseudomonas* sp. ID1. Carbohydr Polym 117:1028–1034

Caruso C, Rizzo C, Mangano S, Poli A, Di Donato P, Finore I, Nicolaus B, Di Marco G, Michaud L, Lo Giudice A (2018) Production and biotechnological potential of extracellular polymeric substances from sponge-associated Antarctic bacteria. Appl Environ Microbiol 84:e01624–e01617

Chalasani KB, Russel-Jones GJ, Jain AK, Diwan PV, Jain SK (2007) Effective oral delivery of insulin in animal models using vitamin B12-coated dextran nanoparticles. J Control Rel 122:141–150

Chen B, You W, Huang J, Yu Y, Chen W (2010) Isolation and antioxidant property of the extracellular polysaccharide from *Rhodella reticulata*. World J Microbiol Biotechnol 26:833–840

Chen Y, Mao W, Tao H, Zhu W, Qi X, Chen Y, Li H, Zhao C, Yang Y, Hou Y, Wang C (2011) Structural characterization and antioxidant properties of an exopolysaccharide produced by the mangrove endophytic fungus *Aspergillus* sp. Y16. Bioresour Technol 102(17):8179–8184

Chen YT, Yuan Q, Shan LT, Lin MA, Cheng DQ, Li CY (2013) Antitumor activity of bacterial exopolysaccharides from the endophyte *Bacillus amyloliquefaciens* isolated from *Ophiopogon japonicus*. Oncol Lett 5:1787–1792

Chen F, Cheng-Gang R, Tong Z, Yu-Jia W, Chuan-Chao D (2016) A novel exopolysaccharide elicitor from endophytic fungus *Gilmaniella* sp. AL12 on volatile oils accumulation in *Atractylodes lancea*. Sci Rep 6:34735

Chen M, Li Y, Liu Z, Qu Y, Zhang H, Li D, Zhou J, Xie S, Liu M (2018) Exopolysaccharides from a *Codonopsis pilosula* endophyte activate macrophages and inhibit cancer cell proliferation and migration. Thorac Cancer 9(5):630–639

Chikkanna A, Ghosh D, Kishore A (2018) Expression and characterization of a potential exopolysaccharide from a newly isolated halophilic thermotolerant bacteria *Halomonas nitroreducens* strain WB1. Peer J 6:e4684

Colliec JS, Chevolot L, Helley D, Ratiskol J, Bros A, Sinquin C, Roger O, Fischer AM (2001) Characterization, chemical modifications and *in vitro* anticoagulant properties of an exopolysaccharide produced by *Alteromonas infernus*. Biochim Biophys Acta (BBA): General Subjects 1528:141–151

Costa CR, Menolli RA, Osaku EF, Tramontina R, de Melo RH, do Amaral AE, Duarte PA, de Carvalho MM, Smiderle FR, Silva JL, Mello RG (2019) Exopolysaccharides from *Aspergillus terreus*: Production, chemical elucidation and immunoactivity. Int J Biol Macromol 139:654–664

Cunha MA, Santos VA, Calegari GC, Luna WN, Marin SL, Dekker RF, Barbosa-Dekker AM (2019) Structure and biological properties of lasiodiplodan: an uncommon fungal exopolysaccharide of the (1→ 6)-β-D-glucan type. In: Extracellular sugar-based biopolymers matrices. Springer, Cham, pp 409–432

Delattre C, Pierre G, Gardarin C, Traiki M, Elboutachfaiti R, Isogai A (2015) Antioxidant activities of a poly glucuronic acid sodium salt obtained from TEMPO-mediated oxidation of xanthan. Carbohydr Polym 116:34–41

Dvir I, Stark AH, Chayoth R, Madar Z, Arad SM (2009) Hypocholesterolemic effects of nutraceuticals produced from the red microalga *Porphyridium* sp. in rats. Nutrients 1:156–167

El-Newary SA, Ibrahim AY, Asker MS, Mahmoud MG, El Awady ME (2017) Production, characterization and biological activities of acidic exopolysaccharide from marine *Bacillus amyloliquefaciens* 3MS 2017. Asian Pac J Trop Med 10(7):652–662

Fooladi T, Soudi MR, Alimadadi N, Savedoroudi P, Heravi MM (2019) Bioactive exopolysaccharide from *Neopestalotiopsis* sp. strain SKE15: Production, characterization and optimization. Int J Biol Macromol 129:127–139

Freire-Nordi CS, Vieira AA, Nascimento OR (2005) The metal binding capacity of Anabaena spiroides extracellular polysaccharide: an EPR study. Process Biochem 40(6):2215–2224

Freitas F, Alves VD, Pais J, Costa N, Oliveira C, Mafra L, Hilliou L, Oliveira R, Reis MA (2009) Characterization of an extracellular polysaccharide produced by a Pseudomonas strain grown on glycerol. Bioresour Technol 100(2):859–865

Freitas F, Alves VD, Torres CA, Cruz M, Sousa I, Melo MJ, Ramos AM, Reis MA (2011) Fucose-containing exopolysaccharide produced by the newly isolated Enterobacter strain A47 DSM 23139. Carbohydr Polym 83(1):159–165

Fu L, Zhang J, Yang G (2013) Present status and applications of bacterial cellulose-based materials for skin tissue repair. Carbohydr Polym 92:1432–1442

Gugliandolo C, Spanò A, Lentini V, Arena A, Maugeri TL (2013) Antiviral and immunomodulatory effects of a novel bacterial exopolysaccharide of shallow marine vent origin. J Appl Microbiol 116(4):1028–1034

Gugliandolo C, Spanò A, Maugeri T, Poli A, Arena A, Nicolaus B (2015) Role of bacterial exopolysaccharides as agents in counteracting immune disorders induced by. Herpes Virus. Microorganisms 3(3):464

Guo S, Mao W, Li Y, Gu Q, Chen Y, Zhao C, Li N, Wang C, Guo T, Liu X (2013) Preparation, structural characterization and antioxidant activity of an extracellular polysaccharide produced by the fungus *Oidiodendron truncatum* GW. Process Biochem 48(3):539–544

Gupta M, Gupta AK (2004) Hydrogel pullulan nanoparticles encapsulating pBUDLac Z plasmid as an efficient gene delivery carrier. J Control Rel 99:157–166

Hagerstrom H (2003) Polymer gels as pharmaceutical dosage forms, Ph.D. thesis, ActaUniversitatisUpsaliensis, Uppsala

Han J, Cai J, Borjihan W, Ganbold T, Rana TM, Baigude H (2015) Preparation of novel curdlan nanoparticles for intracellular siRNA delivery. Carbohydr Polym 117:324–330

Hassan SW, Ibrahim HA (2017) Production, characterization and valuable applications of exopolysaccharides from marine *Bacillus subtilis* SH1. Polish J Microbiol 66(4):449–461

Hu X, Pang X, Wang PG, Chen M (2019) Isolation and characterization of an antioxidant exopolysaccharide produced by *Bacillus* sp. S-1 from Sichuan Pickles. Carbohydr Polym 204:9–16

Huang M, Wang F, Zhou X, Yang H, Wang Y (2015) Hypoglycemic and hypolipidemic properties of polysaccharides from Enterobacter cloacae Z0206 in KKAy mice. Carbohydr Polym 117:91–98

Ivanova E, Bazaka K, Crawford RJ (2014) Natural polymer biomaterials: advanced applications. In: Dodds F (ed) New functional biomaterials for medicine and healthcare. Woodhead Publishing Series in Biomaterials, 67, Woodhead, Cambridge, pp 32–70

Jiang P, Li J, Han F, Duan G, Lu X, Gu Y, Yu W (2011) Antibiofilm activity of an exopolysaccharide from marine bacterium *Vibrio* sp. QY101. PLoS One 6(4):e18514

Jiao P, Tseng-Crank J, Corneliusen B, Yimam M, Hodges M, Hong M, Maurseth C, Oh M, Kim H, Chu M, Jia Q (2014) Lipase inhibition and antiobesity effect of *Atractylodes lancea*. Planta Med 80(07):577–582

Kambourova M, Mandeva R, Dimova D, Poli A, Nicolaus B, Tommonaro G (2009) Production and characterization of a microbial glucan, synthesized by *Geobacillus tepidamans* V264 isolated from Bulgarian hot spring. Carbohydr Polym 77(2):338–343

Khan T, Park JK, Kwon JH (2007) Functional biopolymers produced by biochemical technology considering applications in food engineering. Korean J Chem Eng 24(5):816–826

Kumar AS, Mody KH (2009) Microbial exopolysaccharides: variety and potential applications. In: Rehm BHA (ed) Microbial production of biopolymers and polymer precursors: applications and perspectives. Caister Academic, Norfolk, pp 229–254

Kumar AS, Mody KH, Jha B (2007) Bacterial exopolysaccharides–a perception. J Basic Microbiol 47:103–117

Kumar D, Saini N, Pandit V, Ali S (2012) An insight to pullulan: a biopolymer in pharmaceutical approaches. Int J Basic Appl Sci 1:202–219

Kumar CG, Mongolla P, Pombala S (2018) Lasiosan, a new exopolysaccharide from *Lasiodiplodia* sp. strain B2 (MTCC 6000): structural characterization and biological evaluation. Process Biochem 72:162–169

Leathers TD, Nunnally MS, Ahlgren JA, Côté GL (2003) Characterization of a novel modified alternan. Carbohydr Polym 54(1):107–113

Li P, Luo C, Sun W, Lu S, Mou Y, Peng Y, Zhou L (2011) In vitro antioxidant activities of polysaccharides from endophytic fungus *Fusarium oxysporum* Dzf17. Afr J Microbiol Res 5 (32):5994–5997

Li P, Sun W, Luo C, Shan T, Mou Y, Lu S et al (2012) In vitro evaluation of antioxidant activities of polysaccharides from the endophytic fungus *Berkleasmium* sp. Dzf12. Afr J Microbiol Res 6:471–477

Li Y, Li Q, Hao D, Jiang D, Luo Y, Liu Y, Zhao Z (2015) Production, purification, and antibiofilm activity of a novel exopolysaccharide from *Arthrobacter* sp. B4. Prep Biochem Biotechnol 45 (2):192–204

Li H, Jiao X, Sun Y, Sun S, Feng Z, Zhou W, Zhu H (2016) The preparation and characterization of a novel sphingan WL from marine *Sphingomonas* sp. WG. Sci Rep 6:37899

Li J, Wu G, Qin C, Chen W, Chen G, Lu W (2019) Structure characterization and otoprotective effects of a new endophytic exopolysaccharide from Saffron. Molecules 24(4):749

Lin M-H, Yang Y-L, Chen Y-P, Hua K-F, Lu C-P, Sheu F, Lin G-H, Tsay S-S, Liang S-M, Wu S-H (2011) A novel exopolysaccharide from the biofilm of *Thermus aquaticus* YT-1 induces the immune response through toll-like receptor 2. J Biol Chem 286(20):17736–17745

Lin SM, Baek CY, Jung JH, Kim WS, Song HY, Lee JH, Ji HJ, Zhi Y, Kang BS, Bahn YS, Seo HS (2020) Antioxidant activities of an exopolysaccharide (deinopol) produced by the extreme radiation-resistant bacterium *Deinococcus radiodurans*. Sci Rep 10(1):1–1

Liu J, Luo J, Ye H, Sun Y, Lu Z, Zeng X (2009) Production, characterization and antioxidant activities in vitro of exopolysaccharides from endophytic bacterium *Paenibacillus polymyxa* EJS-3. Carbohydr Polym 78(2):275–281

Liu J, Luo JG, Ye H, Sun Y, Zeng XX (2012) Preparation, antioxidant and antitumor activities in-vitro of different derivatives of levan from endophytic bacterium *Paenibacillus polymyxa* EJS-3. Food Chem Toxicol 50:767–772

Llamas I, Béjar V, Martínez-Checa F, Martínez-Cánovas MJ, Molina I, Quesada E (2011) *Halomonas stenophila* sp. nov., a halophilic bacterium that produces sulphate exopolysaccharides with biological activity. Int J Syst Evol Microbiol 61(10):2508–2514

Luang-In V, Saengha W, Deeseenthum S (2018) Characterization and bioactivities of a novel exopolysaccharide produced from lactose by *Bacillus tequilensis* PS21 isolated from Thai Milk Kefir. Microbiol Biotechnol Lett 46(1):9–17

Mahapatra S, Banerjee D (2012) Structural elucidation and bioactivity of a novel exopolysaccharide from endophytic *Fusarium solani* SD5. Carbohydr Polym 90:683–689

Mahapatra S, Banerjee D (2013a) Evaluation of in vitro antioxidant potency of exopolysaccharide from endophytic *Fusarium solani* SD5. Int J Biol Macromol 53:62–66

Mahapatra S, Banerjee D (2013b) Optimization of a bioactive exopolysaccharide production from endophytic *Fusarium solani* SD5. Carbohydr Polym 97:627–634

Mahapatra S, Banerjee D (2016) Production and structural elucidation of exopolysaccharide from endophytic *Pestalotiopsis* sp. BC55. Int J Biol Macromol 82:182–191

Mahgoub AM, Mahmoud MG, Selim MS, Awady MEE (2018) Exopolysaccharide from marine *Bacillus velezensis* MHM3 induces apoptosis of human breast cancer MCF-7 cells through a mitochondrial pathway. Asian Pacific J Cancer Prev 19(7):1957

Manca MC, Lama L, Improta R, Esposito E, Gambacorta A, Nicolaus B (1996) Chemical composition of two exopolysaccharides from *Bacillus thermoantarcticus*. Appl Environ Microbiol 62(9):3265–3269

Matricardi P, DiMeo C, Coviello T, Hennink WE, Alhaique F (2013) Interpenetrating polymer networks polysaccharide hydrogels for drug delivery and tissue engineering. Adv Drug Deliv Rev 65:1172–1187

Matsuda M, Yamori T, Naitoh M, Okutani K (2003) Structural revision of sulfated polysaccharide B-1 isolated from a marine pseudomonas species and its cytotoxic activity against human cancer cell lines. Mar Biotechnol 5:13–19

Matsumura K, Nakajima N, Sugai H, Hyon SH (2014) Self-degradation of tissue adhesive based on oxidized dextran and poly-L-lysine. Carbohydr Polym 113:32–38

Misaki A (2004) Elsinan, an extracellular alpha 1, 3: 1, 4 glucan produced by *Elsinoe leucospila*: production, structure, properties and potential food utilization. Foods Food Ingredients J Jpn 209:4

Mocanu G, Mihai D, Dulong V, Picton L, LeCerf D, Moscovici M (2011) Anionic polysaccharide hydrogels with thermosensitive properties. Carbohydr Polym 83:52–59

Mocanu G, Souguir Z, Picton L, LeCerf D (2012) Multi-responsive carboxymethyl polysaccharide cross-linked hydrogels containing Jeffamine side-chains. Carbohydr Polym 89:578–585

Mohamed AA, Hassan AH, Ibrahim I, Sahar WM, Hassan UM, Abdul-Raouf U (2018a) Antimicrobial and anti-tumor activities of exopolysaccharides produced by the biofilm of marine *Halomonas saccharevitans* AB2 isolated from Suez Gulf, Egypt. Egypt J Aquat Biol Fish 22 (5):99–119

Mohamed SS, Amer SK, Selim MS, Rifaat HM (2018b) Characterization and applications of exopolysaccharide produced by marine *Bacillus altitudinis* MSH2014 from Ras Mohamed, Sinai, Egypt. Egypt J Basic Appl Sci 5(3):204–209

Moriello VS, Lama L, Poli A, Gugliandolo C, Maugeri TL, Gambacorta A, Nicolaus B (2003) Production of exopolysaccharides from a thermophilic microorganism isolated from a marine hot spring in flegrean areas. J Ind Microbiol Biotechnol 30(2):95–101

Moscovici M (2015) Present and future medical applications of microbial exopolysaccharides. Front Microbiol 6:1012

Mukherjee A, Das R, Sharma A, Pal A, Paul AK (2017) Production and partial characterization of extracellular polysaccharide from endophytic *Bacillus cereus* RCR 08. Microbiol Commun 10:612–622

Naeye B, Raemdonck K, Remaut K, Sproat B, Demeester J, DeSmedt SC (2010) PEGylation of biodegradable dextran nano gels for siRNA delivery. Eur J Pharm Sci 40:342–351

Nikolova B, Semkova S, Tsoneva I, Antov G, Ivanova J, Vasileva I, Kardaleva P, Stoineva I, Christova N, Nacheva L, Kabaivanova L (2019) Characterization and potential antitumor effect of a heteropolysaccharide produced by the red alga Porphyridium sordidum. Eng Life Sci 19 (12):978–985

Nwodo UU, Green E, Okoh AI (2012) Bacterial exopolysaccharides: functionality and prospects. Int J Mol Sci 13(11):14002–14015

Orlandelli RC, Vasconcelos AFD, Azevedo JL, da Silva MDLC, Pamphile JA (2016) Screening of endophytic sources of exopolysaccharides: Preliminary characterization of crude exopolysaccharide produced by submerged culture of *Diaporthe* sp. JF766998 under different cultivation time. Biochim Open 2:33–40

Ouyang Z, Zhang L, Zhao M, Wang P, Wei Y, Fang J (2012) Identification and quantification of sesquiterpenes and polyacetylenes in *Atractylodes lancea* from various geographical origins using GC-MS analysis. Rev Bras Farm 22(5):957–963

Pan F, Hou K, Li DD, Su TJ, Wu W (2019) Exopolysaccharides from the fungal endophytic *Fusarium* sp. A14 isolated from *Fritillaria unibracteata* Hsiao et KC Hsia and their antioxidant and antiproliferation effects. J Biosci Bioeng 127(2):231–240

Panosyan H, Di Donato P, Poli A, Nicolaus B (2018) Production and characterization of exopolysaccharides by *Geobacillus thermodenitrificans* ArzA-6 and *Geobacillus toebii* ArzA-8 strains isolated from an Armenian geothermal spring. Extremophiles 22(5):725–737

Panthavee W, Noda M, Danshiitsoodol N, Kumagai T, Sugiyama M (2017) Characterization of exopolysaccharides produced by thermophilic lactic acid bacteria isolated from tropical fruits of Thailand. Biol Pharm Bull 40(5):621–629

Pei H, Lin YZ, Rao LY, Wang YX, Niu LN, Wang HM, Yang W (2012) Antiviral activities of exopolysaccharide from one fungi strain isolated from mangrove in Hainan province in vitro. J Hainan Med Univ 18(2):152–154

Peng Q, Li M, Xue F, Liu H (2014) Structure and immunobiological activity of a new polysaccharide from *Bletilla striata*. Carbohydr Polym 107:119–123

Prathyusha AM, Sheela GM, Bramhachari PV (2018) Chemical characterization and antioxidant properties of exopolysaccharides from mangrove filamentous fungi *Fusarium equiseti* ANP2. Biotechnol Rep 19:00277

Prezotti FG, Cury BSF, Evangelista RC (2014) Mucoadhesive beads of gellan gum/pectin intended to controlled delivery of drugs. Carbohydr Polym 113:286–295

Radchenkova N, Vassilev S, Panchev I, Anzelmo G, Tomova I, Nicolaus B, Kambourova M (2013) Production and properties of two novel exopolysaccharides synthesized by a thermophilic bacterium *Aeribacillus pallidus* 418. Appl Biochem Biotechnol 171(1):31–43

Radchenkova N, Vassilev S, Martinov M, Kuncheva M, Panchev I, Vlaev S, Kambourova M (2014) Optimization of the aeration and agitation speed of *Aeribacillus palidus* 418 exopolysaccharide production and the emulsifying properties of the product. Process Biochem 49(4):576–582

Radchenkova N, Panchev I, Vassilev S, Kuncheva M, Dobreva S, Kambourova M (2015) Continuous cultivation of a thermophilic bacterium *Aeribacillus pallidus* 418 for production of an exopolysaccharide applicable in cosmetic creams. J Appl Microbiol 119(5):1301–1309

Radchenkova N, Boyadzhieva I, Atanasova N, Poli A, Finore I, Di Donato P, Kambourova M (2018) Extracellular polymer substance synthesized by a halophilic bacterium *Chromohalobacter canadensis* 28. Appl Microbiol Biotechnol 102(11):4937–4949

Raemdonck K, Naeye B, Hogset A, Demeester J, DeSmedt SC (2010) Prolonged gene silencing by combining siRNA nanogels and photochemical internalization. J Control Release 145:281–288

Raguénès G, Christen R, Guezennec J, Pignet P, Barbier G (1997a) *Vibrio diabolicus* sp. nov., a new polysaccharide-secreting organism isolated from a Deep-Sea hydrothermal vent polychaete annelid, *Alvinella pompejana*. Int J Syst Evol Microbiol 47(4):989–995

Raguénès GHC, Peres A, Ruimy R, Pignet P, Christen R, Loaec M, Rougeaux H, Barbier G, Guezennec JG (1997b) *Alteromonas infernus* sp. nov., a new polysaccharide-producing bacterium isolated from a deep-sea hydrothermal vent. J Appl Microbiol 82(4):422–430

Rairakhwada D, Pal AK, Bhathena ZP, Sahu NP, Jha A, Mukherjee SC (2007) Dietary microbial levan enhances cellular non-specific immunity and survival of common carp (*Cyprinus carpio*) juveniles. Fish Shellfish Immunol 22(5):477–486

Raveendran S, Poulose AC, Yoshida Y, Maekawa T, Kumar DS (2013) Bacterial exopolysaccharide based nanoparticles for sustained drug delivery, cancer chemotherapy and bioimaging. Carbohydr Polym 91(1):22–32

Remminghorst U, Rehm BHA (2009) Microbial production of alginate: biosynthesis and applications. In: Rehm BHA (ed) Microbial production of biopolymers and polymer precursors: applications and perspectives. Caister Academic, Norfolk, pp 13–42

Rokstad AM, Lacik I, de Vos P, Strand BL (2014) Advances in biocompatibility and physico-chemical characterization of microspheres for cell encapsulation. Adv Drug Deliv Rev 6:113–130

Ruiz-Ruiz C, Srivastava GK, Carranza D, Mata JA, Llamas I, Santamaría M, Quesada E, Molina IJ (2011) An exopolysaccharide produced by the novel halophilic bacterium *Halomonas stenophila* strain B100 selectively induces apoptosis in human T leukaemia cells. Appl Microbiol Biotechnol 89(2):345–355

Saldias C, Velasquez L, Quezada C, Leiva A (2015) Physicochemical assessment of dextran-g-poly (ε-caprolactone) micellear nanoaggregates as drug nanocarriers. Carbohydr Polym 117:458–467

Santra HK, Banerjee D (2020) Natural products as fungicide and their role in crop protection. In: Natural bioactive products in sustainable agriculture. Springer, Singapore, pp 131–219

Sarilmiser HK, Oner ET (2014) Investigation of anti-cancer activity of linear and aldehyde activated levan from *Halomonas smyrnensis* AAD6T. Biochem Eng J 92:28–34

Sharma S, Mukkur TKS, Benson HAE, Chen Y (2009) Pharmaceutical aspects of intranasal delivery of vaccines using particulate systems. J Pharm Sci 98:812–843

Silva KA, Juenet M, Meddahi-Pellé A, Letourneur D (2015) Polysaccharide-based strategies for heart tissue engineering. Carbohydr Polym 116:267–277

Squillaci G, Finamore R, Diana P, Restaino OF, Schiraldi C, Arbucci S, Ionata E, La Cara F, Morana A (2016) Production and properties of an exopolysaccharide synthesized by the extreme halophilic archaeon *Haloterrigena turkmenica*. Appl Microbiol Biotechnol 100(2):613–623

Srikanth R, Reddy CHHSS, Siddartha G, Ramaiah MJ, Uppuluri KB (2015) Review on production, characterization and applications of microbial levan. Carbohydr Polym 120:102–114

Sun L, Wang L, Zhou Y (2012) Immunomodulation and antitumor activities of different molecular weight polysaccharides from *Porphyridium cruentum*. Carbohydr Polym 87:1206–1210

Sun ML, Zhao F, Shi M, Zhang XY, Zhou BC, Zhang YZ, Chen XL (2015) Characterization and biotechnological potential analysis of a new exopolysaccharide from the Arctic marine bacterium *Polaribacter* sp. SM1127. Sci Rep 5:18435

Sun K, Chen Y, Niu Q, Zhu W, Wang B, Li P, Ge X (2016) An exopolysaccharide isolated from a coral-associated fungus and its sulfated derivative activates macrophages. Int J Biol Macromol 82:387–394

Survase SA, Saudagar PS, Bajaj IB, Singhal RS (2007) Scleroglucan: fermentative production, downstream processing and applications. Food Technol Biotechnol 45(2):107–118

Sutherland IW (1998) Novel and established applications of microbial polysaccharides. Trends Biotechnol 16:41–46

Sutherland IW (2002) A sticky business. Microbial polysaccharides: current products and future trends. Microbiol Today 29:70–71

Tassa C, Shaw SY, Weissleder R (2011) Dextran-coated iron oxide nanoparticles: a versatile platform for targeted molecular imaging, molecular diagnostic and therapy. Acc Chem Res 44:842–852

Thankam FG, Muthu J (2014) Alginate based hybrid copolymer hydrogels- influence of pore-morphology on cell-material interaction. Carbohydr Polym 112:235–244

Umezawa H, Okami Y, Kurasawa S, Ohnuki T, Ishizuka M, Takeuchi T, Shiio T, Yugari Y (1983) Marinactan, antitumor polysaccharide produced by marine bacteria. J Antibiot 36(5):471–475

Vandamme EJ, Soetaert W (1995) Biotechnical modification of carbohydrates. FEMS Microbiol Rev 16:163–186

Vu B, Chen M, Crawford RJ, Ivanova EP (2009) Bacterial extracellular polysaccharides involved in biofilm formation. Molecules 14:2535–2554

Vuyst DL, Degeest B (1999) Heteropolysaccharides from lactic acid bacteria. FEMS Microbiol Rev 23(2):153–177

Wang L, Zhang H, Yang L, Liang X, Zhang F, Linhardt RJ (2017) Structural characterization and bioactivity of exopolysaccharide synthesized by *Geobacillus* sp. TS3-9 isolated from radioactive radon hot spring. Adv Biotechnol Microbiol 4(2):1–8

Wang Y, Li Y, Li S, Li Q, Fan W, Kiatoukosin L, Chen J (2019) Extracellular polysaccharides of endophytic fungus *Alternaria tenuissima* F1 from *Angelica sinensis*: Production conditions, purification, and antioxidant properties. Int J Biol Macromol 133:172–183

Wang N, Jia G, Wang C, Chen M, Xie F, Nepovinnykh NV, Goff HD, Guo Q (2020) Structural characterisation and immunomodulatory activity of exopolysaccharides from liquid fermentation of *Monascus purpureus* (Hong Qu). Food Hydrocoll 103:105636

Widner B, Behr R, Von-Dollen S, Tang M, Heu T, Sloma A, Sternberg D, DeAngelis PL, Weigel PH, Brown S (2005) Hyaluronic acid production in *Bacillus subtilis*. Appl Environ Microbiol 71:3747–3752

Xiong YW, Ju XY, Li XW, Gong Y, Xu MJ, Zhang CM, Yuan B, Lv ZP, Qin S (2020) Fermentation conditions optimization, purification, and antioxidant activity of exopolysaccharides obtained from the plant growth-promoting endophytic actinobacterium *Glutamicibacter halophytocola* KLBMP 5180. Int J Biol Macromol 153:1176

Yan MX, Mao WJ, Liu X, Wang SY, Xia Z, Cao SJ, Li J, Qin L, Xian HL (2016) Extracellular polysaccharide with novel structure and antioxidant property produced by the deep-sea fungus *Aspergillus versicolor* N2bc. Carbohydr Polym 147:272–281

Yang B, Xiao B, Sun T (2013) Antitumor and immunomodulatory activity of Astragalus membranaceus polysaccharides in H22 tumor-bearing mice. Int J Biol Macromol 62:287–290

Yang J, Li X, Xue Y, Wang N, Liu W (2014) Anti-hepatoma activity and mechanism of corn silk polysaccharides in H22 tumor-bearing mice. Int J Biol Macromol 64:276–280

Ye S, Liu F, Wang J, Wang H, Zhang M (2012) Antioxidant activities of an exopolysaccharide isolated and purified from marine Pseudomonas PF-6. Carbohydr Polym 87(1):764–770

Ye S, Zhang J, Liu Z, Zhang Y, Li J, Li YO (2016) Biosynthesis of selenium rich exopolysaccharide (Se-EPS) by Pseudomonas PT-8 and characterization of its antioxidant activities. Carbohydr Polym 142:230–239

Yildiz YS, Anzelmo G, Ozer T, Radchenkova N, Genc S, Di Donato P, Nicolaus B, Toksoy Oner E, Kambourova M (2014) *Brevibacillus themoruber*: a promising microbial cell factory for exopolysaccharide production. J Appl Microbiol 116(2):314–324

Yim JH, Kim SJ, Ahn SH, Lee CK, Rhie KT, Lee HK (2004) Antiviral effects of sulfated exopolysaccharide from the marine microalga *Gyrodinium impudicum* strain KG03. Mar Biotechnol 6(1):17–25

Yu YJ, Chen Z, Chen PT, Ng IS (2018) Production, characterization and antibacterial activity of exopolysaccharide from a newly isolated *Weissella cibaria* under sucrose effect. J Biosci Bioeng 126(6):769–777

Yu M, Chen M, Gui J, Huang S, Liu Y, Shentu H, He J, Fang Z, Wang W, Zhang Y (2019) Preparation of *Chlorella vulgaris* polysaccharides and their antioxidant activity in vitro and in vivo. Int J Biol Macromol 137:139–150

Yuan W, Hu ZH, Su J, Wu F, Liu ZH, Jin T (2012) Preparation and characterization of recombinant human growth hormone-Zn2+−dextran nanoparticles using aqueous phase –aqueous phase emulsion. Nanomedicine 8:424–427

Yuan J, Zhou J, Li X, Dai C (2016) The Primary mechanism of endophytic fungus *Gilmaniella* sp. AL12 promotion of plant growth and sesquiterpenoid accumulation in *Atractylodes lancea*. Plant Cell Tissue Organ Cult 125:571–584

Zanchetta P, Lagarde N, Guezennec J (2003) A new bone-healing material: a hyaluronic acid-like bacterial exopolysaccharide. Calcif Tissue Int 72(1):74–79

Zeng YJ, Yang HR, Wang HF, Zong MH, Lou WY (2019) Immune enhancement activity of a novel polysaccharide produced by *Dendrobium officinale* endophytic fungus *Fusarium solani* DO7. J Funct Foods 53:266–275

Zhang J, Liu L, Chen F (2019) Production and characterization of exopolysaccharides from *Chlorella zofingiensis* and *Chlorella vulgaris* with anti-colorectal cancer activity. Int J Biol Macromol 134:976–983

Zhang J, Yang B, Chen H (2020) Identification of an endophytic fungus *Pilidiella guizhouensis* isolated from *Eupatorium chinense* L. and its extracellular polysaccharide. Biologia 10:1–9

Zhao D, Jiang J, Du R, Guo S, Ping W, Ling H, Ge J (2019) Purification and characterization of an exopolysaccharide from *Leuconostoc lactis* L2. Int J Biol Macromol 139:1224–1231

Zheng LP, Zou T, Ma YJ, Wang JW, Zhang YQ (2016) Antioxidant and DNA damage protecting activity of exopolysaccharides from the endophytic bacterium *Bacillus cereus* SZ1. Molecules 21(2):174

Zheng YJ, Yang HR, Ou XY, Su HH, Zong MH, Yang JG, Lou WY (2019) Fungal polysaccharide similar with host *Dendrobium officinale* polysaccharide: preparation, structure characteristics and biological activities. Int J Biol Macromol 141:460–470

Zong A, Cao H, Wang F (2012) Anticancer polysaccharides from natural resources: are view of recent research. Carbohydr Polym 90:1395–1410

Microbial Biopolymers: Pharmaceutical, Medical, and Biotechnological Applications

18

Rohit Godbole, Asha Goutam, and Aniket Mali

Abstract

Biopolymers are a heterogeneous, acquiescent group of materials that can be synthesized or obtained from several biological compounds or their building blocks, such as sugars, amino acids, lipids, and nucleic acids, sourced from living organisms such as animals, plants, and microbes. Thus, they are broadly divided as carbohydrates or polysaccharides (e.g., glycogen, starch, and cellulose), proteins or polypeptides, nucleic acids or polynucleotides (e.g., DNA, RNA), lipids, and synthetically derived materials from resins, vegetable oils, and fats. Their unique properties like the possession of diverse functional groups and the ability to adopt precise 3D shapes and structures empower biopolymers to function as bioactive biomolecules capable of playing several biological roles. Biopolymers present some advantages such as a well-defined and more complex structure responsible for their biological activities, biodegradability, inexpensiveness, and renewability along with some distinctive features of smart biopolymers such as sensitivity to light intensity, pH, temperature, humidity, and electromagnetic fields. Microorganisms play a significant role in producing a great variety of biopolymers, ranging from viscous solutions to plastics, such as polysaccharides, polyesters, polyamides, and polyphosphates. The microbial biopolymers can be effectively produced from pure cultures, selective laboratory mutants, or genetically modified microorganisms. They are acquiring a significant place in a wide range of biomedical applications for use as medical, pharmaceutical and

R. Godbole
Sailife Sciences, Pune, Maharashtra, India

A. Goutam
iMED Global Pvt. Ltd, Bangalore, Karnataka, India

A. Mali (✉)
National Centre for Cell Science, Pune, Maharashtra, India

A. Vaishnav, D. K. Choudhary (eds.), *Microbial Polymers*,
https://doi.org/10.1007/978-981-16-0045-6_18

biotechnological materials, cosmetics, food additives, biosensors, absorbents, packaging materials, along with regular industrial applications such as industrial plastics, clothing fabrics, water treatment chemicals, data storage elements, and many more. In this chapter, we specifically reviewed the leading microbial biopolymers and their wide range of specific applications in the field of Medicine, Pharmacy, and Biotechnology.

Keywords

Biopolymers · Microbial biopolymers · Biomedical applications · Pharmaceutical applications · Biotechnological applications · Medical applications

18.1 Introduction to Biopolymers

Due to the wide range of applications for their specific properties and scientific features, polymers have gained an important place in different areas of science and day-to-day life. Polymers are the macromolecules that typically comprised the same or different monomeric subunits, combined in different ways, and are of natural or synthetic origin. Although we are more dependant on synthetic polymers nowadays, there are plenty of health and environmental issues associated with their use (Qureshi et al. 2020). Thus, the quest for biomaterials that can be used as natural polymers is on the rise not only to substitute the polymers of synthetic origin for different purposes but also to lessen the nonrenewable resource's consumption and to curtail the environmental issues associated with them (Mohan et al. 2016). Biopolymers are polymeric biomolecules that are produced by living organisms such as animals, plants, and microbes. In a more specific way, biopolymers can be defined as a heterogeneous, acquiescent group of materials that can be synthesized or obtained from several biological compounds or their building blocks such as sugars, amino acids, lipids, and nucleic acids (Verma et al. 2020). Based on their monomeric units, biopolymers are broadly divided as polysaccharides, polypeptides, and polynucleotides. Popular examples of the biopolymers are carbohydrates, polysaccharides (e.g., glycogen, starch, and cellulose), proteins, peptides, nucleic acids (e.g., DNA, RNA), and lipids. Materials that are produced with the use of synthetic chemistry approaches from available biological materials including resins, vegetable oils, and fats can also be described as biopolymers (Hernández et al. 2014). The repeating monomer units of the biopolymers are generally saccharides, amino acids, or aliphatic polyesters and hence they may possess diverse functional groups that are responsible for their different biological roles (Prameela et al. 2018; Qureshi et al. 2020). Compared to the simpler and more random structures of synthetic polymers, biopolymers adopt precise 3D shapes and structures due to their complex molecular assemblies. This unique and important property empowers biopolymers to function as bioactive biomolecules. The sustainability combined with biodegradability is the prime feature differentiating the biopolymers from polymers derived from fossil fuels (Mohan et al. 2016). In summary, compared to polymers, biopolymers present some advantages such as a well-defined and more complex structure responsible for their biological activities, biodegradability, inexpensiveness, and renewability. On top of this, smart biopolymers exhibit some unique and distinctive features such as sensitivity to light intensity, pH, temperature, humidity, and electromagnetic fields (Pattanashetti et al. 2017). Thus, they are acquiring a significant place in a wide range of industrial applications for use as medical, pharmaceutical and biotechnological materials, cosmetics, food additives, biosensors, absorbents, packaging materials, industrial plastics, clothing fabrics, water treatment chemicals, data storage elements, and many more (Rebelo et al. 2017).

Wool, leather, silk, and cellulose are the biological materials popularly known as traditional commodity polymers. However, there are plenty of uncommon or less used biopolymers owing to their limited or lack of specific required features. In the current era of modern science, particularly in biotechnology, modern biotechnological methods such as recombinant DNA technology, bacterial fermentation, advanced genetic engineering, and nanotechnology have the potential to modify or develop the uncommon and underutilized biopolymers, their derivatives, and blends for the desired new properties or features to optimize their uses for specific applications (Onar 2004). As mentioned earlier, biopolymers are generally developed from sources of natural origin, by synthesizing from different chemicals using biological materials, or by synthesizing from living organisms. Thus, as an umbrella term, Biopolymers cover polymers that fulfill the following criteria:

- *Biobased polymers*—In simple words, these are the polymers derived from the organic materials of living organisms (i.e., Biomass), in whole or in part.
- *Biodegradable polymers*—These are the biopolymers that can be degraded by living organisms or microorganisms like bacteria.
- *Biocompatible polymers*—These are harmless and highly compatible Polymers for use in tissues of humans or other animals.

Due to the feasibility of producing the biopolymers as synthetic, modified, or application-oriented forms, and copious availability of natural sources in the form of plants, animals, and microorganisms (specifically, bacteria and fungi), the bulk production of biopolymers is achievable at industrial scale. One of the novel approaches for biopolymer production is genetically engineered plants. However, a variety of biopolymers can also be produced from another important and leading source; microorganisms. Similarly, the production of animal protein-based biopolymers like gelatine, silk, wool, collagen, and polysaccharides (chitin, chitosan, glycosaminoglycans, and hyaluronic acid) is also an interesting option for the preparation of biopolymers (Mohan et al. 2016; Wankhade 2020). Figure 18.1 depicts the developmental history of a few important biopolymers.

During the development of the biopolymers, the characterization of biopolymers is also a crucial step in which the important properties of interests are scientifically analyzed. The main purposes of such characterization are to develop parameters for the processing and to determine the end-use characteristics of the biopolymers under

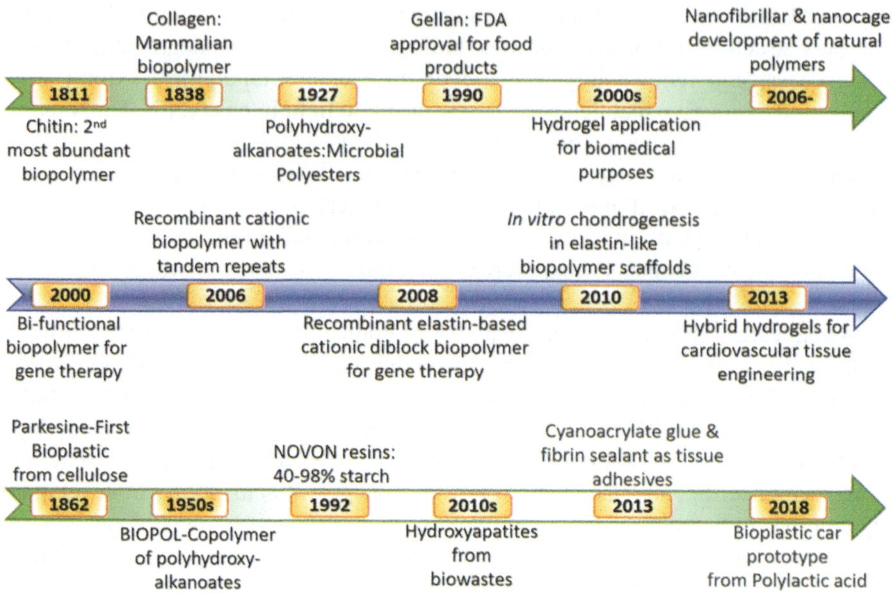

Fig. 18.1 Developmental timeline of some important biopolymers

development. Several techniques such as Fourier Transform Infrared Spectroscopy (FTIR), X-ray diffraction analysis, NMR spectroscopy, Thermal analysis, VIS-NIR spectroscopy, and Scanning Kelvin Probe Microscopy (SKPM) are generally used for the biopolymer characterization (Mohan et al. 2016).

18.2 Classification of Biopolymers

Based on different scales and parameters, biopolymers are classified in different ways (Fig. 18.2).

(a) On the basis of biodegradability these polymers are broadly classified as follows:
   - *Biodegradable Biopolymers*—These polymers offer alternatives to polymers derived from fuels and have been produced from saccharides, polysaccharides, or other organic components in different proportions, which are biodegradable. Bacteria present in the compost, soil, and marine sediment can easily degrade these biopolymers (Mohan et al. 2016). During in vivo use, biodegradable polymers break down and are eliminated from the body subsequently they serve their intended purpose. Due to their abilities to break down, excrete, or absorb, these are nowadays becoming the desired and obvious choice materials for biomedical applications, and due to their wide range of applications they are also referred to as "materials of today," e.g.,

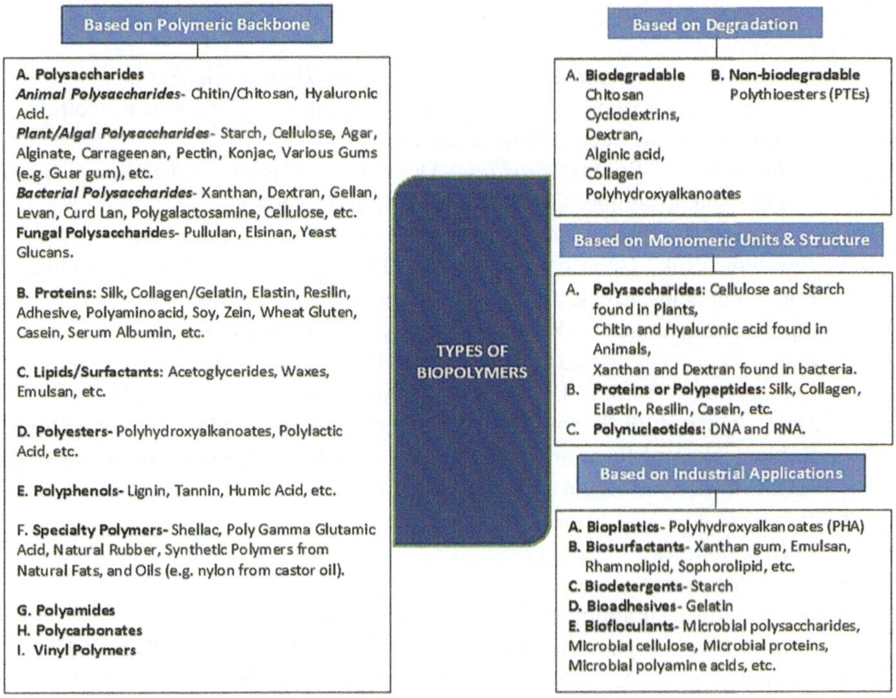

Fig. 18.2 Different ways of classifying the Biopolymers

Chitosan, Cyclodextrins, Dextran, Alginic acid, Collagen, Polyhydroxyalkanoates, etc. (Kunduru et al. 2016).
- *Nonbiodegradable Biopolymers*—These polymers do not break down or degrade after exposure to microorganisms. They unite the benefits of decreased carbon footprint during production with their tenacity towards biodegradation, which facilitates conservative applications requiring durable materials, e.g., Polythioesters (PTEs) (Andreeßen and Steinbüchel 2019).

(b) On the basis of the structure of biopolymer consisting of monomers as repeating units, they are mainly classified into three classes:
- *Polysaccharides*—These polymers are carbohydrate structures generally formed by linearly attached monosaccharides, e.g., Cellulose and Starch from plants, Chitin and Hyaluronic acid found in animals, Xanthan and Dextran found in bacteria, etc.
- *Proteins or Polypeptides*—These are polymers made up of short or long chains of amino acids, e.g., Silk, collagen, Elastin, Resilin, casein, etc.
- *Polynucleotides*—These are long polymers composed of 13 or more nucleotide monomers, e.g., DNA and RNA.

(c) On the basis of the backbone present in the polymer structure, biopolymers are also grouped as follows (Yadav et al. 2015; Mohan et al. 2016):

- *Polysaccharides*
  - *Animal Polysaccharides*—Chitin/Chitosan, Hyaluronic Acid.
  - *Plant/Algal Polysaccharides*—Starch, Alginate, Carrageenan, Cellulose, Konjac, Agar, Pectin, Guar gum, etc.
  - *Bacterial Polysaccharides*—Dextran, Gellan, Xanthan gum, Polygalactosamine, Cellulose, Curd Lan, Levan, etc.
  - *Fungal Polysaccharides*—Elsinan, Yeast Glucans, Pullulan.
- *Proteins*—Silk, Gelatin or Collagen, Elastin, Casein, Resilin, Adhesive, Polyaminoacid, Soy, Zein, Wheat Gluten, Serum Albumin, etc.
- *Lipids or Surfactants*—Waxes, Acetoglycerides, Emulsan, etc.
- *Polyesters*—Polyhydroxyalkanoates, Polylactic Acid, etc.
- *Polyphenols*—Tannin, Lignin, Humic Acid, etc.
- *Specialty or High-Performance Polymers*—Shellac, Poly-γ-Glutamic Acid, Rubber, Natural Fats, and Oils-derived polymers; synthetic in origin (e.g., castor oil-derived nylon).
- *Polyamides*
- *Polycarbonates*
- *Vinyl Polymers*

(d) Based on industrial applications, biopolymers can be classified as:

- *Bioplastics*—Polyhydroxyalkanoates (PHA) (Saharan and Ankita 2012).
- *Biosurfactants*—Xanthan gum, Emulsan, Rhamnolipid, Sophorolipid, etc. (Salek and Gutierrez 2016).
- *Biodetergents*—Starch (Mohan et al. 2016).
- *Bioadhesives*—Gelatin (Mohan et al. 2016).
- *Biofloculants*—Microbial polysaccharides, Microbial cellulose, Microbial proteins, Microbial polyamine acids, etc. (Rebah et al. 2018).

From the above classification, it is inferred that the majority of biopolymers are sourced from a wide-ranging group of microorganisms. In this chapter, we will discuss in detail the microbial biopolymers and their applications in the field of Pharmacy, Medical, and Biotechnology.

18.3 Microbial Production of Biopolymers

Microorganisms play a significant role in producing a great variety of biopolymers, ranging from viscous solutions to plastics, such as polysaccharides, polyesters, polyamides, and polyphosphates. Microbial Biopolymers are synthesized by living microorganisms using enzymes having the capability of processivity that connect various structural blocks like saccharides and amino acids or fatty acids possessing hydroxy functional groups, in order to produce molecules with higher molecular mass. The physical properties of these biopolymers are decided by their compositions and molecular masses. Microbes synthesize these polymers either as a response to various environmental factors, via their defense mechanism, or as a storage material. These polymers are effectively produced from pure cultures, selective laboratory mutants, or genetically modified microorganisms (Moradali and Rehm 2020; Verma et al. 2020). The biopolymers having promising applications in the fields of tissue engineering and drug delivery should be commercially valuable and custom-made, and their production can be easily carried out by genetically modified biopolymer-producing microorganisms. Nowadays the bacterial biopolymers have gained a great reputation in a wide range of biomedical applications (Liguori et al. 2016; Verma et al. 2020). For a large scale, microbial production of biopolymers, typical nutrients and precise environmental settings are required. Thus, Fermentation or Chemical Polymerization followed by Fermentation are the commonly used techniques or methods for the production of these polymers (Mohan et al. 2016). Microbial production of biopolymers has several advantages over other methods of biopolymer productions such as the viability of high heterologous expression, ease of growth to achieve high cell density, cost-effective substrate utilization, and selective production of the desired biopolymers (Verma et al. 2020). The microbial biopolymers obtained from a broad array of microorganisms can be classified into various categories based on their chemical structures; as described in Table 18.1.

The synthesis of these biopolymers has two important aspects: (a) Mechanism of synthesis by microorganisms and (b) Commercial Production.

(a) ***Mechanism of synthesis by microorganisms:*** The bacterial polymer biosynthesis is governed by signaling pathways. As a response to an external stimulus, various transcriptional factors activate promoters controlling the expression of functionally active genes resulting in the synthesis of biopolymers. However, bacteria also possess various enzymes that can be used for the synthesis of high molecular weight biopolymers. Polymers with a molecular weight of more than 100 kDa (e.g., Polysaccharides) cannot be synthesized by chemical synthesis (Moradali and Rehm 2020). An example of such an enzyme is Levansucrase, which acts on sucrose to produce Levan (Ananthalakshmy and Gunasekaran 1999).
(b) ***Commercial production:*** Commercial synthesis is required as many of these biopolymers have shown great potential in fields of Medicines, biotechnology, food sciences, and much more. The cost of production and yield become driving factors rather than the survival and thriving of a microbe itself. For example, exopolysachharides (EPS) are microbial biopolymers that are derived from the microbial synthesis of a variety of intracellular or extracellular polysaccharides. Specific genes code for enzymes required for biosynthesis, assembly, and transport of the completed polymers. The production is influenced by various physical and chemical factors such as temperature, fermentation period, media composition, and carbon source (Vijayendra and Shamala 2014). Alteration in these factors can have an impact on the yield and quality of the polymer produced. Biopolymers are widely produced through fermentation. Where necessary media with nutrients and carbon source is prepared followed by

Table 18.1 List of biopolymers with their chemical structure and microbial sources

Microbial biopolymers	Chemical structure	Microbial sources	Ref.
Bacterial biopolymers			
Polyhydroxy Alkonates (PHA), and Poly-β-hydroxybutyrate (PHB)	Polyesters that accumulate in the cytoplasm as energy reserves	Gram-positive and Gram-negative bacteria • *Pseudomonas putida* • *Ralstonia eutropha* • *Alcaligenes latus* • *Aeromonas hydrophila* • Archaea	U.S. Congress (2000); Favaro et al. (2019); Moradali and Rehm (2020)
Polylactic acid (PLA)	An optical isomer of L-lactic acid	• *Lactobacillus delbrueckii* • *Lactobacillus. leichmanii* • *Lactobacillus amylophilus* • *Lactobacillus rhamnosus* • *Lactobacillus bulgaricus*	Abdel-Fattah et al. (2012); Vijayendra and Shamala (2014)
Bacterial cellulose	A polysaccharide made up of a linear chain of β-(1 → 4)-linked glucose monomeric units	• *Escherichia coli* • *Salmonella enterica* • *Sarcina* species • *Agrobacterium* species • *Rhizobium* species • *Pseudomonas fluorescens* • *Komagataeibacter hansenii*	Tsouko et al. (2015); Moradali and Rehm (2020)
Xanthan Gum	Complex copolymer with five different sugars	• Xanthomonas campestris	U.S. Congress (2000); Song et al. (2019)
Gellan	Complex polysaccharides having a glucose-glucuronic acid-glucose-rhamnose sequence	• *Sphingomonas paucimobilis* (formerly known as *Pseudomonas elodea*)	U.S. Congress (2000)
Dextran	Water-soluble Polysaccharide ranging from 3 to 2000 kDa	• *Leuconostoc mesenteroides*	Moscovici (2015); Kunduru et al. (2016)

(continued)

Table 18.1 (continued)

Microbial biopolymers	Chemical structure	Microbial sources	Ref.
Levan	Neutral homopolysaccharide of fructose	• *Bacillus polymyxa PTCC1020* • *Bacillus subtilis* • *Aerobacter levanicum* • *Erwinia herbicola* • *Streptococcus salivarius* • *Zymomonas mobilis*	Vijayendra and Shamala (2014)
Pullulan	Linear polymer with three glucose units	• *Aureobasidium pullulans*	Akaraonye et al. (2010)
Alginate	(1,4)-linked β-D-mannuronate (M) and its C-5 epimer α-L-guluronate (G) present in the copolymer as the building block residues	• *Pseudomonas aeruginosa* • More *Pseudomonas* species • *Azotobacter* species	Szekalska et al. (2016); Wróblewska-Krepsztul et al. (2019)
Hyaluronate	Unbranched and negatively charged heteropolymer	• *Streptococcus* (Pa) belonging to Group A • *Bacillus cereus* G9241 (PT) • *Pasteurella multocida* (PT) • *Lactococcus lactis* • *Streptococcus equi* (NPT) • *Bacillus subtilis* • *Corynebacterium glutamicum* (NPT) • *Escherichia coli* (NPT)	Güngör et al. (2019); Moradali and Rehm (2020)
Poly-γ-Glutamic acid (γ-PGA)	Water-soluble, anionic, biodegradable amino acid homopolymers	• *Bacillus anthracis* (PT) • *Fusobacterium nucleatum* (PT) • *Streptococcus epidermidis* (PT) • *Bacillus licheniformis* • *Bacillus subtilis* • *Escherichia coli* (NPT)	Kunduru et al. (2016); Favaro et al. (2019)

(continued)

Table 18.1 (continued)

Microbial biopolymers	Chemical structure	Microbial sources	Ref.
Polyphosphate (PolyP)	Orthophosphate residue homopolymer	• *Citrobacter freundii*	Wang et al. (2018c)
Curdlan	–	• *Agrobacterium* species	Zhan et al. (2012)
Fungal biopolymers			
Chitin	A biopolymer made up of arbitrarily disseminated *N*-acetyl-D-glucosamine (*N*-GlcNAc) monomeric units	The fungal classes of • Basidiomycetes • Ascomycetes • Zygomycetes • Deuteromycetes	Elsoud and El Kady (2019)
Chitosan	A biopolymer composed of β-(1-4)-linked-D-glucosamine (GlcN) monomeric units arranged linearly	Several fungal classes: • *Basidiomycetes* • *Ascomycetes* • *Zygomycetes* • *Deuteromycetes*	Elsoud and El Kady (2019)
Fungal Polysaccharides (Gums): Intrapolysaccharides (IPS), Exopolysaccharides (EPS)	Biopolymers having long chains of polysaccharides possessing sugars or derivatives of sugars as monomeric units, in repeated or branched fashion	Produced by several fungal strains • *Agrocybe cylindracea* (EPS) • *Aspergillus terreus* (EPS) • *Lentinus giganteus* (EPS) • *Fomitopsis pinicola* (IPS) • *Ganoderma lucidum* (IPS) • *Termitomyces albuminosus*, etc	Luft et al. (2020)
Fungal Glucans: β-Glucan	Biopolymers containing glucose monomeric units that are connected with α or β glycosidic bonds and possess highly variable structures, found in cell walls of fungi	• *Saccharomyces cerevisiae*	Luft et al. (2020)
Fungal Glucan: Scleroglucan		• *Sclerotium rolfsii*	
Fungal Glucan: Pullulan		• *Aureobasidium pullulans*	
Fungal Glycolipids	They consist of a carbohydrate moiety linked to fatty acids	Produced by members of the phyla Ascomycota, Basidiomycota, Zygomycota • *Ustilago scitaminea NBRC 32730* • *Ustilago maydis*	Abdel-Mawgoud and Stephanopoulos (2018); Luft et al. (2020)

(continued)

Table 18.1 (continued)

Microbial biopolymers	Chemical structure	Microbial sources	Ref.
Fungal Lipopeptides	Comprised of a hydrophobic tail, in the form of a fatty acid connected to a hydrophilic head among amino acids ranging from 4 to 12	Produced by diverse fungal strains • *Fusarium* spp. BS8 • *Penicillium chrysogenum* SNP5	Luft et al. (2020)
Algal biopolymers			
Alginates	(1,4)-linked β-D-mannuronate (M) and its C-5 epimer α-L-guluronate (G) present in the copolymer as the building block residues	Sourced from brown seaweeds, like • *Laminaria hyperborea* • *Laminaria japonica* • *Laminaria digitata* • *Macrocystis pyrifera* • *Ascophyllum nodosum*	Lee and Mooney (2012)
Carrageenan	A sulfated polysaccharide containing alternate residues of galactose creating a linear sequence of monomeric units	Red algae of the • *Kappaphycus* genus • *Eucheuma* genus	Rinaudo (2008)
Fucoidans	A sulfated polysaccharide containing 1N3-linked α-L-fucopyranosyl residues in the form of a chain of the polymer	Many species of brown algae like • *Fucus vesiculosus* • *Sargassum stenophyllum*	Li et al. (2008)
Ulvans	A sulfated polysaccharide containing a chain of different sugars or derivatives of sugars as monomeric units, linked together by alpha and beta linkages	Source: Green algae belonging to • *Ulva* genus • *Enteromorpha* genus	Lahaye and Robic (2007)
Agarose	A linear biopolymer consists of the alternately arranged simple sugar-residues, linked by glycosidic bonds	• Red algae (Red algae cell wall)	Usov (1998)

Notes: *PT* Pathogenic, *NPT* Nonpathogenic

fermentation. The polymer then is extracted by cell disruption followed by a suitable extraction process such as centrifugation (U.S. Congress 2000). Glucose, fructose, arabinose, and maltose, etc., are commonly used as a carbon source for Pullulan production. Substituting these sources with jaggery can reduce the cost of production of Pullulan (Anil Kumar et al. 2007). Gellan, polysaccharide synthesized in the log phase shows varying yields depending upon strain used. A carotenoid free mutant of *S. paucimobolis* ATCC 31461 produced 13% more gellan (13.2 g/L) in comparison with the parent strain (11.89 g/L). Polyhydroxyalkanoates (PHAs) are produced by downstream processing where recovery of PHA from cells is very important. Traditionally cell lysis was carried out by chlorine compounds which were replaced by the cell lytic enzyme of *Microbispora* sp. Thus, making the process environment friendly and less expensive. Large scale production of these polymers require bioreactors such as continuous stir tank reactors, continuous flow packed bed reactor, etc. The choice of fermenter depends upon the viscosity and fluid properties of the polymer produced. The use of fermenters fitted with different types of agitators such as hydrofoil, gasfoil, intermig, and scaba agitators provide better homogenization, air, and heat transfer and are also energy efficient thus effective quality and economics (Vijayendra and Shamala 2014). Commercial aspects of biopolymer production have gained importance only because of the tremendous potential of these in the fields of medicines, food, packaging, and biotechnology.

18.4 Microbial Biopolymers and Their Biomedical Applications

As mentioned earlier, a wide range of microorganisms can serve as one of the important sources for the production and innovation of the biopolymers capable of diversified applications in various fields, with their improved or modified properties. A brief introduction to some of these polymers with their few important biomedical applications is given in this section.

18.4.1 Polyhydroxyalkanoates (PHAs)

Several stress conditions like abundant carbon accessibility and depletion in nutrition under stress conditions such as high carbon (C) availability and limitations of nutrients rich in N, K, Mg, O, etc., may initiate many microorganisms to produce PHAs. PHAs are polyester thermoplastics, found in the cytoplasm of a cell and have similar properties to that of synthetic plastics. Their unique properties, for instance, the degradability by living organisms, compatibility with biological tissues, and nontoxic nature bestow them with desirable properties or features ideally required for the applications in the field of biomedicine (Ray and Kalia 2017). PHAs are optically active and biodegradable polymers that are produced via microbial fermentation process followed by microbial cell lysing (Barouti et al. 2017; Narancic et al.

2020). PHAs have extremely good biocompatibility and hence have the potential for several medical applications in wound care, personal hygiene products, and tissue engineering (U.S. Congress 2000). PHAs are renewable, biocompatible, biodegradable, and optically active. PHA-based drug delivery systems are also being explored. Nearly 300 species manufacture PHAs. The PHAs are commonly synthesized by the known pathway in which two acetyl-CoA moieties are condensed. These moieties can be further converted to hydroxybutyryl CoA and ultimately Poly-β-hydroxybutyrate (PHB) (Vijayendra and Shamala 2014). PHB polymer also finds several applications from surgery to treatment in a variety of diseases. The principal advantage is of being the nontoxic and biodegradable nature of the polymer. Due to the properties such as compatibility and ability to support cell growth, PHB has found its application in tissue engineering scaffolds and cell growth. It is an ideal biopolymer intended for the cell growth of several cells, such as fibroblasts, osteoblasts, and smooth muscle cells, due to its ability of PHB to get metabolized in the biosynthetic pathway of cells. PHB has the potential to act as a biodegradable carrier for the drug (Narancic et al. 2020) or a substitute for implants. The 4-hydroxy butyrate (HB), a degradation product of PHB, has shown pharmaceutical activity and is suitable for treating the long-lasting sleep disorder-narcolepsy. Due to the structural similarity with the gamma γ-aminobutyric acid (GABA), it reduces narcolepsy by interacting with the GABA receptor, which is a muscle tone regulator, and it also acts as a chemical messenger (neurotransmitter) in the central nervous system of mammals (Watanabe et al. 2002; Wani et al. 2016). The PHA derivative 3-hydroxybutyrate methyl easter has potential against Alzheimer's disease, as it inhibits mitochondrial damage (Zhang et al. 2013). PHAs are used in stomach wall patches (Löbler et al. 2002), reagents are used in bovine tuberculosis skin tests as PHA-beads (Chen et al. 2014; Parlane et al. 2016), and nano/micro adhesives for bioimaging (Jahns et al. 2008). However, with all these applications, the industrial-scale production of PHAs is still having limitations (Singh et al. 2015).

18.4.2 Polylactic Acid (PLA)

L-lactic acid, an optically active monomer is derived from the fermentation of *Lactobacilli*. Most applications of PLA copolymers are from therapeutic classes. Devices made from PLA-Poly glycolic acid (PGA) are extensively used in controlled release devices. Polylactide structures have found their applications in surgery, as they have a high degradation rate. Copolymers of PLA and PGA exhibit various advantages to their monomer counterparts concerning biocompatibility and nontoxicity (Onar 2004). PLA is extensively studied in the field of nanoparticles. However, many of the nanoparticulate systems are in combination with synthetic polymers as well. PLA-Poly lactic-*co*-glycolic acid (PLGA) engineered nanoparticles have been studied for the delivery of anticancer drugs. BIND-014 is one such example where nanoparticle has a core, which is polylactic and hydrophobic and a hydrophilic outer shell made up of polyethylene glycol, having a prostate-specific membrane antigen (PSMA). This delivery system showed better targeting of drug docetaxel (Von Hoff et al. 2016; Narancic et al. 2020). Drugs like cisplatin and paclitaxel are also studied (Tian et al. 2017). Various drugs when used with these systems showed better targeting, better response, or diminished toxicity. For example, curcumin-nanoencapsulation in Poly lactic-*co*-glycolic acid (PLGA)/PEG/Folic acid conjugated nanoparticles showed a decrease in cervical cancer in xenografts mice models (Thulasidasan et al. 2017). PLA can also be conjugated with an amphiphilic block copolymer to create micellar structures. Thermoresponsive PLA gels were also studied for the delivery of methotrexate where increased stability of the gel against mucosal clearance and the controlled release of the drug in mice models was achieved even though the mechanism of thermoresponse remained unclear (Zhang et al. 2014). PLA/PLGA nanomaterials have shown extensive applications in drug delivery, organ targeting, and are generally biocompatible. The stability of loaded drugs due to, the reaction between the carboxyl group of polymer and N-terminus of the drug (especially protein-based drugs), and erosion of nanoparticles can pose challenges (Narancic et al. 2020).

18.4.3 Bacterial Cellulose (BC)

Bacterial cellulose, an extracellular polysaccharide is produced by diverse bacteria. As compared to vegetal cellulose, bacterial cellulose shows a high degree of medical applications. But BC lacks functionalities to trigger cell attachments or have a very slow degradation rate. To overcome these drawbacks they are subjected to various in situ and ex situ modifications including the addition of chitosan, gelatin, PHB, etc., in the culture medium. Such modifications can alter the physical-mechanical properties of BC and open up a new range of applications (Keshk and Nada 2016). BC with altered porosity is used in tissue engineering, as higher pore size helps muscle cells in growth (Lv et al. 2016). BC/Polyvinyl alcohol (PVA) and BC/Hyaluronic Acid (HA) composites are used in the artificial cornea as they have higher visible light transmittance than plain BC (Jia et al. 2018). BCs have extensively studied in the field of wound healing. BC with modified topography and impregnated vaccarin has shown an increase in collagen deposition on the wound bed. Various nanocomposites such as BC/TiO_2, BC/Ag, and BC/ZnO have shown antimicrobial action against *Pseudomonas aeruginosa*, *S. aureus,* and *E. coli.* Systems of poloxamer loaded BC are used in the treatment of infected wounds. *Komagataeibacter hansenii* produced cellulose has Generally Recognized As Safe (GRAS) status and is used in biomedical dressings and implants (Alkhatib et al. 2017; Gorgieva and Trček 2019). With all these applications, BC production costs are still higher and 30% of the cost is attributed to carbon sources. Thus, future efforts are needed for making the BC production industry transferable.

18.4.4 Kefiran

Kefiran is an extracellular polysaccharide containing several hydroxyl groups, sourced from numerous *Lactobacillus* species and microbes found in kefir grains, for instance, *L. kefir, L. kefiranofaciens*, *L. parakefir*, *L. kefirgranum*, and *L. delbrueckii* sub-spp. *Bulgaricus* (Micheli et al. 1999). The use of kefiran as carrier material was patented in 2012. It is used as a film-forming polymer and gelling agent. It has also shown antibacterial and antitumor properties (Vijayendra and Shamala 2014). Kefiran also shows antioxidant activity which can be further enhanced by the addition of chitosan. Kefiran has been investigated for its applications in Nanotechnology as well where Kefiran-based nanocomposites films are being investigated as a biomatrix. Kefiran with glycerol and sorbitol as plasticizers has also been used to produce edible films. These films are extensively hydrophilic and to overcome this drawback oleic acid is added as a hydrophobic component. Kefiran has shown immunomodulatory as well as hypotensive effects (Moradi and Kalanpour 2019). Kefiran has shown to be effective in inhibiting allergic reactions and inflammation. It shows anti-inflammatory action through mast cell degranulation (Furuno and Nakanishi 2012). The immunomodulatory effect of kefiran could be used for the therapy of intestinal pathologies. This is due to the ability of kefiran to modify the balance of immune cells in the intestinal mucosa (Medrano et al. 2011). Kefiran also shows antiproliferative properties which makes it a good anticancer candidate. Kefiran produced by *L. kefiranaciens* showed activity against hepatocellular carcinoma and cervical carcinoma cells (Elsayed et al. 2017).

18.4.5 Levan

A variety of microbes such as algae, fungi, and bacteria, and some plant species can naturally produce levan (Laws et al. 2001). It is an exopolysaccharide having a branched structure where the core chain is composed of the D-fructofuranosyl residues which are linked by β (2 → 6) linkages, while at the branches of the chain, the same residues are linked by β (2 → 1) linkages (Arvidson et al. 2006). Several microbes of the genera *Saccharomyces*, *Halomonas, Acetobacter*, *Streptococcus*, *Xanthomonas, Pseudomonas, Zymomonas*, and *Bacillus* can secrete an enzyme, levansucrase, that is capable of carrying out the trans-fructosylation reaction on the substrate derived from the sucrose in order to yield the microbial levan (Srikanth et al. 2015b). Levan is a versatile polymer showing outstanding potential in the biomedical field. Levan also exhibits antimicrobial, anti-hyperglycemic, anti-HIV, and anti-inflammatory properties and also used for the slow release of insulin. Upon acid hydrolysis in the stomach, it produces di-fructo-furanose, fructose, and fructo-oligosaccharides which show prominent prebiotic effects (Vijayendra and Shamala 2014; Srikanth et al. 2015b). Thus, it is one of the important biopolymers having the potential for several biomedical applications.

18.4.6 Dextran

The dextran was first isolated by Louis Pasture from wine in 1961, and it was one of his great contributions in the field of science. It is produced by an enzyme dextransucrase, which is an extracellular enzyme of bacterial origin, capable of producing α-glucan in the form of dextran from sucrose. This enzyme is usually produced by numerous bacteria of mesophilic and thermophilic genera: *Lactobacillus, Streptococcus*, *Gluconobacter*, and *Leuconostoc*. The simple but unique biopolymeric nature, bacterial origin as a neutral exopolysaccharide composed of glucose monomeric units having significant biocompatibility and biodegradability, makes it an extremely desirable candidate for use in nanosized medicines, nano-drug carriers, and biomaterial-based imaging or biosensor applications (Banerjee and Bandopadhyay 2016). Dextran is used as a macromolecular carrier system for drug-polymer conjugates. The neutral pH and high solubility in water make it suitable for crosslinking. Dextran-based hydrogels are investigated for organ-specific drug delivery and angiogenic potential in particular tissues (Sun and Mao 2012). Dextran in combination with amylase, amylopectin, chitosan, pullulan, chitin, cyclodextrin, chondroitin sulfate, levan, and carrageenans is prepared to achieve colon-targeted drug delivery. Dextran-phospholipid micelles as amphiphilic drug carriers have also been evaluated. Dextran crosslinked FeO_2 nanoparticles have potential as a diagnostic biosensor (Banerjee and Bandopadhyay 2016). The antiulcer and anticoagulant activities are also reported for dextran sulfates, which are derivatives of the dextran obtained from the chemical modifications (U.S. Congress 2000). The hemoglobin-dextran conjugate is used as a substitute for blood with no renal side effects. This conjugate has a high binding and releases affinity for oxygen and the ability to escape into the extravascular spaces, resulting in the decreased risk of edema in the tissue (Wang et al. 2014). Many dextran-based patents are getting filed in recent years. Use of microbeads with biodegradability and enhanced drug absorption properties for cancer treatment, dextran as a cryoprotectant, prebiotic compositions, dextran-based polymer tissue adhesives, and production methods of oligodextrans which are useful in the treatment disorders related to the cholesterol and gastrointestinal tract are examples of few such patents. With such a broad array of potential roles in the biomedical field, only a select few have made it to a large scale (U.S. Congress 2000).

18.4.7 Pullulan

The fermentation of black yeast, like *Aureobasidium pullulans,* yields a nonionic polysaccharide, known as Pullulan. Due to its general properties like water solubility, biodegradability, it has been widely exploited in the food and pharmaceutical industries. On the other hand, owing to its unique characteristics, such as nontoxicity, non-immunogenicity, noncarcinogenic, and non-mutagenic nature, it has been greatly explored in recent days for several applications in the fields of medicine, pharmacy, and biotechnology namely, delivery of genes, targeting drug delivery, tissue engineering, wound healing, and similarly for some applications in the field of diagnosis such as nuclear medicine imaging procedures (Kumar et al. 2012). Pullulan has been used as hydrogels for tissue engineering, where Pullulan gels with more than 90% of water content were prepared. Pullulan is studied as a diluent and film-forming agent in fast-dissolving tablets. Nanogels prepared from the hydrophobic Pullulan-based Nanogels, capable of self-assembling have shown greater colloidal stability. Magnetic nanoparticles formed with coatings of Pullulan acetate for treatment of hypothermia has also been studied (Ferreira et al. 2011; Kumar et al. 2012). Pullulan-silver nanoparticle composites nanoparticles have been studied for antibacterial applications (Shahidul Islam et al. 2011). Pullulan being a water-soluble polymer has been investigated for its potential as blood or plasma substitutes (Shingel and Petrov 2002). Pullulan was also subjected to a large number of chemical modifications such as esterification, etherification, sulfation, amidification, and copolymerization so as to increase its potential in biomedical applications (Kumar et al. 2012).

18.4.8 Alginates (ALGs)

These biopolymers are a class of natural anionic polysaccharides found in the cell walls of the brown algae, such as *Laminaria hyperborea*, *Macrocystis pyrifera*, *Ascophyllum nodosum*, and some bacterial genera including *Pseudomonas* and *Azotobacter*. The "Alginates" term is generally mentioned to alginic acid and alginic acid-derived salts and derivatives (Remminghorst and Rehm 2006; Sachan et al. 2009; Szekalska et al. 2016). However, alginates obtained from bacteria have not received GRAS status as of now (Moradali and Rehm 2020). ALGs are linear biopolymers linear in a structure and made up of 1,4-linked β-D-mannuronic acid and 1,4 α-L-guluronic acid monomeric units organized in homogenous or heterogenous block-like patterns (Szekalska et al. 2016). ALG-based biomaterials are used in several fields such as therapeutic delivery of drugs, tissue engineering, dental impression materials, and dressings of wounds. Alginate gels are crossed linked with Ca^{2++} ions and are used for tissue engineering studies. Sodium and magnesium alginates are used in anti-reflux suspensions or tablets. Magnesium and potassium alginates are used in the dental impression material. Alginate gel viscosity is dependent on alginate itself, along with calcium source and sequestrant. Various cells can be encapsulated and used for specific treatments. Examples of such alginate encapsulated cells are Hepatocytes (Liver failure), Parathyroid cells (Hypocalcemia), and Langerhans islets (Diabetes). Propylene glycol alginate is found useful as a suspension stabilizing agent, plasticizer, binder, and emulsifying agent (Sudha 2017). Based on the several discoveries related to ALG over the last decades, it is apparent that these natural multifunctional polymers emerged as desirable compounds in the biomedical and pharmaceutical fields.

18.4.9 Hyaluronic Acid (HA)/Hyaluronate

The primary source of the naturally occurring HA is tissues of vertebrates; however, it is also produced by several bacteria as an exopolysaccharide. It is a heteropolymer carrying a negative charge and unbranched structure composed of β-D-glucuronic acid and β-D-*N*-acetylglucosamine monomeric units arranged alternately and joined by β (1,4) and β (1,3) glycosidic bonds. Bacterial hyaluronates obtained from some bacteria, such as *S. equi*, have been approved the GRAS status and thus extensively utilized in the formulations of ophthalmics, cosmetics, topically applied preparations, and viscosupplements for various therapeutic purposes (e.g., in the treatment osteoarthritis of the knee) (Moradali and Rehm 2020). Although the use of HA poses significant handling challenges such as the long time required for solubilization and high viscosity, it is widely used for conjugation. HA has carboxylic functionality which can be used for direct conjugation without the need for a linker. It can be used for targeted delivery towards the HA receptor (Tammi et al. 2002; Kunduru et al. 2016). As HA plays important role in various regulatory processes in many organisms, it has vast spectra of applications in the biomedical field. It can be used as a disease indicator in the case of liver cirrhosis, arthritis, and tumors. It is also useful in ear and eye surgeries. In eye surgery, it is used in glaucoma surgery, retinal reattachment, and also used as protective for corneal tissue. Wherein in the case of ear surgery, it is used as a scaffold material. Owing to its ability to stimulate wound healing, it is also used in wound healing and scar control (Onar 2004).

18.4.10 Poly-γ-Glutamic Acid (γ-PGA)

The γ-PGA is a microbial biopolymer produced by several *Bacillus* species in the form of uncommon, negatively charged polyamide, consisting of D- and L-glutamic acid monomers joined by amide bonds, conferring unique properties to γ-PGA such as biodegradability, edibility, aqueous solubility, and nontoxicity. Due to these characteristics, it has been widely used as biodegradable plastics, flocculants, biological adhesive, and food additives (Lee et al. 2014). Its unique features like negatively charged chelating cations and high hygroscopic nature made it potentially useful in developing hydrogels, blends for pharmaceutical, and bioremediation uses. It has been also explored for its commercial applications as an anti-scaling agent, dispersant, thickening agent, food additive, wetting agent, and drug delivery system (Moradali and Rehm 2020). These are protease-resistant and are also investigated as surgical adhesives and hemostatic agents (Kunduru et al. 2016). In summary, for about eight decades several scientists not only explored the bacterial development of γ-PGA but also explored its potential and commercial applications in the fields of medicine, pharmacy, and food (Lee et al. 2014).

18.4.11 Polyphosphates (PolyPs)

These are phosphate chains in the condensed form where several inorganic phosphates, ranging from three to a few hundred, with a negative charge and multiple high-energy anhydride bonds, that help PolyPs to function as energy storage material. Besides functioning as phosphate storage, they also contribute to metabolic processes as an energy source, buffering solutions, metal chelating activities, and DNA uptake processes (Moreno and Docampo 2013; Moradali and Rehm 2020). Due to their ability to regulate cell signaling, they can affect the life-cycle of bacteria by influencing bacterial cell viability, cell growth, ability to survive against stress, and their harmfulness or disease-causing capacity (Wang et al. 2018a). The development of virulence factors results due to a clear correlation with polyP metabolism. PolyPs are extremely negatively charged and have the ability to react with other positively charged polymers (Alginates and Hyaluronate) or inorganic cations (e.g., Ca^{2+}, Zn^{2+}, Mg^{2+}, Na^{+}, K^{+}, and Fe^{2+}) or organic molecules such as peptides, polyamine, or amino acids. Due to these unique characteristics, they are useful for the development of hydrogels and nanoparticulate carriers for the biomineralization of bones and additional pharmaceutical purposes, where they form different hydrogels of varying strengths. Such hydrogels of desired material properties act as smart biomaterials for a variety of biomedical applications (Ackermann et al. 2018; Wang et al. 2018b).

18.4.12 Chitin and Chitosan

Although the chief sources for large industrial-scale manufacture of these biopolymers are shrimp and crab shell wastes, various classes of fungi including *Ascomycetes, Deuteromycetes*, *Zygomycetes, Phycomycetes, and Basidiomycetes* are also capable of the same. Not all but cell walls of the mentioned fungal species possess these biopolymers. Both these biopolymers share structural similarities with that of cellulose, consisting of the hundreds of β (1,4) linked D-glucose units. Generally, both these polymers exist as copolymers rather than as homopolymers (Akila 2014; Rahman Bhuiyan et al. 2014; Elsoud and El Kady 2019). Both these biopolymers have some unique properties like compatibility with biological tissues, degradability by living organisms, therapeutic potential against bacteria, nontoxic potential, and biocompatibility, biodegradability, strong antibacterial potential, nontoxicity, and increased moisture absorption which make them appropriate for versatile applications in the fields of medicine, pharmacy and biotechnology. Thus, they have been not only extensively explored in numerous devices for medical use in tissue or organ-replacement and body-functions-related activities. Chitosan alone or as derivatives also reported to have the potential to support the processes of tissue engineering as biomaterials. There are some other biological properties of these biopolymers that make them pharmacologically effective in the treatment of cancer, pain, bleeding, cholesterolemia, microbial infections, and oxidative stress. Interestingly their biological properties are generally influenced by their physicochemical properties including molecular weight, deacetylation, and moisture level, etc. (Aranaz et al. 2009; Islam et al. 2017). It is important to note that in the continuous developments, the industrial-scale manufacture of these biopolymers from fungi has captivated the consideration of several fields throughout the world, owing to their significant merits compared to presently available techniques.

After this brief introduction, we have summarized several important applications of the prime microbial biopolymers in the fields of Medicine, Pharmacy, and Biotechnology in Table 18.2.

Table 18.2 Leading microbial biopolymers with their important pharmaceutical, medical, and biotechnological applications

Microbial biopolymer	Pharmaceutical applications	Medical applications	Biotechnological applications	Ref.
Polyhydroxy Alkonates (PHA), and Poly-β-hydroxybutyrate (PHB)	• Controlled Release DDS • PHA-based nanoparticle DDS • PEGlytaed PHA Nanoparticles as DDS in Cancer • PHB-RGD-based Nanoparticles as DDS in Cancer • As ingredients useful in products of personal care and related packagings • Antibiotic loaded PHB nano- and microspheres • Biodegradable cosmetic containers • HB compounds resulting from PHB degradation serve as pharmacologic agents • HB—Useful in treating the AWS, sleep disorder, chronic and catatonic SCZ, atypical psychosis, CBS, chronic distress, PD, vascular failure, cancers, exposure to radiations, and other nervous system-related disorders • PHA beads—Vaccine (e.g., Hepatitis C vaccine) delivery agent	• PHA and PHB for Surgical sutures • PHA and PHB for Bone plates • PHA and PHB for Wound care • PHB—applicability to reconstructive surgery • PHB—applicability to the manufacture of vascular grafts, pericardial patches, heart valves for circulatory system-related applications • PHB—useful in the development of pericardial substitute • PHB—useful in the development of vascular prostheses • PHB—Useful in bone healing due to the piezoelectric property • PHB—in Dental and Cardio-vascular implants • PHA—Stomach wall patch • PHA—Peripheral nerve guide • PHA beads—Nano/ microdevices for bioimaging	• PHB—in tissue engineering scaffolds • PHB—as a suitable biomaterial for different cell-growths such as FI, bone-forming cells, chondrocytes, HUVECs, and RbAOSMC • PHA—useful in protein chips • PHA—useful in the purification of proteins • PHA—useful in clinical diagnostics for immobilizing the Abs • PHA beads—Tuberculin skin test reagent for diagnosis of bovine tuberculosis • PHA beads—Recombinant protein production • PHA beads—Protein purification • PHA beads—Fluorescence-activated cell sorting based diagnostics • PHB hydroxyhexanoate—Scaffold for cartilage tissue engineering	U.S. Congress (2000); Rodríguez-Contreras and Manero (2016); Wani et al. (2016); Ray and Kalia (2017); Moradi and Kalanpour (2019); Moradali and Rehm (2020)

		in medical approaches • PHB valerate—Sutures • PHB valerate—Myocardial Patch • PHB valerate—Bone Regeneration • FN and ALG coated PHB Fibers—Fibronectin and alginate coated PHB-fibers for treating the injuries to the spinal cord	• PHB hydroxyhexanoate—as a biomaterial for SMC-Scaffolds useful in TE	
Polylactic acid (PLA)	• PLA and PLGA-based Nanomedicine for Potential DDS • PLA-PGA copolymers-based Controlled Release DDS (Devices)—for several drugs and biomolecules • PLGA and PLA-based nanoparticles—for intracellular delivery of drugs • PLA-PGA copolymers-based microsphere or microcapsules, pellets • PLA-PGA Copolymer—as prostheses with porous coatings capable to prolong the release of several bioactive compounds including GFs • PLGA and PLA-based	• PLA-PGA copolymers-based implants and hollow fibers • PLA Sutures for Surgery • PLA/PGA-based copolymers in staples, sutures, reinforcement materials, and clips • PLA/PGA derived copolymers in experimental orthopedic surgery • PLA composites reinforced with Ca and PO_4^{3-} glass fibers for bone medical implants • PLA-chitosan composite—biomaterial as tissue scaffolds for bones • PLA-based copolymeric SR devices in CMF surgery • PLA-based copolymers	• PLGA and PLA-based nanoparticles—for intracellular delivery genes • Chitosan and PLGA microspheres—for controlled release of entrapped antigens • PLA/Pluronic blend films—for controlled release of proteins • PLA composite-based scaffolds for efficient Bone-TE-related processes • PGA-PLA tubular scaffolds for ICC transplanted as EOU • Poly-L-lactic acid for successful management of soft tissue augmentation • PLA-based scaffolds for cultured osteoblasts that can grow and mineralize	Onar (2004); Pawar et al. (2014); Saini et al. (2016); Nofar et al. (2019); Narancic et al. (2020)

(continued)

Table 18.2 (continued)

Microbial biopolymer	Pharmaceutical applications	Medical applications	Biotechnological applications	Ref.
	polymers as effective antimicrobial polymers • PLA-coated titanium for better dental implants	developed by compression molding in the development of plates and screws for orthopedic medical applications • PLA-based Scaffold biomaterials for stimulating the in vivo cartilage regeneration • PLA-PGA copolymers-based prostheses	• PEG-PLA-PLGA-based Three-Dimensional Fibrous Scaffolds for Cell Culture	
Bacterial Cellulose (BC)	• BC/Octenidin/Poloxamer hybrid system as Controlled Release DDS for Octenidin • BC/CMC/Methotrexate for topical treatment of psoriasis • BC/PHEMA Hydrogel matrice for Wound healing • BC/iron oxide nanoparticles for use in Blood vessels • BC/ALG hybrid matrix as Controlled DDS for Cancer Drugs (e.g., Doxorubicin) • BC membranes for implantable drug delivery device • BC-based composites as Sustained DDS	• Composites developed from the nanocrystals of BC and Fibers of Re-CT as a potential biomaterial for Sutures • BC/Octenidin/Poloxamer hybrid applicable system as an innovative treatment-option for wound infections • BC with modified topography for Wound dressing • Vaccarin impregnated on BC for Wound dressing • TEMPO-Oxidized BC with Ag nanoparticles for Wound dressing • BC/ZnO nanocomposite for Wound dressing	• BC/PHEMA Hydrogel matrice for Rat mesenchymal stem cells (rMSCs) proliferation • BC/PHEMA Hydrogel matrice for Tissue replacement • BC with tuned porosity for Tissue engineering • BC/urinary bladder matrix for improved proliferation and adhesion of RPEC • BC matrix for enzyme immobilization • Multiple layer BC matrixes for Bioprocess—biotransformations • BC-based nanocomposite	Cacicedo et al. (2016); Ullah et al. (2016); Picheth et al. (2017); Gorgieva and Trček (2019); Moradali and Rehm (2020)

	• BC-based composites having antibacterial activity • BC-based composites for a slower release of Insulin • BC-based composites for Drug Delivery with lowered/ controlled permeation rate	• BC/TiO_2 nanocomposite for Wound dressing • BC/ε-PLL nanocomposite for Wound dressings • BC/Ag nanoparticle composite for Wound dressings • SMN-zein nanoparticle/ BC nanocomposite for Wound dressings • BC/PVA composite as an Artificial cornea • Tubular BC materials were used as microvessels prosthesis	for inside cell activities by penetrating cell membranes with nanofibrils • Hydrophilic BC for (bacterial) cell immobilization • BC/PEG scaffolds for wound dressing or tissue engineering applications • Cellulose-based nanocomposites for enhanced endothelial cell adhesion and angiogenesis	
Xanthan Gum (XG)	• XG in the preparation of SODF (e.g., Tablets for TDD) • XG in the preparation of Liquid Oral Dosage Forms (e.g., Reconstitutable suspensions) • XG in Ophthalmic Drug Delivery (e.g., Eye drops, Eye gels) • XG in Buccal Drug Delivery (e.g., Double-layer buccal patches developed from XG with mucoadhesion property) • XG in Advanced Drug	• XG for prevention of intra-abdominal adhesion following abdominal surgery • XG use in Wound Healing • XG in Dermal Patches as a Drug release retarding polymer • Xanthan-PVA Microsponges for Non-Fickian release behavior	• Xanthan-based Hydrogels developed from XG for human fibroblasts viability, in vitro • Xanthan-CTS-*g*-GMA-based hydrogels appropriate for the proliferation of FI • MNPs developed from Xanthan for the greater in vitro proliferation of FI • MNPs developed from Xanthan for in vitro neuronal differentiation of ESCs (In vitro) • Xanthan-polypyrrole hydrogel for enhanced	Petri (2015); Kumar et al. (2018); Singhvi et al. (2019); Song et al. (2019)

(continued)

Table 18.2 (continued)

Microbial biopolymer	Pharmaceutical applications	Medical applications	Biotechnological applications	Ref.
	Delivery (e.g., Microspheres) • XG in Brain Drug Delivery (e.g., Brain-targeting Nanoemulgel with Mucoadhesion property) • XG in Nasal Drug Delivery (e.g., XG-based Nasal Gel with pH-controlled in situ application) • XG in Cosmetics (e.g., bioadhesive hydrogel for skin application) • XG-Lignin based Hydrogel for a slower release of the Drug • Xanthan—Hydrogel for pH-controlled release • Xanthan-Chitosan for Liposomes' coatings • XG use in Toothpaste		fibroblasts proliferation • Xanthan-hydroxyapatite hydrogel for improved alkaline phosphatase activity and viability of osteoblasts	
Gellan Gum (GG)	• Gellan gum-based in situ gels as DDS • Gellan gum-based Microparticles as DDS • Gellan gum-based Microcapsules as DDS • GG-Polyacrylic acid-based Tablets for Gastric Delivery of Drugs • GG-Polymethacrylamide-	• GG-MA-laponite based novel Nanocomposite hydrogels as innovative wound dressing material • GG Spongy-like hydrogels as Cell encapsulating systems for Cell Therapy • GG-GRGDS Hydrogels for Cell Therapy in Spinal	• Hydrogels developed from GG for the TE of bones and related applications • Glycidyl-Methacrylated-GG-based Beads in Bone Tissue Engineering • GG/Gold Nanorods Nanohydrogels for Bone Tissue Engineering • GG/Alginate 3D scaffolds	Oliveira et al. (2010); Osmałek et al. (2014); Spera (2018); Ahmad et al. (2019); Palumbo et al. (2020)

	based Tablets as Sustained DDS • GG-Mal/Sericin-based Hydrogels for improved bioavailability of Drugs • GG-Sericin-based Nanocomposites for Cancer Therapy • GG-PEG-Chitosan-based Hydrogels for Wound Treatment • Immediate release gellan tablets • Gellan beads gelated by Al^{3+} and crosslinked by glutaraldehyde as DDS • Carboxymethyl-GG-based Hydrogels for improved bioavailability of drugs • Gellan granules as DDS • GG-glucosamine Films for Cancer Therapy • Hydrogels prepared by conjugation of 2-(2-amino ethyldisulfanyl) nicotinic acid with GG for In situ Topical Vaginal Application • In situ gelling floating gellan system as DDS • GG-CHPTAC Nanoparticles for Topical	cord Injuries • GG/HA Spongy-like hydrogels as Cell encapsulating systems for Wound healing • Hydrogels developed from Methacrylated-GG to regenerate the Nucleus pulposus • GG/SPD or SPM Hydrogels as Cell encapsulating systems for Neuronal regeneration • GG-GRGDS Hydrogels as Cell encapsulation system for cell therapy in Spinal cord injuries	for Bone Tissue Engineering • GG Hydrogels as Cell encapsulating systems for Chondrogenic differentiation • GG Hydrogels as Cell encapsulating systems for Cartilage Tissue Engineering • GG/Gly and HNTs Hydrogels as Cell encapsulating systems for Soft tissue Engineering • GG/PANi Spongy-like hydrogels as Cell encapsulating system for Skeleton muscle Tissue Engineering • GG/PEGDA 3D printed scaffold as Cell encapsulating system for Tissue Engineering • Methacrylated-GG with collagen Hydrogels as Cell encapsulating system for 3D-printed living cells scaffolds	

(continued)

Table 18.2 (continued)

Microbial biopolymer	Pharmaceutical applications	Medical applications	Biotechnological applications	Ref.
	Cutaneous Applications • GG-based Sustained delivery ophthalmic system • GG/BTA/CHO Nanohydrogels for Skin-Topical Applications • Cetyl-GG and GG-Sorbitan mono-oleate Micelles for Disease-Specific Treatments • UTFs developed from Glycerol and GG for use as Mucoadhesion-based DDS • GG/Chitosan Nanocomplexes as Oral DDS • GG-CTS based Spongy-like Complexes for Nasal Drug Delivery • GG/K-carrageenan and GG-Carbopol 943P-based hydrogels for In situ Ophthalmic applications • Microparticles developed from GG, Modified Pectin, and nanofillers for stomach-specific drug delivery • GG/PVA Electrospun mats as Gastroretentive DDS • Hydrogels developed from Poloxamer 407 and GG			

	for In situ periodontal applications of drugs			
Dextran	• Dextran-based hydrogels with thermoresponsive properties as DDS • Dextran-based hydrogels containing moieties for immobilizing the drugs to a single polymer chain • Dextran-based hydrogels as DDS for antifungal molecules (e.g., Amphotericin B) to prevent the growth and proliferation of fungi • Dextran-based hydrogels as injectable in situ crosslinking networks for the delivery of Antifungal Drugs (e.g., Amphotericin B) • Glycidyl methacrylated dextran for Controlled Delivery of BMP-2 in Wound healing application • Chitosan-Dextran-based DDS for the Controlled Delivery of Doxorubicin as an Anticancer Drug in various types of cancers • Biodegradable albumin and dextran sulfate microbeads for improved	• Dextran allyl isocyanate–ethylamine with PEG diacrylate-based hydrogels as instructive scaffolds to initiate the new blood vessels development and regeneration of the skin in full-thickness burn-wounds • Injectable hydrogels made from dextran for the delivery of biomolecules and cells • Succinyl chitosan and dextran aldehyde-based hydrogels to reduce intra-abdominal adhesions • Hydrogels formed by CMDX-ADH with DX-CHO for prevention of abdominal adhesion • Gamma radiation stable dextran solutions for increasing efficiency of blood separation • Electrospun dextran fibers and devices useful as Immediate drug release devices and Bandages • Dextran-based polymer tissue adhesive for	• Dextran-based hydrogels containing moieties capable of interacting with biomolecules • A dextran-based hydrogel encapsulating hESC for enhancing vasculogenesis • Apatit blends containing a dextran/pullulan mixture for bone tissue formation to regenerate damaged bone tissue • Acetylated dextran Hepatocyte for Controlled Delivery of growth factor fragment in Myocardial infarction (MI) related Tissue Engineering Applications • Methacrylated dextran for Controlled Delivery of VEGF in Ischemia-related Tissue Engineering Applications • Chitosan–polyethyleneimine for Controlled delivery of BMP-2 gene for the repair of bone defect during Bone Tissue Engineering • Dextran-peptide hybrid	Sun and Mao (2012); Maia et al. (2014); Wang et al. (2014); Moscovici (2015); Banerjee and Bandopadhyay (2016); Kirschning et al. (2018); Miao et al. (2018)

(continued)

Table 18.2 (continued)

Microbial biopolymer	Pharmaceutical applications	Medical applications	Biotechnological applications	Ref.
	anticancer drug adsorptivity • Dextran-based Prebiotic Compositions	ophthalmic procedures, tissue repair, antiadhesive applications, drug delivery, and as the fistula or punctum-sealing plug • Dextran-based Diagnosis method for CD206 expressing cell Disorders	for efficient gene delivery • Dextrin, Dextran, Isomaltooligosaccharide, and their derivatives-based Cryprotective Agents • Dextran-based Serum-free (Protein-free or Low-protein) mammalian cell culture medium	
Levan	• Levan as an edible coating material for foods and pharma products • Chemically modified levan derivatives as anti-AIDS molecules • Levans (Amorphous and Microcrystalline) with Antitumor Activity • Levans having Bifidogenic effects • Levans with Hyperglycelic Inhibitor Properties • Levan-based Nanoparticles with Anti-inflammatory Properties • Nonionic Levans with Antihyperlipidemic and hypocholesterolemic properties	• Levan with a molecular weight between 3000 and 100,000 Da as Plasma Substitute • Levans as therapeutic agents in dental caries • Levan as an anti-clotting agent in the heart-surgeries • Levan as an anti-clotting factor for treating the angiographic restenosis • Levans (film-forming) in Wound Healing • Levan-based films and adhesives for medical applications • Levan as blood plasma volume expanders • Levans—useful in preventing shock from hemorrhage, burns, and	• Levans as potential polymers in the Regenerative Medicine Applications • Levans as promising polymer materials for developing Cells Encapsulating Scaffolds in Tissue Engineering • Levan-based polymer films for TE and regenerative medicine applications • PEG/Levan two-phase liquid system for the Purification of Biological Materials (e.g., Bovine Serum Albumin) • Levan-based Nanocomposites for several Biomedical Uses • Levans as reducing agents to obtain different	Srikanth et al. (2015a); Avsar et al. (2016); Tomulescu et al. (2016); Gomes et al. (2018); González-Garcinuño et al. (2018)

	• Levorotatory Levans (or its derivatives) having Antiviral effects • Levan as prolongator of drug activity • Levans as Coating Materials in DDS • Levans as Tablet Binders • Levans as drug carrier of the nanoscale size range for peptides and proteins • Levans as color and flavor enhancers in capsule and tablet manufacturing • Levans having immunomodulatory properties on the macrophages, B-lymphocytes, and T-lymphocytes • Levans as formulation aid in cosmetic products • Levan-based gels as DDS • Levans as effective Emulsifying Agents and Encapsulating Agent	surgery • Levans—useful in reducing the risk of thrombosis and embolism	nanoparticle systems • Levan-based Nanoparticles to deliver Peptides and Protein Drugs	
Pullulan	• Pullulan-based Microspheres for pH and temp-dependent controlled drug release • Drug loaded Pullulan for Liver targeting	• Acetylated Pullulan-coated superparamagnetic iron oxide nanoparticles in the Hyperthermic Treatment for Cancer • Pullulan as potential	• Carboxymethylated pullulan conjugated with heparin for TE processes • Pullulan/Interferon (recombinant protein) covalent conjugate for	Mishra et al. (2011); Kumar et al. (2012); Prajapati et al. (2013); Singh et al. (2017); Tabasum et al. (2018)

(continued)

Table 18.2 (continued)

Microbial biopolymer	Pharmaceutical applications	Medical applications	Biotechnological applications	Ref.
	• Innovative Pullulan/ Doxorubicin-bioconjugates for passive targeting of tumors • Amphiphilic-based Nanogels as carriers for Nucleic Acids, Proteins, Peptides, and other molecules lipophilic molecules • Hydrophobized pullulan conjugates as TDDS for metronidazole, nicotinic acid, sulfathiazole, mitoxantrone, or epirubicin • Hydrophobized pullulan conjugates as polymeric nanocarriers in cancer chemotherapy • Hydrophobized pullulan conjugates as polymeric nanocarriers in protein stabilization • Hydrophobized pullulan conjugates as polymeric nanocarriers in artificial vaccines • pH-responsive, thermoresponsive, and photoresponsive nanogels as	blood plasma substitute or plasma expander • Chemically modified Pullulan-CHO derivatives-based Nanogels for the delivery into the Quantum dots cells for Bioimaging or Medical Imaging • Pullulan-SUconjugates for stimulating the insulin production • Pullulan folate-based Nanogels in the Photodynamic Therapy • Pullulan/PVA/silver hybrid antibacterial nanofibers for Medical Applications • *Pullulan/HLC* Hydrogels as efficient *Soft Tissue Fillers*, with reduced inflammation, improved cell adhesion property • *Pullulan/Chondroitin Sulfate* composite hydrogels as Injectable Cell Delivery Carrier in Cartilage Tissue Engineering • Pullulan/Pemulan blended Hydrogels for	preserving its antiviral and immunoregulatory activity and enhancing its liver accumulation • Heparin-conjugated pullulan for improved VEC-proliferation, and suppressed SMC-proliferation in TE • Pullulan/Glutaraldehyde-based polymeric nanogels as a DNA carrier system for improving gene loading efficiency, controlled release properties, biocompatibility, and enhanced stability • Hydrophobized pullulan conjugates as artificial molecular chaperones along with β-CD or as hybrid nanogels • Conjugate of pullulan derivative and plasmid DNA with Zn^{2+} coordination for Gene Delivery Applications • Pullulan-Dextran-based Scaffolds for Bone Tissue Engineering • Pullulan-Protamine based Gene Delivery Vector	

	DDS • Cholesterol-Pullulan conjugated nanogels as a new vehicle for Orodigestive Vaccine Therapy • Pullulan as FFA for the development of OTF (e.g., Rapid dissolving films of cetirizine hydrochloride) • Pullulan—Use as a mucoadhesive ophthalmic vehicle	Superficial Vaginal Infections Treatment • *Hyaluronic acid* grafted *pullulan* polymers for the formation of novel biocompatible Wound Healing Films	• Carboxyl pullulan/HLC/BDDE-based Novel Hydrogels for Tissue Engineering • Pullulan/cellulose acetate-based Hydrogels for Bone Tissue Engineering	
Alginates	• AA as a binder or disintegrant in the Manufacturing of Tablets • AA in Modified-Release Drug Products • AA for masking the unpleasant taste of drugs in the formulation • AA as a thickening agent, Stabilizing Agent, Viscosity modifier, and Thickener • NaALG as Suspending agent, viscosity increasing agent, and Thickener • NaALG as a disintegrating agent, binding agent, Stabilizing agent, and release-modifying agent in Manufacturing of Tables • NaALG as a Disintegrant	• Oxidized-NaALG Hydrogels for healing the Corneal Wounds • Thermal Polymerizable ALG-glycidyl methacrylate Hydrogel for TE in the repair of Myocardium • Alginate/antacid anti-reflux preparations as floating gels or rafts (Neutral in pH) in treating the GORD • Calcium Alginate Fibers for Wound Healing • Antibiotic or Antiseptic-infused ALG-Gelatin-based Sponges for Wound Healing • NaALG as protective coatings for protecting the injured skin or mucous membrane from irritating	• ALG-*g*-PEG Microspheres for Specific Cell Microencapsulation • Sulfated ALG Microspheres for Reduction in Secretion of Inflammatory Cytokines and Improvement in Biocompatibility • Phosphorylated ALG Gels for TE-related Mineralization and Chelation processes • Galactosylated ALG Microcapsules as mechanically stable and selectively permeable Cell-carrier systems • *α*-Cyclodextrin-ALG conjugate Beads for Immobilization of Bacteria,	Jain and Bar-Shalom (2014); Szekalska et al. (2016); Usman et al. (2017); Fernando et al. (2019)

(continued)

Table 18.2 (continued)

Microbial biopolymer	Pharmaceutical applications	Medical applications	Biotechnological applications	Ref.
	and Diluent in Capsule Formation • Ammonium Alginate as Color-diluting agent, Emulsifying agent, Film-forming agent, and Moisturizing agent • CaALG as a disintegrating agent in Tablet Manufacturing • PG-ALG as Stabilizer, Emulsifying, Suspending and Viscosity increasing agent • Oxidized-NaALG Beads as Sustained Oral DDS • NaALG–MAS composite films for modifying drug release from the Tablets • NaALG-*co*-polyacrylamide Hydrogel as Sustained-release gastroretentive Drug Carrier • Starch-*g*-poly(acrylic acid)-NaALG Hydrogel Beads as pH-sensitive matrices for the Oral Drug Delivery • β-Cyclodextrin-ALG conjugate Gels as Controlled DDS	skin situations • Strong-gelling NaALG formulation in Overweight/ Obesity Management • NaALG and K-ALG as Dental impression material and Denture fixatives • Gauze-Swab-Dressings soaked with NaALG formulation are for application in abdominal surgery procedures to stop the bleeding • ALG Dressings for Tooth cavities, Fistulas, and Sinuses • Oxidized sodium alginate conjugated PhE-NH2 polymeric nanoparticles for biological imaging applications	• PG-ALG Microparticles as sustained-release-systems for encapsulated-Proteins • ALG for purification of enzymes (e.g., pectinase, lipase, etc.) in Precipitation, Partitioning, and Chromatographic separation techniques • Nanosized Calcium ALG beads as Carriers for sensitive biomolecules like DNA • ALG containing nanofibers as tissue engineering scaffolds for Bone, Cartilage, and Skin • PLGA-chitosan/ PLGAalginate nanoparticles for seeding HUCMSCs • DNA-loaded Chitosan-ALG-Dextran Sulfate Nanoparticles are as Gene Transfection Vectors	

	• ALG oligosaccharides with biological activity as Therapeutic Compounds useful in the Treatment of Various Diseases (e.g., ALG oligosaccharides; oligomannuronate (HS971) for Alzheimer's disease and neurodegenerative diseases)			
Hyaluronic Acid (HA) or Hyaluronate	• HA-drug conjugates as Prodrugs with enhanced Physicochemical Characteristics and Therapeutic effectiveness (e.g., HA-Exendin, HA-Curcumin Conjugates) • Liposomes with modified-surface properties developed from HA-PL conjugates as drug carriers • More safe and effective Anticancer Therapeutics in the form of Liposomes developed by HA-coatings for the targeted delivery of siRNA, doxorubicin, gemcitabine, and docetaxel, via CD44 Cell Receptors • HA conjugated with Flt1 peptide to develop Nanoparticles as NGPDDS for DEXA in Asthma-	• HA-based soft matrices in RM applications such as Differentiation, Expansion, and Implantation of Stem Cells for clinical use • HA hydrogels developed by redox-initiation for heart muscle repairing • Photocrosslinked HA-hydrogels for the engineering of heart valves (Valvular Engineering) • HA-hydrogels as Cell-Free Therapies for delivery of encapsulated molecules, proteins and stimulation of natural healing processes • HA-based Polymeric-films embedded for the Extended Vitamin E in the Skin Wound-Management, • Hydrogels prepared from	• PEGDA-crosslinked thiolated HA and gelatin for Cell Therapy • HA-Hydrogels encapsulating eEPCs (murine) were encapsulated into to create a Bioartificial Stem Cell Niche • HA-Hydrogels encapsulating NPCs for targeted delivery to the brain affected by cerebral infarction • Nanoporous HA-Hydrogel Microparticles as a release system for BMP-2 to stimulate the robust cartilage-specific ECM production • HA-rich Hydrogel Matrices for the generation of Novel Cell Types and Cell aggregates,	Burdick and Prestwich (2011); Highley et al. (2016); Fallacara et al. (2018); Tiwari and Bahadur (2019)

(continued)

Table 18.2 (continued)

Microbial biopolymer	Pharmaceutical applications	Medical applications	Biotechnological applications	Ref.
	management • CTS Nanoparticles with HA-coatings for the delivery of BMZ in the Topical Treatment of Atopic Dermatitis • HA-based Nanoparticles with surface modification strategies to target cancer • HA microspheres and Microparticles for improved mucoadhesion and release characteristics • Chemically modified HA derivatives-based Hydrogels in the form of polymeric-3D-networks for Topical delivery of Anti-inflammatory, Antibacterial drugs, Antibodies, and Proteins • HA as an API in several Eye-drop formulations • Native HA solutions as Artificial tears • HA as a Moisturizing-API in Cosmetics	linear HA and its chemical derivatives as 3D polymeric networks for Topical Delivery of Cells • HA hydrogels for Intraocular, Intratympanic, Intraarticular, and Dermal delivery of Therapeutics • Drug-infused Films, Hydrogels, and Wound Dressings developed from HA, and its derivatives for Wound healing Applications in Dermatology, Ophthalmology, Otolaryngology, Rhinology, and Odontology • HA-based Viscosurgical Devices useful for Eye tissue protection and lubrication, replacement of lost vitreous fluid, and management of eye treatments • HA as Biomaterial to develop Dermal Fillers	• Modified HA macromers as encapsulation and release systems for DNA delivery • HA hydrogels for controlling the Differentiation of Entrapped Stem Cells • HA-Hydrogels for Cell-based Therapies and Research • HA and its synthetic derivatives-based 3D Scaffolds for in vitro Cell culture activities • The HA-3D Cell Culture for studying or investigating the mechanisms of Anti-cancer Drugs • Methacrylated HA-based 3D Culture Models for in vitro evaluation of Glial Scar formation • A printable-ink in the form of a hydrogel developed by double crosslinking of HA for Biomedicine applications	
Chitin (CT) and Chitosan (CTS)	• CTS-based Implants in the form of Drug-releasing scaffolds invented by WST	• CT and CTS-based Scaffolds and Polymeric membranes for Wound-	• CT or CTS polymeric blends as strong Osteoconductive Scaffold-	Anitha et al. (2014); Zargar et al. (2015); Elieh-Ali-Komi

	• CT and CTS Membranes, CTS/Pectin, CTS/ALG Crosslinked Scaffolds in Drug Delivery Applications • CT and CTS use in Skin- and Hair-care Products (Cosmetics) • A pH-sensitive DDS consisting of a Glutaraldehyde-crosslinked CTS-GHN • CS-PEG-ALG Microspheres for LMW-Heparin delivery • CS-CMC Microcapsules as Protective matrix against the Stomach's Acidic pH during Oral administration of Proteins and Drugs • CTS-based TrDD devices with chitosan have been developed for the permeation-controlled delivery • CTS Gel for Intratumoral Administration • CTS-based Beads, Films, Microcapsules, Coated Tablets as DDS • CTS-based Sustained DDS (e.g., per-oral Sustained-Release tablets)	Healing applications • CT and CTS Films, Sponges, Composites, Hydrogels as Wound Dressings in Wound Healing Applications • CTS-based Contact lenses, Surgical Sutures, BDM, and ABV • CTS as Effective Hemostatic Agent when formulated as Microspheres, Bandages, Dressings, or Absorbable Sponges • CT as a promising role in the development of an Artificial Kidney Membrane • CT Scaffolds and Membranes to treat Patients with Deep Burns, Wounds • CT Membranes for Deep burns, Orthopedic Trauma, and Ulcer conditions • CT-based products like Finely Divided Powder, Nonwoven Fabrics, Porous beads, Lyophilized Soft fleeces or Gels, Gauges, Laminated Sheets, and Transparent Films for Medical Applications • CTS–Heparin	Matrices in TE • Heparin modified CTS-PLAGA sintered scaffold for bone TE applications • CT and CTS-based composites in Cartilage Tissue Engineering (Cartilage Regeneration) • ALG/CTS-Polyelectrolyte Complex, CTS/HA composite fibers, and CT composites for improved cell attachment in Tendon/Ligament Tissue Engineering • CT and CTS as Skin Substitute Materials in Skin/Dermal Tissue Engineering • CT and CTS Membranes/Scaffolds for Hepatic Regeneration in Liver Tissue Engineering • CTS Fibers, Crosslinked CTS Films, Tri-composite CT/CTS/Gelatin Films in Nerve Regeneration Applications • CTS scaffolds for Periodontal Bone Regeneration • CTS-CaALG Capsules to	and Hamblin (2016); Islam et al. (2017)

(continued)

Table 18.2 (continued)

Microbial biopolymer	Pharmaceutical applications	Medical applications	Biotechnological applications	Ref.
		Cyanoborohydride surface as a very promising Catheter • CTS-coated PIHCA Nanoparticles for Intravenous Delivery of Small Interfering RNA (siRNA)	develop an Artificial Pancreas for the treatment of Diabetes Mellitus • CT for Non-viral Gene Delivery Systems	

Notes: *DDS* Drug Delivery System, *PEG* Polyethylene Glycol, *RGD* Arg-Gly-Asp-oligopeptide, *HB* 4-hydroxy butyrate, *AWS* Alcohol Withdrawal Syndrome, *SCZ* schizophrenia, *CBS* Chronic Brain Syndrome, *PD* Parkinson's Disease, *FN* Fibronectin, *ALG* Alginates, *FI* Fibroblasts, *HUVECs* Human Umbilical Vein Endothelial Cells, *RbAOSMC* Rabbit Aortic Smooth Muscle Cells, *Abs* Antibodies, *SMC* Smooth Muscle Cells, *TE* Tissue Engineering, *PGA* Poly Glycolic Acid, *SR* Self-reinforced, *CMF* Cranio-maxillofacial, *PLGA* Poly Lactic-*co*-Glycolic Acid, *GFs* Growth factors, *ICC* Intestinal Crypt Cells, *EOU* Epithelial Organoid Units, *TEMPO* 2,2,6,6-Tetramethylpiperidinyloxy, *ZnO* Zink Oxide, TiO_2: Titanium Dioxide, *ε-PLL* poly-L-Lysine, *Ag* Silver, *SMN* Silymarin, *CMC* Carboxymethyl Cellulose, *PHEMA* Polyhydroxyethylmethacrylate, *Re-CT* Regenerated Chitin, *RPEC* Retinal Pigment Epithelium Cells, *ALG* Alginate, *PVA* Polyvinyl alcohol, *SODF* Solid Oral Dosage Forms, *TDD* Targeted Drug Delivery, *CTS-g-GMA* Chitosan-*g*-Glycidyl Methacrylate, *MNPs* Magnetic Nanoparticles, *ESCs* Embryonic Stell Cells, *GRGDS* a fibronectin-mimetic peptide, *CHPTAC* *N*-(3-chloro-2-hydroxypropyl)-trimethyl ammonium chloride, *BTA* Butyric Acid, *CHO* Cholesterol, *UTFs* Ultra-thin Films, *SU* Sulfonylurea, *VEC* Vascular Endothelial Cells, *β-CD* β-cyclodextrin, *FFA* Film-Forming Agent, *OTF* Oral Thin Films, *Gly* Glycine, *SPD* spermidine trihydrochloride, *SPM* spermine tetrahydrochloride, *HA* hyaluronic acid, *PANi* polyaniline, *MA* Methacrylate, *hESC* human embryonic stem cells, *DX-CHO* aldehyde-modified dextran, *CMDX-ADH* Hydrazide-Modified Carboxymethyldextran, *BMP-2* Bone Morphogenetic Protein-2, *VEGF* Vascular Endothelial Growth Factor, *HLC* Human-like Collagen, *BDDE* 1,4butanediol diglycidyl ether (BDDE), *AA* Alginic Acid, *CaALG* Calcium Alginate, *PG-ALG* Propylene Glycol Alginate, *NaALG* Sodium Alginate, *GORD* gastro-Oesophageal Reflux Disease, *HUCMSCs* Human Umbilical Cord Mesenchymal Stem Cells, *PL* Phospholipids, *NGPDDS* Next-Generation Pulmonary Drug Delivery System, *DEXA* Dexamethasone, *BMZ* Betamethasone, *API* Active Pharmaceutical Ingredient, *RM* Regenerative Medicine, *MAS* Magnesium Aluminum Silicate, *eEPCs* Embryonic Endothelial Progenitor Cells, *NPCs* Neural Progenitor Cell, *WSP* Wet Spinning Tehnique, *ECM* Extracellular Matrix, *GHN* Gelatin Hybrid Network, *LMW* Low Molecular Weight, *BDM* Blood Dialysis Membranes, *ABV* Artificial Blood Vessels, *PLAGA* poly(lactic-*co*-glycolic)acid, *TrDD* Transdermal-Drug Delivery, *PIHCA* Polyisohexylcyanoacrylate

References

Abdel-Fattah AM, Gamal-Eldeen AM, Helmy WA, Esawy MA (2012) Antitumor and antioxidant activities of levan and its derivative from the isolate Bacillus subtilis NRC1aza. Carbohydr Polym 89:314–322. https://doi.org/10.1016/j.carbpol.2012.02.041

Abdel-Mawgoud MA, Stephanopoulos G (2018) Simple glycolipids of microbes: chemistry, biological activity and metabolic engineering. Synth Syst Biotechnol 3:3–19. https://doi.org/10.1016/j.synbio.2017.12.001

Ackermann M, Wang X, Wang S et al (2018) Collagen-inducing biologization of prosthetic material for hernia repair: polypropylene meshes coated with polyP/collagen. J Biomed Mater Res B Appl Biomater 106:2109–2121. https://doi.org/10.1002/jbm.b.34016

Ahmad S, Ahmad M, Manzoor K et al (2019) A review on latest innovations in natural gums based hydrogels: preparations & applications. Int J Biol Macromol 136:870–890. https://doi.org/10.1016/j.ijbiomac.2019.06.113

Akaraonye E, Keshavarz T, Roy I (2010) Production of polyhydroxyalkanoates: the future green materials of choice. J Chem Technol Biotechnol 85:732–743. https://doi.org/10.1002/jctb.2392

Akila RM (2014) Fermentative production of fungal chitosan, a versatile biopolymer (perspectives and its applications). Adv Appl Sci Res 5:157–170

Alkhatib Y, Dewaldt M, Moritz S et al (2017) Controlled extended octenidine release from a bacterial nanocellulose/Poloxamer hybrid system. Eur J Pharm Biopharm 112:164–176. https://doi.org/10.1016/j.ejpb.2016.11.025

Ananthalakshmy VK, Gunasekaran P (1999) Optimization of levan production by Zymomonas mobilis. Braz Arch Biol Technol 42:291–298. https://doi.org/10.1590/s1516-89131999000300004

Andreeßen C, Steinbüchel A (2019) Recent developments in non-biodegradable biopolymers: precursors, production processes, and future perspectives. Appl Microbiol Biotechnol 103:143–157

Anil Kumar PK, Shamala TR, Kshama L et al (2007) Bacterial synthesis of poly(hydroxybutyrate-co-hydroxyvalerate) using carbohydrate-rich mahua (Madhuca sp.) flowers. J Appl Microbiol 103:204–209. https://doi.org/10.1111/j.1365-2672.2006.03221.x

Anitha A, Sowmya S, Kumar PTS et al (2014) Chitin and chitosan in selected biomedical applications. Prog Polym Sci 39:1644–1667. https://doi.org/10.1016/j.progpolymsci.2014.02.008

Aranaz I, Mengibar M, Harris R et al (2009) Functional characterization of chitin and chitosan. Curr Chem Biol 3:203–230. https://doi.org/10.2174/187231309788166415

Arvidson SA, Rinehart BT, Gadala-Maria F (2006) Concentration regimes of solutions of levan polysaccharide from Bacillus sp. Carbohydr Polym 65:144–149. https://doi.org/10.1016/j.carbpol.2005.12.039

Avsar G, Gunduz O, Oner ET (2016) Sulfated levan as a polymer for biomedical applications. New Biotechnol 33:S78. https://doi.org/10.1016/j.nbt.2016.06.988

Banerjee A, Bandopadhyay R (2016) Use of Dextran nanoparticle: a paradigm shift in bacterial exopolysaccharide based biomedical applications. Int J Biol Macromol 87:295–301. https://doi.org/10.1016/j.ijbiomac.2016.02.059

Barouti G, Jaffredo CG, Guillaume SM (2017) Advances in drug delivery systems based on synthetic poly(hydroxybutyrate) (co)polymers. Prog Polym Sci 73:1–31. https://doi.org/10.1016/j.progpolymsci.2017.05.002

Burdick JA, Prestwich GD (2011) Hyaluronic acid hydrogels for biomedical applications. Adv Mater 23:H41–H56. https://doi.org/10.1002/adma.201003963

Cacicedo ML, Castro MC, Servetas I et al (2016) Progress in bacterial cellulose matrices for biotechnological applications. Bioresour Technol 213:172–180. https://doi.org/10.1016/j.biortech.2016.02.071

Chen S, Parlane NA, Lee J et al (2014) New skin test for detection of bovine tuberculosis on the basis of antigen-displaying polyester inclusions produced by recombinant Escherichia coli. Appl Environ Microbiol 80:2526–2535. https://doi.org/10.1128/AEM.04168-13
Elieh-Ali-Komi D, Hamblin MR (2016) Chitin and chitosan: production and application of versatile biomedical nanomaterials. Int J Adv Res 4:411
Elsayed EA, Farooq M, Dailin D et al (2017) In vitro and in vivo biological screening of kefiran polysaccharide produced by Lactobacillus kefiranofaciens. Biomed Res 28:594–600
Elsoud MMA, El Kady EM (2019) Current trends in fungal biosynthesis of chitin and chitosan. Bull Natl Res Cent 43:59
Fallacara A, Baldini E, Manfredini S, Vertuani S (2018) Hyaluronic acid in the third millennium. Polymers (Basel) 10:701. https://doi.org/10.3390/polym10070701
Favaro L, Basaglia M, Casella S (2019) Improving polyhydroxyalkanoate production from inexpensive carbon sources by genetic approaches: a review. Biofuels Bioprod Biorefin 13:208–227. https://doi.org/10.1002/bbb.1944
Fernando IPS, Lee WW, Han EJ, Ahn G (2019) Alginate-based nanomaterials: fabrication techniques, properties, and applications. Chem Eng J 391:123823. https://doi.org/10.1016/j.cej.2019.123823
Ferreira SA, Coutinho PJG, Gama FM (2011) Synthesis and characterization of self-assembled nanogels made of pullulan. Materials (Basel) 4:601–620. https://doi.org/10.3390/ma4040601
Furuno T, Nakanishi M (2012) Kefiran suppresses antigen-induced mast cell activation. Biol Pharm Bull 35:178–183. https://doi.org/10.1248/bpb.35.178
Gomes TD, Caridade SG, Sousa MP et al (2018) Adhesive free-standing multilayer films containing sulfated levan for biomedical applications. Acta Biomater 69:183–195. https://doi.org/10.1016/j.actbio.2018.01.027
González-Garcinuño Á, Tabernero A, Domínguez Á et al (2018) Levan and levansucrases: polymer, enzyme, micro-organisms and biomedical applications. Biocatal Biotransformation 36:233–244. https://doi.org/10.1080/10242422.2017.1314467
Gorgieva S, Trček J (2019) Bacterial cellulose: production, modification and perspectives in biomedical applications. Nanomaterials 9:1352
Güngör G, Gedikli S, Toptaş Y et al (2019) Bacterial hyaluronic acid production through an alternative extraction method and its characterization. J Chem Technol Biotechnol 94:1843–1852. https://doi.org/10.1002/jctb.5957
Hernández N, Williams RC, Cochran EW (2014) The battle for the "green" polymer. Different approaches for biopolymer synthesis: bioadvantaged vs. bioreplacement. Org Biomol Chem 12:2834–2849. https://doi.org/10.1039/c3ob42339e
Highley CB, Prestwich GD, Burdick JA (2016) Recent advances in hyaluronic acid hydrogels for biomedical applications. Curr Opin Biotechnol 40:35–40. https://doi.org/10.1016/j.copbio.2016.02.008
Islam S, Bhuiyan MAR, Islam MN (2017) Chitin and chitosan: structure, properties and applications in biomedical engineering. J Polym Environ 25:854–866. https://doi.org/10.1007/s10924-016-0865-5
Jahns AC, Haverkamp RG, Rehm BHA (2008) Multifunctional inorganic-binding beads self-assembled inside engineered bacteria. Bioconjug Chem 19:2072–2080. https://doi.org/10.1021/bc8001979
Jain D, Bar-Shalom D (2014) Alginate drug delivery systems: application in context of pharmaceutical and biomedical research. Drug Dev Ind Pharm 40:1576–1584. https://doi.org/10.3109/03639045.2014.917657
Jia Y, Zhu W, Zheng M et al (2018) Bacterial cellulose/hyaluronic acid composite hydrogels with improved viscoelastic properties and good thermodynamic stability. Plast Rubber Compos 47:165–175. https://doi.org/10.1080/14658011.2018.1447340
Keshk SMAS, Nada AMA (2016) Heterogeneous derivatization of bacterial and plant cellulose. Biosci Biotechnol Res Asia 1:39–42

Kirschning A, Dibbert N, Dräger G (2018) Chemical functionalization of polysaccharides—towards biocompatible hydrogels for biomedical applications. Chemistry 24:1231. https://doi.org/10.1002/chem.201701906

Kumar D, Saini N, Pandit V, Ali S (2012) An insight to pullulan: a biopolymer in pharmaceutical approaches. Int J Basic Appl Sci 1:202–219. https://doi.org/10.14419/ijbas.v1i3.101

Kumar A, Rao KM, Han SS (2018) Application of xanthan gum as polysaccharide in tissue engineering: a review. Carbohydr Polym 180:128–144. https://doi.org/10.1016/j.carbpol.2017.10.009

Kunduru KR, Basu A, Domb AJ (2016) Biodegradable polymers: medical applications. Encycl Polym Sci Technol 1–22

Lahaye M, Robic A (2007) Structure and function properties of Ulvan, a polysaccharide from green seaweeds. Biomacromolecules 8:1765–1774. https://doi.org/10.1021/bm061185q

Laws A, Gu Y, Marshall V (2001) Biosynthesis, characterisation, and design of bacterial exopolysaccharides from lactic acid bacteria. Biotechnol Adv 19:597–625. https://doi.org/10.1016/S0734-9750(01)00084-2

Lee KY, Mooney DJ (2012) Alginate: properties and biomedical applications. Prog Polym Sci 37:106–126. https://doi.org/10.1016/j.progpolymsci.2011.06.003

Lee N, Go T, Lee S et al (2014) In vitro evaluation of new functional properties of poly-c-glutamic acid produced by Bacillus subtilis D7. Saudi J Biol Sci 21:153–158. https://doi.org/10.1016/j.sjbs.2013.09.004

Li B, Lu F, Wei X, Zhao R (2008) Fucoidan: structure and bioactivity. Molecules 13:1671–1695. https://doi.org/10.3390/molecules13081671

Liguori R, Ventorino V, Pepe O, Faraco V (2016) Bioreactors for lignocellulose conversion into fermentable sugars for production of high added value products. Appl Microbiol Biotechnol 100:597–611. https://doi.org/10.1007/s00253-015-7125-9

Löbler M, Saß M, Kunze C et al (2002) Biomaterial implants induce the inflammation marker CRP at the site of implantation. J Biomed Mater Res 61:165–167. https://doi.org/10.1002/jbm.10155

Luft L, Confortin TC, Todero I et al (2020) An overview of fungal biopolymers: bioemulsifiers and biosurfactants compounds production. Crit Rev Biotechnol 0:1–22. https://doi.org/10.1080/07388551.2020.1805405

Lv X, Yang J, Feng C et al (2016) Bacterial cellulose-based biomimetic nanofibrous scaffold with muscle cells for hollow organ tissue engineering. ACS Biomater Sci Eng 2:19–29. https://doi.org/10.1021/acsbiomaterials.5b00259

Maia J, Evangelista MB, Gil H, Ferreira L (2014) Dextran-based materials for biomedical applications. In: Gil MH (ed) Carbohydrates applications in medicine. Research Signpost, pp 31–53

Medrano M, Racedo SM, Rolny IS et al (2011) Oral administration of kefiran induces changes in the balance of immune cells in a murine model. J Agric Food Chem 59:5299–5404. https://doi.org/10.1021/jf1049968

Miao T, Wang J, Zeng Y et al (2018) Polysaccharide-based controlled release systems for therapeutics delivery and tissue engineering: from bench to bedside. Adv Sci 5:1700513. https://doi.org/10.1002/advs.201700513

Micheli L, Uccelletti D, Palleschi C, Crescenzi V (1999) Isolation and characterisation of a ropy Lactobacillus strain producing the exopolysaccharide kefiran. Appl Microbiol Biotechnol 53:69–74. https://doi.org/10.1007/s002530051616

Mishra B, Suneetha V, Rath K (2011) The role of microbial pullulan, a biopolymer in pharmaceutical approaches: a review. J Appl Pharm Sci 1:45–50

Mohan S, Oluwafemi OS, Kalarikkal N, Thomas S, Songca SP (2016) Biopolymers–application in nanoscience and nanotechnology. In: Recent advances in biopolymers. IntechOpen, pp 47–66

Moradali MF, Rehm BHA (2020) Bacterial biopolymers: from pathogenesis to advanced materials. Nat Rev Microbiol 18:195–210. https://doi.org/10.1038/s41579-019-0313-3

Moradi Z, Kalanpour N (2019) Kefiran, a branched polysaccharide: preparation, properties and applications: Kefiran, a branched polysaccharide: preparation, properties and applications: a review. Carbohydr Polym 223:115100. https://doi.org/10.1016/j.carbpol.2019.115100

Moreno SNJ, Docampo R (2013) Polyphosphate and its diverse functions in host cells and pathogens. PLoS Pathog 9:e1003230. https://doi.org/10.1371/journal.ppat.1003230

Moscovici M (2015) Present and future medical applications of microbial exopolysaccharides. Front Microbiol 6:1012. https://doi.org/10.3389/fmicb.2015.01012

Narancic T, Cerrone F, Beagan N, O'Connor KE (2020) Recent advances in bioplastics: application and biodegradation. Polymers (Basel) 12:920. https://doi.org/10.3390/POLYM12040920

Nofar M, Sacligil D, Carreau PJ et al (2019) Poly (lactic acid) blends: processing, properties and applications. Int J Biol Macromol 125:307–360. https://doi.org/10.1016/j.ijbiomac.2018.12.002

Oliveira JT, Martins L, Picciochi R et al (2010) Gellan gum: a new biomaterial for cartilage tissue engineering applications. J Biomed Mater Res A 93:852–863. https://doi.org/10.1002/jbm.a.32574

Onar N (2004) Usage of biopolymers in medical applications. In: 3rd Indo-Czech textile research conference

Osmałek T, Froelich A, Tasarek S (2014) Application of gellan gum in pharmacy and medicine. Int J Pharm 466:328–340. https://doi.org/10.1016/j.ijpharm.2014.03.038

Palumbo FS, Federico S, Pitarresi G et al (2020) Gellan gum-based delivery systems of therapeutic agents and cells. Carbohydr Polym 229:115430. https://doi.org/10.1016/j.carbpol.2019.115430

Parlane NA, Chen S, Jones GJ et al (2016) Display of antigens on polyester inclusions lowers the antigen concentration required for a bovine tuberculosis skin test. Clin Vaccine Immunol 23:19–26. https://doi.org/10.1128/CVI.00462-15

Pattanashetti NA, Heggannavar GB, Kariduraganavar MY (2017) Smart biopolymers and their biomedical applications. Proc Manuf 12:263–279. https://doi.org/10.1016/j.promfg.2017.08.030

Pawar RP, Tekale SU, Shisodia SU et al (2014) Biomedical applications of poly (lactic acid). Rec Pat Regen Med 4:40–51

Petri DFS (2015) Xanthan gum: a versatile biopolymer for biomedical and technological applications. J Appl Polym Sci 132. https://doi.org/10.1002/app.42035

Picheth GF, Pirich CL, Sierakowski MR et al (2017) Bacterial cellulose in biomedical applications: a review. Int J Biol Macromol 104:97–106. https://doi.org/10.1016/j.ijbiomac.2017.05.171

Prajapati VD, Jani GK, Khanda SM (2013) Pullulan: an exopolysaccharide and its various applications. Carbohydr Polym 95:540–549. https://doi.org/10.1016/j.carbpol.2013.02.082

Prameela K, Mohan CM, Ramakrishna C (2018) Biopolymers for food design: consumer-friendly natural ingredients. In: Biopolymers for food design. Academic, pp 1–32

Qureshi D, Nayak SK, Anis A, Ray SS, Kim D, Nguyen TTH, Pal K (2020) Introduction of biopolymers: food and biomedical applications. In: Biopolymer-based formulations. Elsevier Inc., pp 1–45

Rahman Bhuiyan MA, Shaid A, Khan MA (2014) Cationization of cotton fiber by chitosan and its dyeing with reactive dye without salt. Chem Mater Eng 2:96–100. https://doi.org/10.13189/cme.2014.020402

Ray S, Kalia VC (2017) Biomedical applications of polyhydroxyalkanoates. Indian J Microbiol 57:261–269. https://doi.org/10.1007/s12088-017-0651-7

Rebah FB, Mnif W, Siddeeg SM (2018) Microbial flocculants as an alternative to synthetic polymers for wastewater treatment: a review. Symmetry (Basel) 10:556. https://doi.org/10.3390/sym10110556

Rebelo R, Fernandes M, Fangueiro R (2017) Biopolymers in medical implants: a brief review. Proc Eng 200:236–243. https://doi.org/10.1016/j.proeng.2017.07.034

Remminghorst U, Rehm BHA (2006) Bacterial alginates: from biosynthesis to applications. Biotechnol Lett 28:1701–1712. https://doi.org/10.1007/s10529-006-9156-x

Rinaudo M (2008) Main properties and current applications of some polysaccharides as biomaterials. Polym Int 57:397–430. https://doi.org/10.1002/pi.2378

Rodríguez-Contreras AMSM-CFJG, Manero JM (2016) Modification of titanium surfaces by adding antibiotic-loaded PHB spheres and PEG for biomedical applications. J Mater Sci Mater Med 27:124

Sachan N, Pushkar S, Jha A, Bhattcharya A (2009) Sodium alginate: the wonder polymer for controlled drug delivery. J Pharm Res 2:1191–1199

Saharan B, Ankita SD (2012) Bioplastics-for sustainable development: a review. Int J Microb Resour Technol 1:11–23

Saini P, Arora M, Kumar MNVR (2016) Poly(lactic acid) blends in biomedical applications. Adv Drug Deliv Rev 107:47–59. https://doi.org/10.1016/j.addr.2016.06.014

Salek K, Gutierrez T (2016) Surface-active biopolymers from marine bacteria for potential biotechnological applications. AIMS Microbiol 2:92–107. https://doi.org/10.3934/microbiol.2016.2.92

Shahidul Islam M, Ashraful Islam Molla M, Sarker M et al (2011) Fabrication of pllulan/silver nanoparticle composite nanospheres using electrospray technique for antibacterial applications. Int J Basic Appl Sci 11:36–40

Shingel KI, Petrov PT (2002) Behavior of γ-ray-irradiated pullulan in aqueous solutions of cationic (cetyltrimethylammonium hydroxide) and anionic (sodium dodecyl sulfate) surfactants. Colloid Polym Sci 280:176–182. https://doi.org/10.1007/s00396-001-0599-2

Singh M, Kumar P, Ray S, Kalia VC (2015) Challenges and opportunities for customizing polyhydroxyalkanoates. Indian J Microbiol 55:235–249. https://doi.org/10.1007/s12088-015-0528-6

Singh RS, Kaur N, Rana V, Kennedy JF (2017) Pullulan: a novel molecule for biomedical applications. Carbohydr Polym 171:102–121. https://doi.org/10.1016/j.carbpol.2017.04.089

Singhvi G, Hans N, Shiva N, Kumar Dubey S (2019) Xanthan gum in drug delivery applications. In: Natural polysaccharides in drug delivery and biomedical applications. Academic, pp 121–144

Song Z, Zhang Y, Shao H et al (2019) Effect of xanthan gum on the prevention of intra-abdominal adhesion in rats. Int J Biol Macromol 126:531–538. https://doi.org/10.1016/j.ijbiomac.2018.12.233

Spera R (2018) Gellan gum for tissue engineering applications: a mini review. Biomed J Sci Tech Res 7:5785–5786. https://doi.org/10.26717/bjstr.2018.07.001474

Srikanth R, Reddy CHSSS, Siddartha G et al (2015a) Review on production, characterization and applications of microbial levan. Carbohydr Polym 120:102–114. https://doi.org/10.1016/j.carbpol.2014.12.003

Srikanth R, Siddartha G, Reddy CHSSS et al (2015b) Antioxidant and anti-inflammatory levan produced from Acetobacter xylinum NCIM2526 and its statistical optimization. Carbohydr Polym 123:8–16. https://doi.org/10.1016/j.carbpol.2014.12.079

Sudha PN (2017) Industrial applications of marine biopolymers. CRC Press, Taylor & Francis Group

Sun G, Mao JJ (2012) Engineering dextran-based scaffolds for drug delivery and tissue repair. Nanomedicine 7:1771–1784. https://doi.org/10.2217/nnm.12.149

Szekalska M, Puciłowska A, Szymańska E et al (2016) Alginate: current use and future perspectives in pharmaceutical and biomedical applications. Int J Polym Sci 2016:7697031. https://doi.org/10.1155/2016/7697031

Tabasum S, Noreen A, Maqsood MF et al (2018) A review on versatile applications of blends and composites of pullulan with natural and synthetic polymers. Int J Biol Macromol 120:603–632. https://doi.org/10.1016/j.ijbiomac.2018.07.154

Tammi MI, Day AJ, Turley EA (2002) Hyaluronan and homeostasis: a balancing act. J Biol Chem 277:4581–4584. https://doi.org/10.1074/jbc.R100037200

Thulasidasan AKT, Retnakumari AP, Shankar M et al (2017) Folic acid conjugation improves the bioavailability and chemosensitizing efficacy of curcumin-encapsulated PLGA-PEG nanoparticles towards paclitaxel chemotherapy. Oncotarget 8:107374. https://doi.org/10.18632/oncotarget.22376

Tian J, Min Y, Rodgers Z et al (2017) Co-delivery of paclitaxel and cisplatin with biocompatible PLGA-PEG nanoparticles enhances chemoradiotherapy in non-small cell lung cancer models. J Mater Chem B 5:6049–6057. https://doi.org/10.1039/c7tb01370a
Tiwari S, Bahadur P (2019) Modified hyaluronic acid based materials for biomedical applications. Int J Biol Macromol 121:556–571. https://doi.org/10.1016/j.ijbiomac.2018.10.049
Tomulescu C, Stoica R, Sevcenco C et al (2016) Levan—a mini review. Sci Bull Ser F Biotechnol 20:309–317
Tsouko E, Kourmentza C, Ladakis D et al (2015) Bacterial cellulose production from industrial waste and by-product streams. Int J Mol Sci 16:14832–14849. https://doi.org/10.3390/ijms160714832
U.S. Congress O of TA (2000) Technical overview of biopolymer field. Biopolym Mak Mater Nature's W 39:19–50
Ullah H, Wahid F, Santos HA, Khan T (2016) Advances in biomedical and pharmaceutical applications of functional bacterial cellulose-based nanocomposites. Carbohydr Polym 150:330–352. https://doi.org/10.1016/j.carbpol.2016.05.029
Usman A, Khalid S, Usman A et al (2017) Algal polysaccharides, novel application, and outlook. In: Algae based polymers, blends, and composites. Elsevier, pp 115–153
Usov AI (1998) Structural analysis of red seaweed galactans of agar and carrageenan groups. Food Hydrocoll 12:301–308. https://doi.org/10.1016/S0268-005X(98)00018-6
Verma ML, Kumar S, Jeslin J, Dubey NK (2020) Microbial production of biopolymers with potential biotechnological applications. In: Biopolymer-based formulations. Elsevier Inc., pp 105–137
Vijayendra SVN, Shamala TR (2014) Film forming microbial biopolymers for commercial applications—a review. Crit Rev Biotechnol 34:338–357. https://doi.org/10.3109/07388551.2013.798254
Von Hoff DD, Mita MM, Ramanathan RK et al (2016) Phase I study of PSMA-targeted docetaxel-containing nanoparticle BIND-014 in patients with advanced solid tumors. Clin Cancer Res 22:3157–3163. https://doi.org/10.1158/1078-0432.CCR-15-2548
Wang S, Yuan F, Chen G et al (2014) Dextran-based thermo-responsive hemoglobin-polymer conjugates with oxygen-carrying capacity. RSC Adv 4:52940–52948. https://doi.org/10.1039/c4ra06397j
Wang L, Yan J, Wise MJ et al (2018a) Distribution patterns of polyphosphate metabolism pathway and its relationships with bacterial durability and virulence. Front Microbiol 9:782. https://doi.org/10.3389/fmicb.2018.00782
Wang X, Schröder HC, Müller WEG (2018b) Amorphous polyphosphate, a smart bioinspired nano-/bio-material for bone and cartilage regeneration: towards a new paradigm in tissue engineering. J Mater Chem B 6:2385–2412. https://doi.org/10.1039/c8tb00241j
Wang X, Wang X, Hui K et al (2018c) Highly effective polyphosphate synthesis, phosphate removal, and concentration using engineered environmental bacteria based on a simple solo medium-copy plasmid strategy. Environ Sci Technol 52:214–222. https://doi.org/10.1021/acs.est.7b04532
Wani SJ, Shaikh SS, Sayyed RZ (2016) Microbial biopolymers in biomedical field. MOJ Cell Sci Rep 3:65–67. https://doi.org/10.15406/mojcsr.2016.03.00055
Wankhade V (2020) Animal-derived biopolymers in food and biomedical technology. In: Biopolymer-based formulations. Elsevier Inc., pp 139–152
Watanabe M, Maemura K, Kanbara K et al (2002) GABA and GABA receptors in the central nervous system and other organs. Int Rev Cytol 213:1–47. https://doi.org/10.1016/S0074-7696(02)13011-7
Wróblewska-Krepsztul J, Rydzkowski T, Michalska-Pożoga I, Thakur VK (2019) Biopolymers for biomedical and pharmaceutical applications: recent advances and overview of alginate electrospinning. Nanomaterials 9:404. https://doi.org/10.3390/nano9030404

Yadav P, Yadav H, Shah VG et al (2015) Biomedical biopolymers, their origin and evolution in biomedical sciences: a systematic review. J Clin Diagn Res 9:21–25. https://doi.org/10.7860/JCDR/2015/13907.6565
Zargar V, Asghari M, Dashti A (2015) A review on chitin and chitosan polymers: structure, chemistry, solubility, derivatives, and applications. ChemBioEng Rev 2:204–226. https://doi.org/10.1002/cben.201400025
Zhan XB, Lin CC, Zhang HT (2012) Recent advances in curdlan biosynthesis, biotechnological production, and applications. Appl Microbiol Biotechnol 93:525–531. https://doi.org/10.1007/s00253-011-3740-2
Zhang J, Cao Q, Li S et al (2013) 3-Hydroxybutyrate methyl ester as a potential drug against Alzheimer's disease via mitochondria protection mechanism. Biomaterials 34:7552–7562. https://doi.org/10.1016/j.biomaterials.2013.06.043
Zhang X, Zheng Y, Wang Z et al (2014) Methotrexate-loaded PLGA nanobubbles for ultrasound imaging and Synergistic Targeted therapy of residual tumor during HIFU ablation. Biomaterials 35:5148–5161. https://doi.org/10.1016/j.biomaterials.2014.02.036

Mycobacterium Biofilms Synthesis, Ultrastructure, and Their Perspectives in Drug Tolerance, Environment, and Medicine

19

Kundan Kumar Chaubey, Mohd. Abdullah, Saurabh Gupta, Manthena Navabharath, and Shoor Vir Singh

Abstract

Bacteria protect their cells from harsh environmental conditions by naturally forming biofilms. Biofilms consist of bacterial cells, exopolysaccharides, extracellular protein as well as extracellular DNA regulated by different genetic and environmental factors include medium, nutrient availability, matrix properties, pH, temperature, organic material, etc. During biofilm synthesis, bacteria use quorum sensing activity for cell-to-cell signaling for biofilm development as well as biofilm maintenance. In mycobacteria, glycopeptidolipids are supportive in biofilms formation and surface attachment. Mycolic acids of acid-fast bacteria contribute to the formation of the mycobacterial envelope with overall structure and functions. It also provides a permeability barrier to resist the microbes from many common therapeutic agents like antibiotics and disinfectants. Several reports showed that drug-tolerant biofilms formed by mycobacterial species were resistant to antibiotics and disinfectants in vitro. Therefore, in this time we have to focus to develop new strategies to combat biofilm formation in mycobacterium and other pathogenic bacterial species which give drug tolerance, host defense, and protection from adverse environmental conditions.

Keywords

Biofilm · Mycobacteria · Environmental factor · Exopolysaccharides · Extracellular · DNA · Extracellular protein · Quorum sensing · Drug tolerance

K. K. Chaubey (✉) · M. Abdullah · S. Gupta · M. Navabharath · S. V. Singh
Department of Biotechnology, Institute of Applied Sciences and Humanities, GLA University, Mathura, Uttar Pradesh, India
e-mail: kundan.chaubey@gla.ac.in

A. Vaishnav, D. K. Choudhary (eds.), *Microbial Polymers*,
https://doi.org/10.1007/978-981-16-0045-6_19

19.1 Introduction

Biofilms synthesized by bacteria contribute to persistent chronic infections raising serious health concern worldwide by their capabilities to tolerate drugs (antibiotics, disinfectants, etc.), host defense systems, and other external or environmental stresses (Costerton et al. 1999). Biofilm synthesis is multistep, well-regulated process such as (I) adsorption of molecules (macro and micro) to host surfaces, (II) adhesion of bacteria on host surface triggers the discharge of extracellular polymeric substances (EPS), and (III) Bacterial colonization and biofilm maturation (Fig. 19.1).

Biofilms save the microorganism from altered pH, nutrients scarcity, osmolarity, mechanical, and shearing forces (Fux et al. 2005; McCarty et al. 2012) as well as access of bacterial biofilm communities by antibiotics and various cells responsible for the host's immunity (Stewart and William Costerton 2001). So, the matrix of biofilms provides additional power of resistance to bacterial pathogens to tolerate the unfavorable and harsh conditions as well as resistance to various antibiotics which open the way for escalation of pathogenic infections like multidrug resistant (MDR), extensively drug resistant (EDR), and totally drug resistant (TDR). Biofilm synthesis in mycobacteria mostly similar to other biofilm communities, though, some of the mycobacterial species may develop biofilms on surfaces and air-media interface (Ojha et al. 2008). The pathway of synthesis of biofilms is depicted in Fig. 19.2. Mostly surface fimbriae or pili are found absent in mycobacteria, so some mycobacterial proteins play a potential role in accumulation as well as in attachment of cells of mycobacterial species (Menozzi et al. 1996). Exopolysaccharide formed in

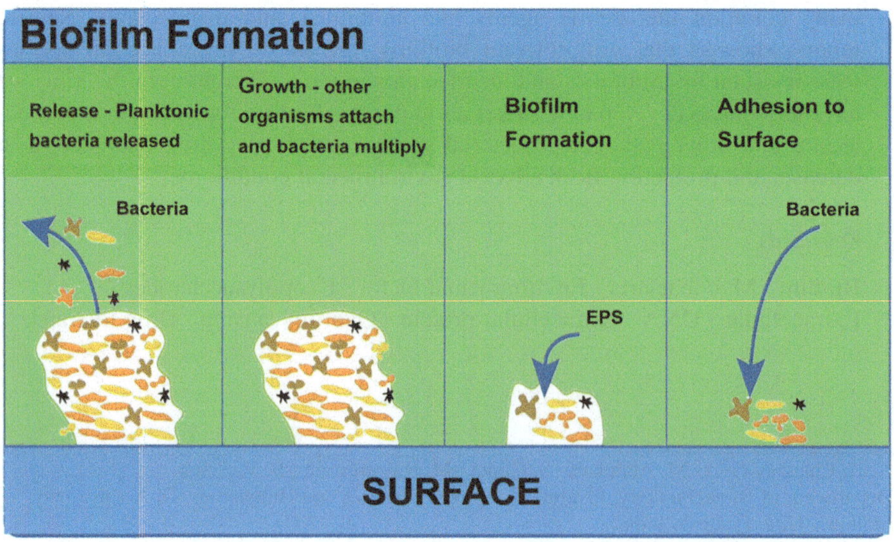

Fig. 19.1 Bacterial adhesion, biofilm formation, bacterial growth, and release of planktonic bacteria

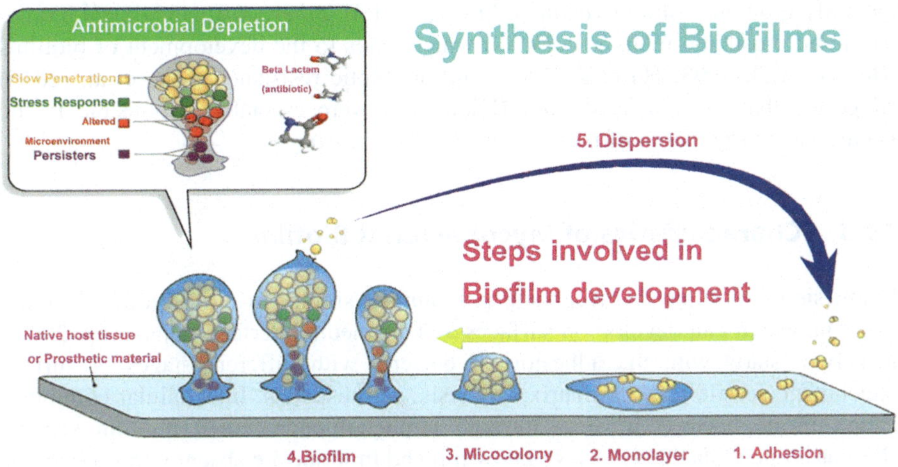

Fig. 19.2 Pathway of synthesis of biofilms

various biofilm-producing bacteria gives the ability to attach diverse surfaces and development of biofilms are found absent in mycobacteria (Zamora et al. 2007; Ojha et al. 2015). Initially, glycopeptidolipids play a crucial role in surface attachment during biofilm synthesis by *Mycobacterium smegmatis* (Recht and Kolter 2001). In nonmotile mycobacteria, shorter chain of mycolic acids play important role in the synthesis of biofilm structure (Ojha et al. 2008). A potential permeability barrier of Mycolic acids gives resistance in mycobacteria to antibiotics (Zambrano and Kolter 2005) and essential role in sliding motility related to biofilm spreading (Recht et al. 2000; Recht and Kolter 2001). Esteban et al. (2008) and Ojha et al. (2015) reported about GroEL1 chaperone which is responsible for the biofilm development in *M. smegmatis*. In various studies, several mycobacterial species have been found to form antibiotic and disinfectant resistant biofilms using genetically controlled mechanisms where metabolic state and activation of resistance genes play a crucial role. So, observations of this study make us focused to develop new strategies to combat biofilm synthesis in mycobacterium and other pathogenic bacteria for their survival in adverse environmental conditions.

19.2 History of Mycobacterial Biofilms

Costerton et al. (1978) given their initial observations on the modern concept of biofilm and in the next decade Wallace Jr. (1987) and Schulze-Robbecke and Fischeder (1989) discussed about environmental mycobacterial biofilms. Esteban and Marta (2018) have discussed the phenomenon of mycobacterial cells forming "aggregates" or "pellicles." Subsequently, Ojha et al. (2008) described *M. tuberculosis* forming "pellicles" in vitro/liquid media and Esteban and Marta (2018) given similar descriptions for avian bacilli and others that mycobacteria naturally grow in biofilm structures. Infection of *M. tuberculosis* in clinical biomaterial and prosthetic joints is common which leads to the development of biofilms (Berbari et al. 1998; Ha et al. 2005) and antibiotic resistance. Ojha et al. (2015) suggested that mycolic acids and DNA, both were crucially responsible in the synthesis and regulation of biofilms in *M. tuberculosis*.

19.3 Characteristics of Mycobacterial Biofilm

Synthesis of biofilms in mycobacteria is almost similar to other biofilm-forming bacteria, which can develop on surfaces and air-media interface (Ojha et al. 2008). Mostly it starts with the adhesion of bacteria with different stages of surface attachment, sessile growth, matrix synthesis, and dispersal. Intercellular communication inside bacteria governs through a phenomenon called quorum sensing (Richards and Ojha 2014). As we know pili and fimbriae are absent in mycobacteria and certain proteins potentially responsible for the aggregation and attachment of mycobacteria to other cells (Menozzi et al. 1996). Some of the bacteria produce exopolysaccharide components of the extracellular matrix which help in the attachment to different surfaces and development of biofilms; however, it is absent in mycobacterial species (Zamora et al. 2007; Zambrano and Kolter 2005; Ojha et al. 2015). It has established that nutrients, ions (Ca^{2+}, Zn^{2+}, and Mg^{2+}), and various carbon sources (glucose and peptone) effect and regularize the bacterial behavior and synthesis of biofilms (Carter et al. 2003). Esteban et al. (2008) reported that rapidly growing mycobacteria (RGM) use tap water as the nutrient source to develop their biofilm. Though biofilms synthesized by *Mycobacterium avium* complex organisms were among the first mycobacterial biofilms ever described (Carter et al. 2003). Many non-tuberculosis mycobacteria (NTM) species forming heterospecies biofilms in environmental specimens (Falkinham III 2002, 2009). In nonmotile mycobacteria, shorter-chain mycolic acids may form a hydrophobic extracellular matrix and play a crucial role in biofilm synthesis (Ojha et al. 2005). Zambrano and Kolter (2005) reported high resistance of mycobacteria to antibiotics and disinfectants is attributed mainly to mycolic acids present in the cell wall, which help provide a permeability barrier.

19.4 Ultrastructure of Biofilm

A group of sessile microbial communities is called microbial biofilm which is attached with substratum and surrounded in the self-produced pool of noncrystalline extracellular polymeric matrix (Hoiby et al. 2011). Bacterial biofilms protect bacteria and make them advance to save from the diverse environmental stresses like desiccation, antimicrobials attack by the immune system, and ingestion by protozoa (Wilkins et al. 2014). Some bacterial species use quorum sensing (QS) for communication within cells of biofilms to regulate the specific gene's expression and coordinate the disassembly of the biofilm community (Solano et al. 2014).

19.5 Resistance to Antibiotics

Biofilm producing bacteria raise antibiotic resistance which contributes to chronic infections because it helps in slow or incomplete penetration of the antibiotics into the biofilm (Kumon et al. 1994; Mah and O'Toole 2001). Multicellular nature of biofilm communities is the main key factor of antibiotics resistance. Extracellular polymeric substances (EPS) produced by mycobacterial cells hold them together and develop multicellular consortia for heterogeneous environment in biofilms that initiates the biofilm to function as a multicellular system. Here, signaling between the cells helps in biofilm maturation into complex structures which is called quorum sensing (QS) process (Davies et al. 1998). This nature of multicellularity of bacterial communities forming biofilm cause antibiotics resistance; if we interfere with the complex structure and integrity of biofilms that increase the efficacy of antibiotics and the host defenses which leads to quick treatment of this persistent pathogenicity. Complexities developed by biofilm formation causing treatment complication in the series of human infections on various biological implants (heart and urinary catheters, replacement of heart valves, and joint implants) (Singh et al. 2000). EPS produced during biofilm formation reduces the effect of antibiotics via diffusion–reaction inhibition phenomenon. Antibiotic activity may be chelated by complex formed between EPS and antibiotics and may be degraded through enzymatically based reactions (Oubekka et al. 2002). Biofilms producing bacterial communities only can survive under antibiotics stress when they are in stationary phase and viable-but-nonculturable state (Li et al. 2014). Biofilms possess most of the cells in the stationary phase which have the decreased antibiotics susceptibility, among them at least 1% becomes tolerant to antibiotics (Maisonneuve and Gerdes 2014) and older biofilms show higher tolerance to antibiotics (Monzón et al. 2002). Uptake of resistance genes by horizontal gene transfer is the other mechanism of antibiotic resistance of biofilms communities (Mah 2012). Various studies reported that in vitro biofilms production by mycobacterial species was found resistant to antibiotics or disinfectants (Greendyke and Byrd 2008; Ortiz-Perez et al. 2011). Greendyke and Byrd (2008) reported that metabolic state and activation of resistance genes are crucial for the progress in mycobacterial antibiotics resistance.

19.6 Mycobacterial Biofilms in the Environment

The capacity of mycobacteria to live in biofilms presents these creatures with numerous favorable circumstances over the planktonic type of growth, including, for instance, protection from ecological animosities, and mycobacteria that grow in biofilms have associated with natural sources (Schulze-Robbecke and Fischeder 1989). Numerous reports of ecological examinations have shown the presence of these mycobacteria, particularly in water system (Falkinham III 2002). As of late, *Mycobacterium chimaera* has arisen as a significant nosocomial microbe related to defilement of heater-cooler units for heart medical procedure all through the world (Kohler et al. 2015), and extraordinary purification measures have been important to eliminate mycobacterial biofilms from these devices (Garvey et al. 2016). It has as of late proposed that rising numbers of NTM secluded from clinical examples could be associated with changes in disinfectants utilized in plumbing frameworks, specifically, the change from chlorine to chloramine in water frameworks in the USA (Falkinham III et al. 2016). A fascinating issue is the connection between NTM and free-living amoebae (Vaerewijck et al. 2005). It is notable that mycobacteria can be intracellular parasites, which may clarify the likely connection between NTM and amoeba in water and biofilm samples (Marciano-Cabral et al. 2010).

19.7 Mycobacterial Biofilms in Medicine: Clinical Implications

Since the idea arose, biofilms have been perceived as a particularly significant pathogenic factor in human diseases (Hall-Stoodley and Stoodley 2005). Even though acute infection is for the most part brought about by planktonic creatures, the pathogenesis of the chronic disease is firmly connected with biofilm development. Also, the developing significance of biomaterials, for example, catheters or prostheses, in present-day medication has added to the importance of biofilms and their administration in human infection (Patino et al. 2008). The clinical pertinence of mycobacterial biofilms can be dissected in two gatherings: biofilms in NTM infection and biofilms in tuberculosis. These two gatherings will be investigated underneath.

19.7.1 Nontuberculous Mycobacterial Disease

The range of infections brought about by NTM is wide, yet practically every one of them share a typical trademark: they are normally chronic. One of the most widely recognized disorders brought about by NTM is a respiratory disease (Esteban et al. 2012), caused either by gradually developing mycobacteria, for example, M. avium complex or quickly developing species, for example, M. abscessus. Respiratory sickness typically influences patients with inclining conditions, for example, old tuberculosis scars, silicosis, bullae, and other lung cavities where NTM can build up a biofilm that initially colonizes the host, and later causes intrusive illness. Mycobacterial biofilms have as of late been distinguished in histological samples from lung cavitary disease (Fennelly et al. 2016). Test data shows the significance of biofilm advancement in the potential of *M. avium* to attack bronchial epithelial cells (Esteban and Marta 2018).

Another significant biofilm-related class of diseases, and the most distinctive, are those related to biomaterial. NTM has been accounted for as the reason for various conditions, the most usually isolated organism being RGM. These organism act as aggressive microorganisms and have been portrayed in a wide range of device-related diseases. Among these, catheter-related bacteremia has been accounted for in various populaces, including malignant growth patients (Esteban and Marta 2018). Other devices that can be contaminated with NTM are prosthetic joints. The most well-known causative agent is RGM, yet gradually developing mycobacteria have additionally been secluded, though seldom. Other biomaterial-related disease incorporates abdominal mesh infection, transplant-related keratitis, pacemaker infection, mammoplasty infection, prosthetic valve endocarditis, and others. Practically these conditions require embed expulsion since antibiotic treatment alone is powerless for eradicating the sessile organisms (Esteban and Marta 2018).

19.7.2 Mycobacterium Tuberculosis Disease

Seemingly one of the most intriguing discoveries with regard to ongoing years in the field of mycobacterial biofilm research is the revelation that M. tuberculosis can build up a biofilm. The primary reports originated from instances of tuberculosis infection related to clinical biomaterial, prosthetic joints specifically. Clinical perceptions prompted that biomaterial expulsion was important to deal with these infections, regardless of whether the M. tuberculosis strain was sensitive in vitro to the antibiotics used. The trouble in treating these diseases could be because of the way that biofilm is a rooted system of antibiotic resistance (Esteban and Marta 2018). Although, the job of biofilms in the pathogenesis of tuberculosis stays hazy. This pathogenic process is tricky, and includes, in particular, intracellular endurance and host defense evasion mechanisms. It has been recommended that the significance of biofilms in this infection is because of their involvement in the process of caseous necrosis and cavitation formation in lung tissue, a site where *M. tuberculosis* could form a biofilm (Esteban and Marta 2018).

19.8 Infection Associated with Biofilm

Around 80% of chronic and intermittent microbial contaminations in the human body are because of bacterial biofilm. Microbial cells inside biofilms have indicated 10–1000 times a larger number of antibiotics resistance than the planktonic cells (Mah 2012). Biofilm is formed in assorted ecological niches, including freshwater waterways, rocks, deep-sea vents, and hydrothermal hot springs. Biofilm-related contaminations can be extensively divided into two kinds. The biofilms might be formed on the abiotic surfaces, particularly diseases related to inhabiting clinical gadgets (Donlan and Costerton 2002) and local biofilm diseases of host tissue (Burmolle et al. 2010). Urinary tract and bloodstream diseases caused by biofilm first formed on clinical inserts, for example, heart valves, catheters, contact lenses, joint prostheses, intrauterine gadgets, and dental unit. These diseases can be treated by the expulsion of the inserts which expanding the expense of the treatment as well as it gets risky for patients (Costerton et al. 2005). Host tissue related biofilm diseases are often chronic, including lung infections of cystic fibrosis patients, chronic osteomyelitis, chronic prostatitis, chronic rhinosinusitis, chronic otitis media, chronic wounds, recurrent urinary tract infection, endocarditis, periodontitis, and dental caries (Burmolle et al. 2010).

19.9 Biofilm Formation by *Mycobacterium smegmatis*

Mycobacterium smegmatis (Msm) is a rapidly growing, nonpathogenic species of mycobacteria that is frequently utilized as a model organism to study the physiology of mycobacteria. Recht et al. showed that transposon mutants in the glycol-peptido-lipids (GPL) biosynthesis pathway are weakened for colonization and biofilms synthesis (Chakraborty and Kumar 2019). GPLs are a significant part of the myco-bacterial cell wall, and it was seen that they play a significant function in the early attachment of mycobacterial cells to the base like Polyvinyl chloride (PVC). The same group later on showed that acetylation of GPL is likewise significant in determining the state morphology and formation of biofilms on PVC plates. Subsequently, mycolic acids, another significant segment of the mycobacterial cell divider, was embroiled in pellicle biofilm development. Some of the modern studies have exhibit estrated that in the pellicle, the mycobacterial cells are encapsulated in self-created EPS. Thus, mycobacterial pellicles are viewed as a type of biofilms (Chakraborty and Kumar 2019). The part of monomeromycolyl diacylglycerol (MMDAG) in cell-to cell connection and biofilm development was separately demonstrated in a different reported study (Purdy et al. 2013). In this study, a transposon mutant library was made and examined for defects in colony morphology. It was seen that transposon insertion in the Lsr2 (a histone-like protein in mycobacteria) prompts smooth, wet, and round colonies, contrasting to the dry, harsh, and rugose colonies of the parent Msm strain. This strain was also undermined in biofilm formation. Further investigation recommended that this transposon mutant strain has a comparable level of GPLs and mycolic acid however is insufficient in the MMDAGs (Chen et al. 2006). The part of Lsr2 in biofilm formation was established by another study which showed that Lsr2 mutant of Msm has expanded sliding motility, decreased surface hydrophobicity, and is lessened for pellicle development (Arora et al. 2008). The function of Lsr2 in the diverse genetically programmed phases of biofilm formation in Msm was exemplified by Yang et al. in 2017, where they observe the engagement of various genes in the individual phases of biofilm formation—from connection with the base to intercellular conglomeration followed by maturation of the pellicle architecture (Chakraborty and Kumar 2019).

19.10 Biofilm in Nontuberculosis Mycobacteria (NTM)

The genus Mycobacterium incorporates astounding human microbes, for example, Mycobacterium tuberculosis and Mycobacterium leprae, the two individuals from the M. tuberculosis complex (MTC), and an enormous gathering of nontuberculous mycobacteria (NTM). The NTM bunch includes more than 172 different species with distinct virulence features (http://www.bacterio.net/mycobacterium.html). NTM ordinarily displays saprophytic, commensally, and symbiotic behaviors (Faria et al. 2015). Albeit generally nonpathogenic, NTM are significant environmental adaptable microbes of humans and animals, including poultry and fish. The NTM are universal in nature sharing with human and animals a wide assortment of living spaces (Faria et al. 2015). The effect of NTM infection has been especially grave in immune-compromised individuals being related to life-threatening diseases in AIDS and transplanted patients. Yet, a rise in the occurrence of pulmonary diseases and healthcare-associated infections (HAI) in strong immune population featured the significance of NTM on human well-being. Clinical devices related diseases, one batch of HAI, normally connected with bacterial biofilm multiplication on these materials, have been portrayed for NTM. The universal idea of NTM even permits their diligence inside biofilms on other medical unit surfaces, for example, water pipes. Biofilm constancy inside medical care units speaks about a danger to human well-being since it favors the beginning and spread of HAI (Faria et al. 2015).

19.10.1 *Mycobacterium avium*

M. avium is competent for infecting people (Young et al. 1986), domestic animals (Thoen et al. 1981), and birds (Grange et al. 1990) and is found richly in different ecological niches, for example, water bodies and soil. Chakraborty and Kumar (2019) shown the presence of *M. avium* biofilms in the drinking water delivery system. *M. avium* infection is hard to treat and requires long treatment with numerous medications. Development of *M. avium* biofilms in vivo could elucidate the necessity of such a delayed treatment. Development of colony biofilms has been read broadly for *M. avium*. Sliding motility is subject to the presence of glycopeptidolipids (or GPLs are a class of amphiphilic molecules restricted in the cell envelope) and assumes a critical function in colony biofilm formation in *M. avium* (Chakraborty and Kumar 2019). Noteworthily, Yamazaki et al. noticed a connection between *M. avium* biofilm development and its ability to colonize the bronchial and bronchiolar mucosa. *M. avium* strains, namely MAC104, MAC101, and MAC A5 could attack and infect the bronchiolar epithelial cells. It might also be noted, these strains are competent enough to form biofilms in vitro. Despite that isogenic mutant clones of MAC A5 (namely 5G4, 6H9, and 9B5), that are attenuated to form biofilms could not attack and infect the bronchiolar epithelial cells. Moreover, it was seen that these mutant strains were improbable to infect mice lung while the MAC A5 was competent for infecting mice lungs, spleen, and liver. Moreover, M. avium biofilms or their supernatants were fit for instigating TNF-α production when matched up with their planktonic counterparts. This enormous TNF-α production in counter of M. avium biofilms prompts apoptotic cell demise of the macrophages (Chakraborty and Kumar 2019).

19.10.2 *Mycobacterium abscessus*

M. abscessus (Mab) is a rapid developing NTM that causes a wide array of human diseases, including those of lung, skin, soft tissue, ocular and central nervous system, and so on. Mab is arising as a major microorganism related with cystic fibrosis (Harris and Kenna 2014). In an elegant study, Qvist et al. (2015) shown the presence of Mab microcolonies encompassed by extracellular matrix in meager intra-alveolar walls. Microcolonies are microscopic communities of ~50 cells that suddenly aggregate and could nucleate the development of a biofilm. In a thought-provoking, another investigation by Fennelly et al. demonstrated the presence of Mab biofilms in the lung cavity of a patient with chronic obstructive pulmonary disease utilizing M. abscessus specific PNA-FISH probes. They further demonstrated that ~2% of Mabs in the lung cavity were in the biofilms and enclosed in the extracellular network (Fennelly et al. 2016). Interestingly, microbes living in biofilms display drug resilience.

19.10.3 *Mycobacterium fortuitum* and *Mycobacterium chelonae*

Other significant rapidly growing human pathogens are *M. fortuitum* and *M. chelonae*. These are astute human microorganisms that basically infect individuals with a compromised immune system or those enduring chronic illnesses. These species are known to prompt postsurgical infection and could form biofilms in eye or skin tissues. These species rapidly form biofilms within 48 h (Chakraborty and Kumar 2019). The biofilm formation is sturdy and does not rely on the nature of the substrate (Chakraborty and Kumar 2019). Critically, these biofilms are biocide resistant. The biofilms of *M. fortuitum* are impermeable to many antibiotics, including ciprofloxacin, and hence interpretating that permeability of biofilm for antibiotics might be responsible for antimicrobial drug resistance. Interestingly, these biofilms contain eDNA and a combination of antibiotics with DNase was more successful in distort the biofilms and killing biofilm inhabitants than antibiotics alone (Chakraborty and Kumar 2019).

19.10.4 *Mycobacterium ulcerans*

M. ulcerans causes buruli ulcer in humans. Buruli ulcer is the third most basic infection caused by a mycobacterial microorganism after tuberculosis and Leprosy. M. ulcerans is a moderate growing environmental bacterium that is competent enough to form biofilms on the salivary glands of the aquatic insect Naucoriscimicoides and on aquatic plants (Chakraborty and Kumar 2019). In an exquisite study, Marsollier et al. (2007) portrayed the biofilms of *M. ulcerans*. SEM examination uncovered that in the biopsy sample from Buruli ulcers patients and those segregated from infected mice mycobacterial cells were arranged into discrete bacterial groups encompassed in the extracellular matrix proposing the formation of biofilms in vivo (Chakraborty and Kumar 2019).

19.10.5 *Mycobacterium marinum*

M. marinum is a moderate developing bacterium that causes infection in fish and periodically infects people. Significantly, *M. marinum* could develop biofilms within 14 days on diversified abiotic surfaces. However, silicon surface yielded the most elevated levels of biofilm production. Hall-Stoodly also studied the ultrastructure of these biofilms and saw that the cording of mycobacterial cells advanced during the later period of biofilm development. This cording phenotype was concealed by oleic albumin dextrose catalase (OADC) supplement (Hall-Stoodley et al. 2006). Lipooligosaccharides are cell wall components and play a significant function in cell motility. M. marinum mutants not competent for forming lipooligosaccharides were defective in biofilm formation.

19.11 Conclusion and Future Prospective

This chapter reports the bacterial biofilms synthesis to protect bacteria from harsh environmental conditions, antibiotics, and disinfectants. Glycopeptidolipids in mycobacteria showed potential role in surface attachment during biofilm formation. Several mycobacterial species, including *M. tuberculosis*, have been found to form drug-tolerant biofilms in vitro. These observations will help to develop new strategies to combat biofilm formation in mycobacterium and other pathogenic bacteria for their survival in adverse environmental conditions.

Acknowledgments Authors are very thankful to DBT, New Delhi and NRDC, New Delhi for financial assistance.

References

Arora K, Whiteford DC, Lau-Bonilla D, Davitt CM, Dahl JL (2008) Inactivation of lsr2 results in a hypermotile phenotype in Mycobacterium smegmatis. J Bacteriol 190:4291–4300. https://doi.org/10.1128/JB.00023-08

Berbari EF, Hanssen AD, Duffy MC, Steckelberg JM, Osmon DR (1998) Prosthetic joint infection due to *Mycobacterium tuberculosis*: a case series and review of the literature. Am J Orthop 27:219–227

Burmolle M, Thomsen TR, Fazli M, Dige I, Christensen L, Homøe P et al (2010) Biofilms in chronic infections—a matter of opportunity—monospecies biofilms in multispecies infections. FEMS Immunol Med Microbiol 59:324–336

Carter G, Wu M, Drummond DC, Bermudez LE (2003) Characterization of biofilm formation by clinical isolates of Mycobacterium avium. J Med Microbiol 52:747–752. https://doi.org/10.1099/jmm.0.05224-0

Chakraborty P, Kumar A (2019) The extracellular matrix of mycobacterial biofilms: could we shorten the treatment of mycobacterial infections? Microb Cell 6(2):105

Chen JM, German GJ, Alexander DC, Ren H, Tan T, Liu J (2006) Roles of Lsr2 in colony morphology and biofilm formation of Mycobacterium smegmatis. J Bacteriol 188:633–641. https://doi.org/10.1128/JB.188.2.633-641.2006

Costerton JW, Gessey GC, Cheng KJ (1978) How bacteria stick. Sci Am 238:86–95. https://doi.org/10.1038/scientificamerican0178-86

Costerton JW, Stewart PS, Greenberg EP (1999) Bacterial biofilms: a common cause of persistent infections. Science 284:1318–1322

Costerton JW, Montanaro L, Arciola CR (2005) Biofilm in implant infections: its production and regulation. Int J Artif Organs 28:1062–1068

Davies DG, Parsek MR, Pearson JP, Iglewski BH, Costerton JW, Greenberg EP (1998) The involvement of cell-to-cell signals in the development of a bacterial biofilm. Science 280:295–298

Donlan RM, Costerton JW (2002) Biofilms: survival mechanisms of clinically relevant microorganisms. Clin Microbiol Rev 15(2):167–193

Esteban J, Marta G (2018) Mycobacterium biofilms. Front Microbiol 8:2651

Esteban J, Martín-de-Hijas NS, Kinnari TJ, Ayala G, Fernández-Roblas R, Gadea I (2008) Biofilm development by potentially pathogenic non-pigmented rapidly growing mycobacteria. BMC Microbiol 8:184

Esteban J, García-Pedrazuela M, Muñoz-Egea MC, Alcaide F (2012) Current treatment of nontuberculous mycobacteriosis: an update. Expert Opin Pharmacother 13:967–986. https://doi.org/10.1517/14656566.2012.677824

Falkinham JO III (2002) Nontuberculous mycobacteria in the environment. Clin Chest Med 23:529–551. https://doi.org/10.1016/S0272-5231(02)00014-X

Falkinham JO III (2009) Surrounded by mycobacteria: nontuberculous mycobacteria in the human environment. J Appl Microbiol 107:356–367. https://doi.org/10.1111/j.1365-2672.2009.04161.x

Falkinham JO III, Williams MD, Kwait R, Lande L (2016) Methylobacterium spp. as an indicator for the presence or absence of Mycobacterium spp. Int J Mycobacteriol 5:240–243. https://doi.org/10.1016/j.ijmyco.2016.03.001

Faria S, Joao I, Jordao L (2015) General overview on nontuberculous mycobacteria, biofilms, and human infection. J Pathog 2015:809014

Fennelly KP, Ojano-Dirain C, Yang Q, Liu L, Lu L, Progulske-Fox A et al (2016) Biofilm formation by Mycobacterium abscessus in a lung cavity. Am J Respir Crit Care Med 193:692–693. https://doi.org/10.1164/rccm.201508-1586IM

Fux CA, Costerton JW, Stewart PS, Stoodley P (2005) Survival strategies of infectious biofilms. Trends Microbiol 13:34–40

Garvey MI, Ashford R, Bradley CW, Bradley CR, Martin TA, Walker J et al (2016) Decontamination of heater-cooler units associated with contamination by atypical mycobacteria. J Hosp Infect 93:229–234. https://doi.org/10.1016/j.jhin.2016.02.007

Grange JM, Yates MD, Boughton E (1990) The avian tubercle bacillus and its relatives. J Appl Bacteriol 68(5):411–431

Greendyke R, Byrd TF (2008) Differential antibiotic susceptibility of Mycobacterium abscessus variants in biofilms and macrophages compared to that of planktonic bacteria. Antimicrob Agents Chemother 52:2019–2026

Ha KY, Chung YG, Ryoo SJ (2005) Adherence and biofilm formation of Staphylococcus epidermidis and Mycobacterium tuberculosis on various spinal implants. Spine 30:38–43

Hall-Stoodley L, Stoodley P (2005) Biofilm formation and dispersal and the transmission of human pathogens. Trends Microbiol 13:7–10. https://doi.org/10.1016/j.tim.2004.11.004

Hall-Stoodley L, Brun OS, Polshyna G, Barker LP (2006) Mycobacterium marinum biofilm formation reveals cording morphology. FEMS Microbiol Lett 257:43–49. https://doi.org/10.1111/j.1574-6968.2006.00143.x

Harris KA, Kenna DT (2014) Mycobacterium abscessus infection in cystic fibrosis: molecular typing and clinical outcomes. J Med Microbiol 63:1241–1246. https://doi.org/10.1099/jmm.0.077164-0

Hoiby N, Ciofu O, Johansen HK, Song ZJ, Moser C, Jensen PØ et al (2011) The clinical impact of bacterial biofilms. Int J Oral Sci 3:55

Kohler P, Kuster SP, Bloemberg G, Schulthess B, Frank M, Tanner FC et al (2015) Healthcare-associated prosthetic heart valve, aortic vascular graft, avium in Acanthamoeba lenticulata. Appl Environ Microbiol 79:3185–3192. https://doi.org/10.1128/AEM.03823-12

Kumon H, Tomochika K, Matunaga T, Ogawa M, Ohmori H (1994) A sandwich cup method for the penetration assay of antimicrobial agents through Pseudomonas exopolysaccharides. Microbiol Immunol 38:615–619

Li L, Mendis N, Trigui H, Oliver JD, Faucher SP (2014) The importance of the viable but nonculturable state in human bacterial pathogens. Front Microbiol 5:258

Mah T-F (2012) Biofilm-specific antibiotic resistance. Future Microbiol 7:1061–1072

Mah T-F, O'Toole GA (2001) Mechanisms of biofilm resistance to antimicrobial agents. Trends Microbiol 9:34–39

Maisonneuve E, Gerdes K (2014) Molecular mechanisms underlying bacterial persisters. Cell 157:539–548

Marciano-Cabral F, Jamerson M, Kaneshiro ES (2010) Free-living amoebae, Legionella and Mycobacterium in tap water supplied by a municipal drinking water utility in the USA. J Water Health 8:71–82. https://doi.org/10.2166/wh.2009.129

Marsollier L, Brodin P, Jackson M, Kordulakova J, Tafelmeyer P, Carbonnelle E et al (2007) Impact of Mycobacterium ulcerans biofilm on transmissibility to ecological niches and Buruli ulcer pathogenesis. PLoS Pathog 3:e62. https://doi.org/10.1371/journal.ppat.0030062

McCarty SM, Cochrane CA, Clegg PD, Percival SL (2012) The role of endogenous and exogenous enzymes in chronic wounds: a focus on the implications of aberrant levels of both host and bacterial proteases in wound healing. Wound Repair Regen 20:125–136

Menozzi FD, Rouse JH, Alavi M, Laude-Sharp M, Muller J, Bischoff R et al (1996) Identification of a heparin-binding hemagglutinin present in mycobacteria. J Exp Med 184:993–1001. https://doi.org/10.1084/jem.184.3.993

Monzón M, Oteiza C, Leiva J, Lamata M, Amorena B (2002) Biofilm testing of Staphylococcus epidermidis clinical isolates: low performance of vancomycin in relation to other antibiotics. Diagn Microbiol Infect Dis 44:319–324

Ojha A, Anand M, Bhatt A, Kremer L, Jacobs WR Jr, Hatfull GF (2005) GroEL1: a dedicated chaperone involved in mycolic acid biosynthesis during biofilm formation in mycobacteria. Cell 123:861–873. https://doi.org/10.1016/j.cell.2005.09.012

Ojha AK, Baughn AD, Sambandan D, Hsu T, Trivelli X, Guerardel Y et al (2008) Growth of Mycobacterium tuberculosis biofilms containing free mycolic acids and harbouring drug-tolerant bacteria. Mol Microbiol 69:164–174

Ojha AK, Jacobs WR Jr, Hatfull GF (2015) Genetic dissection of mycobacterial biofilms. Methods Mol Biol 1285:215–226

Ortiz-Perez A, Martin-de-Hijas N, Alonso-Rodriguez N, Molina-Manso D, Fernandez-Roblas R, Esteban J (2011) Importance of antibiotic penetration in the antimicrobial resistance of biofilm formed by non-pigmented rapidly growing mycobacteria against amikacin, ciprofloxacin and clarithromycin. Enferm Infecc Microbiol Clin 29:79–84

Oubekka SD, Briandet R, Fontaine-Aupart MP, Steenkeste K (2002) Correlative time resolved fluorescence microscopy to assess antibiotic diffusion reaction in biofilms. Antimicrob Agents Chemother 56:3349–3358

Patino S, Alamo L, Cimino M, Casart Y, Bartoli F, Garcia MJ et al (2008) Autofluorescence of mycobacteria as a tool for detection of Mycobacterium tuberculosis. J Clin Microbiol 46:3296–3302. https://doi.org/10.1128/JCM.02183-07

Purdy GE, Pacheco S, Turk J, Hsu FF (2013) Characterization of mycobacterial triacylglycerols and monomeromycolyl diacylglycerols from Mycobacterium smegmatis biofilm by electrospray ionization multiple-stage and high-resolution mass spectrometry. Anal Bioanal Chem 405:7415–7426. https://doi.org/10.1007/s00216-013-7179-4

Qvist T, Eickhardt S, Kragh KN, Andersen CB, Iversen M, Hoiby N et al (2015) Chronic pulmonary disease with Mycobacterium abscessus complex is a biofilm infection. Eur Respir J 46:1823–1826. https://doi.org/10.1183/13993003.01102-2015

Recht J, Kolter R (2001) Glycopeptidolipid acetylation affects sliding motility and biofilm formation in Mycobacterium smegmatis. J Bacteriol 183:5718–5724
Recht J, Martinez A, Torello S, Kolter R (2000) Genetic analysis of sliding motility in Mycobacterium smegmatis. J Bacteriol 182:4348–4351
Richards JP, Ojha AK (2014) Mycobacterial biofilms. Microbiol Spectr 2:MGM2-0004-2013. https://doi.org/10.1128/microbiolspec.MGM2-0004-2013
Schulze-Robbecke R, Fischeder R (1989) Mycobacteria in biofilms. Zentralbl Hyg Umweltmed 188:385–390
Singh PK, Schaefer AL, Parsek MR, Moninger TO, Welsh MJ, Greenberg EP (2000) Quorum-sensing signals indicate that cystic fibrosis lungs are infected with bacterial biofilms. Nature 407:762–764
Solano C, Echeverz M, Lasa I (2014) Biofilm dispersion and quorum sensing. Curr Opin Microbiol 18:96–104
Stewart PS, William Costerton J (2001) Antibiotic resistance of bacteria in biofilms. Lancet 358:135–138
Thoen CO, Karlson AG, Himes EM (1981) Mycobacterial infections in animals. Rev Infect Dis 3:960–972
Vaerewijck MJ, Huys G, Palomino JC, Swings J, Portaels F (2005) Mycobacteria in drinking water distribution systems: ecology and significance for human health. FEMS Microbiol Rev 29:911–934. https://doi.org/10.1016/j.femsre.2005.02.001
Wallace RJ Jr (1987) Nontuberculous mycobacteria and water: a love affair with increasing clinical importance. Infect Dis Clin North Am 1:677–686
Wilkins M, Hall-Stoodley L, Allan RN, Faust SN (2014) New approaches to the treatment of biofilm-related infections. J Infect 69:S47–S52
Yang Y, Thomas J, Li Y, Vilchèze C, Derbyshire KM, Jacobs WR Jr et al (2017) Defining a temporal order of genetic requirements for development of mycobacterial biofilms. Mol Microb 105:794–809
Young LS, Inderlied CB, Berlin OG, Gottlieb MS (1986) Mycobacterial infections in AIDS patients, with an emphasis on the Mycobacterium avium complex. Rev Infect Dis 8:1024–1033
Zambrano MM, Kolter R (2005) Mycobacterial biofilms: a greasy way to hold it together. Cell 123:762–764
Zamora N, Esteban J, Kinnari TJ, Celdran A, Granizo JJ, Zafra C (2007) In-vitro evaluation of the adhesion to polypropylene sutures of nonpigmented, rapidly growing mycobacteria. Clin Microbiol Infect 13:902–907

A Comprehensive Review on Different Microbial-Derived Pigments and Their Multipurpose Activities

20

Archana S. Rao, Sidhartha Pratim Deka, Sunil S. More, Ajay Nair, Veena S. More, and K. S. Ananthjaraju

Abstract

Microbial pigments are colored secondary metabolites produced by various kinds of microorganisms. These pigments vary in color from red, green, to yellow, etc. They typify the organism by giving it the color, help in photosynthesis, and play vital roles in pathogenicity. Some of the more important microbial pigments are carotenoids, flavonoids (anthocyanins), and some tetrapyrroles (chlorophylls, phycobiliproteins). They are majorly found in various species of bacteria and fungi. In the recent past, these pigments have been found to be useful beyond their inherent biological relevance. Due to the growing need for natural colorants and dyes in various industries, these microbial pigments are being extensively sought after; in particular, in the food industry as food colorants, in the textile industry as natural dyes, in the medical industry as anticancerous compounds and immunosuppressants, etc. Here, in this chapter, we have outlined what microbial pigments are, their types, their characteristic occurrences along with their multi-faceted commercial application. Factors that improve the production of these pigments and the methods adopted for their extraction are also discussed in this chapter. The need for an eco-friendly, but sustainable industry demands that we substitute synthetic coloring agents with these biological pigments.

A. S. Rao · S. P. Deka · S. S. More (✉) · A. Nair
School of Basic and Applied Sciences, Dayananda Sagar University, Bangalore, India
e-mail: drsunil@dsu.edu.in

V. S. More
Department of Biotechnology, Sapthagiri College of Engineering, Bangalore, India

K. S. Ananthjaraju
Department of Chemistry, Dayananda Sagar College of Engineering, Bangalore, India

A. Vaishnav, D. K. Choudhary (eds.), *Microbial Polymers*,
https://doi.org/10.1007/978-981-16-0045-6_20

Keywords

Microbial pigments · Secondary metabolites · Synthetic dye · Industrial applications

20.1 Pigments, an Introduction

Pigments are any type of coloring agent or compound that imparts its color onto substances. They may originate from a lot of sources, most of them being synthetic. These pigments may be utilized in many areas, such as food coloring, or as dyes for clothes, and similar industries. Since, the dawn of civilization, pigments have been utilized for the purposes of painting, dyeing clothes, as food coloring, etc. In the past, these pigments were made from natural sources such as crushed plant matter like leaves, flowers, and fruits. However, as time passed and the industrial revolution took over, the use of natural pigments declined.

Artificial pigments have retained their use up till the modern age, as their synthesis can be done in a laboratory using chemicals, and thus is quite viable when compared to growing, collecting, and crushing up plants. These pigments also provided a range of colors not readily available from the natural world. However, these pigments have had adverse side effects ranging from pollution of environment, toxicity, and allergic reactions to humans. However, carcinogenic effects on those who come in direct contact with these pigments, on a daily basis, via colorants in edible food, or dyes of clothing seem to be the most notable of all the adverse impacts. Consequentially, it has become imperative that an alternate source of pigments and dyes be found.

This is where microorganisms presented themselves as potential alternate sources for pigments. Microbes produce an array of colored components, such as the green colored photosynthetic apparatus, or are themselves colored. These colors arise from biological pigments produced by these organisms. Pigments like Chlorophyll, Anthraquinone, β-carotene, etc., have specific biological roles in the organisms that produce them. For example, β-carotene (and α-carotene) impart yellow, orange, or reddish color depending on its concentration. However, in addition, carotenes also protect the organism from the harmful effects of UV radiation, as well as having antioxidant properties.

20.2 Microbial Pigments

Considering their eco-friendly demeanor, microorganisms have become the most sought-after alternatives for several synthetic pigments. Various species of bacteria, fungi, and algae are prime candidates for the isolation of these natural pigments. These pigments vary in color from yellow, green, red, and blue and various mixtures and shades in between. Indeed, all the colors of the rainbow can be generated from these pigments (Fig. 20.1).

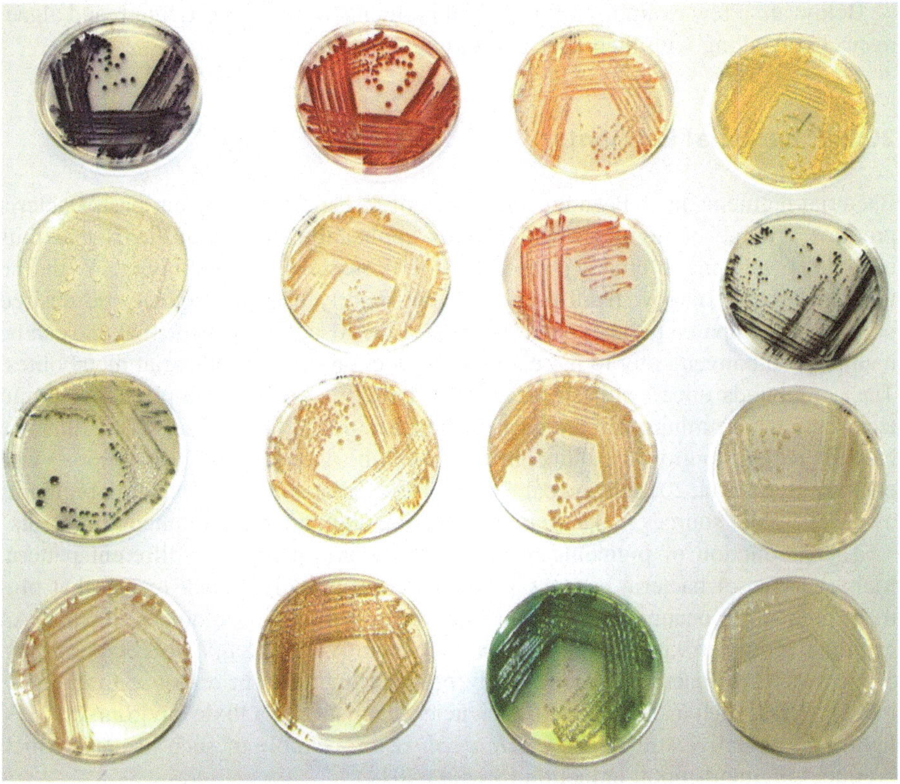

Fig. 20.1 Different pigments produced by microorganisms in culture plates. Source: Exploringtheinvisible.com C-MOULD: living paints

They have many potential roles as coloring agents for clothes and food and have many medicinal properties.

Below are a few examples of microbial pigments from bacteria, fungi, and algae, respectively along with their uses and biological roles.

20.2.1 Bacterial Pigments

Bacterial pigments have been an active area of research in the recent past. Bacteria produce a range of pigments which finds application in various fields and thus became a valuable promising resource for the scientific community. With the inherent ability to grow readily and fast, it is fairly easy to produce the desired product, as compared to other chemical processes. Also, their genetic makeup can be manipulated conveniently to increase the production of such industrial metabolites. The compounds once synthesized can also be extracted by a simple liquid-liquid technique which minimizes the cost of the operation. Plus, the substrates that the bacteria require to grow are quite cheap and can be used for bulk production (Venil et al. 2013; Saviola 2014; Rao et al. 2017). These advantages of producing pigments using microbial sources make them a promising alternative for synthetic pigments.

The production of pigments by different bacteria displays a different pattern. Many species of bacteria produce pigments constitutively, which means that pigment production occurs at a constant rate, regardless of physiological demand. An example of bacteria producing pigments constitutively is *Staphyloccocus aureus.* Some species of bacteria however only produce pigments in response to environmental stress, such as nutritional deficiencies, exposure to a toxic environment, UV light, and oxidative stress. *Escherichia coli is a*n example of such bacteria which produces pigments upon induction (Saviola 2014).

Bacteria produce various range of pigments which finds different applications. Below are a few examples of important pigments that are produced by bacterial species and their roles (Table 20.1).

Pigments such as anthraquinone, melanin, indigoidine also have many commercial uses as pharmaceuticals, antioxidants, and bioleaching, respectively.

20.2.1.1 Pyocyanin

Pyocyanin (Fig. 20.2) is a blue-green pigment synthesized by bacteria. It is a secondary metabolite, which has the property to oxidize and reduce other molecules. It is primarily used by the bacterium as a toxin, to kill other microorganisms that compete against it.

Commercially pyocyanin is extracted from *Pseudomonas aeruginosa.* The biological role of this pigment that is produced by *P. aeruginosa* is to cause cystic fibrosis in mammals. This toxin is responsible for the destruction of mammalian lung tissue. In blood, it can exist in the zwitterion state (a molecule that contains equal number of positive and negative functional groups) and can thus easily cross the cell membrane and disrupt cell activity.

Pyocyanin acts by inactivating the enzyme catalase. It does so by reducing the gene expression of catalase as well as directly targeting the enzyme itself. This in turn increases the concentration of hydrogen peroxide. These H_2O_2 oxidize the

Table 20.1 Bacterial pigments

Pigment	Color	Isolated from	Production	Isolation	Application	Reference
Pyocyanin	Blue/ green	*Pseudomonas aeruginosa*	Solid state fermentation	Percolation; solid-liquid extraction	Antimicrobial agent	Hassan and Fridovich (1980)
Astaxanthin	Red/ orange	*Agrobacterium aurantiacum*	Submerged fermentation	Liquid-liquid extraction	Antioxidant	Ambati et al. (2014)
Staphyloxanthin	Golden/ yellow	*Staphylococcus aureus*	Submerged or solid state fermentation	Liquid-liquid extraction	Antibiotic	Liu et al. (2005)
Violacein	Purple	*Chromobacterium violaceum*	Submerged fermentation	Maceration; solid-liquid extraction and crystallization	Antitumor, antimicrobial, and antiparasitic agent	Andrighetti-Fröhner et al. (2003)
Prodigiosin	Red	*Serratia marcescens*	Submerged or solid state fermentation	Liquid-liquid extraction	Antibiotic	Feng et al. (2015)

Pyocyanin, which in turn convert molecular oxygen into superoxide radicals by oxidizing NADPH to $NADP^+$. This depletes the NADPH substrates and also the superoxide species inhibits cell pathways such as cytokines. This eventually leads to the inactivity of the cell and finally death (Hassan and Fridovich 1980).

20.2.1.2 Astaxanthin

Astaxanthin (Fig. 20.3) is a part of a broad family of pigments known as carotenoids. It is a keto-carotenoid containing both hydroxyl (–OH) and a keto (=O) group and is comprised of five carbon precursors. Astaxanthin is a red-orange pigment, as a consequence of extended chains of conjugated double bonds at the center of the compound. Due to these double bonds at the center, Astaxanthin also has an antioxidant property. Thus, it has also been used as a dietary supplement for humans, animals, and in aquaculture.

Astaxanthin is produced in bacteria such as *Agrobacterium aurantiacum,* as well as in many species of yeast like *Xanthophyllomyces dendrorhous* (Barredo et al. 2017). It is generally found in aquatic organisms, responsible for their red color. The pigment is transferred from species to species through the food chain. Primarily, algae are natural sources of astaxanthin in marine ecosystems. The pigment is then transferred to the primary consumers of algae, such as salmon, red trout, flamingos, and crustaceans like shrimps and crabs. Consequentially, following consumption, the primary consumers themselves take on a reddish hue. The concentration of the pigment determines the strength of the color, which varies on both species and individual levels, with diet and habitat influences playing an equally important role.

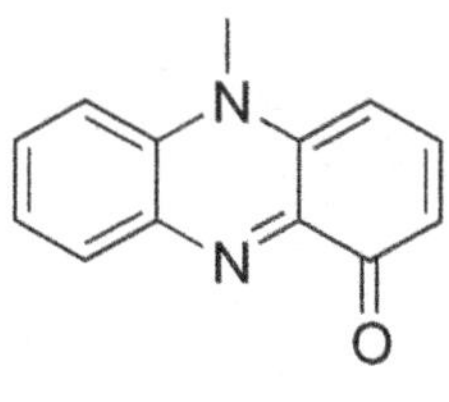

Fig. 20.2 Structure of pyocyanin. (Source: https://pubchem.ncbi.nlm.nih.gov/compound/Pyocyanine)

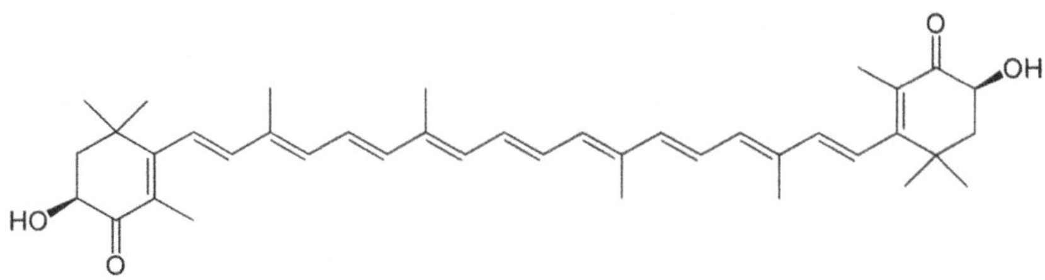

Fig. 20.3 Structure of astaxanthin. (Source: https://pubchem.ncbi.nlm.nih.gov/compound/Astaxanthin)

In terms of usage, the pigment serves as an antioxidant and a coloring agent. Artificially, astaxanthin can also be used as an animal feed to impart the red-orange color to farm-raised salmon and to chicken for the color of the egg yolks (Ambati et al. 2014).

20.2.1.3 Staphyloxanthin

Staphyloxanthin (Fig. 20.4) is a golden colored pigment which is classified under carotenoid pigments. It is produced exclusively by some strains of *S. aureus*. Biochemically, the biosynthesis of staphyloxanthin is quite similar to that of human cholesterol. The pigment function principally as virulence factors. This pigment is majorly an antioxidant, which helps in counteracting reactive oxygen species produced by the host as a defense mechanism against the organism.

Research has conclusively demonstrated how staphyloxanthin improves *S. aureus's* pathogenicity by comparing a mutant strain lacking staphyloxanthin was compared against a normal pigmented wild type strain. The experiment clearly showed the role of pigment in survival of the organism in presence of hydrogen peroxide and human neutrophils. The pigmented strains can cause wounds lingered with pus, whereas, the wound caused by nonpigmented strain healed quickly. This shows the importance of pigment in pathogenicity (Liu et al. 2005).

20.2.1.4 Violacein

Violacein (Fig. 20.5) is a purple pigment that is synthesized by a large number of bacteria such as *Chromobacterium violaceum*. It is known to have high antibiotic activity, such as a strong antibacterial action against *S. aureus* and other gram-positive pathogens. It has further shown to have antiviral, antifungal properties, and

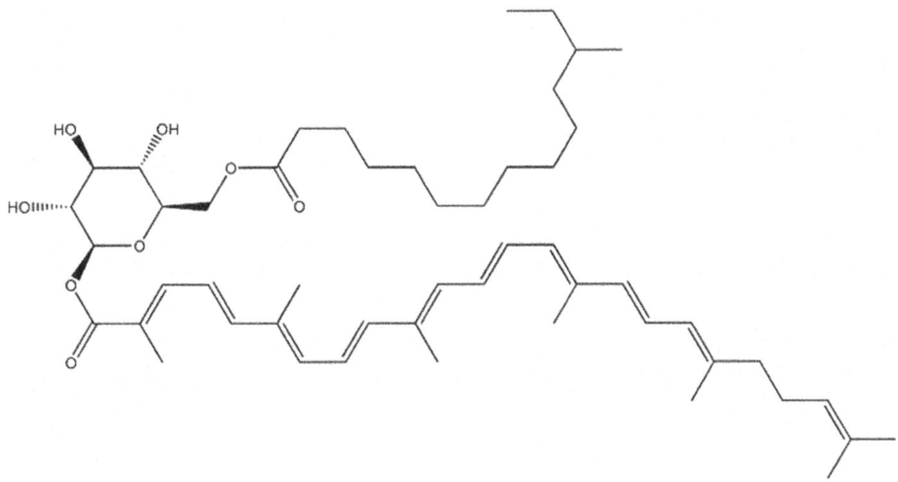

Fig. 20.4 Structure of staphyloxanthin. (Source: https://pubchem.ncbi.nlm.nih.gov/compound/Staphyloxanthin)

most importantly it has been shown to play a vital role as a cytotoxic anticancerous agent.

Violacein is a bis-indole pigment comprised of two indole rings. The bis-indole structure is a type of secondary metabolite and is associated with biofilm production. Furthermore, the production of this compound by the bacterium is regulated by quorum sensing. Violacein is synthesized by the enzymatic condensation of two tryptophan molecules, further requiring the involvement of five other proteins. However, the commercial production of this pigment has been difficult, and thus synthesis is currently being pursued through genetic engineering and synthetic biology (Andrighetti-Fröhner et al. 2003).

20.2.1.5 Prodigiosin

Prodigiosin (Fig. 20.6) is a red colored pigment produced as a secondary metabolite by many gram-negative bacteria. Production of this pigment via organisms such as *Serratia marcescens* is of particular interest to industry. The biological activity for prodigiosin range from properties such as antimalarial, antifungal, and an immuno-suppressant. It has been found to be very effective against *Borrelia burgdorferi* (which is the causative organism of Lyme disease) in the stationary phase of its growth cycle (Feng et al. 2015).

Prodigiosin also possesses the capacity to trigger apoptosis in tumor or cancerous cells. The exact pathway for this is immensely complex and not fully elucidated. Thus, prodigiosin is a pigment that has undergone extreme investigation in the preclinical phase for the treatment of pancreatic cancer (Williamson et al. 2008).

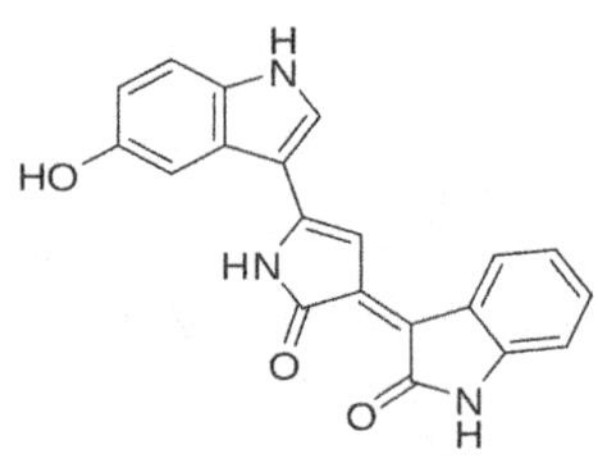

Fig. 20.5 Structure of violacein. (Source: https://pubchem.ncbi.nlm.nih.gov/compound/Violacein)

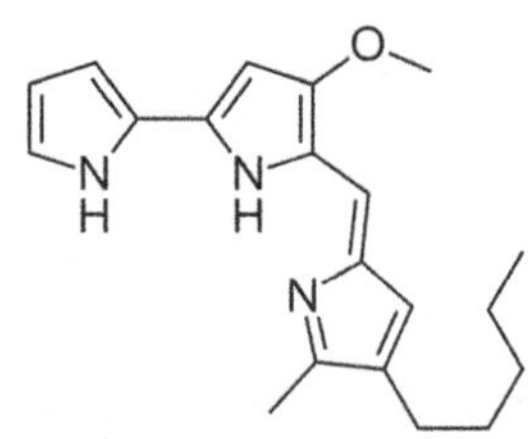

Fig. 20.6 Structure of prodigiosin. (Source: https://pubchem.ncbi.nlm.nih.gov/compound/Prodigiosin)

20.2.2 Fungal Pigments

Fungi are diverse in nature with an even more diverse array of pigments produced by them. Most fungal pigments are secondary metabolites. They may be produced when environmental conditions are unfavorable and generally produced under nutrient scarcity. These pigments have found intense use in the commercial market in a variety of industries. Fungal culturing is quite simple and large-scale production of compounds from fungi is a very feasible idea for industries (Mukherjee et al. 2017).

Below are a few fungal pigments and their various roles (Table 20.2).

Additionally, fungal pigments such as asperversin exhibit antifungal properties and anthraquinone can be used as an antioxidant.

20.2.2.1 Riboflavin

Riboflavin (Fig. 20.7) is also known as vitamin B_2 and is naturally produced as a pigment by several microorganisms. However, three species are employed industrially, as they produce higher levels of pigment. These are the fungi *Eremothecium gossypii* which is also known as *Ashbya gossypii,* yeast *Candida famata,* and the bacteria *Bacillus subtilis.* Among these, *A. gossypii* produces the highest level of riboflavin.

Riboflavin is generally a yellow pigment and has an additional property that protects the spores of *A. gossypii* from ultraviolet light. Even in its natural state, the organism overproduces this pigment thus made it a suitable candidate for industrial production of riboflavin.

Due to its specific interaction with UV light, Riboflavin appears fluorescent. This is why it is used in chemical blend tanks or bioreactors to detect leaks. Medically it is also used to treat Corneal ectasia. Riboflavin can also be taken as dietary supplements to combat vitamin deficiencies.

A. gossypii is one of many organisms that industrially produce riboflavin. These organisms are generally genetically modified to increase their riboflavin yield. In *A. gossypii,* the concentration of riboflavin is so high, that the mycelium where it accumulates is seen as reddish-brown. The mycelium eventually bursts, releasing colored crystals (Stahmann et al. 2000).

20.2.2.2 β-Carotene

β-carotene (Fig. 20.8) is an abundant pigment found in various fungal species. It is red/orange in color and as most carotenoids have antioxidant properties. β-carotene functions as an intermediary product in many fungi, towards the formation of retinal, which is produced through the symmetrical oxidative cleavage of β-carotene (Mata-Gómez et al. 2014; Biesalski et al. 2007).

β-carotene can be commercially used for food dyes, as it is naturally responsible for the coloration of many fruits and vegetables. It is also believed to have a role in the prevention of cancer and to decrease the risk of contracting cataract.

Molds are generally utilized for the industrial production of β-carotene. *Blakeslea trispora* is one of the species that is readily cultivated on large scale and the

Table 20.2 Fungal pigments

Pigment	Color	Isolated from	Production	Isolation	Application	Reference
Riboflavin	Yellow	*Ashbya gossypii*	Solid state fermentation	Liquid-liquid extraction	Treatment for corneal ectasia	Stahmann et al. (2000)
β-carotene	Red/ orange	*Blakeslea trispora*	Solid state fermentation	Percolation; solid-liquid extraction	Antioxidant	Mata-Gómez et al. (2014)
Naphthoquinone	Yellow or red	*Fusarium verticillioides*	Solid state fermentation	Liquid-liquid extraction	Antimalarial, anticancerous, and antibacterial properties	Boonyapranai et al. (2008)
Lycopene	Red	*Blakeslea trispora*	Co-fermentation; submerged fermentation	Crystallization and filtration	Antioxidant	Olempska-Beer (2006)
Benzoquinone	Yellowish	*Fusarium*	Solid state fermentation	Liquid-liquid extraction	Anticancer agent	Zheng et al. (2017)

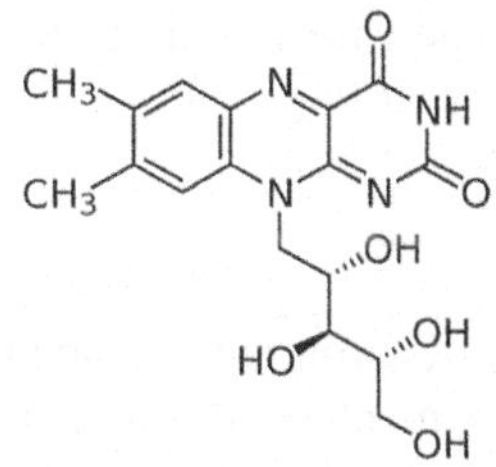

Fig. 20.7 Structure of riboflavin. (Source: https://pubchem.ncbi.nlm.nih.gov/compound/Riboflavin)

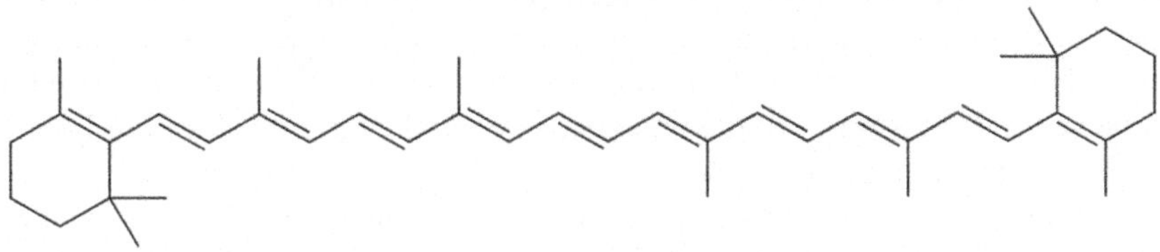

Fig. 20.8 Structure of β-carotene. (Source: https://pubchem.ncbi.nlm.nih.gov/compound/beta-Carotene)

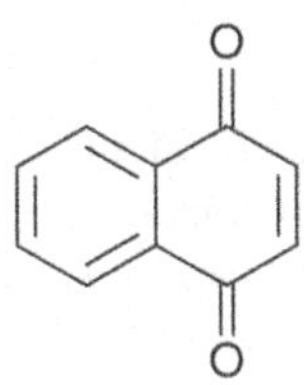

Fig. 20.9 Structure of naphthoquinone. (Source: https://pubchem.ncbi.nlm.nih.gov/compound/1_4-Naphthoquinone)

carotenoids are then extracted from the inoculum through drying or pressing (Mata-Gómez et al. 2014).

20.2.2.3 Naphtoquinone

Naphtoquinones (Fig. 20.9) are collective sets of secondary metabolites produced by many species of fungi such as *Fusarium verticillioides*, *Monascus* sp., *Ophiocordyceps unilateralis*, etc. These are all derivatives of naphthoquinone and have shown to possess strong antimalarial activity, anticancerous, and antibacterial properties.

Naphthoquinone pigments are generally yellow in color but show up as red under acidic conditions, and purple under alkaline conditions. Currently, naphthoquinones have a wide range of applications, as dyes in foods, in cosmetics, and also in pharmaceutical manufacturing processes. They also have shown to improve symptoms of tuberculosis and increase immunity in conjunction with chemotherapeutic drugs (Boonyapranai et al. 2008).

20.2.2.4 Lycopene

Lycopene (Fig. 20.10) is a type of carotenoid, which has bright red coloration and is produced in many natural red pigmented fruits and vegetables, most notably tomato. It is also biosynthesized by many microorganisms such as fungi, bacteria, and algae. It functions as an intermediary molecule as it leads directly to the production of β-carotene and other carotenoids.

Blakeslea trispora is the most abundant producer of Lycopene. Once processed by the enzyme lycopene cyclase, lycopene produces β-carotene. But for the development of lycopene at an industrial level, *B. tripsora* is grown in a medium which inhibits lycopene cyclase and so prevents the lycopene being converted into β-carotene. The zygospores of the organism contain the largest amount of lycopene, from where the compound is isolated.

For humans, lycopene serves purposes such as an antioxidant to maintain a healthy heart and also to lower risks of certain types of cancers. Although it can be found in dietary supplements, it is most effective when taken in the form of lycopene-rich foods (Olempska-Beer 2006).

20.2.2.5 Benzoquinone

Benzoquinone (Fig. 20.11) is a newly identified pigment from *Fusarium* species. The pigment showed to have an anticancerous effect against breast cancer cell lines. The pigment compound, termed as compound VI, was considered for its potential therapeutic use in the treatment of human breast cancer (Zheng et al. 2017).

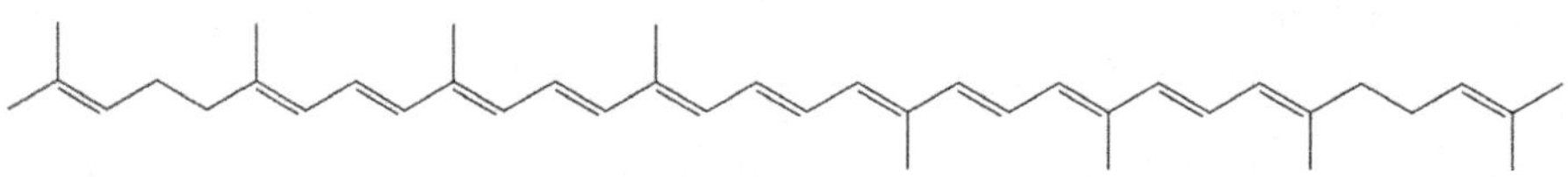

Fig. 20.10 Structure of lycopene. (Source: https://pubchem.ncbi.nlm.nih.gov/compound/Lycopene)

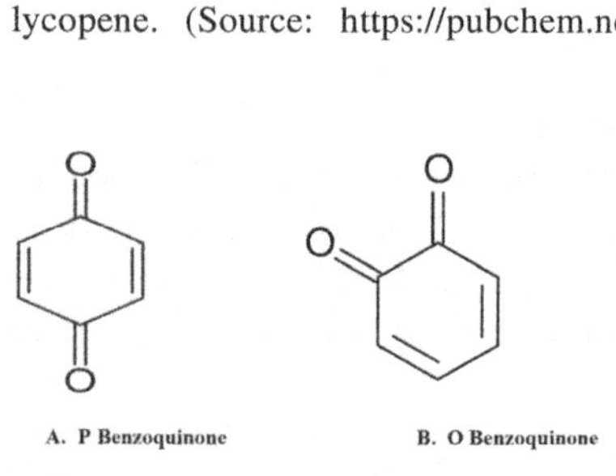

Fig. 20.11 Structure of benzoquinone (**a**, **b**). (**a**) P Benzoquinone. (**b**) O Benzoquinone. (Source: https://pubchem.ncbi.nlm.nih.gov/compound/1_4-Benzoquinone and https://pubchem.ncbi.nlm.nih.gov/compound/1_2-Benzoquinone)

20.2.3 Algal Pigments

Algae are photosynthetic organisms and generally grow in aquatic habitats. They are usually found in the multicellular organization, but microalgal forms are known to equally exist. Algae in general exhibit a variety of pigments such as chlorophylls, phycobiliproteins, and carotenoids and exhibit colors from green, yellow, brown, and red. Most algal pigments are produced constitutively throughout their lifetime as they play important roles in the metabolic processes of the organism. For example, chlorophylls play an integral role in the structure of the lamellae present in the chloroplast and play a key role in photosynthesis.

Species of microalgae can be readily exploited as a source for natural dyes and pigments. These pigments are utilized in numerous ways in industries such as pharmaceuticals, medicine, cosmetics, food, etc. (Algal pigments—Robertson's Laboratory www.wordpress.clarku.edu).

Below are some algal pigments and their various roles (Table 20.3).

20.2.3.1 Chlorophyll

Microalgae are capable of carrying out photosynthesis and thus produce the green pigment known as chlorophyll which imparts a green color onto the organism. Chlorophyll (Fig. 20.12) is an important pigment responsible for being a vital piece in the process by which algae harness solar energy to synthesize their own food.

Additionally, chlorophyll has multiple commercial uses. It can be used as an additive in the pharmaceutical industry. It is readily employed as a natural food coloring agent. It also has medicinal uses as an antioxidant and as anti-mutagentic properties. For industrial production of chlorophyll from algae species, *Chlorella* and *Spirulina* are generally utilized (Hosikian et al. 2010).

20.2.3.2 Fucoxanthin

Fucoxanthin (Fig. 20.13) is a marine carotenoid that is found in microalgae such as brown seaweeds, macroalgae, and diatoms. It is one of the most abundant carotenoids found in nature and is responsible for the red brownish coloration of the organism. As a carotenoid, it has a strong antioxidant activity and plays a significant role in the photosynthetic process.

Fucoxanthin has multifaceted health benefits for humans. It has strong anti-obesity and antidiabetic effects. It has also shown antitumorigenic activity and can arrest the cell cycle as well as induce apoptosis. It has the potential to also be used as an anti-inflammatory agent as well as an antioxidant. Additionally, it can also be used to treat illnesses of the liver as it shows certain hepatoprotective effects.

Fucoxanthin can be commercially extracted from species such as *Undaria pinnatifida, Laminaria japonica, Phaeodactylum tricornutum,* and *Cylindrotheca closterium* (Karpiński and Adamczak 2019).

Table 20.3 Algal pigments

Pigment	Color	Isolated from	Production	Isolation	Application	Reference
Chlorophyll	Green	*Chlorella* and *Spirulina*	Open air and photobioreactor system	Maceration and organic solvent extraction	Antioxidant	Hosikian et al. (2010)
Fucoxanthin	Red/ brownish	*Phaecodactylum tricornutum*	Solid state fermentation	Pressurized liquid extraction	Anti-obesity effect, antidiabetic effect	Karpiński and Adamczak (2019)
Lutein	Yellow	*Dunaliella* sp.	Photoautotrophic and heterotrophic cultivation	Accelerated solvent extraction	Pharmaceuticals	Sun et al. (2015)
Phycocyanin	Light blue	*Spirulina platensis*	Submerged fermentation	Decoction water extraction	Diagnostic agent, anticancer agent	Liu et al. (2016)
Phycoerythrin	Red	*Phyllophora antarctica*	Submerged fermentation	Masceration, gel filtration, and anion exchange techniques	Cosmetics, food colorant	Fleurence (2003)

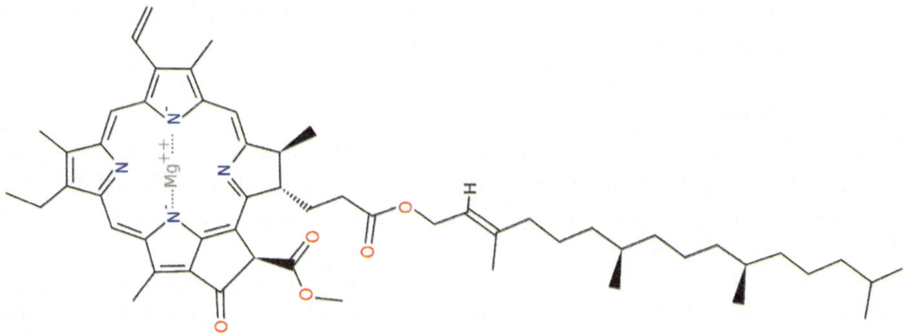

Fig. 20.12 Structure of chlorophyll. (Source: https://pubchem.ncbi.nlm.nih.gov/compound/Chlorophyll)

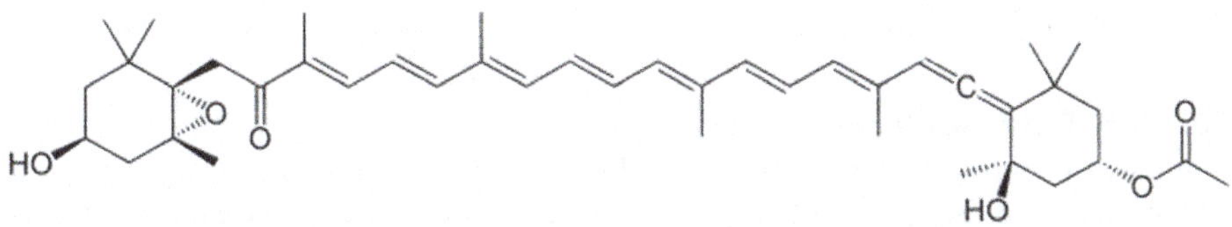

Fig. 20.13 Structure of fucoxanthin. (Source: https://pubchem.ncbi.nlm.nih.gov/compound/Fucoxanthin)

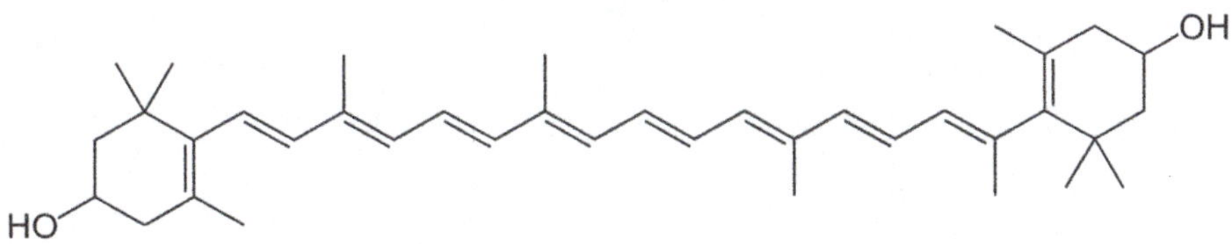

Fig. 20.14 Structure of lutein. (Source: https://pubchem.ncbi.nlm.nih.gov/compound/Lutein)

20.2.3.3 Lutein

Lutein (Fig. 20.14) is classified as a Xanthophyll under carotenoids. It is a yellow pigmented compound and contributes heavily to the Xanthophyll cycle. While the majority of Lutein in nature is present in plants and green leafy vegetables, microalgae such as *Chlorella* which also readily produce Lutein as a component required for photosynthesis. It also acts as a strong antioxidant. Lutein is present in small amounts in the macula of the retina of the human eye. Its precise function has not yet been determined.

Lutein's strong antioxidant properties make it a much sought-after pigment with regards to human health. Dietary intake of Lutein also prevents the early onset of atherosclerosis (accumulation of fats in and on the arterial walls). It is also known to have remedial effects on cataracts (clouding of the eye lens that affects vision). Most importantly, it has shown to help in the treatment of Macular degeneration, a degenerative disease of the human eye. Lutein is commercially extracted from *Chlorella* and *Dunaliella* sp. (Sun et al. 2015).

20.2.3.4 Phycocyanin

Phycocyanin (Fig. 20.15) is a pigment-protein which acts as an accessory to chlorophyll during the process of photosynthesis. It has a characteristic light blue color and is synthesized in many species of microalgae such as *Spirulina platensis, Anabaena,* and other blue-green algae.

Phycocyanin has many uses in the industry. In biotechnology, it is used as a fluorescent probe, in various detection assays. It is used both as a cosmetic and food colorant. In the medical field, it can be used to improve anemia and has hepatorenal (liver and kidney) protective activity. It is a strong antioxidant, has anti-inflammatory as well as antitumorigenic activity.

S. platensis is the best-suited microorganism for the commercial production of Phycocyanin (Liu et al. 2016).

20.2.3.5 Phycoerythrin

Similar to Phycocyanin, Phycoerythrin (Fig. 20.16) is also an accessory protein to Chlorophyll and thus plays an important role in light harvesting and photosynthesis. It is a red colored pigment and is primarily synthesized by red algae species such as *Porphyridium cruentum* as well as cryptophytes. It, along with Phycocyanin,

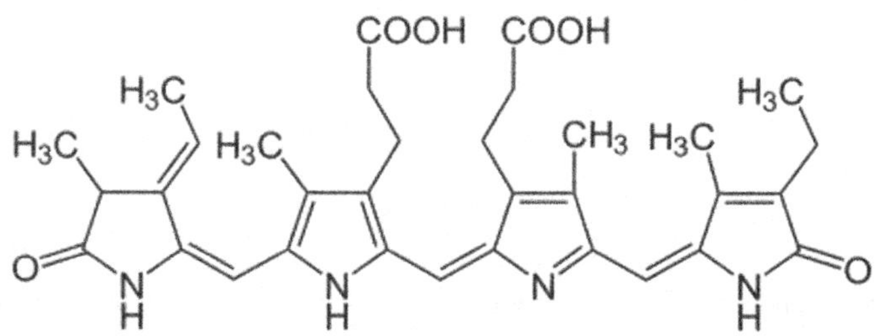

Fig. 20.15 Structure of phycocyanin. (Source: https://pubchem.ncbi.nlm.nih.gov/substance/8011185)

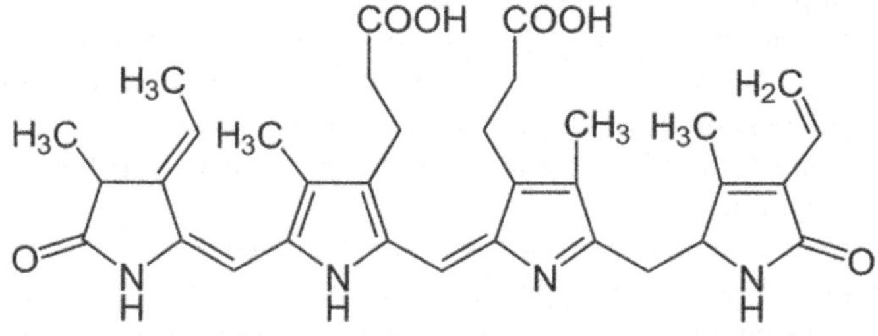

Fig. 20.16 Structure of phycoerythrin. (Source: https://pubchem.ncbi.nlm.nih.gov/substance/315492484)

belongs to a group of proteins known as Phycobiliproteins, which collectively have a light harvesting role during photosynthesis.

Phycoerythrin has shown the potential to be used as a coloring agent in food as well as a dye in cosmetics. It can be used in the field of biotechnology as a fluorescent probe in flow cytometry, microscopy, and immune assays. It can be employed as an internal marker in electrophoretic techniques and size gel exclusion chromatography. It has shown antiparasitic and antitumor activities. The pigment can induce programed cell death in tumor cells and its natural fluorescence helps track the progress of the pigment inside the tumor cells.

Commonly, red algae such as *Phyllophora antarctica* and *P. palmate* are used during industrial isolation of phycoerythrin (Fleurence 2003).

20.3 Production of Microbial Pigments

When any compound is required to be isolated from microbial sources there are certain methodologies to be followed. Production of pigments from microorganisms is no different from identifying the desired microbe to culturing it in vitro and allowing for the organism to synthesize the pigment in desirable quantities. The pigment is then to be separated from culture and analyzed which is finally made available for commercial use. These processes occur in sequential order and are illustrated as such in the flow chart below (Fig. 20.17).

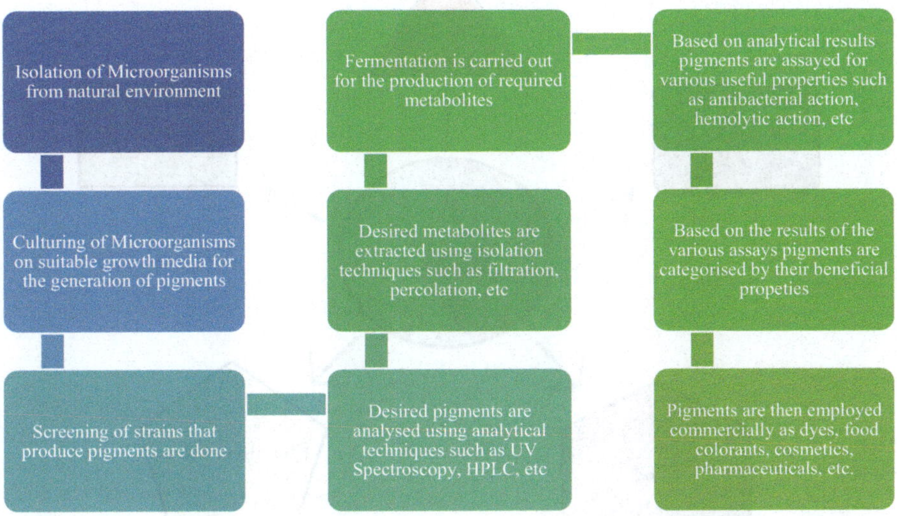

Fig. 20.17 General protocol for isolation of pigments

20.4 Applications of Microbial Pigments

In the next section, we will see the impact microbial pigments have had commercially, in various industries (Fig. 20.18).

20.4.1 Textile Industry

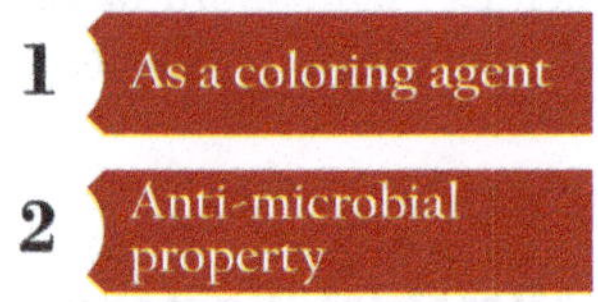

Textile is a gigantic industry, providing items of clothing to more than seven billion people worldwide. Fashion and apparel design has become one of the largest conglomerates in the last 100 years. And more recently, it has taken steps to go down an eco-friendlier route. Since the dawn of the industry, hundreds of synthetic dyes had been developed for the purpose of adding color to clothes. However, there quite a lot of problems were later discovered in the use of these synthetic dyes. A large amount of these dyes are lost as effluents during the dyeing and finishing process. Due to their synthetic nature, these dyes accumulate and persist in the environment. Since they are highly stable against light, water, detergents, and

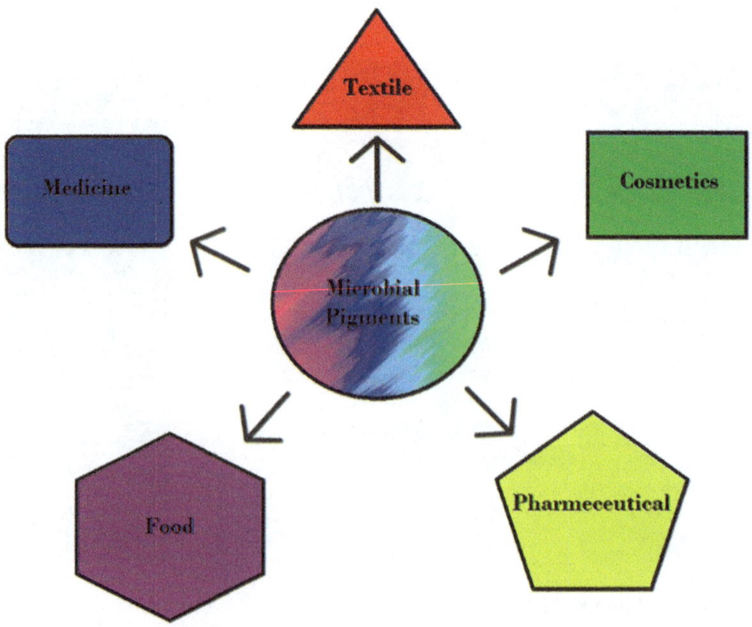

Fig. 20.18 Application of microbial pigments in different fields

chemicals, they end up as pollutants that cannot be degraded and end up being toxic to the natural flora and fauna. In addition to this, synthetic dyes have also been found to be toxic to the human body. Many have been found to have a carcinogenic effect, being the cause of skin, kidney, liver, and bladder cancer.

Microbial pigments are reliable, biodegradable, and natural sources for dyes. Species of bacteria such as *Vibrio* produce Prodigiosin which can dye wool, nylon, acrylics, and silk (Rao et al. 2017). Pigments from *S. marcescens* can be used with tamarind as a mordant to color acrylic, polyester microfibers, polyester, silk, and cotton.

Fungal species such as *F. oxysporum* can be used to dye wool fabrics (Nagia and El-Mohamedy 2007). Quite recently it was reported that wet blue goat nappa skin can be dyed using the pigment from *Penicillium minioluteum* (Sudha et al. 2016). Another species of fungi known as *Talaromyces verruculosus* produces a red pigment that shows an adequate color tone for cotton without any cytotoxic effects (Chadni et al. 2017).

Microbial pigments exhibit the property of producing different tones in different textiles. For example, the pigment from a bacterium known as *Janthinobacterium lividum* (Shirata et al. 2000) produces a blueish purple color tone on silk, cotton, and wool, while producing a dark blue in nylon and vinylon. Deep blue and red pigments from *Streptomyces* strains NP2 and NP4 demonstrate changes in the dyeing ability with respect to the material used; polyacrylamide and acrylic fibers stained vibrantly, while cotton and cellulosic fibers stained weakly (Kramar et al. 2014). The dyeing ability of a yellow pigment from *Thermomyces* shows a higher affinity to silk fabric, as compared to others such as cotton and wool (Poorniammal et al. 2013).

But adding color to the fabrics is just one of the several properties held by microbial pigments. Microbial dyed textiles also showed antimicrobial properties. Fabrics dyed with the prodiginines from *Vibrio* sp. showed antibacterial properties against *Staphylococcus aureus* and *Escherichia coli* (Dufossé 2009).

Considering its nontoxic nature and the added health benefits microbial pigments can be a suitable replacement for synthetic dyes in textiles. These pigments are extensively available and are relatively cost-effective to produce. They do not pollute the environment, something that adds further to the market appeal (Rao et al. 2017; Venil et al. 2014).

20.4.2 Food Industry

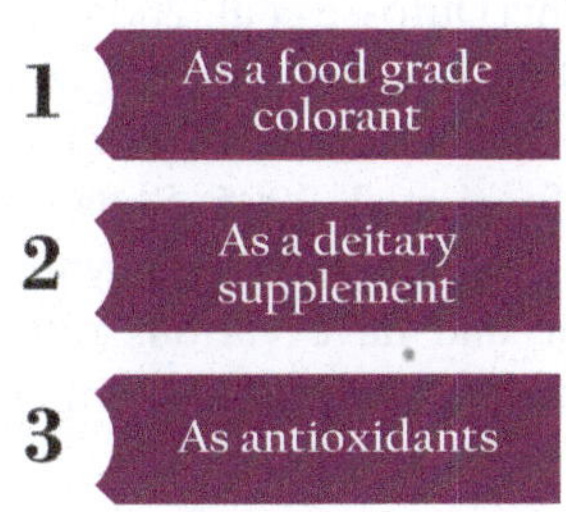

Food is, it is needless to say, one of the most important necessities of life. However, the art of cooking has changed over the past. Merely, making the food taste good is not sufficient, it needs to look attractive too. To achieve the same, various dyes are employed as an additive, either synthetic or natural. In recent years, the industry has slowly moved towards natural food colorants, as it was found that the synthetic ones had many adverse health benefits associated with them; carcinogenicity being one among them.

Microbes are a good source of natural colorants due to the wide array of pigments they collectively produce. The production of these pigments on a commercial scale is relatively cheap and with a high yield. Various species of fungi and bacteria can be used for this purpose. Species such as astaxanthin from *Xanthophyllomyces dendrorhous* (Ramirez et al. 2000), Riboflavin from *Ashbya gossypii* (Dufossé 2018), Carotene from *Streptomyces* sp., and lycopene from *Erwinia uredevora* can be added to food to increase its appeal (Dufossé 2017a, b).

Bacterial species such as *Bradyrhizobium* sp. produce canthaxanthin, which is an orange to deep pink colored carotenoid and is an approved food colorant that is used in salmon and poultry feed (Lorquin et al. 1997). Being a carotenoid, canthaxanthin has antioxidant properties which serve as an additional health benefit along with it being a colorant. Prodigiosin from *S. marcescens* is a red pigment with antibiotic, antimalarial, and antineoplastic activity and it has been successfully been applied to yogurt, milk, and carbonated drinks (Dufossé 2017a, b, 2018; Bennett and Bentley 2000). *Chromobacterium violaceum* produces the purple pigment called violacein. It has antifungal, antibacterial, antitumor properties and can also be used as a food colorant (Duran et al. 2012).

Fungal species such as basidiomycetous yeast produce the red-orange pigment astaxanthin which is an approved coloring agent, used in the fish and animal food industry (Guedes et al. 2011; Zuluaga et al. 2017). β-carotene from species such as *Blakeslea trispora, Fusarium sporotrichioides, Mucor circinelloides,* etc., is a red-orange pigment with antioxidant and anticancer properties and is used as a food colorant for many food items ranging from red to yellow in color (Terao 1989; Dufossé 2018). Lycopene from *Fusarium* sp. is a bright red pigment and a carotenoid with the potential to attenuate persistent diseases such as some types of cancer and coronary heart disease. It is also used as a coloring agent for meat in some countries (Di Mascio et al. 1989).

Algae also contribute as food colorants. Phycocyanin from *Spirulina* is a blue pigment which is used as a natural coloring agent "Lina Blue" in foods and beverages and is also found in ice creams and sweets (Eriksen 2008). Lutein harvested from *Chlorella* is a yellow pigment and has antioxidant property and is readily used as a food colorant (Dufossé et al. 2005; Chen et al. 2016). Phycoerythrin from *Porphyridium cruentum* and other microalgae shows antioxidant, antitumor, and immunoregulatory activity and is currently in the developmental stage of being used as a food colorant (Dufossé et al. 2005; Sonani et al. 2014).

Microbial pigments are a good alternative to synthetic food dyes as microbes are found almost in every niche and have various roles in nature. They are already affiliated with food and are responsible for the fermentation of food products. They are a better alternative than plants as their availability does not depend on seasons, high yield per hectare, and a very straightforward downstreaming process. A considerable amount of research has been done to lower production and processing costs, to increase stability and shelf life to compete with synthetic colors.

Additionally, these pigments can also have a large impact on improving the quality of life. They impart many crucial health benefits. Pigments such as violacein, carotenoids, anthocyanins, and naphthoquinone have been shown to have strong antioxidant properties.

Many microbial pigments have also shown antitumor properties. Scytonemin, a green-yellow pigment, which is produced by aquatic cyanobacteria has been shown to have such an ability. It inhibits the action of a protein that is responsible for the regulation of the cell cycle, thereby effectively stopping cancer cells from proliferating (Stevenson et al. 2002). Prodigiosin, produced by species such as *S marcescens* and *Pseudomoalteromonas rubra,* shows apoptotic effect against human cervical carcinoma (Kavitha et al. 2010).

Antibiotic activity is another property these pigments may impart along with coloration. A pigment obtained from an endophytic fungus was shown to be more potent than the commercially available drug Streptomycin. It had strong antibacterial activity against *Klebsiella pneumoniae, Staphylococcus aureus, Salmonella typhi,* and *Vibrio cholera* (Visalakchi and Muthumary 2010). Violacein also inhibits bacterial growth as well as exhibiting antifungal, anti-protozoan, and antiviral properties (Lopes et al. 2009; Nakamura et al. 2003; Sen et al. 2019).

20.4.3 Cosmetic Industry

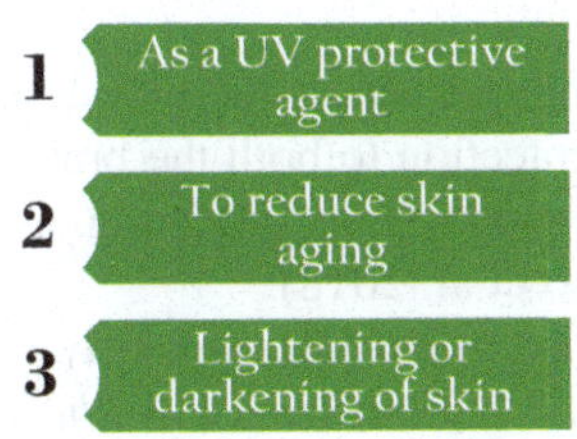

Cosmetics occupy a large part in the day-to-day life of a majority of people. Whether it be skincare creams, ointments, or sunscreen, cosmetics are used quite profusely. Synthetic ingredients such as chemicals in cosmetics however have led to adverse side effects on the human population ranging from mild symptoms such as headaches, skin irritation to pigmentation disorders and allergic reactions. However, symptoms can also be quite harmful, such as lead poisoning, damage to reproductive organs, lung diseases, and may even cause skin cancer.

Society has seen a gradual shift to natural or organic cosmetics in recent years, as these compounds are generally healthier and cause little to no damage to the human body. Certain brands such as Juicy Beauty, Innisfree have many cosmetic or skincare products that are synthesized using natural components, such as plants and plant bioproducts. In fact, the latter of the two brands have taken the "organic" label one step further, by significantly reducing the use of nonbiodegradable packaging material, and rarely selling its products with additional outer packaging. Cosmetic products from India like Mamaearth, Biotique, and age-old cream Vicco turmeric uses natural ingredients in their products. The market appeal of natural products thus has only increased over the past years.

However, the use of plants on an industrial scale is not very cost-effective, and require large areas for cultivation and extraction can be difficult and a lengthy procedure. An alternative therefore is the use of microbial bioproducts as ingredients for cosmetics. Microbes are easily cultured and are readily available from the environment. They take up much less place as compared to plants and their yield is significantly higher. Microbial pigments are a good type of secondary metabolite, as they can also act as a natural colorant for cosmetics and have many additional properties that may be beneficial to humans.

Astaxanthin is a red pigment produced by bacterial species such as *Paracocus, Agrobacterium aurantiacum* and fungal species such as *Thraustrochytrids, Rhodotorula,* and *Phaffia rhodozyma*. It is the most potent antioxidant available in nature and is significantly more effective than Vitamin E. Studies on this pigment has shown to increase skin moisture and moisture retention, elasticity along with promoting smoothness of skin, decreasing fine lines and wrinkles (Ambati et al. 2014).

Astaxanthin also naturally has UV light projection which can help human skin against damage from the sun, like dark spots on the skin. It can also reduce multiple free radicals at the same time and projects the skin against oxidation damage and inflammation. Additionally, it has been shown to increase blood flow, which helps the skin to have better circulation and in turn increased cell turnover number (Tominaga et al. 2012).

Astaxanthin has a rare ability to cross the blood–brain barrier and blood–retinal barrier and can thus offer protection to both the brain and the eyes. Additionally, it can support joint health, reduce signs of aging as well as reduce the risk of Alzheimer's disease (Galasso et al. 2018).

Fucoxanthin is another pigment that is readily produced by many species of brown algae or Phaeophytes. A diatom known as *Phaeodactylum tricornutum* can be used for the commercial production of fucoxanthin. It has strong antioxidant properties, along with the ability to brighten skin. It does so by inhibiting the melanogenesis pathway in melanoma and UVB induced skin pigmentation. This anti-pigmentary activity can be induced by the application of the pigment on the skin (topically) or by taking in orally (Karpiński and Adamczak 2019).

Lutein is a pigment synthesized by many species of algae, bacteria, and fungi that show antioxidant and antiaging properties similar to astaxanthin and fucoxanthin.

β-carotene is a carotenoid and an important pigment that can be obtained from microorganisms. Industrially it can be obtained from bacterial species such as *Blakeslea trispora,* fungal species like *Phycomyces blakesleeanus,* and algal species such as *Dunaliella salina*. It has a large demand in the cosmetic industry due to its large contribution to human health and skincare (Gupta et al. 2019).

β-carotene has been shown to determine epithelial growth, prevent excessive exfoliation of the skin, and accelerate healing of skin damage. It also prevents the development of neoplastic changes and protects against the harmful effects of UV. As a carotenoid, it can also help in the prevention of cancer due to its antioxidant properties (Grune et al. 2010).

Skin aging due to exposure to light for prolonged amounts of time and photooxidation of skin are lifelong problems most people try to avoid. Carotenoids have been proven to be successful in prevention and also in the treatment of age-related changes of skin. The antiaging properties are achieved by increasing skin density, neutralization of singlet oxygen, and by generation of retinoic acid which is mediated by provitamin A. Additionally carotenoids can soothe sunburn too.

It has been shown that utilizing β-carotene in a dose of 30 mg/day for 10 weeks leads to an increase in the yellow pigmentation of the skin and reduction in erythema. β-carotene can also be applied topically as a type of sunscreen to reduce the risk of sunburn, as it can increase the capacity to absorb sunlight through the skin. Cosmetics utilizing β-carotene actually act as a type of UV filter due to its ability to block UV radiation (Grune et al. 2010).

And finally, β-carotene can be applied externally to ease skin roughness or dryness of the epidermis. It also helps in skin discoloration, spots, or uneven coloring. It is able to accomplish this due to its capacity to penetrate the epidermis and then convert into a compound known as retinyl palmitate. This compound is characterized by optimal skin absorption (Biesalski and Obermueller-Jevic 2001).

Melanin is another important pigment that is present as biological pigment in mammalian skin, hair, eyes, ears, and the nervous system. It is also produced by birds, squids, insects, plants, and many microorganisms. One important source of industrial melanin is from the bacterium *Streptomyces* in a newly isolated strain named as *Streptomyces glaucescens* NEAE-H. It can produce a large amount of the pigment melanin (El-Ahmady El-Naggar and El-Ewasy 2017).

Melanin is used as a skin-conditioning agent, as well as an antioxidant, antiaging, and skin-protecting agent such as a sunscreen. However, more importantly, it can be introduced into a number of skincare products and that can induce skin coloring or tinting, which essentially does the opposite of skin whitening. It also exhibits anti-inflammatory and immunoregulatory activities.

Various other pigments such as chlorophyll, lycopene, zeaxanthin, riboflavin, etc., can be utilized in cosmetic products, due to their antioxidant properties and the ability to promote better skincare (Rao et al. 2017; Gupta et al. 2019; Tominaga et al. 2012).

20.4.4 Pharmaceuticals and Medicine

The field of medicine has accomplished huge goals in the improvement of human life. Medical treatments in response to ailments started all the way back in 431 BCE during the ancient Greek civilization. Hippocrates is known as the first physician and treated his patients with a combination of physical therapy and natural medicine. Over the years, medicine has evolved along with society and had come up with many new ways to combat various ailments and diseases. However, things are far from perfect and medicine cannot always keep up when new diseases and pathogens keep arising.

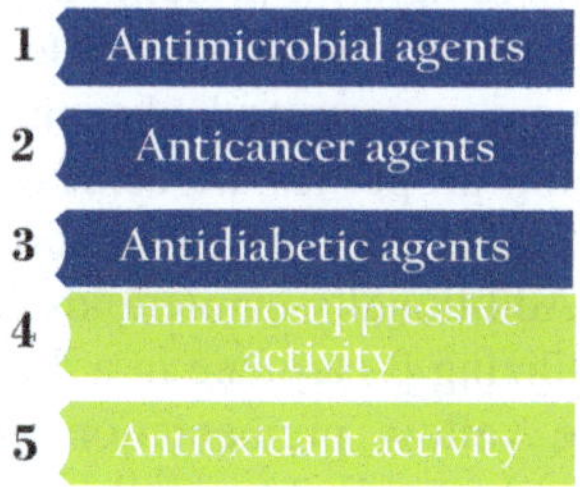

There are many treatments and drugs available to treat diseases, but a majority of them are man-made and based on synthetic chemicals. These synthetic drugs do help in the treatment of diseases, but when taken in excess or in some cases even in measured doses, cause side effects, that deteriorates a person's well-being. They may also accumulate in the body and cause toxicity and sometimes may even contribute to cancer.

In the last couple of years, medicine has seen a shift back to its basics. People nowadays prefer natural medicine over a synthetic drug. And so the need has arisen to look for natural sources of obtaining remedies from. But it is not only because of a shift in trend. It is also the knowledge that since all disease-causing pathogenic microorganisms arise from nature, it would be wise to assume that nature also has a way to deal with them.

One important source of such natural remedies is from microorganisms themselves. Microorganisms generally produce compounds that can neutralize or even terminate one another. One such type of compounds is their pigments. Microbial pigments inherently have many properties that can help in against other microbes. They also additionally contain many extra health benefits that affect human health.

20.4.4.1 Antimicrobial Activity

Microbial pigments have antimicrobial properties that can affect a wide range of pathogens. Pigments like pyocyanin and pyorubin which can be industrially extracted from *P. aeruginosa* have shown to be significantly effective against species of *Citrobacter* which are the causative organisms for urinary tract and wound infection. Various pigments produced from *Micrococcus luteus* KF532949 showed good antimicrobial action against other wound associated pathogens such as *Staphylococcus* sp., *Klebsiella* sp., *and Pseudomonas* sp. (Umadevi and Krishnaveni 2013).

Upon successive encounters with synthetic drugs, pathogens may evolve to become drug-resistant. This is how multidrug-resistant bacteria have arisen. However, natural pigments from microorganisms also develop and adapt to fight against drug-resistant species. *Streptomyces hygroscopicus* synthesizes a pigment that has shown strong antimicrobial activity against certain drug-resistant pathogens like methicillin and vancomycin-resistant strains of *Staphylococcus aureus.* It additionally shows antimicrobial action against β lactamase producing strains of *Escherichia coli, P. aureginosa,* and *Klebsiella* sp. (Berlanga et al. 2000; Selvameenal et al. 2009; Rao et al. 2017).

Fungal pigments also serve as active antimicrobial agents. A pigment produced by *Monascus ruber* CCT 3802 demonstrated antimicrobial activity against common foodborne bacteria *S. aureus, E. coli,* and *Salmonella enteritidis.* Another endophytic fungus known as *Monodictys castaneae* produced a pigment that could inhibit human pathogenic bacteria such as *S. aureus, K. pneumoniae,* and *Vibrio cholera* (Vendruscolo et al. 2013; Visalakchi and Muthumary 2010).

Algae are also a good source of pigments that show strong antimicrobial activity. Fucoxanthin is a carotenoid that is produced by brown algae and diatoms. Fucoxanthin was tested for its antimicrobial action strength against 20 species of bacteria. Its strongest action was against the following species: *Streptococcus agalactiae, S. epidermidis,* and *S. aureus* (Karpiński and Adamczak 2019). Another algal species *Oscillatoria* produces a pigment known as phycocyanin. This pigment also showed strong antimicrobial activity, in a study conducted in 2016. It showed to be effective against human pathogens *Klebsiella* sp., *Pseudomonas* sp., and *S. aureus* (Shanmugam et al. 2017). The algal species of *Gracilaria* produces phycoerythrin that was tested for antimicrobial activity. It showed results against *P. aureginosa* and *S. aureus* (Sudhakar et al. 2015).

Antifungal properties are also evident in certain microbial pigments. A yellow pigment synthesized by a *Bascillus* sp. which was isolated from a lichen (*Dirinaria agealita*) showed promising antifungal activity against three species of fungus: *Rhizoctonia solani, Fusarium oxysporum,* and *Sclerotium rolfsii.* It is interesting to note here that it was reported that the activity of the pigment extract was different near each fungus (Dawoud et al. 2020). Another pigment isolated from a fungus, *Aspergillus nishimurae* demonstrated antimicrobial action against other fungi, such as *Aspergillus fumigatus, Aspergillus niger, Candida albicans,* and *Cryptococcus neoformans* (Hosoe et al. 2011).

Pseudomonas and *Streptomyces* species produce certain phenazine compounds which have shown promising antiviral activities (Schneemann et al. 2011). The pigment violacein also has strong antiviral properties against herpes simplex virus, poliovirus, and simian rotavirus SA II (Durán et al. 2007). Certain quinone derivative pigments such as anthraquinones, napthaquinones, and benzoquinones are also known to have antiviral properties (Koyama 2006; Gessler et al. 2013).

Nostoc spongiaeforme TISTR 8169 is a freshwater cyanobacteria. It can synthesize and release a violent pigment known as nostocine A. In a study conducted in 2003, it was found that nostocine A, extracted from *N. spongiaeforme* TISTR 8169 acts as a growth inhibitor for other microalgae (Hirata et al. 2003).

In addition to the ones already listed, pigments such as carotenoids, melanins, quinones, monascins, flavins, violacein, and indigo have also been reported to have adequate antimicrobial activity (Rao et al. 2017).

All the abovementioned pigments act similarly to commercially available antimicrobial agents. These pigments might act as micro-biostatic agents that arrest growth and stop the increasing numbers of the target organism. This is generally achieved by inhibiting certain biological pathways related to cell proliferation or they may act as micro-biocidal agents that completely kills the target organism population. This is generally achieved by the inhibition of cell wall synthesis and thereby inadvertently lysing the cells of the organism.

Thus, these pigments are suitable candidates to be used as novel drugs and may one day replace the synthetic drugs commercially available at the moment.

20.4.4.2 Antioxidant Activity

Free radicals are a class of molecules that are highly volatile and react with organic cellular composition because they contain unpaired electrons. These molecules are generated as by-products of the biological metabolic processes of organism, including human beings. These reactive species can cause a lot of damage to cells of the human body, including cellular components such as DNA, proteins, and cell membranes. The damage is done by removing electrons through the process of oxidation. This phenomenon is known as oxidative stress.

Increase in free radicals of the body in turn increases the chances of occurrence of chronic diseases such as cancer, diabetes, cardiovascular, and autoimmune disorders. In order to avoid this, antioxidants can be employed. Antioxidants neutralize free radicals, by donating electrons to pair with the reactive species (Rao et al. 2017).

Microorganisms also generate free radicals and have their own defense mechanism against them. Carotenoids are a very important class of pigments that all have very strong antioxidant properties. They are very diverse groups of natural pigments, both structurally and functionally. They are ubiquitously found in all organisms capable of performing photosynthesis. They are precursors of retinol or vitamin A, but they also play a role in photoprotection. Their most important function is as efficient physical and chemical quenchers of singlet oxygen species ($^1O_2$) as well as potent scavengers of other reactive oxygen species. Carotenoids and some of their metabolites are thought to have a protective role in numerous free radical-mediated disorders (Fiedor and Burda 2014; Rao et al. 2017).

Carotenoids are industrially extracted from many microorganisms. Some of them are *Blakelsea trispora* (β-carotene), *Phaffia rhodozyma* (Astaxanthin), *Muriellopis* sp. (Lutein), *Dietzia natronolimnaea* (Canthaxanthin), etc. . .

Anthraquinones from endophytic fungus *Stemphylium lycopersici* (Li et al. 2017) and melanin from *Streptomyces glaucescens* NEAE-H (El-Ahmady El-Naggar and El-Ewasy 2017 were shown to have antioxidant properties. Xanthomonadin pigment is a yellow pigment synthesized by *Xanthomonas* bacteria (Tuli et al. 2015) has been shown to have antioxidant activity and photodamage protection. These bacteria are naturally plant pathogen, but their pigment could be used for human benefit (Rao et al. 2017).

Antioxidant supplements are commonly used in the pharmaceutical industry. Microbial sources of these compounds could significantly cut down on the production and extraction costs and increase the yield of the desired pigments. As these pigments are already in use in the treatment of oxidative stress, adapting the microbial versions of the pigments into commercially usable drugs is an economically viable option.

20.4.4.3 Anticancer Agents

Cancer is one of the major diseases that have been the bane of human beings for centuries. The phenomenon of cells of certain tissues start growing without restraint and out of control causing tumors dates all the way back to Egypt in the year 1500 BCE. There are several different types of cancers; some of the most common types are breast cancer, prostate cancer, skin cancer, lung cancer, leukemia, etc.

Cancer therapeutics has been an extensive area of research from a long time. There are not many compounds in existence that can readily cure cancer, but some can help prevent and reduce the risk of developing it. One such class of compounds are biological pigments. And a good chunk of such pigments are synthesized by microbial species.

A large number of microbial pigments possess anticancer activity. Consequentially many efforts were taken to use these pigments as anticancer agents thus laying the foundation for continued research in the said direction. A gram-negative bacteria, *Pseudoalteromonas* sp. 1020R which was isolated from the Pacific coast of Japan (Wang et al. 2012) produces a pigment, prodigiosin, and three other similar compounds. These pigments exhibited varying degrees of cytotoxicity against U937 leukemia cells. It was observed the pigments could trigger cell death, with typical features of apoptosis.

Streptomyces glaucescens NEAE-H produces the pigment melanin naturally. Upon testing this pigment for its biological properties, anticancer property was recorded. It was shown that the pigment has a potent cytotoxic activity against HFB4 skin cancer cell lines (El-Ahmady El-Naggar and El-Ewasy 2017).

Certain pigments extracted from the fungus *Monascus* have demonstrated various anticancer activity against different types of cancer cell lines. Pigments such as monascin showed anticancer activity against mouse melanoma cells. Ankaflavin showed anti-proliferative activity against Hep G2 and A549 human cancer cell lines. Monaphilone A and B showed an inhibitory effect against Hep-2 human laryngeal cancer cell lines (Feng et al. 2012).

Three bianthraquinone derivatives from the mangrove endophytic fungus *Alternaria* sp. ZJ9-6B were shown to have moderate cytotoxic activity towards MDA-MB-435 and MCF-7 cancer cells (Huang et al. 2011).

Of particular interest was the anticancer agent, prodigiosin, which has been tested against more than 60 cancer cell lines for its anticancer properties. Results showed that it had a strong anticancer activity due to the presence of multiple cellular targets (Darshan and Manonmani 2015).

It is evident from the above instances that microbial pigments possess the potential to be widely used as therapeutic agents in the treatment of various types of cancers.

20.4.4.4 Immunosuppressive Activity

Pseudoalteromonas dentrificans produces a red pigment known as cycloprodigioson. This compound, which is stable under several physiochemical conditions, exhibits immunosuppressive activity. It does so by inhibition of proliferation of T-cells and PMA (Phorbol 12-myristate 13-acetate) stimulated Jurkat cells (Kawauchi et al. 1997). At levels that were not harmful to the cells, the pigment inhibited the following processes: T-cell mediated immune functions, mixed lymphocyte response, local graft vs. host reaction, and T-dependent antibody response (Han et al. 1998).

A similar pigment, undecylprodiogiosin displayed the ability to inhibit the peripheral human T and B lymphocytes as well as to elicit the inhibition of retinoblastoma protein phosphorylation, by the inhibition of cyclin-dependent kinase-2 and cyclin-dependent kinase-4 in human lymphocytes (Songia et al. 1997).

Tambajamines are a group of alkaloids which are natural products of microbes and are structurally related to prodiginines. These compounds are also known to have immunosuppressive action (Stafsnes and Bruheim 2013).

20.4.4.5 Antidiabetic Activity

Certain microorganisms are shown to have antidiabetic action. *Monascus*, a red mold, ferments *Dioscorea* plant species and was found to show the delaying of the onset of diabetes. This occurs through the simultaneous action of antioxidant effects, protection of pancreatic β-cells, and control of hyperglycemia by decreasing blood glucose and serum-free fatty acid levels. Studies performed on Streptozotocin-induced diabetic rats provided promising results (Shi et al. 2011).

Ethanol extracts from *Monascus* fermented durian seeds have demonstrated antidiabetic action by the action of inhibiting α-glucosidase (Srianta et al. 2013).

The above phenomenon occurs with respect to *Monascus* due to the fact that the class of pigments known as monascins are known to block the phosphorylation of PPAR-γ (a transcription factor) by the enzyme p-JN Kinase, which results in antidiabetic activity (Lee et al. 2011).

20.4.4.6 Anti-adipogenic Activity

Adipogenesis is the process by which adipose cells and tissues are generated. Adipose tissues are responsible for the storage of fats and are unhealthy for the human body in large quantities. Pigments such as monascin and ankaflavin assist in and boost the delipidation of mature adipose cells. They accomplish this by promoting the release of glycerol, and the process is increased by about 113.2% by monascin and 278.3% by ankaflavin. They reduce the downregulation activity of HR-LPL (heparin-releasable lipoprotein lipase). This allows for the lipase to act on the lipid molecules to a greater extent than normal with a reported increase of 45.3% by monascin and 58.1% by ankaflavin (Jou et al. 2010).

High-fat diet which has been supplemented with monascin pigment derivatives along with 4-phenylbutylamine (PBA) and 2-(*p*-toyly) ethylamine have exhibited inhibitory action against adipogenic differentiation in 3T3-L1cells (Choe et al. 2012).

20.4.4.7 Anti-atherosclerosis Activity

Research investigating the heart aorta of hamsters demonstrated that the pigment ankaflavin contributes towards the prevention of accumulation of fatty liver and lipid plaque and the pigment monascin prevents the accumulation of enhanced high-density lipoprotein cholesterol. It was additionally observed that ankaflavin can act as a potential hypolipidemic agent (Lee et al. 2013).

Astaxanthin, isolated from different sources has shown to improve oxidative stress, inflammation, increase lipid metabolism, and glucose metabolism. This leads to it demonstrating preventive actions against atherosclerotic cardiovascular diseases. In clinical trials, it has shown to inhibit low-density lipoprotein (LDL) oxidation and increase high-density lipoprotein (HDL)-cholesterol and adiponectin levels (Kishimoto et al. 2016).

20.4.4.8 Anti-inflammatory Activity

Monascin, the yellow pigment produced by *Monascus,* has shown an anti-inflammatory effect. It achieves this by the inhibition of the inflammatory signal pathways of protein kinase C (PKC) and JNK (c-Jun N-terminal kinase) phosphorylation in a C2C12 cell model (Lee et al. 2011).

Monascin present in the *Monascus* fermented dioscorea also has demonstrated anti-inflammatory property in Streptozotocin-induced diabetic rats. This was possible as the pigment reduced the inflammatory cytokine TNF-α levels and enhancing IL-2 cytokine expression (Shi et al. 2011).

20.4.4.9 Antimalarial Activity

The violet pigment extracted from the bacteria *Chromobacterium violaceum* known as violacein has potent antimalarial activity against the protozoan *Plasmodium falciparum* strains, in vitro (Lopes et al. 2009).

Prodiginine derivatives such as cycloprodigiosin (Kim et al. 1999), prodigiosin, undecylprodigiosin, hepltylprodigiosin, and metacylcoprodigiosin (Stankovic et al. 2014) have all shown antimalarial activity against *P. falciparum* (Lazaro et al. 2002).

A study from 1993 had also reported that melanin extracted from a *Bacillus thuringiensis* subsp. *israelensis* had shown mosquitocidal activity (Liu et al. 1993).

20.4.4.10 Anti-tuberculosis Activity

The pigment violacein isolated from *Janthinobacterium* sp. Ant5-2 and flexirubin isolated from the *Flavobacterium* sp. Ant342 exhibited the inhibition of *Mycobacterium tuberculosis* growth. Both of those bacterial species reside in the Antarctic (Mojib et al. 2010).

20.4.4.11 Anti-HIV Activity

Various compounds were extracted from different species of fungus known as *Phoma* and were tested and found to have anti-HIV properties. It was observed that they could inhibit the activity of the inherent enzyme integrase (Rai 2009).

Clinical studies are also underway to assess the activity of violacein on AIDS-related lymphoma (Durán et al. 2016).

20.4.4.12 Anti-Alzheimeric Activity

Some phycobiliproteins isolated from red algae and cyanobacteria species have demonstrated some anti-Alzheimeric properties (Sonani et al. 2016).

20.4.4.13 Anti-hypertensive Activity

Monascus fermented Dioscorea when administered orally and at a low dosage were shown to decrease systolic and diastolic blood pressures in hypertensive rats (Wu et al. 2009).

20.4.4.14 Antiulcerogenic Activity

The purple pigment violacein has been shown to inhibit gastric damage which causes ulcers. It works in the presence of β-cyclodextrin (βCD) inclusion complex in molar ratios of 1:1 and 1:2 (Durán et al. 2003).

A pigment isolated from *Chryseobacterium* and *Flavobacterium*, known as flexirubin, is used in the treatment of chronic skin disease, eczema, gastric ulcers, etc. (Kim 2013; Venil et al. 2014).

20.5 Other Applications

20.5.1 Cytotoxic Activity

In a study performed in 2007 two brominated yellow pigments bromoalterochromide A and bromoalterochromide A' were isolated from a marine bacterial strain *Pseudoalteromonas maricaloris* KMM 636^T. These pigments showed a cytotoxic effect on the developing eggs of a species of sea urchin (Speitling et al. 2007).

Another study from 2009 demonstrated the highlighted cytotoxic activity of a deep blue pigment that was extracted from strains of a bacteria known as *Rheinheimera*. The cytotoxic activity was reported against a species of brine shrimp (Grossart et al. 2009).

20.5.2 Antifouling Activity

Biological fouling is the undesirable accumulation of organisms on surfaces, generally moist, which can then lead to corrosion. The organisms responsible for fouling are generally algae, diatoms, small animals, and plants. This phenomenon is generally observed on the hull of ships, reservoirs of untreated water.

Pseudoalteromonas species which are pigmented, isolated from marine plants and animals, were found to be effective against common fouling organisms such as invertebrate larvae of *Hydroides elegans* and *Balanus amphirite,* algal spores of *Ulva lactuca* and *Polysiphonia* sp., diatoms, bactreia, and fungi (Holmström et al. 2002).

Prodigiosin producing *S. marcescens* CMST07 also reported similar antifouling activity, which was able to inhibit marine fouling bacteria like *Alteromonas* sp. and *Gallionella* sp. Additionally, the pigment prodigiosin was found to impede cyanobacterial adhesion on glass surfaces (Priya et al. 2013).

20.5.3 Algicidal Activity

Prodigiosin isolated from a marine bacterium *Hahella chejuensis* demonstrated potent algicidal activity. It showed a near-complete inhibition of algal activity against the red-tide dinoflagellate *Chochlodinium polykrikoides* (Jeong et al. 2005).

Another pigment known as xylinden, with a blue-green color, isolated from a fungal species *Chlorociboria aeruginosa,* was seen to be able to control the growth of *Chlorella* (Sakaki et al. 2002).

20.5.4 Insecticidal Activity

Naphthoquinone-derived pigments such as javanicin, fusarubin, and similarly related compounds have been reported to display insecticidal activity (Margalith 1992; Gessler et al. 2013).

20.5.5 Herbicidal Activity

Many species of fungi such as *Phoma exigua* var. *exigua, Phoma foveata, Phoma glomerata, P. herbarum, P. macdonaldii, P. macrostoma, P. multirostrata, P. proboscis, P. sorghina,* and *P. tracheiphila* produce anthraquinones and some derivatives possess strong herbicidal activities (Rai 2009).

A study performed in 2011 managed to isolate a pigment (anhydropseudophlegmacin-9,10-quinone-3′-amino-8′-*O*-methyl ether) from the fungal plant pathogen *Phoma herbarum* FGCC#54. This compound showed potential herbicidal activity against notable weeds *Hyptis suaveolens, Lantana camara, Parthenium hysterophorus,* and *Sida acuta* (Quereshi et al. 2011).

20.5.6 Antiparasitic Activity

Chromobacterium violaceum isolated pigment, violacein demonstrated in laboratory experiments that it is capable of antiparasitic activity. It has strong trypanocide properties which can inhibit the growth of the euglenoid parasite *Trypanosoma cruzi* (Haun et al. 1992). Violacein was also reported to inhibit the growth of chloroquine-sensitive and -resistant strains of *Plasmodium falciparum* (Lopes et al. 2009).

20.5.7 Antiprotozoal Activity

Janthinobacterium lividum and *Chromobacterium violaceum* produce the pigment violacein which was reported to have antiprotozoal activity. It could efficiently inhibit the cultures of nanoflagellates *Spumella* sp. and *Orchromonas* sp. At higher concentrations, it resulted in the complete reduction of cells. This indicates why the nanoflagellates refrain from ingesting the abovementioned bacteria (Matz et al. 2004). Prodigiosin, are pigment, isolated from *Serratia* also have shown to possess antiprotozoal activity (Williams and Quadri 1980).

20.5.8 Antileishmanial Activity

Violacein was observed to be able to reduce the viability and inhibit the growth of parasite *Leishmania amazonensis* which is the causative organism for leishmaniasis (Leon et al. 2001). Prodigiosin isolated from *S. marcescens* variant SM365 has also shown evidence to be able to lyse the organism *Leishmania chagasi* (Moraes et al. 2008).

20.5.9 Antinematodal Activity

P. fluorescens produces a class of compounds, known as phenazines, which were found to suppress the hatching of the eggs and to enhance the juvenile mortality of the root-knot nematode *Meloidogyne incognita,* in vitro (Sankari et al. 2014).

20.5.10 Fluorescent Probes

Phycoerythrins and other phycobiliproteins are pigments derived from cyanobacteria which are readily used as fluorescent probes in various biotechnological assays and are commercially available as conjugated activity (Chakdar and Pabbi 2012).

20.6 The Road Ahead and Challenges

Microbes as a source of color has several advantages. They are easily synthesized and controlled, have very simple modes of extraction and are eco-friendly. They have great potential considering their wide usability in various industries such as textile, food, pharmaceuticals, medicine as well as in various other commercial aspects. Microorganism is suitable for large-scale production of desired compounds, given the ease with which they can be cultured with minimum resources.

Microorganisms utilized for the large-scale production of pigments (or any other desired compound) can be tweaked using genetic engineering to facilitate a larger yield of the products. In addition to this, the use of genomic manipulation can also make microorganism a common host that is able to synthesize pigments not only from bacteria, fungi, and algae but also from higher organisms such as plants and animals. This will elevate the significance of microbial pigments to newer heights.

It is said that "there are more microbial species on earth than there are stars in the night sky." And according to an estimate from the year 2016, there are about one trillion microbial species on earth, and only 0.1% of them have yet been discovered. This fact alone states that we have only scratched the surface in terms of the industrial potential microorganisms possess. In particular, as far as the usability of microbial pigments as an alternative to chemical dyes is concerned, we can assume, that there are numerous more pigments that the microbial community can offer us, and we have only observed a mere handful of them. Going forward, this particular topic of research and study is only going to grow and flourish to a greater extent.

However, no scientific endeavor can be made without its due challenges. One of the crucial obstacles standing in the way of utilizing microbial pigments in a large scale is commercialization. In the market, preference is still given to synthetic colors. The synthetic products are cheaper in comparison to the natural ones, as synthesis of artificial dyes out of chemicals is very cost-effective.

Contamination of the microbial cultures posts one of the major obstacles to using microbes as potential sources for pigments. One contaminated culture results in a large scale loss of time and assets. The remedy for this however is simple. The individuals working with the microorganisms need to be trained by professionals to following proper practices and maintain a sterile environment for the culturing of microorganisms.

Many microbial pigments are not stable under normal environmental factors such as temperature, light, pH, etc. These compounds are rendered inactive when exposed to these conditions and thus have a short shelf life. As such, they need to be stabilized by employing various techniques to make these pigments industrially useful. Employing such techniques however increases the market value of these compounds, thereby losing the competition with their synthetic counterparts.

20.7 Conclusion

There is no denying the impact color has had on human life and civilization. We have implemented the use of various colors since the dawn of mankind. From vibrant cave paintings created by the early humans during the Stone Age, by using crushed berries and leaves. Over the years color has become a symbol for many things and has also been associated psychologically with human emotion.

Since the last couple of decades, synthetic dyes have been the primary source of color. Chemicals were developed to produce various hues and pigments and they are utilized in many different aspects of human civilization such as clothing, food, and even in medicine. However, the continuous use of these chemicals (pigmented or not) proved to have many adverse effects on humans as well as the rest of the environment.

So the search began for natural sources of components that could substitute the toxic chemicals, now prevalent in many industries of civilization. And one such source of natural, beneficial compounds were pigments isolated from microorganisms. These pigments played important roles in the organism's life cycle. But served equally useful functions outside of their hosts.

Pigments isolated from bacteria such as *S. marcescens* and fungi like *F. oxysporum* can be used readily in the textile industry to dye fabrics and fibers. Astaxanthin from *X. dendrorhous* and Riboflavin from *A. gossypii* are useful food colorants and have additional health benefits such as antioxidant activity. Fucoxanthin from certain diatoms though a strong antioxidant but can be also used in cosmetics as it is observed to lighten skin. In contrast, melanin from *S. glaucescens* NEAE-H is shown to darken skin color.

Microbial pigments play an immensely large role in the pharmaceutical and medical field as a substitute for synthetic medicine. They have many important properties that are beneficial to humans, such as antimicrobial activity, anticancer activity, immunosuppressive activity to name a few. They can be used in the treatment of some diseases and also incorporated into drugs for the betterment of human health.

Despite their various uses, microbial pigments have only come into use in recent times. Their applications are vast and there is much more yet to be known about them. Research on this subject will only continue to grow, despite the technical challenges. The future for microbial pigments looks promising as more microbes are being discovered and many more biological compounds produced by them are explored.

References

Ambati RR, Phang SM, Ravi S, Aswathanarayana RG (2014) Astaxanthin: sources, extraction, stability, biological activities and its commercial applications—a review. Mar Drugs 12 (1):128–152. https://doi.org/10.3390/md12010128

Andrighetti-Fröhner CR, Antonio RV, Creczynski-Pasa TB, Barardi CRM, Simões CMO (2003) Cytotoxicity and potential antiviral evaluation of violacein produced by *Chromobacterium violaceum*. Mem Inst Oswaldo Cruz 98(6):843–848. https://doi.org/10.1590/s0074-02762003000600023

Barredo JL, Garcia-Estrada C, Kosalkova K, Barreiro C (2017) Biosynthesis of Astaxanthin as a main carotenoid in the Heterobasidiomycetous Yeast *Xanthophyllomyces dendrorhous*. J Fungi 3(3):44. https://doi.org/10.3390/jof3030044

Bennett JW, Bentley R (2000) Seeing red: the story of prodigiosin. Adv Appl Microbiol 47:1–32. https://doi.org/10.1016/S0065-2164(00)47000-0

Berlanga M, Ruiz N, Borrell J, Montero T, Vinas M (2000) Role of the outer membrane in the accumulation of quinolones by *Serratia marcescens*. Can J Microbiol 46:716–722. https://doi.org/10.1139/w00-052

Biesalski HK, Obermueller-Jevic UC (2001) UV light, beta-carotene and human skin—beneficial and potentially harmful effects. Arch Biochem Biophys 389(1):1–6. https://doi.org/10.1006/abbi.2001.2313

Biesalski HK, Chichili GR, Frank J, von Lintig J, Nohr D (2007) Conversion of β-carotene to retinal pigment. Vitam Horm 75:117–130. https://doi.org/10.1016/S0083-6729(06)75005-1

Boonyapranai K, Tungpradit R, Lhieochaiphant S, Phutrakul S (2008) Optimization of submerged culture for the production of naphthoquinones pigment by *Fusarium verticillioides*. Warasan Khana Witthayasat Maha Witthayalai Chiang Mai 35(3):457–466

Chadni Z, Rahaman MH, Jerin I, Hoque KMF, Reza MA (2017) Extraction and optimisation of red pigment production as secondary metabolites from *Talaromyces verruculosus* and its potential use in textile industries. Mycology 8:48–57. https://doi.org/10.1080/21501203.2017.1302013

Chakdar H, Pabbi S (2012) Extraction and purification of Phycoerythrin from Anabaena variabilis (CCC421). Phykos 42:25–31

Chen C-Y, Jesisca HC, Lee D-J, Chang C-H, Chang J-S (2016) Production, extraction and stabilization of lutein from microalga Chlorella sorokiniana MB-1. Bioresour Technol 200:500–505. https://doi.org/10.1016/j.biortech.2015.10.071

Choe D, Woo S, Lee J, Shin CS (2012) Evaluation of the amine derivatives of Monascus pigment with anti-obesity activities. Food Chem 134(1):315–323. https://doi.org/10.1016/j.foodchem.2012.02.149

Darshan N, Manonmani HK (2015) Prodigiosin and its potential applications. J Food Sci Technol 52:5393–5407. https://doi.org/10.1007/s13197-015-1740-4

Dawoud TM, Alharbi NS, Theruvinthalakal AM, Thekkangil A, Kadaikunnan S, Khaled JM, Almanaa TN, Sankar K, Innasimuthu GM, Alanzi KF, Rajaram SK (2020) Characterization and antifungal activity of the yellow pigment produced by a Bacillus sp. DBS4 isolated from the lichen *Dirinaria agealita*. Saudi J Biol Sci 27(5):1403–1411. https://doi.org/10.1016/j.sjbs.2019.11.031

Di Mascio PD, Kaiser S, Sies H (1989) Lycopene as the most efficient biological carotenoid singlet oxygen quencher. Arch Biochem Biophys 274(2):532–538. https://doi.org/10.1016/0003-9861(89)90467-0

Dufossé L (2009) Pigments, microbial. In: Schaechter M (ed) Encyclopedia of Microbiology, Third edn. Elsevier/Academic Press, New York, NY, pp 457–471

Dufossé L (2017a) Chapter 4: Current carotenoid production using microorganisms. In: Bio-pigmentation and biotechnological implementations, 1st edn. Wiley, pp 87–106

Dufossé L (2017b) Red colourants from filamentous fungi: are they ready for the food industry? J Food Compos Anal 69:156–161. https://doi.org/10.1016/j.jfca.2017.11.002

Dufossé L (2018) Chapter 4: Microbial pigments from bacteria, yeasts, fungi, and microalgae for the food and feed industries. In: Handbook of food bioengineering, 1st edn. Academic, pp 113–132. https://doi.org/10.1016/B978-0-12-811518-3.00004-1

Dufossé L, Galaup P, Yaron A, Arad SM, Blanc P, Murthy KNC, Ravishankar GA (2005) Microorganisms and microalgae as sources of pigments for food use: a scientific oddity or an industrial reality? Trends Food Sci Technol 16:389–406. https://doi.org/10.1016/j.tifs.2005.02.006

Durán N, Justo GZ, Melo PS, De Azevedo MBM, Brito ARMS, Almeida ABA, Haun M (2003) Evaluation of the antiulcerogenic activity of violacein and its modulation by the inclusion complexation with β-cyclodextrin. Can J Physiol Pharmacol 81(4):387–396. https://doi.org/10.1139/y03-033

Durán N, Justo GZ, Ferreira CV, Melo PS, Cordi L, Martins D (2007) Violacein: properties and biological activities. Biotechnol Appl Biochem 48(Pt 3):127–133. https://doi.org/10.1042/BA20070115

Duran M, Ponezi AN, Faljoni-Alario A, Teixeira MFS, Justo GZ, Duran N (2012) Potential applications of violacein: a microbial pigment. Med Chem Res 21:1524–1532. https://doi.org/10.1007/s00044-011-9654-9

Durán N, Justo GZ, Durán M, Brocchi M, Cordi L, Tasic L, Castro GR, Nakazato G (2016) Advances in Chromobacterium violaceum and properties of violacein-Its main secondary metabolite: A review. Biotechnol Adv 34(5):1030–1045. https://doi.org/10.1016/j.biotechadv.2016.06.003

El-Ahmady El-Naggar N, El-Ewasy SM (2017) Bioproduction, characterization, anticancer and antioxidant activities of extracellular melanin pigment produced by newly isolated microbial cell factories *Streptomyces glaucescens* NEAE-H. Sci Rep 7:42129. https://doi.org/10.1038/srep42129

Eriksen NT (2008) The technology of microalgal culturing. Biotechnol Lett 30(9):1525–1536

Feng Y, Shao Y, Chen F (2012) Monascus pigments. Appl Microbiol Biotechnol 96:1421–1440. https://doi.org/10.1007/s00253-012-4504-3

Feng J, Shi W, Zhang S, Zhang Y (2015) Identification of new compounds with high activity against stationary phase *Borrelia burgdorferi* from the NCI compound collection. Emerg Microbes Infect 4(6):e31. https://doi.org/10.1038/emi.2015.31

Fiedor J, Burda K (2014) Potential role of carotenoids as antioxidants in human health and disease. Nutrients 6(2):466–488. https://doi.org/10.3390/nu6020466

Fleurence J (2003) R-Phycoerythrin from red macroalgae: strategies for extraction and potential application in biotechnological area. University of Nantes

Galasso C, Orefice I, Pellone P, Cirino P, Miele R, Ianora A, Brunet C, Sansone C (2018) On the neuroprotective role of astaxanthin: new perspectives? Mar Drugs 16(8):247. https://doi.org/10.3390/md16080247

Gessler NN, Belozerskaya TA, Barashkova AS (2013) Fungal anthraquinones. Appl Biochem Microbiol 49(2):85. https://doi.org/10.1134/S000368381302004X

Grossart H-P, Thorwest M, Plitzko I, Brinkhoff T, Simon M, Zeeck A (2009) Production of a blue pigment (Glaukothalin) by marine Rheinheimera spp. Int J Microbiol 2009:701735. https://doi.org/10.1155/2009/701735

Grune T, Lietz G, Palou A, Ross AC, Stahl W, Tang G, Thurnham D, Yin SA, Biesalski HK (2010) β-Carotene is an important vitamin a source for humans. J Nutr 140(12):2268S–2285S. https://doi.org/10.3945/jn.109.119024

Guedes AC, Amaro HM, Malcata FX (2011) Microalgae as sources of carotenoids. Mar Drugs 9:625–644. https://doi.org/10.3390/md9040625

Gupta PL, Rajput M, Oza T, Trivedi U, Sanghvi G (2019) Eminence of microbial products in cosmetic industry. Nat Prod Bioprospect 9(4):267. https://doi.org/10.1007/s13659-019-0215-0

Han SB, Kim HM, Kim YH, Lee CW, Jang ES, Son KH, Kim SU, Kim YK (1998) T-cell specific immunosuppression by prodigiosin isolated from *Serratia marcescens*. Int J Immunopharmacol 20(1–3):1–13. https://doi.org/10.1016/s0192-0561(97)00062-3

Hassan HM, Fridovich I (1980) Mechanism of the antibiotic action pyocyanin. J Bacteriol 141(1):156–163. https://doi.org/10.1128/JB.141.1.156-163.1980

Haun M, Pereira MF, Hoffmann ME, Joyas A, Campos V, Filardi LD, De Castro SL, Duran N (1992) Bacterial chemistry. VI. Biological activities and cytotoxicity of 1,3-dihydro-2H-indol-2-one derivatives. Biol Res 25(1):21–25

Hirata K, Susilaningsih D, Mahakhant A, Miyamoto K, Yoshitomi S, Iwabe O, Polchai J (2003) Bioactivities of nostocine a produced by a freshwater cyanobacterium *Nostoc spongiaeforme* TISTR 8169. J Biosci Bioeng 95(5):512–517. https://doi.org/10.1016/S1389-1723(03)80053-1

Holmström C, Egan S, Franks A, McCloy S, Kjelleberg S (2002) Antifouling activities expressed by marine surface associated Pseudoalteromonas species. FEMS Microbiol Ecol 41(1):47–58. https://doi.org/10.1111/j.1574-6941.2002.tb00965.x

Hosikian A, Lim S, Halim R, Danquah MK (2010) Chlorophyll extraction from microalgae: a review on the process engineering aspects. Int J Chem Eng 2010:391632. https://doi.org/10.1155/2010/391632

Hosoe T, Mori N, Kamano K, Itabashi T, Yaguchi T, Kawai K (2011) A new antifungal yellow pigment from *Aspergillus nishimurae*. J Antibiot (Tokyo) 64(2):211–212. https://doi.org/10.1038/ja.2010.132

Huang CH, Pan JH, Chen B, Yu M, Huang HB, Zhu X (2011) Three bianthraquinone derivatives from the mangrove endophytic fungus Alternaria sp. ZJ9-6B from the South China Sea. Mar Drugs 9:832–843

Jeong H, Yim JH, Lee C, Choi S-H, Park YK, Yoon SH, Hur C-G, Kang H-Y, Kim D, Lee HH, Park KH, Park S-H, Park H-S, Lee H-K, Oh TK, Kim JF (2005) Genomic blueprint of Hahella chejuensis, a marine microbe producing an algicidal agent. Nucleic Acids Res 33 (22):7066–7073. https://doi.org/10.1093/nar/gki1016

Jou PC, Ho B-Y, Hsu T-W, Pan T-M (2010) The effect of Monascus secondary polyketide metabolites, monascin and ankaflavin, on adipogenesis and lipolysis activity in 3T3-L1. J Agric Food Chem 58(24):12703–12709. https://doi.org/10.1021/jf103121c

Karpiński TM, Adamczak A (2019) Fucoxanthin-an antibacterial carotenoid. Antioxidants (Basel) 8(8):239. https://doi.org/10.3390/antiox8080239

Kavitha R, Aiswarya S, Ratnawali CMG (2010) Anticancer activity of red pigment from *Serratia marcescens* in human cervix carcinoma. Int J ChemTech Res 2:784–787

Kawauchi K, Shibutani K, Yagisawa H, Kamata H, Nakatsuji S, Anzai H, Yokoyama Y, Ikegami Y, Moriyama Y, Hirata H (1997) A possible immunosuppressant, cycloprodigiosin hydrochloride, obtained from *Pseudoalteromonas denitrificans*. Biochem Biophys Res Commun 237 (3):543–547. https://doi.org/10.1006/bbrc.1997.7186

Kim S (2013) Marine biomaterials: characterization, isolation and applications. CRC Press, New York, pp 1–787

Kim HS, Hayashi M, Shibata Y, Wataya Y, Mitamura T, Horii T, Kawauchi K, Hirata H, Tsuboi S, Moriyama Y (1999) Cycloprodigiosin hydrochloride obtained from Pseudoalteromonas denitrificans is a potent antimalarial agent. Biol Pharm Bull 22(5):532–534. https://doi.org/10.1248/bpb.22.532

Kishimoto Y, Yoshida H, Kondo K (2016) Potential anti-atherosclerotic properties of astaxanthin. Mar Drugs 14(2):35. https://doi.org/10.3390/md14020035

Koyama J (2006) Anti-infective quinone derivatives of recent patents. Recent Pat Antiinfect Drug Discov 1(1):113–125. https://doi.org/10.2174/157489106775244073

Kramar A, Llic-Tomic T, Petkovic M, Radulovic N, Kostic M, Jocic D, Nikodinovic-Runic J (2014) Crude bacterial extracts of two new Streptomyces sp. isolates as bio-colorants for textile dyeing. World. J Microbiol Biotechnol 30:2231–2240. https://doi.org/10.1007/s11274-014-1644-x

Lazaro JEH, Nitcheu J, Predicala RZ, Mangalindan GC, Nesslany F, Marzin D, Concepcion GP, Diquet B (2002) Heptyl prodigiosin, a bacterial metabolite, is antimalarial in vivo and non-mutagenic in vitro. J Nat Toxins 11(4):367–377

Lee B-H, Hsu W-H, Liao T-H, Pan T-M (2011) The Monascus metabolite monascin against TNF-α-induced insulin resistance via suppressing PPAR-γ phosphorylation in C2C12 myotubes. Food Chem Toxicol 49(10):2609–2617. https://doi.org/10.1016/j.fct.2011.07.005

Lee ME, Aswani A, Han AS, Tomlin CJ, Dueber JE (2013) Expression level optimization of a multi enzyme pathway in the absence of a high throughput assay. Nucleic Acid Res 41:10668–10678. https://doi.org/10.1093/nar/gkt809

Leon LL, Miranda CC, De Souza AO, Durán N (2001) Antileishmanial activity of the violacein extracted from Chromobacterium violaceum. J Antimicrob Chemother 48(3):449–450. https://doi.org/10.1093/jac/48.3.449

Liu YT, Sui MJ, Ji DD, Wu IH, Chou CC, Chen CC (1993) Protection from ultraviolet irradiation by melanin of mosquitocidal activity of *Bacillus thuringiensis* var. *israelensis*. J Invertebr Pathol 62(2):131–136. https://doi.org/10.1006/jipa.1993.1088

Liu GY, Essex A, Buchanan JT, Datta V, Hoffman HM, Bastian JF, Fierer J, Nizet V (2005) *Staphylococcus aureus* golden pigment impairs neutrophil killing and promotes virulence through its antioxidant activity. J Exp Med 202(2):209–215. https://doi.org/10.1084/jem.20050846

Liu Q, Huang Y, Zhang R, Cai T, Cai Y (2016) Medical application of *Spirulina platensis* derived C-phycocyanin. Evid Based Complement Alternat Med 2016:7803846. https://doi.org/10.1155/2016/7803846

Li F, Xue F, Yu X (2017) GC-MS, FTIR and Raman analysis of antioxidant components of red pigments from Stemphylium lycopersici. Curr Microbiol 74:532–539

Lopes SCP, Blanco YC, Justo GZ, Nogueira PA, Rodrigues FLS, Goelnitz U, Wunderlich G, Facchini G, Brocchi M, Duran N, Costa FTM (2009) Violacein extracted from *Chromobacterium violaceum* inhibits Plasmodium growth in vitro and in vivo. Antimicrob Agents Chemother 53:2149–2152. https://doi.org/10.1128/AAC.00693-08

Lorquin J, Molouba F, Dreyfus BL (1997) Identification of the carotenoid pigment canthaxanthin from photosynthetic Bradyrhizobium strains. Appl Environ Microbiol 63:1151–1154

Margalith PZ (1992) Pigment microbiology. Chapman & Hall, London, pp 1–156

Mata-Gómez LC, Montañez JC, Méndez-Zavala A, Aguilar CN (2014) Biotechnological production of carotenoids by yeasts: an overview. Microb Cell Fact 13:12. https://doi.org/10.1186/1475-2859-13-12

Matz C, Deines P, Boenigk J, Arndt H, Eberl L, Kjelleberg S, Jürgens K (2004) Impact of violacein-producing bacteria on survival and feeding of bacterivorous nanoflagellates. Appl Environ Microbiol 70(3):1593–1599. https://doi.org/10.1128/aem.70.3.1593-1599.2004

Mojib N, Philpott R, Huang JP, Niederweis M, Bej AK (2010) Antimycobacterial activity in vitro of pigments isolated from Antarctic bacteria. Antonie Van Leeuwenhoek 98(4):531–540. https://doi.org/10.1007/s10482-010-9470-0

Moraes CS, Seabra SH, Castro DP, Brazil RP, De Souza W, Garcia ES, Azambuja P (2008) Leishmania (Leishmania) chagasi interactions with Serratia marcescens: ultrastructural studies, lysis and carbohydrate effects. Exp Parasitol 118(4):561–568. https://doi.org/10.1016/j.exppara.2007.11.015

Mukherjee G, Mishra T, Deshmukh SK (2017) Fungal pigments: an overview. In: Developments in fungal biology and applied mycology. pp 525–541. https://doi.org/10.1007/978-981-10-4768-8_26

Nagia FA, El-Mohamedy RSR (2007) Dyeing of wool with natural anthraquinone dyes from *Fusarium oxysporum*. Dyes Pigments 75:550–555. https://doi.org/10.1016/j.dyepig.2006.07.002

Nakamura Y, Asada C, Sawada T (2003) Production of antibacterial violet pigment by psychrotropic bacterium RT102 strain. Biotechnol Bioprocess Eng 8:37–40

Olempska-Beer Z (2006) Lycopene from *Blakslea Trispora* Chemical and Technical Assessment (CTA) Office of Food Additive Safety, Center for Food Safety and Applied Nutrition, U.S. Food and Drug Administration, College Park, Maryland, USA. Corpus ID: 87294035

Poorniammal, Parthiban M, Gunasekaran S, Murugesan R, Thilagavathy R (2013) Natural dye production from Thermomyces sp. fungi for textile application. Indian J Fiber Textile Res 38:276–279

Priya KA, Satheesh S, Ashokkumar B, Varalakshmi P, Selvakumar G, Sivakumar N (2013) Antifouling activity of prodigiosin from estuarine isolate of *Serratia marcescens* CMST 07. In: Microbiological research in agroecosystem management. Springer India, pp 11–21. https://doi.org/10.1007/978-81-322-1087-0_2

Quereshi S, Khan NA, Pandey AK (2011) Anthraquinone pigment with herbicidal potential from Phoma herbarum FGCC#54. Chem Nat Compd 47:521

Rai M (2009) Advances in fungal biotechnology. I. K. International Pvt Ltd, New Delhi, p 545

Ramirez J, Nunez ML, Valdivia R (2000) Increased astaxanthin productionby a *Phaffia rhodozyma* mutant grown on date juice from *Yucca fillifera*. J Ind Microbiol Biotechnol 24:187–190. https://doi.org/10.1038/sj.jim.2900792

Rao MPN, Xiao M, Li W-J (2017) Fungal and bacterial pigments: secondary metabolites with wide applications. Front Microbiol 8:1113. https://doi.org/10.3389/fmicb.2017.01113

Sakaki T, Shibata M, Mukai K, Sakai M, Wakamatsu K, Miyauchi S (2002) *Chlorociboria aeruginosa* pigment as algicide. Jap Kokai Tokkyo Koho 2002291493

Sankari MK, Jonathan EI, Ardhanareeswaran N (2014) Isolation of phenazine and its activity against root-knot nematode, *Meloidogyne incognita*. Indian J Nematol 43:180–183

Saviola B (2014) Pigments and pathogenesis. J Mycobact Dis 4:5, 2161–1068. https://doi.org/10.4172/2155-9880.1000168

Schneemann I, Kunz AL, Wiese J, Imhoff JF (2011) Genetic approach for the fast discovery of phenazine producing bacteria. Mar Drugs 9(5):772–789. https://doi.org/10.3390/md9050772

Selvameenal L, Radhakrishnan M, Balagurunathan R (2009) Antibiotic pigment from desert soil actinomycetes; biological activity, purification and chemical screening. Indian J Pharm Sci 71:499–504. https://doi.org/10.4103/0250-474X.58174

Sen T, Barrow CJ, Deshmukh SK (2019) Microbial Pigments in the Food Industry-Challenges and the Way Forward. Front Nutr 6:7

Shanmugam A, Sigamani S, Venkatachalam H, Jayaraman JD, Ramamurthy D (2017) Antibacterial activity of extracted phycocyanin from Oscillatoria sp. J Appl Pharm Sci 7(3):62–67. https://doi.org/10.7324/JAPS.2017.70310

Shi Y-C, Liao J-W, Pan T-M (2011) Antihypertriglyceridemia and anti-inflammatory activities of monascus-fermented dioscorea in streptozotocin-induced diabetic rats. Exp Diabetes Res 2011:710635. https://doi.org/10.1155/2011/710635

Shirata A, Tsukamoto T, Yasui H, Hayasaka T, Hayasaka S, Kojima A, Kato H (2000) Isolation of bacteria producing bluish-purple pigment and use for dyeing. Jpn Agric Res Q 34:131–140

Sonani RR, Singh NK, Kumar J, Thakar D, Madamwar D (2014) Concurrent purification and antioxidant activity of phycobiliproteins from Lyngbya sp. A09DM: an antioxidant and anti-aging potential of phycoerythrin in *Caenorhabditis elegans*. Process Biochem 49:1757–1766. https://doi.org/10.1016/j.procbio.2014.06.022

Sonani RR, Rastogi RR, Patel R, Madamwar D (2016) Recent advances in production, purification and applications of phycobiliproteins. World J Biol Chem 7(1):100–109. https://doi.org/10.4331/wjbc.v7.i1.100

Songia S, Mortellaro A, Taverna S, Fornasiero C, Scheiber EA, Erba E, Colotta F, Mantovani A, Isetta AM, Golay J (1997) Characterization of the new immunosuppressive drug undecylprodigiosin in human lymphocytes: retinoblastoma protein, cyclin-dependent kinase-2, and cyclin-dependent kinase-4 as molecular targets. J Immunol 158(8):3987–3995

Speitling M, Smetanina OF, Kuznetsova TA, Laatsch H (2007) Bromoalterochromides A and A', unprecedented chromopeptides from a marine *Pseudoalteromonas maricaloris* strain KMM 636T. J Antibiot (Tokyo) 60(1):36–42. https://doi.org/10.1038/ja.2007.5

Srianta I, Kusumawati N, Nugerahani I, Artanti N, Xu GR (2013) In vitro-glucosidase inhibitory activity of Monascus-fermented durian seed extracts. Int Food Res J 20:533–536

Stafsnes MH, Bruheim P (2013) Pigmented marine heterotrophic bacteria. In: Kim S (ed) Marine biomaterials: characterization, isolation and applications. CRC Press, Taylor & Francis Group, London, pp 117–148

Stahmann K-P, Revuelta JL, Seulberger H (2000) Three biotechnical processes using *Ashbya gossypii*, *Candida famata*, or *Bacillus subtilis* compete with chemical riboflavin production. Appl Microbiol Biotechnol 53(5):509–516. https://doi.org/10.1007/s002530051649

Stankovic N, Senerovic L, Ilic-Tomic T, Vasiljevic B, Nikodinovic-Runic J (2014) Properties and applications of undecylprodigiosin and other bacterial prodigiosins. Appl Microbiol Biotechnol 98(9):3841–3858. https://doi.org/10.1007/s00253-014-5590-1

Stevenson CS, Capper EA, Roshak AK, Marquez B, Grace K, Gerwick WH, Jacobs RS, Marshall LA (2002) Scytonemin—a marine natural product inhibitor of kinases key in hyperproliferative inflammatory diseases. Inflamm Res 51:112–114

Sudha, Gupta C, Aggarwal S (2016) Dyeing wet blue goat nappa skin with a microbial colorant obtained from *Penicillium minioluteum*. J Clean Prod 127:585–590. https://doi.org/10.1016/j.jclepro.2016.03.043

Sudhakar MP, Arunkumar K, Perumal K (2015) A preliminary study on antioxidant and antimicrobial activities of phycoerythrin from different Gracilaria species. National Institute of Ocean Technology. Seaweed Res Util 38(1):49–54

Sun Z, Zhou Z-G, Li T, Jiang Y (2015) Microalgae as a source of lutein: chemistry, biosynthesis, and carotenogenesis. Adv Biochem Eng Biotechnol 153:37–58. https://doi.org/10.1007/10_2015_331

Terao J (1989) Antioxidant activity of b-carotene-related carotenoids in solution. Lipids 24:659–661. https://doi.org/10.1007/BF02535085

Tominaga K, Hongo N, Karato M, Yamashita E (2012) Cosmetic benefits of astaxanthin on humans subjects. Acta Biochim Pol 59(1):43–47

Tuli HS, Chaudhary P, Beniwal V, Sharma AK (2015) Microbial pigments as natural color sources: current trends and future perspectives. J Food Sci Technol 52:4669–4678. https://doi.org/10.1007/s13197-014-1601-6

Umadevi K, Krishnaveni M (2013) Antibacterial activity of pigment produced from *Micrococcus luteus* KF532949. Int J Chem Anal Sci 4:149–152. https://doi.org/10.1016/j.ijcas.2013.08.008

Vendruscolo F, Müller BL, Moritz DE, Oliveira D, Schmidell W, Ninow JL (2013) Thermal stability of natural pigments produced by Monascus ruber in submerged fermentation. Biocatal Agric Biotechnol 2(3):278–284

Venil CK, Zakaria ZA, Ahmad WA (2013) Bacterial pigments and their applications. Process Biochem 48:1065–1079. https://doi.org/10.1016/j.procbio.2013.06.006

Venil CK, Aruldass C, Dufossé L, Zakaria ZA (2014) Current perspective on bacterial pigments: emerging sustainable compounds with coloring and biological properties for the industry-an incisive evaluation. RSC Adv 4(74):39523–39529. https://doi.org/10.1039/c4ra06162d

Visalakchi S, Muthumary J (2010) Antimicrobial activity of the new endophytic Monodictys castaneae SVJM139 pigment and its optimization. Afr J Microbiol Res 3(9):550–556

Wang Y, Nakajima A, Hosokawa K, Soliev AB, Osaka I, Arakawa R, Enomoto K (2012) Cytotoxic prodigiosin family pigments from Pseudoalteromonas sp. 1020R isolated from the Pacific coast of Japan. Biosci Biotechnol Biochem 76:1229–1232. https://doi.org/10.1271/bbb.110984

Williams RP, Quadri SM (1980) The pigments of Serratia. In: Graevenitz AV, Rubin SJ (eds) The genus Serratia. CRC Press Inc., Boca Raton, pp 31–75

Williamson NR, Fineran P, Gristwood T, Chawrai SR (2008) Anticancer and immunosuppressive properties of bacterial prodiginines. Future Microbiol 2(6):605–618. https://doi.org/10.2217/17460913.2.6.605

Wu C-L, Lee C-L, Pan T-M (2009) Red mold dioscorea has a greater antihypertensive effect than traditional red mold rice in spontaneously hypertensive rats. J Agric Food Chem 57 (11):5035–5041. https://doi.org/10.1021/jf900349v

Zheng L, Cai Y, Zhou L, Huang P, Ren X, Zuo A, Meng X, Xu M, Liao X (2017) Benzoquinone from *Fusarium* pigment inhibits the proliferation of estrogen receptor-positive MCF-7 cells through the NF-κB pathway via estrogen receptor signaling. Int J Mol Med 39(1):39–46. https://doi.org/10.3892/ijmm.2016.2811

Zuluaga M, Gregnanin G, Cencetti C, Di Meo C, Gueguen V, Letourneur D, Meddahi-Pellé A, Pavon-Djavid G, Matricardi P (2017) PVA/Dextran hydrogel patches as delivery system of antioxidant astaxanthin: a cardiovascular approach. Biomed Mater 13:015020. https://doi.org/10.1088/1748-605X/aa8a86

Websites

https://pubchem.ncbi.nlm.nih.gov/compound/1_2-Benzoquinone
https://pubchem.ncbi.nlm.nih.gov/compound/1_4-Benzoquinone
https://pubchem.ncbi.nlm.nih.gov/compound/1_4-Naphthoquinone
https://pubchem.ncbi.nlm.nih.gov/compound/Astaxanthin
https://pubchem.ncbi.nlm.nih.gov/compound/beta-Carotene
https://pubchem.ncbi.nlm.nih.gov/compound/Chlorophyll
https://pubchem.ncbi.nlm.nih.gov/compound/Fucoxanthin
https://pubchem.ncbi.nlm.nih.gov/compound/Lutein
https://pubchem.ncbi.nlm.nih.gov/compound/Lycopene

https://pubchem.ncbi.nlm.nih.gov/compound/Prodigiosin
https://pubchem.ncbi.nlm.nih.gov/compound/Pyocyanine
https://pubchem.ncbi.nlm.nih.gov/compound/Riboflavin
https://pubchem.ncbi.nlm.nih.gov/compound/Staphyloxanthin
https://pubchem.ncbi.nlm.nih.gov/compound/Violacein
https://pubchem.ncbi.nlm.nih.gov/substance/315492484
https://pubchem.ncbi.nlm.nih.gov/substance/8011185
https://wordpress.clarku.edu/debrobertson/laboratory-protocols/algal-pigments/

Microbial Polysaccharides with Potential Industrial Applications: Diversity, Synthesis, and Their Applications

21

Himadri Tanaya Behera, Abhik Mojumdar, Chiranjib Mohapatra, and Lopamudra Ray

Abstract

Polysaccharides are nontoxic, biodegradable, and natural carbohydrate polymers virtually produced by almost all microbes, covering the surface of the cell and serve as the intracellular energy stores, and play a vital aspect in many biological pathways such as response to the infection, signal transduction, adhesion, and acts as a buffer against changing environments. They are mostly present as extracellular capsules, slime creations, or sheaths. Bioactive polysaccharides were proven to show many immune modulatory and health-promoting effects in diseases like cancer and some inflammatory diseases. Thus, substantial research has been carried out on revealing the mechanism behind the biological activity by structural and functional analysis. Apart from their applications in the health sectors, these polysaccharides have huge pertinence in the food, cosmetic, pharmaceutical industries as thickeners, probiotics, bioadhesives, and gelling agents. In environmental sectors, they are used as biosorbent and bioflocculant. In this chapter, several production processes of EPSs using a cost-effective biomass, commercial applications of the EPSs, and general mechanisms of EPS production are briefly described.

Keywords

Exopolysaccharides · Alginate · Dextrans · Gallans · Biosynthesis pathway

H. T. Behera · A. Mojumdar · C. Mohapatra
School of Biotechnology, KIIT Deemed to be University, Bhubaneswar, Odisha, India

L. Ray (✉)
School of Biotechnology, KIIT Deemed to be University, Bhubaneswar, Odisha, India

School of Law, KIIT Deemed to be University, Bhubaneswar, Odisha, India
e-mail: lopamudra.ray@kls.ac.in; lray@kiitbiotech.ac.in

A. Vaishnav, D. K. Choudhary (eds.), *Microbial Polymers*,
https://doi.org/10.1007/978-981-16-0045-6_21

21.1 Microbial Polysaccharides

Polysaccharides can be extracted from algae, higher-order plants, and also from bacterial and fungal sources; these bioactive polysaccharides can be recovered from fermentation culture broths. However, for the credible polysaccharide production at industrial level, microbial sources are more preferred over plants and algae due to their rapid production abilities under restrained fermentation conditions. Bacterial/fungal production of the polysaccharides can be achieved only within few days or weeks, whereas a minimum of 3–6 months is required for recovery of polysaccharides from the plant source and another main difficulty of using the plant source is its geographic and seasonal variations (Donot et al. 2012). A report estimated that approximately 3.9 billion US dollars will be controlled by algal and plant-based polysaccharides like alginate, galactomannans, carrageenan, starch and pectin, etc., by 2012 (Phillips 2012). Thus, overriding these plant and algal sourced polysaccharides by the microbial equivalent needs an innovative and considerable approach for the development of microbial extracellular polysaccharides (Exopolysaccharides, EPSs) with industrial significances.

Microbial polysaccharides are carbohydrate polymers with high molecular weight and are present as lipopolysaccharides (LPS) at the outer membrane mainly determining the immunogenic properties. Sometimes it may be enveloped as capsular polysaccharides (CPS) creating a distinct outer layer (Capsule) loosely connected to cell exterior (Cuthbertson et al. 2009). CPSs have a direct role in adherence and pathogen resistance to some specific as well as nonspecific host immunity (Taylor and Roberts 2005). EPSs have a multifunctional role in a microorganism, which includes adhesion (Taylor and Roberts 2005), formation of biofilms (Mann and Wozniak 2012), cell-to-cell interaction, stability of the cell from extreme environments (Suresh Kumar et al. 2007).

Polysaccharides can be categorized based on the presence of several repeating monomeric residues, i.e., homopolysaccharides and heteropolysaccharides (Sutherland 1982). Only one type of monomeric units bonded together to form linear chains (curdlan, pullulan, cellulose) or ramified chains (dextrose) are classified as homopolysaccharides. Two or more types of different monosaccharides composed of three to eight residues (xanthan and gellan) are classified as heteropolysaccharides (Purama et al. 2009; Sutherland 1972). Bacteria are the main source of EPS producer and some most commonly used polysaccharides like xanthan, dextran, gellan, curdlan are produced by species of *Xanthomonas*, *Leuconostoc*, *Sphingomonas*, *Alcaligens*. Dextran was the first commercial microbial polysaccharides approved for use in the food industry (Sarwat et al. 2008). Diversity in the structure of the bacterial polysaccharides can be confirmed by the presence of different sugars present on the polysaccharides.

Thus, more studies should be carried out for the optimization of the EPS production process with the formulation of new and improved functional food ingredients with nutritional as well as economic benefits.

21.2 General Polysaccharide Structure and Physical Properties

According to the nature of monomeric repeats, the EPS can be divided as homo and heteropolysaccharides. Either these polymers are attached to the outer surface of the membrane or are remain exposed to the extracellular space. The behaviors of the EPSs are affected by the nature and composition of the branching residues and points. While a minimum of 100 unique monomers comprise the microbial polysaccharides and many branching sites are still to be deciphered. Most industrial heteropolysaccharides have a regular structure of two to eight repeating units consisting of two to four different monomers (Sutherland 2001). Simple sugars (Pyranose) like D-glucose, L-fructose, D-galactose, L-rhamnose, and D-mannose are the very common types of monomers. The presence of pentose sugars is very common in the eukaryotic species than the prokaryotic species (Sutherland 2001). Varying degrees of acyl substitution is observed in microbial EPS and random chains of acetate esters and pyruvate ketals are mostly common. But Alginate is one of the exceptions to this standard generalizations, consisting of repeating units of heavily acetylated D-mannuronic and L-guluronic units in the main chain (Sutherland 2001). Different combinations of sugar monomers with their linkages, branching chains, and some organic modifications are responsible factors for the wide variety of possible conformations.

Thick hydrocolloid suspensions and gels are formed by many microbial polysaccharides making them interesting for industrial uses. Again this property can be modified by acetylation or any other chemical modification of cations and solvated cations (Sutherland 1994). Charge and flexibility of the chains are also influenced by critical factors like pH and temperature. Several condensation linkages in polysaccharides with the difference in the sequence of the monomer provide EPS with a broad range of various feasible shapes and architecture. A complex entanglement is observed by the association of high molecular weight polymer chains contributing to its observed physical properties (Atkins 1986).

21.3 Common Metabolic Precursors

Varied pools of precursor molecules are required for the exopolysaccharide synthesis and most usual precursors are energy-rich nucleoside diphosphate-modified monosaccharide. These monosaccharide diphosphates are the main primary unit serving as the main energy source for its synthesis (Steenbergen and Vimr 2003). Several mechanisms like epimerization, decarboxylation, reduction, oxidation, and rearrangement drive various interconversions of the sugar.

Aside from the apparent requirement of the activated monosaccharide component, EPSs often are decorated with modified sugars or any non-sugar component. Acetyl and Acyl groups on the EPS are added by Acetyl CoA and acyl CoA derivatives respectively (Leitão and Sá-Correia 1993; Bossio et al. 1996).

Amino acids, esters, acyl like organic acids may be present on the EPS as a non-sugar component. A lipid acceptor is assumed to be required for assembly during the EPSs synthesis at the cell membrane. By standard enzymatic phosphorylation reaction, ATP derived phosphate groups may be added to the EPS. Sulfate groups can be added to EPS by amino acid, but the mechanism is not well accepted and characterized.

21.4 Common Analytical Techniques

For the structural investigations of the polysaccharides, the polymer needs to be digested via acid hydrolysis or by recombinant enzymes (Mischnick 1998). The mixtures obtained after hydrolysis may be analyzed by methods like high performance liquid chromatography (HPLC), thin layer chromatography (TLC), and gas-liquid chromatography (GLC). However, much information regarding the primary structures may be lost during the hydrolysis process. Therefore, this should be carried out meticulously not to modify the monomeric units during hydrolysis. Gel filtration chromatography (GFC), viscometric, and light scattering analysis can be used for the determination of the size of the side groups. The location of side chains and linkages can be determined by hydrolysis after methanolysis step (Chaplin 1982). Information regarding the glycosidic bonds can be figured out by periodate oxidation. Stepwise digestions may be opted by using residue-specific hydrolases for the study of size and identity of resulting fragments. Compositional and structural analysis in both native and modified polymers can be studied by nuclear magnetic resonance (NMR), ^{13}C, and cross-polarization. By optical rotation method the glycosidic linkage conformation can be determined (Duus et al. 2000). Regarding glycoprotein analysis, matrix-associated laser desorption or ionization mass spectroscopy (MALDI) can be opted for the quickest analysis of the carbohydrates and linkages (Harvey 1999).

21.5 Commercial Microbial Polysaccharide and Its Commercial Applications

21.5.1 Alginate

Alginate is the salt form of alginic acid and consists of polymeric units of D-mannopyranosyluronic acid and L-glucopyranosyluronic residues linked via 1,4-linkage without any branching. Origin of the term alginate comes from the algin, as for many years these were used to emanate from marine algae only. However, in 1964 a strain *Pseudomonas* was reported to produce an extracellular capsular polysaccharide which was almost identical to the algae-derived alginate (Linker and Jones 1964). Later it was found that alginate production is common among many bacteria like nitrogen-fixing *Azotobacter vinelandii*, *A. chroococcum* (Cote and Krull 1988), *A. beijerinckii* (Taylor 1998), opportunistic human pathogen *Pseudomonas aeruginosa* (Evans and Linker 1973), nonpathogenic *P. mendocina* (Hacking et al. 1983), many plant pathogens of *Pseudomonas* species (Fett et al.

Table 21.1 Applications of commercially produced alginate (Rehm and Valla 1997)

Applications	Functions of alginate
Water treatment	In the flocculation process, increases the size of the aggregate
Pharmaceutical and biotechnology industry	Cell immobilization, oesophageal reflux treatment, dental-impression material
Production of ceramics	Reduced rates of drying surface
Textile printing	Fixation, an increase in brightness, and color yield
Can sealing	Cord formation into slits of can lids
Welding rod production	Covering materials for welding of rods, improving bendability
Paper and board treatments	Improvement of surface uniformity
Creaming of latex	Functions as a concentrating agent of natural latex during extraction from rubber plants

1986). Other than alginate, these bacteria produce various types of EPS, depending on the strain types, carbon, and nitrogen source. Alginates can be produced on a large scale by the bacterial fermentation process, but only the significant source of the alginate is still marine algae.

The structural variations of the bacterial alginates are determined by the arrangement of L-guluronate and D-mannuronate residues. The soluble and non-gelling properties of the sequence of poly-β-D-mannuronatecan be modified to form a rigid gel structure by the introduction of α-L-guluronate sequence into the sequence chain. Alginate with good gelling properties can be achieved by growing the bacteria under favorable conditions for biosynthesis with a suitable ratio of guluronate and mannuronate sequences. In bacteria, alginates are synthesized via synthase-dependent pathway. In bacteria, poly-β-D-mannuronate is first assembled and epimerization of few D-mannuronate units to L-guluronate (Larsen and Haug 1971; Pindar and Bucke 1975) occurs by extracellular polymannuronic acid C-5 epimerase from *A. vinelandii* (Ertesvåg et al. 1994). *O*-acetyl units joined to bacterial sugar units affect the structural and property relationship, as the C-5 epimerase cannot act on the mannuronate residues containing *O*-acetyl groups (Davidson et al. 1977).

Recently, alginate is being used in various industries for the production of brown seaweeds. Based on the gelling, viscosifying, and suspension-stabilizing qualities alginate is used in many food industries (McCouch et al. 2002; Gacesa 1988). Besides their use in the food industry, these have been used as a wound dressing material, drug delivery and tissue engineering, and cell encapsulation. Besides this, some of the other applications of the alginates are depicted in Table 21.1.

21.5.2 Dextrans

Dextran, defined as the extracellular homopolysaccharide with α-(1-6)-linked D-glucopyranosyl residues with various branching points by α-(1-2), α-(1-3), or α-(1-4) linkages (Mark et al. 1987). For many years dextrose has been identified as a polymeric unit of D-glucose, and hence the origin of the name. Due to the broad

Table 21.2 Dextrans produced by different LAB isolated from different food sources (Kothari et al. 2014)

Microorganisms	Food source	Linkages
Lactobacillus satsumensis NRRL B-59839	Kefir grains	44% of α-(1-3), 37% of α-(1-6)
Peddiococcus pentosaceus CRAG3	Ripened cucumber	75% α-(1-6), 25% α-(1-3)
Leuconostoc mesenteroides AA1	Ripened cabbage	α-(1-6) linkages only
Weissella cibaria CMGDEX3	Cabbage	96.6% α-(1-6), 3.4% α-(1-3)
Leuconostoc mesenteroides CMG713	Grapes	α-(1-6) linkage
Weissella confuse Cab3	Ripened cabbage	97% of α-(1-6), 3% of α-(1-3)
Weissella cibaria JAG8	Apple peels	93% of α-(1-6), 7% of α-(1-3)

range of utilizations in the food, pharmaceutical, chemical, and biomaterial industries, this polysaccharide has gained importance. Dextrans are abundantly produce by lactic acid bacteria (LAB) (Cerning 1990). Four specific categories of α-D-glucan produced by LAB are: (1) mutant produced by *Streptococcus* species, consisting majority of α (1-3) linkages (2) reuteran produced only by *Lactobacillus reuteri*, consisting of α-(1-4) linkages (3) alternant found only in *L. mesenteroides*, containing alternating α-(1-3) and α-(1-6) linkages (Monchois et al. 1999) (4) dextran produced extensively by LAB genera, e.g., *Weissella, Leuconostoc, Lactobacillus, Pediococcus,* and *Streptococcus* consisting of α-(1-6) linkages in the main backbone and α-(1-2), α-(1-3), α-(1-4) branched linkages (Kothari and Goyal 2013). Each type of dextran produced by different microbial strains is structurally different. In foods, like maple syrup, honey, crystalline sugar, sauerkraut juice dextran exists naturally in very small amounts. A bacterial enzyme dextransucrase acts on the substrate sucrose to synthesize dextran. To date, NRRL B-512F is the commercial dextran produced by *Leuconostoc mesenteroides* (Table 21.2).

Dextran is synthesized from sucrose by dextransucrase, releasing a fructose molecule via a double displacement reaction mechanism. In the early stage of the reaction, α-(1-2) glycosidic bond of the sucrose splits with a formation of an intermediate glucosyl enzyme and release of a fructose molecule. Glucosyl unit of the intermediate is attached via β-glycosidic linkages to the catalytic nucleophile. In the second stage of the reaction, the glucosyl unit is altered to the nonreducing sugar molecule of a growing chain, reforming the glycosidic bond.

Since the 1950s, dextran has been used as food ingredients due to its gelling, viscosifying, texturing, and emulsifying properties. The application of dextrans may depend on its chemical and physicochemical properties such as chain length; high molecular weight polysaccharides dissolved well in water; and improves the properties for food product formulation. Dextran possesses a high demand as an attractive food ingredient, making a present consumer flow towards choosing additive-free food options (Table 21.3).

Apart from the application of commercial dextrans in the food processing industry, dextran also has various usages in the medicinal industry, photo film, petroleum, textile, fine chemicals, and cosmetic industries.

Table 21.3 Applications of Dextran in food industries

Applications	Reference
Bakery	Katina et al. (2009)
Improves crumb texture, the volume of the loaf, the softness of the bread, mouth-feel, and freshness	
Fermented dairy products	Mende et al. (2013)
Improves the viscosity and creaminess, reduces syneresis	
Ice cream	Naessens et al. (2005)
Cryoprotectant	
Confectionary	Maina et al. (2011)
Increase the retention of moisture content and viscosity, prevents the crystallization of sugar and used as gelling agents in jelly candies and gums	
Cheesemaking	Agrawal and Hassan (2008)
Enhances the water-binding properties and moisture content of the nonfat substances	
Prebiotics	Das et al. (2014)
Functional food	
Frozen and dried food	Bhavani and Nisha (2010)
As a preservative, protection from the damage caused due to oxidation and chemical damage	

Antithrombotic effect exerted by dextran is owing to its binding abilities with erythrocyte membranes, platelets, and vascular endothelium which increases the electronegativity and reduces the erythrocyte aggregation. In some eye drops also, dextran is used as a lubricant. In the photographic industry, the qualities of the silver emulsions are improved by the addition of dextran biopolymers. In cosmetic industries, dextran is used as a moisturizer and thickener by forming complex salts with anionic surfactants forming a film on hair and skin having moisturizing effects. Various reports are there suggesting the anti-inflammatory effects of the dextran (Islam et al. 2008; Farombi et al. 2013). In the laboratory, some size exclusion chromatography matrices like Sephadex contain dextran. In a fluorescent microscope, dextran is used to visualize the early endosome. Dextrans used as a stabilizing coating protects the metal nanoparticles from oxidation and improving its biocompatibility (Bhavani and Nisha 2010).

Dextran with its derivatives possess a wide range of usages in technical, biological, pharmaceutical, and food industries (Bhavani and Nisha 2010). Hence, the industrial production of dextran using low-cost substrates like agricultural and food wastes should be highly encouraged.

21.5.3 Gellan

Since the early 1980s, gellan was introduced as an alternative to the agar. Gellan is an anionic heteropolysaccharide composed of repeating units of

Table 21.4 Potential food applications of gellan (Omoto et al. 1999; Guohua 2002)

Major food areas	Products
Jams and jellies	Bakery filling reduced-calorie jams
Water-based gels	Dessert gels
Pet foods	Gelled pet foods and canned pet foods
Dairy products	Milkshake, ice cream, yogurt
Fabricated foods	Meats, vegetables
Confectionery	Pectin jellies, fillings, marshmallow, starch jellies
Puddings	Pie-fillings, instant desserts, canned puddings

D-glucopyranosyluronic acid (GlcpA), D-glucopyranosyl (Glcp), and L-rhamnopyranosyl (Rhap) residues. Bacterium like Sphingomonas *elodea, Pseudomonas elodea*, and Auromonas *elodea* synthesizes gellan. A soft and elastic gel is formed by the native gel, whereas a more rigid with brittle gel is formed by the sulfated gellan (Sandford and Baird 1983). The gel-forming ability of the gellan requires cations (divalent or monovalent) and the presence of any chelating agents like EDTA which prevents the formation of the gel. The rate of acyl group substitution, the cation types and its concentration. The concentration of the polysaccharide regulates the gelling property of the polysaccharide. However, gellan is used as a substitute to agar, but a lower concentration of gellan (0.5–0.8%) with magnesium is required for making microbiological plate media than agar used (1.5%). When agar is used for the gel preparation, heating is necessary to dissolve the dried gel, whereas no gel formation occurs when gellan is dissolved in water (Sandford 1979).

Compared to any other microbial polysaccharides, in a relatively short period, gellan has achieved a huge commercial success and is marketed as a substitute to agar due to its high purity, perspicuity, and its ability to form gels at a lower concentration. In the year 1992, the FDA approved gellan to be used as a food additive in frostings, jams, jellies, fillings, as stabilizing and thickening agents (Anon 1992). Some potential food application of the gellan is shown in Table 21.4.

In pharmaceutical industries for oral (Kubo et al. 2003), buccal (Remuñán-López et al. 1998), rectal (El-Kamel and El-Khatib 2006), nasal (Cao et al. 2009), and ophthalmic (Kesavan et al. 2010) administrations various forms of formulations like solid, semi-solid, and liquid use the gellan-based material. In the tissue regeneration process, bone repair, delivery of genes, or in the synthesis of biosensor gellan containing materials are being studied. Some reports regarding the pharmaceutical application of gellan containing formulations are depicted in Table 21.5.

21.5.4 Welan

Alcaligenes species produces welan by aerobic fermentation process (Kuo et al. 1986), which is structurally almost identical to the backbone sequence of the gellan. The backbone of the welan contains an extra side group, α-L-rhannopyranosyl or α-L-mannopyranosyl joined via (1,3-) linkages to β-D-glucopyranosyl residues. The

Table 21.5 Pharmaceutical formulations based on Gellan gum (Osmałek et al. 2014)

Type of formulations	Application
Oral formulations	
Gellan granules	Sympathomimetic
Gellan beads	Antibiotic
In situ gelling system	Antipyretic, analgesic
Gellan capsules and beads	Inhibitor of phosphodiesterase
Ophthalmic formulations	
Gelling ophthalmic solution	Anti-glaucoma
Soluble bioadhesive ocular insert	Antibiotic
Albumin nanoparticles with gellan	Anti-glaucoma
Gelling ophthalmic nano-emulsion	Fungal keratitis
Nasal formulations	
Nasal gel	Epithelial uptake testing, nausea, motion sickness
Intranasal microparticle	Antiemetic
Nasal formulation containing thiolated gellan	Prevention of motion sickness disorder
Gellan microspheres	Erectile dysfunction

distributions of the side chains are random and nearly 66% of the side chains contain rhamnose units. Sometimes, *O*-acyl groups may present on the repeat units. Welan is not soluble in isopropanol solutions and this property is used for retrieving the polymer from the culture. They keep up the viscosity even at a higher temperature (135 °C) with a half-life of 900 min, whereas at this temperature 0.4% xanthan solution has a viscosity half-life nearly equal to zero (Sandford 1979). However, the viscosity of the welan solution is reduced upon the addition of salt.

The elevated heat tolerance capacity and great salt compatibility of the welan gum makes it a possible candidate for the natural gas and oil recovery. In oil-industry drilling actions and hydraulic fracturing tasks, welan gum is used extensively instead of xanthan gum. In the formulation of the insulating material, welan-ethylene glycol compositions are used.

21.5.5 Pullulan

In 1959, Pullulan, the water-soluble extracellular α-D-glucan was reported first and produced by *Aureobasidium pullunana* (Bender et al. 1959). Some other microbial sources of Pullulan are *Tremella mesenterica*, *Cytaria harioti*, *Cytaria darwinni*, *Cryphonectira parasitica*, *Teloschistes flavicans*, and *Rhodototula bacarum*. This polymer is composed of α (1-6) linked maltotriose residues joined from the reducing end of one residue to the nonreducing point of the adjacent polysaccharide units. Nearly, 5–7% of the chains contain maltotetraose (Catley and Whelan 1971; Sandford 1979).

Table 21.6 Various industrial applications of pullulan (Jones et al. 2017; Singh et al. 2008)

	Applications
Food industry	Replacement of starch in solid and liquid food
	Increase the shelf-life of food
	Functions as prebiotics, enhance the growth of beneficial bacteria
	Used in beverages and sauces as a low-viscosity filler
	Inhibits the fungal growth in food
	Used as a stabilizer, binder, adhere nuts to cookies
	Employed in plant fertilizers, seed coatings, and binder for tobacco
	Substitute cholesterol or fatty acid and stabilize the fatty emulsion
	Incorporated into the foods designed for diabetics and impaired glucose tolerance patients
	Used as a packaging material for nuts, confectionaries, vegetables, meats, dried foods
Pharmaceutical industry	Used as a denture adhesive
	Used in sugar layer of pharmaceutical tablets, pills to prevent the brownish color change
	Helps in sustained release of the drug formulations
	Tissue engineering applications of heparin conjugated carboxymethyl Pullulan
	Explored as a potential alternative to plasma
	Used as a component of cosmetic and makeup lotions, powders, shampoos, tooth powders, and face packs
Miscellaneous industry	Can be formed into good resembling nylon or rayon fibers
	Used as an adhesive agent in paper coatings
	Improves the activity and retention of enzyme immobilized with Pullulan gel

The flexibility in the pullulan structure and its increased solubility are due to the regular α-(1-6) linkages. In its powdered form, it is white, non-hygroscopic, easily dissolved in hot/cold water, nontoxic, biodegradable, and also biocompatible. Except for dimethylformamide and dimethylsulfoxide, it is not soluble in the organic solvents. In aqueous solution, pullulan is not able to form a gel, but the viscosity of the solution increases which is again proportional to the molecular weight of the pullulan. In the dried form the polymer possesses excellent adhesive property, making it suitable for agglomerated products. Various reports regarding the food and pharmaceutical applications of the water-soluble biopolymer pullulan have been reviewed (Table 21.6).

Despite various valuable applications of Pullulan, the major drawback of the application of the polymer is its production price, which is approximately three times more than the dextran and xanthan. Much research is focused on its improvement in production such as improved fermentation condition (Schuster et al. 1993), Pullulan synthesis using immobilized *A. pullulans* cells (Mulchandani et al. 1989), bioreactor system (Gibbs and Seviour 1992), recovering pullulans from fermentation broths (Yamasaki et al. 1993) for a high yield of pullulan improvement in production strains, mainly the reduction of the melanin production by genetic engineering strategies could open a new avenue for Pullulan utilization.

21.5.6 Scleroglucan

Scleroglucan, the neutral, branched homopolysaccharide produced by fungi of genus sclerotium. This polymer is composed of (1-3) joined β-D-glucopyranosyl units, with the presence of one D-glycopyranyl residue joined by β (1-6) linkages at each third sugar (Singh and Gross 2001). In its triple-helical conformation, it shows the highest stability. Aggregation of the helices is prevented when the D-glucopyranosyl residue project to the outward of the helix, resulting in the insolubility of the polymer. At a temperature above 90 °C or pH 12 (Bo et al. 1987) the helix form changes to coil form and at a temperature below 8 °C, a weak gel is formed (Mahajan and Kimerling 2013). Viscosity remains constant at temperature 10–90° C. During shear scleroglucan shows pseudoplastic flow behavior.

For enhancing the oil recovery scleroglucan act as a mobility control agent (Moorhouse et al. 1977; Pirri et al. 1996). An immune response can be stimulated and repression of some forms of cancer occurs by the administration of scleroglucan. It also possesses antitumor, antiviral, and antimicrobial activity (Singh et al. 1974; Jong and Donovick 1989). Due to its specific rheological properties, it can be used as adhesive, painting inks, watercolors, and liquid animal feed composition (Halleck 1969). Several Japanese patents are there describing the use of this polysaccharide in quality improvement of frozen foods (Coviello et al. 2005), steamed foods, rice crackers, bakery products, and Japanese cakes (Coviello et al. 2005). In the pharmaceutical industry, scleroglucan may be used as a coating material for tablets (Sheth and Lachman 1967), laxative (Duc 1982). The chemical/physical cross-linking of scleroglucan helps in sustained release and controlled delivery systems, which represent an attractive challenge for the future with a wide potentiality of scleroglucan in several industries.

21.5.7 Curdlan

Curdlan is a linear homopolysaccharide containing D-glucan units joined by β-(1-3)-linkages, synthesized principally by *Alcaligenes faecalis* var. myxogenes, and also by *Agrobacterium radiobacter*, *Rhizobium trifolii*, *Rhizobium meliloti*, and *Rhizobium niftal* (Chien et al. 2019). On heating above 55 °C, curdlan is able to form a gel in aqueous neutral or dilute alkaline solutions, but upon cooling the gel form dissociates and is referred to as "low-set" or thermoreversible gel. On the other hand, an aqueous solution of curdlan on heating to 80–145 °C form a "high-set" thermoreversible gel. Under acidic conditions, the polysaccharides undergo the acid-catalyzed cleavage.

Owing to the remarkable rheological properties, and its physical stabilities, curdlan can be applied in the food industry either by adding a very low level directly to the existing food system or by creating a newer type of food products by using a higher level of curdlan. The curdlan molecule is capable of absorbing water up to a 100 times of its weight helping in the development of a moisture food product; this is due to the helical conformation of the curdlan molecule. Mouth-filling property of fat-containing products can be mimicked by using hydrated and heated forms of this polysaccharide. Elasticity and strength of the packaged noodles can be increased by the use of curdlan. Apart from its application in the food industry, due to its physical and chemical properties, curdlan shows a potential application in the biomedical and pharmaceutical industries. Sulfated derivatives of the curdlan show anti-AIDS or antiviral activity (Jagodzinski et al. 1994). Some of the derivatives of curdlan show a great activity against blastomere infection (Yu et al. 2011) and this ideal polymer also can act as a therapeutic factor for the treatment of malaria (McIntosh et al. 2005). A class of immune cells such as macrophages (Kataoka et al. 2002), monocytes, and natural killer cells (Zhan et al. 2012) is also generated by this polymer. Sulfated curdlan has high fucose content, making it useful for the treatment of viral diseases, cancer, and other inflammatory diseases. Curdlan drug-impregnated gels stay intact in the rectum and allow a local diffusion of drugs, compared with other rectal suppository systems (Kanke et al. 1995). Encapsulation of cancer chemotherapy drug doxorubicin with curdlan increases the stability of the drug under frozen conditions (Subedi et al. 2009). Heavy metals like Pb(II), Cd(II), Cu(II), and Mn(II) can be adsorbed by curdlan in combination with the activated carbon (Moon and Lee 2005). Therefore, a focused attempt should be pointed towards the production of curdlan at cost-effective rates to meet industrial demands.

21.5.8 Xanthan Gum

Xanthan is an anionic group of heteropolysaccharide, synthesized by bacteria in the genus *Xanthomonas*; *X. campestris.* This commercially important xanthan gum composed of repeating units of D-glucopyranosyl units having side chains consisting of D-mannopyranosyl and D-glucopyranosyluronic acid residues. At the internal O-6 position, D-mannosy unit is acetylated. The viscosity of the xanthan gum shows very well pseudoplastic behavior. Highly viscous solution is obtained at relatively lower concentrations of the biopolymer and this is maintained over a wide range of pH and temperature. The stability of the viscosity at elevated temperature is maintained by the addition of mono- and divalent cations (Holzwarth 1976). Similarly, at pH 1.5–3, the stability of the viscosity solution is maintained by the addition of salt (Sandford 1979). This naturally occurring polymer is fully biodegradable. Though the structural backbone of the xanthan gum is almost similar to the cellulose, it shows resistance towards cellulase, which may be due to the presence of trisaccharide side chains. However, fungal cellulase enzymes cleave the xanthan gum only in its disordered form but xanthan gum in its helix conformation shows resistance towards fungal cellulase. Various applications of the xanthan gum are reviewed, particularly in the food, cosmetics, and pharmaceutical applications (Table 21.7).

Table 21.7 Various industrial applications of the xanthan gum (Katzbauer 1998; Sanderson 1981)

	Applications
Food industry	Used as stabilizer and thickening agents
	Contributes pourability, pumpability to the dressings and sauces
	Prevents thickness to the soups and gravies
	Helps in adjusting the flow behavior of the prepared and semi-prepared foods
	Increase shelf-life and retains the moisture content of the bakery product
	Control batter rheology in the prepared cake mixes
	Improves the product stability of the frozen foods from repeated freeze-thaw cycles
	Application as a stabilizer in toppings, desserts with some other hydrocolloids
	Gives the mouth-feel a full-bodied taste and releases good flavor in beverages
	Supply moistness in low-fat muffins and low-fat frankfurters
	Used as a stabilizer in liquid animal feed
Cosmetics and pharmaceutical industries	Used in the toothpaste to allow easy extrusion from the tube and keeps a stable stand of toothpaste on the brush
	Used as stabilizer and thickener in personal care products like face creams, eye contour gel, etc.
	Prevents separation of insoluble ingredients, e.g., barium sulfate in X-ray contrast media
Other applications	Used in carpet printing pastes, formulations in ink, and clay coating form in paper and printing industries
	Blended in herbicides and insecticides to improve agrochemical applications to the plants

21.6 EPS Production Using Low-Cost Biomass Resource

Fermentation technology is a very widely used method for EPS production. Depending on the microbial strain used the fermentation process may vary (batch, fed-batch, and continuous processes). During EPS production, significant variations in the rheological properties may happen, resulting in a highly viscous and non-Newtonian broth. This change causes severe problems in mixing, transfer of heat, a supply of oxygen, and cause inconstancy in the nature and value of the final product. Whatever may be the scale of production, carbon to nitrogen ratio have to be maintained and the substrates used generally are the cost-effective available form.

21.6.1 Cost-Effective Biomass Resources

During the process of fermentation, nearly 30% of the total cost is generally devoted to the medium used. The media normally used in this process are costly such as yeast extract, peptone, and salts. So measures should be taken during the selection of media alternatives that can be cheap and compatible with the end product of fermentation. Recent research has moved towards the use of multicomponent feedstock systems to optimize the cost-effectiveness of the process and the use of low-cost alternative media such as wastewater generated from Olive mill (OMW), syrups, molasses, etc. (Sutherland 1972; Nicolaus et al. 2010). Lately, the by-products and wastes of several industries and as well as from the agricultural practices have been used frequently as nutrients during the industrial fermentation process.

21.6.1.1 Syrups and Molasses

Due to high sucrose and nutrient content, syrup and molasses have been applied frequently as the medium for the mass production of the commercially available polysaccharides like pullulan (Leathers and Gupta 1994), xanthan (Lopez et al. 2001), dextran (Vedyashkina et al. 2005), scleroglucan, levan (de Oliveira et al. 2007), and gellan (Göksungur et al. 2004). Other advantages that make it more suitable for this process include its availability and low cost. The low yield of pullulan and scleroglucan from the crude molasses need preprocessing to increase the yield amount. Increasing production of pullulan (Roukas 1998) and levan (Santos et al. 2005) has been reported after acid pretreatment of sugar beet molasses. Treatment of acid along with carbon has improved the yield amount of pullulan and levan, which might be due to its ability to remove heavy metals and colored substances. It has also been reported that for the elimination of iron, zinc, and some other mixtures from molasses, treatment with tricalcium phosphate ($Ca_3(PO_4)_2$), TCP is an efficient method. These metals are known to replace calcium during the reaction. Potassium ferrocyanide ($K_3[Fe(CN)_6]$) was also employed for removing heavy metals (Yin et al. 2002). Sugarcane syrup and crude sugar beet molasses resulted in a very low yield of levan with *Paenibacillus polymyxa*. Therefore, pretreatments like cane syrup and beet molasses along with peptone had to pass through gel filtration and ion exchange columns in order to raise the yield of the levan when compared to the sucrose. Five different physical and chemical pretreatment methods and their combined treatments, such as clarification, pH change, sulphuric acid, TCP, and treatment with activated carbon were employed for the synthesis of levan from halophilic *Halomonas* species. Sugar beet molasses often treated with TCP persuaded by sulphuric acid and carbon treatment ultimately leads to the production of the highest amount of levans. At the same time, treatment of sugar beet molasses including aeration, acid, activated carbon, $K_3[Fe(CN)_6]$ treatments, and anion or cation-exchange chromatography resulted in lower yields of xanthan from *X. campestris,* whereas in case of the untreated molasses significant level of production was noticed. Crude sugarcane with the help of microbial culture of S. *paucimobilis* resulted in the highest production of gellan. Coconut water and sugarcane juice have also been used as an alternative source for the production of EPS with *Lactobacillus confuses* cultures (Öner 2013).

21.6.1.2 Sugarbeet Pulp (SBP)

During the extraction process, the fibrous material persists as a by-product after the collection of sugar. Cellulose, hemicellulose, and pectin are the major constituents of sugar beet pulp. It is frequently used in the cattle industry, used as livestock feed. On the other hand, it has been used as a new and renewable source for the microbial production of EPS (Hutnan et al. 2000).

Change in composition and physical properties occurs due to its pretreatment such as autoclaving at 122–136 °C for 1 h leading to increased solubility and swelling of pectin and arabinans (Guillon et al. 1992). Sudden evaporation of ammonia due to the pretreatment like Ammonia pressurization depressurization (APD) reported to bursting of sugar beet pulp due to the increased hydrolysis of cellulose (Foster et al. 2001).

0.89 g xanthan/gram of SBP production was reported from microbial fermentation of sugar beet pulp by *X. campestris* (Yoo and Harcum 1999). It has also been reported that reduced particle size as a result of autoclaving increases its availability for microbial fermentation and production of EPS.

21.6.1.3 Olive Mill Wastewater (OMW)

A dark-colored fluid mainly composed of water released from the olive mill, waters produced from machinery cooling, and the remnant of the fruit is recognized as olive mill wastewater (OMW). It is the by-product produced during the manufacturing of Olive oil. Chemically it consists of Proteins, lipids, and carbohydrates as well as various other organic compounds which include monoaromatic and polyaromatic compounds (Aguilera et al. 2008). Because of its high carbon to nitrogen ratio, it has been used as the source material for microbial polymer production. Usually, it is helpful in the production of pullulan and xanthan (Ramos-Cormenzana et al. 1995; Lopez and Ramos-Cormenzana 1996). Before the microbial fermentation process collected OMW must be passed through filtration and diluted with distilled water/ saline water to prevent the effect of phenol. By the above approach, OMW has been used to produce some metal-binding EPS by *Paenibacillus jamilae*. The only concern associated with OMW is its decreasing concentration due to dilution. However, without dilution, β-glucan can be produced by the fungus *Botryosphaeria rhodina*. This process involved centrifugation and steam sterilization rather than dilution. Other than EPS, it has been utilized for the manufacturing of other biopolymers such as polyhydroxyalkanoates (PHAs) (Cerrone et al. 2010).

21.6.1.4 Cheese Whey

During the preparation of cheese, whey is a vital by-product produced. Mainly its nutrient composition depends on milk and from which it is derived which in turn is dependent on the processing methods of milk. Chemically 70% of the total contents solids of whey mainly consists of lactose. Despite having a variety of supplements and growth factors, the production of EPS is completely dependent on the lactose utilization capacity of the microorganism. *S. paucimobilis*can produces gellan only from the glucose, sweet cheese whey, and lactose as a substrate in media (Pollock 1993). Highly viscous gellan was obtained when the discussed strain utilizes lactose as substrate.

21.6.1.5 Pomace

Waste products from grape pomace have a significant contribution to agricultural industries. Processing of grapes in the wine industries leads to the production of grape pomace as the by-product. These by-products are often used as animal diets as well as for the production of ethanol by fermentation and distillation process. Grape pomace is mainly composed of sugar units and a low amount of proteins. The pullulan produced has also contained chemical compounds similar to that of Grape pomace. Through solid-state fermentation, xanthan has been produced by *X. campestris*, using grape, peels of tangerine, and solid remains of apple as substrate (Stredansky and Conti 1999). These substrates have been pretreated through an alkaline solution for neutralizing the effect of organic acids before adding to the fermentation media. Xanthan yields usually depend on the sugar content of the substrate, i.e., pomace as well as absorption extent of the solid material.

21.6.1.6 Lignocellulosic Biomass

For microbial systems, lignocellulosic biomass is a significant, cost-effective source with the ability to produce EPS on a large scale. Microbial cells are able to utilize the lignocellulosic source via endoglucanases or cellobiose for the production of biopolymers. The filamentous fungi *S. rolfsii* and other medicinal fungi are best suited for utilizing lignocellulosic source for EPS production due to their ability for metabolizing various pentose sugars like xylose and arabinose. Schizophyllan and arabinoxylan were produced by *Schizophyllum commune* from corn fiber pretreated with alkaline H_2O_2 (Leathers et al. 2006). Cornstalks used by *Lentinus squarrosulus* for the EPS production by solid-state fermentation and 5 g/L EPS was obtained from the fermentation medium (Isikhuemhen et al. 2012). Acid hydrolysates of wood were used for the production of EPS from bacterial culture. A high amount of polyhydroxyalkanoates (PHAs) were produced by the bacterial strain *Brevundimonas vesicularis* and *Sphingopyxis macrogoltabida* using acid hydrolyzed sawdust (Devi et al. 2012). Different residues generated from the biomass used for the production of EPS are corn steep, hydrolyzed potato starch, peach pulp, peat hydrolysates, spent grain and spent sulfite liquors, etc. (Cheng et al. 2011).

21.7 Biosynthesis Pathways of Microbial Polysaccharides

21.7.1 General Maneuvering for the Engineering of Bacterial Polysaccharides

Polysaccharides synthesized by bacteria possess a wide range of utilizations in the food processing industries and are used as, emulsifying agents, gelling agents, stabilizers, and viscofiers. Owing to these beneficial properties, various investigations were conducted to genetically modify the organisms to develop unique variants of polysaccharides for augmentating the production. The supposed objectives for the manipulation are the composition, molecular weight, and sequence of comonomers.

Contingent upon the purpose, the engineering methods can be subdivided into many different categories. One aim is an increase in the volumetric productivity for the cost-effective production of several exopolysaccharides (EPS). This type of study mostly focuses on enhancing the carbon flux towards the polymer by increasing the EPS precursor molecules. However, this method was proved to be favorable only for some EPS producer organisms, and unsuccessful in some cases may be due to alteration of the multi-domain protein complexes, which are involved in the polymerization and secretion processes (Thorne et al. 2000; Schatschneider et al. 2013; Videira et al. 2000; Wu et al. 2014; Van Kranenburg et al. 1999). Single gene knockouts were also employed for the enhancement in the product yield and also for the alteration in the chemical structure of the EPS (Gaytán et al. 2012; Nunez et al. 2009). But the genetic engineering strategy, along with enhancing the productivity it may reduce the viscosifying properties of the EPS. So rather than EPS bioengineering, optimization of the process parameters is a more promising alternative for EPS producers.

Another strategy aims at the alteration of the molecular structure, changing the behavior and characteristics of the polymer. These tailor-made variant productions can be based on the addition or deletion of monomeric sugar units from the side chain. The rheological behavior of the polymer can be controlled via acetylation and pyruvylation of polymers (Donati and Paoletti 2009; Hassler and Doherty 1990). The degree of pyruvylation and acetylation has a reversed effect on viscosity. Higher viscosity results in a higher rate of pyruvylation, and decreased viscosity of the EPS results in a higher degree of acetylation. The viscosity of the EPS can be adjusted by following these general rules.

Altered composition of the repeat units can be resulted by shift of whole gene clusters towards substitute host strains (Pollock et al. 1997; Van Kranenburg et al. 1999). However, these heterologous strategies may result in the lower production of foreign polymers (Stingele et al. 1999). Presently, four well-known mechanisms are there for producing EPS in bacteria, which are briefly discussed below.

21.7.1.1 Production of Exopolysaccharide via Wzx/Wzy-Dependent Pathway

In this pathway, at the cytoplasmic face of the membrane, the individual repeating residues joined to an undecaprenyl diphosphate (C55) anchor and are put together by different glycosyltransferases. These then translocated across the cytoplasmic membrane via a flippase protein called Wzx protein. Their polymerization in the periplasmic space occurs in the next step by Wzy (polymerase) protein (Cuthbertson et al. 2009; Islam and Lam 2014; Morona et al. 2009). Transport of these polymerized repeats to the exterior of the cell from the periplasmic space is based on proteins assigned to the outer surface polysaccharide export (OPX) and the enzyme polysaccharide co-polymerase (PCP). The polysaccharides which are assembled through the Wzx/Wzy pathway show a distinctly assorted kind of sugar motif and are categorized as heteropolymers (e.g., xanthan, sphingan, succinoglycan, colonic acid). The Wzx and Wzy genes present within the extracellular polysaccharide operons of the strains using this pathway.

21.7.1.2 The ABC Transporter Pathway

This pathway mainly synthesizes capsular polysaccharide (CPS) which does not represent the exopolysaccharides as they are linked to the cell surface. In the ABC transporter pathway glycosyltransferases (GTs) assemble the synthesized CPS at the inner side of the membrane. A homopolymer results when operons containing only a single GT unit is involved, and a heteropolymer when more than one GT's are involved. These ABC transporter proteins are approximately similar to OPX and PCP proteins in the Wzx/Wzy pathway of polysaccharide synthesis (Cuthbertson et al. 2009; Morona et al. 2009). One of the key difference in the Wzx/Wzy and the ABC transporter pathways are the presence phosphatidylglycerol (conserved glycolipid) and a poly-2-keto-3-deoxyoctulosonic acid (Kdo) linker in CPSs, synthesized via ABC transporter pathway at the reducing end (Willis et al. 2013; Willis and Whitfield 2013).

21.7.1.3 Productions of Exopolysaccharide via Synthase-Dependent Pathways

In this pathway, the entire polymer strand across the membrane and cell wall secreted are not dependent on flippase activity for the translocation process. An envelope-spanning multiprotein complex subunit often performs the polymerization and translocation process (Rehm 2010). A cluster of homopolymers, requiring only a single type of sugar precursor molecules is synthesized via this pathway. For example, Biosynthesis of curdlan, consisting of only β-(1-3)-linked glucose; Biosynthesis of cellulose, consisting of β-(1-4)-linked glucose units.

Heteropolysaccharides like Alginate and Hyaluronic acids are synthesized following the synthase-dependent pathway. Alginates are exopolysaccharides consisting of continuous (1-4)-linked β-D-mannuronic acid (M-block) and α-L-guluronic acid (G-blocks) or arranged as alternating MG-blocks (Rehm and Valla 1997; Gutsche et al. 2006). Hyaluronic acid polymerization and secretion are catalyzed by a single enzyme, hyaluronan synthase, and consisting of glucuronic acid and *N*-acetyl-glucosamine (GlcNAc) (Liu et al. 2011).

21.7.1.4 Extracellular Synthesized Polysaccharides

Some polysaccharides like dextran and levan, are synthesized extracellularly. Synthesis occurs via glycosyltransferase, and after synthesis, they are covalently attached to the exterior of the cell surface.

21.7.2 Bioengineering Strategies Towards Tailor-Made Exopolysaccharide

Tailor-made variants of the polysaccharides synthesized by microbes can be produced by overexpression of single genes or complete operons or by domain shuffling of GTs. The increasing perspicacity into the mechanisms, gene functions, and proteins involved in different biosynthetic pathways with the growing demand for techniques unfolding the metabolic networks have promising potential for the production of engineered polysaccharides. NGS approach unravels the novel EPS encoding gene clusters and enhances the understanding of the mechanism of the pathway of the EPS synthesis along with its diversity and modification. EPS biosynthesis can be induced by synthetic biology approaches, to separate from the natural regulatory networks. Artificial repeat units can be designed by a targeted shuffling of glycosyltransferases, involved in the repeat unit's assembly process. A different era of bacterial polysaccharides might be made by the increased understanding of the regulatory mechanism behind EPS biosynthesis obtained through a combination of bioinformatics and synthetic biology approaches.

Apart from this, in vivo and in vitro modifications to EPS can be introduced, by altering EPS modifying enzymes through protein engineering approaches. Alginate showing an extent of material properties can be produced by epimerases of *A. vinelandii* (Martinsen et al. 1989; Mørch et al. 2007; de Reese et al. 2015). Tailored variants of alginates with a large area of applications can be obtained by including other monomers along with guluronic acid and mannuronic acid. Molecular weight and glycosidic bond selectivity of the homopolymers like curdlan and cellulose can be modified by the targeted modifications of the synthases enzymes involved. Similarly, the regular stringent order of cellulose β-(1-4)-glycosidic linkages can be altered by modifying the orientation of UDP-glucose.

Short-chain fructo-oligosaccharides and oligolevans produced from sucrose yielding approximately 65% with productivity of 96 g/L/h. This method is based on an endo-inulase and immobilized levansucrase (Tian et al. 2014). Several low-molecular-weight oligo- or polysaccharides having high applicability in the food and medicine industry by several carbohydrate-modifying enzymes. Rapid and specific screening approaches for the identification of novel structures of polysaccharides will build upon the foundation of EPS yield and industrialization.

21.8 Conclusion

As discussed throughout the chapter, the microbial polysaccharides have a diverse role in the food, pharmaceutical, and textile industries. However, the industrial production of the EPS is dependent on the functional novelty and the cost of production, therefore a template-based synthesis of polysaccharides should be opted in the future to limit the relationship between chain length, its distribution, and function. Polysaccharides, when mixed with a biopolymer system, by physical or chemical cross-linking with salt or protein, can provide biocompatibility, biodegradability, and stability. Moreover, the approach towards the employment of functional polysaccharides in vitro is becoming the most attractive direction. Following the improved analytical and genetic tools and screening methods, the existing gap can be filled with new avenues in this field.

References

Agrawal P, Hassan AN (2008) Characteristics of reduced fat Cheddar cheese made from ultrafiltered milk with an exopolysaccharide-producing culture. J Dairy Res 75(2):182

Aguilera M, Quesada MT, del Águila VG, Morillo JA, Rivadeneyra MA, Ramos-Cormenzana A, Monteoliva-Sánchez M (2008) Characterisation of Paenibacillus jamilae strains that produce exopolysaccharide during growth on and detoxification of olive mill wastewaters. Bioresour Technol 99(13):5640–5644

Anon (1992) 21 CFR 172-Food additives permitted for direct addition to food for human consumption: gellan gum. Fed Regist 57(228):55444–55445

Atkins E (1986) Biomolecular structures of naturally occurring carbohydrate polymers. Int J Biol Macromol 8(6):323–329

Bender H, Lehmann J, Wallenfels K (1959) Pullulan, ein extracelluläres Glucan von Pullularia pullulans. Biochim Biophys Acta 36(2):309–316

Bhavani AL, Nisha J (2010) Dextran—the polysaccharide with versatile uses. Int J Pharm Biol Sci 1 (4):569–573

Bo S, Milas M, Rinaudo M (1987) Behaviour of scleroglucan in aqueous solution containing sodium hydroxide. Int J Biol Macromol 9(3):153–157

Bossio J, Semino C, de Iannino Iñón N, Dankert M (1996) The in vitro biosynthesis of the exopolysaccharide produced by Rhizobium leguminosarum bv. trifolii, strain NA 30. Cell Mol Biol (Noisy-le-Grand) 42(5):737–758

Cao S-L, Ren X-W, Zhang Q-Z, Chen E, Xu F, Chen J, Liu L-C, Jiang X-G (2009) In situ gel based on gellan gum as new carrier for nasal administration of mometasone furoate. Int J Pharm 365 (1–2):109–115

Catley B, Whelan WJ (1971) Observations on the structure of pullulan. Arch Biochem Biophys 143 (1):138–142

Cerning J (1990) Exocellular polysaccharides produced by lactic acid bacteria. FEMS Microbiol Rev 7(1–2):113–130

Cerrone F, Sanchez-Peinado MDM, Juárez-Jimenez B, González-López J, Pozo C (2010) Biological treatment of two-phase olive mill wastewater (TPOMW, alpeorujo): polyhydroxyalkanoates (PHAs) production by Azotobacter strains. J Microbiol Biotechnol 20 (3):594–601

Chaplin MF (1982) A rapid and sensitive method for the analysis of carbohydrate components in glycoproteins using gas-liquid chromatography. Anal Biochem 123(2):336–341

Cheng K-C, Demirci A, Catchmark JM (2011) Pullulan: biosynthesis, production, and applications. Appl Microbiol Biotechnol 92(1):29

Chien C-Y, Enomoto Y, Iwata T (2019) Synthesis of C2-regioselectively substituted curdlan acetate propionate and the effect of C2 substituent on their properties. ACS Sustain Chem Eng 7 (11):9857–9864

Cote GL, Krull LH (1988) Characterization of the exocellular polysaccharides from Azotobacter chroococcum. Carbohydr Res 181:143–152

Coviello T, Palleschi A, Grassi M, Matricardi P, Bocchinfuso G, Alhaique F (2005) Scleroglucan: a versatile polysaccharide for modified drug delivery. Molecules 10(1):6–33

Cuthbertson L, Mainprize IL, Naismith JH, Whitfield C (2009) Pivotal roles of the outer membrane polysaccharide export and polysaccharide copolymerase protein families in export of extracellular polysaccharides in gram-negative bacteria. Microbiol Mol Biol Rev 73(1):155–177

Das D, Baruah R, Goyal A (2014) A food additive with prebiotic properties of an α-d-glucan from Lactobacillus plantarum DM5. Int J Biol Macromol 69:20–26
Davidson I, Sutherland I, Lawson C (1977) Localization of O-acetyl groups of bacterial alginate. Microbiology 98(2):603–606
de Oliveira MR, da Silva RSSF, Buzato JB, Celligoi MAPC (2007) Study of levan production by Zymomonas mobilis using regional low-cost carbohydrate sources. Biochem Eng J 37 (2):177–183
de Reese J, Sperl N, Schmid J, Sieber V, Plank J (2015) Effect of biotechnologically modified alginates on LDH structures. Bioinspir Biomimet Nanobiomater 4(3):174–186
Devi ES, Vijayendra S, Shamala T (2012) Exploration of rice bran, an agro-industry residue, for the production of intra-and extra-cellular polymers by Sinorhizobium meliloti MTCC 100. Biocatal Agric Biotechnol 1(1):80–84
Donati I, Paoletti S (2009) Material properties of alginates. In: Alginates: biology and applications. Springer, pp 1–53
Donot F, Fontana A, Baccou J, Schorr-Galindo S (2012) Microbial exopolysaccharides: main examples of synthesis, excretion, genetics and extraction. Carbohydr Polym 87(2):951–962
Duc ANC (1982) Glucosylglucans and their use in gastroenterology, especially in the treatment of colon disorders. Eur Patent 45
Duus JØ, Gotfredsen CH, Bock K (2000) Carbohydrate structural determination by NMR spectroscopy: modern methods and limitations. Chem Rev 100(12):4589–4614
El-Kamel A, El-Khatib M (2006) Thermally reversible in situ gelling carbamazepine liquid suppository. Drug Deliv 13(2):143–148
Ertesvåg H, Doseth B, Larsen B, Skjåk-Bræk G, Valla S (1994) Cloning and expression of an Azotobacter vinelandii mannuronan C-5-epimerase gene. J Bacteriol 176(10):2846–2853
Evans LR, Linker A (1973) Production and characterization of the slime polysaccharide of Pseudomonas aeruginosa. J Bacteriol 116(2):915–924
Farombi EO, Adedara IA, Ajayi BO, Ayepola OR, Egbeme EE (2013) Kolaviron, a natural antioxidant and anti-inflammatory phytochemical prevents dextran sulphate sodium-induced colitis in rats. Basic Clin Pharmacol Toxicol 113(1):49–55
Fett WF, Osman SF, Fishman ML, Siebles T (1986) Alginate production by plant-pathogenic pseudomonads. Appl Environ Microbiol 52(3):466–473
Foster BL, Dale BE, Doran-Peterson JB (2001) Enzymatic hydrolysis of ammonia-treated sugar beet pulp. In: Twenty-second symposium on biotechnology for fuels and chemicals. Springer, pp 269–282
Gacesa P (1988) Alginates. Carbohydr Polym 8(3):161–182
Gaytán I, Pena C, Núñez C, Córdova MS, Espín G, Galindo E (2012) Azotobacter vinelandii lacking the Na+-NQR activity: a potential source for producing alginates with improved properties and at high yield. World J Microbiol Biotechnol 28(8):2731–2740
Gibbs P, Seviour R (1992) Influence of bioreactor design on exopolysaccharide production by Aureobasidium pullulans. Biotechnol Lett 14(6):491–494
Göksungur Y, Uçan A, Güvenç U (2004) Production of pullulan from beet molasses and synthetic medium by Aureobasidium pullulans. Turk J Biol 28(1):23–30
Guillon F, Barry JL, Thibault JF (1992) Effect of autoclaving sugar-beet fibre on its physico-chemical properties and its in-vitro degradation by human faecal bacteria. J Sci Food Agric 60 (1):69–79
Guohua H (2002) The speciality of gellan gum and its application in food industry. China Food Addit 6
Gutsche J, Remminghorst U, Rehm BH (2006) Biochemical analysis of alginate biosynthesis protein AlgX from Pseudomonas aeruginosa: purification of an AlgX-MucD (AlgY) protein complex. Biochimie 88(3–4):245–251
Hacking A, Taylor I, Jarman T, Govan J (1983) Alginate biosynthesis by Pseudomonas mendocina. Microbiology 129(11):3473–3480
Halleck FE (1969) Paint composition containing polysaccharides. Google Patents

Harvey DJ (1999) Matrix-assisted laser desorption/ionization mass spectrometry of carbohydrates. Mass Spectrom Rev 18(6):349–450
Hassler RA, Doherty DH (1990) Genetic engineering of polysaccharide structure: production of variants of xanthan gum in Xanthomonas campestris. Biotechnol Prog 6(3):182–187
Holzwarth G (1976) Conformation of the extracellular polysaccharide of Xanthomonas campestris. Biochemistry 15(19):4333–4339
Hutnan M, Drtil M, Mrafkova L (2000) Anaerobic biodegradation of sugar beet pulp. Biodegradation 11(4):203–211
Isikhuemhen OS, Mikiashvili NA, Adenipekun CO, Ohimain EI, Shahbazi G (2012) The tropical white rot fungus, Lentinus squarrosulus Mont.: lignocellulolytic enzymes activities and sugar release from cornstalks under solid state fermentation. World J Microbiol Biotechnol 28 (5):1961–1966
Islam ST, Lam JS (2014) Synthesis of bacterial polysaccharides via the Wzx/Wzy-dependent pathway. Can J Microbiol 60(11):697–716
Islam M, Murata T, Fujisawa M, Nagasaka R, Ushio H, Bari A, Hori M, Ozaki H (2008) Anti-inflammatory effects of phytosteryl ferulates in colitis induced by dextran sulphate sodium in mice. Br J Pharmacol 154(4):812–824
Jagodzinski PP, Wiaderkiewicz R, Kurzawski G, Kloczewiak M, Nakashima H, Hyjek E, Yamamoto N, Uryu T, Kaneko Y, Posner MR (1994) Mechanism of the inhibitory effect of curdlan sulfate on HIV-1 infection in vitro. Virology 202(2):735–745
Jones M, Huynh T, Dekiwadia C, Daver F, John S (2017) Mycelium composites: a review of engineering characteristics and growth kinetics. J Bionanosci 11(4):241–257
Jong S-C, Donovick R (1989) Antitumor and antiviral substances from fungi. In: Advances in applied microbiology, vol 34. Elsevier, pp 183–262
Kanke M, Tanabe E, Katayama H, Koda Y, Yoshitomi H (1995) Application of curdlan to controlled drug delivery. III. Drug release from sustained release suppositories in vitro. Biol Pharm Bull 18(8):1154–1158
Kataoka K, Muta T, Yamazaki S, Takeshige K (2002) Activation of macrophages by linear (1→ 3)-β-D-glucans implications for the recognition of fungi by innate immunity. J Biol Chem 277 (39):36825–36831
Katina K, Maina NH, Juvonen R, Flander L, Johansson L, Virkki L, Tenkanen M, Laitila A (2009) In situ production and analysis of Weissella confusa dextran in wheat sourdough. Food Microbiol 26(7):734–743
Katzbauer B (1998) Properties and applications of xanthan gum. Polym Degrad Stab 59 (1–3):81–84
Kesavan K, Nath G, Pandit J (2010) Preparation and in vitro antibacterial evaluation of gatifloxacin mucoadhesive gellan system. DARU: Journal of Faculty of Pharmacy, Tehran University of Medical Sciences 18(4):237
Kothari D, Goyal A (2013) Structural characterization of enzymatically synthesized dextran and oligosaccharides from Leuconostoc mesenteroides NRRL B-1426 dextransucrase. Biochemistry (Moscow) 78(10):1164–1170
Kothari D, Das D, Patel S, Goyal A (2014) Dextran and food application. In: Polysaccharides. Springer
Kubo W, Miyazaki S, Attwood D (2003) Oral sustained delivery of paracetamol from in situ-gelling gellan and sodium alginate formulations. Int J Pharm 258(1–2):55–64
Kuo M-S, Mort AJ, Dell A (1986) Identification and location of L-glycerate, an unusual acyl substituent in gellan gum. Carbohydr Res 156:173–187
Larsen B, Haug A (1971) Part I. Composition and structure of alginate produced by Azotobacter vinelandii (Lipman). Carbohydr Res 17:287–296
Leathers TD, Gupta SC (1994) Production of pullulan from fuel ethanol byproducts by Aureobasidium sp. strain NRRl Y-12,974. Biotechnol Lett 16(11):1163–1166
Leathers TD, Nunnally MS, Price NP (2006) Co-production of schizophyllan and arabinoxylan from corn fiber. Biotechnol Lett 28(9):623–626

Leitão J, Sá-Correia I (1993) Manipulation of Pseudomonas aeruginosa alginate pathway by varying the level of biosynthetic enzymes and growth temperature. J Appl Bacteriol 74 (4):452–459
Linker A, Jones RS (1964) A polysaccharide resembling alginic acid from a Pseudomonas micro-organism. Nature 204(4954):187–188
Liu L, Liu Y, Li J, Du G, Chen J (2011) Microbial production of hyaluronic acid: current state, challenges, and perspectives. Microb Cell Fact 10(1):99
Lopez M, Ramos-Cormenzana A (1996) Xanthan production from olive-mill wastewaters. Int Biodeterior Biodegradation 38(3–4):263–270
Lopez M, Moreno J, Ramos-Cormenzana A (2001) Xanthomonas campestris strain selection for xanthan production from olive mill wastewaters. Water Res 35(7):1828–1830
Mahajan S, Kimerling L (2013) Concise encyclopedia of semiconducting materials & related technologies. Elsevier
Maina NH, Virkki L, Pynnönen H, Maaheimo H, Tenkanen M (2011) Structural analysis of enzyme-resistant isomaltooligosaccharides reveals the elongation of α-(1→ 3)-linked branches in Weissella confusa dextran. Biomacromolecules 12(2):409–418
Mann EE, Wozniak DJ (2012) Pseudomonas biofilm matrix composition and niche biology. FEMS Microbiol Rev 36(4):893–916
Mark HF, Bikales N, Overberger CG, Menges G (1987) Encyclopedia of polymer science and engineering, vol 11: photographic applications to polyesters, elastomeric
Martinsen A, Skjåk-Bræk G, Smidsrød O (1989) Alginate as immobilization material: I. Correlation between chemical and physical properties of alginate gel beads. Biotechnol Bioeng 33(1):79–89
McCouch SR, Teytelman L, Xu Y, Lobos KB, Clare K, Walton M, Fu B, Maghirang R, Li Z, Xing Y (2002) Development and mapping of 2240 new SSR markers for rice (Oryza sativa L.) (supplement). DNA Res 9(6):257–279
McIntosh M, Stone B, Stanisich V (2005) Curdlan and other bacterial (1→ 3)-β-D-glucans. Appl Microbiol Biotechnol 68(2):163–173
Mende S, Peter M, Bartels K, Dong T, Rohm H, Jaros D (2013) Concentration dependent effects of dextran on the physical properties of acid milk gels. Carbohydr Polym 98(2):1389–1396
Mischnick P (1998) Determination of the substituent distribution of O-(2-hydroxypropyl) inulin by methylation analysis. Starch-Stärke 50(1):33–36
Monchois V, Willemot R-M, Monsan P (1999) Glucansucrases: mechanism of action and structure–function relationships. FEMS Microbiol Rev 23(2):131–151
Moon C-J, Lee J-H (2005) Use of curdlan and activated carbon composed adsorbents for heavy metal removal. Process Biochem 40(3–4):1279–1283
Moorhouse R, Walkinshaw M, Arnott S (1977) Extracellular microbial polysaccharides. In: ACS symposium series, vol 90. p 911
Mørch Ý, Donati I, Strand B, Skjåk-Bræk G (2007) Molecular engineering as an approach to design new functional properties of alginate. Biomacromolecules 8(9):2809–2814
Morona R, Purins L, Tocilj A, Matte A, Cygler M (2009) Sequence-structure relationships in polysaccharide co-polymerase (PCP) proteins. Trends Biochem Sci 34(2):78–84
Mulchandani A, Luong JH, LeDuy A (1989) Biosynthesis of pullulan using immobilized Aureobasidium pullulans cells. Biotechnol Bioeng 33(3):306–312
Naessens M, Cerdobbel A, Soetaert W, Vandamme EJ (2005) Leuconostoc dextransucrase and dextran: production, properties and applications. J Chem Technol Biotechnol 80(8):845–860
Nicolaus B, Kambourova M, Oner ET (2010) Exopolysaccharides from extremophiles: from fundamentals to biotechnology. Environ Technol 31(10):1145–1158
Nunez C, Bogachev AV, Guzman G, Tello I, Guzman J, Espín G (2009) The Na+-translocating NADH: ubiquinone oxidoreductase of Azotobacter vinelandii negatively regulates alginate synthesis. Microbiology 155(1):249–256
Omoto T, Uno Y, Asai I (1999) The latest technologies for the application of gellan gum. In: Physical chemistry and industrial application of gellan gum. Springer, pp 123–126

Öner ET (2013) Microbial production of extracellular polysaccharides from biomass. In: Pretreatment techniques for biofuels and biorefineries. Springer, pp 35–56
Osmałek T, Froelich A, Tasarek S (2014) Application of gellan gum in pharmacy and medicine. Int J Pharm 466(1–2):328–340
Phillips K (2012) Demand for healthier foods stirs up growth in hydrocolloids. Chemical Week. http://www.accessmylibrary.com/article-1G1–167346654/demand-healthier-foodsstirs.html. Accessed 20
Pindar DF, Bucke C (1975) The biosynthesis of alginic acid by Azotobacter vinelandii. Biochem J 152(3):617–622
Pirri R, Gadioux J, Riveno R (1996) Scleroglucan gel applied to the oil industry. Google Patents
Pollock TJ (1993) Gellan-related polysaccharides and the genus Sphingomonas. Microbiology 139 (8):1939–1945
Pollock T, Mikolajczak M, Yamazaki M, Thorne L, Armentrout R (1997) Production of xanthan gum by Sphingomonas bacteria carrying genes from Xanthomonas campestris. J Ind Microbiol Biotechnol 19(2):92–97
Purama RK, Goswami P, Khan AT, Goyal A (2009) Structural analysis and properties of dextran produced by Leuconostoc mesenteroides NRRL B-640. Carbohydr Polym 76(1):30–35
Ramos-Cormenzana A, Monteoliva-Sanchez M, Lopez M (1995) Bioremediation of alpechin. Int Biodeterior Biodegradation 35(1–3):249–268
Rehm BH (2010) Bacterial polymers: biosynthesis, modifications and applications. Nat Rev Microbiol 8(8):578–592
Rehm B, Valla S (1997) Bacterial alginates: biosynthesis and applications. Appl Microbiol Biotechnol 48(3):281–288
Remuñán-López C, Portero A, Vila-Jato JL, Alonso MJ (1998) Design and evaluation of chitosan/ ethylcellulose mucoadhesive bilayered devices for buccal drug delivery. J Control Release 55 (2–3):143–152
Roukas T (1998) Pretreatment of beet molasses to increase pullulan production. Process Biochem 33(8):805–810
Sanderson GR (1981) Applications of xanthan gum. Br Polym J 13(2):71–75
Sandford PA (1979) Exocellular, microbial polysaccharides. In: Advances in carbohydrate chemistry and biochemistry, vol 36. Elsevier, pp 265–313
Sandford PA, Baird J (1983) Industrial utilization of polysaccharides. In: The polysaccharides. Elsevier, pp 411–490
Santos M, Rodrigues A, Teixeira JA (2005) Production of dextran and fructose from carob pod extract and cheese whey by Leuconostoc mesenteroides NRRL B512 (f). Biochem Eng J 25 (1):1–6
Sarwat F, Qader SAU, Aman A, Ahmed N (2008) Production & characterization of a unique dextran from an indigenous Leuconostoc mesenteroides CMG713. Int J Biol Sci 4(6):379
Schatschneider S, Persicke M, Watt SA, Hublik G, Pühler A, Niehaus K, Vorhölter F-J (2013) Establishment, in silico analysis, and experimental verification of a large-scale metabolic network of the xanthan producing Xanthomonas campestris pv. campestris strain B100. J Biotechnol 167(2):123–134
Schuster R, Wenzig E, Mersmann A (1993) Production of the fungal exopolysaccharide pullulan by batch-wise and continuous fermentation. Appl Microbiol Biotechnol 39(2):155–158
Sheth P, Lachman L (1967) The coating of tablets. Fr Patent 1(480):874
Singh SK, Gross RA (2001) Overview: introduction to polysaccharides, agroproteins, and poly (amino acids). ACS Publications
Singh PP, Whistler RL, Tokuzen R, Nakahara W (1974) Scleroglucan, an antitumor polysaccharide from Sclerotium glucanicum. Carbohydr Res 37(1):245–247
Singh RS, Saini GK, Kennedy JF (2008) Pullulan: microbial sources, production and applications. Carbohydr Polym 73(4):515–531

Steenbergen SM, Vimr ER (2003) Functional relationships of the sialyltransferases involved in expression of the polysialic acid capsules of Escherichia coli K1 and K92 and Neisseria meningitidis groups B or C. J Biol Chem 278(17):15349–15359

Stingele F, Vincent SJ, Faber EJ, Newell JW, Kamerling JP, Neeser JR (1999) Introduction of the exopolysaccharide gene cluster from Streptococcus thermophilus Sfi6 into Lactococcus lactis MG1363: production and characterization of an altered polysaccharide. Mol Microbiol 32 (6):1287–1295

Stredansky M, Conti E (1999) Xanthan production by solid state fermentation. Process Biochem 34 (6–7):581–587

Subedi RK, Kang KW, Choi H-K (2009) Preparation and characterization of solid lipid nanoparticles loaded with doxorubicin. Eur J Pharm Sci 37(3–4):508–513

Suresh Kumar A, Mody K, Jha B (2007) Bacterial exopolysaccharides—a perception. J Basic Microbiol 47(2):103–117

Sutherland I (1972) Bacterial exopolysaccharides. In: Advances in microbial physiology, vol 8. Elsevier, pp 143–213

Sutherland I (1982) Biosynthesis of microbial exopolysaccharides. In: Advances in microbial physiology, vol 23. Elsevier, pp 79–150

Sutherland IW (1994) Structure-function relationships in microbial exopolysaccharides. Biotechnol Adv 12(2):393–448

Sutherland IW (2001) Exopolysaccharides in biofilms, flocs and related structures. Water Sci Technol 43(6):77–86

Taylor CM (1998) Glycopeptides and glycoproteins: focus on the glycosidic linkage. Tetrahedron 54(38):11317–11362

Taylor CM, Roberts IS (2005) Capsular polysaccharides and their role in virulence. In: Concepts in bacterial virulence, vol 12. Karger Publishers, pp 55–66

Thorne L, Mikolajczak M, Armentrout R, Pollock T (2000) Increasing the yield and viscosity of exopolysaccharides secreted by Sphingomonas by augmentation of chromosomal genes with multiple copies of cloned biosynthetic genes. J Ind Microbiol Biotechnol 25(1):49–57

Tian F, Karboune S, Hill A (2014) Synthesis of fructooligosaccharides and oligolevans by the combined use of levansucrase and endo-inulinase in one-step bi-enzymatic system. Innovative Food Sci Emerg Technol 22:230–238

Van Kranenburg R, Boels IC, Kleerebezem M, De Vos WM (1999) Genetics and engineering of microbial exopolysaccharides for food: approaches for the production of existing and novel polysaccharides. Curr Opin Biotechnol 10(5):498–504

Vedyashkina T, Revin V, Gogotov I (2005) Optimizing the conditions of dextran synthesis by the bacterium Leuconostoc mesenteroides grown in a molasses-containing medium. Appl Biochem Microbiol 41(4):361–364

Videira PA, Cortes LL, Fialho AM, Sá-Correia I (2000) Identification of the pgmG gene, encoding a bifunctional protein with phosphoglucomutase and phosphomannomutase activities, in the gellan gum-producing strain Sphingomonas paucimobilis ATCC 31461. Appl Environ Microbiol 66(5):2252–2258

Willis LM, Whitfield C (2013) Structure, biosynthesis, and function of bacterial capsular polysaccharides synthesized by ABC transporter-dependent pathways. Carbohydr Res 378:35–44

Willis LM, Stupak J, Richards MR, Lowary TL, Li J, Whitfield C (2013) Conserved glycolipid termini in capsular polysaccharides synthesized by ATP-binding cassette transporter-dependent pathways in Gram-negative pathogens. Proc Natl Acad Sci 110(19):7868–7873

Wu Q, Tun HM, Leung FC-C, Shah NP (2014) Genomic insights into high exopolysaccharide-producing dairy starter bacterium Streptococcus thermophilus ASCC 1275. Sci Rep 4(1):1–8

Yamasaki H, Lee M-S, Tanaka T, Nakanishi K (1993) Improvement of performance for cross-flow membrane filtration of pullulan broth. Appl Microbiol Biotechnol 39(1):21–25

Yin X, Calderin L, Stott M, Sayer M (2002) Density functional study of structural, electronic and vibrational properties of Mg-and Zn-doped tricalcium phosphate biomaterials. Biomaterials 23 (20):4155–4163

Yoo SD, Harcum SW (1999) Xanthan gum production from waste sugar beet pulp. Bioresour Technol 70(1):105–109

Yu L, Wu J, Liu J, Zhan X, Zheng Z, Lin CC (2011) Enhanced curdlan production in Agrobacterium sp. ATCC 31749 by addition of low-polyphosphates. Biotechnol Bioprocess Eng 16(1):34–41

Zhan X-B, Lin C-C, Zhang H-T (2012) Recent advances in curdlan biosynthesis, biotechnological production, and applications. Appl Microbiol Biotechnol 93(2):525–531

Eco-friendly Microbial Biopolymers: Recent Development, Biodegradation, and Applications

22

Chandrabose Selvaraj and Sanjeev Kumar Singh

Abstract

Microorganism has the capability to synthesis a wide range of biopolymers that take in a diverse biological function and possess powerful antimicrobial properties, and thus most of the biopolymers are widely employed in medical and food industries. Microbial biopolymers such as exopolysaccharides are the most widely used material and act as a key role in several fields, including the fabric industry, dairy, and cosmetics. Hence, a more honest apprehension of the underlying process involved in biopolymer production and its regulation under certain conditions is required. Recent improvements such as protein engineering are used to raise their activity and, especially, the antimicrobial polymers are effectively fighting against pathogens and have got considerable attention in the research study. Certain models of polymer are significantly affected in the metabolic and protein engineering approach to improve the turnout and their highlights are discussed in this chapter. Along with the recent developments, the metabolic pathway derived production of biopolymers and applications were also talked about thoroughly.

Keywords

Microbial polymer · Polysaccharides · Xanthan · Biodegradation · Biodegradable polymers

C. Selvaraj (✉) · S. K. Singh (✉)
Department of Bioinformatics, Computer-Aided Drug Design, and Molecular Modeling Lab, Science Block, Alagappa University, Karaikudi, Tamil Nadu, India
e-mail: selnikraj@bioclues.org; sksingh@alagappauniversity.ac.in

A. Vaishnav, D. K. Choudhary (eds.), *Microbial Polymers*,
https://doi.org/10.1007/978-981-16-0045-6_22

22.1 Introduction

Biopolymers are produced by a variety of natural origins, including plants, animals, and microorganisms; generally, they are water-soluble and easily degradable with natural and chemical solvents. Biopolymers have novel and unique physicochemical properties due to their wide diversity of structure and physical attributes. Hence, it is often applied in a broad scope of applications including, food and waste management industries. It can as well be utilized as potential coagulants, suspending agents, binders, stabilizers, and lubes. Recently it emerged as an important source of several polymeric materials and gradually turned into an economic competitor for natural sources such as algae and other plants. The genetically modified microbes may produce the new exopolysaccharides due to its novel superior properties which will open a fresh window of industrial applications and also significantly increase their demand. Biopolymers are a kind of exopolysaccharides that plays a key function in different roles in microbial cells, based on this it can be classified into three major characters such as extracellular polysaccharides, intracellular polymers, and structural elements of cell structures such as lipopolysaccharides and teichoic acids. Founded on the microbial system the exopolysaccharides are formed as a prominent capsule around the cell, and thus it becomes one of the cell wall components, while others form slimes and can diffuse away into the liquid phase during the fermentation (Singha 2012).

Most of the polysaccharides are widely used in starch-based industries. The plant-derived polysaccharides such as gum Arabic and guar gum are the major constituents in the market. Xanthan gum is one of the microbial exopolysaccharides from *Xanthomonas campestris* (Singha 2012). It is likewise applied to synthesis nanomaterials that have significant bonding properties with fabrics hence it helps in the production of wears with several properties like antimicrobial effect, fire retardancy, water, oil and dust repellent, UV protection, and enhanced electric conductivity. Microbial polysaccharides are attracting the high-interest level; especially, in the manufacture of medical dressing fabrics. For instance, the aubasidan and rodexman fibers show significant protective and therapeutic function, and hence it is highly used for the preparation of unique dressing for wounds and shows significant operative response including anti-inflammatory and healing effect without any side effects (Palza 2015). This high market value of the exopolysaccharides attracts the researchers to produce the exopolysaccharides and it is used in green technology to better the surroundings. It will perhaps go for microbes as a renewable resource (Ciechańska 2004). Some of the polymers derived from microbes play a significant role in the textile world. For example, the water-soluble microbial polymers Pullsn derived from *Aureobasidium pullulans* is used as the water repellant in the fabric industry. It can be completely soluble in water and partially soluble in organic solvents (Sena et al. 2006). So, it can receive the application in the manufacture of water-resistant raincoats and covers (Babu et al. 2013).

The polymers produced from the soil bacterium are linear polysaccharides and are hydrophilic in nature so it can play as a water-binding factor. Another polymer from microbes is alginate fibers, widely used in wound healing due to its special mannuronic acid's ability (Goh et al. 2012). It also has high moisture absorption and it can be used in sports apparel for absorbing sweat and used as flame retardants (Gowri et al. 2010). These fabrics are used in various applications along with clothing, institutional draperies, commercial carpet, military garments, and bedding (Tian et al. 2013). Microbial cellulose is the widely used natural polymer in textile industries which is produced by *Acetobacter xylinum* (Tyagi and Suresh 2013). Owing to their water-holding ability and a high crystallinity index biopolymer is widely used in wound healing (Retegi et al. 2010; Ashjaran et al. 2013). Xanthan gum is another example of microbial exopolysaccharides that improve the development of more efficient fermentation processes. Rheology is one of the most important factors of the polymer which is highly influenced by their structure and composition. This structural composition of microbial polymers significantly depends on several factors, including the nature of the substrate, microbial species, and other fermentation conditions. This chapter discusses the production of microbial production and recent advances in the development of the eco-friendly polymer, in various applications and emphasizes the important factors affecting the production and processing of microbial polymers.

22.2 Types of Biopolymers

22.2.1 Pullulan

The pullulan is a well-known polymer derived from *Aureobasidium pullularia* which shows significant activities for enzymes like amylase, fructosyltransferase, invertase, and glucose oxidase and proteolytic enzymes. It plays a critical role in various industries. Pullulans are extracellular water-soluble polysaccharides with a molecular weight ranging from 10,000 to 400,000 mainly containing starch and other sugar sources. The chemistry of glucan-based pullulans consists of α-maltotriose units polymerized linearly through 1-6-linkage on the terminal glucose. Microbial polysaccharides have been used for a wide range of applications including food and drinks, the process of low-calorie foods and hygienic preparation, and soil stabilizers (Singh et al. 2008). It highly resembles the styrene in gloss, which has more hardness and transparency and much more elastic. The oxygen permeability of the pullulans varies based on the thickness of the pullulans, for instance, the oxygen permeability of 0.01 mm thick pullulans films varies from 0.6 to 2 compared with cellophane and polypropylenes. Based on the biodegradability of the polymer pullulans are significantly used in food packing films coating food containers of perishable fruit and vegetables (Iran 2009). Compared to bacteria and yeast, fungal cells contain less nucleic acid and high levels of proteins and the filamentous nature of fungal mycelium facilitates the formation of single-cell proteins (SCP).

22.2.2 Poly-β-Hydroxybutyrate

The polymeric compound poly-β-hydroxybutyrate (PHB) is produced by several bacterial species and serves as a strong polymer as an energy source (Katircloglu et al. 2003; Aslim et al. 2002; Seidi and Alemzadeh 2003). Recently, it has received high attention in utilizing PHB as polymeric materials in the drug delivery system (Seidi and Alemzadeh 2003). The lack of fatty acid-containing lipids is soluble in chloroform and similar solvents. The PHB production on high-level yield is processed while the nitrogen is exhausted from the medium, and here the synthesis depends on the concentration of O_2 (Katircloglu et al. 2003; Aslim et al. 2002). The culture of *Ralsotonia eutropha* along with *Lactobacillus delbrueckii* with programmed control of nitrogen and carbon in simple batch fermentation produce the high yield of PHB. Currently, it is used as biodegradable thermoplastic in the trade name of Bipol at Billingham and UK (Jung et al. 2005; Ganduri et al. 2005). It is also used in a variety of applications and acts as a potent substitute for plastic (Valappil et al. 2007). The species such as *Alcaligen euthrophus* and Reuthropha are the best choice of the organism for PHB production which reduces about 70–80% of its biomass, where glucose is used as substrate (Li et al. 2007). Several microbial organisms are currently being processed by optimizing fermentation to produce efficient PHB production at a lower cost (Jang et al. 2009; Shih et al. 2007). The methods used in the optimization process require the accurate and rapid quantification of PHB in microbial cells. Several techniques like IR spectrometry, gas chromatography, gravimetry, and high-pressure liquid chromatography are used in various steps of optimization processes (Monteil-Rivera et al. 2007; Seidi and Alemzadeh 2003; Taguchi et al. 2001). The growth and investigation of PHB production have highly influenced the utilization of synthetic and natural sugar sources such as whey, starch, and molasses which were used as a substrate in the fructose synthetic medium.

22.2.3 Cellulose and Its Derivatives

In the year 1838, Payen discovered the principle structural constituent of the cell walls and found the cellulose is a major part of most of the plants. About 95% are present in the leaves and stalks and it was observed that 50% in wood. Owing to the ubiquity of the plant kingdom cellulose is the major source. Structurally cellulose is linear polymers and connected with β (1, 4)-glycosidic bonds. The homopolymer of glucose is not water-soluble due to inner and outer molecular hydrogen bonding hence the chemical composition is modified either with suitable solvents. Commercially it can be produced as cellulose ethers including carboxymethyl cellulose (CMC), hydroxyethyl cellulose (HEC), ethyl hydroxyethyl cellulose (EHEC), and methyl hydroxyethyl cellulose (MHEC).

22.2.4 Chitin and Pectin

Chitin is another type of naturally occurring polysaccharide found in fungi, yeast, and in the cytoskeletons of insects and crustaceans. The focused view of the structural chemistry of chitin shows it is composed of p- (1->4) 2-acetamido-2-deoxy-D-glucose repeat units which are deacetylated. The deacetylation-based degree is different from species to species and it varies from 8% to 15%. Most forms of the chitin and its derivatives are used in several biomedical and pharmaceutical applications. Chitosan is the most diluted form of chitin and has been manufacturing in the form of a tablet and used as a binder, lubricant, and disintegrating agents (Felt et al. 1998; Singla and Chawla 2001). Next to chitin, pectin present in the cell wall and that surrounding growing and dividing plant cells and acting as an essential component in the growth and ripening process (Ridley et al. 2001). The composure of pectin includes the α-1,4-linked D-galacturonic acid units that are used at large levels with ethyl cellulose, in the drug delivery system (Shirwaikar et al. 2008).

22.2.5 Bacterial Biopolymers

Among the microbes, especially the bacteria hold high potential to convert the carbon source into a wide range of polymers with various chemical and physical properties. Though the bacteria significantly synthesis only a few intracellular polymers and play a role in various functions in a wide range of prokaryotes,it can be specific for certain bacterial taxa that can be involved in the synthesis of polymers. The polymers synthesized from the bacteria are classified into four types, and they are named as polysaccharides, polyesters, polyamides, and inorganic polyanhydrides including polyphosphates (Ryder et al. 2007; Pham et al. 2004). These polymers play a unique role in various biological functions. It acts as a potent material that provides a substantial advantage under certain environmental conditions. Most of the bacteria have been used for the large-scale production of polymers through medium- to large-scale fermentations and it is estimated that 2000–100,000 tons. The polymers derived from the other natural sources have advantages due to their unique features such as production from renewable sources and their biodegradability and biocompatibility. The basic properties of the microbial polymers are their biodegradable nature; hence, it can be used in many applications in commodity products and are attractive because of the longing to evade the use of oil-based polymers and accumulate in the environment. When the polymers are exposed in the environment they are fully degraded and mineralized into CO_2 and H_2O; this conversion is catalyzed by depolymerase and hydrolases which attack the backbone of the microbial polymers and leads to the formation of lower molecular-mass products and are easily degradable and also they can be taken by the other microorganism and used as carbon and energy source (Chen 2009).

22.2.6 Polysaccharides

The bacterial polymers are classified into exopolysaccharides, capsular polysaccharides, and intracellular polysaccharides. Exopolysaccharides are either secreted by cell wall-associated enzymes. The capsular polymers such as K30 antigen and the intracellular polysaccharides (glycogen) are synthesized by the other inner cell membrane and other cytosolic enzymes. Based on the structural organization polysaccharides are further classified into repeat unit polymers and non-repeating polymers (Whitfield 2006; Valla et al. 2009; Remminghorst and Rehm 2006). The polymers are different in both structural and chemical composition because they require different enzymes and proteins for the organization, and hence these are reflected in the varied organizations of the biosynthesis of the gene cluster. Both the exo and capsular polysaccharides are produced by the gene cluster and are extensively involved in the regulation of transcription and also in signal transduction pathways and quorum sensing (Hickman and Harwood 2008; Leech et al. 2008; Weber et al. 2006; Sakuragi and Kolter 2007). The induction of these exopolysaccharides is highly associated with the establishment of the biofilm production and acting as an important factor (Hall-Stoodley et al. 2004). The nucleotide associated sugar molecules like ADP; diphosphate sugar derivatives like uDP-*N*-acetylglucosamine are often used as precursors for the synthesis of exopolysaccharides. Besides, the enzymes like pyrophosphorylase and dehydrogenases are the polymer-specific enzymes that are highly required for the synthesis of activated polymer precursor, which is the first step of the polymer synthesis and have been targeted by metabolic engineering to enhance the polymer production (Ruffing and Chen 2006). The secretion and polymerization of the exopolysaccharides significantly affect the carbon flux and affect the formation of a high molecular weight exopolymer. The synthesis of the catalytic unit is habitually contained in the cytoplasmic membrane and their associated enzymes/proteins that are mandatory for the translation of polymers across the cell membrane.

22.2.7 Exopolysaccharides

A wide range of bacteria and some archaea synthesize exopolysaccharides and based on their chemical composition, subunits, and molecular weight, exopolysaccharides are significantly used in various applications in industrial and medical fields. The best examples of exopolysaccharides are dextran and xanthan, which are commercially produced on a large scale. Alginate is another exopolysaccharide that also has potential applications. Dextrans are water-soluble and behave as Newtonian fluids and viscosity nature changes based on the functional concentration, temperature, and average molecular mass. The acid hydrolysis of the dextrans produces the fractions of defined molecular masses and the low immunogenicity of the dextrans is often used in both clinical and pharmaceutical applications. For instance, dextrans act as blood plasma extenders in chromatography media (Leathers 2005). The hydrolase and glycoside superfamily enzymes such as dextransucrase and glucan sucrase are acting as key enzymes involved in the synthesis of dextran. The secreted dextran with a molecular weight of 160 kDa is significantly anchored on the cell wall (van Hijum et al. 2006). Glucansucrase substantially catalyzes the hydrolysis of the glycosidic bond in the glucose transfer mechanism and regulates the growing, reducing end of the covalently linked glucan chain (Robyt et al. 2008). The gene dsrS encoded the enzyme dextransucrase expression and the hyaluronic acid is encoded by HA synthase (HasA) which is involved in polymerization and secretion (Brown and Pummill 2008). Most of the exopolysaccharides are polymerized and secreted by multiprotein complexes and this complex-mediated synthesis is categorized as two general pathways one is shown by the Wzy-dependent polymerization and secretion mechanism which are involved in xanthan biosynthesis. Alginate is widely used in medical applications and significantly linked to their stabilizing gelling properties and has the potential to retain water. Seaweeds and genetically engineered bacteria *Azotobacter vinelandii* and Pseudomonas fluorescens are used for the commercial production of alginates. The in vitro synthesis of alginates involves the complete envelope and provides confirmation that it is multiprotein complex, and the outer membrane has coordinated polymerization and secretion (Rehm 2009).

Gum Arabic is synthesized by *Acacia Senegal* and *Acacia Seyal* trees. These species naturally grow in the semi-arid and sub-Saharan regions of Africa and are promoted in other countries including Europe and the US. These countries are having GA markets importing 40 kTn/year, while Japan is the largest GA production (2 kTn/year) in Asian countries. Several studies have reported the molecular structure of the GA and revealed that it has a group of macromolecules with a higher proportion of carbohydrates mainly composed of D-glucose and L-arabinose units and a low proportion of protein. The chemical composition is highly influenced by the origin, climate, age of the tree, and processing conditions. The chemistry and composure of GA may depend on its origin, climate, harvest season, tree age, and processing conditions such as spray drying. (Al-Assaf et al. 2007; Flindt et al. 2005; Hassan et al. 2007).

22.2.8 Capsular Polysaccharides

Capsular polysaccharides play a major role in bacterial virulence. They are attached to the cell and also function as the surface antigen. The biosynthesis and assembly of capsular polysaccharides are extensively studied with *E. coli* and pathogenic bacteria *Klebsiella pneumonia* which contributes to the pathogen's ability to evade phagocytosis by macrophage cells (Yamanaka et al. 2008). Capsular polymers are considered as valuable materials and the biosynthesis of these polymers could identify the potential targets for the treatment of infection. Hence, the exopolysaccharides and their derivatives are serving as potential drug molecules. It proved the repeat unit of polysaccharides biosynthesis and understanding of the characterization of their biosynthetic pathways significantly contributed towards understanding the biosynthesis of other commercially valuable polysaccharides. Several repeat units of these polymers are highly required for the lipid carrier undecaprenyl phosphate for translocation of this unit to cross the cell membrane and Wzy-dependent pathway where it undergoes polymerization and involved in the secretion of oligosaccharides (Cuthbertson et al. 2009; Ford et al. 2009; Whitfield and Larue 2008). Table 22.1 shows various biopolymers and sources. The previous reports state that lipopolysaccharides such as O antigen polymerization and this secretion are widely distributed Wzy-dependent biosynthesis pathways (Raetz and Whitfield 2002). However, the assembly of the K30 antigen is an example of repeat unit polysaccharides which are involved in sugar-phosphate transfer from the nucleotide molecule to undecaprenyl phosphate sugar and catalyzed the initiating glycosyltransferase. These glucose transfers are also catalyzed by the monofunctional glycosyltransferase and provide the potential drug target for the drug development process (Barreras et al. 2008). At the side of the cytoplasmic membrane, the polymerization occurs and requires the transfer of repeat unit which is mediated by Wzx a putative transport protein that significantly catalyzed the transfer of nascent polymer from its undecaprenyl phosphate to lipid-linked repeat unit (Collins and Derrick 2007). Wzx is substantially involved in the polymerization and controls the length of the nascent polymer chain via periplasm to the outer membrane. The polysaccharide co-polymerase (PCP) family members share similar structural features including transmembrane domain and extended periplasmic domains which allow assembly of PCP into oligomers and define interaction with Wzy promoters subsequently control the polymer chain length (Collins and Derrick 2007). The recent crystal structure of the PCP and Wza from *E. coli* has been used to study the insight of structural requirements for translocation of the nascent polysaccharides from periplasm via the outer membrane (Ford et al. 2009; Collins and Derrick 2007).

22.2.9 Polyamides and Polyesters

Polyamides are a distinct group of biopolymers synthesized by non-ribosomally with three major distinct components including extracellular polyamides, poly-γ-glutamate (PgA), and ε-poly-L-lysine which have significant antibacterial activities. PgA is also synthesized from *Bacillus anthracis* and plays a crucial role in virulence mechanism with immunogenicity hence it triggers a humoral immune response in the infected host. The released PgA serves as a potent source of nitrogen and carbon and has been associated as a water-binding module in the biofilm matrix. Hence, the bacterial polymers act as potent renewable, nontoxic, and biodegradable. Polyhydroxyalkanoates is another type of bacterial polymers also known as bioplastics, which accumulate as a carbon source when growth is limited due to starvation of the other micronutrients. The excessive accumulation is regulated by the biosynthesis genes (Kessler and Witholt 2001). The hydrophobic PHA core is mostly enclosed by protein and convoluted in PHA metabolism (Grage et al. 2009; Jendrossek 2009). It can vary significantly in their chemical composition; about 150 known chemical constituents are present in it; hence, it is commercially

Table 22.1 Different bacterial polymers and its source and applications

Bacterial type	Polymer class	Primary structure	Main component	Application
Bacteria and archaea	Glycogen	α-(1,6)-branched α-(1,4)-linked homopolymer	Glucose	NA
Pseudomonas spp. and *Azotobacter* spp.	Alginate	β-(1,4)-linked non-repeating heteropolymer	Mannuronic acid and guluronic acid	Used as a potent biomaterial for tissue scaffold and drug delivery
Agrobacterium spp., *Rhizobium* spp., and *Cellulomonas* spp.	Curdlan	β-(1,3)-linked homopolymer	Glucose	Agrobacterium spp., Rhizobium spp., and Cellulomonas spp.
Leuconostoc spp. and *Streptococcus* spp.	Dextran	α-(1,2)/α-(1,3)/α-(1,4)-branched α-(1,6)-linked homopolymer	Glucose	Used as a food additive (thickener or a gelling agent)
Alphaproteobacteria, Betaproteobacteria, Gammaproteobacteria, and Gram-positive bacteria	Cellulose	β-(1,4)-linked homopolymer	D-Glucose	Used in food processing (nata de coco) and wound dressing
Escherichia coli	K30 antigen	β-(1,2)-linked repeating heteropolymer consisting of tetrasaccharides units	Mannose, galactose, and glucuronate	NA
Sphingomonas spp.	Gellan	Heteropolymer with tetrasaccharides units	Glucose, rhamnose, and glucuronate	Culture media additive, food additive (for example, as a gelling agent) or for encapsulation

significant as a renewable and biodegradable alternative to oil-based plastics. Owing to a high melting point (50–180 °C), it can be used in various applications to replace the oil-based materials, for example, used in the production of the film, fiber, molded products, coating, and adhesives (Andrade 2003).

22.2.10 Polyanhydrides

Polyanhydride is inorganic Polyphosphates found in all the living cells, especially in bacteria, it can form intracellular storage particles and also form a membrane-anchored complex with low molecular weight polyhydroxy butyrate which facilitated the uptake of DNA and irons (Kameda 2001). However, it can affect several factors including bacterial physiology, biofilm formation, and pathogenicity. These biodegradable polyphosphates are used in various industrial applications including flavor enhancing agents. Initially, it was produced by the degradation of rock phosphate, recently it can be produced on a large scale via bacterial polyphosphates, but it is not economically feasible. Most of the polyphosphates are commercially used in water treatment due to the ability of the accumulation of polyphosphate. Polyphosphate kinase (PPK) is the key enzyme that plays an important role in biosynthesis. In addition to biosynthesis, the PPK is also involved in catalyzing the reverse reaction phosphorylating ADP to produce ATP (Zhu et al. 2005). The crystal structure of PPK from *E. coli* has been used to study the reaction mechanism of PPK, the first step of the phosphate synthesis is the autophosphorylation of a conserved histidine residue. Phosphate synthesis is the autophosphorylation of a conserved histidine residue, which probably acts as a nucleophile that attacks the γ-phosphate group of ATPs via the phosphodiester bond. Whereas the enzyme PPK is isolated from *P. aeruginosa* which catalyzes polyphosphate from GTP and prefers GDP to ADP as a phosphate group acceptor (Ishige et al. 2002). This GDP kinase activity extends for the formation of GTP subsequently affects the biosynthesis of the alginate precursor (Rao et al. 2009).

22.3 Biosynthesis of Microbial Biopolymers

Polyhydroxyalkanoates (PHA) is one of the biopolymers produced by microorganisms during stress conditions. Structurally, they are made of (R)-3-hydroxy fatty acid molecules and act as storage molecules that provide potential energy and carbon source during stress conditions. This stored globular nature of the molecules is produced in the form of the phospholipid-protein layer with a polyester core (Lutke-Eversloh et al. 2002; Marang et al. 2018). Among the various biopolymers, PHA plays a key role in industries and is often preferred for large-scale production because of its biocompatibility and eco-friendly nature (Shah et al. 2008). PHA is composed of about 600–35,000 monomeric units of R-hydroxy fatty acids and around 250 PHA molecules have been identified to date, and several studies were ongoing to enhance the physicochemical properties of the polymers

(Zinn and Hany 2005). Recently, genetically modified organisms are created to produce biopolymers with modified functional groups including ethyl, methyl, propyl, and pentyl (Escapa et al. 2011). These modified properties of the polymer may enhance their activity in several applications ranging from biodegradable backing to medical packing and can be used in pharmaceutical applications to treat several diseases including AIDS and cancer (De Roo et al. 2002; Rai et al. 2011).

Most of the biopolymers are replacing the existing synthetic polymers due to their biodegradability and its association with thermoplastic and elastomers. The toxicity of the polymeric compounds is extremely diluted by using agro-industrial debris and the application of PHA in the medical subject area which should be vastly untainted to meet the requirements (Koller 2018; Peptu and Kowalczuk 2018). During the stress conditions, PHA granules are produced in the form of energy granules stored in the cytosol of the organism and act as essential nutrient components in the growth medium and cover vast application in industries than other types of polymers. It is a thermoplastic polyester and substitutes for synthetic polymers. Several types of bacteria including gram-negative and gram-positive, photosynthetic, anaerobic have the capability for the accumulation of intercellular PHA for their energy and carbon source. Generally, it can be produced by fermentation of lipids as linear polyester molecules consist of three repetitive units of hydroxy alkanoic acid with varying lengths of carbon groups and help for the sporulation of various microbial species (Rawte and Mavinkurve 2001). The production of PHA molecules under stress conditions or insufficient nutrient elements like nitrogen, sulfur, oxygen, and magnesium, the PHA molecules act as an energy source. The polymers produced by *E. coli* and *Azotobacter vinelandii* and B. subtilis exist in the microsome and mitochondria in the form of calcium-containing molecules. The maximum production of polymers occurred in the medium when it was supplemented with a sufficient carbon source with rice bran and soy molasses (Aljuraifani et al. 2018). In another approach, the bacterial strains are isolated from the soil and organic debris has been practiced to produce bioplastic. The physicochemical properties including biodegradability and its mechanical properties are assisting PHB as an effective substitute for synthetic petroleum plastics (Jacquel et al. 2008; Kunasundari and Sudesh 2011). These changes are mostly borne away by two central steps to the release and solubilization of polyhydroxyalkanoates and are extensively produced from cellular biomass (Jacquel et al. 2008).

Several solvents including methanol and ethanol are successfully used for the extraction of PHA from its cellular mass. Some of the microbial origins of polymers from *Ralstonia eutropha*, *Alcaligenes latus*, and *Pseudomonas oleovarans* is effectively derived from their cellular mass (Verlinden et al. 2007; Pei et al. 2011). The ability of the microbial production is enhanced by both growth and non-growth mediated PHA components and used in the significant progress of polymeric composites in productions (Mohapatra et al. 2017). Bacillus sp. are effectively used to produce the PHA under environmental conditions, the enzymes present in the Bacillus sp. are significantly suppressing the production of PHA molecules. Besides, most of the bacterial sp. has the capability to synthesis PHA during both log phase and stationary phase (Borah et al. 2002). The production of PHA is associated with microbial biosynthetic pathways and varies according to the species involved in the process of accumulation (Chen 2010). Till date about eight biosynthetic pathways are identified to produce PHA; among them, three are well-studied to produce PHA monomers. The conversion of carbohydrate to glucose or fructose to form homopolymers for the production of PHA and another study, PHA was synthesized via PHA synthase where two thiol groups are involved; one thiol group form bond with loading site and another one bonds with elongation site. The genetically modified organism has generally enhanced production, among them *E. coli* is the standard organism (Nielsen et al. 2017). Some of the *E. coli* strains use lactose as the primary source of carbon precursor for the PHA production. Hence, the economically valuable milk proteins are essentially used as a potent source of carbon and are utilized by modulating PHA production via pha operon. Some of the strains such as DH5, XL1-blue are not able to use lactose as a carbon source; hence, the genetically modified *E. coli* is often used in the production of PHA which uses lactose as a carbon source for large-scale production. The pha operon-modified organisms such as *A. latus* and *Cupriavidus necator* species are efficient organism for PHA production with dry cell mass weight of 119.5 g (Ahn et al. 2000). Another bacterial strain *Microbacterium* sp. WA81 enhances the PHA production in batch fermentation. The adaptation of biodegradable polymers in the plastic industries is increasing due to the negative impacts of synthetic polymers on the environment. Recently, microbial polymers are widely used in polymeric industries and supplemented with inexpensive and resourceful carbon and nitrogen, where the bacteria can successfully renovate the carbon source into water and CO_2 via enzymatic reaction (Farrin 2005; Vroman and Tighzert 2009). There are three enzymatic reactions involved in the chemolithoautotrophic bacteria. The conversion of substrate acetyl coenzyme A into acetoacetyl-Coa is the initial step which occurs via condensation of b ketothiolase (PhbA) enzymes, then these molecules undergo reduction reaction by stereospecific acetoacetyl-CoA reductase (PhbB) to R-(II)-3-hydroxybutyryl-CoA. Subsequently, the reduced constituents are polymerized to form a PHA molecule in the presence of PHB synthase (Santhanam and Sasidharan 2010). The microorganism *R. eutropha, B. megaterium, Pseudomonas* sp. *recombinant E. coli,* and *Enterobactor aerogenes* are used to manage the biopolymers (Fig. 22.1).

22.4 Types of Antimicrobial Groups Incorporated in Polymers

Due to the presence of teichoic acids and phospholipids, both gram-positive and gram-negative bacterial cells have a net charge on the surface of the cell wall (Muñoz-Bonilla and Fernández-García 2012; Timofeeva and Kleshcheva 2011). These negatively charged cell surfaces interact significantly with polymers which bear cationic functional groups and are having potent antimicrobial properties. Among the different functionalities, quaternary ammonium/phosphonium groups are the most explored functionalities with efficient antimicrobial properties used for the synthesis of cationic polymers and this positive charge is vital for the

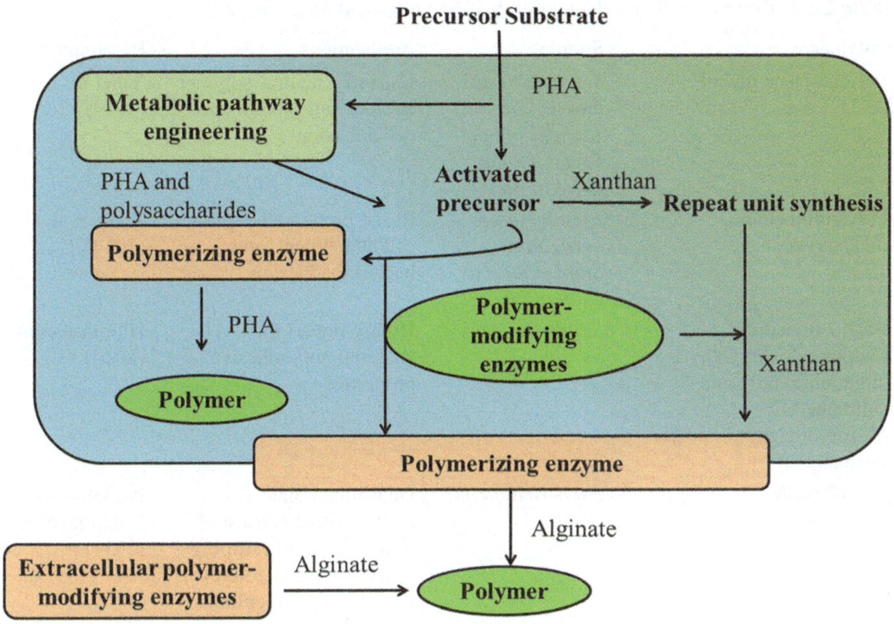

Fig. 22.1 The strategies to produce modified biopolymers with altered properties

interaction of neighboring bacterial cells. The cationic polymers are effectively interacting with bacterial cells and destroying them; hence, they have the potential of preventing their proliferation. Several studies have been reported that using cationic polymers, for example, Munoz-Bonilla and Fernandez-Garcia distinguish the polymers which have aromatic or heterocyclic structures, these structures are mostly acquired by chemical modification with polystyrene or poly(vinylpyridine). However, the polymer containing pyridinium group is extensively used which is obtained by quaternization of poly(4-vinylpyridine) (Tiller et al. 2001; Tiller et al. 2002; Park 2004; Li and Shen 2000).

Aromatic group-containing polymers show excellent antimicrobial activity and have been reported with imidazole derivatives (Anderson and Long 2010). These imidazole groups are the important part of histidine and related compounds hence it has been used in both uncharged and quaternized to form imidazolium salt which shows potential antimicrobial activity (Soykan et al. 2005). Another large group is acrylic and methacrylic antimicrobial polymers that have been employed owing to a large number of monomers such as 2-(dimethylamino)ethyl methacrylate (DMAEMA), which result in the formation of different types of polymers (homo and copolymers) with various structural properties (Gottenbos et al. 2001; Gottenbos et al. 2003). Cationic conjugated polymers are used to differentiate the alteration of single or double bonds within the backbone of each functional group to provide the additional properties of the polymer, especially to enhance their low solubility. Poly (phenylene ethylene) (PPE)-based conjugated polymers is the most extensively used

Table 22.2 Commercially used antimicrobial polymers and their source

Polymer	Source	Applications	Reference
Poly(ethylene glycol)	*Staphylococcus aureus* *Escherichia coli* *Pseudomonas aeruginosa*	Used as a neutral polymer and prevent protein and cell adhesion	Yu et al. (2014)
Poly(sulfobetaine methacrylate)	*Pseudomonas aeruginosa* *Staphylococcus epidermidis*	Resist protein adsorption, cell attachment, and bacterial adhesion	Zhou et al. (2014)
Poly[3-dimethyl (methacryloyloxyethyl) ammonium propane sulfonate-b-2-(diisopropylamino) ethyl methacrylate]	*Staphylococcus aureus*	Highly impart the bacterial anti-adhesive properties	Hippiusa et al. (2014)
Poly(2-methyl-2-oxazoline)	*Escherichia coli*	Dual-functional antimicrobial surface of poly(L-lysine)-graft-poly (2-methyl-2-oxazoline)-quaternary ammonium	Pidhatika and Rakhmatullina (2014)
Polyphenols	*Streptococcus mitis* *Fusobacterium nucleatu* *Porphyromonas gingivalis*	Effectively inhibits the periodontal bacteria	Shahzad et al. (2015)

in different fabricated polyelectrolytes which show effective white light-activated biocides (Corbitt et al. 2009; Xing et al. 2009). Polysiloxanes are another interesting group that show significant antimicrobial activity obtained from quaternary ammonium salt side-chain group have enhanced activity with main-chain flexibility and the contact of the biocide groups than other types of the polymers which tolerates the concentration of active group at the bacterial cell membrane thus improving the efficacy (Sauvet 2003; Mizerska 2009). The polymer has high-density functional groups and nonlinear architecture are known as hyperbranched polymers that are modified with substitution of cationic and hydrophobic groups to optimize the antimicrobial activity (Gao et al. 2007; Pasquier 2007). Though the presence of large polydispersity polymers significantly inhibits the systematic interaction with the cell membrane (Abid et al. 2010). Table 22.2 represents the commercially used antimicrobial polymers and their source.

22.5 Halogen Containing Polymers

The binding of one or more nitrogen atoms in halogen (*N*-halamines) is extensively used as antimicrobial activity due to their number of qualities including more activity against a broad range of microorganisms, enhanced stability, regenerability, and safety to both human and environment. Structurally, halamines have one or more nitrogen atoms that are directly bonded with halogen and the antimicrobial activity via interaction of cellular receptors with positive halogen atoms. Then it undergoes an oxidation reaction, the positively charged halogen atom effectively inhibits the enzymatic reactions in the cell subsequently leading to cell death. *N*-halamine having N-Cl or N-Br moieties that show potent antimicrobial activity against a wide range of microorganisms and the halogen atom of the *N*-halamine are used as disinfecting agents (Sun and Sun 2004). Several studies have been reported that the halamine groups containing polymers are more stable and achieve long-term storage of antimicrobial chlorine (Sun et al. 2001). Recently, the study carried out by Chen and Sun (2006) describes the series of 3-alkyl-5-5-dimethylhydantoin derivatives by reacting with dimethyl hydantoin with different alkyl chain length (C-2- to C-22).

22.6 Factors Affecting the Production and Purification of Biopolymers

Numbers of factors are influencing the production of microbial polymers. The factors like temperature, the composition of culture medium and culture conditions, aeration and agitation are necessary for every bioprocess; among them, rheology of the biopolymers is the most important factor which causes the major impact on various physical properties including heat transfer mechanism mass, oxygen which leads the success of bioprocess. Owing to the secretion of biopolymers and the viscosity of the culture fluid it can dramatically increase during cultivation. Thus, the basic knowledge of rheology is essential for the understanding of the production and for the successful purification of microbial polymers.

22.7 The Pathway Involved in Microbial Polymer

Based on the biological functions the polymers especially polysaccharides are classified into intracellular storage polysaccharides and capsular polysaccharides which are intricately linked to the cell surface and exopolysaccharides that are vital for biofilm formation and pathogenicity. Generally, four different pathways such as Wzx/Wzy-dependent pathway, ATP-binding cassette transporter dependent pathway, synthase-dependent pathway, and extracellular pathway are involved in the production of carbohydrates and polymers with single sucrose as a precursor. Precursor molecules undergo stepwise elongation and are recognized by various enzymatic transformations. In the Wzx/Wzy-dependent pathway, the repeating individuals are linked to a undecaprenol diphosphate anchor and are assembled by

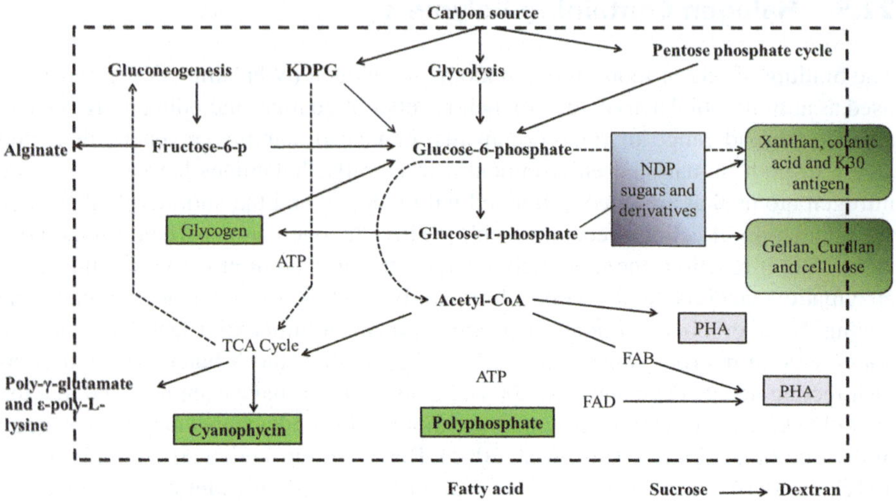

Fig. 22.2 The central metabolic pathway of bacteria in polymer biosynthesis

several glycosyltransferases and located to the cytoplasm via cell membrane by Wzx proteins then it undergoes polymerization at periplasmic space by the Wzy proteins and exported to the cell surface (Cuthbertson et al. 2009; Morona et al. 2009; Islam and Lam 2014). Finally, all the polysaccharides are assembled by the Wzx/Wzy. In the second pathway, exopolysaccharides are synthesized by ABC transporter-dependent pathway, these polysaccharides are linked to the cell surface, are synthesized via ABC transporter-dependent pathways, are assembled by the action of GT at the cytoplasmic face, and are also involved in the assembly and action of GT-containing operon (Whitney and Howell 2013) (Fig. 22.2).

22.8 Application of Biopolymers

22.8.1 Biopolymers for Water Retention

Several studies have reported that the impact of water availability and its optimal condition for the microenvironment reside in the biological soil crusts. Biopolymers especially exopolysaccharides have the capability in sustaining the water for longer periods of time in the soil environment (Potts 2001). Owing to their hygroscopicity these polymers cause a significant impact on hydraulic conductivity of the soil via reducing water infiltration subsequently leads the surface sealing and pore closing (Malam-Issa et al. 2009; Fischer et al. 2010), and also enhance the water adsorption thus providing the spongy nature to the soil (Rossi et al. 2012). In plants, these exopolysaccharides are acting as a water reservoir and are involved in the conduction of water to the root when the soil has a low moisture content (Carminati et al. 2011). Deng et al. (2015) reported that the 2D pore structure for the understanding of the structural architecture also measures the effects of microbial exopolysaccharides on the soil microstructure geometry. The impact of cyanobacterial exopolysaccharides on the soil microenvironment limits the water evaporation thus giving high water potential to the soil and possibly enabling the dehydration tolerance mechanism (Adessi et al. 2018). In another study, Colica et al. (2014) reported the impact of exopolysaccharides on the hydraulic conductivity and moisture-retaining capacity of the physically disrupted cyanobacterial in the soil microenvironment. The results indicated the impact of exopolysaccharides on maintaining the crust structure and water-retaining capacity in the soil and reducing water infiltration (Lan et al. 2010). The polymers produced by Pseudomonas sp. are significantly holding the water content and slowly dry compared to the unamended soil matrix and implying the role of EPS to enhance the survival of the desiccation. Xanthan and dextran derived from *Xanthomonas campestris* and *Leuconostoc mesenteroides* respectively have shown the potent water holding capacity. The glucose diffusion function of the water potential through exopolysaccharides revealed that the diffusion of glucose is slower in both the exopolysaccharides compared to water. Compared to dextran the glucose diffusion is higher in xanthan due to their water holding capacity and the structure of the polysaccharide matrixes (Chenu 1993; Chenu and Roberson 1996).

22.8.2 Biopolymers for Soil Adhesion

Soil has several minerals and is composed of a lot of pores. The rearrangement of these particles highly influences soil fertility, erodibility, and plant germination (Bronick and Lal 2005). In the aggregated soil, it was observed that more water content is retained in interstitial space, whereas the larger inter aggregate pores highly enhance the gas exchange (Or et al. 2007; Morales et al. 2010). Hence, the intermediate water content is highly required for the plant roots and soil communities. Exopolysaccharides play a key role in soil aggregation by absorption of these polymers on the surface of the minerals which form bond between them and increase the adhesion of the soil particles to the plant rooting and stability in the soil (Gajic et al. 2010; Park et al. 2007; Bomfeti et al. 2011). The microorganism like *Rhizobium leguminosarum* and *R. tropici* strains have shown to produce large amounts of rhizobial exopolysaccharides which significantly enhance the soil stability and enhancing the soil aggregation (Kaci et al. 2005; Santaella et al. 2008). The role of rhizobial exopolysaccharides are highly promoting the stabilization and fertility of the soil and enhance the soil aggregation and stability via adhesion of *P. putida* on soil under the stress condition (Kohler et al. 2006).

22.8.3 Role of Biopolymers for Nutrient Accumulation and Vegetative Growth

Exopolysaccharides effectively improved soil fertility and promoted the formation of mineral aggregation (Kögel-Knabner et al. 2008). The physicochemical properties of the soil significantly control the nutrient accumulation of the soil and also it can trap, bind, and concentrate the nutrients and organic materials and provides a major source of carbon for the microbial community in cells (Rodríguez-Caballero et al. 2013). Owing to the nature of aggregation, exopolysaccharides significantly enhance the level of water retention in the soil environment leading to improving plant germination. Microorganisms such as Bacillus, Pseudomonas, Pantoe strains are commonly used to produce exopolysaccharides and have been directly used in the soil for improving the yield of crops (Cipriano et al. 2016). Cipriano et al. (2016) reported the 54 Pseudomonas strains on lettuce growth which increased the plant biomass and increased the impact of the rhizosphere bacterial community. Besides, several studies have reported that the various microbial species including *Burkholderia cepacia*, *Bacillus subtilis*, and Pseudomonas sp. are also improving plant growth and crop yields (Stamenković et al. 2018). The strain of *R. tropici* was used as one of the authorized microorganism for the production of rhizobial inoculant and acts as a model species for the production of exopolysaccharides and is applied to coat seeds to enhance the germination rates with low water content (Larson et al. 2012). Kaci et al. (2005) reported the impact of microbial polymer from Rhizobium strains KYGT207 which shows increased root-adhering properties and increases the root dry mass ratio. The polymers produced from cyanobacterial sp. increase the nutrient uptake and yield in chickpea (Bidyarani et al. 2016). Manjunath et al. (2016) demonstrated the improvement of soil nutrients for the good yield of okra plants. Several other studies also reported the impact of various polymers especially exopolysaccharides from a variety of microorganisms for enhanced soil aggregation, vegetative growth, and protection under stress conditions (Costa et al. 2018).

22.8.4 Biopolymers for Heavy Metal Sorption

Owing to the nature of eco-friendly and safe alternatives, biopolymers are used for immobilization of heavy metals in the soil environment (Sardar et al. 2018). Exopolysaccharides are effectively binding with heavy metal ions and forming hydrophilic bonds with negatively charged polysaccharides and functional groups like carboxyl and hydroxyl groups. These hydrophobic interactions significantly improved the embedding of heavy metals onto the exopolysaccharides (Hou et al. 2013). The factors including ratio of carbohydrate and protein, pH, and ionic strength of the medium are remarkably affecting the organometal complexation process and related to the properties of the heavy metal species (Ding et al. 2018). Causse et al. (2016) reported that the impact of xanthan is adsorbing of Cu and showing remarkable changes in Cu bond length and conformational changes in xanthan structures. Pérez et al. (2008) reported the adsorption of heavy metals such as Pd, Cd, Ni, and Cu by exopolysaccharides synthesized from *P. jamilae* which enhances the binding affinity of the exopolysaccharides. The bacterial strains such as *P. vulgaris* AR6 and AR8 show potent Cr adsorbing capability, whereas the strains of *K. pneumoniae*, *B. firmus,* and mycobacterium are significantly adsorbing Cr ions (Bennett et al. 2013). Several studies have reported the impact of different microbial exopolysaccharides on various heavy metals including cyanobacteria (Freire-Nordi et al. 2005), Pseudomonas sp. (Lau et al. 2005), and M. xanthus (Merroun et al. 2003).

22.8.5 Role of Biopolymers in Soil Stability and Soil Structure

Biopolymers have been demonstrated as potent substitutes for the synthetic oil-based polymer which effectively stabilizes the additives because of their cost-effective nature and very fewer undesirable influences on the environment. The structural difference of exopolysaccharides highly influences the nature of physico-chemical and biological properties and enhances the functionality and cost-effectiveness and biodegradable nature compared to the synthetic polymers, and hence it can be widely used as an effective alternative in various applications (Larson et al. 2012). The adhesive nature of biopolymers play a vital role in the mechanism of protection against biofilm formation; hence, the characteristic feature of exopolysaccharides has been investigated as potential soil engineering agents (Larson et al. 2012). The result reported the evaluation of biopolymers produced by *R. tropici* with more stability to control erosion and suspended solids transport and stabilization of heavy metals in the soil (Larson et al. 2012). Castellane et al. (2015) reported that the microbial polymer from R. tropici strains shows effective soil-stabilizing agents for various industrial applications and reported the impact of biofilms and exopolysaccharides in promoting soil erosion resistance (Gerbersdorf et al. 2008). Other bio-mediated soil improvement techniques including bio grouting and carbonate precipitation are used for the enhancement of strength and stiffness of granular soils (van Paassen et al. 2010). Various microbial polymers from microorganisms especially from bacteria are used in various fields including construction and geotechnical engineering and these microbial polymers significantly increase the interparticle interaction and improving and strengthening soils (Ringelberg et al. 2014). The microbial polymers are widely used for soil strengthening and interact with the charged surface of the soil and form electrostatic bonds by which it strengthens the soil; this effect is primarily dependent on the type of soil, hydration level, and mixing method (Chang et al. 2015). The other microbial polymers such as gellan gum and agar gum are also used for the strengthening of the clayey and sandy soils (Chang et al. 2016). And the results revealed that both the microbial polymers show significant strengthening effects. The gel-type polymers are worked in best with pure soils and evaluate the strengthening potential of the biopolymers (Chang et al. 2016).

22.8.6 Biomedical Applications

Recently biopolymers have a great interest in the biomedical field including tissue engineering pharmaceutical carriers and medical devices (Meinel et al. 2009). Gelatin is common and widely used in medicine for dressing wounds as an adhesive agent. Porous gelatins are used for holding drugs or nutrients to be supplied to the wound for healing (Vlierberghe et al. 2007). PLGA-based scaffolds are extensively used in biomedical engineering and the drug delivery system (Zhao et al. 2016). MWCNT-incorporated electrospun nanofibers show a significant application in many aspects of tissue engineering (Reddy et al. 2015). The cross-linking of the biopolymers highly improves the properties of biomaterials and causes undesirable changes to the polymers. Glutaraldehydes are widely used as a cross-linking agent, but it is more difficult to handle (Othman 2014).

22.8.7 Food Industry

In the food industry, biopolymers are widely used for food coating and packaging materials. Replacing the oil-based packaging biopolymers shows more advantages due to their sustainable and improved technical properties. In the encapsulation process, they provide enhanced product shelf-life and reduce the carbon footprint-related packaging (Mohy Eldin et al. 2012). In the food industry, these biopolymers are used in three different areas such as food coating, food packaging, and encapsulation. The most commercially used microbial polymers are universally used in food packaging and the biodegradable polyester like starch, PLA, and PHA are processed by conventional equipment and are used in several applications. Starch and PKA are the most attractive types of polymers due to their physicochemical property, excellent transparency, and good water resistance. The improvement of their barrier and thermal properties is the major challenge of these molecules and they can perform like polyethylene terephthalate (PET) and other materials such as Zein and chitosan show significant potential as gas barriers. The high rigidity and the difficulty to manage these materials in conventional methods are the primary drawbacks. The hydrophilic nature of the biopolymers significantly affected their use as end products. Renewable polymers are also employed in the food encapsulation process. Previously, it has been described as technology to protect the sensibility of the subject matter and also release of defined methods for wrapping solids and small capsules, which can put out their content under a stressful environment (Champagne and Fustier 2007).

The number of novel technologies is utilized in the pharmaceutical sector with increasing interest in edible films and coatings using polymers. Glycerol, acetylated monoglycerides, and polyethylene glycol are mostly applied to alter the mechanical properties of the film coating and making remarkable changes to the barrier properties of the picture. Though it causes a bunch of advantages, the major advantage of coating can be used as a vehicle for integrating both natural and chemical compounds, including antimicrobial agents, antioxidants, and enzymes like probiotics, minerals, and vitamins. It also can be ingested with food products by which it enhances the safety, nutritional, and sensory attributes (Shit and Shah 2014). Chitosan is one of the widely used antimicrobial polymers significantly used in the process of food packaging and to preserve the food against the broad spectrum of microorganisms. This incorporation of antimicrobial compounds with edible films of biopolymers offers a unique approach to enhance the shelf-life of the food products and safety. Lysozyme is another important and often used antimicrobial enzyme used in food packaging (Chaudhary et al. 2010). Owing to the hydrophilic nature starch has a high sensitivity to relative humidity and significantly reduces the plasticizers and enhances the flexibility of the matrix (Myllärinen et al. 2002).

22.9 Biodegradation of Microbial Polymers

The conversion of organic substances into small fragments with very less molecular weight by certain species of microorganisms is known as biodegradation that can be further converted into carbon dioxide and water. This biodegradation of polymers is highly influenced by the physicochemical properties of the polymer and the efficiency achieved by the microorganism is causally related to the key properties of the molecular weight and crystallinity of the polymers. Exo-enzymes present in the outside of the cell are effectively involved in polymer degradation and have a wide range of reactivity including oxidation and hydrolysis. This action on polymers can be generally described as depolymerization and effectively degrades the polymers into smaller, simple molecules that can take in the microbial cell (Kumar et al. 2011; Bhardwaj et al. 2012).

Several steps are involved in the polymer biodegradation mechanism where the microorganism breaks down the macromolecules into small fragments via biochemical transformation. Biodegradation may cause alteration of the polymer properties including the digestion of microbial enzymes and reduction in molecular weight. These degradable particles are again digested into nontoxic to the environment. Generally, microorganism forms catalytic enzymes for biodegradation via different enzymatic activity and bond cleavage.

There are four basic steps involved in biodegradation; first, it alters the chemical and physical properties, then the materials and the bond present in the polymers are cleaved by enzymatic reaction and the uptake of these molecules by microorganisms in the form of minerals. Both anaerobic and aerobic conditions favor the mineralization of polymers, in the aerobic conditions biodegradation of polymers leads to the formation of carbon dioxides and water while under the anaerobic condition the polymers are degraded into carbon dioxide, methane, and water (Singh and sharma 2008). Starch and cellulose-based plastics are biodegradable and are easily composting via microbial enzymes and can reduce landfilling and solve waste management. The degradation of high molecular weight synthetic polymers is degraded into lower molecular oligomers which can be utilized by microorganisms via biological activity (Sivan 2011; Song et al. 2009; Leja and Lewandowicz 2010;

Kumar et al. 2011). The chemical groups of the polymers have significantly attacked the polycaprolactone and oil-based polymers (Leja and Lewandowicz 2010; Song et al. 2009). Hence, the enzymatic reactions are used to control pollution and develop eco-friendly surroundings.

22.10 Concluding Remarks

The degrading capacity of the microorganism and the investigation of proficient bacteria for the large-scale production and these microorganisms can degrade both organic and inorganic materials and plastic polymers. Hence, to improve the strains to improve the quantity and quality of the polymers are produced, the effort must be made to explore the microorganism to enhance the production of polymers. Owing to increasing environmental issues and limitations it will act as a potential alternative for the sustainable fraction and with increased activity of the general system. Bacteria have remained as an ideal source of tailor-made polymers which provide a space for the production of polymers hence the understating of the polysaccharides or polymers in structure and function could be used for the development of potent microbial polymers with applications in food industries and medicine.

Acknowledgments The authors CS and SKS acknowledge the RUSA-Phase 2.0 grant sanctioned No: F.24-51/2014-U, Policy (TNmulti-Gen), Dept. of Edn, Govt. of India, Dt.09.10.2018.

References

Abid C, Chattopadhyay S, Mazumdar N, Singh H (2010) Synthesis and characterization of quaternary ammonium PEGDA dendritic copolymer networks for water disinfection. J Appl Polym Sci 116:1640–1649. https://doi.org/10.1002/app.31510

Adessi A, de Carvalho RC, De Philippis R, Branquinho C, da Silva JM (2018) Microbial extracellular polymeric substances improve water retention in dryland biological soil crusts. Soil Biol Biochem 116:67–69. https://doi.org/10.1007/s42832-019-0017-7

Ahn WS, Park SJ, Lee SY (2000) Production of poly (3-Hydroxybutyrate) by fed-batch culture of recombinant Escherichia coli with a highly concentrated whey solution. Appl Environ Microbiol 66:3624–3627. https://doi.org/10.1128/aem.66.8.3624-3627.2000

Al-Assaf S, Phillips GO, Aoki H, Sasaki Y (2007) Characterization and properties of Acacia senegal (L.) Willd. var. senegal with enhanced properties (Acacia (sen) SUPER GUM(TM)): part 1—controlled maturation of Acacia senegal var. senegal to increase viscoelasticity, produce a hydrogel form and convert a poor into a good emulsifier. Food Hydrocoll 21:319–328. https://doi.org/10.1016/j.foodhyd.2006.04.011

Aljuraifani AA, Berekaa MM, Ghazwani AA (2018) Bacterial biopolymer (polyhydroxyalkanoate) production from low-cost sustainable sources. Microbiol Open 2018:e755. https://doi.org/10.1002/mbo3.755

Anderson EB, Long TE (2010) Imidazole-and imidazolium-containing polymers for biology and material science applications. Polymer 51:2447–2454

Andrade AL (2003) Plastics and the environment. In: Andrady A (ed) Wiley Interscience, Hoboken, pp 3–76

Ashjaran A, Yazdanshenas ME, Rashidi A, Khajavi R, Rezaee A (2013) Overview of bio nanofabric from bacterial cellulose. J Text Inst 104:121–131. https://doi.org/10.1080/00405000.2012.703796
Aslim B, Nuryuksek D, Beyatli Y (2002) Determination of PHB growth quantities of certain bacillus species isolated from soil. Turk Electron J Biotechnol (Special Issue):24
Babu RP, O'Connor K, Seeram R (2013) Current progress on bio-based polymers and their future trends. Prog Biomater 2:1–16. https://doi.org/10.1186/2194-0517-2-8
Barreras M, Salinas SR, Abdian PL, Kampel MA, Ielpi L (2008) Structure and mechanism of GumK, a membrane-associated glucuronosyltransferase. J Biol Chem 283:25027–25035. https://doi.org/10.1074/jbc.M801227200
Bennett RM, Cordero PRF, Bautista GS, Dedeles GR (2013) Reduction of hexavalent chromium using fungi and bacteria isolated from contaminated soil and water samples. Chem Ecol 29:320–328. https://doi.org/10.1080/02757540.2013.770478
Bhardwaj H, Gupta R, Tiwari A (2012) Microbial population associated with plastic degradation. Sci Rep 1(2):1–4. https://doi.org/10.4172/scientificreports.272
Bidyarani N, Prasanna R, Babu S, Hossain F, Saxena AK (2016) Enhancement of plant growth and yields in chickpea (Cicer arietinum L.) through novel cyanobacterial and biofilmed inoculants. Microbiol Res 188:97–105. https://doi.org/10.1016/j.micres.2016.04.005
Bomfeti CA, Florentino AF, Guimarães AP, Cardoso PG, Guerreiro MC, Moreira FMS (2011) Exopolysaccharides produced by the symbiotic nitrogen-fixing bacteria of leguminosae. Rev Bras Ciênc Solo 35:657–671. https://doi.org/10.1590/S0100-06832011000300001
Borah B, Thakur PS, Nigam JN (2002) The influence of nutritional and environmental conditions on the accumulation of poly-beta-hydroxybutyrate in Bacillus mycoides RLJ B-017. J Appl Microbiol 92(4):776–783
Bronick CJ, Lal R (2005) Soil structure and management: a review. Geoderma 124:3–22. https://doi.org/10.1016/j.geoderma.2004.03.005
Brown SH, Pummill PE (2008) Recombinant production of hyaluronic acid. Curr Pharm Biotechnol 9:239–241. https://doi.org/10.2174/138920108785161488
Carminati A, Schneider CL, Moradi AB, Zarebanadkouki M, Vetterlein D, Vogel HJ, Hildebrandt A, Weller U, Schüler L, Oswald SE (2011) How the rhizosphere may favor water availability to roots. Vadose Zone J 10:988–998. https://doi.org/10.1071/FP13330
Castellane TCL, Persona MR, Campanharo JC, Lemos EGM (2015) Production of exopolysaccharide from rhizobia with potential biotechnological and bioremediation applications. Int J Biol Macromol 74:515–522. https://doi.org/10.1016/j.ijbiomac.2015.01.007
Causse B, Spadini L, Sarret G, Faure A, Travelet C, Madern D, Delolme C (2016) Xanthan exopolysaccharide: Cu2+ complexes affected from the pH-dependent conformational state; Implications for environmentally relevant biopolymers. Environ Sci Technol 50:3477–3485. https://doi.org/10.1021/acs.est.5b03141
Champagne CP, Fustier P (2007) Microencapsulation for the improved delivery of bioactive compounds into foods. Curr Opin Biotechnol 18(2):184–190. https://doi.org/10.1016/j.copbio.2007.03.001
Chang I, Im J, Prasidhi AK, Cho GC (2015) Effects of Xanthan gum biopolymer on soil strengthening. Constr Build Mater 74:65–72. https://doi.org/10.1016/j.conbuildmat.2014.10.026
Chang I, Im J, Cho GC (2016) Geotechnical engineering behaviors of gellan gum biopolymer treated sand. Can Geotech J 53:1658–1670
Chaudhary D, Dong Y, Kar KK (2010) Hydrophilic plasticized biopolymers: morphological influence on physical properties. Mater Lett 64:872–875. https://doi.org/10.1016/j.matlet.2010.01.048
Chen GQ (2009) A microbial polyhydroxyalkanoates (PHA) based bio- and materials industry. Chem Soc Rev 38:2434–2446. https://doi.org/10.1039/B812677C

Chen GQ (2010) Plastics completely synthesized by bacteria: polyhydroxyalkanoates. In: Chen GQ (ed) Plastics from bacteria: natural functions and applications, vol 14. Springer, Berlin/Heidelberg, pp 17–37

Chen Z, Sun Y (2006) N-halamine-based antimicrobial additives for polymers: preparation, characterization and antimicrobial activity. Ind Eng Chem Res 45(8):2634–2640. https://doi.org/10.1021/ie060088a

Chenu C (1993) Clay—or sand—polysaccharides associations as models for the interface between microorganisms and soil: water-related properties and microstructure. Geoderma 1993 (56):143–156. https://doi.org/10.1016/0016-7061(93)90106-U

Chenu C, Roberson EB (1996) Diffusion of glucose in microbial extracellular polysaccharide as affected by water potential. Soil Biol Biochem 28:877–884. https://doi.org/10.1016/0038-0717(96)00070-3

Ciechańska D (2004) Multifunctional bacterial cellulose/chitosan composite materials for medical applications. Fibres Text East Eur 12:48

Cipriano MA, Lupatini M, Lopes-Santos L, da Silva MJ, Roesch LF, Destéfano SA, Freitas SS, Kuramae EE (2016) Lettuce and rhizosphere microbiome responses to growth promoting Pseudomonas species under field conditions. FEMS Microbiol Ecol 92:fiw197

Colica G, Li H, Rossi F, Li D, Liu Y, De Philippis R (2014) Microbial secreted exopolysaccharides affect the hydrological behavior of induced biological soil crusts in desert sandy soils. Soil Biol Biochem 68:62–70. https://doi.org/10.1016/j.soilbio.2013.09.017

Collins RF, Derrick JP (2007) Wza: a new structural paradigm for outer membrane secretory proteins? Trends Microbiol 15:96–100

Corbitt TS, Ji E, Ding L, Ista LK (2009) Light and dark biocidal activity of cationic poly (arylene ethynylene) conjugated polyelectrolytes. Photochem Photobiol Sci 8(7):998–1005. https://doi.org/10.1039/b902646k

Costa OY, Raaijmakers JM, Kuramae EE (2018) Microbial extracellular polymeric substances: ecological function and impact on soil aggregation. Front Microbiol 9:1636. https://doi.org/10.3389/fmicb.2018.01636

Cuthbertson L, Mainprize IL, Naismith JH, Whitfield C (2009) Pivotal roles of the outer membrane polysaccharide export and polysaccharide copolymerase protein families in export of extracellular polysaccharides in Gram-negative bacteria. Microbiol Mol Biol Rev 73:155–177. https://doi.org/10.1128/MMBR.00024-08

De Roo G, Kellerhals MB, Ren Q, Witholt B, Kessler B (2002) Production of chiral R-3-hydroxyalkanoic acids and R-3-hydroxyalkanoic acid methylesters via hydrolytic degradation of polyhydroxyalkanoate synthesized by Pseudomonads. Biotechnol Bioeng 77:717–722. https://doi.org/10.1002/bit.10139

Deng JZ, Orner EP, Chau JF, Anderson EM, Kadilak AL, Rubinstein RL, Bouchillon GM, Goodwin RA, Gage DJ, Shor LM (2015) Synergistic effects of soil microstructure and bacterial EPS on drying rate in emulated soil micromodels. Soil Biol Biochem 83:116–124

Ding P, Song W, Yang Z, Jian J (2018) Influence of Zn(II) stress induction on component variation and sorption performance of extracellular polymeric substances (EPS) from Bacillus vallismortis. Bioprocess Biosyst Eng 41:781–791. https://doi.org/10.1007/s00449-018-1911-6

Escapa IF, Morales V, Martino VP, Pollet E, Averous L, Garcıa JL, Prieto MA (2011) Disruption of b-oxidation pathway in Pseudomonas putida KT2442 to produce new functionalized PHAs with thioester groups. Appl Microbiol Biotechnol 89:1583–1598. https://doi.org/10.1007/s00253-011-3099-4

Farrin J (2005) Biodegradable plastics from natural resources. Institute of Technology, Rochester

Felt O, Buri P, Gurny R (1998) Chitosan: a unique polysaccharide for drug delivery. Drug Dev Ind Pharm 24(11):979–993. https://doi.org/10.3109/03639049809089942

Fischer T, Veste M, Wiehe W, Lange P (2010) Water repellency and pore clogging at early successional stages of microbiotic crusts on inland dunes, Brandenburg, NE Germany. Catena 80:47–52. https://doi.org/10.1016/j.catena.2009.08.009

Flindt C, Alassaf S, Phillips G, Williams P (2005) Studies on acacia exudate gums. Part V. Structural features of Acacia seyal. Food Hydrocoll. 19(4), 687–701. https://doi.org/10.1016/j.foodhyd.2004.09.006

Ford RC, Brunkan-LaMontagne AL, Collins RF, Clarke BR, Harris R, Naismith JH, Whitfield C (2009) Structure-function relationships of the outer membrane translocon Wza investigated by cryo-electron microscopy and mutagenesis. J Struct Biol 166(2):172–182. https://doi.org/10.1016/j.jsb.2009.02.005

Freire-Nordi CS, Vieira AAH, Nakaie CR, Nascimento OR (2005) Effect of polysaccharide capsule of the microalgae Staurastrum iversenii var. americanum on diffusion of charged and uncharged molecules, using EPR technique. Braz J Phys 36:75–82. https://doi.org/10.1590/S0103-97332006000100013

Gajic B, Durovic N, Dugalic G (2010) Composition and stability of soil aggregates in fluvisols under forest, meadows, and 100 years of conventional tillage. J Plant Nutr Soil Sci 2010 (173):502–509. https://doi.org/10.1002/jpln.200700368

Ganduri VSRK, Ghosh S, Patnaik PR (2005) Mixing control as a device to increase PHB production in batch fermentations with co-cultures of lactobacillus Delbrueckii and Ralstonia Eutropha. Process Biochem 40:257

Gao B, Zhang X, Zhu Y (2007) Studies on the preparation and antibacterial properties of quaternized polyethyleneimine. J Biomater Sci Polym Ed 18(5):531–544. https://doi.org/10.1163/156856207780852523

Gerbersdorf SU, Manz W, Paterson DM (2008) The engineering potential of natural benthic bacterial assemblages in terms of the erosion resistance of sediments. FEMS Microbiol Ecol 66:282–294. https://doi.org/10.1111/j.1574-6941.2008.00586.x

Goh CH, Heng PWS, Chan LW (2012) Alginates as a useful natural polymer for microencapsulation and therapeutic applications. Carbohydr Polym 88:1–12. https://doi.org/10.1016/j.carbpol.2011.11.012

Gottenbos B et al (2001) Antimicrobial effects of positively charged surfaces on adhering gram-positive and gram-negative bacteria. J Antimicrob Chemother 48(1):7–13. https://doi.org/10.1093/jac/48.1.7

Gottenbos B, Grijpma DW, van der Mei HC (2003) Positively charged biomaterials exert antimicrobial effects on gram-negative bacilli in rats. Biomaterials 24(16):2707–2710. https://doi.org/10.1093/jac/48.1.7

Gowri S, Almeida L, Amorim T, Carneiro N, Souto AP, Esteves MF (2010) Polymer nanocomposites for multifunctional finishing of textiles—a review. Text Res J 80:1290–1306. https://doi.org/10.1177/0040517509357652

Grage K, Jahns AC, Parlane N, Palanisamy R, Rasiah IA, Atwood JA, Rehm BHA (2009) Bacterial polyhydroxyalkanoate granules: biogenesis, structure, and potential use as nano-/micro-beads in biotechnological and biomedical applications. Biomacromolecules 10:660–669. https://doi.org/10.1021/bm801394s

Hall-Stoodley L, Costerton JW, Stoodley P (2004) Bacterial biofilms: from the natural environment to infectious diseases. Nat Rev Microbiol 2:95–108. https://doi.org/10.1038/nrmicro821

Hassan F, Shah AA, Hameed A, Ahmed S (2007) Synergistic effect of photo and chemical treatment on the rate of biodegradation of low density polyethylene by Fusarium sp. AF4. J Appl Polym Sci 105:1466–1470. https://doi.org/10.1002/app.26328

Hickman JW, Harwood CS (2008) Identification of FleQ from Pseudomonas aeruginosa as a c-di-GMP responsive transcription factor. Mol Microbiol 69:376–389. https://doi.org/10.1111/j.1365-2958.2008.06281.x

Hippiusa C, Bütünb V, Erel-Goktepec I (2014) Bacterial anti-adhesive properties of a monolayer of zwitterionic block copolymer micelles. Mater Sci Eng C 41:354–362. https://doi.org/10.1016/j.actbio.2016.04.033

Hou W, Ma Z, Sun L, Han M, Lu J, Li Z, Mohamad OA, Wei G (2013) Extracellular polymeric substances from copper-tolerance Sinorhizobium meliloti immobilizeCu2+. J Hazard Mat 261:614–620. https://doi.org/10.1016/j.jhazmat.2013.06.043

Iran A (2009) The study on microbial polymers: pullulan and PHB. Iran J Chem Chem Eng 28:13–21

Ishige K, Zhang H, Kornberg A (2002) Polyphosphate kinase (PPK2), a potent, polyphosphate-driven generator of GTP. Proc Natl Acad Sci U S A 99:16684–16688. https://doi.org/10.1073/pnas.262655299

Islam ST, Lam JS (2014) Synthesis of bacterial polysaccharides via the Wzx/Wzy-dependent pathway. Can J Microbiol 60:697–716. https://doi.org/10.1139/cjm-2014-0595

Jacquel N, Lo CW, Wei YH (2008) Isolation and purification of bacterial poly (3-hydroxyalkanoates). Biochem Eng J 39(1):15–27. https://doi.org/10.3144/expresspolymlett.2011.60

Jang JY, Kim MS, Jeong HM, Shin CM (2009) Graphite oxide/poly(methyl methacrylate) nanocomposites prepared by a novel method utilizing macroazoinitiator. Compos Sci Technol 69:186–191. https://doi.org/10.1016/j.compscitech.2008.09.039

Jendrossek D (2009) Polyhydroxyalkanoate granules are complex subcellular organelles (carbonosomes). J Bacteriol 191:3195–3202. https://doi.org/10.1128/JB.01723-08

Jung L, Phyo KH, Kim KC, Park HK, Kim IG (2005) Spontaneous liberation of intracellular polyhydroxy-butyrate granules in Escherchia coli. Res Microbiol 156:865. https://doi.org/10.1111/j.1365-2672.2007.03335.x

Kaci Y, Heyraud A, Barakat M, Heulin T (2005) Isolation and identification of an EPS-producing Rhizobium strain from arid soil (Algeria): characterization of its EPS and the effect of inoculation on wheat rhizosphere soil structure. Res Microbiol 156:522–531. https://doi.org/10.1016/j.resmic.2005.01.012

Kameda A (2001) A novel ATP regeneration system using polyphosphate-AMP phosphotransferase and polyphosphate kinase. J Biosci Bioeng 91:557–5563. https://doi.org/10.1263/jbb.91.557

Katircloglu H, Aslim B, Nuryuksek D, Mercan N, Beyatli Y (2003) Production of poly-B hydroxbutyrate (PHB) and differentiation of putative bacillus mutant strains by SDS-PAGE of total cell protein. Afr J Biotechnol 2(6):147

Kessler B, Witholt B (2001) Factors involved in the regulatory network of polyhydroxyalkanoate metabolism. J Biotechnol 86:97–104. https://doi.org/10.1016/s0168-1656(00)00404-1

Kögel-Knabner I, Guggenberger G, Kleber M, Kandeler E, Kalbitz K, Scheu S, Eusterhues K, Leinweber P (2008) Organo-mineral associations in temperate soils: integrating biology, mineralogy, and organic matter chemistry. J Plant Nutr Soil Sci 171:61–82. https://doi.org/10.1002/jpln.200700048

Kohler J, Caravaca F, Carrasco L, Roldán A (2006) Contribution of Pseudomonas mendocina and Glomus intraradices to aggregate stabilization and promotion of biological fertility in rhizosphere soil of lettuce plants under field conditions. Soil Use Manag 22:298–304. https://doi.org/10.1111/j.1475-2743.2006.00041.x

Koller M (2018) Biodegradable and biocompatible polyhydroxy-alkanoates (PHA): auspicious microbial macromolecules for pharmaceutical and therapeutic applications. Molecules 23 (2):362. https://doi.org/10.3390/molecules23020362

Kumar AA, Karthick K, Arumugam KP (2011) Biodegradable polymers and its applications. Int J Biosci Biochem Bioinform 1(3):173–176

Kunasundari B, Sudesh K (2011) Isolation and recovery of microbial polyhydroxyalkanoates. Express Polym Lett 5(7):620–634. https://doi.org/10.3144/EXPRESSPOLYMLETT.2011.60

Lan SB, Hu CX, Rao BQ, Wu L, Zhang DL, Liu YD (2010) Non-rainfall water sources in the topsoil and their changes during formation of man-made algal crusts at the eastern edge of Qubqi Desert, Inner Mongolia. Sci China Life Sci 53:1135–1141. https://doi.org/10.1007/s11427-010-4049-6

Larson SL, Newman JK, Griggs CS, Beverly M, Nestler C (2012) Biopolymers as an alternative to petroleum-based polymers for soil modification; ERDCTR-12-8, U.S. Army Corps of Engineers Report; Engineer Research and Development Center, Vicksburg

Lau TC, Wu XA, Chua H, Qian PY, Wong PK (2005) Effect of exopolysaccharides on the adsorption of metal ions by Pseudomonas sp. CU-1. Water Sci Technol 52:63–68. https://doi.org/10.2166/wst.2005.0182

Leathers TD (2005) In: Steinbüchel A, Marchessault RH (eds) Biopolymers for medical and pharmaceutical applications. Wiley-VCH, Weinheim, pp 537–560

Leech AJ, Sprinkle A, Wood L, Wozniak DJ, Ohman DE (2008) The NtrC family regulator AlgB, which controls alginate biosynthesis in mucoid Pseudomonas aeruginosa, binds directly to the algD promoter. J Bacteriol 190:581–589. https://doi.org/10.1128/JB.01307-07

Leja K, Lewandowicz G (2010) Polymer biodegradation and biodegradable polymers—a review. Pol J Environ Stud 19(2):255–266

Li G, Shen J (2000) A study of pyridinium-type functional polymers. IV. Behavioral features of the antibacterial activity of insoluble pyridinium-type polymers. J Appl Polym Sci 78:676–684. https://doi.org/10.1002/1097-4628(20001017)78:3<676::aid-app240<3.0.co;2-e

Li H, Chi Z, Wang X, Duan X, Ma L, Gao L (2007) Purification and characterization of extracellular amylase from the marine yeast Aureobasidium pullulans N13d and its raw potato starch digestion. Enzym Microb Technol 40:1006

Lutke-Eversloh T, Fischer A, Remminghorst U, Kawada J, Marchessault RH, Bogershausen A, Kalwei M, Eckert H, Reichelt R, Liu SJ, Steinbuchel A (2002) Biosynthesis of novel thermoplastic polythioesters by engineered E. coli. Nat Mater 1:236–240. https://doi.org/10.1038/nmat773

Malam-Issa O, Defarge C, Trichet J, Valentin C, Rajot JL (2009) Microbiotic soil crusts in the Sahel of Western Niger and their influence on soil porosity and water dynamics. Catena 77:48–55. https://doi.org/10.1016/j.catena.2008.12.013f

Manjunath M, Kanchan A, Ranjan K, Venkatachalam S, Prasanna R, Ramakrishnan B, Hossain F, Nain L, Shivay YS, Rai AB, Singh B (2016) Beneficial cyanobacteria and eubacteria synergistically enhance bioavailability of soil nutrients and yield of okra. Heliyon 2:e00066. https://doi.org/10.1016/j.heliyon.2016.e00066

Marang L, van Loosdrecht MCM, Kleerebezem R (2018) Enrichment of PHA-producing bacteria under continuous substrate supply. Nat Biotechnol 41:55–61. https://doi.org/10.1016/j.nbt.2017.12.001

Meinel AJ, Kubow KE, Klotzsch E, Fuentes M, Smith ML, Vogel V, Merkle HP, Meinel L (2009) Optimization strategies for electrospun silk fibroin tissue engineering scaffolds. Biomaterials 30:3058–3067. https://doi.org/10.1016/j.biomaterials.2009.01.054

Merroun ML, Chekroun K, BArias JM, Gonzalez-Munoz MT (2003) Lanthanum fixation by Myxococcus xanthus: cellular location and extracellular polysaccharide observation. Chemosphere 52:113–120

Mizerska U (2009) Polysiloxane cationic biocides with imidazolium salt (ImS) groups, synthesis and antibacterial properties. Eur Polym J 45(3):779–787. https://doi.org/10.1007/s10904-011-9485-7

Mohapatra S, Maity S, Dash HR, Das S, Pattnaik S, Rath CC, Samantaray D (2017) Bacillus and biopolymer: prospects and challenges. Biochem Biophys Rep 12:206–213. https://doi.org/10.1016/j.bbrep.2017.10.001

Mohy Eldin MS, Soliman EA, Hashem AI, Tamer TM (2012) Biopolymer modifications for biomedical applications. In Theophanides T (ed) Infrared spectroscopy—life and biomedical sciences. InTech. https://doi.org/10.5772/35489

Monteil-Rivera F, Betancourt A, Tra HV, Yezza A, Hawari J (2007) Use of headspace solid phase micro extraction for the quantification of poly(3-hydroxybutyrate) in microbial cells. J Chromatogr A 1154:4

Morales VL, Parlange JY, Steenhuis TS (2010) Are preferential flow paths perpetuated by microbial activity in the soil matrix? A review. J Hydrol 393:29–36. https://doi.org/10.1016/j.jhydrol.2009.12.048

Morona R, Purins L, Tocilj A, Matte A, Cygler M (2009) Sequence-structure relationships in polysaccharide co-polymerase (PCP) proteins. Trends Biochem Sci 34:78–84. https://doi.org/10.1016/j.tibs.2008.11.001

Muñoz-Bonilla A, Fernández-García M (2012) Polymeric materials with antimicrobial activity. Prog Polym Sci 37(2):281–339. https://doi.org/10.1016/j.progpolymsci.2011.08.005

Myllärinen P, Partanen R, Seppälä J, Forsella P (2002) Effect of glycerol on behaviour of amylose and amylopectin films. Carbohydr Polym 50:355–361. https://doi.org/10.1016/S0144-8617(02)00042-5

Nielsen C, Rahman A, Rehman AU, Marie Walsh K, Miller CD (2017) Food waste conversion to microbial Polyhydroxyalkanoates. Microb Biotechnol 10(6):1338–1352. https://doi.org/10.1111/1751-7915.12776

Or D, Smets BF, Wraith JM, Dechesne A, Friedman SP (2007) Physical constraints affecting bacterial habitats and activity in unsaturated porous media—a review. Adv Water Resour 30:1505–1527. https://doi.org/10.1016/j.advwatres.2006.05.025

Othman SH (2014) Bio-nanocomposite materials for food packaging applications: types of biopolymer and nano-sized filler. Agric Agric Sci Proc 2:296–303. https://doi.org/10.1016/j.aaspro.2014.11.042

Palza H (2015) Antimicrobial polymers with metal nanoparticles. Int J Mol Sci 2015 (16):2099–2116. https://doi.org/10.3390/ijms16012099

Park ES (2004) Antibacterial activities of polystyrene-block-poly(4-vinyl pyridine) and poly(styrene-random-4-vinyl pyridine). Eur Polym J 40(12):2819–2822. https://doi.org/10.1016/j.eurpolymj.2004.07.025

Park EJ, Sul WJ, Smucker AJ (2007) Glucose additions to aggregates subjected to drying/wetting cycles promote carbon sequestration and aggregate stability. Soil Biol Biochem 39:2758–2768. https://doi.org/10.1016/j.soilbio.2007.06.007

Pasquier N (2007) From multifunctionalized poly (ethylene imine)s toward antimicrobial coatings. Biomacromolecules 8(9):2874–2882. https://doi.org/10.1021/bm070353r

Pei L, Schmidt M, Wei W (2011) Conversion of biomass into bioplastics and their potential environmental impacts. In: Elnashar M (ed) Biotechnology of biopolymers. InTech, pp 57–74. https://doi.org/10.5772/18042

Peptu C, Kowalczuk M (2018) Biomass-derived polyhydroxy-alkanoates: biomedical applications. In: Popa VI, Volf I (eds) Biomass as renewable raw material to obtain bioproducts of high-tech value. Elsevier, Oxford, pp 271–313

Pérez JA, García-Ribera R, Quesada T, Aguilera M, Ramos-Cormenzana A, Monteoliva-Sánchez M (2008) Biosorption of heavy metals by the exopolysaccharide produced by Paenibacillus jamilae. World J Microbiol Biotechnol 24:2699–2704. https://doi.org/10.1007/s11274-008-9800-9

Pham TH, Webb JS, Rehm BHA (2004) The role of polyhydroxyalkanoate biosynthesis by *Pseudomonas aeruginosa* in rhamnolipid and alginate production as well as stress tolerance and biofilm formation. Microbiology 150:3405–3413. https://doi.org/10.1099/mic.0.27357-0

Pidhatika B, Rakhmatullina E (2014) The synthesis of polymeric dual-functional antimicrobial surface based on poly(2-methyl-2-oxazoline). Indones J Biotechnol 19:12–22

Potts M (2001) Desiccation tolerance: a simple process? Trends Microbiol 9:553–559. https://doi.org/10.1016/s0966-842x(01)02231-4

Raetz CR, Whitfield C (2002) Lipopolysaccharide endotoxins. Annu Rev Biochem 71:635–700. https://doi.org/10.1146/annurev.biochem.71.110601.135414

Rai R, Keshavarz T, Roether JA, Boccaccini AR, Roy I (2011) Medium chain length polyhydroxyalkanoates, promising new biomedical materials for the future. Mater Sci Eng R Rep 72:29–47. https://doi.org/10.1016/j.mser.2010.11.002

Rao NN, Gomez-Garcia MR, Kornberg A (2009) Inorganic polyphosphate: essential for growth and survival. Annu Rev Biochem 78:605–647. https://doi.org/10.1146/annurev.biochem.77.083007.093039

Rawte T, Mavinkurve S (2001) Biodegradable plastics bacterial polyhydroxy alkanoates. Indian J Microbiol 47:233–245

Reddy N, Reddy R, Jiang Q (2015) Crosslinking biopolymers for biomedical applications. Trends Biotechnol 33:362–369. https://doi.org/10.1016/j.tibtech.2015.03.008

Rehm BHA (2009) In: Rehm BHA (ed) Microbiology monographs, alginates: biology and applications. Springer, Berlin, pp 55–72

Remminghorst U, Rehm BHA (2006) Bacterial alginates: from biosynthesis to applications. Biotechnol Lett 28:1701–1712. https://doi.org/10.1007/s10529-006-9156-x

Retegi A, Gobilondo N, Pena G, Zuluaga R, Casro C (2010) Bacterial cellulose films with controlled microstructure-mechanical property relationships. Cellulose 17:661–669. https://doi.org/10.1007/s10570-009-9389-7

Ridley BL, O'Neill MA, Mohnen D (2001) Pectins: structure, biosynthesis and oligogalacturonide related signaling. Phytochemistry 57:929–967. https://doi.org/10.1016/s0031-9422(01)00113-3

Ringelberg DB, Cole DM, Foley KL, Ruidaz-Santiago CM, Reynolds CM (2014) Compressive strength of soils amended with a bacterial succinoglycan: effects of soluble salts and organic matter. Can Geotech J 51:747–757. https://doi.org/10.1021/bk-2019-1323.ch013

Robyt JF, Yoon SH, Mukerjea R (2008) Dextransucrase and the mechanism for dextran biosynthesis. Carbohydr Res 343:3039–3048. https://doi.org/10.1016/j.carres.2008.09.012

Rodríguez-Caballero E, Cantón Y, Chamizo S, Lázaro R, Escudero A (2013) Soil loss and runoff in semiarid ecosystems: a complex interaction between biological soil crusts, microtopography, and hydrological drivers. Ecosystems 16:529–546. https://doi.org/10.1007/s10021-012-9626-z

Rossi F, Potrafka RM, Garcia-Pichel F, De Philippis R (2012) The role of the exopolysaccharides in enhancing hydraulic conductivity of biological soil crusts. Soil Biol Biochem 46:33–40. https://doi.org/10.1016/j.soilbio.2011.10.016

Ruffing A, Chen RR (2006) Metabolic engineering of microbes for oligosaccharide and polysaccharide synthesis. Microb Cell Factories 5:25. https://doi.org/10.1186/1475-2859-5-25

Ryder C, Byrd M, Wozniak DJ (2007) Role of polysaccharides in *Pseudomonas aeruginosa* biofilm development. Curr Opin Microbiol 10:644–648. https://doi.org/10.1016/j.mib.2007.09.010

Sakuragi Y, Kolter R (2007) Quorum-sensing regulation of the biofilm matrix genes (pel) of Pseudomonas aeruginosa. J Bacteriol 189:5383–5386. https://doi.org/10.1128/JB.00137-07

Santaella C, Schue M, Berge O, Heulin T, Achouak W (2008) The exopolysaccharide of Rhizobium sp. YAS34 is not necessary for biofilm formation on Arabidopsis thaliana and Brassica napus roots but contributes to root colonization. Environ Microbiol 10:2150–2163. https://doi.org/10.1111/j.1462-2920.2008.01650.x

Santhanam A, Sasidharan S (2010) Microbial production of polyhydroxyalkanoates (PHA) from Alcaligens spp. and Pseudomonas oleovorans using different carbon sources. Afr J Biotechnol 9:3144–3150

Sardar UR, Bhargavi E, Devi I, Bhunia B, Tiwari ON (2018) Advances in exopolysaccharides based bioremediation of heavy metals in soil and water: a critical review. Carbohydr Polym 199:353–364. https://doi.org/10.1016/j.carbpol.2018.07.037

Sauvet G (2003) Amphiphilic block and statistical siloxane copolymers with antimicrobial activity. J Polym Sci A Polym Chem 41(19):2939–2948. https://doi.org/10.1021/acs.biomac.8b00685

Seidi A, Alemzadeh I (2003) Poly hydroxyl butyrate production. In: 8th chemical engineering congress, Mashad, I.R. IRAN, pp 115–120

Sena RF, Costelli MC, Gibson LH, Coughlin RW (2006) Enhanced production of pullulan by two strains of A. pullulans with different concentrations of soybean oil in sucrose solution in batch fermentations. Braz J Chem Eng 23:507–515. https://doi.org/10.1590/S0104-66322006000400008

Shah AA, Hasan F, Hameed A, Ahmed S (2008) Biological degradation of plastics: a comprehensive review. Biotechnol Adv 26:246–265. https://doi.org/10.1016/j.biotechadv.2007.12.005

Shahzad M, Millhouse E, Culshaw S, Edwards CA, Ramage G, Combet E (2015) Selected dietary (poly)phenols inhibit periodontal pathogen growth and biofilm formation. Food Funct 6:719–729. https://doi.org/10.1039/c4fo01087f

Shih WJ, Chen YH, Shih CJ, Hon MH, Wang MC (2007) Structural and morphological studies on poly(3-hydroxybutyrate acid) (PHB)/chitosan drug releasing microspheres prepared by both single and double emulsion processes. J Alloys Compd 826:434–435. https://doi.org/10.4103/2277-9175.188490

Shirwaikar A, Shirwaikar A, Prabu SL, Kumar GA (2008) Herbal excipients in novel drug delivery systems. Indian J Pharm Sci 70:415–422. https://doi.org/10.4103/0250-474X.44587

Shit SC, Shah PM (2014) Edible polymers: challenges and opportunities. J Polym 4:1–13. https://doi.org/10.1155/2014/427259

Singh B, Sharma N (2008) Mechanistic implications of plastic degradation. Polym Degrad Stab 93:561–584. https://doi.org/10.1016/j.polymdegradstab.2007.11.008

Singh RS, Saini GK, Kennedy JF (2008) Pullulan: microbial sources, production and applications. Carbohydr Polym 73:515. https://doi.org/10.1016/j.carbpol.2008.01.003

Singha TK (2012) Microbial extracellular polymeric substances: production, isolation and applications. IOSR J Pharm 2:276–281

Singla AK, Chawla M (2001) Chitosan: some pharmaceutical and biological aspects—an update. J Pharm Pharmacol 53(8):1047–1067. https://doi.org/10.1211/0022357011776441

Sivan A (2011) New perspectives in plastic biodegradation. Curr Opin Biotechnol 22:422–426. https://doi.org/10.12691/jaem-5-1-2

Song JH, Murphy RJ, Narayan R, Davies GBH (2009) Biodegradable and compostable alternatives to conventional plastics. Philos Trans R Soc Biol 364:2127–2139. https://doi.org/10.1098/rstb.2008.0289

Soykan C, Coşkun R, Delibaş A (2005) Microbial screening of copolymers of N-vinylimidazole with phenacyl methacrylate: synthesis and monomer reactivity ratios. J Macromol Sci A 42 (12):1603–1619. https://doi.org/10.1080/10601320500246693

Stamenković S, Beškoski V, Karabegović I, Lazić M, Nikolić N (2018) Microbial fertilizers: a comprehensive review of current findings and future perspectives. Span J Agric Res 16:09–01. https://doi.org/10.5424/sjar/2018161-12117

Sun Y, Sun G (2004) Novel refreshable N-halamine polymeric biocides: N-chlorination of aromatic polyamides. Ind Eng Chem Res 43(17):5015–5020. https://doi.org/10.1021/ie030846m

Sun Y et al (2001) Novel refreshable N-halamine polymeric biocides containing imidazolidin-4-one derivatives. J Polym Sci A Polym Chem 39(18):3073–3084. https://doi.org/10.1002/pola.1288

Taguchi S, Maehara A, Takase K, Nakahara M, Nakamura H, Doi Y (2001) Analysis of mutational effects of a polyhydroxybutyrate (PHB) polymerase on bacterial PHB accumulation using an in vivo assay system. FEMS Microbiol Lett 198:65. https://doi.org/10.1111/j.1574-6968.2001.tb10620.x

Tian GX, Ji Q, Quan FY, Xia YZ (2013) Influence of calcium ion on thermal degradation and flame retardance behaviour of alginate fiber. Adv Mater Res 631:447–451. https://doi.org/10.4028/www.scientific.net/AMR.631-632.447

Tiller JC, Liao C, Lewis K, Kilbanov (2001) Designing surfaces that kill bacteria on contact. Proc Natl Acad Sci U S A 98(11):5981–5985. https://doi.org/10.1073/pnas.111143098

Tiller JC, Lee SB, Lewis K, Kilbanov AM (2002) Polymer surfaces derivatized with poly(vinyl-N-hexylpyridinium) kill airborne and waterborne bacteria. Biotechnol Bioeng 79(4):465–471. https://doi.org/10.1002/bit.10299

Timofeeva L, Kleshcheva N (2011) Antimicrobial polymers: mechanism of action, factors of activity, and applications. Appl Microbiol Biotechnol 89(3):475–492. https://doi.org/10.1007/s00253-010-2920-9

Tyagi N, Suresh S (2013) Isolation and characterization of cellulose producing bacterial strain from orange pulp. Adv Mater Res 626:475–479. https://doi.org/10.4028/www.scientific.net/AMR.626.475

Valappil SP, Misra SK, Boccaccini AR, Keshavarz T, Bucke C, Roy I (2007) Large-scale production and efficient recovery of PHB with desirable material properties, from the newly characterized Bacillus Cereus SPV. J Biotechnol 132:251. https://doi.org/10.1016/j.jbiotec.2007.03.013. Epub 2007 Apr 22

Valla S, Ertesvag H, Tonouchi N, Fjaervig E (2009) In: Rehm BHA (ed) Microbial production of biopolymers and polymer precursors: applications and perspectives. Caister Academic, Norwich, pp 43–78

van Hijum SA, Kralj S, Ozimek LK, Dijkhuizen L, van Geel-Schutten IG (2006) Structure-function relationships of glucansucrase and fructansucrase enzymes from lactic acid bacteria. Microbiol Mol Biol Rev 70:157–176. https://doi.org/10.1128/MMBR.70.1.157-176.2006

van Paassen LA, Ghose R, van der Linden TJ, van der Star WR, van Loosdrecht MC (2010) Quantifying biomediated ground improvement by ureolysis: large-scale biogrout experiment. J Geotech Geoenviron Eng 136:1721–1728. https://doi.org/10.1007/s13762-018-1821-9

Verlinden RAJ, Hill DJ, Kenward MA (2007) Bacterial synthesis of biodegradable polyhydroxyalkanoates. J Appl Microbiol 102(6):1437–1449. https://doi.org/10.1111/j.1365-2672.2007.03335.x

Vlierberghe SV, Cnudde V, Dubruel P, Masschaele U, Cosijns A, Paepe ID, Jacobs PJS, Hoorebeke LV, Remon JP, Schacht E (2007) Porous gelatin hydrogels: 1. Cryogenic formation and structure analysis. Biomacromolecules 8:331–337. https://doi.org/10.1021/bm060684o

Vroman I, Tighzert L (2009) Biodegradable polymers. Materials 2:307–344. https://doi.org/10.3390/ma2020307

Weber H, Pesavento C, Possling A, Tischendorf G, Hengge R (2006) Cyclic-di-GMP-mediated signalling within the σS network of Escherichia coli. Mol Microbiol 62:1014–1034. https://doi.org/10.1111/j.1365-2958.2006.05440.x

Whitfield C (2006) Biosynthesis and assembly of capsular polysaccharides in Escherichia coli. Annu Rev Biochem 75:39–68. https://doi.org/10.1146/annurev.biochem.75.103004.142545

Whitfield C, Larue K (2008) Stop and go: regulation of chain length in the biosynthesis of bacterial polysaccharides. Nat Struct Mol Biol 15:121–123. https://doi.org/10.1038/nsmb0208-121

Whitney JC, Howell PL (2013). Synthase-dependent exopolysaccharide secretion in Gram-negative bacteria. Trends Microbiol 21:63–72. https://doi.org/10.1016/j.tim.2012.10.001

Xing C, Xu Q, Tang H, Liu L, Wang S (2009) Conjugated polymer/porphyrin complexes for efficient energy transfer and improving light-activated antibacterial activity. J Am Chem Soc 131(36):13117–13124. https://doi.org/10.1021/ja904492x

Yamanaka K, Maruyama C, Takagi H, Hamano Y (2008) ε-poly-l-lysine dispersity is controlled by a highly unusual nonribosomal peptide synthetase. Nat Chem Biol 4:766–772. https://doi.org/10.1038/nchembio.125

Yu K, Mei Y, Hadjesfandiari N, Kizhakkedathu JN (2014) Engineering biomaterials surfaces to modulate the host response. Colloids Surf B Biointerfaces 124:69–79. https://doi.org/10.1016/j.colsurfb.2014.08.009

Zhao W, Li J, Jin K, Liu W, Qiu X, Li C (2016) Fabrication of functional PLGA-based electrospun scaffolds and their applications in biomedical engineering. Mater Sci Eng C 59:1181–1194. https://doi.org/10.1016/j.msec.2015.11.026

Zhou R, Ren PF, Yang HC, Xu ZK (2014) Fabrication of antifouling membrane surface by poly (sulfobetaine methacrylate)/polydopamine co-deposition. J Membr Sci 466:18–25. https://doi.org/10.1016/j.memsci.2014.04.032

Zhu Y, Huang W, Lee SS, Xu W (2005) Crystal structure of a polyphosphate kinase and its implications for polyphosphate synthesis. EMBO Rep 6:681–687. https://doi.org/10.1038/sj.embor.7400448

Zinn M, Hany R (2005) Tailored material properties of polyhydroxyalkanoates through biosynthesis and chemical modification. Adv Eng Mater 7:408–411. https://doi.org/10.1002/adem.200500053

Part IV

Advances in Microbial Polymers

Microbial Biopolymers as an Alternative Construction Binder

23

Gye-Chun Cho, Ilhan Chang, and Jooyoung Im

Abstract

With an increasing need for civil infrastructures owing to increasing populations and growing concerns on climate change, the need for sustainable construction material has become ever more present. Worldwide sustainable development has led to various technologies and materials for various engineering practices. Among such developments, the use of biopolymers in the field of geotechnical engineering has shown great promise. Biopolymers are naturally derived organic polymers that have shown a significant engineering performance. As a new material with a low carbon footprint, the use of biopolymer as a sustainable material has been increasing in recent years. In this chapter, the engineering performances of biopolymers in geotechnical engineering practices, methods of implementation, and limitations and future prospects of this material will be investigated.

Keywords

Biopolymer · Xanthan gum · Gellan gum · Biopolymer-based soil treatment (BPST)

G.-C. Cho (✉) · J. Im
Department of Civil and Environmental Engineering, Korea Advanced Institute of Science and Technology (KAIST), Daejeon, Republic of Korea
e-mail: gyechun@kaist.edu; jooyoungim@kaist.ac.kr

I. Chang
Department of Civil Systems Engineering, Ajou University, Suwon-si, Republic of Korea
e-mail: ilhanchang@ajou.ac.kr

A. Vaishnav, D. K. Choudhary (eds.), *Microbial Polymers*,
https://doi.org/10.1007/978-981-16-0045-6_23

23.1 Introduction

The increasing population and economic growth worldwide have led to the increased development of civil infrastructures. Coupled with the growing concerns for climate change, the need for sustainable construction materials has become ever more present. In the field of geotechnical engineering, there are two major forms of soil improvement, mechanical stabilization (e.g., compaction, vibration, anchors) and chemical stabilization (i.e., cement, asphalt, and gypsum) (Makusa 2012).

Chemical stabilization makes use of particle cementation using chemical processes to enhance the strength characteristics of the soils (Han 2015; Makusa 2012). Several commonly used materials for such purposes are cement, lime, fly ash, and asphalt. However, various environmental concerns have been expressed about the use of such materials, particularly the use of cement, which has been identified as a major contributor to greenhouse gases (Karol 2003). Cement has been the most widely used and accepted method in the field of construction and geotechnical engineering since the 1960s (Chu et al. 2009). However, as the production and use of one ton of cement lead to the emission of approximately one ton of CO_2 (Worrell et al. 2001), it has been found that the cement industry has a large impact on climate change, accounting for approximately 5–8% of global CO_2 emissions (United States Geological Survey (USGS) 2018; Metz et al. 2005). As the problems and side effects of climate change have become clearer, development and research into sustainable alternative construction materials have resulted in new technologies, such as unconventional binders, waste reuse, and biological approaches.

23.1.1 Alternative Binders

In the late twentieth century, research began on chemically synthesized polymers as an alternative method of soil enhancement. One such polymer that has been widely tested in the field of geotechnical engineering is polyacrylamide (PAM) (Malik and Letey 1991). PAM is a linear polymer that is synthesized from acrylamide ($CH_2CHCONH_2$) monomers, and it is classified as a nontoxic, hydrophilic polymer with a negatively charged surface (Kulicke et al. 1982; Malik and Letey 1991; Laird 1997). In the field of geotechnical engineering, PAM has been used for soil strengthening purposes and to help reduce soil erosion (Fox and Bryan 1992; Orts et al. 2007).

Although its polymer form is known to be nontoxic, its monomer, acrylamide, is a toxic substance that is a suspected carcinogen (Tareke et al. 2002). As the production of PAM still results in trace elements of acrylamides (Woodrow et al. 2008), and concerns exist regarding the depolymerization of PAM into acrylamides (Prajapat and Gogate 2016), the use of PAM has been greatly limited.

23.1.2 Waste Reusage

Another form of sustainable development is the reuse of waste materials such as blast furnace slag, fly ash, or demolition wastes as aggregates. Such materials are either used as additives in more conventional binders such as cement, as performance enhancers, or they are used in geopolymerization processes (Abubakar and Baharudin 2012; Bakharev 2005; Oh et al. 2010). The geopolymerization process involves the use of alkali-activated silicon and aluminum ions in either soils or waste products (i.e., blast furnace slag and fly ash) to produce a cementation effect between the soil particles (Duxson et al. 2007; Kim et al. 2006). When these waste materials are subjected to an alkaline solution, the excess oxide ions (O^{2-}), resulting from the reaction between the hydroxyl ions (OH^-) and the free water (H_2O), form monomers with the silicon and aluminum ions found in the waste products, which in turn condense around the soil particles and create a cementation effect (Hench 1998).

However, several factors limit the widespread use of these geopolymerized waste materials. The geopolymerization process generally involves heating (above 60 °C) for sufficient geopolymerization to occur, and as such, this technology is difficult to apply in the field (Rovnaník 2010; van Jaarsveld et al. 2002). Additionally, studies have found that geopolymers are susceptible to water, with a significant decrease in strength when saturated, and are especially sensitive to acidic solutions as the acid results in the depolymerization of the aluminosilicate polymers (Bakharev 2005).

23.1.3 Biological Approaches

23.1.3.1 Microbial Induced Calcite Precipitation

Among biological approaches to soil improvement, the use of microbes for soil cementing (i.e., bio-mineralization) has been widely studied. Microbial induced calcite precipitation (MICP) is the most recognized method of soil treatment, and it makes use of microorganisms that precipitate calcium carbonate, such as *Sporosarcina pasteurii* and *Bacillus pasteurii*, through biological processes to cement the soil particles together (DeJong et al. 2006; Whiffin et al. 2007).

The strengthening mechanism of MICP involves the precipitation of calcium carbonate through urea hydrolysis (Stockes-Fischer et al. 1999). Urea is supplied to ureolytic bacteria that convert urea into ammonium and carbonate, and the carbonate derived from this biological process then reacts with calcium ions supplied to the system to form calcium carbonate precipitates (Fujita et al. 2000). The calcium carbonate precipitates then bind to the soil grains and create a cementitious effect that enhances the strength and stiffness of the soil (Mortensen et al. 2011).

Although this method is a novel approach to ground improvement, studies have shown that its use in the field has several limitations. As the method used in the MICP process involves the use of a living organism, the conditions of the organism's survival greatly affect the performance. Studies have shown that the use of MICP is most applicable to coarse-grained soils, and due to several factors (i.e., limitations to food supply and unsuitable environment for bacterial growth) the use of MICP in fine-grained soils is greatly limited (Rong et al. 2011). Moreover, the use of MICP results in a byproduct of ammonium chloride in the soil, which generally requires soil flushing to remove (Pham et al. 2013).

23.1.3.2 Biopolymers

Biopolymers are defined as organic polymers that are derived from biological processes. They are generally composed of repeated units of monomers bonded together into a larger formation. Biopolymers can be categorized as polynucleotides (i.e., RNA and DNA), polypeptides (i.e., amino acids), and polysaccharides. Among these sub-categories, polysaccharides are the most widely used biopolymer in various industries because of their unique and diverse properties (US National Library of Medicine 2011; Kalia and Averous 2011).

Polysaccharides are polymers made up of carbohydrate monomers with the properties of the polysaccharides, which vary with the overall structure and chemical composition of the polymers. They are widely found in nature because of their roles in key biological processes, such as skeletal structure formation, water binding mechanisms, and reserve substance assimilation (Belitz et al. 2009). Owing to their unique properties, such biopolymers have been used as thickening agents, stabilizers, sweeteners, and gel-forming agents for food, agriculture, cosmetics, medical, and pharmaceutical purposes (Saha and Bhattacharya 2010; Lorenzo et al. 2012; Van de Velde and Kiekens 2002).

For geotechnical purposes, biopolymers, such as natural bitumen, straw, and sticky rice, were used in ancient civilizations. In ancient Chinese civilizations, sticky rice was used as a mortar with river sand for various geotechnical purposes (FuWei et al. 2009). Studies have shown that biopolymers can help stabilize the soil by acting as a cementing binder for soil particles (Akbulut and Cabalar 2014; Chang et al. 2015a; Cho and Chang 2018).

23.2 Common Biopolymers Used in Civil and Construction Engineering Practices

23.2.1 Xanthan Gum

Xanthan gum is a polysaccharide biopolymer that consists of a repeated structure of two glucose, two mannose, and one glucuronic acid (Garcıa-Ochoa et al. 2000; Milas and Rinaudo 1979). It is naturally produced from the microbe *Xanthomonas campestris* and is either in a helix or random coil formation (Butler 2016). When xanthan gum is exposed to water, it absorbs the water to form a viscous solution with a high degree of stability in a wide range of temperature, pH, and electrolyte conditions (Chang et al. 2015a).

Xanthan gum is mostly used for its highly viscose and thickening properties (Garcıa-Ochoa et al. 2000). It has often been used in the food industry because of its high stability with temperature and other food ingredients (Garcıa-Ochoa et al.

2000). It has also been used in industries as a suspending agent for drilling purposes as a mud thickener (Becker et al. 1998).

23.2.2 Gellan Gum

Gellan gum is produced during the fermentation process of the microbe *Sphingomonas elodea* (Giavasis et al. 2000). Its structure consists of a repeated chain of two D-glucose monomers with one D-glucuronic acid and one L-rhamnose residue (Stokke et al. 1993).

Gellan gum has a temperature-dependent structure that restrains the gellan molecules during the heating and cooling of the gel solution to form a double helix, resulting in a stiff hydrogel (i.e., thermogelation) (Chang et al. 2015b). At high temperatures, the gellan molecules show single helix structures, while at lower temperatures, the gellan molecules exhibit a double helix structure (Giavasis et al. 2000). Gellan gum has been used in the food and medical industries (Oliveira et al. 2010; Imeson 1992).

23.2.3 Starch

Starch is one of the most commonly found biopolymers and is mainly derived from organic foodstuffs such as seeds, grains, and roots. Starch is generally composed of two main polysaccharides: amylose and amylopectin. The properties of starch will differ based on the different ratios of amylose to amylopectin, along with the size and structure of the polymers (Mishra and Rai 2006). Different sources of starch have different ratios and structures of amylose to amylopectin.

Starches are widely used in a variety of industries such as food, textiles, cosmetics, and pharmaceuticals (Eliasson 2004). They are mainly used as disintegrants and diluents, strengtheners, and adhesives (Rodrigues and Emeje 2012; Nachtergaele 1989; Vishnuvarthanan and Rajeswari 2013).

23.2.4 Beta-Glucan

Beta-glucan is a biopolymer that mainly consists of D-glucose monomers joined together through glycosidic bonds (Bacic et al. 2009). It is a biopolymer commonly found in items such as cellulose, bran, yeasts, fungi, and bacteria (Misaki et al. 1990; Zhu et al. 2015; Lazaridou and Biliaderis 2007). Beta-glucan is mostly used in the cosmetic industry as a moisturizer (Kogan et al. 2005; Pillai et al. 2005).

23.2.5 Guar Gum

Guar gum is a polysaccharide extracted from the seeds of the plant *Cyamopsis tetragonoloba*. It has a main backbone of beta-D-mannopyronose with 1,4 linkages with side branches of alpha-D-galactose monomers (Risica et al. 2005). It has the significant characteristic of rapid hydration even in cold water and is capable of producing high viscosities even at lower concentrations (Mudgil et al. 2014).

Guar gum has mostly been used in the food industry because of its stabilizing, emulsifying, and thickening properties (Mudgil et al. 2014; Whistler and Hymowitz 1979). It has also been used as a flocculant, foam stabilizer, and aid in filtration and water treatment and as an additive in pharmaceutical applications (Khullar et al. 1998; Krishanaiah et al. 2003; Chourasia and Jain 2004; Krishnaiah et al. 2001).

23.2.6 Casein

Casein is a phosphorous protein biopolymer that is most commonly found in mammalian milk. Casein is a hydrophobic biopolymer found in a suspension of particles called a casein micelle (Ruhsing Pan et al. 1999). The structure of the micelles is held together by calcium phosphate particles that interact directly with the surface of the casein particles (Fox and McSweeney 2003). Due to its hydrophobic properties, casein has been used as an additive in food, industrial paints, adhesives, plastics, and various medical practices (Semenova et al. 2010; Rose 2000).

23.3 Geotechnical Engineering Behaviors of BPST

23.3.1 Microscopic Interaction Between Biopolymers and Soil Particles

23.3.1.1 Biopolymers: Coarse Particles

Biopolymers have a high specific area with electrical charges, which enable them to interact with other biopolymers, ions, and water (Dickinson et al. 2003). In the case of biopolymers in coarse-grained soils, the surface charge along the surface of the coarse grains can be considered negligible for interaction purposes with the biopolymers. Therefore, biopolymers heavily interact with water when saturated, and when dehydrated, they interact with other biopolymers to form biopolymer films (Chang and Cho 2014; Chang et al. 2016).

The SEM images in Fig. 23.1 show how biopolymers create interconnecting biopolymer films between soil particles. The biopolymers are shown to coagulate around the sand particles, encompassing the soils in a very thin layer of a biopolymer film.

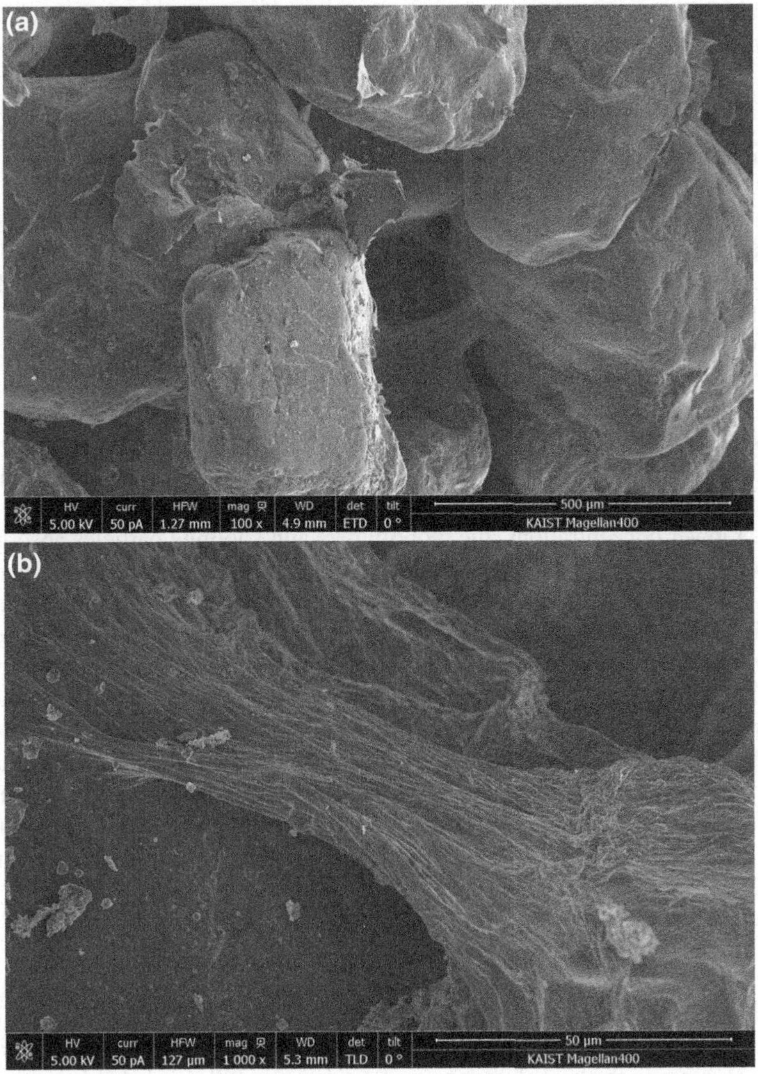

Fig. 23.1 SEM images of 1% gellan gum-treated sands: (**a**) soil-biopolymer mixtures and (**b**) biopolymer film (Chang et al. 2016)

23.3.1.2 Biopolymers: Clay Particles

In the case of biopolymers in clayey soils, as both the biopolymers and clay particles have surface charges, there is a more direct interaction between the biopolymers and clay particles than in the case of the coarse-grained soils. As most biopolymers are negatively charged due to the OH^- groups found in the carbohydrate chains, they are mostly found to interact with the positively charged edges of the clay particles, with some biopolymers forming connections between the kaolinite coagulates (Chang and Cho 2019). This mechanism is shown in Fig. 23.2.

23.3.2 Soil Consistency and Electrical Sensitivity

When we consider the effects of biopolymer-based soil treatment (BPST) on soil consistency and electrical sensitivity, studies have found that, as biopolymers are not electrically neutral, they have varying degrees of effects depending on the soil characteristics and the concentrations of biopolymers present in the soil.

Liquid limit (LL) tests performed with a xanthan gum BPST of kaolinite-sands and bentonite-sands are shown in Fig. 23.3 (Chang et al. 2018b). In Fig. 23.3a it is shown that at lower biopolymer concentrations, the addition of biopolymers increased the LL of the kaolinite, and after a certain concentration was achieved, the LL of the biopolymer kaolinite mixture was reduced. This mechanism is explained by the fact that as biopolymers are hydrophilic and are known to absorb water, at lower concentrations, the biopolymers and kaolinite particles absorb water separately, thereby increasing the water-holding capacity of the soil and increasing the LL. At higher concentrations, due to the closer proximity of the xanthan gum and kaolinite particles, it is expected that there is a facilitation of the interaction between the kaolinite particles and the xanthan gum biopolymers, thereby reducing the LL of the soil. This indicates that, in the case of xanthan gum, the tendency to interact with water molecules is greater than that of kaolinite particles.

In the case of xanthan gum BPST with bentonite soils, Fig. 23.3b shows that with an increase in the biopolymer concentration, the LL of the soil decreased slightly. As the surface charge characteristics of bentonite soils are significantly larger than those of kaolinite, it is believed that xanthan gum directly interacts with the bentonite particles even at lower biopolymer concentrations. The neutralization of the bentonite's surface charge with its interactions with the xanthan gum particles results in a slight decrease in the LL of the soil.

With regard to the soil classification changes with the addition of biopolymers, Fig. 23.4 shows the effects that xanthan gum BPST has on the soil classification based on electrical sensitivity (Jang and Santamarina 2016). The results indicate that with an increase in the biopolymer concentrations, the soils tended to have a reduced electrical sensitivity and increased plasticity (Chang et al. 2018b).

23.3.3 Strengthening Parameters

23.3.3.1 Unconfined Compressive Strength (UCS)

Previously, it was mentioned that biopolymer soil interactions were dependent on the soil type (i.e., coarse-grained or fine-grained soils) due to the surface charge characteristics of the biopolymers and soils. This difference in behavior is also observed when we examine the strengthening capabilities of biopolymers.

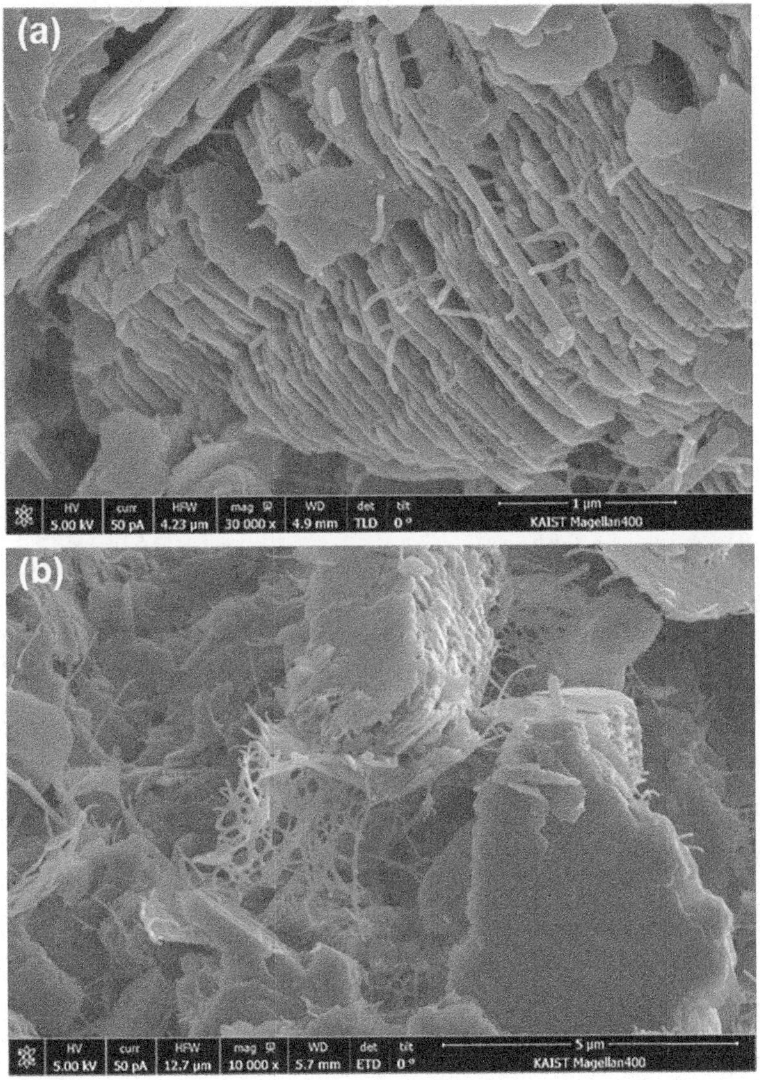

Fig. 23.2 SEM images of 1% gellan gum-treated kaolinite (**a**) biopolymers' interaction on the edges of the kaolinite particles (**b**) formation of biopolymer connections between the kaolinite coagulates (Chang and Cho 2019)

In Fig. 23.5, the various strengthening capabilities of different biopolymers with coarse and fine soils are shown. It is seen that on average, the biopolymer-treated fine soils display a higher strengthening efficiency than that of coarse-grained soils. As the electrical surface charges between the biopolymers and the fine soils allow for a more direct method of interaction, the overall strengthening capabilities are improved. It should also be noted that the biopolymer type and concentration play a large role in the strengthening behavior of the soils.

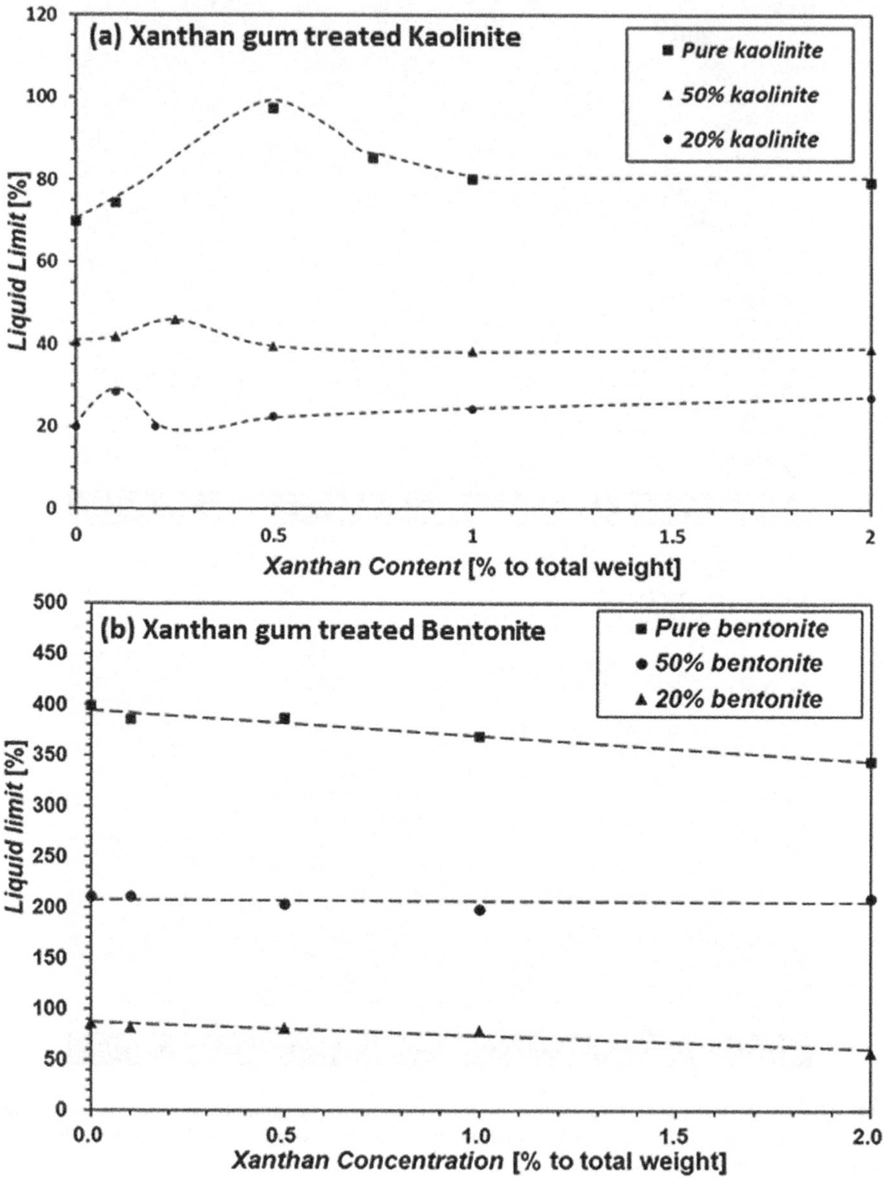

Fig. 23.3 Liquid limit tests of xanthan gum-treated (**a**) kaolinite-sand mixtures and (**b**) bentonite-sand mixtures (Chang et al. 2018b)

Moreover, as biopolymers are known to be hydrophilic materials, they are highly sensitive to the presence of water, and their strengthening capabilities have been shown to be directly related to the water content of the soils (Fig. 23.6) (Chang et al. 2015b).

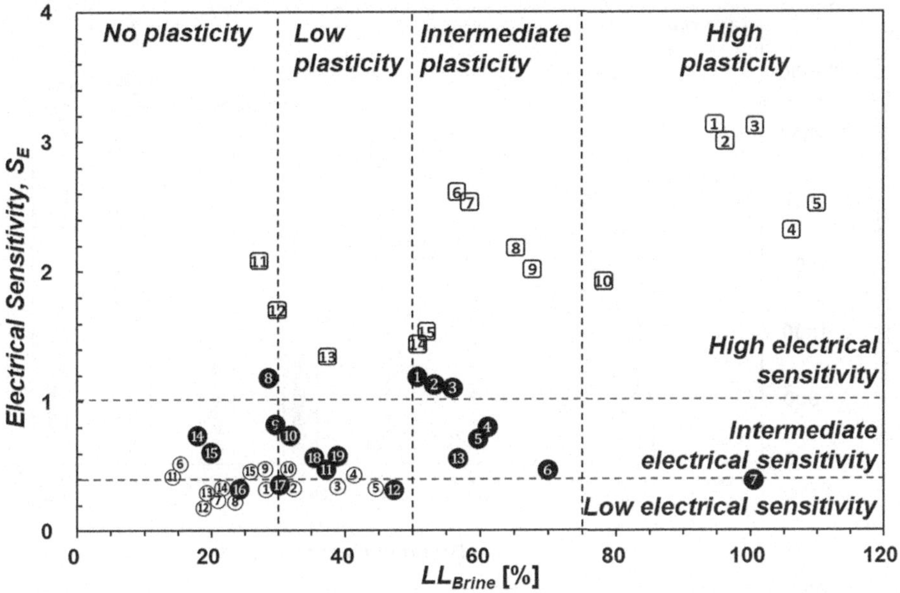

	0	0.1	0.2	0.25	0.5	0.75	1	2	5
Pure Kaolinite	❶	❷			❸	❹	❺	❻	❼
50% Kaolinite	❽	❾		❿	⓫		⓬	⓭	
20% Kaolinite	⓮	⓯	⓰		⓱		⓲	⓳	
Pure Illite	①	②			③		④	⑤	
50% Illite	⑥	⑦			⑧		⑨	⑩	
20% Illite	⑪	⑫			⑬		⑭	⑮	
Pure Bentonite	[1]	[2]			[3]		[4]	[5]	
50% Bentonite	[6]	[7]			[8]		[9]	[10]	
20% Bentonite	[11]	[12]			[13]		[14]	[15]	

Fig. 23.4 Effects of xanthan gum biopolymer treatment on the soil classification based on electrical sensitivity (Chang et al. 2018b)

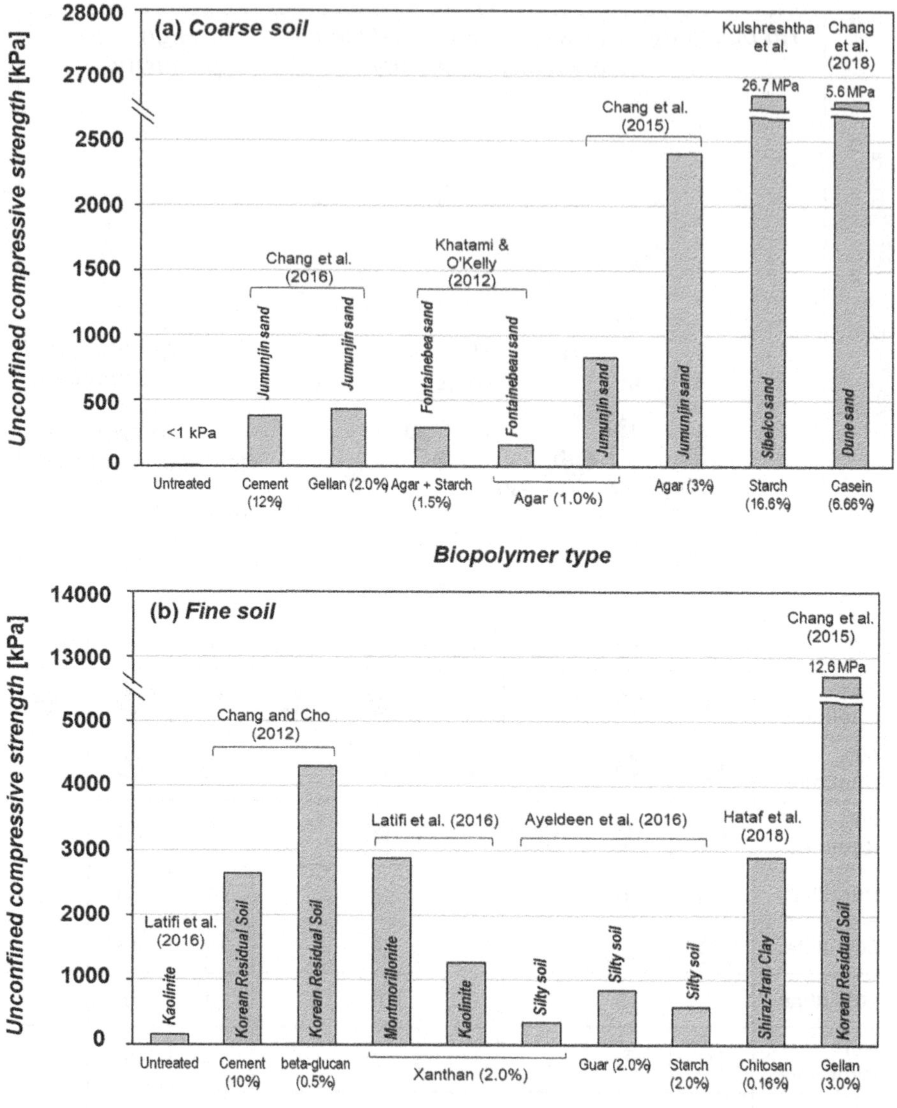

Fig. 23.5 Unconfined compressive strengths of biopolymer-treated (**a**) coarse soils and (**b**) fine soils (Chang et al. 2015b, 2020; Khatami and O'Kelly 2012; Fatehi et al. 2018; Kulshreshtha et al. 2017; Chang and Cho 2012; Ayeldeen et al. 2016; Latifi et al. 2016; Hataf et al. 2018)

23.3.3.2 Interparticle Cohesion

In the case of coarse-grained soils, the BPST encompass the soil particles while forming biopolymer-film connections between them. This mechanism results in an increase in soil cohesion. With fine-grained soils, the charge characteristics of

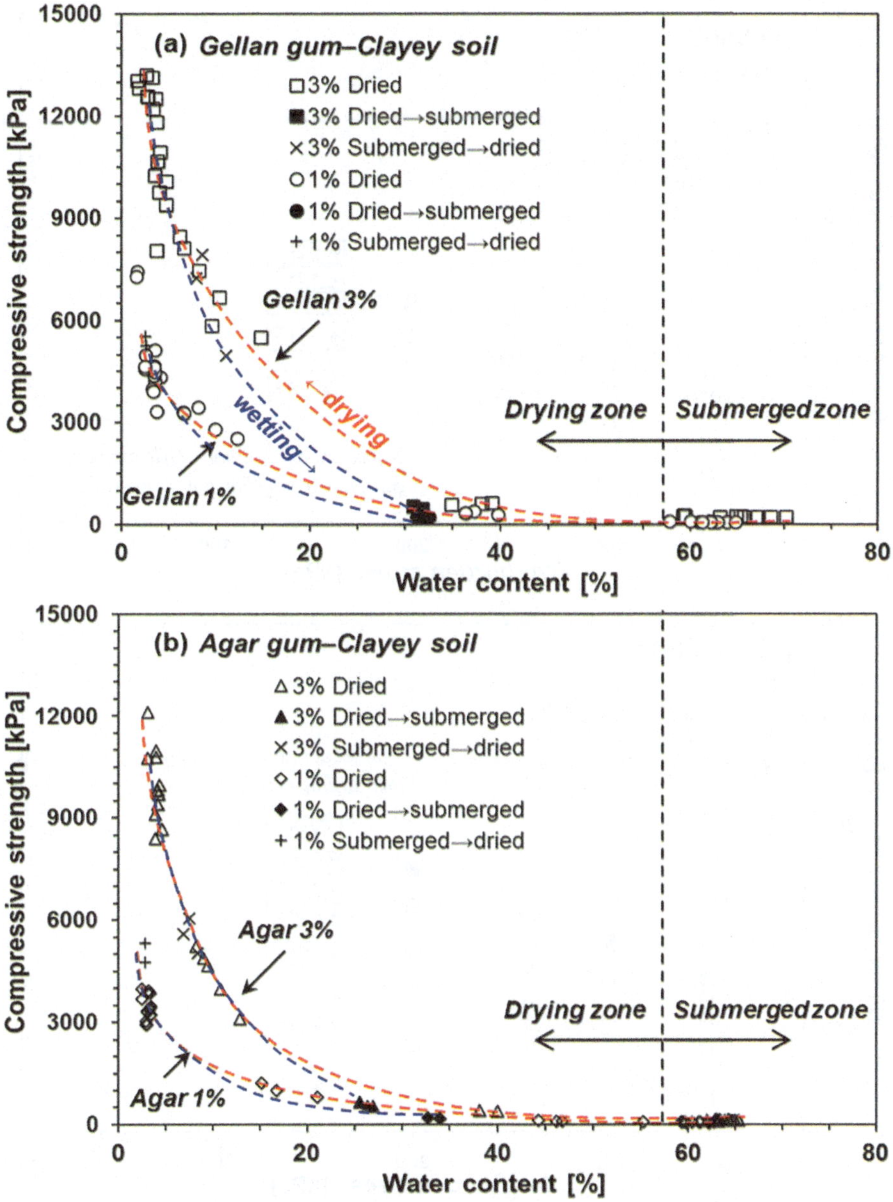

Fig. 23.6 Uniaxial compressive strength variations of (**a**) gellan gum and (**b**) agar gum-treated clayey soils at different water contents (Chang et al. 2015b)

biopolymers interact directly with the surface charge of the clay particles through hydrogen bonding. With this bonding mechanism, the cohesion of the soil is enhanced. Direct shear tests have shown this trend with biopolymer-treated sands

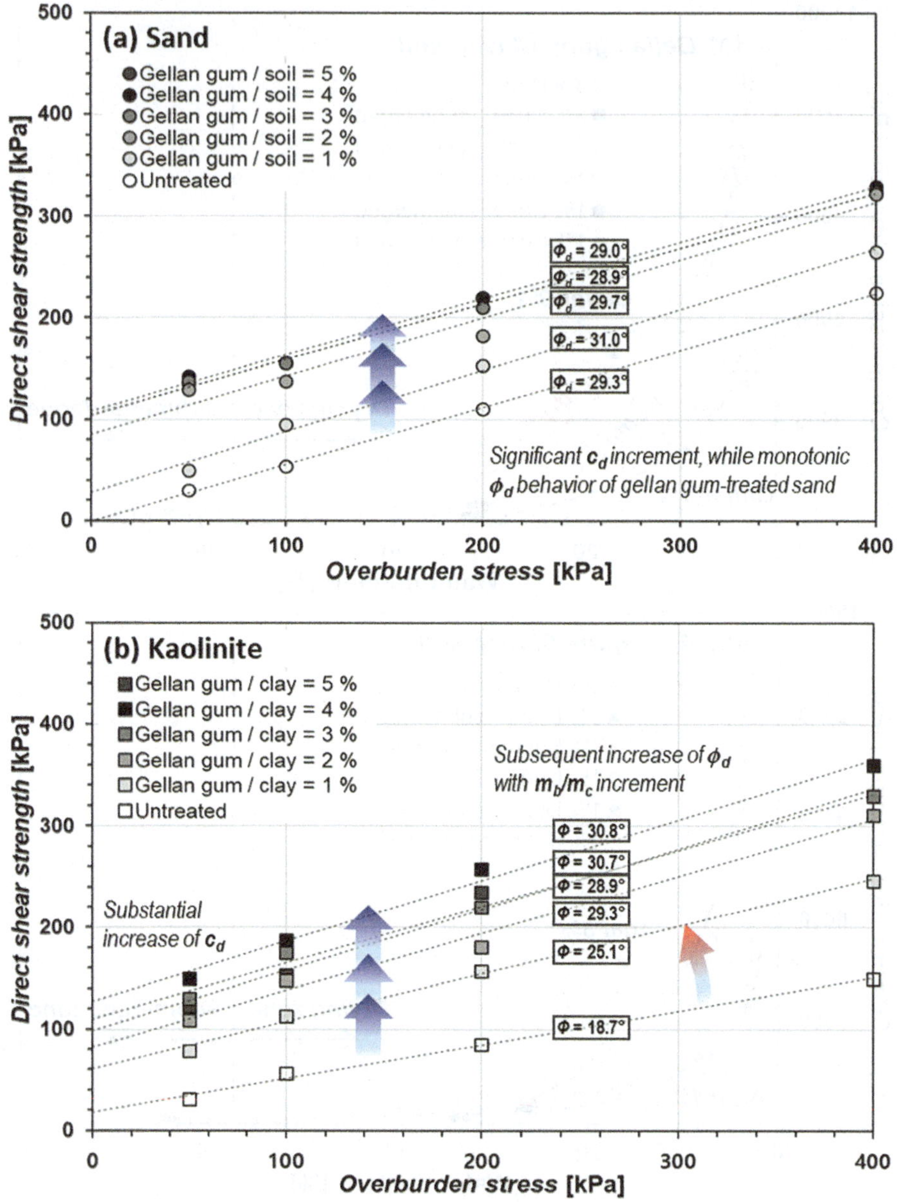

Fig. 23.7 Direct shear results for gellan gum-treated (**a**) sand and (**b**) kaolinite (Chang and Cho 2019)

and kaolinite samples, Fig. 23.7 (Chang and Cho 2019). As shown, there is a definitive increase in the cohesion for both sandy soils and clayey soils. It should be noted that the increase in cohesion can vary greatly depending on the biopolymer used for BPST.

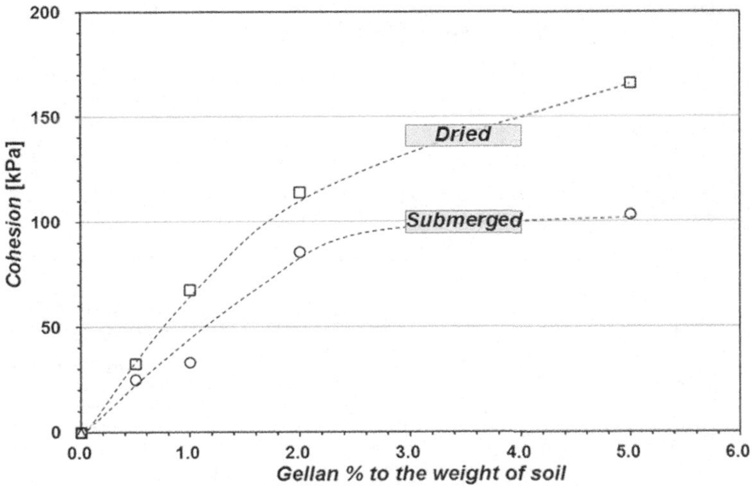

Fig. 23.8 Increase in cohesion with dehydration of gellan gum-treated sands (Chang et al. 2016)

The use of biopolymers in soils has also been known to vary based on the degree of hydration of the biopolymer–soil mixtures. As biopolymers are generally hydrophilic in nature, they tend to easily absorb water, and with the absorption of water, there is a decrease in the strengthening efficiency of BPST. Conversely, with the dehydration of the biopolymers, there is a large increase in the strengthening capabilities of the biopolymer (Chang et al. 2016). This effect is shown in Fig. 23.8, in which the cohesion of the gellan gum-treated sands increases when completely dried.

The overall strengthening capabilities of BPST on soil cohesion shows that the increase in cohesion is directly related to the concentration of the biopolymers used in the soil (Fig. 23.9) (Chang and Cho 2019). The efficiency of the strengthening mechanism will differ based on the soil and biopolymer types used in the soil treatment process.

23.3.3.3 Dilatancy and Interparticle Friction Angle

With regard to the effects of biopolymers on the dilatancy and interparticle friction angle of the soil, the results of direct shear tests shown in Fig. 23.10 show the effects of gellan gum treatment on the sand. With an increase in the biopolymer concentration, the direct shear tests showed an increase in the dilative properties of the soil (Chang and Cho 2019). This behavior was shown to be especially true for soils with clay particles (Fig. 23.11).

Clay soils are known to have low friction angles owing to the shape of their particles. With a platelet-like particle shape, its dilative properties, and thereby its friction angles, are low. However, BPST increases the friction angle of these clay particles (Chang and Cho 2019). It is believed that, owing to the electrical surface charges of both the biopolymers and clay particles, there is a high degree of

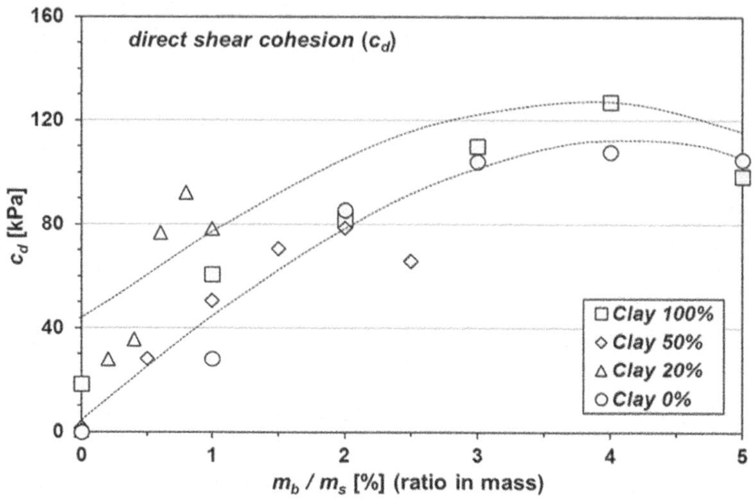

Fig. 23.9 Increase in the cohesion of gellan gum-treated sands from direct shear tests (Chang and Cho 2019)

interaction between them. With a strong connection between the biopolymers and the clay particles, the biopolymers help bridge and connect the clay particles into a bulk conglomerate. These conglomerates then act as large coarse-grained particles exhibiting larger dilatancies and friction angles. This behavior is illustrated in Fig. 23.12.

In the case of coarse-grained soils, BPST does not seem to have a dominant effect on the friction angle. However, with dehydration, the biopolymer connections between the soil particles strengthen enough to overcome the force generated by the dilatancy of the soil. As such, under dry conditions, there is an increase in the friction angle, as shown in Fig. 23.13, as the biopolymer-treated coarse-grained soils form larger conglomerates (Chang et al. 2016).

23.3.4 Hydraulic Conductivity

The use of BPST has also been known to decrease the hydraulic conductivity of soils. The behavior of the biopolymers with regard to the hydraulic conductivity (Fig. 23.14) shows that up to a certain biopolymer concentration, the hydraulic conductivity decreases at an exponential rate and then levels off (Chang et al. 2016). This behavior can be explained by the fact that as the biopolymers take up more and more space within the pores of the soil (e.g., pore-clogging), the hydraulic conductivity decreases rapidly. After a certain concentration of biopolymers is present in the soil, the pore spaces within the soil are completely filled up with the biopolymers, and any additional increase in the biopolymer concentration does not significantly decrease the hydraulic conductivity of the soils.

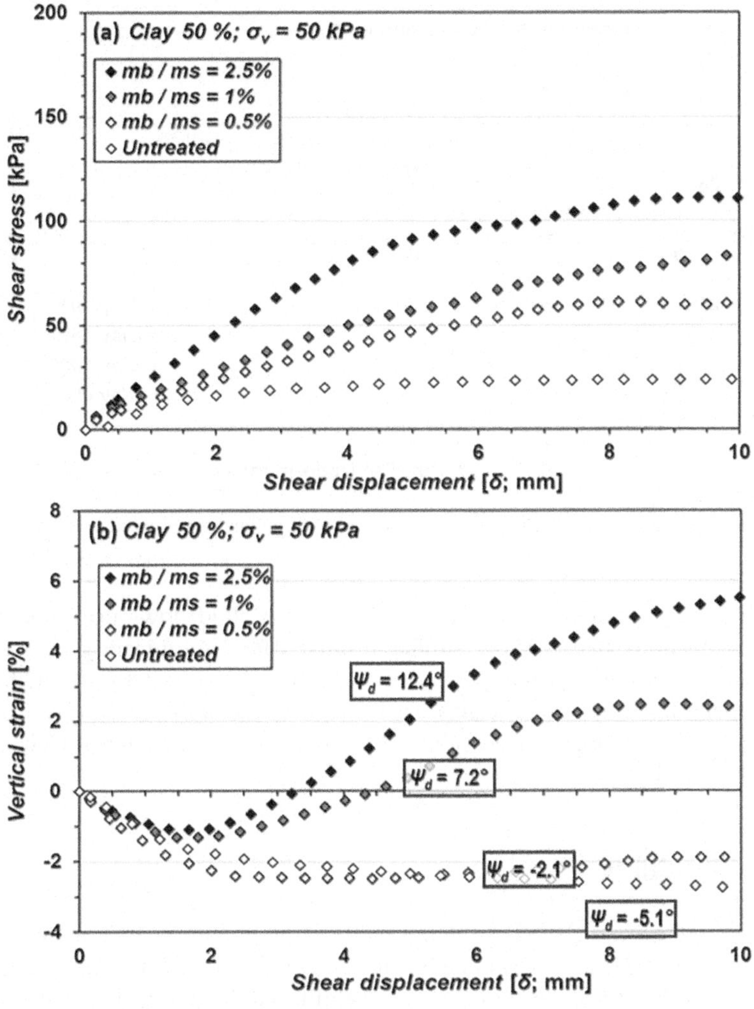

Fig. 23.10 Direct shear behaviors of gellan gum-treated sand (**a**) Stress–strain curves (**b**) Dilatancy (Chang and Cho 2019)

23.3.5 Erosion Behavior

The use of biopolymer has been shown to increase the cohesion of soils and facilitates the formation of larger soil conglomerates, where BPST shows sufficient soil erosion reduction effects. The effects of BPST on sandy soils (Fig. 23.2) show how the addition of biopolymers to the soil reduces the cumulative erosion that the soil exhibits to rainfall to nearly negligible levels (Chang et al. 2015c) (Fig. 23.15).

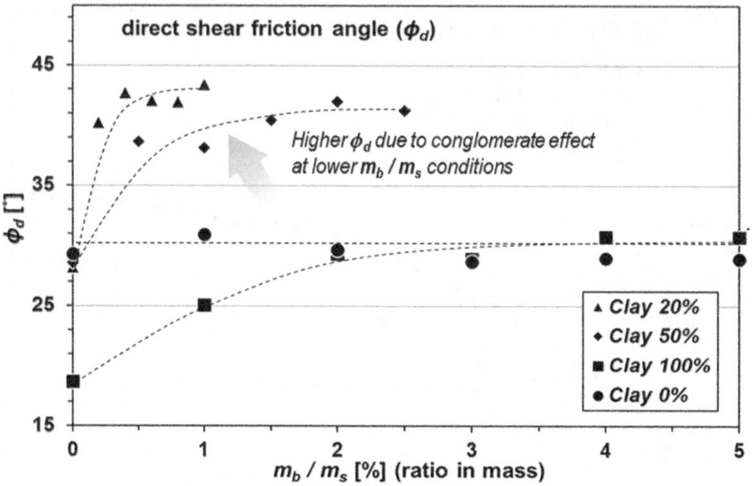

Fig. 23.11 Changes in the friction angle of the soil with gellan gum treatment (Chang and Cho 2019)

Additionally, erosion conditions tested at higher fluid flows using the erosion function apparatus (EFA) show similar results: BPST significantly increases the resistance of soils to erosion (Kwon et al. 2020). As shown by the addition of biopolymers, there is a significant improvement in the soil's erosion resistance to fluid flow, showing signs of erosion at flow rates higher than 1 m/s (Kwon et al. 2020) (Fig. 23.16).

23.3.6 Durability

One major concern for BPST is the durability of biopolymers. As a material that is highly sensitive to water, seasonal changes and repeated wetting and drying cycles may hinder and degrade the engineering capabilities of biopolymers in soils. Experimental tests (Fig. 23.17) showed that with repeated drying and wetting cycles, the strength and stiffness of the biopolymers show a slight degradation with each loading cycle (Chang et al. 2017).

23.3.7 Vegetation Growth

As the majority of biopolymers are carbohydrate polymers, biopolymers are non-toxic to the environment. In addition, due to the high water-holding capability and natural origin of biopolymers, BPST has shown optimistic effects in the promotion of vegetation growth. In Fig. 23.18, we can see the effects of biopolymers on germination rate of oats and the survival efficiency of oat sprouts under severe drought conditions (Tran et al. 2019; Chang et al. 2015c). The results show that

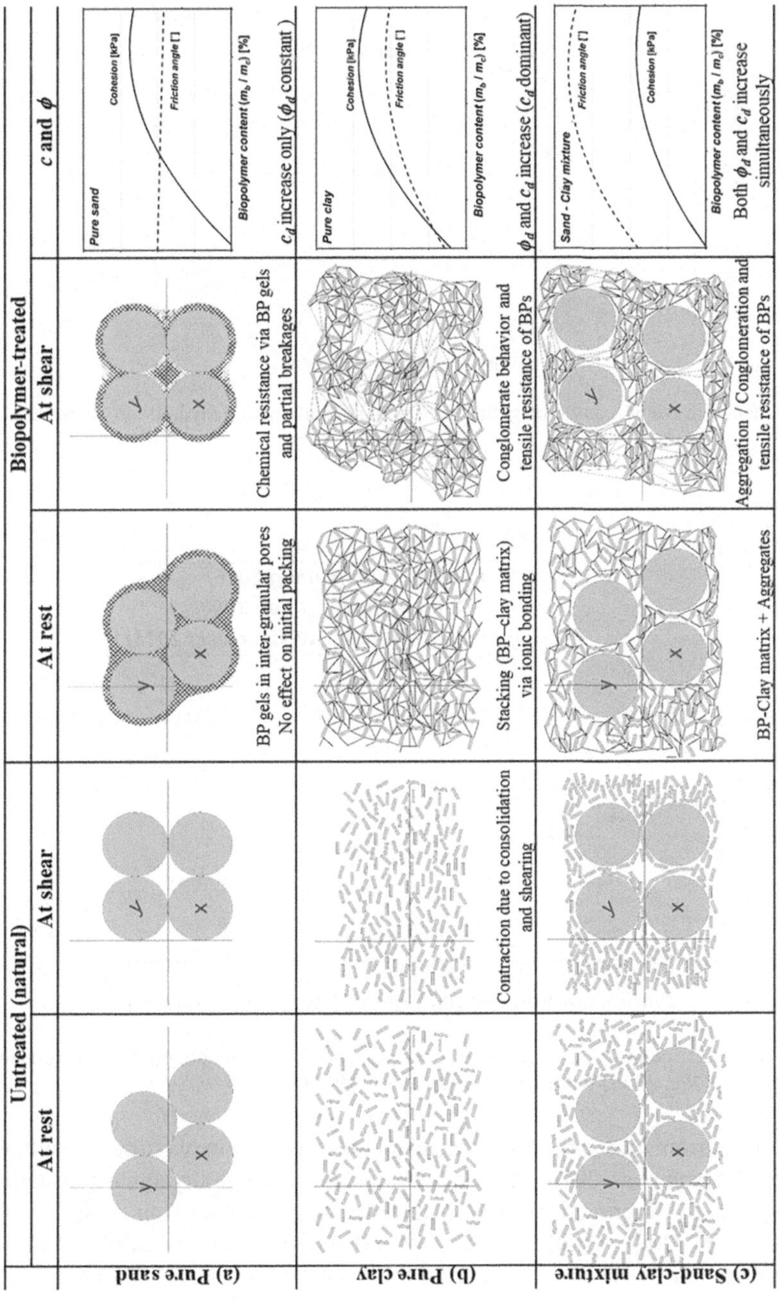

Fig. 23.12 Schematic model of the shearing behavior of gellan gum-treated coarse and fine-grained soils (Chang and Cho 2019)

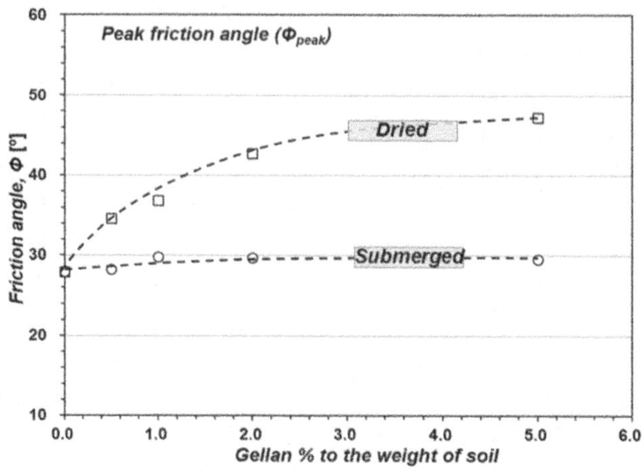

Fig. 23.13 Change in friction angle of gellan gum-treated sands with dehydration (Chang et al. 2016)

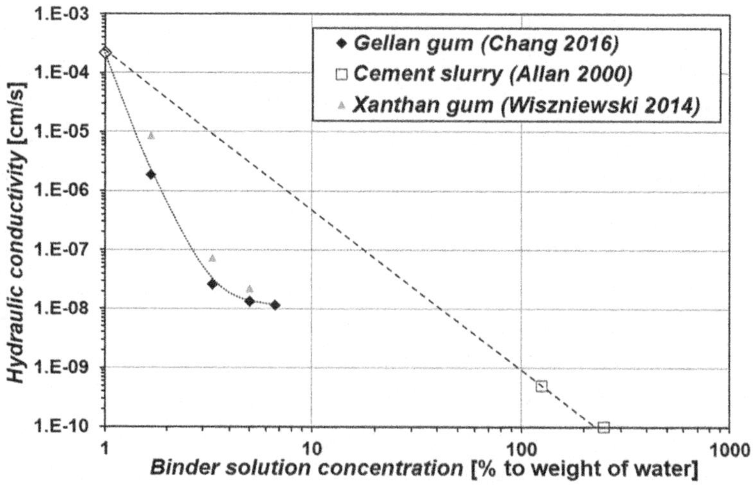

Fig. 23.14 Hydraulic conductivity reduction comparison between biopolymers and cement (Chang et al. 2016; Allan 2000; Wiszniewski and Cabalar 2014)

with BPST, the germination rate of a plant can increase to close to two times than that of the control (Larson et al. 2013). Moreover, the plant survivability against drought conditions is greatly improved by the addition of biopolymers (Tran et al. 2019; Larson et al. 2013). However, it should be noted that a different type of biopolymer and plant species may react differently from the results shown in Fig. 23.18.

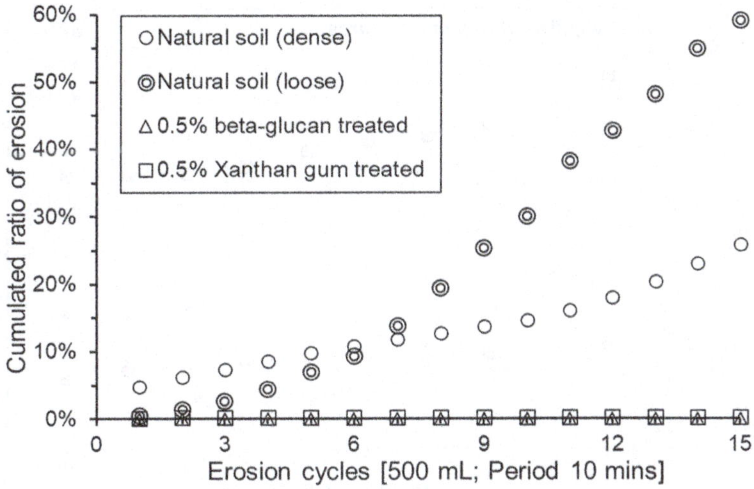

Fig. 23.15 Erosion response of biopolymer-treated and untreated sandy soils to repeated precipitation simulations (Chang et al. 2015c)

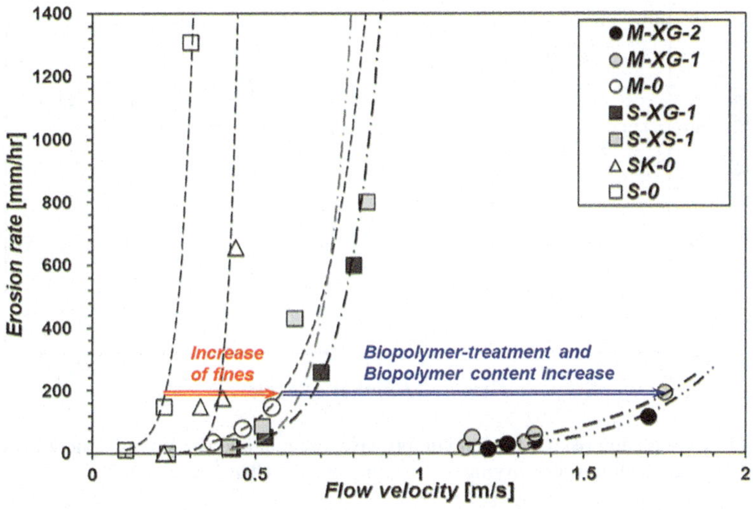

Fig. 23.16 EFA test results for xanthan gum (XG), xanthan-starch (XS), and untreated silica sand (S), silica 90%–kaolinite 10% (SK), and soft marine soil (M) at biopolymer contents (biopolymer to soil ratio in mass) of 0, 1, and 2% (Kwon et al. 2020)

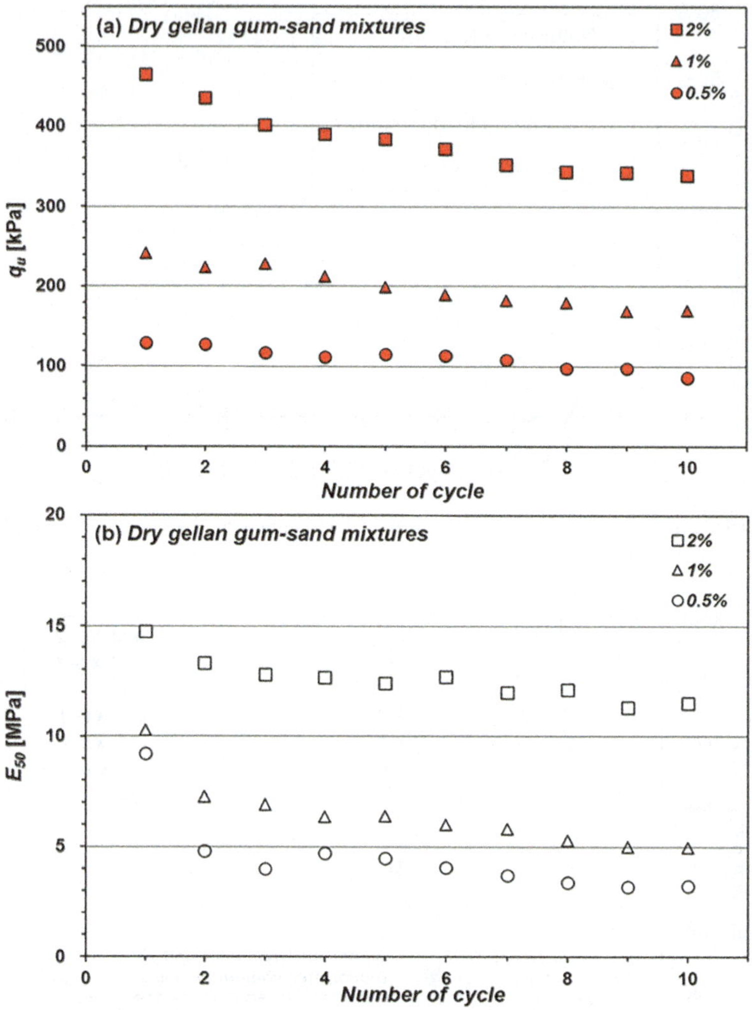

Fig. 23.17 Degradation of (**a**) uniaxial compressive strength and (**b**) Young's modulus of gellan gum-treated sands with repeated drying and wetting cycles (Chang et al. 2017)

23.4 BPST Implementation in Geotechnical Engineering Practices

23.4.1 Implementation Methods

The implementation methods of BPST in the field have not been thoroughly studied. However, the following three methods—spraying, grouting, and direct mixing—are regarded to be the most applicable form for BPST practical implementation.

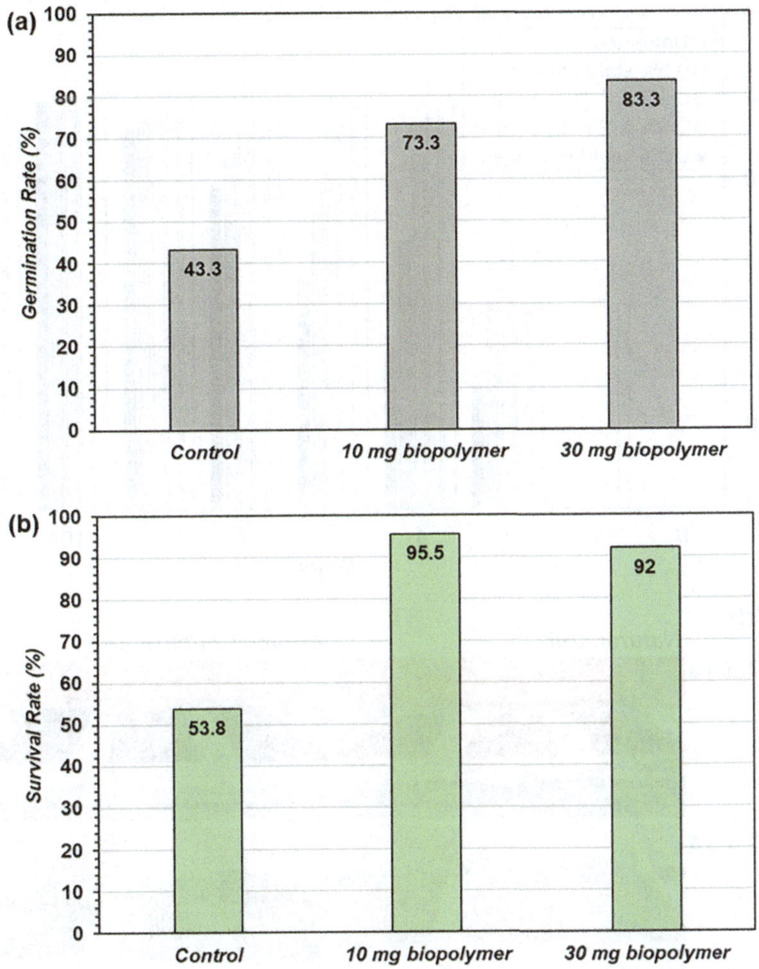

Fig. 23.18 Effects of biopolymers on the (**a**) germination rate, (**b**) survival rate during drought conditions, (**c**) wilting rate during drought conditions, and (**d**) vegetation growth promotion (Larson et al. 2013; Tran et al. 2019; Chang et al. 2015c)

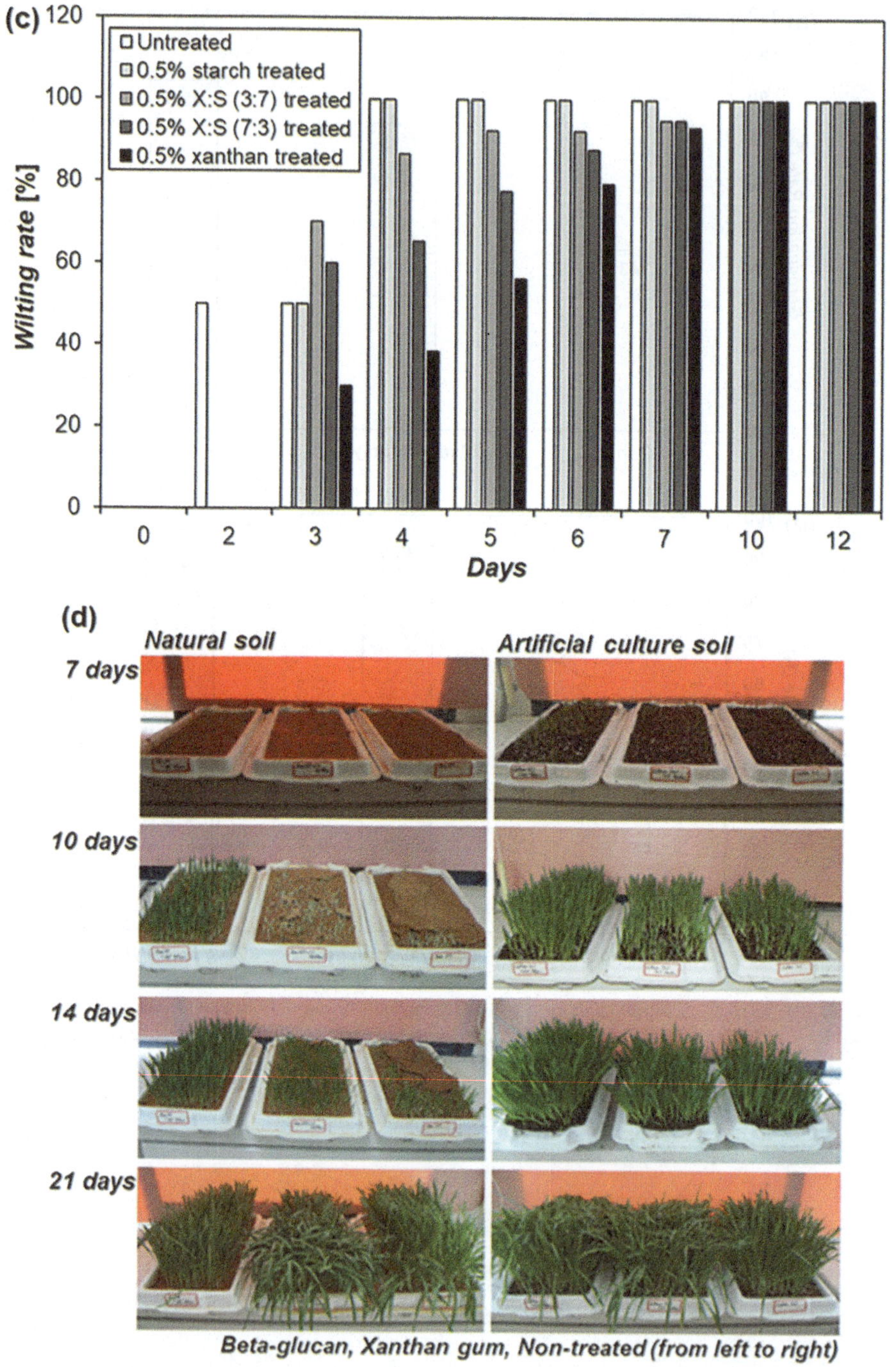

Fig. 23.18 (continued)

Fig. 23.19 Wet spraying method in which a biopolymer solution is mixed directly with the soil before spraying (**a**) In-situ wet mixing system (**b**) Biopolymer-soil mixture spraying

Fig. 23.20 Procedure for biopolymer implementation using the dry spraying method (**a**) Test site overview (**b**) Site soil supply via pneumatic pressure (**c**) Biopolymer solution preparation (**d**) Dry soil and biopolymer solution spraying (**e**) Multiple spraying nozzle (**f**) Completion of BPST on embankment surfaces (Chang et al. 2020)

23.4.1.1 Spraying: Wet and Dry

In the case of spraying, there are two considered methods: (1) wet spraying (Fig. 23.19) and (2) dry spraying (Fig. 23.20). In the case of wet spraying, either the biopolymer solution or a biopolymer solution mixed with soil is directly sprayed on the applicable area.

Wet Spraying

When spraying only the biopolymer solution, a device such as a hydroseeder can be used to apply the solution on the soil, and the infiltration of the biopolymer can be relied on to treat the soil. When the biopolymer solution is mixed with the soil before spraying, a very viscous and fluid mixture is necessary for reliable spraying. Additionally, heavy-duty pumps are required to push the mixture through the spraying nozzle (Fig. 23.19).

Dry Spraying

In the case of dry spraying (Fig. 23.20), a pneumatic compressor is used as the transport mechanism to deliver the soil to the site (Fig. 23.20b). A relatively dry soil is used to prevent clogging in the pipes, and a biopolymer solution is pumped (Fig. 23.20c) into the main pipe at approximately 2–5 m behind the exit nozzle (Fig. 23.20d). The rate of biopolymer pumping must be regulated to achieve the desired mixing concentrations.

23.4.1.2 Injection: Grouting

Biopolymer injection is a possible implementation method that can be used for hydraulic conductivity control purposes. Grouting is a method in which a material is injected into the voids in the ground through the use of fluid pressure. To utilize biopolymers for grouting purposes, a careful method of injection will be necessary, as the use of biopolymers can drastically increase the viscosity of a fluid. Moreover, the initial mixing concentrations of the injected biopolymer will need to be considered as the engineering properties of biopolymer-treated soils will differ based on the mixing concentrations (Fig. 23.21).

23.4.1.3 In Situ Soil Mixing and Compaction

The last possible methods of BPST implementation are (1) direct mixing of the biopolymer into the soil (like deep cement mixing; DCM) or (2) mixing and compacting biopolymer-treated in-situ soil at the desired location (e.g. pavement using site soil) (Fig. 23.22). However, in-situ mixing and compaction method may require specialized equipment for biopolymers as a sufficient kneading of the biopolymer-treated soils maximizes the engineering performance. Additionally, as biopolymer-treated soils react almost immediately and form a sticky viscous mixture, a sufficiently powerful mixer will be required (Fig. 23.22).

23.4.2 Erosion Control

Several erosion control tests using biopolymers have been performed under in situ conditions. In one case study performed by Larson et al. (2013), a biopolymer solution was sprayed over a berm to test the changes in erosion resistance. The biopolymers were applied using a hydroseeder with a single spray, double spray, and double spray at a two-foot depth (Larson et al. 2013). The results of observations over a 6-month period showed that the use of biopolymers sufficiently reduced the erosion of the soil, in particular the surface roughness of the slopes (Fig. 23.23) (Larson et al. 2013).

Fig. 23.21 Schematic diagram of biopolymer grouting methodology

Fig. 23.22 Direct mixing and compaction procedure (**a**) Site overview (**b**) Site cleaning and trimming (**c**) Biopolymer-site soil mixing (**d**) Biopolymer-site soil mixture distribution (**e**) Compaction (**f**) Completion (Chang et al. 2020)

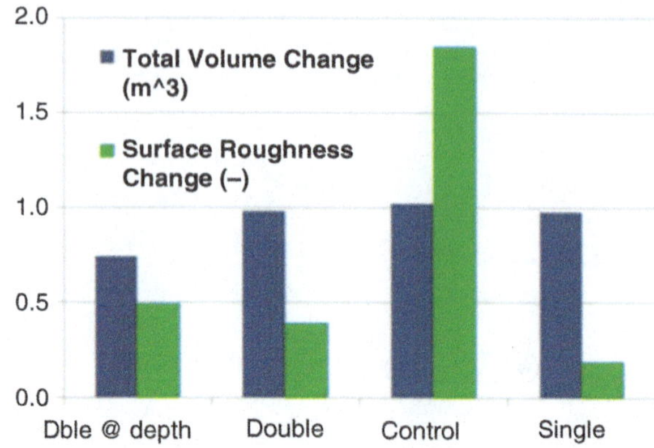

Fig. 23.23 Performance of biopolymer hydroseed spraying on erosion control of a berm (Larson et al. 2013)

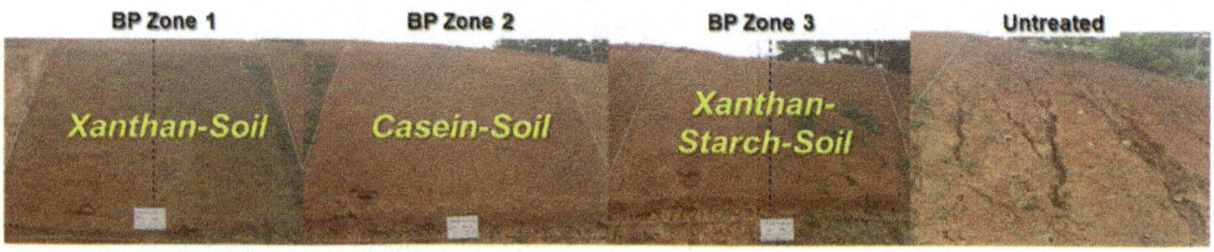

Fig. 23.24 Reduction of surface erosion on target slope after heavy rainfall

In another case study, a dry mixing method (Fig. 23.20) was attempted to assess the erosion resistance of a biopolymer-treated slope (Chang et al. 2020). The target soil of the slope, silty sand, was transported using an air compressor, and a biopolymer solution was mixed with the soil close to the exit nozzle. The surface erosion, shown in Fig. 23.24, was significantly reduced in the cases with biopolymer treatment.

Both biopolymer spraying and dry spraying methods of application have been shown to be viable for the purposes of erosion control.

23.4.3 Grouting Control and Injection

The grouting and injection of biopolymers into the soil can help to immediately reduce the transport mechanisms within the soil due to the high viscosity of the biopolymers. In one study conducted by Khatami and O'Kelly (2018), biopolymers were used for grouting purposes to inhibit the bleeding of particulate grouts such as granulated blast furnace slags (GGBS) (Khatami and O'Kelly 2018). In this case, the grouting material was prepared as a solution and added to a biopolymer solution, as the direct addition of a grouting powder resulted in an uneven mixing of the solution.

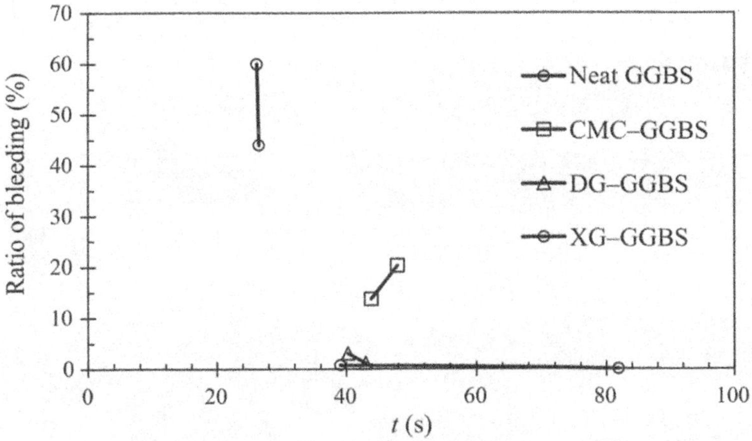

Fig. 23.25 Reduction of GGBS bleeding using carboxymethylcellulose (CMC), diutan gum (DG), and xanthan gum (XG) biopolymers (Khatami and O'Kelly 2018)

The results, shown in Fig. 23.25, showed that the use of biopolymers greatly reduced the effects of bleeding of the grout material (Khatami and O'Kelly 2018).

23.4.4 Vegetation Promotion and Degraded Site Recovery

As biopolymers have shown sufficient vegetation enhancement properties, a site application has been attempted to verifiy the vegetation growth promotion effect of BPST in the field. In a particular study, a biopolymer solution was mixed with cultivated soil and natural sandy soils to test its vegetation enhancement capabilities (Chang et al. 2020). The method used for implementation was the wet spraying method (Fig. 23.19), in which the biopolymer solution was mixed with the target soil before spraying was commenced. The results from this test showed that the use of biopolymers had a large effect on the germination rate of the plants (Chang et al. 2020). Additionally, it was observed that the wet spraying method was an adequate method for BPST implementation for vegetation growth promotion (Fig. 23.26).

23.5 Future Prospects of BPST

23.5.1 Economic Feasibility

The price per unit of biopolymers is significantly higher than that of conventional materials, such as cement. However, biopolymers are used at significantly lower concentrations than are conventional binders. If we compare the prices of one of the more commercially available biopolymers, xanthan gum, with cement (Table 23.1), the prices per unit are 2300 USD/ton and 123.5 USD/ton for xanthan gum and cement, respectively. If we consider that xanthan gum biopolymer can be used at

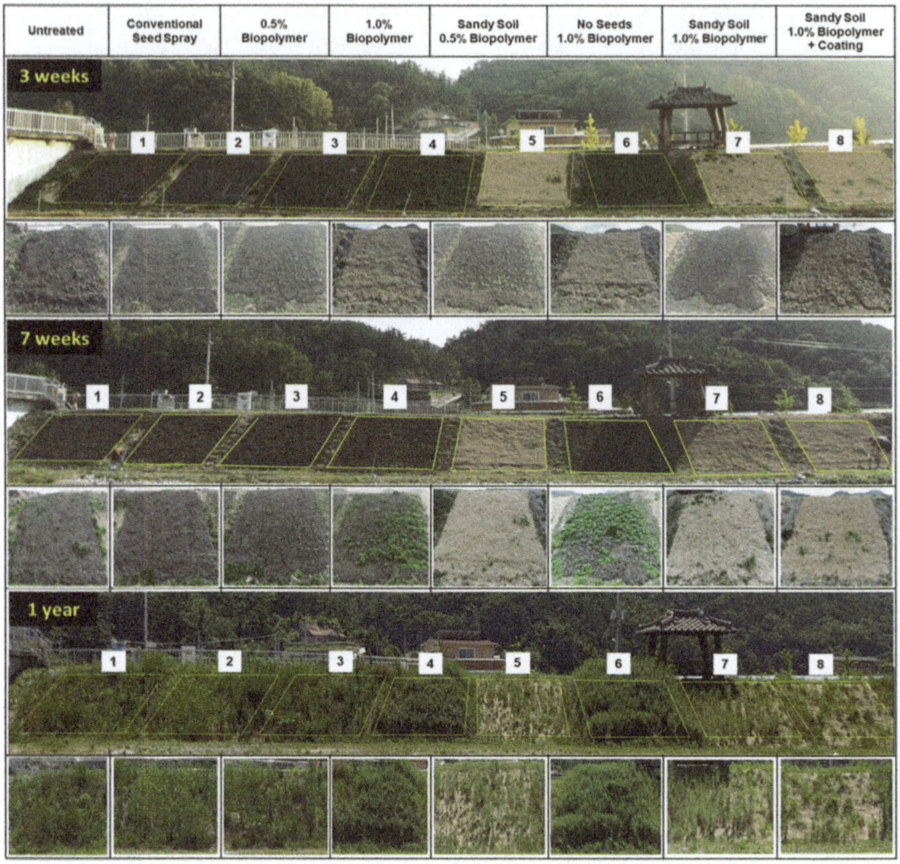

Fig. 23.26 Biopolymer-treated soils for vegetation growth (Chang et al. 2020)

Table 23.1 Economic feasibility of xanthan gum and beta-glucan BPST in comparison with cement-based ground improvement

	Cement	Xanthan gum	Beta-glucan
Market price of material per ton	123.5 USD[a]	2300 USD[b]	50,000 USD[b]
Required amount for 1 ton of soil treatment	100 kg (10% to soil mass)	5 kg (0.5% to soil mass)	5 kg (0.5% to soil mass)
Material price for 1 ton of soil treatment	12.34 USD	11.5 USD	250 USD

[a]Price of cement in the United States, in 2019 (www.statista.com)
[b]Average price found in 2020 (www.alibaba.com)

concentrations of 0.5% to the soil mass, while the typical cement treatment uses approximately 10% of cement to the soil mass, the prices per unit of treated soil are 11.5 USD/ton and 12.35 USD/ton of soil treated using xanthan gum and cement,

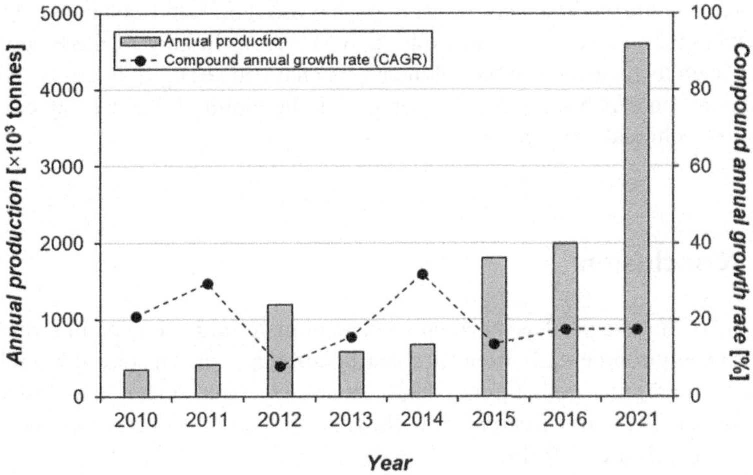

Fig. 23.27 Global biopolymer market (status and growth) (Chang et al. 2020)

respectively. This shows that the price of xanthan gum is comparable to that of conventional soil binders.

However, it should be noted that the cost of the biopolymer treatment of soils is highly dependent on the biopolymer used for the treatment. If we compare xanthan gum and beta-glucan, we can see from Table 23.1 that there is an extremely large disparity between their prices. However, as biopolymer use has been increasing over the past few years, it is expected that the overall price of biopolymers will drop to levels comparable to that of xanthan gum treatment (Chang et al. 2020) (Fig. 23.27).

23.5.2 Limitations and Challenges

Although BPST has shown several beneficial engineering properties in soils, the use of biopolymers in geotechnical engineering has also been limited by several challenges. The use of biopolymers in soils has always resulted in a certain degree of inconsistency. Owing to the highly hydrophilic nature of biopolymers, their engineering capabilities are highly reliant on the moisture content within the soil. Studies have shown that with biopolymer usage, the saturated strength can decrease to 10% of the dry strength (Chang et al. 2015b). With such a large variation in their engineering performance, their use and applicability in the long term are called into question.

There are several methods that can be used to overcome this limitation, such as the use of hydrophobic biopolymers. In one study, the use of casein proteins, a hydrophobic protein commonly found in mammalian milk, was used as a binder in soils (Chang et al. 2018a). The results of this study showed that with this method, the saturated strength of the soils was greatly enhanced (i.e., up to 650 kPa), and although there was a decrease in strength from the dry to saturated soils, the decrease was not as significant as for previously studied biopolymers (Chang et al. 2020).

Another viable method may be through the alteration and enhancement of the biopolymers themselves using methods such as a cross-linking of the biopolymers. Through such techniques, the biopolymer's mechanical strength, ductility, and even water absorption capabilities can be altered to fit the required engineering criteria for various geotechnical applications.

23.6 Conclusion

Biopolymers that are used as a soil enhancement material have been introduced as a sustainable, environmentally friendly construction material. The use of biopolymers, such as xanthan gum, gellan gum, starch, guar gum, beta-glucan, and casein, have been studied by various researchers worldwide. The current findings of biopolymer behaviors in soils are as follows:

- Biopolymer interactions—The surface charge characteristics of biopolymers allow for the biopolymer molecules to interact with (a) other biopolymers, (b) water, and (c) surface charges on fine soils. In the case of coarse soils, the biopolymer molecules interact with other biopolymer molecules to create a biopolymer film that encompasses the soil particles. In the case of fine soils, the biopolymer molecules interact with the surface charges on the fine soils and other biopolymer molecules to create a net interaction between the soil particles.
- Soil consistency—Biopolymers have water absorption capabilities that can alter the LL and PL of different soils. The addition of biopolymers provides the soils a higher plasticity and lower electrical sensitivity.
- Strength—Biopolymers are capable of significantly enhancing the strength of soils mostly through an increase in cohesion. The biopolymers either encompass the soil particles or interact directly with the fine particles to create a large conglomerate. These conglomerates tend to increase the dilatancy and friction angle of the soils. The efficiency of the strengthening capability is largely dependent on the presence of water.
- Erosion—As biopolymers enhance interparticle interactions, biopolymer-treated soils have been shown to have significant resistance to erosion. The highly viscous nature and binding properties of biopolymers have been shown to inhibit soil particle movement.
- Hydraulic conductivity—Biopolymers are generally hydrophilic and have water absorption capabilities. When biopolymers are present in the pore spaces of soil, they can absorb water and fill the pore spaces within the soil. Through this mechanism, the use of biopolymers can greatly reduce the hydraulic conductivity of the soil.
- Vegetation—The water-holding capabilities of biopolymers, along with their natural origin, have been shown to enhance the germination, growth rate, and drought survival rate of plants. Different combinations of plants and biopolymer types may show varying results.

As biopolymer-based soil treatment (BPST) has various beneficial engineering properties, along with the fact that biopolymer use is a sustainable and environmentally friendly approach to construction activities, its use as a future engineering material shows great promise. However, to fully incorporate BPST technology into current engineering practices, the limitations of biopolymer use need to be addressed.

Acknowledgments The research described in this chapter was financially supported by the Korea Ministry of Land, Infrastructure and Transport (MOLIT) as the Water Management Research Program (20AWMP-B114119-05).

References

Abubakar AU, Baharudin KS (2012) Properties of concrete using tanjung bin power plant coal bottom ash and fly ash. Int J Sust Constr Eng Technol 3(2):56–69

Akbulut N, Cabalar AF (2014) Effects of biopolymer on some geotechnical properties of a sand. In: New frontiers in geotechnical engineering—geotechnical special publications 243. ASCE, Shanghai, pp 28–37. https://doi.org/10.1061/9780784413456.004

Allan ML (2000) Materials characterization of superplasticized cement-sand grout. Cem Concr Res 30(6):937–942

Ayeldeen MK, Negm AM, El Sawwaf MA (2016) Evaluating the physical characteristics of biopolymer/soil mixtures. Arab J Geosci 9(5):1–13. https://doi.org/10.1007/s12517-016-2366-1

Bacic A, Fincher GB, Stone BA (2009) Chemistry, biochemistry, and biology of (1-3)-[beta]-glucans and related polysaccharides, 1st edn. Academic, Amsterdam

Bakharev T (2005) Resistance of geopolymer materials to acid attack. Cem Concr Res 35 (4):658–670. https://doi.org/10.1016/j.cemconres.2004.06.005

Becker A, Katzen F, Pühler A, Ielpi L (1998) Xanthan gum biosynthesis and application: a biochemical /genetic perspective. Appl Microbiol Biotechnol 50(2):145–152. https://doi.org/10.1007/s002530051269

Belitz H-D, Grosch W, Schieberle P (2009) Food chemistry. 4th revised and extended edition. Springer, Leipzig

Butler M (2016) Xanthan gum: applications and research studies. Food science and technology. Nova Science Publisher's, Inc., New York

Chang I, Cho G-C (2012) Strengthening of Korean residual soil with β-1,3/1,6-glucan biopolymer. Constr Build Mater 30(0):30–35. https://doi.org/10.1016/j.conbuildmat.2011.11.030

Chang I, Cho G-C (2014) Geotechnical behavior of a beta-1,3/1,6-glucan biopolymer-treated residual soil. Geomech Eng 7(6):633–647. https://doi.org/10.12989/gae.2014.7.6.633

Chang I, Cho G-C (2019) Shear strength behavior and parameters of microbial gellan gum-treated soils: from sand to clay. Acta Geotech 14(2):361–375. https://doi.org/10.1007/s11440-018-0641-x

Chang I, Im J, Prasidhi AK, Cho G-C (2015a) Effects of Xanthan gum biopolymer on soil strengthening. Constr Build Mater 74(0):65–72. https://doi.org/10.1016/j.conbuildmat.2014.10.026

Chang I, Prasidhi AK, Im J, Cho G-C (2015b) Soil strengthening using thermo-gelation biopolymers. Constr Build Mater 77(0):430–438. https://doi.org/10.1016/j.conbuildmat.2014.12.116

Chang I, Prasidhi AK, Im J, Shin H-D, Cho G-C (2015c) Soil treatment using microbial biopolymers for anti-desertification purposes. Geoderma 253–254(0):39–47. https://doi.org/10.1016/j.geoderma.2015.04.006

Chang I, Im J, Cho G-C (2016) Geotechnical engineering behaviors of gellan gum biopolymer treated sand. Can Geotech J 53(10):1658–1670. https://doi.org/10.1139/cgj-2015-0475
Chang I, Im J, Lee S-W, Cho G-C (2017) Strength durability of gellan gum biopolymer-treated Korean sand with cyclic wetting and drying. Constr Build Mater 143:210–221. https://doi.org/10.1016/j.conbuildmat.2017.02.061
Chang I, Im J, Chung M-K, Cho G-C (2018a) Bovine casein as a new soil strengthening binder from dairy wastes. Constr Build Mater 160:1–9. https://doi.org/10.1016/j.conbuildmat.2017.11.009
Chang I, Kwon Y-M, Im J, Cho G-C (2018b) Soil consistency and inter-particle characteristics of xanthan gum biopolymer containing soils with pore-fluid variation. Can Geotech J 56:1206. https://doi.org/10.1139/cgj-2018-0254
Chang I, Lee M, Tran ATP, Lee S, Kwon Y-M, Im J, Cho G-C (2020) Review on biopolymer-based soil treatment (BPST) technology in geotechnical engineering practices. Transport Geotech 24:100385
Cho G-C, Chang I (2018) Soil stabilization and improvement method using biopolymer. United States Patent
Chourasia M, Jain S (2004) Potential of guar gum microspheres for target specific drug release to colon. J Drug Target 12(7):435–442
Chu J, Varaksin S, Klotz U, Mengé P (2009) Construction processes, State-of-the-art report. In: 17th international conference on soil mechanics and geotechnical engineering. Alexandria, Egypt, pp 3006–3135
DeJong JT, Fritzges MB, Nüsslein K (2006) Microbially induced cementation to control sand response to undrained shear. J Geotech Geoenviron Eng 132(11):1381–1392
Dickinson E, Vliet TV, Wageningen Centre for Food Sciences (2003) Food colloids, biopolymers and materials. Special publication, vol 284. Royal Society of Chemistry, Cambridge
Duxson P, Fernández-Jiménez A, Provis JL, Lukey GC, Palomo A, van Deventer JSJ (2007) Geopolymer technology: the current state of the art. J Mater Sci 42(9):2917–2933. https://doi.org/10.1007/s10853-006-0637-z
Eliasson A-C (2004) Starch in food: structure, function and applications. Woodhead Publishing in food science and technology. CRC Press, Boca Raton
Fatehi H, Abtahi SM, Hashemolhosseini H, Hejazi SM (2018) A novel study on using protein based biopolymers in soil strengthening. Constr Build Mater 167:813–821. https://doi.org/10.1016/j.conbuildmat.2018.02.028
Fox D, Bryan RB (1992) Influence of a polyacrylamide soil conditioner on runoff generation and soil erosion: field tests in Baringo District, Kenya. Soil Technol 5(2):101–119
Fox PF, McSweeney PLH (2003) Advanced dairy chemistry: 1 proteins. Kluwer Academic/Plenum, New York
Fujita Y, Ferris FG, Lawson RD, Colwell FS, Smith RW (2000) Calcium carbonate precipitation by ureolytic subsurface bacteria. Geomicrobiol J 17:305–318
FuWei Y, BingJian Z, ChangChu P, YuYao Z (2009) Traditional mortar represented by sticky rice lime mortar—one of the great inventions in ancient China. Sci China Ser E: Technol Sci 52 (6):1641–1647
Garcıa-Ochoa F, Santos V, Casas J, Gomez E (2000) Xanthan gum: production, recovery, and properties. Biotechnol Adv 18(7):549–579
Giavasis I, Harvey LM, McNeil B (2000) Gellan gum. Crit Rev Biotechnol 20(3):177–211
Han J (2015) Principles and practice of ground improvement. Wiley, Hoboken
Hataf N, Ghadir P, Ranjbar N (2018) Investigation of soil stabilization using chitosan biopolymer. J Clean Prod 170:1493–1500
Hench LL (1998) Sol-gel silica. Properties, processing and technology transfer. Noyes Publications
Imeson A (1992) Thickening and gelling agents for food, 1st edn. Blackie, London, New York
Jang J, Santamarina JC (2016) Fines classification based on sensitivity to pore-fluid chemistry. J Geotech Geoenviron Eng 142(4). https://doi.org/10.1061/(ASCE)GT.1943-5606.0001420
Kalia S, Averous L (2011) Biopolymers: biomedical and environmental applications. Wiley

Karol RH (2003) Chemical grouting and soil stabilization. Civil and environmental engineering, vol 12, 3rd edn, Revised, and Expanded edn. Marcel Dekker, Inc., New York

Khatami H, O'Kelly B (2012) Improving mechanical properties of sand using biopolymers. J Geotech Geoenviron Eng 139(8):1402–1406. https://doi.org/10.1061/(ASCE)GT.1943-5606.0000861

Khatami H, O'Kelly BC (2018) Prevention of bleeding of particulate grouts using biopolymers. Constr Build Mater 192:202

Khullar P, Khar RK, Agarwal SP (1998) Evaluation of guar gum in the preparation of sustained-release matrix tablets. Drug Dev Ind Pharm 24(11):1095–1099

Kim D, Lai HT, Chilingar GV, Yen TF (2006) Geopolymer formation and its unique properties. Environ Geol 51(1):103–111

Kogan G, Stasko A, Bauerova K, Polovka M, Soltes L, Brezova V, Navarova J, Mihalova D (2005) Antioxidant properties of yeast (1 -> 3)-beta-D-glucan studied by electron paramagnetic resonance spectroscopy and its activity in the adjuvant arthritis. Carbohydr Polym 61(1):18–28

Krishanaiah Y, Latha K, Rao LN, Karthikeyan R, Bhaskar P, Satyanarayana V (2003) Development of colon targeted oral guar gum matrix tablets of Albendazole for the treatment of helminthiasis. Indian J Pharm Sci 65(4):378

Krishnaiah Y, Seetha Devi A, Nageswara Rao L, Bhaskar Reddy P, Karthikeyan R, Satyanarayana V (2001) Guar gum as a carrier for colon specific delivery; influence of metronidazole and tinidazole on in vitro release of albendazole from guar gum matrix tablets. J Pharm Sci 4 (3):235–243

Kulicke WM, Kniewske R, Klein J (1982) Preparation, characterization, solution properties and rheological behaviour of polyacrylamide. Prog Polym Sci 8(4):373–468

Kulshreshtha Y, Schlangen E, Jonkers H, Vardon P, Van Paassen L (2017) CoRncrete: a corn starch based building material. Constr Build Mater 154:411–423

Kwon Y-M, Ham S-M, Kwon T-H, Cho G-C, Chang I (2020) Surface-erosion behavior of biopolymer-treated soils assessed by EFA. Geotech Lett 10:1–7

Laird DA (1997) Bonding between polyacrylamide and clay mineral surfaces. Soil Sci 162 (11):826–832. https://doi.org/10.1097/00010694-199711000-00006

Larson S, Newman JK, O'Connor G, Griggs C, Martin A, Nijak G, Lord E, Duggar R, Leeson A (2013) Biopolymer as an alternative to petroleum-based polymers to control soil erosion: Iowa Army ammunition plant. Environmental Security Technology Certification Program Office (DOD), Arlington

Latifi N, Horpibulsuk S, Meehan CL, Abd Majid MZ, Tahir MM, Mohamad ET (2016) Improvement of problematic soils with biopolymer—an environmentally friendly soil stabilizer. J Mater Civ Eng 29(2):04016204

Lazaridou A, Biliaderis CG (2007) Molecular aspects of cereal beta-glucan functionality: physical properties, technological applications and physiological effects. J Cereal Sci 46(2):101–118

Lorenzo G, Zaritzky N, Califano A (2012) Rheological analysis of emulsion-filled gels based on high acyl gellan gum. Food Hydrocoll 30:672–680

Makusa GP (2012) Soil stabilization methods and materials. Lulea, Sweden: sn

Malik M, Letey J (1991) Adsorption of polyacrylamide and polysaccharide polymers on soil materials. Soil Sci Soc Am J 55(2):380–383. https://doi.org/10.2136/sssaj1991.03615995005500020014x

Metz B, Davidson O, Coninck HD, Loos M, Meyer L (2005) Carbon dioxide capture and storage. Intergovernmental Panel on Climate Change, New York

Milas M, Rinaudo M (1979) Conformational investigation on the bacterial polysaccharide xanthan. Carbohydr Res 76(1):189–196. https://doi.org/10.1016/0008-6215(79)80017-8

Misaki A, Sone Y, Mitsuhashi M, Miyake T (1990) Beta-D-glucan, and its production and uses. United States Patent

Mishra S, Rai T (2006) Morphology and functional properties of corn, potato and tapioca starches. Food Hydrocoll 20(5):557–566

Mortensen B, Haber M, DeJong J, Caslake L, Nelson D (2011) Effects of environmental factors on microbial induced calcium carbonate precipitation. J Appl Microbiol 111(2):338–349

Mudgil D, Barak S, Khatkar BS (2014) Guar gum: processing, properties and food applications—a review. J Food Sci Technol 51(3):409–418

Nachtergaele W (1989) The benefits of cationic starches for the paper industry. Starch—Stärke 41 (1):27–31. https://doi.org/10.1002/star.19890410108

Oh JE, Monteiro PJM, Jun SS, Choi S, Clark SM (2010) The evolution of strength and crystalline phases for alkali-activated ground blast furnace slag and fly ash-based geopolymers. Cem Concr Res 40(2):189–196

Oliveira JT, Martins L, Picciochi R, Malafaya PB, Sousa RA, Neves NM, Mano JF, Reis RL (2010) Gellan gum: a new biomaterial for cartilage tissue engineering applications. J Biomed Mater Res A 93(3):852–863. https://doi.org/10.1002/jbm.a.32574

Orts W, Roa-Espinosa A, Sojka R, Glenn G, Imam S, Erlacher K, Pedersen J (2007) Use of synthetic polymers and biopolymers for soil stabilization in agricultural, construction, and military applications. J Mater Civ Eng 19(1):58–66. https://doi.org/10.1061/(ASCE)0899-1561(2007)19:1(58)

Pham V, Paassen LV, Nakano A, Kanayama M, Heimovaara T (2013) Microbially induced carbonate precipitation (MICP) by denitrification as ground improvement method—process control in sand column experiments. EGU General Assembly 15

Pillai R, Redmond M, Roding J (2005) Anti-wrinkle therapy: significant new findings in the non-invasive cosmetic treatment of skin wrinkles with beta-glucan. Int J Cosmet Sci 27(5):292

Prajapat AL, Gogate PR (2016) Intensified depolymerization of aqueous polyacrylamide solution using combined processes based on hydrodynamic cavitation, ozone, ultraviolet light and hydrogen peroxide. Ultrason Sonochem 31:371–382

Risica D, Dentini M, Crescenzi V (2005) Guar gum methyl ethers. Part I. Synthesis and macromolecular characterization. Polymer 46(26):12247–12255

Rodrigues A, Emeje M (2012) Recent applications of starch derivatives in nanodrug delivery. Carbohydr Polym 87(2):987–994. https://doi.org/10.1016/j.carbpol.2011.09.044

Rong H, Qian C, Wang R (2011) A cementation method of loose particles based on microbe-based cement. Sci China Technol Sci 54(7):1722–1729

Rose RK (2000) Effects of an anticariogenic casein phosphopeptide on calcium diffusion in streptococcal model dental plaques. Arch Oral Biol 45(7):569–575. https://doi.org/10.1016/S0003-9969(00)00017-0

Rovnaník P (2010) Effect of curing temperature on the development of hard structure of metakaolin-based geopolymer. Constr Build Mater 24(7):1176–1183. https://doi.org/10.1016/j.conbuildmat.2009.12.023

Ruhsing Pan J, Huang C, Chen S, Chung Y-C (1999) Evaluation of a modified chitosan biopolymer for coagulation of colloidal particles. Colloids Surf A Physicochem Eng Asp 147(3):359–364. https://doi.org/10.1016/S0927-7757(98)00588-3

Saha D, Bhattacharya S (2010) Hydrocolloids as thickening and gelling agents in food: a critical review. J Food Sci Technol 47(6):587–597

Semenova MG, Dickinson E, Burlakova EB, Zaikov GE (2010) Biopolymers in food colloids: thermodynamics and molecular interactions. Brill, Leiden, Boston

Stockes-Fischer S, Galinat JK, Ban SS (1999) Microbiological precipitation of CaCO3. Soil Biol Biochem 31(11):1563–1571

Stokke BT, Elgsaeter A, Kitamura S (1993) Macrocyclization of polysaccharides visualized by electron microscopy. Int J Biol Macromol 15(1):63–68. https://doi.org/10.1016/S0141-8130(05)80090-X

Tareke E, Rydberg P, Karlsson P, Eriksson C, Tornqvist M (2002) Analysis of acrylamide, a carcinogen formed in heated foodstuffs. J Agric Food Chem 50(17):4998–5006

Tran ATP, Chang I, Cho G-C (2019) Soil water retention and vegetation survivability improvement using microbial biopolymers in drylands. Geomech Eng 17(5):475–483. https://doi.org/10.12989/gae.2019.17.5.475

United States Geological Survey (USGS) (2018) Cement statistics and information. https://www.usgs.gov/centers/nmic/cement-statistics-and-information. Accessed 2019.06.07
US National Library of Medicine (2011) Polynucleotides
Van de Velde K, Kiekens P (2002) Biopolymers: overview of several properties and consequences on their applications. Polym Test 21(4):433–442. https://doi.org/10.1016/S0142-9418(01)00107-6
van Jaarsveld JGS, van Deventer JSJ, Lukey GC (2002) The effect of composition and temperature on the properties of fly ash- and kaolinite-based geopolymers. Chem Eng J 89(1–3):63–73. https://doi.org/10.1016/S1385-8947(02)00025-6
Vishnuvarthanan M, Rajeswari N (2013) Additives for enhancing the drying properties of adhesives for corrugated boards. Alex Eng J 52(1):137–140. https://doi.org/10.1016/j.aej.2012.10.002
Whiffin VS, van Paassen LA, Harkes MP (2007) Microbial carbonate precipitation as a soil improvement technique. Geomicrobiol J 24(5):417–423. https://doi.org/10.1080/01490450701436505
Whistler RL, Hymowitz T (1979) Guar: agronomy, production, industrial use, and nutrition. Purdue University Press
Wiszniewski M, Cabalar AF (2014) Hydraulic conductivity of a biopolymer treated sand. In: New frontiers in geotechnical engineering. pp 19–27
Woodrow J, Seiber J, Miller G (2008) Acrylamide release resulting from sunlight irradiation of aqueous polyacrylamide/iron mixtures. J Agric Food Chem 56(8):2773–2770
Worrell E, Price L, Martin N, Hendriks C, Meida LO (2001) Carbon dioxide emissions from the global cement industry. Annu Rev Energy Environ 26(1):303–329. https://doi.org/10.1146/annurev.energy.26.1.303
Zhu F, Du B, Bian Z, Xu B (2015) Beta-glucans from edible and medicinal mushrooms: characteristics, physicochemical and biological activities. J Food Compos Anal 41:165–173

Genetic Engineering Approaches for High-End Application of Biopolymers: Advances and Future Prospects

24

Saurabh Gupta, Kundan Kumar Chaubey, Vishal Khandelwal, Tarubala Sharma, and Shoor Vir Singh

Abstract

The role of genetic engineering in the development of novel biopolymers has a massive impact on various innovative scopes in biomedical research and also have other industrial applications. Genetic manipulation of non-replica microorganisms is still challenging due to organism-specific variations that hinder universal central dogma or molecular genetics tools, translatable knowledge of intracellular biochemical pathways and their regulatory mechanisms. A better understanding of the essential processes involved in the biosynthesis of microbial polymer and their regulation has created a starting point for metabolic and protein-engineering approaches to improve economic-production efficiency with highly applicable biopolymer properties. On the basis of molecular genetics, sequence, folded structure, molecular weight, and stereochemistry of protein polymers advocate a view for the synthesis of protein-based genetically engineered biopolymers. Genetically engineered microbial polymers have a huge area of research over the last 15 years and advances like CRISPR, genetic engineering-driven coassembly of synthetic bacterial cells and magnetic nanoparticles for capturing heavy metals, cyanobacterial extracellular polymeric substances (EPS), etc., come into sight as an alternative to address biotechnological and biomedical challenges and also boosting biomaterial research. Here, we encapsulated the key aspects of biopolymers (microbial polymers) production and give a brief outline about the approaches used for biopolymer synthesis and their advances that can lead to increased use of biopolymers as valuable renewable products.

S. Gupta (✉) · K. K. Chaubey · V. Khandelwal · T. Sharma · S. V. Singh
Department of Biotechnology, GLA University, Mathura, Uttar Pradesh, India
e-mail: saurabh.gupta@gla.ac.in

A. Vaishnav, D. K. Choudhary (eds.), *Microbial Polymers*,
https://doi.org/10.1007/978-981-16-0045-6_24

Keywords

Genetic engineering · Biotechnology · Microbial polymers · CRISPR · Mutagenesis · Extracellular polymeric substances

24.1 Introduction

For hundreds of years, we have adopted applied biotechnology techniques, such as selectively breeding animals and fermentation. Later, in the early twentieth century studies revealed how microorganisms accomplish commercially advantageous methods which leads to substantial development of vaccines and antibiotics against many diseases (Jain 1996; Lee et al. 2012). Nevertheless, new vaccines are greatly demanded to effectively control new and reemerging pathogenic species. Globally, the researchers took a large leap forward with the invention and development of recombinant DNA technology (RDT) or genetic engineering in the 1970s (Svab et al. 1990; Gomes et al. 2012). The area of genetic engineering proceeds to expedite with modern announcements and applications predicted to help the economy throughout the twenty-first century (Minkenberg et al. 2017; Velazquez-Campoy 2018). Genetic engineering has an intervention in the field of agricultural and industrial biotechnology including different types of plants, animals, and microorganisms (Velazquez-Campoy 2018). In pharmaceuticals, human insulin gene was spliced to synthesize artificial insulin (humulin), using cells of *E. coli* bacteria (Baeshen et al. 2014). Interferons, tissue plasminogen activator (used in blood clots), and urokinase which is functional to kill specific viruses and cancer cells is an outcome of genetic engineering (Takaoka et al. 2013).

Gene targeting also known as replacement strategy is based on homologous recombination to modify an endogenous gene and can be used to delete or add coding region of the gene (exons) and introduce point mutations (modify individual base pairs) (Gerlai 2016). The recent introduction of third-generation CRISPR/Cas RNA-based gene targeting provides a flexible approach for genome editing, epigenetic modulation, genome imaging, and target a protein to the locus of interest using RNA molecules (Geel et al. 2018; Withanage and Daniil 2020). Many researchers now exploit this platform as a new era of gene targeting and are revolutionizing biomedical sciences.

Naturally, biopolymers are synthesized either chemically from a biological material or completely biosynthesized by living organisms and researchers have been involved for many years to boost upon and develop genetically engineered polymers with exceptional properties such as strength, toughness, and adaptability (Yadav et al. 2015; Zhang et al. 2019). There are a number of genetically engineered biopolymers developed, but the main mechanism and the idea show relativity among each other (Kopecek 2003; Gomes et al. 2012; Santos et al. 2019). The procedure of cloning of genes is the main and initiating step and using of cloning vectors such as plasmids, bacteriophage, cosmids, Ti plasmids, yeast cloning vectors, and so on (Khan 2009). The selected characteristic of a polymer is carried out by its responsible gene which is located in the genome of its origin. Using genomic editing tools such as CRISPR/Cas platform, the genes are modified in a genome compatible with its effect (Minkenberg et al. 2017; Gerlai 2016). Target genome selection, plasmids fixation using restriction enzymes and ligases, insertion of designed plasmids into the host cell leading to regulation and multiplication of gene as well as to produce compatible target proteins (https://www.ncbi.nlm.nih.gov/books/NBK21498/). Researches confirmed various genetically engineered based biopolymer outputs which are involved in a variety of applications such as drug delivery, nanoparticle coverings, macromolecular carriers, hydrogels, and tissue engineering, (Kopecek 2003; Gerlai 2016; Santos et al. 2019; Zhang et al. 2019).

24.2 Advancement in Genetically Engineered Biopolymers

24.2.1 Second-Generation Biopolymers

The second-generation genetically engineered biopolymers production technology developed by Chinese scientists in the United States uses microalgae gene technology to produce expensive genetic technology biological products at low cost. SRT-type spider silk material (first product), which fills the gap of traditional genetic engineering. It has very good strength, toughness, thermoplasticity, and is a protein matrix with spider silk (Khan 2009). It is used in bulletproof jackets, parachutes, airbags, bone regeneration brackets, 3D printing, and high-end fabrics. At the same time, this technology has environmental advantages, it can absorb carbon emissions and produce bio-oil as a by-product.

24.2.2 Genetically Engineered Proteins for Tissue Engineering

In 2012, Gomes and his group have established a technology to observe the properties and applicability of chimeric protein-based biopolymers synthesized through recombinant DNA technology. The recombinant proteins were designed with the cDNA of natural origin (Silk, Chitosan, etc.) combining them with the cDNA with bioactive proteins. Genetically engineered proteins with a combination of natural and biological domains consisted of a variety of chemical domains that allows cell adhesion, increased mechanical properties, antimicrobial ability, etc. The applications of these biopolymers (chimeric proteins) are likely suggested to be involved in drug delivery, gene delivery, tissue transplantation, cell encapsulation, and coatings (Gomes et al. 2012).

24.2.3 Genetically Engineered Elastin-Based Biopolymers

Recently some good research groups working on biomaterials or biopolymers have designed genetically engineered Elastin like Recombinants (ELRs) produced using Elastin like polypeptides (ELPs) biosequences found in nature leading to apply them in a variety of biomedical applications (Santos et al. 2019; Xiaowen et al. 2018). Recently, protein-based polymers, a new class of biopolymers called the silk-elastin-proteins (Silk-ELR co-recombinamer covered stents) produced by electro-spinning that combine the flexibility and elasticity of elastin with the high-tensile strength of silk fibroin (Putzu et al. 2019).

In drug delivery studies, different uses of ELRs were observed in different processes of drug delivery such as soluble drug delivery patterns when combined with therapeutic agents, Pharmacokinetic enhancers, fusion protein expression when combined with drugs and chemical conjugation (Santos et al. 2019). The local delivery of ELR conjugates into tumor tissues can be enhanced by therapeutic hyperthermia by taking advantage of tumor vasculature permeability and perfusion (Langer and Tirrell 2004). Besides, the ELR-based nanoparticle formation and conjugation with drugs and therapeutics to form drug nanocarriers were observed using different strategies such as metal ion-induced self-assembly which allows polypeptide assembly and interchain crosslinking. ELR-based macromolecular hydrogels have the potential to be used in tissue engineering such as skeletal system and vascular soft tissues engineering. Bicomponent with ELRs such as ELR-collagen, ELR-silicon has potential effects of involving in tissue engineering (Santos et al. 2019). Genetically engineered Spider silk material has also been used as a modern triboelectric biomaterial. Triboelectric materials are polymers which can generate electron flow by collision leading to generate electricity (Ferrari and Cappello 1997; Withanage and Daniil 2020).

24.2.4 Genetically Engineered Human Osteoblasts Biopolymers

Tissue engineering has been fast-growing fields of biomedical sciences and technologies so that the researchers have developed human osteoblasts biopolymers using genetic engineering procedure for the in vitro and in vivo regeneration and development of functionally active tissues. In this experiment, a cellular model relevant to the in vitro interpretation, i.e., Human osteosarcoma-derived cell line named as Saos-2 was genetically modified and osteoblastic phenotype was observed in Saos-2 (Rodan et al. 1987; Morelli et al. 2007). Incorporating the BMP-2 (Bone Morphogenic Protein) peptide in genetically engineered biomaterials accelerates osteogenic differentiation of mesenchymal cells and maintained an absolute proliferation possibility in vitro.

24.2.5 "Bacterial Builders" Produce Functional Biopolymers

Wyss Institute researchers have developed a modern, adequate, and measurable strategy for the construction of biopolymer using bacterial strains (*E. coli*) as bio-factories for protein structures (known as amyloids). Amyloids are fibrous structures that consist of nanoscale features and are occurring naturally in a normal environment. They were exactly suggested to be an indication of Alzheimer's and Parkinson's diseases caused by protein missfolding. "Functional amyloids" were found to be attaining advantageous aspects in the preservation of organisms and their connections with surfaces, including insects and bacteria (McAlpine 2016; Withanage and Daniil 2020). There is a wide range of applications of amyloid biomaterials such as tissue engineering, water purification, vaccine production, and so on. Existing methods to the production of amyloid proteins depend on extracted amyloids from natural sources. Besides, it was hard to differentiate amyloid leading to the applications and the purification was also an expensive and time-consuming procedure after recombinant amyloid production in *E. coli*. Neel Joshi, Ph.D., has designed a modernized, vacuum filtration technique that allows agile separation of amyloid fibers from *E. coli* cultures after genetic engineering of amyloid protein fibers using *E. coli* as an expression host. Besides, it does not require any specially designed equipment and sidesteps leading to acquire slow purification steps. This developmental technology provides an extraordinary interpretation of current approaches to protein-based materials fabrication and also simplifies and enhances protein crops compared to previous techniques (Withanage and Daniil 2020).

24.3 Approaches for the Production of Microbial Polymers

Bacterial species competently convert different carbon moieties into a wide range of biopolymers with varying material and chemical properties. Although bacterial species synthesize a range of extracellular polymers that they can synthesize is vast but only a few are intracellular polymers. Some of these above polymers provide the same function in most of the prokaryotes, whereas other biopolymers can be specific for certain bacterial taxa and play distinctive biological functions (Anderson et al. 1990; Rehm 2010). Four important classes of polymers produced by bacterial species are polysaccharides, polyamides, polyesters, and polyphosphates (inorganic polyanhydrides). Various major biosynthesis pathways towards the synthesis of the various polymer precursors from intermediates of central metabolism are summarized (Table 24.1; Fig. 24.1). Various strategies like metabolic-engineering targets to increase the production of tailor-made biopolymers, production of bacterial strains with modified and improved yields of biopolymers (in vivo methods) (Fig. 24.2). In vitro production using polymerizing enzymes or genetically engineered enzymes targeted to specific substrates can lead to the formation of new biopolymers with altered material properties that improve their performance (Rehm 2010). And secondly, biopolymers can also be upgraded by exposing them to

Table 24.1 Classes of microbial polymers and their characteristics

Bio-polymer type (localization)	Precursors	Primary structure	Enzyme involved in metabolic engineering	Producer organism	Industrial applications
Polysaccharides					
Alginate (extracellular)	GDP-mannuronic acid	β-(1,4)-linked non-repeating heteropolymer	Glycosyl transferase	*Pseudomonas* spp. and *Azotobacter* spp.	Biomaterial (application in drug delivery and tissue scaffold)
Glycogen (intracellular)	ADP–glucose	α-(1,6)-branched α-(1,4)-linked homopolymer	Glycogen synthase	Bacteria and archaea	In cancer theranostics (in vivo study)
Xanthan	UDP–glucose, GDP–mannose and UDP–glucuronate	β-(1,4)-linked repeating heteropolymer consisting of pentasaccharide units	Xanthan polymerase	*Xanthomonas* spp.	Food additive (as a thickener or an emulsifier)
Dextran (extracellular)	Saccharose	α-(1,2)/α-(1,3)/α-(1,4)-branched α-(1,6)-linked homopolymer	Dextran sucrase	*Lenconostoc* spp. and *Streptococcus* spp.	Blood plasma extender and chromatography media
Gellan (extracellular)	UDP–glucose, dTDP-rhamnose and UDP–glucuronate	β-(1,3)-linked repeating heteropolymer consisting of tetrasaccharide units	Gellan synthase	*Sphingomonas* spp.	Culture media additive, food additive (as a gelling agent) or for encapsulation
Colanic acid (extracellular)	GDP–L-fucose, UDP–D-glucose, UDP–D-galactose and UDP–D-glucuronate	β-(1,4)-linked repeating heteropolymer consisting of hexasaccharide units	Colanic acid polymerase	*Escherichia coli*, *Shigella* spp., *Salmonella* spp. and *Enterobacter* spp.	As a cationic delivery vehicles

K30 antigen (capsular)	UDP–D-glucose, UDP–D-galactose and UDP–D-glucuronate	β-(1,2)-linked repeating heteropolymer consisting of tetrasaccharide units	Polysaccharide polymerase	*Escherichia coli*	K30 carbohydrate polymer binds with Wzi in-vitro and Wzi is a novel outer-membrane lectin that assists in the formation of the bacterial capsule via direct interaction with capsular polysaccharides
Cellulose (extracellular)	UDP–D-glucose	β-(1,4)-linked homopolymer	Cellulose synthase	Alphaproteobacteria, Betaproteobacteria, Gamma-proteobacteria and Gram-positive bacteria	Food, diaphragms of acoustic transducers and wound dressing
Hyaluronic acid (extracellular)	UDP—D-glucuronate and UDP—*N*-acetyl glucosamine	β-(1,4)-linked repeating heteropolymer consisting of disaccharide units	Hyaluronan synthase	*Streptococcus* spp. and *Pasteurella multocida*	Cosmetics, drug delivery and tissue repair
Polyamides					
Cyanophycin granule peptide (intracellular)	(β-spartate arginine) 3-phosphate, ATP, L-arginine and L-aspartate	Repeating heteropolymer consisting of dipeptide units	Cyanophycin synthetase	Cyanobacteria, *Acinetobacter* spp. and *Desulfito bacterium* spp.	Dispersant and water softener (after removal of arginyl residues)
Poly-γ-glutamate (extracellular or capsular)	(Glutamate) *n*-phosphate, ATP and glutamate	Homopolymer	Poly-γ-glutamate synthetase	*Bacillus* spp. and a few Gram-positive bacteria, the Gram-negative bacterium *Fusobacterium nucleatum* and the archaea *Natronococcus occultus* and *Natrialba aegyptiaca*	Replacement of polyacrylate, thickener, humectant, drug delivery and cosmetics

(continued)

Table 24.1 (continued)

Bio-polymer type (localization)	Precursors	Primary structure	Enzyme involved in metabolic engineering	Producer organism	Industrial applications
ε-poly-L-lysine (extracellular)	L-lysine, ATP and L-lysine-AMP	Homopolymer	ε-poly-L-lysine synthetase	*Streptomyces albulus* subsp. *lysinopolymerus*	Feed preservative and, when cross-linked, adsorbent (in medicine)
Polyester and polyanhydrides					
Polyhydroxy-alkanoates (intracellular)	(*R*)-3-hydroxyacyl CoA	Heteropolymer	Polyhydroxy-alkanoate synthase	Bacteria and archaea	Bioplastic, biomaterial and matrices for displaying or binding proteins
Polyphosphate (intracellular)	ATP	Homopolymer	Polyphosphate kinase	Bacteria and archaea	Replacement of ATP in enzymatic synthesis and flavour enhancer

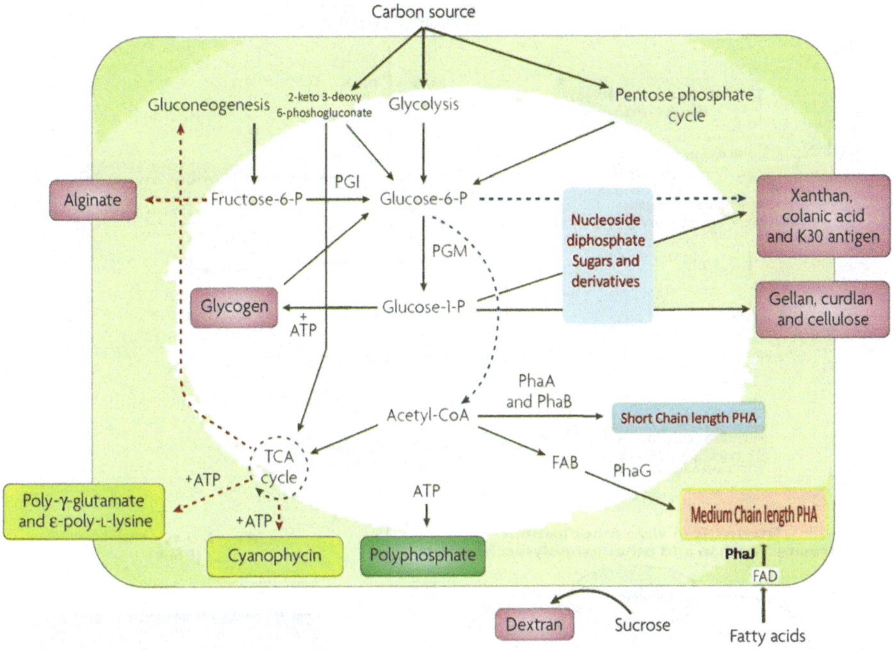

Fig. 24.1 Microbial polymer biosynthesis metabolic pathways towards the synthesis of the various polymer precursors are summarized. Dashed lines indicate multiple enzymatic steps. Solid lines represent the linking of primary metabolic pathways with intermediates (required for the synthesis of biopolymer) or direct enzyme catalyzed alterations towards the immediate biopolymer precursor. Polysaccharides are shown in purple boxes, polyesters are shown as short and medium chain length PHAs in boxes, polyamides and inorganic polyanhydride are shown in yellow and green boxes, respectively. *FAB* fatty acid de novo biosynthesis, *FAD* fatty acid β-oxidation, *PGI* phosphoglucoisomerase, *PGM* phosphorglucomutase, and *Pha* polyhydroxy-alkanoate

polymerizing enzymes or by chemical modifications like poly-hydroxy-alkanoate (Fig. 24.2).

24.4 Discussion and Conclusion

In relevance to the genetic engineering towards the improvement of biopolymers leading to different applications are increasingly boosting up research scope as we have discussed in the introduction. We have discussed and described several important research outcomes which were preceded by leading researchers. Different techniques of genetic engineering have been used leading to the development of novel biopolymers and to improve and modulate the properties of existing materials. Plasmid gene delivery, genetic transformation, genetic transduction, gene

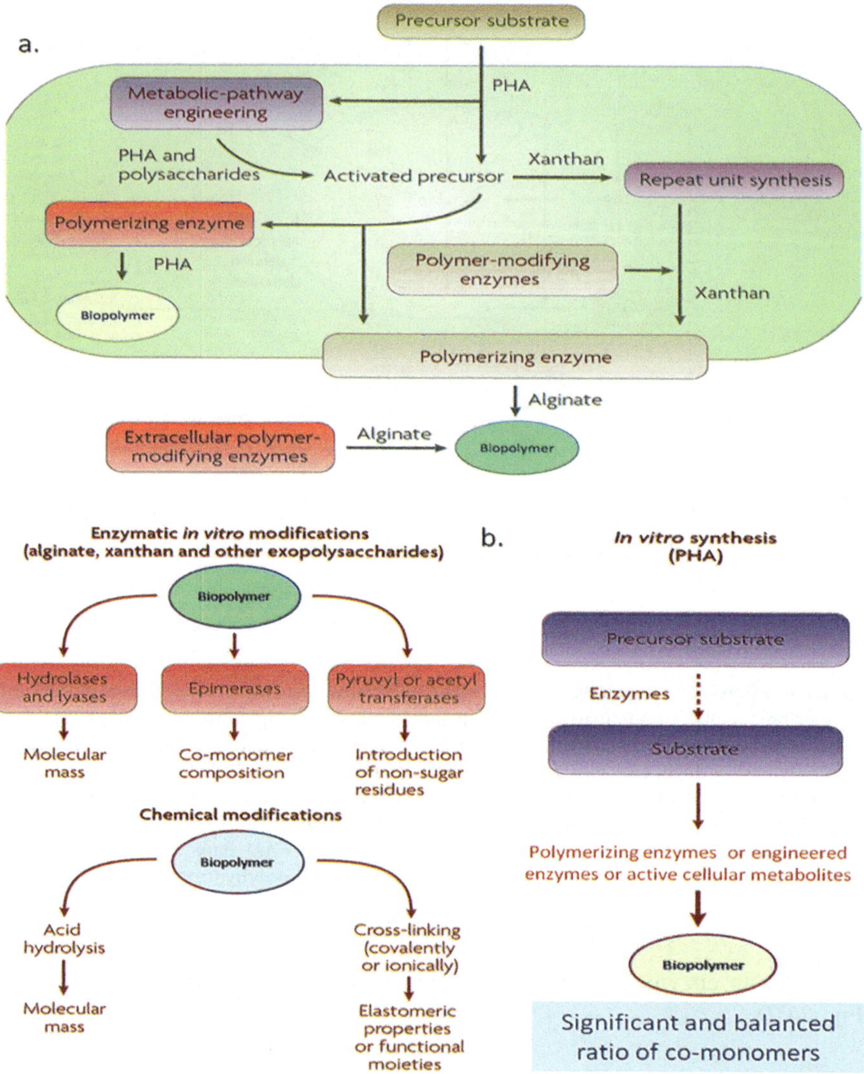

Fig. 24.2 Approaches for the synthesis of modified biopolymers with altered properties to enhance their performance. (**a**) Represent metabolic-engineering targets to increase the production yield of biopolymers using production strains (in vivo production of biopolymers). (**b**) In vitro synthesis using engineered or polymerizing enzymes targeted to specific substrates that can lead to the formation of new biopolymers and secondly, it can be upgraded by exposing to chemical or enzymatic modifications

transfection using different types of expression plasmids and viral vectors or cosmids were observed leading to genetically modify the materials. There are a wide range of genetic modifications that have been done according to the recent research references. The develop biomedical properties as well as improving chemical and physical properties leading to expand the applicable capacity are highlighted remarks along with the development of biomaterials with genetic engineering technologies. Genetically engineered biopolymers were suggested to have an enormous impact on biomedical applications such as drug delivery, tissue engineering, vaccine production, biological visualizations, and so on. Physical and chemical characteristics improved genetically engineered biopolymers allow them to be involved in water purification, using as Nano-generators and more. There is another type of secondary applicability of genetically engineered biomaterials. The use of genetically engineered proteins in the production of hydrogels, films, matrices, pharmaceutical materials, and different types of physiological and anatomical implants can be able to provide extraordinary materials with excellent multiple properties leading to widen the range of applicability.

24.5 Future Prospects

In terms of the genetically engineered materials or polymers, the research scope is still undergoing since the biopolymers have great potentialities to be applied to provide solutions for global problems as well as biological, medical, and chemical applications. Genetically engineered proteins (chimeric proteins); CRISPR/Cas RNA-based gene targeting provides a flexible approach for genome editing and imaging and epigenetic modulation along with targeting a protein to the locus of interest using RNA; cyanobacterial extracellular polymeric substances (EPS), coassembly of synthetic bacterial cells and magnetic nanoparticles (MNPs) for capturing heavy metals can be developed and used as novel biopolymers with biomedical applications as discussed. For the successful exploitation of cyanobacterial EPS, it is important to strategically design the biopolymers, either by genetically engineered the bacterial strains or by chemical modification of the polymers. This requires a better understanding of the extracellular polymeric substances biosynthetic pathways and their relationship with cellular metabolism, as well as available polymer functionalization chemistries.

These novel techniques can be developed in the near future with such properties of biocompatibility, biodegradability, potential energy harvesting capabilities, thermal properties, optical properties, and carcinogenic detection potentialities, and so on. Tissue engineering for bone tissue regeneration and differentiation methods have already been developed and the applications of these genetically engineered materials will be available in industrial clinical settings in the near future. In relevant to these discussions, genetic engineering techniques will be playing a major role in the development of future biopolymers.

References

Anderson AJ, Haywood GW, Dawes EA (1990) Biosynthesis and composition of bacterial poly (hydroxyalkanoates). Int J Biol Macromol 12:102–105

Baeshen NA, Baeshen MN, Sheikh A, Bora RS, Ahmed MM, Ramadan HA et al (2014) Cell factories for insulin production. Microb Cell Fact 13:141

Ferrari FA, Cappello J (1997) Biosynthesis of protein polymers. In: McGrath K, Kaplan D (eds) Protein-based materials. Bioengineering of materials. Birkhäuser, Boston, pp 37–60

Geel TM, Ruiters MHJ, Cool RH, Halby L, Voshart DC, Andrade Ruiz L et al (2018) The past and presence of gene targeting: from chemicals and DNA via proteins to RNA. Philos Trans R Soc Lond B Biol Sci 373:20170077

Gerlai R (2016) Gene targeting using homologous recombination in embryonic stem cells: the future for behavior genetics? Front Genet 7:43

Gomes S, Leonor IB, Mano JF, Reis RL, Kaplan DL (2012) Natural and genetically engineered proteins for tissue engineering. Prog Polym Sci 37:1–17

Jain RK (1996) Delivery of molecular medicine to solid tumors. Science 271:1079–1080

Khan KH (2009) Vectors used in gene manipulation-a retrospective. Adv Biotechnol J 9:1–18

Kopecek J (2003) Smart and genetically engineered biomaterials and drug delivery systems. Eur J Pharm Sci 20:1–16

Langer R, Tirrell DA (2004) Designing materials for biology and medicine. Nature 428:487–492

Lee NH, Lee JA, Park SY, Song CS, Choi IS, Lee JB (2012) A review of vaccine development and research for industry animals in Korea. Clin Exp Vaccine Res 1:18–34

McAlpine KJ (2016) Bacterial builders produce functional biomaterials. Wyss Institute

Minkenberg B, Wheatley M, Yang Y (2017) CRISPR/Cas9-Enabled multiplex genome editing and its application. Prog Mol Biol Transl Sci 149:111–132

Morelli C, Barbanti-Brodano G, Ciannilli A, Campioni K, Tognon M (2007) Cell morphology, markers, spreading, and proliferation on orthopaedic biomaterials. An innovative cellular model for the "in vitro" study. J Biomed Mater Res A 83:178–183

Putzu M, Causa F, Parente M, González de Torre I, Rodriguez-Cabello JC, Netti PA (2019) Silk-ELR co-recombinamer covered stents obtained by electrospinning. Regen Biomater 6:21–28

Rehm BH (2010) Bacterial polymers: biosynthesis, modifications and applications. Nat Rev Microbiol 8:578–592

Rodan SB, Imai Y, Thiede MA, Wesolowski G, Thompson D (1987) Characterization of a human osteosarcoma cell line (Saos-2) with osteoblastic properties. Cancer Res 47:4961–4965

Santos M, Serrano-Dúcar S, Gonzalez-Valdivieso J, Vallejo R, Girotti A, Cuadrado P et al (2019) Genetically engineered elastin-based biomaterials for biomedical applications. Curr Med Chem 26(40):7117–7146

Svab Z, Hajdukiewicz P, Maliga P (1990) Stable transformation of plastids in higher plants. Proc Natl Acad Sci U S A 87:8526–8530

Takaoka A, Hayakawa S, Yanai H, Stoiber D, Negishi H, Kikuchi H et al (2013) Integration of interferon-alpha/beta signalling to p53 responses in tumour suppression and antiviral defence. Nature 424:516–523

Velazquez-Campoy A (2018) Biotechnology in the 21st century: challenges and opportunities. An R Acad Nac Med 135:169–173

Withanage SR, Daniil K (2020) Genetically engineered biomaterials. Adv Genet Eng 9:161

Xiaowen L, Wang C, Liu Z (2018) Protein-engineered biomaterials for cancer theranostics. Adv Healthc Mater 7:1800913

Yadav P, Yadav H, Shah VG, Shah G, Dhaka G (2015) Biomedical biopolymers, their origin and evolution in biomedical sciences: a systematic review. J Clin Diagn Res 9:ZE21–ZE25

Zhang L, Xiang Z, Zhao G, Wu Z, Cui H (2019) Functionalized genetic engineered silk-based biomaterials and their applications. Chin J Biotechnol 35:956–971

Microbial Pigments: Secondary Metabolites with Multifaceted Roles

25

Himadri Tanaya Behera, Abhik Mojumdar, Suchismita Nivedita, and Lopamudra Ray

Abstract

Pigments are compounds capable of absorbing light and are responsible for displaying a specific color. Several ores, insects, plants, microbes are the source of the natural pigments. However, microorganisms have enormous potential in producing pigments due to its stability, availability, productivity, easy downstream processing, and yield. Pigments like melanins, quinines melanins, flavins, violacein, monascins, and carotenoids have been produced by microorganisms. Microbial pigments have gained more interest in present research not just due to their utilization as colorants in industries like food, plastic, textile, paint, printing, and paper, but it also possess several biological properties such as antimicrobial, antioxidant, anticancer, and various other bio-pharmacological activities that have been discussed in this chapter. This chapter will lead to the exploration of microorganisms producing pigments, their application in various fields as well as their ultimatums, and future prospective for various applications.

Keywords

Microorganisms · Pigments · Industries · Medicinal

H. T. Behera · A. Mojumdar · S. Nivedita
School of Biotechnology, KIIT Deemed to be University, Bhubaneswar, Odisha, India

L. Ray (✉)
School of Biotechnology, KIIT Deemed to be University, Bhubaneswar, Odisha, India

School of Law, KIIT Deemed to be University, Bhubaneswar, Odisha, India
e-mail: lopamudra.ray@kls.ac.in; lray@kiitbiotech.ac.in

A. Vaishnav, D. K. Choudhary (eds.), *Microbial Polymers*,
https://doi.org/10.1007/978-981-16-0045-6_25

25.1 Introduction

Natural sources of colors have kindled a keen interest due to its nontoxicity and biodegradability (Cristea and Vilarem 2006). Nature originated pigments like flavonoids, anthraquinone, synthesized by plants and animals are used as the traditional source of colorants. In the present scenario, with the increasing awareness of human safety, the world is favoring the practice of eco-friendly and biodegradable substitute. Natural pigments can be obtained from various sources like several ores, insects, microbes, and plants; however, microbes have enormous possibility for the production of pigments produced naturally due to its stability, availability, productivity, easy downstream processing, and yield (Joshi and Attri 2014). Various reports are there specifying the synthesis and utilization of these bacterial pigments (Usmani et al. 2020; Venil and Lakshmanaperumalsamy 2009; Ahmad et al. 2012). Astaxanthin, bacteriochlorophylls, β-carotene, carotenoids, canthaxanthin, flavins, glaukothalin, indigoids, melanins, monascins, pheomelanin, phenazines, prodigiosin, phycocyanin, quinoneprecursors, violacein, xanthomonadin, etc., are certain microbial pigmented molecules that are produced as by-products by various microbes (Ramesh et al. 2019). Despite the use of microbial pigments in industries like cosmetics, food, plastic, textile, paint, printing, and paper, it also exhibits antioxidant, antimicrobial, cytotoxic, anticancer, antitumor, antimalarial, and antifouling activities (Ramesh et al. 2019).

25.1.1 Brief History of Pigments

Nature is a rich source of colors and the origin of colors from the biological material is often known to be bio-colors. The addition of colors like saffron, turmeric, some vegetable dyes to processed food is a very old procedure. Since the Indus period (2600–1900 BC), dyeing was used in the cloth industry. The ancient record for the natural dye was in China (Siva 2007). People of Aztech and Maya culture, in central and North America, used cochineal dye. Dyes like woad, indigo, madder, Brazilwood, and dark purple were known by the fourth century (Gulrajani 1992). Before 2500 BC, henna was used as a dye and saffron was also mentioned in the Bible. In the Nara period of Japan in the eigth century, Shosoin text contains the reference of the use of colored soybean and adzuki-bean cakes. With the amelioration of civilization and education, the multiple functions of the natural pigments were also greatly impulsed (Krishnamurthy et al. 2002). Nature originated pigments were the only source of colorant for coloring the natural fibers like wool, cotton, and silk, before the onset of synthetic pigments. The use of natural pigments gradually decreased since 1856, after the introduction of commercial synthetic dyes by Perkin (Walford 1980). The use of natural dyes has almost completely replaced by synthetic dyes by the twentieth century. In 1932, Warburg and Christian fractioned a yellow-colored dye from yeast extract that was later elucidated as riboflavin by Karner and Kuhn (Revuelta et al. 2016). In the early 1970s, bacteriorhodopsin, a purple pigment was discovered from *Halobacterium* (Baxter et al. 2014). In 1964, thermorubin, a red colored pigment was screened from moderately thermophilic soil actinomycetes *Thermoactinomyces antibioticus* (Giddings and Newman 2015). However, mass production of synthetic compounds has become cheaper than natural dye due to the advancement in organic chemistry. Recent usages of artificial pigments are in the paper production, leather tanning and textile industry, in agricultural research, photoelectrochemical cells, and food technology, and hair colorings.

25.1.2 Why Natural Pigments over Synthetic Pigments?

The use of synthetic dyes is badly chided by the consumers in the US by the 1960s, the inhibition of these food additives was spread out widely by environmental activities (Koes et al. 1994). Biodegradability, nontoxicity, noncarcinogenicity characteristics of the natural pigments and colorants were highlighted and campaigned these properties as a sales tool. Presently, synthetic colors were interpreted as impurities and this trend has been encouraged (Krishnamurthy et al. 2002). For today's production lines mildly toxic products and natural raw materials are favored as the starting material of pigment production. Some precursors or the end product of the synthesized dye possess carcinogenic effect when disposed to the ecosystem. But on the other hand, natural pigments when disposed to the ecosystem display some advantageous biological activities thus increasing the marketability of the products. So it is imperative for the investigation of these veral natural sources of dyes and colorants with their applications. In food, cosmetics, and textile applications, some industries are now able to produce some microbial pigments (Dufossé 2016). Some pigments produced by microorganisms are listed in Table 25.1. Since ancient times, natural products like isoprenoid, alkaloids, and flavonoids are being used as colorants, fragrances, and flavors. The scarce fortuity and the huge price of the natural pigments have led to the more use and importance of the synthetic colorants in food and other industries. Nevertheless, currently due to the toxicological side effects of some synthetic pigments, the demand of natural products is increasing among the customers (Sowbhagya and Chitra 2010). The limited chemical diversity with the structural complexities of the synthetic libraries and the great success of the natural pigments in several industries has opened a huge demand for the screening of uncharacterized new bacteria for the production of new pigments with their applications (Li and Vederas 2009).

25.2 Ecology of Pigmented Microorganisms

A plethora of researches are published reporting the isolation of pigmented microorganisms, dispersed in various geographical habitats ranging from polar region to tropical region and aerial to deep-sea areas. Microorganisms tolerate the harsh conditions of their geographical regions by producing pigments. Pigmented bacteria like *Stenotrophomonas* and yeast-like *Rhodotorula* can enter the coastal environments, from the discharges of hospitals and domestic sewages accepting the

Table 25.1 Pigment synthesizing microorganisms (Malik et al. 2012)

Microorganisms	Molecules	Pigment color
Bacteria		
Staphylococcus aureus	Zeaxanthin, Staphyloxanthin	Yellow, Golden
Pseudomonas aeruginosa	Phycocyanin	Blue-green
Bacillus		Brown
Xanthomonas oryzae	Xanthomonadin	Yellow
Agrobacterium aurantiacum	Astaxanthin	Pink-red
Janthinobacterium lividum	Violacein	Purple
Cornybacterium michigannise		Grayish to creamish
Chromobacterium violaceum	Violacein	Purple
Rhodococcusmaris		Bluish-red
Brevibacterium sp.		Orange-yellow
Haloferax alexandrinus	Canthaxanthin	Dark red
Achromobacter		Creamy
Cornybacterium insidiosum		Greyish to creamish
Actinomycetes		
Sterptomyces echinoruber	Rubrolone	Red
Streptoverticillium rubrireticuli	Prodigiosin	Red
Yeast		
Rhodotorula sp., *Rhodotorula glutinis*	Torularhodin	Orange-red
Cryptococcus sp.		Red
Phaffia rhodozyma	Astaxanthin	Pink-red
Yarrowia lipolytica		Brown
Algae		
Hematococcus	Canthaxanthin	
Dunaliella salina	β-carotene	Red
Chlorococcum	Lutein	
Fungi		
Penicillium oxalicum	Anthraquinone	Red
Penicilllum cyclopium		Orange
Blakeslea trispora	β-carotene	Cream
Penicillumn algeovensis		Yellow
Fusarium sporotrichioides	Lycopene	Red
Monascus sp.	Ankaflavin	Yellow
Helminthosporium avenae		Bronze
Cordyceps unilateralis	Naphtoquinone	Deep-blood red
Neurospora crassa, Mucor circinelloides	β-carotene	Yellow-orange
Pacilomyces farinosus	Anthraquinone	Red
Ashbya gossypi	Riboflavin	Yellow

marine environment. The pigmented bacteria can be broadly divided into two categories: (1) true marine pigmented bacteria (2) robust pigmented bacteria (primary origin is the terrestrial environment and then can also exists in a coastal habitat). Though in the terrestrial environment, the occurrences of the pigmented bacteria (PB) are more common, the PB isolated from the marine region is also attaining much consideration owing to their diverse bioactive pigmented compounds.

Current researches have been deflected to explore the novel chromogenic compounds with biotechnological and industrial applications from the marine environment. The distribution of the PB in the marine region varies according to geographical and nutritional status. In terms of abundance, the diversification of marine heterotrophic bacteria (MHB) is more when related to pigmented heterotrophic bacteria (PHB). Depending on the site of sampling, seasonal variation, and nutritional availability the colony-forming units (CFU) of the PHB differ. In air-water interfaces (Hermansson et al. 1987), ice centers (Zhang et al. 2008), surface microlayer of the sea and bottom waters (Agogué et al. 2005), salt lakes (Khanafari et al. 2010), glaciers (Miteva et al. 2004), deep-sea hydrothermal outlets (Yurkov et al. 1999), deep hot springs regions the occurrence of PBs are more frequent. Also, from lava caves certain PB communities have been isolated (Ramesh et al. 2019). Isolation of pigments from different sources such as marine sediments, seawater, freshwater, mussel, seagrass, algal mats, sea cucumber (Ramesh et al. 2017), sponge, marine solar saltern, lagoon, microbial mats in Antarctica lakes, sea ice (Nedashkovskaya and Ludwig 2015), oil-contaminated soil, nonsaline alkaline groundwater have been reported.

According to the environments, several pigmented microorganisms exhibit polyextremophilic characteristics, such as dimorphic (*Metschnikowia laotica*), extreme halophilic (*Salinibacter*), thermophilic (*Thermus*), xerophilic (*Penicillium purpurogenum*), psychrophilic (*Kocuria polaris*), alkaliphilic (*Microbacterium arborescens*), radioresistance (*Deinococcus grandis*), polyextremophile (*Halorubrum*), and color mimic (*Celluophaga lytica*) (Ramesh et al. 2019). Different authors also reviewed the distribution of different species of microalgae in various environments (Zachleder 1998; Proschold and Leliaert 2007; Likens 2010; Takaichi 2011; Roy et al. 2011). However, the factors responsible for the restricted distribution of these microorganisms in their environments are still doubtful.

25.3 Sources of Pigments

Microorganisms are the most flexible instruments of biotechnology for creating different molecules including pigments, enzymes, and organic acids and antibiotics. A broad variety of microorganisms including bacteria, yeast, molds, and algae are able to produce pigments. This group of microorganisms must satisfy several criteria, such as (1) able to use a varied range of C and N sources; (2) able to tolerate a wide variety of utmost conditions like pH, temperature, the concentration of minerals, etc.; (3) nontoxic and nonpathogenic; (4) a good amount of pigment yield; and (5) easier separation from its cell mass.

25.3.1 Bacteria

A diverse range of pigments are produced by bacteria such as actinorhodin, pyocyanin, carotenoids, melanin, canthaxanthin, prodigiosin, violacein, zeaxanthin, etc., and possesses certain advantages over fungi and other organisms such as a reduced life cycle making it easy to engineer genetically (Venil et al. 2013, 2014). PBs can be pervasive and are present in different niches like freshwater, soil, desert sand, and marine samples. In comparison to other bacterial communities, various genera of Actinobacteria such as *Micromonospora, Streptomyces, Actinomadura, Nocardia, Kitasatospora, Microbispora, Actinoplanes, Streptosporangium, Rhodococcus,* and *Thermomonospora* are stated for the production of a broad range of pigments. The genus Streptomyces has been reported to produce the maximum variety of pigments (Narsing Rao et al. 2017). Bacterial pigments offer certain advantages such as it has a broad range of activities, pigment-associated genes can be manipulated easily, simple techniques for extraction with minimum operating costs, etc. (Venil et al. 2014).

Strain improvement of several recognized PBs by different strategies have been reported to enhance the yield. Strain improvement through ultraviolet (UV) mutations have been reported to increase in the production of prodigiosin in *Serratia marcescens* (El-Bialy and Abou El-Nour 2015). Optimization of the culture environment has also been outlined for enhancing the yield in *Duganella* sp. (Wang et al. 2009) and *Bacillus* sp. (Mondal et al. 2015). Recent advancements in genetic engineering techniques have also eased the gene manipulation strategies in bacteria to produce the pigment of interest.

25.3.2 Fungi

Filamentous fungi are gaining industrial importance as it can be easily grown in the laboratory and can also enhance large scale production. A varied range of pigments are produced by filamentous fungi like carotenoids, phenazines, violacein, melanins, etc. (Dufosse et al. 2014). Genera like *Curvularia, Drechslera, Eurotium,* and *Fusarium* were reported to produce pigments like anthraquinones, chrysophanol, cynodontin, tritisporin, naphthaquinones erythroglaucin, flavin, helminthosporin, and dihydroxy naphthalene melanin (Babitha 2009).

Monascus used for the production of ang-kak (red mold rice) has been reported to be the most ancient use of fungal pigment. Various colored pigments such as yellow (monascine) orange (monascorubrine rice), purple (monascorubramine) have been produced by *Monaseus* and are used in different food items around the globe (Dufossé et al. 2005). *Monaseus* also possesses antibacterial, anti-obesity, anti-mutagenic, and antiproliferative activities (Feng et al. 2012).

Beyond 200 species of fungus have been noted for the synthesis of carotenes among which mostly belonging to the zygomycetes; order Mucorales including *Blakeslea, Phycomyces, and Mucor.* Other than Mucorales, basidiomycetes such as *Sporidiobolus, Sclerotinia, Sclerotium, Rhodosporidium,* and *Ustilago,* ascomycetes such as *Penicillium, Aspergillus, Aschersonia,* and *Cercospora* have been described for the synthesis of carotenes (Avalos and Limón 2015). Apart from β-carotene, lycopene is another pigment produced by *Blakeslea trispora* (Dufosse et al. 2014).

25.3.3 Yeast

Several yeasts like *Rhodotorula, Yarrowia lipolytica, Phaffia rhodozyma, Cryptococcus* sp. are good sources of microbial pigments (Joshi et al. 2003). However, the cultural conditions of the *P. rhodozyma* play a vital role in the production of the various types of carotenoid pigments like ϒ-carotene, lycopene, echinenone, phoenicoxanthin, and astaxanthin. *P. rhodozyme* produces a red pigment called Astaxanthin during the exponential phase of the growth (Khan and Bhat 2013). In aquaculture feeds, the use of *P. rhodozyma* as a source of pigment provides a good color to the feeds (Nelis and De Leenheer 1991).

25.3.4 Algae

Various types of pigments like canthaxanthin, astaxanthin, lutein, β-carotene are produced by Algae. *Dunaliella salina*, belonging to class Chlorophyta produces β-carotene under high saline and high light intensity conditions. Apart from main chlorophyll pigments, *Rhodophyta*, another red microalga, are also known to contain both red phycocyanins and blue phycoerythrin pigments. Such red and blue pigments account for approximately 30% of the total algal cell protein (Joshi et al. 2003).

25.4 Types of Pigments

25.4.1 Carotenoids

Carotenoids are tetraterpenoid that was first screened by Heinrich Wilhelm Ferdinand Wackenroder (Wackenroder 1831). There are more than 600 reported carotenoids which are categorised into two classes, namely, (1) xanthophylls containing oxygen and (2) carotenes that contain no oxygen and are purely hydrocarbons (Köcher and Müller 2011). Among them, some of the most considerable carotenoids are α- and β-carotenes, cryptoxanthin, canthxanthin, astaxanthin, lycopene, violaxanthin, lutein, neoxanthin, and zeaxanthin.

Commercially, carotenoids are either produced through chemical synthesis which generates hazardous wastes or extracted from vegetables which has a drawback such as geographical and seasonal variability. However, microbial production of carotenoids utilizes low-cost substrates and do not produce environmental hazards.

25.4.1.1 Astaxanthin

Marine organisms such as microalgae are known for the production of a natural carotenoid pigments, Astaxanthin. This is derived from *Phaffia rohodozoa* and *Haematococcus pluviais,* a high commercial value red colored pigment, used in the aquaculture, feed, and pharmaceutical industries. The red basidiomycetes yeast *Xanthophyllomyces dendrorhous* (Golubev 1995) and the green algae, *Heamatococcus pluvialis* produced a red color natural astaxanthin pigment. *Agrobacterium aurantiacum* has also been found to produce astaxanthin but the study related to it has been done to a lesser extent. Astaxanthin is applied to the skin for protection against sunburn, reduction of wrinkles, and other cosmetic benefits. It is used as a coloring in food for the processing of salmon, crabs, shrimp, chicken, and eggs. Astaxanthin is used in agriculture as a food supplement for chickens that produce eggs. The antioxidant properties of Astaxanthin provide the principal source of the supplement's health claims and benefits, particularly when used to help treat cancer (Hama et al. 2012; Zhou et al. 2019).

25.4.1.2 β-Carotene

β-Carotene is used in diverse types of dietary supplements, including multivitamins, vitamin A and antioxidant formulations. *Blakeslea trispora* and *Dunaliella salin* are potential sources of beta-carotene and show a metabolic role of provitamin-A activity. *Mucor circinelloides* (wild type) *and Phycomyces* are a promising source of beta-carotene.

25.4.1.3 Canthaxanthin

Canthaxanthin is a type of carotenoid, which is naturally found in bacteria, algae, and some groups of fungi. It is one of the primary carotenoids produced by *Bradyrhizobium* strains, which contain orange- and dark pink-pigmented bacteriochlorophyll.

25.4.2 Melanin

Melanins are naturally occurring pigments produced by many plants, animals, and many microorganisms. They are indolic polymers and can be classified as eumelanin, pheomelanin, and allomelanin (Banerjee et al. 2014). Melanin can shield fungi from adverse conditions; hence, this pigment has been documented to function as a fungal protector by neutralizing the oxidants produced in counter to stress. In addition to the scavenging function, many biological functions are exhibited by melanin pigment, including cytotoxic, anti-inflammatory, thermoregulatory, antimicrobial, antiviral, and immunomodulatory functions. Melanins absorb a wide range of electromagnetic spectrum, ultimately providing resistance to UV light and preventing photoinduced damage. A diverse range of microorganisms have been reported for the production of melanin such as *Alteromonas nigrifaciens, Aspergillus fumigates, Colletotrichum lagenarium, Cryptococcus neoformans, Magnaporthe grisea, Paracoccidioides brasiliensis, Shewanella colwelliana, Sporothrix schenckii, Vibrio cholerae,* and species of the *Streptomyces* genus (Langfelder et al. 2003; Soliev et al. 2011).

25.4.3 Prodigiosin

A red pigment known as prodigiosin secreted from *Serratia marcescens* (Boger and Patel 1987). It is a secondary metabolite that is synthesized in the last stages of bacterial development of various other microbes such as *Vibrio gazogenes, Serratia rubidaea, Pseudomonas magneslorubra, Vibrio psychroerythrous, Alteromonas rubra, Rugamonas rubra* and Gram-positive actinomycetes, like *Streptomyces longisporusruber* and *Streptoverticilli um rubrireticuli synthesizes* prodigiosin pigment and/or any derivatives of this molecule (Khanafari et al. 2006). It is a potential therapeutic molecule and possesses antimicrobial, antifungal, antimalarial, immunosuppresser, anticancer, and insecticidal activities (Harris et al. 2004; Kamble and Hiwarale 2012).

25.4.4 Phycocyanin

Phycocyanin is one of the blue pigments synthesized by *Pseudomonas aeruginosa* and cyanobacteria (Hassan and Fridovich 1980) and is made up of two subunits of *N*-methyl-1-hydroxyphenazine. It has been used as a biocontrol agent as it possesses antimicrobial and antifungal activities (Jayaseelan et al. 2014). It is used in medicine, food coloring, and cosmetics as it is a potent anti-inflammatory, hepatoprotective, and radical-scavenging agent which is nontoxic and noncarcinogenic.

25.4.5 Riboflavin

Riboflavin is a yellow-colored water-soluble pigment, synthesized by several microorganisms and also known as Vitamin B2. It is an essential vitamin supplement in the human diet and functions as a structural element of coenzymes flavin mononucleotide (FMN) and flavin adenine dinucleotide (FAD). Both of these coenzymes function as hydrogen carriers in the electron transport system and catalyze nonenzymatic redox reactions. Combination of riboflavin with UV light has been reported to be persuasive in lessening the detrimental pathogens in blood products (Goodrich et al. 2006). Riboflavin is produced by bacterium *Bacillus subtilis* and a group of fungi such as *Ashbya gossypii, Candida famata,* etc. It is used in different food items like baby foods, sauces, processed cheese, soft drinks, milk products, etc.

25.4.6 Violacein

A violet colored pigment violacein is isolated from *Chromobacterium violaceum*, but many other microorganisms such as *Duganella* sp., *Pseudoalteromonas ulvae, Collimonas* sp., *Pseudoalteromonas luteoviolacea, Microbulbifer* sp., *Pseudoalteromonas tunicate,* and *Janthinobacterium lividum* have been reported to produce violacein (Aranda et al. 2011; Lu et al. 2009; Hakvåg et al. 2009; Yada et al. 2008). Violacein is known to have a range of biological activities as antibacterial, antiviral, antileishmanial, antiulcerogenic, anticancer agent, and antibiotics against *Staphylococcus aureus* and other Gram-positive pathogens (Soliev et al. 2011; Durán et al. 2007).

25.5 Applications of Microbial Pigments

25.5.1 Biological Significance

Tentatively, microbial pigments with different colors and shades possess defensive properties to protect the organism from UV irradiation thus protecting and increasing their survivability (Agogué et al. 2005) in their surrounding environmental conditions. Antarctic heterotrophic bacteria can able to withstand cold environments by producing carotenoid pigmentation (Dieser et al. 2010). Some symbiotic or epibiotic pigmented bacteria play a defensive role against the pathogenic and many other predatory organisms, protecting their host organisms (Egan et al. 2002). Extremophilic microorganisms like *Thermus filiformis, Halococcus morrhuae,* and *Halobacterium salinarium* produces C50-carotenoids which helps in the survivability of these extremophiles by stabilizing their cell membrane (Mandelli et al. 2012). Carotenoid pigments can efficiently scavenge toxic radicals like reactive oxygen species (ROS), reactive nitrogen species (RNS), and some other nonbiological radicals (Chew and Park 2004; Mandelli et al. 2012). *Chromobacterium violaceum* produces a violet pigment called violacein that protects the lipid membrane such as egg and soybean phosphatidylcholine liposomes and rat liver microsomes against hydroxyl radical-induced lipid peroxidation (Konzen et al. 2006). Survival and formation of biofilms by bacteria are regulated by bacterial phenazines (Pierson and Pierson 2010). Yellow pigmentation of the *Thermus* strain protects them from the sunlight in the natural thermal areas (Rosenberg 2014).

Precisely, it was apparent that pigmented *Vibrio* strains producing prodigiosin survive around 1000-fold more under UV exposure (324 J/m^2) compared to non-pigmented vibrios (Borić et al. 2011). Several diverse toxic drugs are neutralized by the bacterial melanins and protect the cells (Plonka and Grabacka 2006) from stressful conditions like hyperosmotic stress, high temperature, and starvation like *Vibrio cholerae* (Coyne and al-Harthi 1992). *Janthinobacterium lividum* and *Chromobacterium violaceum* show a self-survival mechanism by producing violacein pigments causing cell death to bacterivores like *Spumella* sp.,

Ochromonas sp., and *Bodosaltans* (Matz et al. 2004). Pyoverdin facilitates iron transport in *P. fluorescens* (Visca et al. 2007). *Roseobacter* strains survive in the environment by producing a blue quinine compound indigoidine (Soliev et al. 2011).

Some researchers are published revealing marine-origin pigmented bacteria show higher resistance towards heavy metals and antibiotics than non-pigmented bacteria (Nair et al. 1992). Neutralization of the compounds containing polyphenols piled in the senescing nodules occurs by the melanin pigments of *Rhizobium* species (Johnson and Schroeder 1995). Fungal melanins stabilize the cells from harmful UV radiation. Endophytic fungi protect themselves from insects and other organisms by producing anthraquinones (Gessler et al. 2013). Diverse organisms produce a natural defensive compound against predators called Tambjamines (Blackman and Li 1994). Photosynthetic bacteria possess bacteriochlorophylls, proteorhodopsin, and bacteriorhodopsin similar to chlorophyll of the plants. Bacteriorhodopsin present in the bacteria helps them in obtaining energy from a low amount of organic matter. In short, microbial pigments displays many functions like antioxidant properties (Konzen et al. 2006), UV absorption (Stafsnes et al. 2010), virulence (Liu and Nizet 2009), protection from radiation (Tong and Lighthart 1997), antibiotic activities (Nakamura et al. 2003), and membrane stabilization (Oren 2011). For taxonomic identification and differentiation of various microbes, pigment traits are often used as a biological marker (Oren 2011). *Claviceps purpurea* produces many colors like yellow, black, orange, and red causing warning signs to predators (Lev-Yadun and Halpern 2007).

25.5.2 Microbial Pigments in Pharmacological Industries

25.5.2.1 Anticancer Potential of Bacterial Pigments

Around the globe, approximately six million people are suffering from cancer each year. Though many drugs for this prominent disease are being used and some are also in the trial phase, the serious side effects of those drugs have also been reported. Apoptosis is one of the regulated process for killing tumors without any side effects and is presently focused as a recent cancer treatment option (Lowe and Lin 2000; McConkey et al. 1996). Various natural products can be obtained from marine source (Cragg and Newman 1999; Schwartsmann 2000). In previous years nearly about 3000 compounds have been screened from marine sources and some of them also trialed in the clinical phase (Schwartsmann et al. 2001). Millions of species producing diverse novel agents with therapeutic properties are found in the ocean (Fenical 1997).

Twelve different bacterial strains were isolated and pigment extracts from them were tested for their cytotoxic activities on HeLa cell lines. The result revealed that the extracts were considerably able to impede the proliferation of the HeLa cell lines in a time and concentration-dependent way (Lin et al. 2005). Various other cell lines tested with the bacterial pigments are depicted in Table 25.2. Anthocyanin, a compound having a notable effect in reducing the prospect of cancer (Kong et al. 2003; Kim et al. 2012; Lazzè et al. 2004). *Serratia* culture broth secreted prodigiosin

Table 25.2 Cell types tested against bacterial pigments for cytotoxicity (Numan et al. 2018)

Type of cells	Source of cell lines	Cell line	Activity of pigments
Hela-derived	Human	Hep2	Inhibits the cell growth
Fetal kidney	Monkey	FR hK-4	Inhibits the cell growth
Lymphoma	Human	K562	Less effective than HL60
Uveal melanoma	Human	92.1	Mildly cytotoxic
Cervical epithelial cells	Human	HeLa	Cause apoptosis
Erythroleukemia	Human	TF 1	Cause cell death
Colorectal adenocarcinoma	Human	HCT116	Observed to be effective
Colorectal adenocarcinoma	Human	HT29	Cause cytotoxicity
Choroidal melanoma	Human	OCM-1	Mildly cytotoxic
Promyelocytic leukemia	Human	HL60	Cause apoptosis
Kidney	Monkey	Vero	Inhibits the cell growth

Table 25.3 Biological application of Prodigiosin and other important pigments (Numan et al. 2018)

Bacterial pigment	Applications	Source
Prodigiosin	Diabetes mellitus, cytotoxic, antiproliferative, leukemia patients	Bacteria
Ant342 (F-YOP)	Chemotherapy of tuberculosis	Bacteria
Anthocyanin	Cancer, immune response, inflammation	Bacteria
Violacein	Antiviral, antiprotozoan, anticancer, and antibacterial activities	Bacteria

pigment which shows antiproliferative and cytotoxic potential when tested against ovarian, colon, lung, and kidney cell lines. In the B-cell lymphocytic leukemia patients, prodigiosin is used (Kim et al. 2003; Ahmad et al. 2012). Apart from prodigiosin, the derivatives and analogs of the prodigiosin have been reviewed (Table 25.3). However, there are many more bacterial pigments with strong cytotoxic activities that are still to be discovered.

25.5.2.2 Antioxidant and Anti-hypersensitivity Activities

Other than common carotenoids such as astaxanthin, lutein, β-carotene, etc., a halophilic bacterium *Micrococcus yunnanensis* strain AOY-1 synthesized rare C_{50} carotenoids like sarcinaxanthin diglucoside, sarcixanthin, sarcixanthin monoglucoside also shows antioxidant activity with IC_{50} values of 74 μM, 54 μM, and 57 μM, respectively (Osawa et al. 2010). Lipid peroxidation caused by hydroxyl radicals can be protected by violacein pigment (Konzen et al. 2006). *Streptomyces mediolani* produces an unusual phenolic carotenoid, 3,3′-dihydroxyisorenieratene shows a powerful antioxidant activity (Sonani et al. 2016). Cyanobacterial pigments like C-phycocyanin, lycopene, lutein, and phycobiliproteins also demonstrate strong antioxidant properties (Sonani et al. 2016).

Systolic and diastolic blood pressures of the hypertensive rats have been revealed to be decreased by the oral administration of low-dose (150 mg/kg) of monascus-fermented dioscorea (Wu et al. 2009).

25.5.2.3 Antimicrobial Activities

Bacterial pigments exhibit antimicrobial activities against both Gram-positive and Gram-negative groups of bacteria. *Chromobacterium violaceum* produces a violet pigment having an antibacterial effect against S. *typhi* and *S. aurous* (August et al. 2000). A very strong antimicrobial activity was reported to be shown against *S. aureus*, *E. coli, S. typhi,* and *K. pneumonia* by a red-colored pigment isolated from *H. alkaliphilus* MSRD1 (Suresh et al. 2015). A study reported the antimicrobial effect of nine different isolated pigments against both Gram-positive and Gram-negative groups of bacteria and concluded that the pigments manifested higher efficacy against Gram-negative pathogenic bacteria than Gram-positive (Rashid et al. 2014). A weak antimicrobial effect was displayed by yellow, red, and orange color pigment produced by *Monascus purpureus* (Kim et al. 2006) but showed a positive effect when tested against *S. aureus* (Chen and Tseng 1989). Violacein is an effective pigment, that shows antimicrobial activity except for *Clostridium welchii* (Durán et al. 2012). *Chromobacterium*, a seawater isolate that produces a yellow pigment called tetrabromopyrrole, shows inhibition against human pathogens, and is also known to exhibit autoinhibition (Andersen et al. 1974). Several bacterial and fungal species biosynthesize phenazine compounds (Schneemann et al. 2011), anthraquinones, quinones (Johnson and Schroeder 1995) exhibiting a broad range of antimicrobial activities (Gessler et al. 2013).

25.5.2.4 Antifungal Activity

Prodiginines like undecylprodigiosin, cycloprodigiosin, and prodigiosin exert antifungal activity against fungi like *Histoplasma*, *Trichophyton*, *Asperigillus*, *Didymella*, *Cryptococcus*, *Sacharomyces*, *Penicillum,* and *Verticillum* (Stankovic et al. 2014). *Pseudoaltermonas tunicate* produces a yellow pigment called tambjamines possessing antifungal properties. Violacein isolated from *Janthinobacterium lividum* shows antifungal activity against white root rot causing phytopathogens *Rosellinia necatrix* (Shirata et al. 1997). *Trichoderma harzianum* and *Curuvularia lunata* synthesizes anthraquinone showing antifungal activities against fungal species (Gessler et al. 2013).

25.5.2.5 Immunosuppressive Activity

In Phorbol 12-myristate 13-acetate induced Jurkat cells, the proliferation of the T cells is inhibited by a red pigment extracted from *Pseudoalteromonas denitrificans* known as cycloprodigiosin hydrochloride. This depicts the immunosuppressive activity of the bacterial pigments (Kawauchi et al. 1997). T-cell mediated immune functions like local graft vs. host reaction, proliferation induced by concanavalin-A are inhibited by nontoxic concentrations of the prodigiosin (Han et al. 1998). Undecylprodigiosin with an IC_{50} value at 3–8 ng/mL inhibits the peripheral human T and B lymphocytes. Retinoblastoma phosphorylation is inhibited by inhibiting cyclin-dependent kinase-2 and cyclin-dependent-kinase-4 by undecylprodigiosin (Songia et al. 1997). Immunosuppressive activities by tambjamine alkaloids and other related proteins are also reported (Kim 2013).

25.5.2.6 Anti-HIV and Anti-Alzheimer Activity

The integrase of HIV is inhibited by compounds isolated from the *Phoma* species (Rai 2009). The effect of violacein on AIDS-related lymphoma has also been investigated (Duran et al. 1996).

Anti-Alzheimer activity is displayed by phycobiliproteins isolated from red algae and cyanobacterial species (Sonani et al. 2016).

25.5.2.7 Anti-lipoperoxidant and Antiulcerogenic Activities

A complex of violacein and β-cyclodextrin (βCD) inclusion (1:2) exhibits fourfold increased anti-lipoperoxidant activity in rat liver than violacin alone. Forty percent inhibition of MDA was observed with an IC_{50} at 125 and 505 μM (Durán et al. 2003).

Violacein in complexation with βCD at 1:1 and 1:2 ratio exhibits inhibition of gastric damage (Durán et al. 2003), chronic skin diseases, gastric ulcers, and eczema are treated by the flexirubin pigments isolated from *Chryseobacterium* and *Flavobacterium* (Kim 2013).

25.5.2.8 Anti-obesity, Anti-adipogenic, and Antidiabetic Activities

Pigments isolated from *Monascus* species are generally derivatives of L-Tryptophan and -L-Leucine ethyl esters inhibit the triglycerides and cholesterol contents (Kim et al. 2010). The decreased accumulation of triglyceride and proliferation of preadipocytes of 3T3-L1 cells along with the suppression of transcription factors like C/EPBα, C/EPBβ, C/EPBδ, and C/EPBϒ observed at a concentration of 8 μg/mL of ankaflavin and monascin pigment (Jou et al. 2010). In C57BL/6 mice serum, a pigment known as TEA (2-(*p*-toyly) ethylamine) decreased the content of cholesterol and low-density lipoprotein (LDL) by 24% and 38%, respectively (Choe et al. 2012).

Delipidation of mature adipocytes and the release of glycerol are promoted by monascin (by 113.2%) and ankaflavin (by 278.3%). Reduction of downregulation of enzyme heparin-releasable lipoprotein lipase (HR-LPL) was reduced by monascin and ankaflavin (Jou et al. 2010). Inhibitory effect against the adipogenic differentiation in the 3T3-L1 cell lines was observed when high-fat diet (HFD), added with the 2.5 μM of amine derivatives of monascus pigments, 4-phenylbutylamine (PBA), and 12.5 μM of 2-(*p*-toylyl) ethylamine (TEA) (Choe et al. 2012).

In diabetic rats lured by streptozotocin, red mold dioscorea fermented by *Monascus* protects the pancreatic β-cells, control hyperglycemia delaying diabetes (Shi et al. 2011). The antidiabetic effect of the Monascins is found to be exerted by preventing the ϒ-phosphorylation by enzyme phosphor-c-Jun N-terminal kinase (p-JNK) (Lee et al. 2011).

25.5.2.9 Herbicidal, Insecticidal, and Algicidal Activities

Several fungal species like *Phomaexigua* var., *P. foveata*, *P. macdonaldii*, *P. proboscis*, *P. herbarum*, *P. sorghina*, *P. tracheiphilla* produces anthraquinones possessing herbicidal activities (Rai 2009). From *P. herbarum*, a pigmented compound called anhydro pseudophlegmacin-9,10-quinone-3′-amino-8′-*O*-methyl ether was isolated, showing herbicidal effect against weeds such as *Lantana camara*, *Sidaacuta*, *Parthenium hysterophorus* (Quereshi et al. 2011).

Insecticidal activities of pigments like javanicin, fusarubin, naphthoquinone, and related pigments are reported (Gessler et al. 2013).

Chlorella growth was found to be controlled by a blue-green compound produced by *Chlorociboria aeruginosa*, known as xylindein (Sakaki et al. 2002).

25.5.2.10 Antiviral, Antimalarial, and Antituberculosis Activities

Promising antiviral activities of *Pseudomonas and Streptomyces* synthesized phenazines have been evaluated (Schneemann et al. 2011). An indicative extent of antiviral properties against some viruses like herpes simplex virus, simian rotavirus SA II, poliovirus was demonstrated by violacein (Durán et al. 2007). Anthraquinones, benzoquinone, and naphthoquinones are familiar quinine compounds demonstrating antiviral properties (Gessler et al. 2013).

Bacillus thuringiensis subsp. *israelensis* synthesizes melanin pigment, with mosquitocidal activity (Liu et al. 1993). Violacein, produced by *Chromobacterium violaceum,* exhibits antimalarial activity against *Plasmodium falciparum* (Lopes et al. 2009).

Inhibition of growth of *Mycobacterium tuberculosis* was demonstrated by flexirubin, isolated from *Flavobacterium* sp. Ant342 with a Minimum Inhibitory Concentration of 10.8 μg/mL. Similarly, violacein isolated from *Janthinobacterium* sp. Ant5-2 shows an inhibitory effect against virulent *M. tuberoculosis* with a Minimum Inhibitory Concentration of 34.4 μg/ml (Mojib et al. 2010).

25.5.2.11 Antiprotozoal and Antiparasitic Activities

Red pigments produced by *Serratia* exhibits antiprotozoal activities (Williams 1980). Freshwater isolates *Chromobacterium violaceum* and *Janthinobacterium lividum* produce violacein, possessing protozoan-killing ability. Destruction of the non-flagellates like *Spumella* and *Ochromonas* sp. was observed at its higher concentration of more than 10 μM (Matz et al. 2004).

The growth of *Trypanosoma cruzi* was inhibited in vitro by the violacein pigments extracted from *Chromobacterium violaceum* (Haun et al. 1992). Violacein at its IC_{50} value of 0.85 ± 0.11 μM hinders the growth of both chloroquine-sensitive as well as resistant strains of *Plasmodium falciparum* (Lopes et al. 2009).

25.5.2.12 Antileishmanial and Antitrypanosomal Activities

Worldwide, more than 12 million people are suffering from fatal and disfiguring Leishmaniasis disease. Though the drug for this was developed 50 years ago, the treatment is not efficiently working due to many toxic side effects. Leon et al. reported violacein showing a considerable antileishmanial activity with an EC_{50} value of approximately 4.3 μmol/L. However, violacein was found to be less effective, when compared with the drug pentamidine. Pentamidine of 16.8 μmol/L can inhibit 100% growth of promastigotes, whereas 460.8 μmol/L is required for achieving 100% inhibition (Leon et al. 2001). Although violacein is less effective than pentamidine, it has no side effects.

Prodigiosin extracted from *S. marcescens* disrupts the function of the mitochondria interfere with the oxidative phosphorylation process, eventually causing the cell death of *Trypanosoma cruzi* (Genes et al. 2011).

25.5.3 Microbial Pigments in Food Industries

In food industries, the use of natural food colorants instead of synthetic colorants to make the foods attractive appeared to be an important goal. The demand for natural colors is required more in the food industry since the artificial color additives exert negative impacts on health following their consumption (Aberoumand 2011). Several natural sources of food-grade colorants with their potentials still need to be explored. Though many colors are available isolated from natural sources, microbial colorants show a very important role as a food colorant due to its easy production, down the streaming process, higher productions through strain engineering, availability of starting materials with no seasonal deviations (Malik et al. 2012).

Some food-grade pigments were extracted from bacteria as mentioned in Table 25.4 and a blue pigment with natural color and excellent stability was isolated from cultured soil bacteria. Based on the research of East China University, these blue pigments tapped into the trend of edible food-grade natural colors (Zhang et al. 2006). However, there is a larger insistence for alternatives to artificial colors like sunset yellow, tartrazine, and quinoline yellow in the food markets. Being environment friendly, bacterial pigments as colorants can fulfill the need for visually appealing colors as well as serve as probiotics adding health benefits to the food products (Nagpal et al. 2011).

In the fish industries, the pink color of the salmon fish was enhanced by microbial colorants. Example of some commercial FDA approved food-grade colors used in the markets are pink-red from *Penicillium oxalicum*, carotene from *Blakesela trispora*, monascus pigment, astaxanthin from *Xanthophyllomyces* (Trispora 2003).

Table 25.4 Pigments used in food industries screened from bacteria (Venil et al. 2013)

Bacteria	Pigments	Color	Application
Photosynthetic bacterium, *Bradyrhizobium* sp., *Halobacterium* sp.	Canthaxanthin	Dark red	Food colorants
Flavobacterium sp., *Paracoccus xanthinifaciens*	Zeaxanthin	Yellow	Strengthens yellow color of the animal skins, acts as additives in the poultry feed, highlights the yolk color of the egg
Agrobacterium aurantiacum, *Paracoccus carotinifaciens*, *Halobacterium salinarium*	Astaxanthin	Pink-red	Food supplement, the natural nutritional component
Streptomyces sp.	Carotenoids	Yellow	The coloration of ornamental fishes, feed additive, food colorants

Still, there is a limitation in the availability of FDA approved colorants in the food industry. Some approved colorants are named by their chemical name canthaxanthin, and some are named by their sources such as fruit or vegetable juice. The use of biotechnological tools can be used for the synthesis of colorants easily. The success rate of the fermentation rely upon its approval by the FDA, market acceptability, and size of the capital investment (Dufossé 2016).

25.5.4 Microbial Pigments in the Textile Industries

Approximately, 130 lakh tons of synthetic dyes costing around U$ 23 billion are produced and consumed in the textile industry. However, synthetic dyes are hazardous to health, limiting its use. So in recent years the production and use of natural colorants are gaining more interest in the textile industry (Mapari et al. 2005). Renewable sources like fossil oil are the main source for the production of coloring agents. Other than the technological advancement and economic efficiency various challenges like environmental toxicity, human health, dependency on nonrenewable sources are faced by commercial producers. Wide varieties of chemically modified pigments can be produced through fermentation and this approach proved to be an important source for the colorant production (Ahmad et al. 2012). Rubramines, flavonoids, quinones, carotenoids are the pigments synthesized by the microorganisms with a lesser amount of residues and higher yield when compared to other producers like plants and animals (Numan et al. 2018). Apart from this, anthraquinone-like compounds is isolated from bacteria with antimicrobial properties and brighter color, proving its potential source as antimicrobial textile material (Numan et al. 2018).

Vibrio sp. produces a bright color pigment prodigiosin (Ahmad et al. 2012), which possesses many applications in diverse fields like textile and pharmaceuticals. Mant silk, acrylics, wools, fibers, and nylons are colored by the prodigiosin dye (Alihosseini et al. 2008). Five types of fabrics including polyester, cotton, silk, polyester microfiber can be colored by pigments extracted from *Serratia marcescens* (Venil et al. 2013). The capability in keeping color of the fabrics even in conditions like crocking/rubbing, perspiration, and washing are exhibited by the colorant isolated from *Janthinobacterium lividum*, used for dyeing of wool, cotton, silk, vinylon, wool, and nylon (Hiroshi 2000). Dying of material by colored pigment is achieved by boiling the material with pigmented cells or by dipping in the pigmented extract. The intensity of the color relies upon the extent of soaking time and the temperature of the dye bath. Violacin (violet-colored pigment) and prodigiosin (red-colored pigment) were reported for dyeing jacquard rayon, pure rayon, silk, and acrylic, respectively. In Asia for making flowers, leaves and geometrical designs on gown-like dress making violacin and prodigiosin are being used (Ahmad et al. 2012). The addition of ethyl acetate to purple and red color or acetone to yellow color improves the tone of the color (Ahmad et al. 2012).

25.6 Conclusion and Future Prospectives

Pigments originated from natural sources such as bacteria, microalgae, and fungi that are endowed to be more relevant and demandable over synthetic pigments. Other than an important source for natural colors, it also possesses broad use in the pharmaceutical and textile industry. However, more onerous attempts are still essential to achieve a cost-effective organic substrate for the growth and higher yield of microbial pigments. More investigations should be aimed to define the mechanism behind the action of pigments isolated from microbes. This will ultimately help in designing novel drug formulations for the regulation of many alarming diseases. In the future, analysis required to be further aimed at the structure of the natural pigments and their structural-functional relationship and these can be beneficial for commercial interest.

References

Aberoumand A (2011) A review article on edible pigments properties and sources as natural biocolorants in foodstuff and food industry. World J Dairy Food Sci 6(1):71–78

Agogué H, Joux F, Obernosterer I, Lebaron P (2005) Resistance of marine bacterioneuston to solar radiation. Appl Environ Microbiol 71(9):5282–5289

Ahmad A, Ahmad WW, Zakaria Z, Yosof N (2012) Applications of bacterial pigments as colorant: the Malaysian perspective. Springer briefs in molecular science, New York 3(7):57–74

Alihosseini F, Ju KS, Lango J, Hammock BD, Sun G (2008) Antibacterial colorants: characterization of prodiginines and their applications on textile materials. Biotechnol Prog 24(3):742–747

Andersen R, Wolfe M, Faulkner D (1974) Autotoxic antibiotic production by a marine Chromobacterium. Mar Biol 27(4):281–285

Aranda S, Montes-Borrego M, Landa BB (2011) Purple-pigmented violacein-producing Duganella spp. inhabit the rhizosphere of wild and cultivated olives in southern Spain. Microb Ecol 62 (2):446–459

August P, Grossman T, Minor C, Draper M, MacNeil I, Pemberton J, Call K, Holt D, Osburne M (2000) Sequence analysis and functional characterization of the violacein biosynthetic pathway from Chromobacterium violaceum. J Mol Microbiol Biotechnol 2(4):513–519

Avalos J, Limón MC (2015) Biological roles of fungal carotenoids. Curr Genet 61(3):309–324

Babitha S (2009) Microbial pigments. In: Biotechnology for agro-industrial residues utilisation. Springer, pp 147–162

Banerjee A, Supakar S, Banerjee R (2014) Melanin from the nitrogen-fixing bacterium Azotobacter chroococcum: a spectroscopic characterization. PLoS One 9(1):e84574

Baxter BK, Gunde-Cimerman N, Oren A (2014) Salty sisters: the women of halophiles. Front Microbiol 5:192

Blackman AJ, Li C (1994) New tambjamine alkaloids from the marine bryozoan Bugula dentata. Aust J Chem 47(8):1625–1629

Boger DL, Patel M (1987) Total synthesis of prodigiosin. Tetrahedron Lett 28(22):2499–2502

Borić M, Danevčič T, Stopar D (2011) Prodigiosin from Vibrio sp. DSM 14379; a new UV-protective pigment. Microb Ecol 62(3):528

Chen M-T, Tseng Y-Y (1989) Efficacy of antimicrobial substances from Monascus metabolites on preservation of meat. In: 35 international congress of meat science and technology, Copenhagen (Denmark), 20–25 Aug 1989, SFI

Chew BP, Park JS (2004) Carotenoid action on the immune response. J Nutr 134(1):257S–261S

Choe D, Lee J, Woo S, Shin CS (2012) Evaluation of the amine derivatives of Monascus pigment with anti-obesity activities. Food Chem 134(1):315–323

Coyne V, al-Harthi L (1992) Induction of melanin biosynthesis in Vibrio cholerae. Appl Environ Microbiol 58(9):2861–2865

Cragg GM, Newman DJ (1999) Discovery and development of antineoplastic agents from natural sources. Cancer Investig 17(2):153–163

Cristea D, Vilarem G (2006) Improving light fastness of natural dyes on cotton yarn. Dyes Pigments 70(3):238–245

Dieser M, Greenwood M, Foreman CM (2010) Carotenoid pigmentation in Antarctic heterotrophic bacteria as a strategy to withstand environmental stresses. Arct Antarct Alp Res 42(4):396–405

Dufossé L (2016) Current and potential natural pigments from microorganisms (bacteria, yeasts, fungi, microalgae). In: Handbook on natural pigments in food and beverages. Woodhead Publishing, pp 337–354

Dufossé L, Galaup P, Yaron A, Arad SM, Blanc P, Murthy KNC, Ravishankar GA (2005) Microorganisms and microalgae as sources of pigments for food use: a scientific oddity or an industrial reality? Trends Food Sci Technol 16(9):389–406

Dufosse L, Fouillaud M, Caro Y, Mapari SA, Sutthiwong N (2014) Filamentous fungi are large-scale producers of pigments and colorants for the food industry. Curr Opin Biotechnol 26:56–61

Duran N, Melo P, Haun M (1996) In vitro evaluation of violacein on AIDS-related lymphoma and human tumor cell lines. In: XXV Annual Meetings of the Brazilian Society of Biochemistry and Molecular Biology, Sociedade Brasileira de Bioquímica e Biologia Molecular SBBq, p 150

Durán N, Justo GZ, Melo PS, De Azevedo MB, Souza Brito AR, Almeida AB, Haun M (2003) Evaluation of the antiulcerogenic activity of violacein and its modulation by the inclusion complexation with β-cyclodextrin. Can J Physiol Pharmacol 81(4):387–396

Durán N, Justo GZ, Ferreira CV, Melo PS, Cordi L, Martins D (2007) Violacein: properties and biological activities. Biotechnol Appl Biochem 48(3):127–133

Durán M, Ponezi AN, Faljoni-Alario A, Teixeira MF, Justo GZ, Durán N (2012) Potential applications of violacein: a microbial pigment. Med Chem Res 21(7):1524–1532

Egan S, James S, Holmström C, Kjelleberg S (2002) Correlation between pigmentation and antifouling compounds produced by Pseudoalteromonas tunicata. Environ Microbiol 4 (8):433–442

El-Bialy HA, Abou El-Nour SA (2015) Physical and chemical stress on Serratia marcescens and studies on prodigiosin pigment production. Ann Microbiol 65(1):59–68

Feng Y, Shao Y, Chen F (2012) Monascus pigments. Appl Microbiol Biotechnol 96(6):1421–1440

Fenical W (1997) New pharmaceuticals from marine organisms. Trends Biotechnol 15(9):339–341

Genes C, Baquero E, Echeverri F, Maya JD, Triana O (2011) Mitochondrial dysfunction in Trypanosoma cruzi: the role of Serratia marcescens prodigiosin in the alternative treatment of Chagas disease. Parasit Vectors 4(1):1–8

Gessler N, Egorova A, Belozerskaya T (2013) Fungal anthraquinones. Appl Biochem Microbiol 49 (2):85–99

Giddings L-A, Newman DJ (2015) Bioactive compounds from terrestrial extremophiles. In: Bioactive compounds from terrestrial extremophiles. Springer, pp 1–75

Golubev WI (1995) Perfect state of Rhodomyces dendrorhous (Phaffia rhodozyma). Yeast 11 (2):101–110

Goodrich RP, Edrich RA, Li J, Seghatchian J (2006) The Mirasol™ PRT system for pathogen reduction of platelets and plasma: an overview of current status and future trends. Transfus Apher Sci 35(1):5–17

Gulrajani M (1992) Introduction to natural dyes. Indian Institute of Technology, New Delhi

Hakvåg S, Fjærvik E, Klinkenberg G, Borgos SEF, Josefsen KD, Ellingsen TE, Zotchev SB (2009) Violacein-producing Collimonas sp. from the sea surface microlayer of costal waters in Trøndelag, Norway. Marine Drugs 7(4):576–588

Hama S, Takahashi K, Inai Y, Shiota K, Sakamoto R, Yamada A, Tsuchiya H, Kanamura K, Yamashita E, Kogure K (2012) Protective effects of topical application of a poorly soluble antioxidant astaxanthin liposomal formulation on ultraviolet-induced skin damage. J Pharm Sci 101(8):2909–2916

Han SB, Kim HM, Kim YH, Lee CW, Jang E-S, Son KH, Kim SU, Kim YK (1998) T-cell specific immunosuppression by prodigiosin isolated from Serratia marcescens. Int J Immunopharmacol 20(1–3):1–13

Harris AK, Williamson NR, Slater H, Cox A, Abbasi S, Foulds I, Simonsen HT, Leeper FJ, Salmond GP (2004) The Serratia gene cluster encoding biosynthesis of the red antibiotic, prodigiosin, shows species-and strain-dependent genome context variation. Microbiology 150 (11):3547–3560

Hassan HM, Fridovich I (1980) Mechanism of the antibiotic action pyocyanine. J Bacteriol 141 (1):156–163

Haun M, Pereira MF, Hoffmann ME, Joyas A, Campos V, Filardi LD, De Castro SL, Duran N (1992) Bacterial chemistry. VI. Biological activities and cytotoxicity of 1, 3-dihydro-2H-indol-2-one derivatives. Biol Res 25(1):21–25

Hermansson M, Jones G, Kjelleberg S (1987) Frequency of antibiotic and heavy metal resistance, pigmentation, and plasmids in bacteria of the marine air-water interface. Appl Environ Microbiol 53(10):2338–2342

Hiroshi K (2000) Isolation of bacteria producing bluish-purple pigment and use for dyeing. Jpn Agric Res Q 34:131–140

Jayaseelan S, Ramaswamy D, Dharmaraj S (2014) Pyocyanin: production, applications, challenges and new insights. World J Microbiol Biotechnol 30(4):1159–1168

Johnson EA, Schroeder WA (1995) Microbial carotenoids. In: Downstream processing biosurfactants carotenoids. Springer, pp 119–178

Joshi VK, Attri D (2014) Characterization and conversion of microbial pigments into water soluble pigments for application in food products. Proc Natl Acad Sci India Sect B Biol Sci 84 (4):1053–1058

Joshi VK, Attri D, Bala A, Bhushan S (2003) Microbial pigments

Jou P-C, Ho B-Y, Hsu Y-W, Pan T-M (2010) The effect of Monascus secondary polyketide metabolites, monascin and ankaflavin, on adipogenesis and lipolysis activity in 3T3-L1. J Agric Food Chem 58(24):12703–12709

Kamble K, Hiwarale V (2012) Prodigiosin production from Serratia marcescens strains obtained from farm soil. Int J Environ Sci 3(1):631–638

Kawauchi K, Shibutani K, Yagisawa H, Kamata H, Nakatsuji SI, Anzai H, Yokoyama Y, Ikegami Y, Moriyama Y, Hirata H (1997) A possible immunosuppressant, cycloprodigiosin hydrochloride, obtained from Pseudoalteromonas denitrificans. Biochem Biophys Res Commun 237(3):543–547

Khan SS, Bhat SV (2013) Microbes: a potential source of food colors. Indian Hortic J 3(1 and 2):51–53

Khanafari A, Assadi MM, Fakhr FA (2006) Review of prodigiosin, pigmentation in Serratia marcescens. Online J Biol Sci 6(1):1–13

Khanafari A, Khavarinejad D, Mashinchian A (2010) Solar salt lake as natural environmental source for extraction halophilic pigments. Iran J Microbiol 2(2):103

Kim S-K (2013) Marine biomaterials: characterization, isolation and applications. CRC Press

Kim H, Han S, Lee C, Lee K, Park S, Kim Y (2003) Use of prodigiosin for treating diabetes mellitus. Google Patents

Kim C, Jung H, Kim JH, Shin CS (2006) Effect of monascus pigment derivatives on the electrophoretic mobility of bacteria, and the cell adsorption and antibacterial activities of pigments. Colloids Surf B: Biointerfaces 47(2):153–159

Kim JH, Kim YO, Jeun J, Choi D-Y, Shin CS (2010) L-Trp and L-Leu-OEt derivatives of the Monascus pigment exert high anti-obesity effects on mice. Biosci Biotechnol Biochem 74 (2):304–308

Kim HW, Kim JB, Cho SM, Chung MN, Lee YM, Chu SM, Che JH, Kim SN, Kim SY, Cho YS (2012) Anthocyanin changes in the Korean purple-fleshed sweet potato, Shinzami, as affected by steaming and baking. Food Chem 130(4):966–972

Köcher S, Müller V (2011) The nature and function of carotenoids in the moderately halophilic bacterium Halobacillus halophilus. In: Halophiles and hypersaline environments. Springer, pp 303–317

Koes RE, Quattrocchio F, Mol JN (1994) The flavonoid biosynthetic pathway in plants: function and evolution. BioEssays 16(2):123–132

Kong J-M, Chia L-S, Goh N-K, Chia T-F, Brouillard R (2003) Analysis and biological activities of anthocyanins. Phytochemistry 64(5):923–933

Konzen M, De Marco D, Cordova CA, Vieira TO, Antônio RV, Creczynski-Pasa TB (2006) Antioxidant properties of violacein: possible relation on its biological function. Bioorg Med Chem 14(24):8307–8313

Krishnamurthy K, Siva R, Senthil T (2002) Natural dye-yielding plants of Shervaroy Hills of Eastern Ghats. In: Proceedings of national seminar on the conservation of the Eastern Ghats. Environment Protection Training and Research Institute, Hyderabad, pp 151–153

Langfelder K, Streibel M, Jahn B, Haase G, Brakhage AA (2003) Biosynthesis of fungal melanins and their importance for human pathogenic fungi. Fungal Genet Biol 38(2):143–158

Lazzè MC, Savio M, Pizzala R, Cazzalini O, Perucca P, Scovassi AI, Stivala LA, Bianchi L (2004) Anthocyanins induce cell cycle perturbations and apoptosis in different human cell lines. Carcinogenesis 25(8):1427–1433

Lee B-H, Hsu W-H, Liao T-H, Pan T-M (2011) The Monascus metabolite monascin against TNF-α-induced insulin resistance via suppressing PPAR-γ phosphorylation in C2C12 myotubes. Food Chem Toxicol 49(10):2609–2617

Leon L, Miranda C, De Souza A, Durán N (2001) Antileishmanial activity of the violacein extracted from Chromobacterium violaceum. J Antimicrob Chemother 48(3):449–450

Lev-Yadun S, Halpern M (2007) Position paper: Ergot (Claviceps purpurea)—an aposematic fungus. Symbiosis

Li JW-H, Vederas JC (2009) Drug discovery and natural products: end of an era or an endless frontier? Science 325(5937):161–165

Likens GE (2010) Biogeochemistry of inland waters. Academic

Lin J, Yan XJ, Zheng L, Ma HH, Chen HM (2005) Cytotoxicity and apoptosis induction of some selected marine bacteria metabolites. J Appl Microbiol 99(6):1373–1382

Liu GY, Nizet V (2009) Color me bad: microbial pigments as virulence factors. Trends Microbiol 17(9):406–413

Liu Y-T, Sui M-J, Ji D-D, Wu I-H, Chou C-C, Chen C-C (1993) Protection from ultraviolet irradiation by melanin of mosquitocidal activity of Bacillus thuringiensis var. israelensis. J Invertebr Pathol 62(2):131–136

Lopes SC, Blanco YC, Justo GZ, Nogueira PA, Rodrigues FL, Goelnitz U, Wunderlich G, Facchini G, Brocchi M, Duran N (2009) Violacein extracted from Chromobacterium violaceum inhibits Plasmodium growth in vitro and in vivo. Antimicrob Agents Chemother 53 (5):2149–2152

Lowe SW, Lin AW (2000) Apoptosis in cancer. Carcinogenesis 21(3):485–495

Lu Y, Wang L, Xue Y, Zhang C, Xing X-H, Lou K, Zhang Z, Li Y, Zhang G, Bi J (2009) Production of violet pigment by a newly isolated psychrotrophic bacterium from a glacier in Xinjiang, China. Biochem Eng J 43(2):135–141

Malik K, Tokkas J, Goyal S (2012) Microbial pigments: a review. Int J Microbial Res Technol 1 (4):361–365

Mandelli F, Miranda VS, Rodrigues E, Mercadante AZ (2012) Identification of carotenoids with high antioxidant capacity produced by extremophile microorganisms. World J Microbiol Biotechnol 28(4):1781–1790

Mapari SA, Nielsen KF, Larsen TO, Frisvad JC, Meyer AS, Thrane U (2005) Exploring fungal biodiversity for the production of water-soluble pigments as potential natural food colorants. Curr Opin Biotechnol 16(2):231–238
Matz C, Deines P, Boenigk J, Arndt H, Eberl L, Kjelleberg S, Jürgens K (2004) Impact of violacein-producing bacteria on survival and feeding of bacterivorous nanoflagellates. Appl Environ Microbiol 70(3):1593–1599
McConkey DJ, Zhivotovsky B, Orrenius S (1996) Apoptosis—molecular mechanisms and biomedical implications. Mol Asp Med 17(1):1–110
Miteva V, Sheridan P, Brenchley J (2004) Phylogenetic and physiological diversity of microorganisms isolated from a deep Greenland glacier ice core. Appl Environ Microbiol 70 (1):202–213
Mojib N, Philpott R, Huang JP, Niederweis M, Bej AK (2010) Antimycobacterial activity in vitro of pigments isolated from Antarctic bacteria. Antonie Van Leeuwenhoek 98(4):531–540
Mondal S, Samantaray D, Mishra B (2015) Optimization of pigment production by a novel Bacillus sp. BBMRH isolated from cow dung. J Pure Appl Microbiol 9:2321–2326
Nagpal N, Munjal N, Chatterjee S (2011) Microbial pigments with health benefits-A mini review. Trends Biosci 4(2):157–160
Nair S, Chandramohan D, Bharathi PL (1992) Differential sensitivity of pigmented and non-pigmented marine bacteria to metals and antibiotics. Water Res 26(4):431–434
Nakamura Y, Asada C, Sawada T (2003) Production of antibacterial violet pigment by psychrotropic bacterium RT102 strain. Biotechnol Bioprocess Eng 8(1):37–40
Narsing Rao MP, Xiao M, Li W-J (2017) Fungal and bacterial pigments: secondary metabolites with wide applications. Front Microbiol 8:1113
Nedashkovskaya OI, Ludwig W (2015) Cyclobacteriaceae fam. nov. Bergey's manual of systematics of archaea and bacteria. pp 1–1
Nelis HJ, De Leenheer A (1991) Microbial sources of carotenoid pigments used in foods and feeds. J Appl Bacteriol 70(3):181–191
Numan M, Bashir S, Mumtaz R, Tayyab S, Rehman NU, Khan AL, Shinwari ZK, Al-Harrasi A (2018) Therapeutic applications of bacterial pigments: a review of current status and future opportunities. 3. Biotech 8(4):207
Oren A (2011) A short history of the symposia on halophilic microorganisms: from Rehovot 1978 to Beijing 2010. In: Halophiles and hypersaline environments. Springer, pp 373–382
Osawa A, Ishii Y, Sasamura N, Morita M, Kasai H, Maoka T, Shindo K (2010) Characterization and antioxidative activities of rare C50 carotenoids-sarcinaxanthin, sarcinaxanthin monoglucoside, and sarcinaxanthin diglucoside-obtained from Micrococcus yunnanensis. J Oleo Sci 59(12):653–659
Pierson LS, Pierson EA (2010) Metabolism and function of phenazines in bacteria: impacts on the behavior of bacteria in the environment and biotechnological processes. Appl Microbiol Biotechnol 86(6):1659–1670
Plonka PM, Grabacka M (2006) Melanin synthesis in microorganisms—biotechnological and medical aspects. Acta Biochim Pol 53(3):429–443
Proschold T, Leliaert F (2007) Systematics of the green algae: conflict of classic and modern approaches. Systematics Association Special Volume 75:123
Quereshi S, Khan NA, Pandey A (2011) Anthraquinone pigment with herbicidal potential from Phoma herbarum FGCC# 54. Chem Nat Compd 47(4):521
Rai M (2009) Advances in fungal biotechnology. IK International Pvt Ltd
Ramesh C, Mohanraju R, Murthy K, Karthick P (2017) Molecular characterization of marine pigmented bacteria showing antibacterial activity. Indian J Geo Mar Sci 46(10):2081–2087
Ramesh C, Vinithkumar NV, Kirubagaran R, Venil CK, Dufossé L (2019) Multifaceted applications of microbial pigments: current knowledge, challenges and future directions for public health implications. Microorganisms 7(7):186
Rashid MM, Fakruddin M, Mazumdar RM, Kaniz F, Chowdhury MA (2014) Anti-bacterial activity of pigments isolated from pigment-forming soil bacteria. J Pharm Res Int 4(8):880–894

Revuelta JL, Ledesma-Amaro R, Jiménez A (2016) Industrial production of vitamin B2 by microbial fermentation. In: Industrial biotechnology of vitamins, biopigments, and antioxidants. pp 17–40

Rosenberg E (2014) The family Chitinophagaceae. In: Rosenberg E, DeLong EF, Lory S, Stackebrandt E, Thompson F (eds) The prokaryotes: other major lineages of bacteria and the archaea. Springer, Berlin, Heidelberg

Roy S, Llewellyn CA, Egeland ES, Johnsen G (2011) Phytoplankton pigments: characterization, chemotaxonomy and applications in oceanography. Cambridge University Press

Sakaki T, Shibata M, Mukai K, Sakai M, Wakamatsu K, Miyauchi S (2002) Chlorociboria aeruginosa pigment as algicide. Jpn Kokai Tokkyo Koho JP 2002291493

Schneemann I, Wiese J, Kunz AL, Imhoff JF (2011) Genetic approach for the fast discovery of phenazine producing bacteria. Marine Drugs 9(5):772–789

Schwartsmann G (2000) Marine organisms and other novel natural sources of new cancer drugs. Ann Oncol 11:235–244

Schwartsmann G, da Rocha AB, Berlinck RG, Jimeno J (2001) Marine organisms as a source of new anticancer agents. Lancet Oncol 2(4):221–225

Shi Y-C, Liao J-W, Pan T-M (2011) Antihypertriglyceridemia and anti-inflammatory activities of Monascus-fermented dioscorea in streptozotocin-induced diabetic rats. Exp Diabetes Res 2011:710635

Shirata A, Tsukamoto T, Yasui H, Kato H, Hayasaka S, Kojima A (1997) Production of bluish-purple pigments by Janthinobacterium lividum isolated from the raw silk and dyeing with them. J Sericult Sci Jpn 66(6):377–385

Siva R (2007) Status of natural dyes and dye-yielding plants in India. Curr Sci 92:916–925

Soliev AB, Hosokawa K, Enomoto K (2011) Bioactive pigments from marine bacteria: applications and physiological roles. Evid Based Complement Alternat Med 2011:670349

Sonani RR, Rastogi RP, Patel R, Madamwar D (2016) Recent advances in production, purification and applications of phycobiliproteins. World J Biol Chem 7(1):100

Songia S, Mortellaro A, Taverna S, Fornasiero C, Scheiber EA, Erba E, Colotta F, Mantovani A, Isetta A-M, Golay J (1997) Characterization of the new immunosuppressive drug undecylprodigiosin in human lymphocytes: retinoblastoma protein, cyclin-dependent kinase-2, and cyclin-dependent kinase-4 as molecular targets. J Immunol 158(8):3987–3995

Sowbhagya H, Chitra V (2010) Enzyme-assisted extraction of flavorings and colorants from plant materials. Crit Rev Food Sci Nutr 50(2):146–161

Stafsnes MH, Josefsen KD, Kildahl-Andersen G, Valla S, Ellingsen TE, Bruheim P (2010) Isolation and characterization of marine pigmented bacteria from Norwegian coastal waters and screening for carotenoids with UVA-blue light absorbing properties. J Microbiol 48(1):16–23

Stankovic N, Senerovic L, Ilic-Tomic T, Vasiljevic B, Nikodinovic-Runic J (2014) Properties and applications of undecylprodigiosin and other bacterial prodigiosins. Appl Microbiol Biotechnol 98(9):3841–3858

Suresh M, Renugadevi B, Brammavidhya S, Iyapparaj P, Anantharaman P (2015) Antibacterial activity of red pigment produced by halolactibacillus alkaliphilus MSRD1—an isolate from seaweed. Appl Biochem Biotechnol 176(1):185–195

Takaichi S (2011) Carotenoids in algae: distributions, biosyntheses and functions. Marine Drugs 9 (6):1101–1118

Tong Y, Lighthart B (1997) Solar radiation is shown to select for pigmented bacteria in the ambient outdoor atmosphere. Photochem Photobiol 65(1):103–106

Trispora B (2003) Application for the approval of lycopene from Blakeslea trispora

Usmani Z, Sharma M, Sudheer S, Gupta VK, Bhat R (2020) Engineered microbes for pigment production using waste biomass. Curr Genomics 21(2):80–95

Venil CK, Lakshmanaperumalsamy P (2009) An insightful overview on microbial pigment, prodigiosin. Electron J Biol 5(3):49–61

Venil CK, Zakaria ZA, Ahmad WA (2013) Bacterial pigments and their applications. Process Biochem 48(7):1065–1079

Venil CK, Aruldass CA, Dufossé L, Zakaria ZA, Ahmad WA (2014) Current perspective on bacterial pigments: emerging sustainable compounds with coloring and biological properties for the industry—an incisive evaluation. RSC Adv 4(74):39523–39529

Visca P, Imperi F, Lamont IL (2007) Pyoverdine siderophores: from biogenesis to biosignificance. Trends Microbiol 15(1):22–30

Wackenroder H (1831) Ueber das Oleum radicis Dauci aetherum, das Carotin, den Carotenzucker und den officinellen succus Dauci; so wie auch über das Mannit, welches in dem Möhrensafte durch eine besondere Art der Gährung gebildet wird. Geigers Magazin der Pharmazie 33:144–172

Walford J (1980) Developments in food colours (No. 614.3 W3)

Wang H, Jiang P, Lu Y, Ruan Z, Jiang R, Xing X-H, Lou K, Wei D (2009) Optimization of culture conditions for violacein production by a new strain of Duganella sp. B2. Biochem Eng J 44 (2–3):119–124

Williams R (1980) The pigment of Serratia. The genus Serratia. pp 31–75

Wu C-L, Lee C-L, Pan T-M (2009) Red mold dioscorea has a greater antihypertensive effect than traditional red mold rice in spontaneously hypertensive rats. J Agric Food Chem 57 (11):5035–5041

Yada S, Wang Y, Zou Y, Nagasaki K, Hosokawa K, Osaka I, Arakawa R, Enomoto K (2008) Isolation and characterization of two groups of novel marine bacteria producing violacein. Mar Biotechnol 10(2):128–132

Yurkov VV, Krieger S, Stackebrandt E, Beatty JT (1999) Citromicrobium bathyomarinum, a novel aerobic bacterium isolated from deep-sea hydrothermal vent plume waters that contains photosynthetic pigment-protein complexes. J Bacteriol 181(15):4517–4525

Zachleder V (1998) van den Hoek, C., Mann, DG, Jahns, HM (ed.): Algae. An introduction to phycology. Photosynthetica 4(35):506–506

Zhang H, Zhan J, Su K, Zhang Y (2006) A kind of potential food additive produced by Streptomyces coelicolor: characteristics of blue pigment and identification of a novel compound, λ-actinorhodin. Food Chem 95(2):186–192

Zhang X, Yao T, Tian L, Xu S, An L (2008) Phylogenetic and physiological diversity of bacteria isolated from Puruogangri ice core. Microb Ecol 55(3):476–488

Zhou P, Li M, Shen B, Yao Z, Bian Q, Ye L, Yu H (2019) Directed coevolution of β-carotene ketolase and hydroxylase and its application in temperature-regulated biosynthesis of astaxanthin. J Agric Food Chem 67(4):1072–1080

Bio-fermentative Production of Xanthan Gum Biopolymer and Its Application in Petroleum Sector

26

Nanthakumar Kuppanan, Paul Jeyaseelan, Shishram Chahar, Subhasis Das, Veeranna Channashettar, and Banwari Lal

Abstract

Currently, industries and consumers are focusing on green products to reduce carbon footprints. Xanthan gum is a natural microbial polysaccharide known as gums or exopolysaccharides (EPS) which has excellent characteristics such as gel formation, thickening, emulsion stabilizing effect, salt tolerance, pH resistance, and stability to various enzymes makes this product industrially an important additive. The xanthan gum is a high molecular weight, pseudoplastic, and water-soluble heteropolysaccharide that consists of polymerized pentasaccharide. This biopolymer is a most important fermentation product widely used in the food, oil, petrochemical, agrochemical, and pharmaceutical industry. The quality of Xanthan gum is influenced by several factors such as type of Xanthomonas strain, batch variability, cultivation media, and downstream. Due to versatile rheological properties, xanthan gum is considered as a nonhazardous, environmental friendly product for the drilling of oil wells. The mixture of xanthan gum and other additives used as water-based drilling fluids in oil well application. This chapter discusses the fermentative production of xanthan gum, various factors affecting its biosynthesis, recovery, and its application in the drilling of oil wells.

Keywords

Xanthan gum · *Xanthomonas campestris* · *Biofermentation* · Oil Well drilling

N. Kuppanan (✉) · P. Jeyaseelan · S. Chahar · S. Das · V. Channashettar · B. Lal (✉)
Environmental and Industrial Biotechnology Division, The Energy and Resources Institute (TERI), India Habitat Centre, New Delhi, India
e-mail: k.nanthakumar@teri.res.in; banwaril@teri.res.in

A. Vaishnav, D. K. Choudhary (eds.), *Microbial Polymers*,
https://doi.org/10.1007/978-981-16-0045-6_26

26.1 Introduction

In recent days natural polysaccharides received attention among industries and consumers due to its biodegradability, nontoxic, and biocompatibility over synthetic or chemicals polymers. The microbial-based polysaccharides such as pullulan, curdlan, scleroglucan, dextran, and xanthan are industrially produced in controlled bio-fermentation conditions for the commercial application. Xanthan gum possesses several advantages over plant and synthetic polysaccharides as it can be easily produced in an industrial bioreactor using low-cost waste materials as carbon sources. Bacterial exopolysaccharides are being used as a gums produced by bacterial strains *Xanthomonas campestris*. This polymer has novel properties to form gels and viscous solutions in the fermentation process (Morris et al. 2012). Owing to specific properties such as high molecular weight, rheological properties, pseudoplastic and water-soluble heteropolysaccharide nature, the xanthan gum is used in food, pharmaceutical, cosmetic industry, and petroleum and oil industry. Due to the branching structure, xanthan gum possesses stability under a wide range of temperature, Ph, and ionic concentration.

High purity xanthan gum is used in the pharmaceutical and food industries. It is expensive and produced from glucose and sucrose fermentation to meet the high purity standards. For the food and drug applications, the downstream process is very intensive which leads to an increase in the production cost whereas for non-food applications, a crude product is produced using less expensive substrates, such as waste agricultural products. In the year 1960, the first industrial production of xanthan gum was carried out and became available commercially (Flores-Candia and Deckwer 1999). The viscosity of xanthan gum decreases with increasing shear rate so that it is easy to operate with mixing and pumping. Due to these unique properties, the xanthan gum is used as a major product in oil well drilling for oil recovery. Xanthan gum polymers act as a stabilizer and facilitate cementing, remove cuttings from well, control formation pressure, and seal permeable formation.

Drill-in fluids are generally water-based formulations which contain Xanthan gum and a combination of additives. This combination is made of a viscosifier, fluid loss additive, and bridging particles. Bio-based xanthan gum along with other suitable additives in the drilling fluid formulation is a very important component of technically viable and cost-effective drilling of oil wells. Since 1887, the development of drilling fluid had undergone several technological changes due to its extended application demand and quality (Chilingarian and Vorabutr 1983). Different type of oil, water, and gas-based drilling fluids are used in oil well drilling.

The oil well drilling operation strongly depends on the performance of drilling fluid which in turn depends on formation damage and wellbore. It accounts for more than 15% of the total oil well drilling cost. Xanthan gum-based drilling fluid occupied a major role due to technical performance, cost-effectiveness, and its effect on health, safety, and environment. Moreover, the xanthan-based drilling fluid demand has been increased due to ever-increasing drilling depth and reach, linked with the geographical location induce the oil industries to find alternative products. This review focuses on the xanthan production, microbial fermentation process, different subtracts, control factors, and finally discusses the potential properties and application in the oil and petroleum industries.

26.2 Structure and Properties of Xanthan Gum

The xanthan gum produced by bacterial strains is classified as a branched heteropolysaccharide which contains a cellulose like backbone with three-unit-long side chains at the C-3 position. Xanthan polymer is composed of glucose, mannose, and glucuronic acid in a ratio of 2:2:1 (Fig. 26.1). Xanthan is made up of repeating pentasaccharide units consisting of two D-mannopyranosyl units, two D-glucopyranosyl units, and one D-glucopyranosyluronic acid. 1-4 linked β-D-glucopyranosyl units are the backbone of xanthan. A trisaccharide side chain with a D-glucuronosyl unit is attached between two D-mannosyl units at the O-3 position. At the O-4 position, the β-D-glucopyranosyluronic acid unit links the terminal β-D-mannopyranosyl unit (Kang and Pettitt 1993).

X-ray diffraction characterization studies revealed that orientated xanthan gum fibers as a right-handed, fivefold helix with a rise per backbone disaccharide residue of 0.94 nm. This description also confirms that the trisaccharide side chain is aligned with the backbone and stabilizes the complete conformation by non-covalent interactions, mainly hydrogen bonding. In solution the side chains wrap around

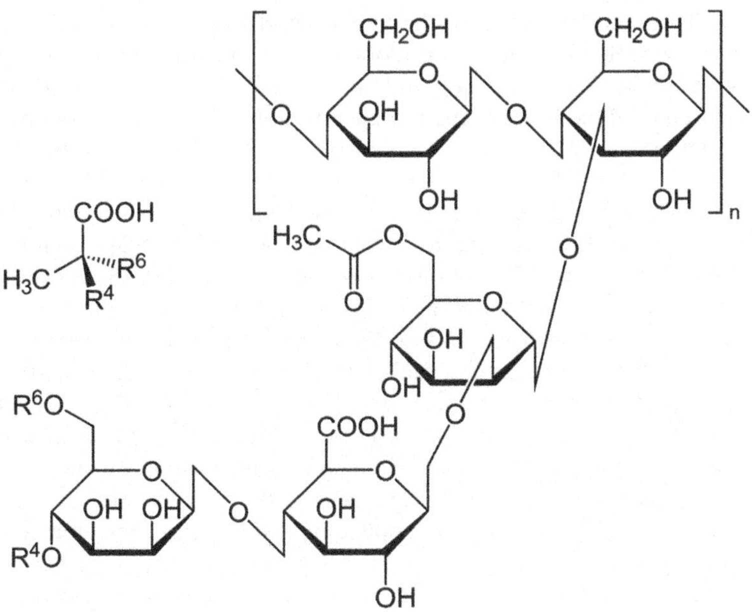

Fig. 26.1 Structure of Xanthan gum

the backbone thereby protecting the labile linkages from attack. This protection is mainly responsible for gum stability under different adverse conditions.

Xanthan is a free flowing cream color powder, soluble in water. It hydrates very quickly if dispersed and possesses excellent thermal stability representing an important class of polymer because of their ability to form gel and viscous solutions that are industrially produced by fermentation techniques (Pelletier et al. 2001; Ruffing and Chen 2006; Morris et al. 2012). The molecular weight distribution ranges from 2×10^6 to 20×10^6 Da. This molecular weight varies with its chains of association. The non-Newtonian behavior of xanthan gum makes it highly viscous, less sensitive for salt, biodegradable, resistant to mechanical degradation, and high stability to acid, alkaline, and heat because of the branched helical backbone (Malik and Sridhar 1992; Whistler and BeMiller 1997). Further, the xanthan is stable and remains unchanged in its viscosity under a wide range of temperatures. Xanthan gum tends to form gels in the presence of divalent cations at higher pH levels (>10). Most of the xanthan gum producing bacteria utilize different activated carbohydrate donors to form the polymer on an acceptor molecule.

26.3 Bio-fermentative Production of Xanthan Gum

The xanthan gum is produced by different species of Gram-negative Xanthomonas bacteria including X. *campestris*, *X. fragaria*, *X. gummisudans*, *X. juglanids*, *X. phaseoli*, and *X. vasculorum* (Kennedy and Bradshaw 1984). Among these *X. campestris* is used in the commercial production of Xanthan gum due to its high yield. The xanthan gum is synthesized by bacterial strains in cell membranes and through the cell-cell signaling or quorum sensing that triggers the formation of biofilm which consists of xanthan gum (Vojnov et al. 2001). The composition and molecular weight of the xanthan chains depends on fermentation conditions and the type of Xanthomonas strains used (Kennedy and Bradshaw 1984; Becker et al. 1998).

The Xanthan gum synthetic pathway can be divided into three parts: (1) Sugar molecule uptake and conversion to nucleotide derivatives. (2) Aggregation of pentasaccharide subunits into isopentyl pyrophosphate carrier. (3) Polymerization of pentasaccharide units and secretion. The biosynthesis of xanthan, as in most polysaccharide-producing bacteria, utilizes various activated carbohydrate donors to form the polymer on an acceptor molecule (Palaniraj and Jayaraman 2011).

The fermentative production of xanthan gum starts with the preparation of X. campestries inoculum from stock cultures which are prepared in a suitable fermentation medium in conventional batch processing using mechanically agitated vessels called fermenter (Fig. 26.2). The aerated bacterial culture that undergoes aerobic process is held at the following operating conditions: temperature 28–30 °C, pH $\sim$ 7, the aeration rate higher than 0.3 (v/v), and the specific power input for agitation higher than 1 kW/m^3. The fermentation process is carried out for predetermined contact time to convert the substrate into the xanthan product. Inoculum preparation includes several stages that require a set of the reactor ranging

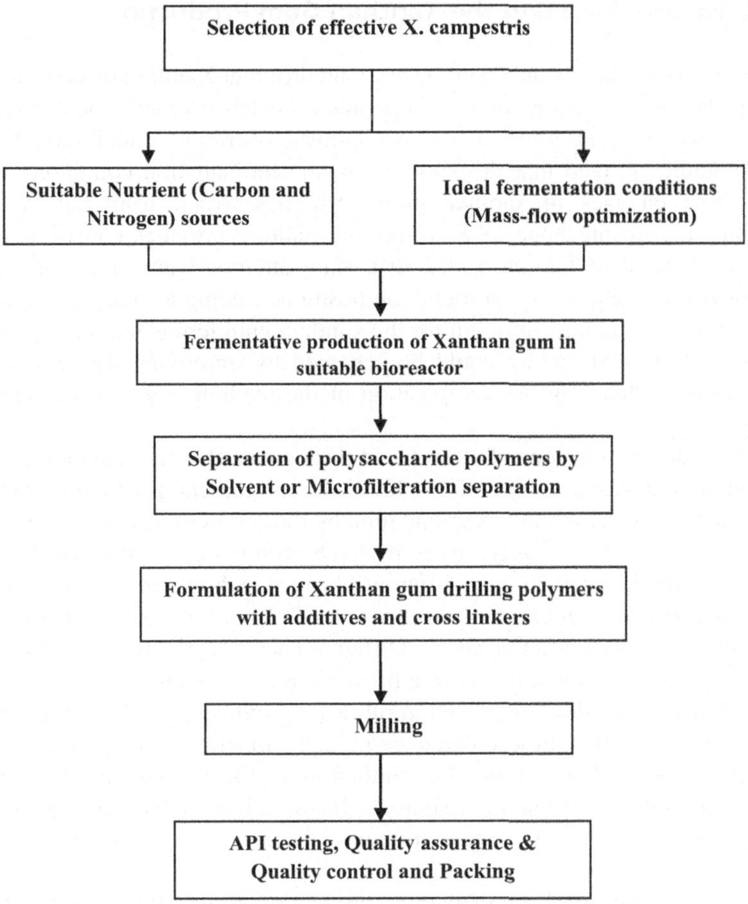

Fig. 26.2 Stages in production of Xanthan gum-based drilling fluid product

from 10 L for the initial seed up to 100 m^3 in the production stage by which the volume is usually maximized by tenfolds (Flickinger and Draw 1999). During the growth phase, the polysaccharide accumulation starts and continues after growth. The pH reduction occurs during the fermentation due to the formation of organic acids. The maintenance of optimum pH 7 helps in maximum xanthan gum yield (Gumus et al. 2010). After product formation in the fermentation stage, multistep downstream processes would perform to separate the xanthan product. In commercial scale xanthan production, the post-fermentation process includes pasteurization of fermented broth followed by separation of xanthan gum by microfiltration or solvent precipitation. The precipitated xanthan gum is further processed in sprayed dry, dewatering, and washing to obtain final xanthan gum. Finally, the xanthan gum is mixed with drilling fluid additives to make final xanthan gum drilling fluid.

26.3.1 Factors Affecting the Xanthan Gum Production

The quality and yield of xanthan gum vary with different Xanthomonas strains, types of fermentation bioreactor, mode of operation—batch or continuous, carbon and nitrogen sources, and physicochemical culture operation conditions. Different research studies proved that the variation in the fermentation conditions leads to characteristic changes in xanthan gum properties. Apart from culture growth conditions, the maintenance of X. campestris culture provides consistency in the production of xanthan (Gumus et al. 2010). Many attempts were reported to optimize operational variables such as nutrient composition, feeding technique, temperature, pH, agitation, and adding antifoam for the xanthan gum fermentation. Improvement of xanthan yield and quality could be achieved by improving the fermenter and process design, changing the composition of the medium, and the media feeding strategy.

The xanthan gum is generally produced through aerobic fermentation process at an optimum temperature range of 27–30 °C. In commercial production, more than 70% of substrate converts into Xanthan gum by the action of X. campestris. In order to enhance the product quality, many research groups have optimized the various operational conditions including different cheaper carbon sources. Most of these studies are carried out in bench-scale experiments. Very few studies tried to upscale the conditions to market the products. Different bacterial strains were also explored to produce quality xanthan gum for application in food, pharmaceutical, and other industrial sectors besides being used in oil drilling industry for the enhancement of crude recovery. Studies showed that even in high temperature or solutions with high concentration of acid, alkali, or salt, xanthan gum still can maintain this property, which is especially important in offshore drilling or other harsh conditions. This high stability and high viscosity nature influence the success of the oil well drilling process.

The present industrial xanthan gum production is usually carried out in the medium that provides a compromise between the concentrations needed for cell growth and xanthan gum formation. Generally, glucose is used as a substrate for the production of xanthan gum in batch production than in continuous production due to its successful results. Due to increasing cost, the alternative raw material has been optimized. Further different cultures would require different media and optimum conditions, many studied on nutrients required for the product side chain variation and optimization in xanthan gum biosynthesis have been reported (Zhang 1999; Garcia-Ochoa et al. 1992, 2000b; Letisse et al. 2001).

The industrial scale production of xanthan is carried out using inexpensive substrates and nutrients. Carbohydrate sources such as sucrose, sugarcane molasses, hydrolyzed starch, corn syrup, hydrolyzed rice, barley and corn flour, acid whey, coconut juice, sugar cane, beet molasses, date juice palm, corn steep liquor, and whey have been successfully used in the production medium (El-Salam et al. 1994; Rosalam and England 2006; Salah et al. 2011). A list of different nonexpensive nutrient sources and its xanthan yield is presented in Table 26.1. Apart from substrate type, the quality of Xanthan gum is also affected by operational factors

Table 26.1 Different nutrient sources explored for the Xanthan gum yield by Xanthomonas sp.

S. no.	Xanthomonas strain	Nutrient source	Xanthan gum yield (g/L)	Reference
1	*X. campestris* NRRL-B-1459	Sucrose	30	De Vuyst et al. (1987)
2	*X. campestris* NRRL B-1459	Agro-industry wastes spent malt grain	32.9–57.1	Miroslav and Elena (1999)
3	*X. campestris*	Apple pomace	28	Stredansky and Conti (1999)
4	*X. campestris*	Grape pomace	10	Stredansky and Conti (1999)
5	*X. campestris*	Glucose	75	Garcia-Ochoa et al. (2000b)
6	*X. campestris* ATCC 1395	Sugar beet molasses	53	Kalogiannis et al. (2003)
7	X. campestris	Cassava bagasse hydrolysate	14	Adnise et al. (2003)
8	*X. campestris* MO-0	Ram horn hydrolysate	25	Kurbanoglu and Kurbanoglu (2007)
9	*X. campestris pv.campestris*	Sugar-beet molasses	*19.8*	Moosavi and Karbassin (2010)
10	*X. campestris NRRL B-1459*	Date palm juice	24.5	Salah et al. (2011)
11	X. campestris PTCC 1473	Milk whey	16.5	Gilani et al. (2011)
12	X. campestris MTCC 2286	Acidified sugarcane molasses	10.3	Murugesan et al. (2012)
13	Xanthomonas campestris NRRL-B 1459 or ATCC 13951	Yeast extract	4.8	Kumaraswamy et al. (2012)
14	M. rouxii NRRL 1894	Molasses	30	Abdel-Aziz et al. (2012)
15	X. campestris PTCC1473	Date extract	11.2	Khosravi et al. (2013)
16	X. campestries 1866	Coconut shell extract	10.3	Nery et al. (2013)
17	X. campestris NRRL B-1459	Extract of spoiled dates and their remains	43.35	Chandrasekaran and Bahkali (2013)
18	X. campestris ATCC 15206	Pulp hydrolysate	20.8	Gowdhaman et al. (2013)
19	*Xanthomonas* NIGEB-K37 strain	Lactose based medium (10%)	14.26	Ramezani et al. (2014)
20	*Xanthomonas* NIGEB-K37 strain	Cheese whey medium	18.4	Ramezani et al. (2014)
21	X. campestris XLM 1521	Whey	12	De Sousa et al. (2014)

(continued)

Table 26.1 (continued)

S. no.	Xanthomonas strain	Nutrient source	Xanthan gum yield (g/L)	Reference
22	X. campestris PTCC1473	Date juice by-products	6.72	Moshaf et al. (2015)
23	X. campestris PTCC1473 and X. pelargonii PTCC1474)	Cheese whey	16.77	Niknezhad et al. (2015)
24	X. campestris pv campestris (b82)	Grape juice concentrate	14.35	Ghashghaei et al. (2016)
25	X. campestris LRELP-1	Kitchen waste hydrolysate	11.73	Li et al. (2016)
26	*X. campestris* WXLB-006	Glycerol	33.9	Wang et al. (2017)
27	X. campestris NCIM 2961	Jackfruit seed powder	51.62	Katherine et al. (2017)
28	*X. campestris (PTCC 1473)*	Broomcorn	8.9	Zahra et al. (2018)
29	X. campestris (BeNa culture collection)	Orange peels	30.19	Mohsin et al. (2018)

such as type of Xanthomonas strain, batch variability, cultivation media, and downstream processing (Garcia-Ochoa et al. 2000a; Desplanques et al. 2012). For commercial production, glucose is the most commonly used substrate for xanthan gum production under batch process (Rosalam and England 2006; Harding et al. 1995). In industrial scale xanthan gum production, Xanthomonas strain is used under submerged fermentation with controlling process variables, such as carbon and nitrogen sources, batch or fed-batch process, pH, temperature, inoculums, airflow rate, agitation, and fermentation time (Garcia-Ochoa et al. 1992; Esgalhado et al. 1995; Umashankar et al. 1996; Liakopoulou et al. 1999; Murugesan et al. 2012; Niknezhad et al. 2015).

According to Amanullah et al. (1988), the rheological complexity of Xanthan fermentations is influenced by mixing phenomena. The study was conducted in a 150 L fermentor which compares the physical and biological performance of four pairs of impellers and the result highlighted that well mixed and oxygenated region increases the capacity for high microbial oxygen uptake rates which govern the Xanthan productivity and quality. Different research group has explored the utilization of various nutrient sources such as soybean and cereal grains. The research findings demonstrated that the production media composition significantly influences the production of xanthan gum. Davidson has observed that limiting magnesium or phosphate resulted in the production of low pyurate containing xanthan gum (Embden–Meyerhop or EMP) (Davidson 1978).

Souw and Demain (1979) have demonstrated that sucrose is an ideal substrate for xanthan gum production and found that succunate and 2-oxoglutarate have stimulatory effects on xanthan gum production in sucrose-based medium. Letisse et al. (2001) performed the fermentation using X. *camprestris* ATCC 13951 and sucrose as the carbon source. They reported that the xanthan gum production yield has been increased by nitrate at 0.79 mmol/g cells h cf. ammonium at 0.52 mmol/g cells h, although the organic acid content (acetate and pyruvate) of the xanthan gum remained constant at 6.0% and 4.6%, respectively. In this study, the ammonium is reported for the biomass accumulation, whereas xanthan gum yields are higher with nitrate as the nitrogen source. The growth initiation is poor for each of the mineral nitrogen sources but could be improved using organic nitrogen (e.g., soybean flour hydrolysate). The xanthan gum-based drilling fluid production involves different stages as highlighted in Fig. 26.3.

Xanthan gum is suitable for tertiary oil recovery and enhanced oil recovery in oil production. Its strong heat resistance properties make it a reliable displacing agent and mobile control agent. Studies show that xanthan gum polymer proved to be suitable for the formulation of fluids which can be used in wellbores with a higher temperature gradient. In this context, many companies realized that technological monitoring is a growing need for organizations competing in dynamic and globalized markets. Further identification of relevant technologies, market niches, partnerships, and innovation and competition movements helps in upgrading products to global market.

26.4 Downstream Separation of Xanthan Gum

There are different researches focused on the standardization of xanthan gum separation from the fermentation medium through various separation processes (Lo et al. 1996; Howell et al. 1996; Yang et al. 1996). Usually, the fermentation broth is pasteurized or sterilized to kill the cells. The thermal treatments also enhance xanthan removal from the cells. Adaptation of ideal treatment conditions (80–100 °C, 10–20 min, pH 6.3–6.9) enhanced xanthan dissolution occurs without thermal degradation and disruption of cells is observed (Garcia-Ochoa et al. 1992). The high viscous nature of xanthan gum is a major problem during the removal of bacterial cell biomass from the fermented broth. The high viscous nature of the compounds causes degradation during centrifugation and heat treatment during separation from cells. The enzymatic lysis of bacterial cells in microbial broth is preferred for the removal of cells, as the structure of the gum is preserved and effective at high viscosity (Suresh and Prasad 2005). The addition of a non-solvent reagent promotes precipitation not only by decreasing the water affinity of the polymer but also by enhancing the binding of the cations, which are present. Thus, xanthan precipitates with lesser amounts of reagents when alcohol and salt are used in combination (Garcia-Ochoa et al. 1992).

The dissolved colloid using methods such as salts, water-miscible non-solvents, and concentration by evaporation. The addition of solvents in the fermentation also influences the recovery process which includes alcohols (methanol, ethanol, isopropanol) and acetone. These are non-solvents for the polysaccharide which

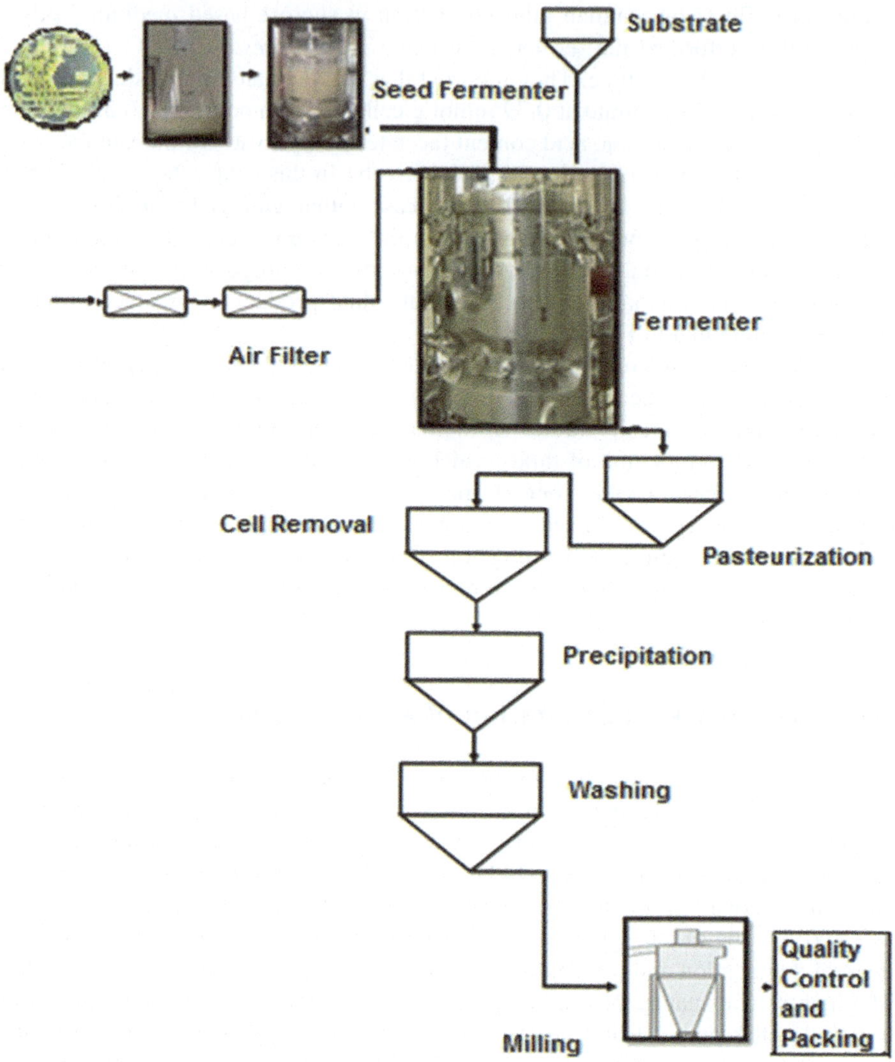

Fig. 26.3 Bio-fermentative production of xanthan gum

can be added to the fermentation broth not only to decrease the solubility until phase separation occurs but also to wash out impurities such as cells, salt, and colored components. Recovery options that have been studied include precipitation with an organic solvent such as ethanol (Zhang and Chen 2010).

Lo et al. (1996) developed a new recovery process that is energy efficient, environmentally benign, and cost-effective using ultrafiltration (UF), an alternative method to alcohol precipitation that recovers xanthan gum from dilute fermentation broth. Even under high-shear-rate conditions, UF did not give any observed adverse effects on the rheological properties and molecular weight of the xanthan polymer. Thus, UF can be used to concentrate xanthan broth from fermentation by a factor of five or higher thereby reducing the amount of alcohol needed for xanthan recovery by at least 80%.

The salt-based precipitation is another xanthan separation process studied. Alcohol is needed per broth volume (Borges et al. 2008). Gumus et al. (2010) precipitated xanthan gum by using 2 volumes of isopropanol per volume of the broth. Addition of salts in sufficient concentration also causes precipitation or complex coacervation due to ion binding of the cations of the added salt to the ionized groups on the polyanion. This leads to charge reversal at the instance when all the available anionic groups are bound to a cation. Polyvalent cations such as calcium, aluminum, and quaternary ammonium salts are especially effective in precipitation.

26.5 Commercial Application of Xanthan Gum

As xanthan showed unique characteristics, xanthan is used in different food applications as a suspending and thickening agent in bakery, beverages, dairy, dressings, pet food, syrups, and toppings. Xanthan gum helps in improving unique food properties such as texturization, viscosity, flavor release, appearance, and water release properties (Palaniraj and Jayaraman 2011). Xanthan gum also controls the rheology of the food product. The industrial application includes oil industries, agricultural, cleaners, coatings, paper, and personal care applications. In agricultural sectors, xanthan product is used as herbicides, fungicides, and insecticides in order to improve the flowability by suspending the solid components and helps in increasing the contact time between pesticides and crop (Flickinger and Draw 1999).

Among various applications, xanthan gum-based drilling fluid occupies a major market demand. Xanthan gum-based product is used for "enhanced oil recovery" (EOR) application. In the petroleum industry, xanthan gum is used in oil drilling fracturing, pipeline cleaning, and work-over and completion. Xanthan gum is also used in polymer flooding as a tertiary oil recovery operation. In this application, polymer-thickened brine is used to drive the slug of the surfactant through porous reservoir rock to mobilize residual oil; the polymer prevents bypassing of the drive water through the surfactant band and ensures good area sweeping (Byong 1996).

The cost of the xanthan product is influenced by several factors including fermentation conditions, nature of additives and copolymers used in its formulation.

26.6 Application of Xanthan Gum in Petroleum Industries

Drilling fluid or drilling mud is used to help the drilling of boreholes while drilling of oil and natural gas wells to explore the petroleum products from the earth. Drilling fluid circulated from the surface, along with drill pipe to the drilling bit, helps to lift the cuttings also controlling subsurface pressures, stabilize the rock, prevent the contamination of subsurface formation hydrocarbon fluids and lubricate the bit

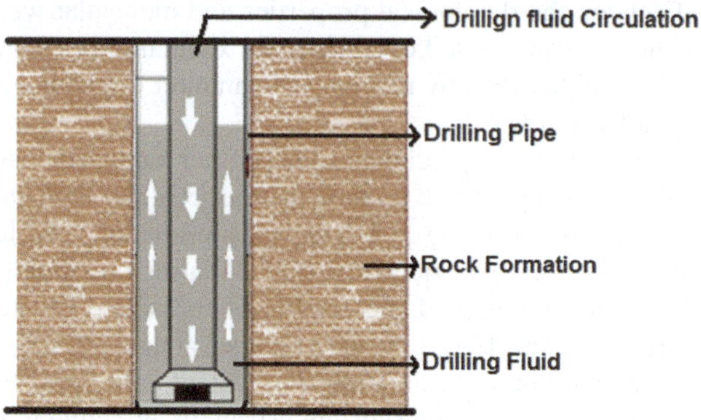

Fig. 26.4 Schematic representation of the oil well drilling process

(Fig. 26.4). Drilling fluid/muds play a key role in oil exploration jobs as the success of geothermal drilling depends on the employment of appropriate drilling fluids.

Drilling fluids is an important component which decides the success of drilling operation, selection of appropriate drilling viscosifier help to eliminate the difficulties of drilling operation, and significant reduction in drilling cost. Moreover, the drilling fluid needs to be effective in a complex subsurface environment under high pressure and high temperature with high rheological properties and filtration properties along with environmental friendly nature.

Major properties of drilling fluid include removing cuttings from the hole, cleaning the hole bottom, cooling and lubricating the bit and drill pipe, preventing cuttings settling in the hole and surface pits, and deposition of impermeable wall cake to prevent fluid loss. Liquid drilling fluids are classified into water-based muds and oil-based muds with water and oil being the continuous phase respectively. Since 1887, the development of drilling fluid underwent several technological changes due to its extended application demand and quality (Chilingarian and Vorabutr 1983). Different types of oil and water-based drilling fluids are used in oil well drilling. Generally oil, water and, gas-based drilling fluids are commonly used and it is classified based on its several aspects such as fluid phase, alkalinity, dispersion, the type of chemicals used in their formulation and the degrees of inhibition.

Initially, the synthetic and oil-based drilling fluids are used and restricted due to its environmental sensitive nature. The most appropriate choice for this purpose is the use of xanthan gum-based drilling fluid. Xanthan gum occupies a prominent place in the market due to its rheological properties that are quite different and unusual, such as high degree of pseudoplasticity, a high viscosity even at low concentrations, stability and compatibility with most metallic salts, excellent solubility, and stability in acidic and alkaline solutions and resistance to degradation at elevated temperatures and various pH levels.

In order to meet the technical advancement, economic concern, and environmental safety, there are several new criterias that have been introduced in addition to the normal drilling fluid function including no environmental damage, not interfering with the normal productivity of the fluid bearing formation and not causing excessive wear of drilling equipment. The drilling operation strongly depends on the performance of the drilling fluid and further the drilling fluid performance is a collective function of several factors including formation damage and wellbore. The other concern is the ever-increasing drilling depth and geographical location also induces the oil industries to find superior alternative products.

The bacterial based xanthan gum along with carboxymethylcellulose (CMC) and polynomic cellulose (PAC) are used in the oil and gas industry for filtration control, enhance the rheological properties of the drilling fluid, reduce the drag, and improve shale stability (Zhang et al. 2001). The xanthan gum is used with starch-based material to reduce the mud fluid loss into the formation and hence to reduce the formation damage (Darley and Gray 1988). Increasing the polymer concentration in KCl drilling fluid increases the consistency index and the flow index (these parameters are used to describe the pseudoplastic behavior of the drilling fluid) (Kok and Alikaya 2005). Blkoor and Fattah (2013) studied the effect of xanthan gum concentration on the properties of water-based drilling fluid filter cake and the formation damage. Along with unique molecular chemistry the xanthan gum also produced in high yields exceeding 40 g/L in batch fermentations in consistent quality, translate into the most successful microbial biopolymer to be marketed. Further, the improvement in the aspect of productivity and properties play a significant role in expanding the industrial use of xanthan gum.

To meet the drilling fluid property many research teams have formulated with suitable bridging particles and copolymers along with xanthan gum. The additives such as barite and bentonite are used to maintain density, rheology, temperature stability, and fluid loss control properties of drilling fluids. However, this imposes several limitations including costly treatment cost and its inability. Some of the vegetable oils considered to be prospective base fluids are rapeseed oil, sunflower oil, palm oil, and groundnut oil. The ester composition within vegetable oils has the intrinsic ability to biodegrade at the presence of "built-in" oxygen in esters. However, esters contain high viscosity properties due to the presence of hydrogen bonding of –OH groups in unsaturated fatty acids. However, hydrogenation process is able to convert double bonds of unsaturated fats into saturated fats thus altering the viscosity properties of vegetable oils to be less viscous. Many findings suggested that the xanthan gum is superior to polyacrylamide, carboxymethylcellulose, and modified starch and plant polysaccharides such as guar gum.

Generally, the water-based drilling fluids are formulations of Xanthan gum and a combination of additives. This combination is made of a viscosifier, fluid loss additive, and bridging particles. The water-based drilling fluids got addition in the market due to its environmental, technical, and safety properties. In global market, the water-based polymers are expected to grow faster due to increasing drilling activities mainly in South America and the Asia Pacific.

The Xanthan gum has the following advantages in oilfield applications.

- Helps in minimized pumping friction in lime freshwater and saltwater muds
- Enhance the drill bit penetration

- Increase the drilling rates in low viscosity and high shear conditions
- Reduce solids buildup in drilling fluids
- Handling high gravel concentrations
- Stabilization of hole-cleaning fluids
- Decreased damage to oil formation
- Decreased maintenance expense
- Reduce the drilling operation cost
- Reduced suspension time
- Provides thixotropic properties
- Provides microbiological stability in water-based formulations
- Complies with requirements of environmentally drilling fluid formulations

Generally, Xanthan gum polymer is cross-linked with additives to meet the API (American Petroleum Institute drilling specification) oil well drilling fluid properties. The most commonly used additives are modified Starch, Polyanionic cellulose (PAC), CarboxymethylCellulose (CMC), Sulphonated asphalts, Polyols (methyl glucoside or polyglycerol), Blend of glucoside and sulfonated asphalt, Ester additives, Lignosulphate (cross-linker), Barite, Lignin, Salts (potassium chloride, potassium hydroxide), Calcium carbonate (bridging agent), and Caustic soda.

API standard parameters for the drilling fluids include borehole stability, fluid loss filtration, penetration rate, temperature stability, density and viscosity, rheological properties, and lubricity. Apart from drilling fluid quality, the success of the oil well drilling also depends on the application of effective drilling fluid. Drilling fluid circulated from the surface, along with drill pipe to the drilling bit, helps to lift the cuttings also controlling subsurface pressures, stabilizes the rock, prevents the contamination of subsurface formation hydrocarbon fluids, and lubricates the bit. Drilling fluid/muds play a key role in oil exploration jobs as the success of geothermal drilling depends on the employment of appropriate drilling fluids.

26.7 Recent Developments and Future Scenarios

Due to their unique properties, the nanoparticles are considered as very good candidates for smart drilling fluid formulation. The combination of conventional polymers and nanotechnology has received much attention due to its enhanced rheological and filtration properties in oil industry engineering. In recent years many researchers have examined the use of various nanoparticles, to formulate drilling fluids with enhanced properties that can withstand high pressure and high temperature (HP/HT) under extreme downhole environments.

Recent studies evidenced the ability of xanthan gum to produce iron-based nanofluids. The xanthan gum is incorporated with nanoparticles such as copper oxide (CuO) and zinc oxide (ZnO) to prepare nanofluids (Xue and Sethi 2012; Ponmani et al. 2014). The choice of adding stable nanofluids in XG aqueous solution into drilling fluid is made to ensure better dispersion of the nanofluids into the drilling mud which influence the thermal, electrical, and rheological properties. The study showed that the increased concentration of nanoparticles enhances the thermal conductivity in both base nanofluid and base drilling fluid. CuO nanofluids in XG water showed greater enhancement in thermal conductivities than ZnO nanofluids in the same system (Ponmani et al. 2014). Many other investigations also focused to explore the efficiency of nanoparticles and natural polymers in enhanced oil recovery from oil wells. In xanthan base fluid, the nanoparticles like CuO and ZnO enhanced the thermal and electrical properties of the drilling fluid. Hence, the nano-enhanced drilling fluid plays an important role in avoiding circulation loss by sealing the formation with both NPs and conventional bulk particles in the fluid system. This method expanded the application of water-based drilling fluid especially in unconventional drilling (Das et al. 2003; Ponmani et al. 2016). Also SiO_2, Fe $(OH)_3$, TiO_2, Al_2O_3 with 0.2 wt.% with Xanthan gum 1–9% improved the solution viscosity (Corredor et al. 2019).

Pandey and Mishra (2011) grafted xanthan gum with methylbis (trimethyl silyloxy) vinylsilane through free radical copolymerization and found that the xanthan crosslinking with vinylsilane monomers enhances its physicochemical and rheological properties. It is also reported that the chemical displacement by xanthan gum as well as xanthan/SiO_2 and xanthan grafted with vinylsilane derivatives were investigated in reservoir environment through rheological performance relative to shearing rate, ionic strength, and thermal stability. The flooding tests showed that grafting of the silica derivative overcome the shortage of xanthan solution in flooding operations relative to the reservoir conditions. The results revealed that the synthesized composite was a promised agent for enhancing oil recovery and profile conformance (Soliman et al. 2020). In this xanthan gum (XG) and nanocomposite combination, the XG/SiO_2 exhibit reasonable heat resistance due to the enrichment of endothermic intramolecular interaction with temperature increasing (El-Hoshoudy et al. 2016; Ye et al. 2013). Moreover, the formation of hydrogen bonding between silanol groups (Si–OH) and carbonyl groups (C=O) (Hu et al. 2018; Maurya and Mandal 2016) of ester in xanthan gum form compacted, rough structure which enhances the salinity and temperature resistance. Various researches demonstrated the advantage of nanoparticle-based drilling fluid additives in order to overcome stern drilling problems. However, there are still challenges that need to be addressed in order to implement the advantage of such advanced technologies in oil and gas industry.

26.8 Conclusion

Xanthan gum is one of the key industrial products and current major markets are propelling in food and oil industries. A recent analysis showed the drilling fluid demand has grown due to technological upgradation in horizontal and unconventional wells in addition to the increase in lateral lengths which require more drilling fluids. As a technological progress, oil well drillers favor the water-based Xanthan gum drilling fluids. Water-based fluids are favored for their lower cost and better environmental benefits.

Recent development in the utilization of effective Xanthomonas strains, low-cost fermentation substrate, advanced downstream separation of xanthan gum leads to the production of inexpensive and superior quality xanthan gum. Conversely, issues like final product cost and high performance need to be addressed through utilization of low-cost fermentation process and raw materials, copolymerization with nano-materials.

References

Abdel-Aziz SM, Hamed HA, Mouafi FE, Gad AS (2012) Acidic pH-shock induces the production of an exopolysaccharide by the fungus Mucor rouxii: utilization of beet-molasses. N Y Sci J 5:52–61

Adnise LW, Soccol CR, Rocha SN, Pandey A (2003) Xanthan gum production from cassava bagasse hydrolysate with Xanthomonas campestris using alternative sources of nitrogen. Appl Biochem Biotechnol 118:305

Amanullah A, Serrano CL, Castro B, Galindo E, Nienow A (1988) The influence of impeller type in pilot scale xanthan fermentations. Biotechnol Bioeng 57:95–108

Becker A, Katzan F, Puhler A, Lelpi L (1998) Xanthan gum biosynthesis and application: a biochemical/genetic perspective. Appl Microbiol Biotechnol 50:145–152

Blkoor SO, Fattah KA (2013) The influence of XC-polymer on drilling fluid filter cake properties and formation damage. J Pet Environ Biotechnol 4:157

Borges CD, Moreira AD, Vendruscolo CT, Ayub MAZ (2008) Influence of agitation and aeration in xanthan production by Xanthomonas campestris pv pruni strain 101. Rev Argent Microbiol 40:81–85

Byong HL (1996) Fundamentals of food biotechnology. VCH Publishers Inc., New York

Chandrasekaran M, Bahkali AH (2013) Valorization of date palm (Phoenix dactylifera) fruit processing by-products and wastes using bioprocess technology—review. Saudi J Biol Sci 20:105–120

Chilingarian GV, Vorabutr P (1983) Drilling and drilling fluids, 2nd edn. Elsevier, Amsterdam, p 767

Corredor LM, Husein MM, Maini BB (2019) Effect of hydrophobic and hydrophilic metal oxide nanoparticles on the performance of xanthan gum solutions for heavy oil recovery. Nanomaterials 9:94

Darley HCH, Gray GR (1988) Composition and properties of drilling fluids. Gulf Publishing, Houston, TX

Das SK, Putra N, Thiesen P, Roetzel W (2003) Temperature dependence of thermal conductivity enhancement for nano-fluids. J Heat Transf 125:567–574

Davidson W (1978) Production of polysaccharide by Xanthomonas comprestris in continuous culture. FEMS Microb Lett 3:347–349

De Sousa Costa LA, Campos MI, Druzian JI, De Oliveira AM, De Oliveira EN (2014) Biosynthesis of xanthan gum from fermenting shrimp shell: yield and apparent viscosity. Int J Polym Sci 2014:1–8

De Vuyst L, Van LJ, Vandamme E (1987) Two-step fermentation process for improved xanthan production by Xanthomonas campestris NRRL-B-1459. J Chem Technol Biotechnol 39:263–273

Desplanques S, Renou F, Grisel M, Malhiac C (2012) Impact of chemical composition of xanthan and acacia gums on the emulsification and stability of oil-in-water emulsions. Food Hydrocoll 27:401–410

El-Hoshoudy A, Desouky S, Betiha M, Alsabagh A (2016) Use of 1-vinyl imidazole based surfmers for preparation of polyacrylamide–SiO2 nanocomposite through aza-Michael addition copolymerization reaction for rock wettability alteration. Fuel 170:161–175

El-Salam A, Fadel MA, Murad HA (1994) Bioconversion of sugarcane molasses into xanthan gum. J Biotechnol 33:103–106
Esgalhado ME, Roseiro JC, Collaço MTA (1995) Interactive effects of pH and temperature on cell growth and polymer production by Xanthomonas campestris. Process Biochem 30:667–671
Flickinger FC, Draw SW (1999) Encyclopedia of bioprocess technology: fermentation, vol 5. pp 2706–2707
Flores-Candia JL, Deckwer WD (1999) Xanthan gum. In: Flickinger MC, Drew SW (eds) Fermentation, biocatalysis, and bioseparation. Wiley, pp 2695–2711
Garcia-Ochoa F, Casas JA, Mohedano AF (1992) Precipitation of xanthan gum. Sep Sci Technol 28:1303–1313
Garcia-Ochoa F, Castro EC, Santos VE (2000a) Oxygen transfer and uptake rates during xanthan gum production. Enzym Microb Technol 27:680–690
Garcia-Ochoa F, Santos JA, Gomez E (2000b) Research review paper of xanthan gum: production recovery and properties. Biotechnol Adv 18:549–579
Ghashghaei T, Soudi MR, Hoseinkhani S (2016) Optimization of xanthan gum production from grape juice concentrate using Plackett Burman design and response surface methodology. Appl Food Biotechnol 3:15–23
Gilani SL, Heydarzadeh HD, Mokhtarian N, Alemian A, Kolaei M (2011) Effect of preparation conditions on xanthan gum production and rheological behavior using cheese whey by Xanthomonas campestris. Aust J Basic Appl Sci 5:855–859
Gowdhaman D, Padmapriya S, Ponnusami V (2013) Effect of citric acid inducer on xanthan gum production from cassava bagasse, a potential agro-industry waste. Res J Pharm, Biol Chem Sci 4:370–374
Gumus T, Demirci AS, Mirik M, Arici M, Aysan Y (2010) Xanthan gum production of Xanthomonas species isolated from different plants. Food Sci Biotechnol 19:201–206
Harding NE, Cleary JM, Luis L (1995) Genetics and biochemistry of xanthan gum production by Xanthomonas campetris. Dlm. In: Hui YH, Khachatourians G (eds) Food technology microorganisms. Wiley–VCH Inc., New York, pp 495–514
Howell JA, Field R, Wu D (1996) Ultrafiltration of high-viscosity solutions: theoretical developments and experimental findings. Chem Eng Sci 51:1405–1415
Hu X, Ke Y, Zhao Y, Lu S, Yu C, Peng F (2018) Synthesis and characterization of a b-cyclodextrin modified polyacrylamide and its rheological properties by hybriding with silica nanoparticles. Coll Surf A Physicochem Eng Asp 548:10–18
Kalogiannis S, Gesthimani L, Maria LK et al (2003) Optimization of xanthan gum production by Xanthomonas campestris grown in molasses. Process Biochem 39:249–256
Kang KS, Pettitt DJ (1993) Xanthan, Gellan, Welan and Rhamsan. In: Whistler RL, Bemiller NJ (eds) Industrial gums: polysaccharides and their derivatives, 3rd edn. Elsevier, pp 341–397
Katherine RF, Muthukumaran C, Sharmila G, Kumar NM, Tamilarasan K, Jaiganesh R (2017) Xanthan gum production using jackfruit-seed-powder-based medium: optimization and characterization. 3 Biotech Springer 7:248
Kennedy JF, Bradshaw IJ (1984) Production properties and applications of xanthan. Prog Ind Microbiol 19:319–371
Khosravi DK, Reyhani FS, Nejad B, Farhadi GBN (2013) Bench scale production of xanthan from date extract by Xanthomonas campestris in submerged fermentation using central composite design. Afr J Biotechnol 10:13520–13527
Kok MV, Alikaya T (2005) Effect of polymers on the rheological properties of KCl/polymer type drilling fluid. Energy Sources 27:405–415
Kumaraswamy M, Khan BA, Rohit K, Purushotham B (2012) Effect of carbon and nitrogen sources on the production of xanthan gum from Xanthomonas campestris isolated from soil. Arch App Sci Res 4:2507–2512
Kurbanoglu EB, Kurbanoglu NI (2007) Ram horn hydrolysate as enhancer of xanthan production in batch culture of Xanthomonas campestris EBK-4 isolate. Process Biochem 42:1146–1149

Letisse F, Chevallereau P, Simon JL, Lindley ND (2001) Kinetic analysis of growth and xanthan gum production with Xanthomonas comprestris on sucrose, using sequentially consumed nitrogen sources. Appl Microbiol Biotechnol 55:417–422

Li P, Li T, Zeng Y, Li X, Jiang X, Wang Y, Zhang Y (2016) Biosynthesis of xanthan gum by Xanthomonas campestris LRELP-1 using kitchen waste as the sole substrate. Carbohydr Polym 151:684–691

Liakopoulou K, Psomas K, Kyriakid A (1999) Xanthan gum production by Xanthomonas campestris w.t. fermentation from chestnut extract. Appl Biochem Biotechnol 82:175–183

Lo YM, Yang ST, Min DB (1996) Kinetic and feasibility studies of ultrafiltration of viscous xanthan gum fermentation broth. J Membr Sci 117:237–249

Malik VS, Sridhar P (1992) Industrial biotechnology. Oxford & IBH Publishing Co., Pvt. Ltd, New Delhi

Maurya NK, Mandal A (2016) Studies on behavior of suspension of silica nanoparticle in aqueous polyacrylamide solution for application in enhanced oil recovery. Pet Sci Technol 34:429–436

Miroslav S, Elena C (1999) Xanthan production by solid state fermentation. Process Biochem 34:581–587

Mohsin A, Zhanga K, Hua J, Rehman SU, Tariq M, Zaman WQ, Khan IM, Yingping Zhuang Y, Guo M (2018) Optimized biosynthesis of xanthan via effective valorization of orange peels using response surface methodology: a kinetic model approach. Carbohydr Polym 181:793–800

Moosavi A, Karbassin A (2010) Bioconversion of sugar-beet molasses into xanthan gum. J Food Process Preserv 34:316–322

Morris ER, Nishinari K, Rinaudo M (2012) Gelation of gellan—a review. Food Hydrocoll 28:373–411

Moshaf S, Hamidi EZ, Azizi MH (2015) Statistical optimization of xanthan gum production and influence of airflow rates in lab-scale fermentor. Appl Food Biotechnol 1:17–24

Murugesan A, Dhevahi B, Gowdhaman D, Bala A, Sathesh P (2012) Production of xanthan employing Xanthomonas campestris using sugarcane molasses. Am J Environ Eng 2:31–34

Nery TBR, Da Cruz AJG, Druzian JI (2013) Use of green coconut shells as an alternative substrate for the production of xanthan gum on different scales of fermentation. Polymeros 23:602–607

Niknezhad SV, Asadollahi AL, Zamani A, Biria D, Doostmohammadi M (2015) Optimization of xanthan gum production using cheese whey and response surface methodology. Food Sci Biotechnol 24:453–460

Palaniraj A, Jayaraman V (2011) Production, recovery and applications of xanthan gum by Xanthomonas campestris. J Food Eng 106:1–12

Pandey S, Mishra SB (2011) Graft copolymerization of ethylacrylate onto xanthan gum, using potassium peroxydisulfate as an initiator. Int J Biol Macromol 49:527–535

Pelletier E, Viebke C, Meadows J, Williams PA (2001) A rheological study of the order–disorder conformational transition of xanthan gum. Biopolymers 59:339–346

Ponmani S, William JKM, Samuel R, Nagarajan R, Sangwai JS (2014) Formation and characterization of thermal and electrical properties of CuO and ZnO nanofluids in xanthangum. Colloids Surf A Physicochem Eng Asp 443:37–43

Ponmani S, Nagarajan R, Sangwai JS (2016) Effect of nanofluids of CuO and ZnO in polyethylene glycol and polyvinylpyrrolidone on the thermal, electrical, and filtration-loss properties of water-based drilling fluids. SPE J 24:405–415

Ramezani A, Jafari M, Goodarzi T, Alavi SM, Salmanian AH, Azin M (2014) Lactose consuming strains of Xanthomonas citri subsp. citri (Xcc) insight into the emergence of natural field resources for xanthan gum production. World J Microbiol Biotechnol 30:1511–1551

Rosalam S, England R (2006) Review of xanthan gum production from unmodified starches by Xanthomonas camprestris sp. Enzym Microb Technol 39:197–207

Ruffing A, Chen RR (2006) Metabolic engineering of microbes for oligosaccharide and polysaccharide synthesis. Microb Cell Factories 5:25

Salah RB, Chaari K, Besbes S, Blecker C, Attia H (2011) Production of xanthan gum from *Xanthomonas campestris* NRRL B-1459 by fermentation of date juice palm by-products (Phoenix dactylifera L.). J Food Process Eng 34:457–474

Soliman AA, El-hoshoudy AN, Attia AM (2020) Assessment of xanthan gum and xanthan-g-silica derivatives as chemical flooding agents and rock wettability modifiers. Oil Gas Sci Technol—Rev IFP Energies Nouv 75:12

Souw P, Demain AL (1979) Nutritional studies on xanthan gum production by Xanthomonas comprestris NRRL B1459. Appl Environ Microbiol 37:1186–1192

Stredansky M, Conti E (1999) Xanthan production by solid state fermentation. Process Biochem 34:581–587

Suresh S, Prasad MS (2005) Technological application of an extracellular cell lytic enzyme in xanthan gum clarification. Braz J Microbiol 36:57–62

Umashankar H, Annadurai G, Chellapandian M, Krishnan MRV (1996) Influence of nutrients on cell growth and xanthan by Xanthomonas campestris. Bioprocess Eng 14:307–309

Vojnov AA, Slater H, Daniels MK, Dow JM (2001) Expression of the gum operon directing xanthan biosynthesis in Xanthomonas campestris and its regulation in planta. Mol Plant-Microbe Interact 14:768–774

Wang Z, Wu Z, Gao MJ, Zhu L, Zhan XB (2017) High production of xanthan gum by a glycerol-tolerant strain *Xanthomonas campestris* WXLB-006. Prep Biochem Biotechnol 47:468–472

Whistler RL, BeMiller JN (1997) Carbohydrate chemistry for food scientists. Eagan Press, St. Paul, pp 117–128

Xue D, Sethi R (2012) Viscoelestic gels of guar and xanthan gum mixtures provide long term stabilization of iron-micro and nanoparticles. J Nanopart Res 14:1239

Yang ST, Lo YM, Min DB (1996) Xanthan gum fermentation by *Xanthomonas campestris* immobilized in a novel centrifugal fibrous-bed bioreactor. Biotechnol Prog 12:630–637

Ye Z, Feng M, Gou S, Liu M, Huang Z, Liu T (2013) Hydrophobically associating acrylamide-based copolymer for chemically enhanced oil recovery. J Appl Polym Sci 130:2901–2911

Zahra S, Maryam N, Farid T, Vahid N (2018) Xanthan gum production from acid hydrolysed broomcorn stem as a sole carbon source by *Xanthomonas campestris*. 3 Biotech 8:296

Zhang G (1999) Chemically defined media for commercial fermentations. Appl Microbiol Biotechnol 51:407–421

Zhang Z, Chen H (2010) Fermentation performance and structure characteristics of xanthan produced by *Xanthomonas campestris* with a glucose/xylose mixture. Appl Biochem Biotechnol 160:1653–1663

Zhang LM, Tan YB, Li ZM (2001) New water-soluble ampholytic polysaccharides for oil field drilling treatment: a preliminary study. Carbohydr Polym 44:255–260

A Comparative Study on Biodegradable Packaging Materials: Current Status and Future Prospects

27

Archana S. Rao, Ajay Nair, Sunil S. More, Arpita Roy, Veena S. More, and K. S. Anantharaju

Abstract

An ever-growing call to contain pollution has led to the search for alternate sources of sustainable industrial raw materials. Recently, biodegradable waste materials have been sought after as a potential alternative. The waste from packaging materials is of great concern as it is nondegradable. Synthetic plastics made with different kinds of polymers such as Expanded polystyrene (EPS), Polyethylene Terephthalate (PETE), Polyethylene (PE), Polypropylene (PP), and Polyethylene terephthalate (PET, are primarily used for packaging materials. These synthetic polymers being lightweight, strong, and economical, find use in equipment protection, civil construction, food packaging and other packages, etc. However, after use, these packaging materials are discarded. And since they are not completely degradable, they accumulate and contaminate the natural ecosystem. Therefore, it is imperative to seek eco-friendly options. Biodegradable materials developed using macrofungi have shown their ability to convert agricultural wastes into biodegradable packaging materials. Mushroom mycelia thus appear to be promising in creating new packaging materials, which are environmentally friendly and economical. Hence, this chapter attempts to compile some of the recent technological advances made in creating packaging materials using mushrooms for a sustainable future.

A. S. Rao · A. Nair · S. S. More (✉) · A. Roy
School of Basic and Applied Sciences, Dayananda Sagar University, Bangalore, Karnataka, India
e-mail: drsunil@dsu.edu.in

V. S. More
Department of Biotechnology, Sapthagiri College of Engineering, Bangalore, Karnataka, India

K. S. Anantharaju
Department of Chemistry, Dayananda Sagar College of Engineering, Bangalore, Karnataka, India

A. Vaishnav, D. K. Choudhary (eds.), *Microbial Polymers*,
https://doi.org/10.1007/978-981-16-0045-6_27

Keywords

Mushrooms · Agricultural wastes · Biodegradable · Plastics · Packaging material

27.1 Introduction

Since the use of natural materials with intrinsic plastic-like properties, synthetic plastics have become perhaps the most important commodity in our everyday life. From laptops that we use to the cell phones, water bottles, chairs, packaging materials to the tiny brush bristles, everything that we can probably lay our eyes on is made of some form of plastic. This extensive dependency on plastics stems from some of the features it is characteristic of, such as being inexpensive, lightweight, durable, corrosion-resistant, in addition to being endowed with thermal and electrical insulation properties.

Globalization has enabled the transport of materials anywhere and everywhere. And the inclusion of digitalization in the industry permits us to access things at our fingertips. Import and export of goods demand good and safe packaging and hence packaging materials play an important role in any industry. Not only this, industries on routine supplies like milk, curd, groceries, and food have been one of the primary consumers of plastics to suffice various needs. Over the past few decades, with the advent of online delivery, plastics have also found their way into our lives in the form of packaging materials to deliver food.

Considering how important packaging materials are to any industry, they have to be managed methodically and categorically. An approach that focuses on either complete refusal towards synthetic plastic usage or minimizes the heavy dependency or at least helps find ways to reuse plastics is paramount. The plastic packaging materials though offer many benefits, they also contribute a significant chunk in the form of hazardous environmental impacts. Hence, the need to find an alternative that not only shares all the qualities present in synthetic plastics but at the same time is eco-friendlier and biodegradable. The eco-friendly substances which can be efficiently employed for the production of packaging materials have thus in the recent past grown in demand. Many materials derived like peanuts, cornstarch, seaweed or cardboard, paper, etc., are being henceforth experimented and used as alternatives for these plastics. Packaging materials made using mushrooms are one such promising eco-friendly substitute to replace the synthetic packaging materials. In this chapter, we discuss the several negative impacts that accompany synthetic packaging material disposal, from there we expand on various available biodegradable alternatives, their respective advantages, and drawbacks. We then deliberate extensively on the promising future mushrooms hold as an emerging packaging material with the challenges associated with it.

27.2 Synthetic Packaging Materials

Synthetics dominate the packaging industry because of the advantages they provide. The packaging materials used widely are synthetic polymers like plastics and are superior to any other packaging materials. Because of qualities like cost-effectiveness, strength to hold the material inside, thermos-stability, being reasonably waterproof, fabrication flexibility, sealability, impermeability to gases, tear and puncture resistance, and being lightweight they hold unparalleled supremacy in terms of widespread usability. Though they are commercially useful, these materials are for single use, so they are naturally a waste afterwards, later entering into the solid waste stream. As they are non-biodegradable, they create many problems ranging from environmental impacts to health concerns. The rate at which these packaging materials are added to the solid waste lot is alarming and demands a substitute to reduce the associated problems.

Common types of packaging materials include plastic, glass, steel, aluminum, paper, cardboard, and wood. Each of these materials will have their applications along with their advantages and disadvantages. Plastic is the most sought after packaging material since it can be used from food packaging to electronic gadgets. The glass packaging is generally used for liquids like medicine, wine, beer, juice, etc. Steel and aluminum are used in certain types of packaging and it can be recycled too. Paper and paperboard boxes are eco-friendly, they can be recycled but still they are not as sturdy as plastics. Wooden packaging materials are one of the oldest but have many limitations such as trapping moisture, carrying pests, and being inherently heavy.

Among all the above mentioned materials, plastic is clearly the most preferred packaging material. Besides serving purposes such as protecting the material inside, these plastic packages are highly customizable to cater to customer needs in terms of design, pattern, etc. Plastics have contrastingly become expensive, from a production point of view. And that when added to its short-term usage, ending up eventually in the waste stream tips the balance towards negative.

Plastics are macromolecular polymers derived mainly from hydrocarbons obtained from crude oil and natural gas (Dahlbo et al. 2017), but it accounts for 2–4% of the overall consumption of crude oil and natural gas (Selke 1990). The plastics used in packaging are thermoplastic organic polymers. These synthetic plastics like polyolefins, polystyrene (PS), polyvinyl chloride (PVC), and some semi-organic polymers like polyamide (PA) and polyesters (PE) are all made up of carbon-carbon bond main chain.

Polymers used substantially in packaging materials are high-density polyethylene (HDPE), medium-density polyethylene (MDPE), low-density polyethylene (LDPE), linear low-density polyethylene (LLDPE), polypropylene (PP), and polyethylene terephthalate (PET), polystyrene. These are different from polyvinyl chloride (PVC) used mainly in pipeline constructions. Polyethylene and polypropylene are mainly used in packaging materials, HDPE is used for making containers, bottles, bags, and wrappers, whereas LDPE is used for bags, containers, agricultural films, etc. PET meets food-grade packaging standards hence is used for bottles and food packages

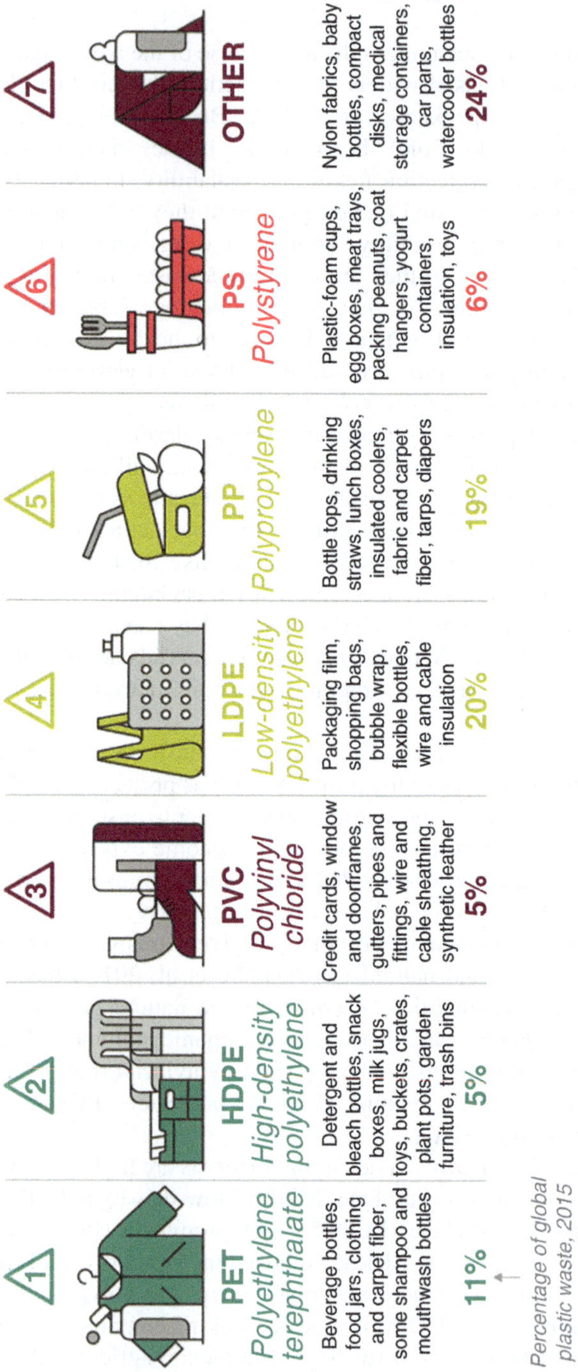

Fig. 27.1 Challenges of recycling plastic

(Fig. 27.1). Being non-biodegradable, plastics can transcend eons (Barnes et al. 2009; Holland et al. 2016). The greenhouse gases (carbon dioxide and methane) thus emitted by the disposal of synthetic plastics can lead to hazardous environmental consequences (Roopesh and Archana 2015).

27.2.1 Hazards of Synthetic Packaging Materials

The global plastic production has increased immensely in the last few decades. A production of 28,600 million metric tons of plastic polymer is expected by 2050 worldwide (Geyer et al. 2017), and by 2026, the production of packaging plastics worth $490 billion is expected.

The chart shows the challenges involved in recycling different types of synthetic plastics. The numbered triangles of different colors represent the various levels in terms of difficulty: green is for easy; pink is for difficult; yellow is for manageable; and maroon is for very difficult.

However, considering the increase in general awareness of how plastics impact nature, they are being banned in many countries. The whole world is looking for alternatives that can replace these single-use plastics. Some of the many problems associated with such single-use packaging materials are discussed below:

27.2.1.1 Unsustainability

The plastics have proven to be an unsustainable packaging material. The raw material for plastic synthesis is petroleum, which is a nonrenewable resource and also an economical commodity of high demand. This clearly shows the unsustainability of plastics (Tudryn et al. 2018). The production, export of plastics from manufacturers to consumers also increases the cost.

27.2.1.2 Disposal

However, with the increase in the production of plastics, there is also been an increase in the hazardous effects that accompany plastic waste disposal. As per a survey carried out by the Central Pollution Control Board (CPCB), in 60 cities in India, the quantum of plastic waste generation has been estimated to be 15,342 tonnes per day, out of which 40% is used by the packaging industries and most remained uncollected (Aryan et al. 2019). The packaging industry is the sector that uses the largest share of plastics when compared to the other plastic materials producing sectors, more than one-third of the plastics produced globally typically ends up in the dump yards. Plastic packaging is widely used for almost every routine use known to man (Dahlbo et al. 2017). Fossil resources are the prime source of raw materials for plastic industries thus the prices of natural and petroleum oils also increase which potentially impacts the plastic market economically.

27.2.1.3 Environmental Pollution

As mentioned earlier, the majority of the plastics produced are single-use products. Though recycling of these plastics is slowly increasing globally, it is again depending on the type of plastics. Polyethylene terephthalate (PET), used in bottles and packaging materials, are highly recycled compared to other types, whereas, expanded polystyrene (EPS) and extruded polystyrene (XPS) are the least recycled plastics.

Though 9–10% of plastics are recycled, upto 30% of plastic produced continues to be in use, leaving 60% to enter the waste stream and eventually terminate in landfills (Geyer et al. 2017). Though plastic waste landfilling was one of the preferred management strategies earlier, now it is considered as the least accepted solution. The plastics undergo both abiotic and biotic degradation producing toxic byproducts including plastic additives like stabilizers, colorants, heavy metals, etc., that later contaminate soil. Contamination of soil affects microbial diversity and soil quality. Methane and carbon dioxide released due to microbial degradation of plastics also contributes to global warming (Biello 2013).

Plastic incineration is a much-suited alternative to landfilling. But fumes released following incineration contain furans, dioxins, polychlorinated biphenyls (PCBs) which further add to the atmospheric pollution. These noxious fumes have deleterious effects on both humans and the environment. The ashes, soot, and powders produced by plastic incineration settles on plants and soil, and thus enters water bodies. The compounds released in the air can combine with rainwater and recontaminate the soil. The groundwater contamination is again a potential problem. Though many plastic waste management strategies are practiced including recycling, incineration, etc., they still have their own fair share of negative impact on nature.

A healthy marine environment is central to a balanced ecosystem. Water pollution with plastic wastes is a serious concern as trillions of plastic wastes are floating on large waterbodies such as oceans. Though these plastics are degraded in a year's time, the degradation products like polystyrene and BPA which are deleterious (Knight 2012) are not completely degraded. The best-known destruction of plastic includes either exposing plastics to UV light to degrade its inherent structural conformation or by grinding them into smaller fragments, termed as microplastics (Danso et al. 2019). These microscopic (size <1 mm), ubiquitous plastic particles were originally found on land and eventually got widely distributed in water bodies. They can be found everywhere from the polar ice to the deep marine world, from human lungs to human stools (Danso et al. 2019). Microplastic fragments in the recent past have been recognized as a potential pollutant threat (Guzzettia et al. 2018). Microplastic, which is essentially a cocktail of polymers and additives, in the marine environment creates hazardous problems for the benthic and pelagic zones of the marine ecosystem. The ingestion of plastic by marine organisms causes entanglement of the gut leading to a slow and painful death (Anthony and Mike 2009). Post-ingestion microplastics cause several detrimental effects such as accumulation in the gut, lungs or kidneys, circulatory systems (Huerta et al. 2016). Some of these adverse effects include negative impacts on reproduction, growth retardation, altered gene expression, and neurotoxicity leading to gut dysbiosis and finally the death of the organism (Fackelmann and Sommer 2019).

Plastics and plastic additives like BPA, Phthalates, dioxins, PAHs, PCBs, nonylphenol, styrene monomer, etc., causes serious health problems in humans

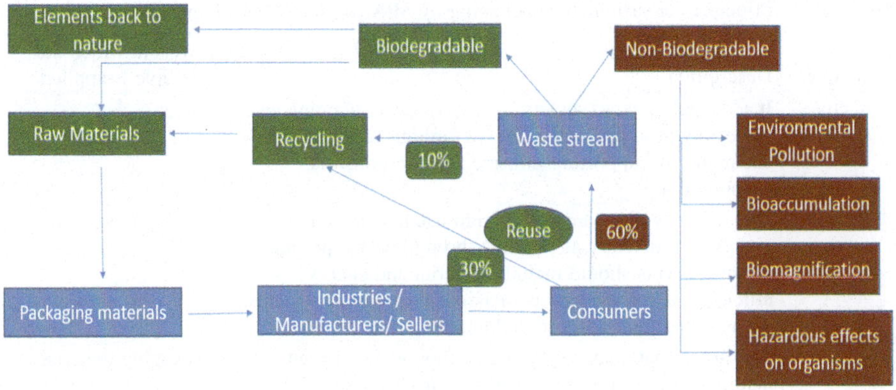

Fig. 27.2 Fate of packaging materials from origin to disposal

like cancer, ovarian disorder, reproductive damage, interference with estrogen, testosterone, and thyroid hormone, etc. (Alabi et al. 2019).

Recently, the biodegradable or semisynthetic form of plastics have started entering the market, giving little hope as far as an eco-friendly substitute to plastic is concerned. There are some polymers containing carbon such as lignin, starch, keratin, cellulose, and gelatin that are compostable or considered environmentally friendly even though they are synthetically produced. But these plastics failed to meet the required standards and did not seem to reduce the problems in an expected manner. Simultaneously, polymerization of such materials can make bioplastic lose its property of compostability. The bioplastic production is also not economical. These limitations therefore did not hold good for bioplastics as potential plastic alternates.

The fate of synthetic packaging materials from its production to disposal is summarized below (Fig. 27.2). As evident beyond reason, there is an immediate need to look for eco-friendly packaging materials.

27.3 Sustainable Packaging Material

The market value of the plastic packaging industry in India is approximately 72 billion USD (www.statistica.com) and our plastic consumption is continuously increasing. The plastic processing industry is estimated to grow to 22 million tonnes (MT) a year by 2020 from 13.4 MT in 2015 and nearly half of this is single-use plastic, according to a Federation of Indian Chambers of Commerce and Industry (FICCI).

The estimated growth rate is alarming and sustainable packaging material production which is environmental friendly thus became the main focus. Sustainable Packaging Alliance (SPA) an Australian organization, defined "sustainable packaging," and highlighted its importance towards social, economic, and environmental

Table 27.1 Principles of sustainable packaging by SPA (Adapted from James et al. 2005)

6. Principle	Description	Levels at which the principle is applied
Effective	It adds real value to society by effectively containing and protecting products as they move through the supply chain and by supporting informed and responsible consumption	Society
Efficient	Packaging systems are designed to use materials and energy as efficiently as possible throughout the product life cycle. This should include material and energy efficiency in interactions with associated support systems such as storage, transport, and handling	Packaging system
Cyclic	Packaging materials are cycled continuously through natural or (industrial) technical systems, minimizing material degradation and/or the use of upgrading additives	Packaging material
Safe	Packaging components do not pose any risks to human health or ecosystems. When in doubt, the precautionary principle applies	Packaging component

safety. SPA also formulated principles to be adhered by sustainable packaging materials (Table 27.1).

SPA also has developed a rapid tool, the Packaging Impact Quick Evaluation Tool (PIQET) to analyze the impact of packaging systems, in material selection, packaging design, packaging performance, and environmental safety (Verghese and Lewis 2005). SPA gives key indicators for sustainable packaging (Table 27.2).

27.3.1 Bioplastics

27.3.1.1 Polylactic Acid Polymer

Polylactic acid (PLA) is a lactic acid-derived polymer (Rafael et al. 2004), obtained by the microbial fermentation of corn starch, maize, cassava, potato, sugarcane, and sugar beet (Selvamurugan and Sivakumar 2019). PLA is a good alternative to using synthetic plastics as it has many advantages: it consumes high carbon dioxide during the production process, has a cost-effective production, it is formed from agricultural wastes (corn starch), it is re-useable and biodegradable, and lastly, the mechanical structure can exploit polystyrene-based polymers. PLA plastics are rigid, stable, transparent, thermostable, and hence highly preferred bioplastic over others. PLA plastics have many applications as packaging material for food, useful in medical, textile, cosmetic, automobile industries, etc. However, there are a few notable disadvantages of using PLA, such as low melting point, penetrable to moisture and oxygen when compared to polyethylene terephthalate, also, the polymer is not resilient enough for packaging materials (Bilal et al. 2017).

Table 27.2 Sustainable Packaging Indicators by SPA (Lewis et al. 2007)

Sustainable packaging principle	Sustainable packaging indicator
Effective	Reduces product waste
	Improves functionality
	Prevents overpackaging
	Reduces business costs
	Achieves satisfactory return on investment (ROI)
Efficient	Improves product/packaging ratio
	Improves efficiency of logistics
	Improves energy efficiency (embodied energy)
	Improves materials efficiency (total amount of material used)
	Improves water efficiency (embodied waster)
	Improves recycled content
	Reduces waste to landfill
Cyclic	Returnable
	Reusable (alternative purpose)
	Recyclable (technically recyclable and system exists for collection and reprocessing)
	Biodegradable
Safe (clean)	Reduces airborne emissions
	Reduces waterborne emissions
	Reduces greenhouse gas emissions
	Reduces toxicity
	Reduces litter impacts

Adapted from Kim and Ruedy (2019)

27.3.1.2 Polyhydroxyalkanoates

Polyhydroxyalkanoates (PHA) are produced from microbial fermentation of plant-based starch. Microbes like *Alcaligens* spp., *Pseudomonas* spp., *Nocardia* spp., etc., are capable of producing PHA under limiting nutrient conditions. The PHA production is intracellular and the cells can accommodate PHA up to 80% of their total weight. PHAs are completely biodegradable by microorganisms under aerobic conditions yielding carbon dioxide and water. It is a safe, nontoxic natural polymer, which can be used in food packaging, medical field, etc.

27.3.1.3 Starch

Starch is made up of the sediments of anhydroglucose containing amylase (25%) and amylopectin (75%) (Bilal et al. 2017). For the first time, the bioplastic blending starch was synthesized from maize and petrochemicals. Later potato, corn, wheat, cassava, and tapioca were also used for bioplastic production (Maulida and Tarigan 2016; Greene 2014). The mixture of starch and petrochemicals is processed, and the slurry is gelatinized which can be set to the desired shape. The advantages of the starch polymer are its transparency, which provides hindrance to oxygen and carbon dioxide. The disadvantages include brittleness, highly hydrophilic in nature, the ability to degrade due to the weak moisture resistance property, weak mechanical properties, and survives without degradation for approximately 40–60 days of duration (Srinivasa and Tharanathan 2007).

27.3.1.4 Cellulose

Cellulose is a polymer of glucose units linked by β 1,4 glycosidic bonds. Being the most abundant carbohydrate available, it is present in plant and algal cell walls. Chemical modification and extraction of natural cellulose give polymers. For bioplastic synthesis, cotton fibers and wood pulp are commonly used. Esterification of cellulose with organic acids gives cellulose esters like cellulose acetate, cellulose propionate, cellulose acetate butyrate, etc. These polymers have application as packaging films, cigarette filters, etc. (Selvamurugan and Sivakumar 2019).

27.3.1.5 Chitin/Chitosan-Based Films

Chitin is an animal-based polysaccharide. The primary source for chitins are invertebrates (Srinivasa and Tharanathan 2007). Approximately 100 million tons of chitins are produced globally. Alike cellulose, chitins are very safe. They are used mainly for food packaging. The advantages of chitin-based plastics are good oxygen and carbon dioxide barrier quality and its durability (Srinivasa and Tharanathan 2007).

27.4 Mushrooms

Mushroom, a macroscopic fungus which is known for its edibility, medicinal uses, colored pigments, toxicity, etc., found its place in the packaging industry because of the innovation from an innovative company named "Ecovative Design." Mushroom is gaining importance for the production of biodegradable, renewable, and cost-effective packaging material. Because of these advantages, mushroom is used in packaging material production, thermal insulation, building materials, protective packaging, etc.

Mushrooms are macroscopic fungi which belong to order *Agaricales*, under the phylum *Basidiomycota*. Generally, edible sporophores like *Agaricus bisporus, Ganoderma lucidum, Inonotus obliquus, Lentinula edodes, Trametes versicolor* (Muszyńska et al. 2018), etc., are called as mushrooms, and nonedible poisonous sporophores like *Inocybe erubescens, Inocybe fastigiata, Amanita phalloids* (Lima et al. 2012), etc., are called as "*toadstool*." The edible mushrooms are a rich source of proteins, fibers, amino acids in particular phenylalanine, threonine, and tyrosine.

27.4.1 Morphology of Fungal Mycelium

All the fungal cell walls are fundamentally similar to each other in terms of structural integrity. The fungal cell wall layer is made up of β-1,3-glucan and chitin linked to each other covalently. The glucan and chitin are bonded in a microfibril basket-like structure around the cell. Mycelium is the vegetative part of the fungus which grows profusely in a branched form called hyphae (Abhijith et al. 2017). The hyphae consist of long slender threadlike structures and multiple hyphae interconnected via a network forms mycelium. Some hyphae form continuous tube-like structures filled with multinucleated cytoplasm called coenocytic hyphae. Organic and inorganic nutrition also affects the mycelial growth within the hyphae network. The addition of optimal quantities of sulfur and nitrogen has been shown to increase the production of branches per millimeter of hyphae (Jones et al. 2017).

27.4.2 Mushroom-Based Packaging Materials

Mushroom mycelium grows easily on any organic substrate, such as biological and agricultural wastes. The vegetative fungal mycelium is composed of frequently branched fibers of hyphae, which feeds on the substrate and entangles it. Mycelia are hyphae or threadlike structures that are made from agricultural wastes mixed with the root-like structures of mushrooms as binding agents. The digested substrate soon gets converted into biodegradable material, which can then be molded into any form. The mycelium foam created can efficiently replace polystyrene form. Mushroom mycelium is biodegradable, light in weight still durable and strong, easy for production, and easy to mold, which makes it a favorite choice for packaging material.

27.4.2.1 Raw Materials

Mushrooms, in particular, of the class Basidiomycota are wood decay fungi (WDF), which are able to degrade cellulose, hemicellulose, lignin by their enzymatic mechanisms (Schwarze et al. 2013), and hence agricultural and biological wastes are generally used as raw materials for the production of mushroom mycelia. Agro-industrial wastes like hemp, wood chips, psyllium husks, cotton hulls, etc., and other industrial wastes can also be used for the same.

27.4.2.2 Production Protocol

Mycelial foam production using agricultural wastes follows different steps as described below (Fig. 27.3).

Step 1: Mycelial Growth

The mushroom hyphae are grown using artificial media and then inoculated onto the substrate (Fig. 27.4). According to Abhijith et al. (2017), the production of mycelia is achieved by the use of organic compounds such as magnesium sulfate ($MgSO_4$), dipotassium phosphate (K_2HPO_4), copper sulfate ($CUSO_4$), and zinc sulfate ($ZnSO_4$). The mineral salt medium along with carbon, nitrogen sources, and inorganic components are used to produce mycelium. Sterilization of the solution is done by autoclaving it at a temperature of 120 °C. The solution is inoculated by adding 10% inoculum in a flask and left for incubation at 28 °C with centrifugation at 150 rpm. The mycelia after washing repeatedly is dried at 80 °C. Once the mycelial

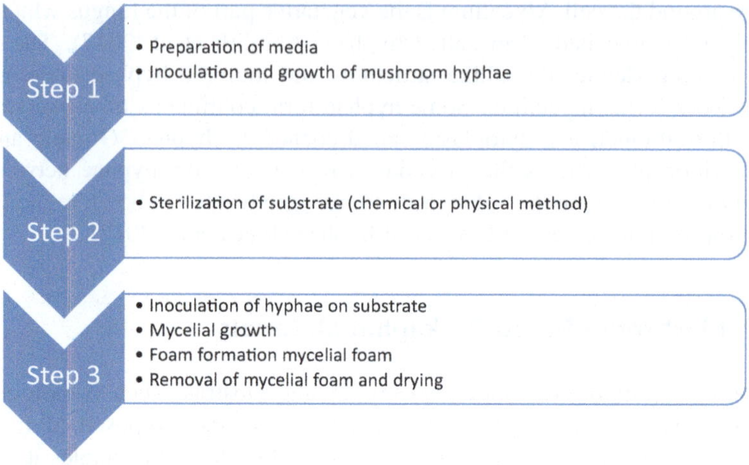

Fig. 27.3 Production protocol for MBFs

Fig. 27.4 Mycelium depicted as a mass of branching threadlike hyphae in a fungus-like bacterial colony

growth turns into the desired shape, the density of the growth of the material is then controlled by dehydration at a temperature less than 60 °C. Dehydrating the mycelium materials is an important step as it deactivates the tissues for any further growth, which eventually hardens the material.

In Kim et al. (2002), a study was conducted that used Potato malt peptone medium (PMP), mushroom complete medium (MCM), and yeast malt extract

(YM) as three different media to cultivate fungi for the sole purpose of the selection of a suitable media that supports the exo-biopolymer synthesis and mycelial growth.

The Composition of the Three Media:

(i) *Potato peptone medium (PMP)*
- Malt extract: 1 g/L.
- Potato dextrose: 24 g/L.
- Peptone: 1 g/L.

(ii) *Mushroom complete medium (MCM)*
- Glucose: 20 g/L.
- Potassium dihydrogen phosphate (KH_2PO_4): 0.46 g/L.
- Dipotassium phosphate (K_2HPO_4): 1 g/L.
- Magnesium sulfate heptahydrate ($MgSO_4 \cdot 7H_20$): 0.5 g/L.
- Peptone: 2 g/L.
- Yeast extract: 2 g/L.

(iii) *Yeast malt extract (YM)*
- Glucose: 10 g/L.
- Malt extract: 3 g/L.
- Peptone: 5 g/L.
- Yeast extract: 3 g/L.

Step 2: Sterilization of Substrate

The raw materials used are sterilized either by using natural antimicrobial agents like cinnamon bark oil, thyme oil, oregano oil, or any other plant product; or by heat sterilization. This eliminates the growth of any other microorganism in the mixture.

Step 3: Mycelial Growth on Substrate

The mushroom mycelium is allowed to grow on an organic substrate or any of the raw materials mentioned above. The filaments invade the substrate and create a fibrous network. It utilizes carbohydrates in the substrate for energy and the growth is very fast. These materials are grown into desirable shape over a short time may be a weeks' time. The mushroom mycelium is not allowed to bear fruiting bodies, so no spore production is seen. The mushroom spores may cause allergies in a few people, so there is no risk associated with mycelium production. When desired growth is achieved, the fungal growth is stopped either by adding antimicrobial agents or by applying heat. This is the base material for mycelial growth which produces foam. This mycelial foam can be used as packaging material.

The agricultural wastes are mixed with mycelium hyphae and then put in the molds of any required size. The whole mixture will be incubated in dark for a week. The mushroom mycelia multiply utilizing the substrate as its nutrition source. As the mycelia grow, the hyphae form a network of fibers throughout the substrate. The continued growth fills the space and produces a solid structure called Mycelium Foam.

This foam is then removed from the mold and dried to stop the further growth of mycelia and spore production. As moisture is a key factor for its growth and reproduction, drying helps in stopping mushroom proliferation.

Mycelium is a natural adhesive that acts to adhere to anything around to create a superdense network of threadlike structures at a low cost. Mushroom-based packaging materials can serve as an alternative to polystyrene foam packaging materials (Abhijith et al. 2017).

The mycelial foam obtained can be used for further fabrication and can be used for making any packaging materials from handbags to faux-leather jackets.

27.4.2.3 Mushroom-Based Foams

Mushroom-based foams (MBFs) can be categorized into two groups, viz., foams and sandwich composites. The MBFs are highly anisotropic, low-density fibrous structures comprising of coexisting fungal biomass and residual substrate in a unique matrix. The solid foam is created because of the impressions left by gas bubbles in the solid matrix. The mycelial foam though does not have such structures, it does share features such as high porosity and slight rigidity. Sandwich composites have an external layer held by mycelial glue growing in the core material. The core is similar to MBF: two layers of lignocellulosic material colonized by mycelium and attached to the core into a single block (Girometta et al. 2019).

27.4.2.4 Properties of MBFs

Density—Though low density of MBFs makes it competitive for packaging material with other synthetic packaging material, it affects the physical-mechanical and thermodynamic properties. Cold or heat pressing affects the final density and helps increasing the quality of the product.

Insulation—MBFs are known to have both thermal (González and González 2015) and acoustic insulation properties (Pelletier et al. 2013).

Fire safety—stable at higher temperatures.

Mechanical properties—compressive strength varies based on fungal species and the protocol followed, still it is competitive with other materials. Tensile strength can be increased by heat pressing and is not affected by either substrate or the species of fungi used. Flexural strength is equally high because of the fungal cell elasticity (Girometta et al. 2019).

27.4.2.5 What Makes MBFs Advantageous?

- **Strong and Biodegradable**—Although the consumer demand for eco-friendly, biodegradable materials have increased, there is still a continued requirement for the materials to be durable and reliable at the same time. Mycelium's ability to satisfy both necessities gives it a serious edge over not just the conventional polystyrene foams but also over other organic plastic alternatives.
- **Moldability**—the foam produced can be molded into any shape thereby requiring a simple protocol for its production. Additionally, because one can grow the exact amount of material needed, one eliminates excess waste generated.

- **Great for small scale business**—The production of mycelium is not dependent on heavy, expensive production machinery. Small scale businesses that may not need large sums of the material can produce at local levels.
- **Cost-competitive**—considering its sustainable nature over its polystyrene counterpart, mycelium inherently becomes cost-competitive as a packaging material.
- **Versatile base material**—Mycelium's foam base offers an extended list of applications like bags, outdoor jackets, insoles, shoes, etc.

27.4.2.6 Multi-Facetted Applications of Mycelium-Based Foams

Although serving as a packaging material is the primary purpose for mycelium-based foams, they can be also used for scaffolding, bags, jackets, furniture, organic plastic production, etc. They have been off late used as raw material for textiles or shoes. While they remain less practical when compared with traditional furniture made out of wood and linen, mycelium-based furniture might last longer given its sustainability.

From footwear, leather alternatives to lampshades, mycelium-based foams could be used for the production. The foam can be shaped to create textiles, clothing and shoes, packages for bottles, etc. Figures 27.5, 27.6, and 27.7 show different materials made out of MBFs.

Greensulate made of water, perlite, flour, and mushroom spores is a fire retardant and can be used as surfboard.

27.5 Future

Since its invention polystyrene foams have served in countless ways, catering for multiple commercial and industrial advancements. They have offered everything from being used as cushioning packages to being incorporated in floatation devices used in rescue missions. Therefore, as a material, it has done its job dutifully to have become the most common household name. But the associated environmental impacts that arise from its inefficient disposal have been a serious troubling issue

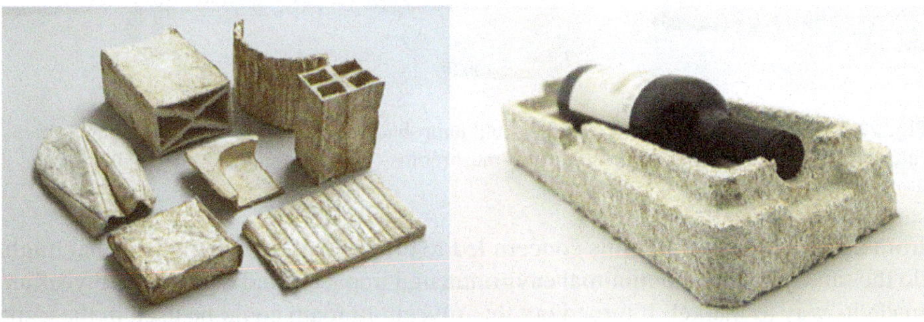

Fig. 27.5 Mycelium-based foams used to create packaging for wine bottles (https://matmatch.com/blog/mycelium-using-mushrooms-to-make-packaging-materials/)

Fig. 27.6 Mycelium panels used to construct the Growing Pavilion. The frame is timber-based, but the panels are made from mushrooms, covered in an organic sealant for sustainability (https://matmatch.com/blog/mycelium-using-mushrooms-to-make-packaging-materials/)

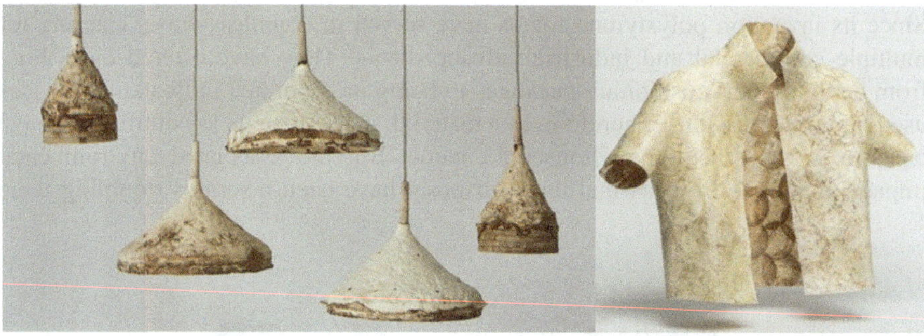

Fig. 27.7 Biodegradable, organic, textiles, and lampshades made out of mushroom-based foams (https://matmatch.com/blog/mycelium-using-mushrooms-to-make-packaging-materials/)

from quite some time. But this concern led to efforts seeking alternatives that might do the same job but with minimal environmental impacts. And eventually, mycelium made its way to market. It turned out that mycelium foam could be used in the same manner as its polystyrene counterpart. And given its lightweight, easy to mold, and easy to dispose trait, mycelium has recently become a far more popular choice, in fulfilling at least some of the major needs for packaging material. Mycelium revolution with organic scaffolding is under research and may come through in near future. Thus, for now, as far as an alternative for plastics is concerned, mycelium holds promise.

There are some challenges in the production of plastic alternatives using mushrooms, like standardization of protocol, maintaining the quality of the product, etc. Secondly, one has to address issues such as the logistics of transportation, the biodegradability of mushrooms is a boon and a bane at the same time. While the material is exceedingly durable, it still breaks down outside. Also, to be used in food packaging it will have to go through rigorous FDA approved quality checks, which could take years to pass. So, will they completely replace plastics? only time will tell. From a sustainable, biodegradable perspective, mycelium appears to perform quite admirably.

27.6 Conclusion

The mycelium-based packaging is composite, easily degradable, and more importantly biodegradable. Mycelium composites utilize agricultural waste material to create eco-friendly packaging materials. The availability of the substrates and the low cost of products makes it efficient and effective unlike polystyrene-based materials. During production, the mycelium composites produce 90% less carbon emission than the Styrofoam packaging materials. The mycelium composites get completely decomposed upon disposal thereby providing rich sources for agricultural decomposers. The mycelium-based composites have the potential to replace polystyrene packaging by engineering it with industrial advancements. The mycelium-based packaging offers a great eco-friendly alternative, in addition to being an immensely cost-effective and sustainable alternative to its petroleum-based packaging materials.

References

Abhijith R, Anagha A, Rejeesh CR (2017) Sustainable packaging application from mycelium to substitute polystyrene: a review. Materials Today: Proc 5(1):2139–2145

Alabi OA, Ologbonjaye KI, Awosolu O, Alalade OE (2019) Public and environmental health effects of plastic wastes disposal: a review. J Toxicol Risk Assess 5(021):1–13. https://doi.org/10.23937/2572-4061.1510021

Anthony LA, Mike AN (2009) Applications and societal benefits of plastics. Biol Sci. https://doi.org/10.1098/rstb.2008.0304

Aryan Y, Yadav P, Samadder SR (2019) Life cycle assessment of the existing and proposed plastic waste management options in India: a case study. J Cleaner Prod 211:1268–1128. https://doi.org/10.1016/j.jclepro.2018.11.236

Barnes DKA, Galgani F, Thompson RC, Barlaz M (2009) Accumulation and fragmentation of plastic debris in global environments. Philos Trans R Soc B Biol Sci 364:1985–1998. https://doi.org/10.1098/rstb.2008.0205

Biello D (2013) Are biodegradable plastics doing more harm than good? Scientific American

Bilal H, Shahzad ASC, Abdullah IH, Khalid MZ, Naseem A (2017) Recent advances on polysaccharides, lipids and protein based edible films and coatings: a review. Int J Biol Macromol 109(1):1095–1107. https://doi.org/10.1016/j.ijbiomac.2017.11.097
Dahlbo H, Poliakova V, Mylläri V, Sahimaa O, Anderson R (2017) Recycling potential of post-consumer plastic packaging waste in Finland. Waste Manag 71:52–61. https://doi.org/10.1016/j.wasman.2017.10.033
Danso D, Chow J, Streit WR (2019) Plastics: environmental and biotechnological perspectives on microbial degradation. Appl Environ Microbiol. https://doi.org/10.1128/AEM.01095-19
Fackelmann G, Sommer S (2019) Microplastics and the gut microbiome: how chronically exposed species may suffer from gut dysbiosis. Mar Pollut Bull 143:193–203. https://doi.org/10.1016/j.marpolbul.2019.04.030
Geyer R, Jambeck JR, Law KL (2017) Production, use, and fate of all plastics ever made. Sci Adv 3 (7):e1700782. https://doi.org/10.1126/sciadv.1700782
Girometta C, Picco AM, Baiguera RM, Daniele D, Babbini S, Cartabia M, Pellegrini M, Savino E (2019) Physico-mechanical and thermodynamic properties of mycelium-based biocomposites: a review. Sustainability 11:281. https://doi.org/10.3390/su11010281
González ME, González DI (2015) Bacterial induced cementation processes and mycelium panel growth from agricultural waste. Key Eng Mater 663:42–49
Greene JP (2014) Biobased and biodegradable polymers. In: Sustainable plastics. John Wiley and Sons, Inc., Hoboken, NJ, pp 71–106
Guzzettia E, Suredab A, Tejadac S, Faggioa C (2018) Microplastic in marine organism: environmental and toxicological effects. Environ Toxicol Pharmcol 64:164–171. https://doi.org/10.1016/j.etap.2018.10.009
Holland ER, Mallory ML, Shutler D (2016) Plastics and other anthropogenic debris in freshwater birds from Canada. Sci Total Environ 571:251–258. https://doi.org/10.1016/j.scitotenv.2016.07.158
Huerta LE, Gertsen H, Gooren H, Peters P, Salánki T, Van Der Ploeg M, Besseling E, Koelmans AA, Geissen V (2016) Microplastics in the terrestrial ecosystem: implications for Lumbricus terrestris (Oligochaeta, Lumbricidae). Environ Sci Technol 50:2685–2691. https://doi.org/10.1021/acs.est.5b05478
James K, Fitzpatrick L, Lewis H, Sonneveld K (2005) Sustainable packaging system development. In: Leal Filho W (ed) Handbook of sustainability research. Peter Lang Scientific Publishing, Frankfurt
Jones MH, Dekiwadia T, Daver CF, John S (2017) Mycelium composites: a review of engineering characteristics and growth kinetics. J Bionanosci 11(4):241–257. https://doi.org/10.1166/jbns.2017.1440
Kim Y, Ruedy D (2019) Mushroom packages. An ecovative approach in packaging industry. In: Dhiman S, Marques J (eds) Handbook of engaged sustainability. Springer Nature Switzerland. https://doi.org/10.1007/978-3-319-53121-2_27-1
Kim SW, Hwang HJ, Park JP, Cho YJ, Song CH, Yun JW (2002) Mycelial growth and exo-biopolymer production by submerged culture of various edible mushrooms under different media. Lett Appl Microbiol 34:56–61
Knight G (2012) Plastic pollution. Raintree Publisher, London
Lewis H, Fitzpatrick L, Verghese K, Sonneveld K, Jordon R (2007) Sustainable packaging
Lima ADL, Costa FR, Garbi Novaes MRC, Percário S (2012) Poisonous mushrooms: a review of the most common intoxications. Nutr Hosp 27(2):402–408. https://doi.org/10.3305/nh.2012.27.2.5328
Maulida MS, Tarigan P (2016) Production of starch based bioplastic from cassava peel reinforced with microcrystalline cellulose Avicel PH101 using sorbitol as plasticizer. J Phys: Conf Ser 710:012012. https://doi.org/10.1088/1742-6596/710/1/012012
Muszyńska B, Grzywacz-Kisielewska A, Kała K, Gdula-Argasińska J (2018) Anti-inflammatory properties of edible mushrooms: a review. Food Chem 243:373–381. https://doi.org/10.1016/j.foodchem.2017.09.149

Pelletier MG, Holt GA, Wanjura JD, Bayer E, McIntyre G (2013) An evaluation study of mycelium based acoustic absorbers grown on agricultural by-product substrates. Ind Crop Prod 51:480–485. Presented at the 22nd International Association of Packaging Research Institutes Symposium

Rafael A, Bruce H, Susan S (2004) An overview of polylactides as packaging materials. Macromol Biosci 4(9):835–864. https://doi.org/10.1002/mabi.200400043. Redefined. http://nbis.org/nbisresources/packaging/sustainable_packaging_guidelines.pdf

Roopesh J, Archana T (2015) Biosynthesis of planet-friendly bioplastics using renewable carbon source. J Environ Health Sci Eng 13:11. https://doi.org/10.1186/s40201-015-0165-3

Schwarze FWMR, Engels J, Mattheck C (2013) Fungal strategies of wood decay in trees. Springer-Verlag, Berlin, Germany, p 185. ISBN 3 540 67205 2

Selke SEM (1990) Packaging and the environment: alternatives, trends and solutions. Technomic Publishing Co., Lancaster, PA

Selvamurugan M, Sivakumar P (2019) Bioplastics—an eco-friendly alternative to petrochemical plastics. Curr World Environ 14(1):49–59. https://doi.org/10.12944/CWE.14.1.07

Srinivasa PC, Tharanathan RN (2007) Chitin/Chitosan—safe, ecofriendly packaging materials with multiple potential uses. Food Rev Int 23(1):53–72. https://doi.org/10.1080/87559120600998163

Tudryn G, Smith LC, Freitag J, Bucinell RB (2018) Processing and morphology impacts on mechanical properties of fungal based biopolymer composites. J Pol Environ 26:1473–1483. https://doi.org/10.1007/s10924-017-1047-9

Verghese K, Lewis H (2005) Sustainable packaging: how do we define and measure it

Environmental Implications of Microbial Bioplastics for a Sustainable Future

28

A. Mbotho, S. I. Magagula, K. M. Moloantoa, J. S. Sefadi, and M. J. Mochane

Abstract

In a world where fast-paced manufacturing and processing of quality, durable, efficient, versatile, and affordable products to meet the demands of an ever-growing population, plastic has dominated many industries. The accumulation of non-biodegradable plastics in the environment has become a global challenge which has facilitated various studies on developing an environmentally friendly strategy to counteract the current plastic pollution predicament. Microbial bioplastics resulting from the interaction of chemical reactions and microorganisms are a promising substitute owing to their biodegradability without the emission of toxic compounds. This chapter explores the synthesis and degradation of these microbial biopolymers. The different types of bioplastics are discussed with emphasis on their impact on the environment as opposed bytheir synthetic counterparts. Additionally, the challenges currently encountered with the production of biopolymers are highlighted, including possible solutions to ensure continuous yet safe and eco-friendly manufacturing of bioplastics without compromising the favorable qualities displayed by non-biodegradable plastics.

A. Mbotho · S. I. Magagula · M. J. Mochane (✉)
Department of Life Sciences, Central University of Technology, Bloemfontein, South Africa
e-mail: mmochane@cut.ac.za

K. M. Moloantoa
Department of Microbial, Biochemical and Food Biotechnology, University of the Free State, Bloemfontein, South Africa

J. S. Sefadi (✉)
Department of Physical and Earth Sciences, Sol Plaatje University, Kimberley, South Africa
e-mail: jeremia.sefadi@spu.ac.za

A. Vaishnav, D. K. Choudhary (eds.), *Microbial Polymers*,
https://doi.org/10.1007/978-981-16-0045-6_28

Keywords

Microbial bioplastics · Non-biodegradable plastics · Biodegradation · Microorganisms · Environment · Pollution

28.1 Introduction

The continuous and excessive use of plastic materials in modern society is evident and nearly unavoidable. Plastic materials have dominated the manufacturing line whereby almost all consumer products produced contain plastic with over 140 million tonnes of plastic produced annually (Caruso 2015) for clothing, electronic devices, cosmetics, home appliances, food and beverage, even transcending to use in the medical application (Ahmed et al. 2018; Muhamad et al. 2015a, b). Plastic is easy to manufacture and its properties enable its formation into various products due to its flexible and moldable nature (Muhamad et al. 2015a, b). Additionally, plastic is durable, lightweight, and resistant to certain environmental conditions including moisture, chemical, corrosion, and thermal resistance (Asrofi and Sujito 2019; Gan and Zhang 2019; Maddah 2016; Mostafa et al. 2018). It comes as no surprise that its popularity in various industries has increased tremendously since its invention in the nineteenth century reaching large-scale production in the twentieth century and ultimately leading to the extensive development of plastic encountered currently in the twenty-first century (Glaser 2019; Krueger et al. 2015).

The massive use coupled with production due to increasing demand for plastic has had a significant impact on the environment (Yang et al. 2018). Not only are synthetic polymers polluting the planet including its water supply but are also associated with increasing temperatures due to global warming as a result of gas emissions which are a consequence of their degradation (Urbanek et al. 2018; Xanthos and Walker 2017). The degradation of synthetic polymers is a problem which has been intensely researched as they are very difficult to degrade naturally, and thereby requiring the use of abiotic means of breaking them down which results in the emission of unfavorable chemicals that affects the environment including all forms of life worldwide. The effects of plastic pollution are also linked to climate change which forms part of the biggest environmental challenges of the twenty-first century according to Perez and coworkers (Pérez et al. 2019).

The intensive search to find solutions to degrade plastic in an environmentally friendly manner remains a mystery. Through ongoing research, various strategies for producing environmentally friendly polymers have been tested including the development of bioplastics which are a promising solution to the global plastic pollution. Bioplastics are an environmentally friendly approach to reduce synthetic plastic as they are produced from renewable sources including polysaccharides (i.e., starch, cellulose, and chitin), lipids (oils), proteins (e.g., gluten and gelatine), plant or microbial polyesters (polyhydroxyalkanoates) and polyesters from bio-derived monomers such as polylactic acid derived from starch (Karamanlioglu et al. 2017). Bioplastic production involves the metabolic activity of microorganisms which depends on the suitable addition of a carbon and energy source that they use for growth and could be sourced from agricultural by-products and starch biopolymers (Trivedi et al. 2016). The benefit of bioplastics is that they are easily biodegradable yielding products such as water, carbon dioxide, and hydrocarbons that are readily assimilated in the environment without producing unfavorable chemicals that promote hazardous conditions (Sidek et al. 2019). The most common carbon sources utilized by microorganisms that produce bioplastics are glucose and sucrose (Rehman et al. 2015; Iram et al. 2019). In addition to carbon sources, a nitrogen source is also required for the development and growth of the microorganisms (e.g., bacteria) involved in the production of biopolymers (Mannina et al. 2019). Examples of nitrogen sources include yeast extract, ammonium chloride, and ammonium sulfate which are readily available and easily broken down by the bacteria to harvest the needed nitrogenous contents of the source (Rehman et al. 2015).

This chapter enlightens on the impact of microbial bioplastics on the environment, the pathways that occur leading to their synthesis and subsequent degradation of bioplastics. The by-products formed from these processes are discussed in comparison to those of synthetic non-biodegradable polymers. This chapter aims to highlight the benefits and challenges encountered due to the use and production of microbial bioplastics in the presence of raw renewable carbon sources as opposed to synthetic polymers which are synthesized artificially from petroleum oils (Shrivastava 2018).

28.2 Microbial Bioplastics

Microbial bioplastics are a special class of biomaterials. Basically, they are polyesters that are produced and accumulated by several bacterial strains, cultured under different nutrient and environmental conditions (Luengo et al. 2003). They are lipid-like in nature and are accumulated intracellularly as storage materials in the form of mobile, amorphous, and liquid granules for the survival of the bacterial microorganisms under stressful conditions (Luengo et al. 2003; Sudesh et al. 2000). The number and size of granules accumulated in each bacterial strain, the monomer composition of the bioplastics, their micromolecular structure, and physicochemical properties depend on the type of producer microorganism (Ha and Cho 2002; Luengo et al. 2003). Furthermore, the physicochemical properties of bioplastics resemble those of petrochemical plastics which is why they are used in replacing these conventional plastics (Luengo et al. 2003). In addition to that, the properties of bioplastics such as biodegradability and biocompatibility have attracted a substantial amount of attention for their use in biotechnological applications (Luengo et al. 2003). However, the major limitation for the use of bioplastics in most industrial applications is their poor mechanical properties. Hence, they are usually incorporated with other materials or biopolymers in order to produce sustainable biocomposites (Raza et al. 2019). Furthermore, due to their light-refracting nature, bioplastic granules appear as light-refracting granules or as electronlucent bodies inside the cells of microorganisms when observed under a scanning and transmission

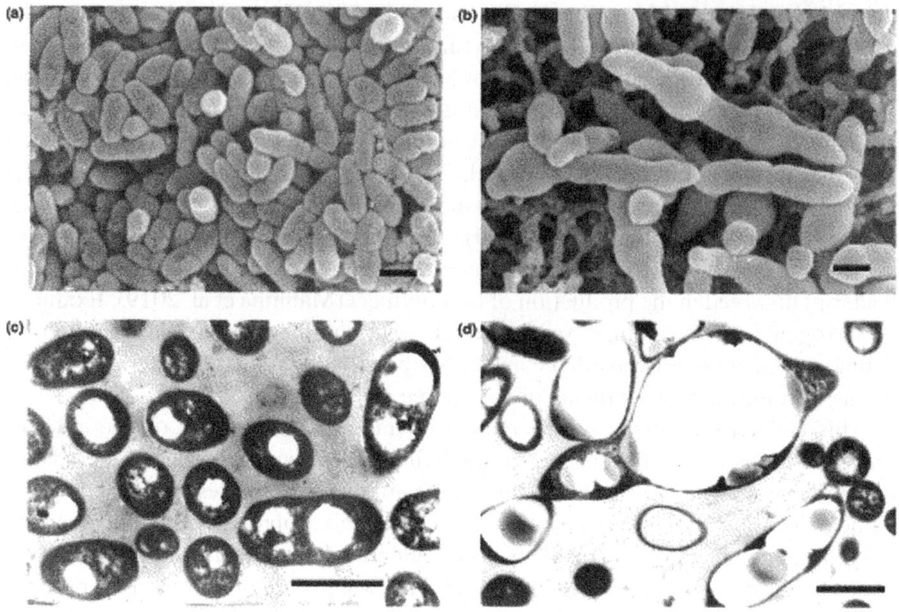

Fig. 28.1 Microphotographs of *P. putida U* (**a**, **c**) and its ΔfadBA *β*-oxidation mutant (**b**, **d**) obtained from scanning (**a**, **b**) and transmission (**c**, **d**) electron microscopy (Luengo et al. 2003). Copyrights Elsevier

electron microscope (see Fig. 28.1). In high-yielding mutant strains, the bioplastic granules can even change the shape of the bacterial microorganism (see also Fig. 28.1) (Luengo et al. 2003).

It has been found that microbes from more than 90 genera synthesize and accumulate bioplastics and these include aerobes, anaerobes, photosynthetic bacteria, archaebacteria, and lower eukaryotes (Sudesh et al. 2000; Zinn et al. 2001). In each bacterial strain, the bioplastics are synthesized from hydroxyl-acyl-CoA derivatives via various metabolic pathways (see Fig. 28.2) (Luengo et al. 2003). The most commonly produced and well-studied microbial bioplastics are polyhydroxybutyrates (PHBs), polyhydroxyalkanoates (PHAs) and their derivatives (Luengo et al. 2003; Sudesh et al. 2000). This is because their physicochemical properties resemble those of conventional petrochemical plastics. However, there are other polyesters that are also produced and accumulated by microorganisms. These polyesters are not produced as much as PHB and PHA because they either require similar biosynthetic enzymes or lack current industrial applications (Luengo et al. 2003).

In spite of the worldwide demand for bioplastics to replace conventional petrochemical plastics, the bulk production of bioplastics is faced with a major challenge. The cost of their production and recovery from bacterial cell cytoplasm is very high (Luengo et al. 2003). However, attempts such as genetic and metabolic engineering have been utilized to produce recombinant organisms such as bacteria, yeasts, or

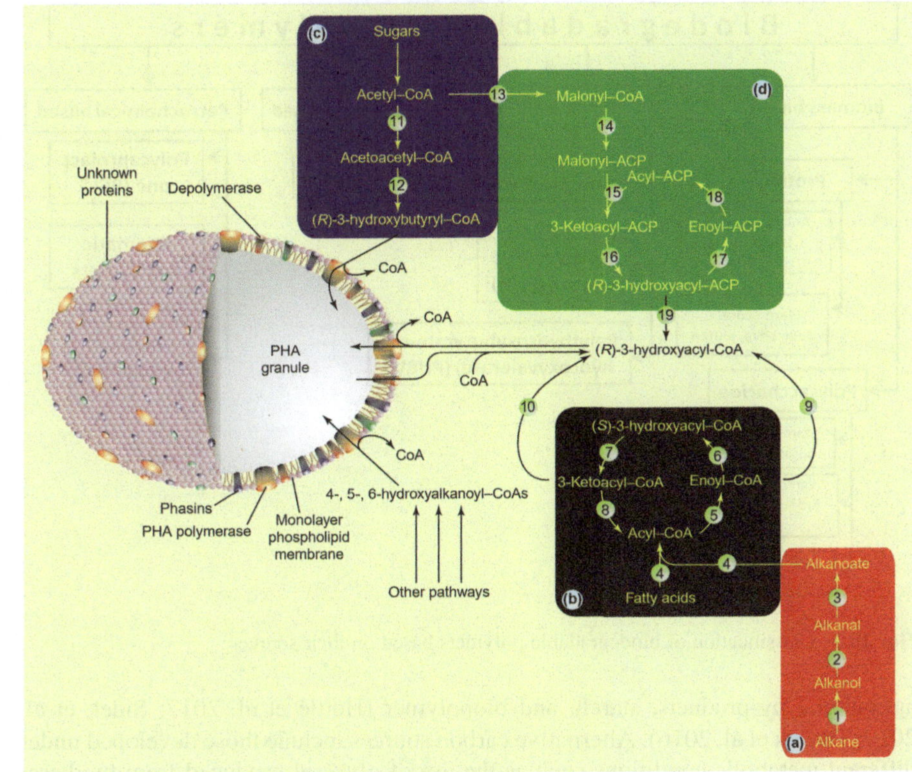

Fig. 28.2 The organization of the PHA granule structure and the interconnections between the various metabolic pathways involved in the biosynthesis and catabolism of PHBs and PHAs. **(a)** Pathway for alkane oxidation. (*1*) Alkane 1-monooxygenase, (*2*) alcohol dehydrogenase, (*3*) aldehyde dehydrogenase, **(b)** β-oxidation of fatty acid. (*4*) acyl–CoA ligase, (*5*) acyl–CoA dehydrogenase, (*6*) enoyl–CoA hydratase, (*7*) 3-hydroxyacyl–CoA dehydrogenase, (*8*) 3-ketothiolase, (*9*) (R)-enoyl–CoA hydratase, (*10*) 3-ketoacyl–CoA reductase. **(c)** Biosynthesis using carbohydrates. (*11*) β-ketothiolase, (*12*) NADPH-dependent acetoacetyl–CoA reductase. **(d)** De novo synthesis of fatty acid. (*13*) acetyl–CoA carboxylase, (*14*) ACP-malonyltransferase, (*15*) 3-ketoacyl-ACP synthase, (*16*) 3-ketoacyl-ACP reductase, (*17*) 3-hydroxyacyl-ACP reductase, (*18*) enoyl-ACP reductase, (*19*) 3-hydroxyacyl-ACP–CoA transacylase (Luengo et al. 2003). Copyrights Elsevier

transgenic plants which are capable of producing and accumulating high yields of bioplastics, and therefore reducing the overall costs (Luengo et al. 2003).

28.3 Different Types of Bioplastics

Many natural, renewable materials have been considered as feedstock in the production of bioplastics. Sources include vegetable fats and oils, corn starch, sugar cane, wood, and algae which are used to derive bioplastics using microorganisms,

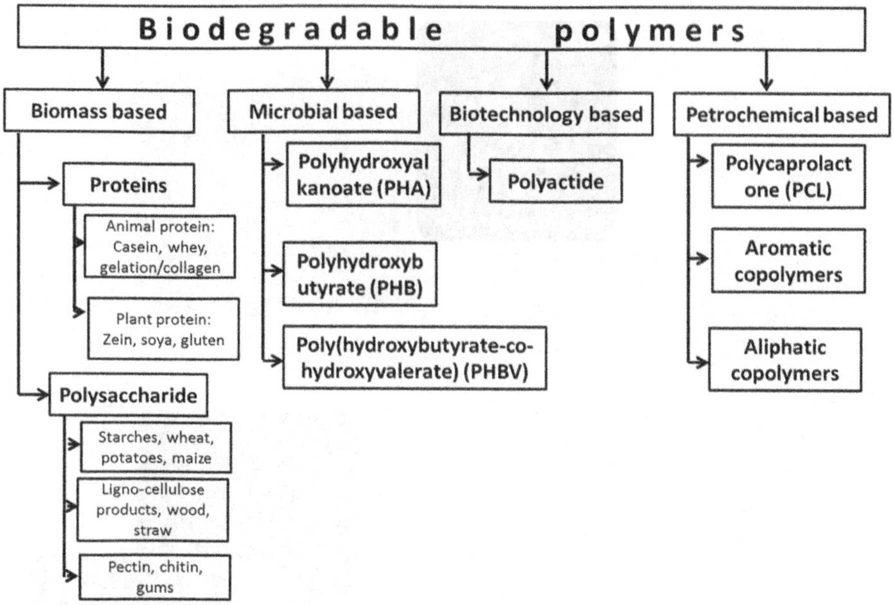

Fig. 28.3 Classification of biodegradable polymers based on their sources

agricultural by-products, starch, and biopolymer (Hottle et al. 2017; Sidek et al. 2019; Trivedi et al. 2016). Alternative carbon sources include those developed under different metabolic conditions such as the use of glycerol produced from biodiesel manufacture to produce P3HB by converting the glycerol to glyceraldehyde-3-phosphate (GAP) with the help of enzymes glycerol kinase (GlpK), glycerol-3-phosphate dehydrogenase (GlpD), and triosephosphate isomerase (TPI) (Zhu et al. 2013). Glycerol use as feedstock is more affordable than the fermentative processes involved in traditional carbon sources. For example, before bioplastics such as PHA can be produced from carbon sources such as carbohydrates, they have to undergo hydrolysis first depending on whether they are monosaccharides, oligosaccharides, or polysaccharides; thereafter they can be fermented to produce PHA which accounts for up to 50% of the entire production cost (Jiang et al. 2016; Zhu et al. 2013). The diverse way which involves the type of pathway used (i.e., pathway I, II, and III) in which carbon sources are metabolized also depends on the type of bioplastic being produced (e.g., PHA, PHB or 3HB and 3 HV) the substrate (glucose, sucrose, or fructose) and the method of fermentation (batch or feed batch) (Mohapatra et al. 2017). The bioplastic can be produced by mixing starch and polyesters to form fillers or blends that are easily biodegradable such as polycaprolactone (Ahmed et al. 2018; Kumar and Thakur 2017). In addition to glucose and fructose, bioplastics can be produced from oil and cellulose (Sukara 2014). Various processes are implemented which ultimately lead to the conversion of the feedstock to a viable bioplastic. Figure 28.3 below shows the classification of different bioplastics.

28.3.1 Starch-Based Bioplastics

In the production of bioplastics, starch is the most commonly used feedstock (Ahmed et al. 2018; Colombo et al. 2016; Hottle et al. 2017; Karamanlioglu et al. 2017; Kumar and Thakur 2017; Mostafa et al. 2020; Muhamad et al. 2015a, b; Sidek et al. 2019). Among the numerous reasons behind its exorbitant use, its widespread availability making it easily accessible, relatively low in cost, as well as its biodegradability are the most noted characteristics of starch as a source of carbon in bioplastic synthesis (More et al. 2017). Starch-based polymers fall under different classifications (see Fig. 28.3) (Kumar and Thakur 2017). Some of the classes include starch-filled polymers and starch-based polymers which include the combination of starch with other material such as polyesters and synthetic polymers (i.e., petroplastics such as polyethylene) (Ilyas et al. 2020) thus forming blends which are readily accessible, cost-effective, and are used in agriculture, engineering, and packaging (Iram et al. 2019). Blends of Starch with other materials are largely used due to their easy degradation by microorganisms such as in the case with blends between starch and polyvinyl alcohol (PVA) (Mathew et al. 2019). Starch-based polymers confer good thermal and mechanical properties including resistance to gas permeation and transparency, these properties are amplified by the blending of starch with PVA which provides tensile strength, chemical resistance, and is fully biodegradable (Zanela et al. 2020). Other Blends include starch with gelatine and starch with cellulose which have the advantage of being natural, renewable, inexpensive, and biodegradable (Jiang et al. 2020).

28.3.1.1 Biopolymers from Gases

The use of greenhouse gases (GHGs) is a plastic production approach where a different substrate in the form of a gas is used as feedstock for the production of biopolymer apart from the traditional sources (Muthuraj and Mekonnen 2018). The approach involves the biological conversion of these GHGs into bioplastics whereby the gases undergo various processes commencing with the capture of the GHG, followed by transport and sequestration to name a few which ultimately lead to their availability and subsequent use as feedstock (Muthuraj and Mekonnen 2018; Pérez et al. 2019). The use of GHGs to produce bioplastics presents great possibilities which would be beneficial for both the environment as the use of these toxic gases would be an effective way to reduce their increasing concentrations in the environment while simultaneously replacing the use of non-biodegradable synthetic polymer use in an affordable manner which will not hinder developments in the industry where plastic manufacturing and use is concerned. A study by Perez et al. (2019) investigated this approach using methane (CH_4) where methanotrophic bacteria including *Methylocystis*, *Methylosinus*, and *Methylocell* were used to degrade CH_4 at different temperatures thus forming bioplastics by two different pathways, namely, the ribulose monophosphate pathway (RuMP) and the serine pathway (REF; REF). The study was able to deduce that methanotrophs are capable of producing PHA of contents ranging from 20 to 60%(wt) utilizing CH_4 as feedstock. The study further highlighted the challenges experienced during this production of

PHA such as the lack of full comprehension in terms of the factors regulating the selective enrichment of methanotrophs for the production of high-performance PHA (Pérez et al. 2019).

28.3.2 Microbial Bioplastics

28.3.2.1 Polyhydroxyalkanoates

Polyhydroxyalkanoates are linear polymers resulting from the fermentation of sugars and lipids by the action of bacteria. PHAs are produced by bacteria as storage for carbon and energy. The bacteria when exposed to conditions of nutrient deficiency will produce PHAs which are then extracted and used in sugar fermentation. PHA has become of tremendous interest due to its properties which are favorable to the environment when compared to those of petrochemical-derived polymers (e.g., polystyrene). PHAs are made from renewable resources, and therefore are completely biodegradable and have many applications including their applications in the medical industry. PHAs have been found to be produced by cyanobacteria under photoautotrophic and chemoheterotrophic conditions utilizing different sources of carbon including acetate and maltose which showed improvement of PHA yields without the need for the addition of other precursors (Singh and Mallick 2017).

The most common polymer used for bioplastic production is polyhydroxybutyrate (PHB). Polyhydroxybutyrate is found in many microorganisms whereby it serves as a carbon and energy source by binding beta-hydroxybutyrate monomers to ester bonds. Examples of bacterial species found to produce PHB include *Vibrio harveyi, Paracoccus species, Micrococcus species, Erythrobactee aquimaris, Halomonas elongata, Bacillus megaterum, Cupriavidus necator, and Halogerax mediterranei.* PHB is similar to conventional plastics in that it is easily mouldable (Bharti and Swetha 2016).

Polylactic acid (PLA) is a bioplastic made from corn and dextrose. PLA is transparent in nature. PLA is of interest due to its properties which closely resemble those of petrochemical-based plastics; additionally, PLA can be produced by utilizing the same production methods used by conventional plastics. PLA is mostly used in film, fiber, cup, and bottle production.

28.4 Synthesis and Degradation of Microbial Bioplastics

The production and destruction of bioplastics much like any other material are regulated by many factors such as environmental factors (moisture, temperature, and pH), nature of substrate used, and the microorganisms involved (i.e., the microbial strain as well as whether it is a pure culture or consortium) (Thakur et al. 2018). All of these factors ultimately determine the type of bioplastic formed, its properties with regards to products required for its synthesis, and which products are formed from its degradation. A study by Mostafa et al. (2020) investigated the effect of temperature, pH, organic compounds (such as glycerine, peptone, urea, yeast extract), and inorganic compounds (such as ammonium chloride, ammonium sulfate, and potassium nitrate) as nitrogen sources in the production of the bioplastic-Poly-3-hydroxybutyrate (PHB) using marine bacterium *Pseudodonghicola xiamenensis*. The study was focused on the dye-based procedure where they found that the highest PHB accumulation (concentration of 15.54 g/L, yield of 38.85% at a production rate of 0.162 g/L/h) was reached after incubation of 96 h at a pH of 7.5, and a temperature of 35 °C in the presence of 4% NaCl with peptone as the preferred nitrogen source (Y. S. Mostafa et al. 2020). The study showed potential for improved production costs for the manufacturing of the biopolymer PHB while simultaneously highlighting the importance of time, temperature, pH, microbial strain (i.e., type, pure or mixed) including its origin (marine environment) and nitrogen source in the synthesis of bioplastics (Carlozzi et al. 2020). Additionally, these parameters and the pathway utilized determine the efficiency and extent of formation (generally, pure cultures have been associated with producing the best yields of polyhydroxyalkanoates (PHAs) whereas mixed cultures are better at adapting to changing conditions and complex substrates) and destruction of the biopolymeric material (Colombo et al. 2016; Mostafa et al. 2020).

The synthesis of bioplastics is observed to occur under conditions of nutrient-limitation in the presence of excess carbon sources (Pérez et al. 2019). The pathway is facilitated by the action of microorganisms whereby they synthesize intracellular or extracellular metabolites as a response to nutritional and environmental stresses. Polyesters are some of the intracellular metabolites synthesized by bacteria under unbalanced conditions of nutrient deficiency (Carlozzi et al. 2019; Monroy and Buitrón 2020). The polyesters are produced as storage material known as polyhydroxyalkanoates (PHAs) (Monroy and Buitrón 2020). The most preferred PHA is polyhydroxybutarate (PHB) due to its properties which closely resemble those of synthetic polymers thus making PHB an interesting candidate in the research to find a replacement for synthetic plastics since PHB is completely biodegradable (Gan and Zhang 2019; Rehman 2015). Figure 28.4 shows the carbon cycle of bioplastics involving the synthesis and biodegradation of bioplastics.

28.5 Parameters Affecting Bioplastic Production

The production of biopolymers involves a number of processes and is affected by many aspects. The substrate or feedstock is but the first factor followed by many more such as the microorganism involved in the synthesis, the environmental conditions including the temperature, pH, as well as the presence or absence of oxygen (aerobic or anaerobic) (Bharti and Swetha 2016). In the production of PHB which is synthesized in the cell, the factors that have control include chemical sources such as oxygen, magnesium, iron, and phosphate in addition to carbon. Fermentation factors such as the ones mentioned above (i.e., temperature and pH) are the physical factors that also regulate PHB production (Bharti and Swetha 2016).

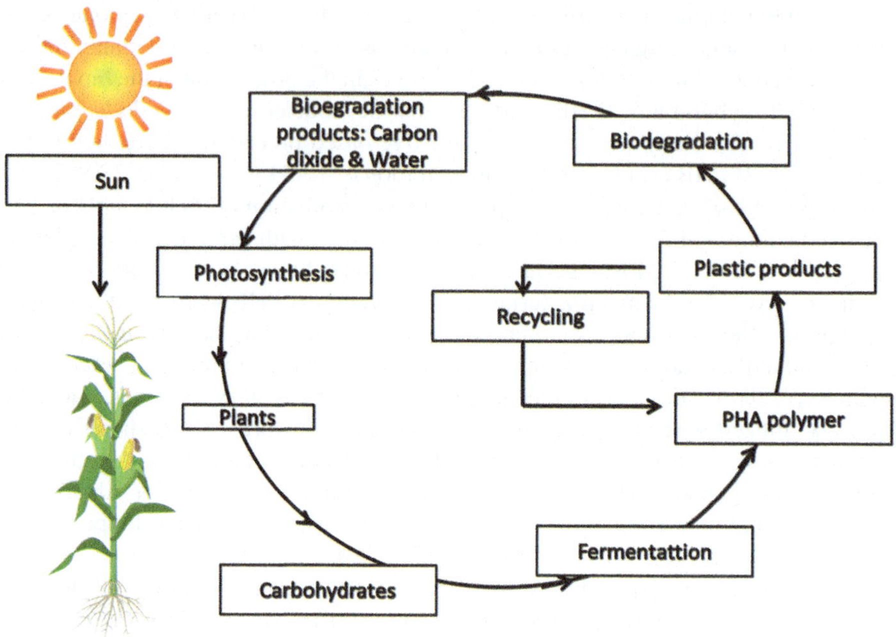

Fig. 28.4 Carbon cycle showing bioplastic synthesis and biodegradation

28.6 Biodegradation of Bioplastics in Different Natural Environments

The utilization of bioplastics has been considered beneficial to the environment, and hence their preference over petrochemical plastics. This is because bioplastics are produced from renewable resources such as crops and agricultural wastes (Emadian et al. 2017). In spite of the benefits associated with the utilization of bioplastics, there is still a need to address their potential negative impact on the environment (Emadian et al. 2017). Hence, in the last few decades, a lot of studies have been focused on the biodegradation of biopolymers in various natural environments such as soil, compost, marine, and other aquatic environments (Emadian et al. 2017). The biodegradation of bioplastics in different environments, environmental conditions, rate of biodegradation, and microorganisms from different microbial communities which are responsible for the biodegradation of bioplastics have been extensively reviewed Emadian et al. (2017). Among the various natural environments, the degree of biodegradability in soil and compost has been extensively studied due to the high microbial diversity present in these environments (Anstey et al. 2014). Even though most plastic wastes are disposed in landfills, the biodegradation of bioplastics in landfills has not yet been investigated (Emadian et al. 2017).

The biodegradability of bioplastics is highly dependent on their physical and chemical structure (Emadian et al. 2017). However, the microbial diversity of the environment in which they are located also plays a significant role in the degree of their biodegradability (Emadian et al. 2017). The biodegradation of bioplastics in various ecosystems is governed by the microorganisms that live there. These microorganisms catalyze the biodegradability of bioplastics via the assistance of responsible enzymes (Emadian et al. 2017). Different fungal and bacterial species that are capable of degrading the bioplastics into end products have been found in environments such as soil, compost, marine water, river water, and other environments. Among the bacterial and fungal species found in these environments, bacterial species such as *Paenibacillus, Pseudomonas, Bacillus,* and *Burkholderia* species and fungal species such as *Aspergillus, Fusarium,* and *Penicillium* species were the most common (Emadian et al. 2017).

Furthermore, the degree of biodegradability of bioplastics has been found to be the highest in environments such as soil and compost (Emadian et al. 2017). In spite of this fact, a substantial amount of plastics in these environments still find their way into water bodies and marine systems. Little is known about the degradability of bioplastics in aquatic systems. Therefore, if care is not taken, the plastics that end up in water bodies and marine systems may end up causing irreversible effects on the fresh water and marine ecosystems by adversely affecting the life of the plant and animal species that live in these environments (Emadian et al. 2017). Even though not proven, this may have adverse effects, in the long run on the health of humans who feed on seafood. The presence of large amounts of plastic waste including bioplastic waste in the marine environment is well known and indisputable. Therefore, there is an urgent need for the investigation of the degree of biodegradability of the bioplastics that end up in aquatic ecosystems (Emadian et al. 2017). However, before this can happen, studies on the number of degrading microorganisms that can be isolated from the marine environment need to be carried out first. This will give an indication of whether the degradability of the bioplastics found in this environment will be high or low.

28.7 Challenges and Environmental Impacts of Microbial Bioplastics

The promising research on bioreduction, production, or remediation of bioplastics is undoubtable; however, there are still challenges which hinder the manufacture and application of these polymers. Firstly, these polymers require large costs for production hence the use of bacteria and GHGs serves as suitable options to save costs while also reducing the carbon footprint in the environment by using these harmful gases to produce biodegradable bioplastics (Ghysels et al. 2018). To counteract the costs, studies have been conducted using affordable renewable sources of carbon such as starch which is available worldwide thus making it the ideal choice. Unfortunately, limitations still prevail as research into starch as a carbon source indicates issues including brittleness, low resistance to heat, poor stability, as well as an increased water/gas permeability, and therefore requiring the addition of plasticizers (i.e., sorbitol) as well as flexibilizers (e.g., glycerol) in order to correct the problem (Jabeen et al. 2015). This solution is redundant especially taking into account that plasticizers are some of the additives which form part of synthetic polymers and are one of the reasons there is a huge demand for the shift from non-biodegradable polymers to biodegradable plastics as some have been found to be carcinogenic (Trivedi et al. 2016; Hottle, et al. 2017; Yang et al. 2018). Moreover, starch-based bioplastics are commonly used bioplastics in the World and can readily be made from home. The excessive use of starch can absorb humidity which facilitates the manufacturing of the drug capsule. Starch, agricultural byproducts, biopolymers, and microorganism are the sources used in the production of bioplastics (Trivedi et al. 2016).

The action of different microorganisms as a response to stresses is the main process behind the synthesis of biopolymers; however, the use of gram-negative bacteria (such as *Ralstonia eutropha*) which are currently the main source of PHAs for biopolymer synthesis has been observed to contain lipopolysaccharide (LPS) endotoxins which are associated with chronic respiratory and gastrointestinal infection. These endotoxins thus limit the number of bacteria that can be used for bioplastic synthesis including limiting the applications that a bioplastic formed under these conditions can be used (Imran et al. 2020; Rehman et al. 2015). The endotoxins induce strong immunogenic reactions which are undesirable for applications involving biomedical sciences as they may cause the abovementioned infections (gastrointestinal and respiratory) including infections associated with water contamination as a result of bacteria such as *Escherichia coli (E.coli)* (Imran et al. 2020). *Bacillus* is missing the LPS layer found in gram-negative bacteria and thus can be used in medical applications; additionally, *Bacillus* can grow at cheaper substrates while producing a large amount of polymeric material (i.e., protein) (Rehman et al. 2015). The absence of lipopolysacharides (LPS) layer in the gram positive bacteria known as Bacillus is useful in biomedical applications and can grow at low cost substrates, while producing large amount of matrix or primary phase called proteins (Rehman et al., 2015). Table 28.1 summarizes the advantages and disadvantages of bioplastics and convetional plastics.

28.8 Future Prospects and Possibilities

In the continuous research for affordable, biodegradable, environmentally friendly, and easily accessible plastic alternatives,one of the solutions is to find more renewable carbon sources to utilize as substrates for polymer production. The research on alternative sources such as glycerol proves to be affordable in processing compared to the fermentative processes associated with bioplastic production using common starches (i.e., glucose, fructose, etc. (Jiang et al. 2020; Zhu et al. 2013)). The development of microbial strains which possess dual capabilities (i.e., recombinant strains) to counteract the need to use different microbes for the synthesis of individual polymers would not only cut processing costs but will also give rise to a wide

Table 28.1 Advantages and disadvantages of bioplastics and conventional plastics

Bioplastics		Conventional Plastics	
Advantages	Disadvantages	Advantages	Disadvantages
They reduce the carbon dioxide footprint (Chen 2014; Reddy et al. 2013; Shamsuddin 2017)	Production costs are high (Bezirhan and Bilgen 2015; Shivam 2016)	Low production cost (Lackner 2015)	Derived from petrochemicals (Lackner 2015)
They are sustainable (Shamsuddin 2017)	Thermally unstable (Jabeen et al. 2015)	They have desirable technological properties (Andrady and Neal 2009; Shamsuddin 2017)	Their recycling is a challenge (Lackner 2015)
They reduce energy efficiency (Bezirhan and Bilgen 2015; Chen 2014; Shamsuddin 2017)	They are challenging to recycle (Bezirhan and Bilgen 2015; El-Kadi 2010)	They are energy efficient (Andrady and Neal 2009)	Non-biodegradable (Lackner 2015)
They are mostly produced from natural feedstock (Bezirhan and Bilgen 2015; Chen 2014; Shamsuddin 2017)	They are brittle (Jabeen et al. 2015; Ilyas et al. 2016)	Thermally recyclable (Lackner 2015)	Combustion products are toxic (Lackner 2015)

range of various polymers that can be produced from numerous carbon sources thus limiting restrictions and widening the scope for polymer applications as new types of polymers with different properties might emerge (Bharti and Swetha 2016).

28.9 Conclusions

The twenty-first century is the age of endless biotechnological advancements which as science progresses so is the development of diverse synthetic materials. Synthetic polymers are at the forefront of developing technologies due to their various impressive properties that make it difficult to discontinue their use as they are detrimental to the environment. The development of bioplastics to replace petroplastics is slowly showing progress with the continuous research by scientists across the globe. It is evident that there are challenges which are restricting the ultimate global production of these plastics including the concerns of potential deforestation should this approach be implemented. At the current moment, production costs are the biggest drawback with the production of bioplastics; therefore, should a practical and affordable way to produce these bioplastics be successful then we may be approaching an era of bioplastic use as opposed to conventional synthetic polymers which have proved to be disadvantageous to the planet as a whole.

References

Ahmed T, Shahid M, Azeem F, Rasul I, Shah AA, Noman M, Hameed A, Manzoor N, Manzoor I, Muhammad S (2018) Biodegradation of plastics: current scenario and future prospects for environmental safety. Environ Sci Pollut Res 25(8):7287–7298

Andrady AL, Neal MA (2009) Applications and societal benefits of plastics. Philos Trans Royal Soc B: Biol Sci 364(1526):1977–1984

Anstey A, Muniyasamy S, Reddy MM, Misra M, Mohanty A (2014) Processability and biodegradability evaluation of composites from poly(butylene succinate) (PBS) bioplastic and biofuel co-products from Ontario. J Polym Environ 22(2):209–218

Asrofi M, Sujito ES (2019) Moisture resistance of sugarcane bagasse cellulose filled tapioca starch biocomposites: effect of cellulose loading. Int J Progr Sci Technol 18(1):01–04

Bezirhan E, Bilgen HD (2015) A review: investigation of bioplastics. J Civil Eng Arch 9:188–192

Bharti S, Swetha G (2016) Need for bioplastics and role of biopolymer PHB: a short review. J Pet Environ Biotechnol 07(272):2

Carlozzi P, Giovannelli A, Traversi ML, Touloupakis E, Di Lorenzo T (2019) Poly-3-hydroxybutyrate and H2 production by Rhodopseudomonas sp. S16-VOGS3 grown in a new generation photobioreactor under single or combined nutrient deficiency. Int J Biol Macromol 135:821–828

Carlozzi P, Di Lorenzo T, Ghanotakis DF, Touloupakis E (2020) Effects of pH, temperature and salinity on P3HB synthesis culturing the marine Rhodovulum sulfidophilum DSM-1374. Appl Microbiol Biotechnol 104(5):2007–2015

Caruso G (2015) Plastic degrading microorganisms as a tool for bioremediation of plastic contamination in aquatic environments. J Pollut Eff Cont 3(3):1–2

Chen Y (2014) Bioplastics and their role in achieving global sustainability. J Chem Pharm Res 6 (1):226–231

Colombo B, Pepè Sciarria T, Reis M, Scaglia B, Adani F (2016) Polyhydroxyalkanoates (PHAs) production from fermented cheese whey by using a mixed microbial culture. Bioresour Technol 218:692–699

El-Kadi S (2010) Bioplastic production from inexpensive sources. ISBN 9783639263725. VDM Verlag Dr. Müller Publishing, Berlin, Germany

Emadian SM, Onay TT, Demirel B (2017) Biodegradation of bioplastics in natural environments. Waste Manag 59:526–536

Gan Z, Zhang H (2019) PMBD: a comprehensive plastics microbial biodegradation database. Database 2019

Ghysels S, Mozumder MSI, De Wever H, Volcke EIP, Garcia-Gonzalez L (2018) Targeted poly (3-hydroxybutyrate-co-3-hydroxyvalerate) bioplastic production from carbon dioxide. Bioresour Technol 249:858–868

Glaser JA (2019) Biological degradation of polymers in the environment. In: Plastics in the environment. IntechOpen. Available from: https://www.intechopen.com/books/plastics-in-the-environment/biological-degradation-of-polymers-in-the-environment

Ha C-S, Cho W-J (2002) Miscibility, properties, and biodegradability of microbial polyester containing blends. Prog Polym Sci 27(4):759–809

Hottle TA, Bilec MM, Landis AE (2017) Biopolymer production and end of life comparisons using life cycle assessment. Resour Conserv Recycl 122:295–306

Ilyas RA, Sapuan S, Sanyang M, Ishak M (2016) Nanocrystalline cellulose reinforced starch-based nanocomposites: a review. In: 5th Postgraduate seminar on natural fiber composites, pp 82–87

Ilyas RA, Sapuan SM, Kadier A, Kalil MS, Ibrahim R, Atikah MSN, Nurazzi NM, Nazrin A, Lee CH, Norrahim MNF, Sari NH (2020) Properties and characterization of PLA, PHA, and other types of biopolymer composites. In: Al-Oqla F, Sapuan SM (eds) Advanced processing, properties, and applications of starch and other bio-based polymers. Elsevier, Amsterdam, The Netherlands, pp 111–138

Imran M, Naeem M, Baig AQ (2020) Topological indices of polyhydroxybutyrate and polycaprolactone. J Inf Optim Sci 41(4):1025–1041

Iram D, Riaz R, Iqbal R (2019) Usage of potential micro-organisms for degradation of plastics. Open J Environ Biol 4(1):007–0015

Jabeen N, Majid I, Nayik GA (2015) Bioplastics and food packaging: a review. Cogent Food Agric 1(1):1117749

Jiang G, Hill DJ, Kowalczuk M, Johnston B, Adamus G, Irorere V, Radecka I (2016) Carbon sources for polyhydroxyalkanoates and an integrated biorefinery. Int J Mol Sci 17(7):1157

Jiang T, Duan Q, Zhu J, Liu H, Yu L (2020) Starch-based biodegradable materials: challenges and opportunities. Adv Ind Eng Polym Res 3(1):8–18

Karamanlioglu M, Preziosi R, Robson GD (2017) Abiotic and biotic environmental degradation of the bioplastic polymer poly (lactic acid): a review. Polym Degrad Stab 137:122–130

Krueger MC, Harms H, Schlosser D (2015) Prospects for microbiological solutions to environmental pollution with plastics. Appl Microbiol Biotechnol 99(21):8857–8874

Kumar S, Thakur KS (2017) Bioplastics-classification, production and their potential food applications. J Hill Agric 8:118–129

Lackner M (2015) Bioplastics Kirk-Othmer encyclopedia of chemical technology. Wiley, New York, pp 1–41

Luengo JM, García B, Sandoval A, Naharro G, Olivera ER (2003) Bioplastics from microorganisms. Curr Opin Microbiol 6(3):251–260

Maddah HA (2016) Polypropylene as a promising plastic: a review. Am J Polym Sci 6(1):1–11

Mannina G, Presti D, Montiel-Jarillo G, Suárez-Ojeda ME (2019) Bioplastic recovery from wastewater: a new protocol for polyhydroxyalkanoates (PHA) extraction from mixed microbial cultures. Bioresour Technol 282:361–369

Mathew S, Jayakumar A, Kumar VP, Mathew J, Radhakrishnan EK (2019) One-step synthesis of eco-friendly boiled rice starch blended polyvinyl alcohol bionanocomposite films decorated with in situ generated silver nanoparticles for food packaging purpose. Int J Biol Macromol 139:475–485

Mohapatra S, Maity S, Dash HR, Das S, Pattnaik S, Rath CC, Samantaray D (2017) Bacillus and biopolymer: prospects and challenges. Biochem Biophys Rep 12:206–213

Monroy I, Buitrón G (2020) Production of polyhydroxybutyrate by pure and mixed cultures of purple non-sulfur bacteria: a review. J Biotechnol 317:39–47

More A, Sen C, Das M (2017) Development of starch-polyvinyl alcohol (PVA) biodegradable film: effect of cross-linking agent and antimicrobials on film characteristics. J Appl Packag Res 9:1–18

Mostafa N, Farag AA, Abo-dief HM, Tayeb AM (2018) Production of biodegradable plastic from agricultural wastes. Arab J Chem 11(4):546–553

Mostafa YS, Alrumman SA, Alamri SA, Otaif KA, Mostafa MS, Alfaify AM (2020) Bioplastic (poly-3-hydroxybutyrate) production by the marine bacterium Pseudodonghicola xiamenensis through date syrup valorization and structural assessment of the biopolymer. Sci Rep 10 (1):8815

Muhamad W, Othman R, Shaharuddin RI, Irani MS (2015a) Microorganism as plastic biodegradation agent towards sustainable environment. Adv Environ Biol 9:8–14

Muhamad WNAW, Othman R, Shaharuddin RI, Hasni MSI (2015b) Microorganism as plastic biodegradation agent towards sustainable environment. Adv Environ Biol 9:8–13

Muthuraj R, Mekonnen T (2018) Recent progress in carbon dioxide (CO_2) as feedstock for sustainable materials development: co-polymers and polymer blends. Polymer 145:348–373

Pérez R, Cantera S, Bordel S, García-Encina PA, Muñoz R (2019) The effect of temperature during culture enrichment on methanotrophic polyhydroxyalkanoate production. Int Biodeterior Biodegradation 140:144–151

Raza ZA, Noor S, Khalil S (2019) Recent developments in the synthesis of poly (hydroxybutyrate) based biocomposites. Biotechnol Prog 35(5):e2855

Reddy RL, Reddy V, Gupta G (2013) Study of bio-plastics as green & sustainable alternative to plastics. Int J Emerg Technol Adv Eng 3(5):76–81
Rehman R (2015) Selection of potent bacterial strain for over-production of PHB by using low cost carbon source for eco-friendly bioplastics. Adv Life Sci 3(1):29–35
Rehman RA, Rao AQ, Ahmed Z, Gul A (2015) Selection of potent bacterial strain for over-production of PHB by using low cost carbon source for eco-friendly bioplastics. Adv Life Sci 3(1):29–35
Shamsuddin I (2017) Bioplastics as better alternative to petroplastics and their role in national sustainability: a review. Adv Biosci Bioeng 5:63
Shivam P (2016) Recent Developments on biodegradable polymers and their future trends. Int Res J Sci Eng 4:17–26
Shrivastava A (2018) Introduction to plastics engineering. William Andrew Publishers
Sidek I, Syed Draman SF, Abdullah S, Anuar N (2019) Current development on bioplastics and its future prospects: an introductory review. INWASCON Technol Mag 1:03–08
Singh AK, Mallick N (2017) Advances in cyanobacterial polyhydroxyalkanoates production. FEMS Microbiol Lett 364(20)
Sudesh K, Abe H, Doi Y (2000) Synthesis, structure and properties of polyhydroxyalkanoates: biological polyesters. Prog Polym Sci 25(10):1503–1555
Sukara E (2014) Potential values of bacterial cellulose for industrial applications. J Selulosa 4 (1):7–16
Thakur S, Chaudhary J, Sharma B, Verma A, Tamulevicius S, Thakur VK (2018) Sustainability of bioplastics: opportunities and challenges. Curr Opin Green Sustain Chem 13:68–75
Trivedi P, Hasan A, Akhtar S, Siddiqui MH, Sayeed U, Khan MKA (2016) Role of microbes in degradation of synthetic plastics and manufacture of bioplastics. J Chem Pharm Res 8 (3):211–216
Urbanek AK, Rymowicz W, Mirończuk AM (2018) Degradation of plastics and plastic-degrading bacteria in cold marine habitats. Appl Microbiol Biotechnol 102(18):7669–7678
Xanthos D, Walker TR (2017) International policies to reduce plastic marine pollution from single-use plastics (plastic bags and microbeads): a review. Mar Pollut Bull 118(1–2):17–26
Yang, S.S., Brandon, A.M., Xing, D.F., Yang, J., Pang, J.W., Criddle, C.S., Ren, N.Q., Wu, W.M. (2018). Progresses in polystyrene biodegradation and prospects for solutions to plastic waste pollution. Paper presented at the IOP Conference Series: EES
Zanela J, Bilck AP, Reis MO, Grossmann MVE, Yamashita F (2020) Modified starches on the properties of extruded biodegradable materials of starch and polyvinyl alcohol. J Polym Environ 28(12):3211–3220
Zhu C, Chiu S, Nakas JP, Nomura CT (2013) Bioplastics from waste glycerol derived from biodiesel industry. J Appl Polym Sci 130(1):1–13
Zinn M, Witholt B, Egli T (2001) Occurrence, synthesis and medical application of bacterial polyhydroxyalkanoate. Adv Drug Deliv Rev 53(1):5–21